Calculus Single Variable

Student Solutions Manual

2nd Edition

Brian E. Blank
Steven G. Krantz
Department of Mathematics
Washington University in St. Louis

Supplement Author
Donald G. Hartig
Mathematics Department
California Polytechnic State University
San Luis Obispo, California

John Wiley & Sons, Inc.

Cover photo: ©Walter Bibikow/JAI/Corbis

Founded in 1807, John Wiley & Sons, Inc. has been a valued source of knowledge and understanding for more than 200 years, helping people around the world meet their needs and fulfill their aspirations. Our company is built on a foundation of principles that include responsibility to the communities we serve and where we live and work. In 2008, we launched a Corporate Citizenship Initiative, a global effort to address the environmental, social, economic, and ethical challenges we face in our business. Among the issues we are addressing are carbon impact, paper specifications and procurement, ethical conduct within our business and among our vendors, and community and charitable support. For more information, please visit our website: www.wiley.com/go/citizenship.

ISBN-13 978-0470-45815-0

Printed in the United States of America

10 9 8 7 6 5 4 3 2 1

Printed and bound by Integrated Book Technology.

Contents

Chapter 1

Basics

1.1 Number Systems

Problems for Practice

1. $2.13 = 213/100$

3. Let $x = 0.232323\ldots$. Then $100x = 23.232323\ldots$. Subtract to obtain $99x = 23$, implying that $x = 23/99$.

5. Let $x = 5.001001001\ldots$. Then $1000x = 5001.001001001\ldots$. Subtract to obtain $999x = 4996$, implying that $x = 4996/999$.

7. Divide 40 into 1.00000 until it terminates to yield a repeating block of zeros: $1/40 = 0.025\overline{0} = 0.025$.

9. Divide 25 into 18.00000 until it terminates to yield a repeating block of zeros: $18/25 = 0.72\overline{0} = 0.72$.

11. Divide 7 into 2.0000000 until the initial remainder (2) repeats yielding a long repeating block: $2/7 = 0.\overline{285714}$.

13. $\{x : 1 \leq x \leq 3\} = [\,1, 3\,]$

15. $\{t : t > 1\} = (\,1, \infty\,)$

17. The number y satisfies the inequality $|y + 4| \leq 10$ when it is 10 units or less from the number -4 on the number line. That is, it must be at, or to the left of, 6 and at, or to the right of, -14: $\{y : |y + 4| \leq 10\} = [\,-14, 6\,]$.

19. Subtract x from each side, and then add 5 to each side of the given inequality to obtain $x < 9$. See the picture.

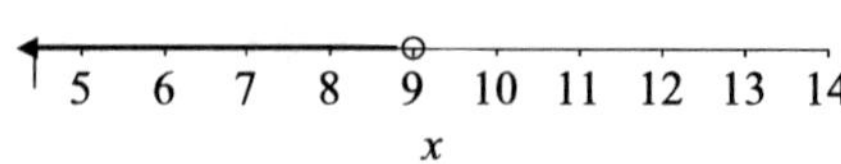

21. Take the square root on both sides. Order is preserved and we get $|t-5| < 3/2$. That is, all numbers that are less than 3/2 units from 5. See the picture.

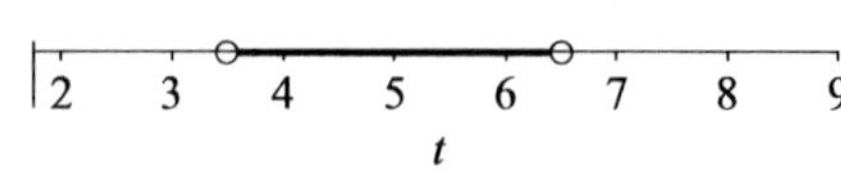

23. Divide both sides by 3 to obtain the inequality $|x+3| \leq 5$. This is all numbers that are less than or equal to 5 units from -3. See the picture.

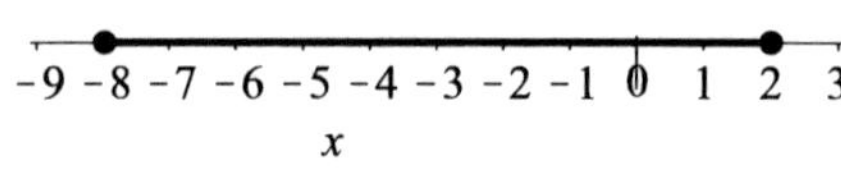

25. The midpoint of the interval is $c = 1$ and its width is 4, so
$$[-1, 3] = \{x : |x - 1| \leq 2\}.$$

27. The midpoint of the interval is $c = 1$ and its width is $2\pi + 2$, so
$$(-\pi, \pi + 2) = \{x : |x - 1| < \pi + 1\}.$$

Further Theory and Practice

29. Rational numbers x and y are of the form $x = m/n$ and $y = p/q$ where m, n, p, and q are integers. Therefore,

$$x + y = m/n + p/q = (m \cdot q + p \cdot n)/(p \cdot q) \quad \text{and} \quad x \cdot y = (m \cdot n)/(p \cdot q)\,.$$

Both x and y are rational because the products and sums of integers are integers.

31. The numbers 22/7 and 3.14 are both rational, π is irrational.

33. One percent of the correct mass is $\epsilon = 0.345$. With $c = 34.5$, the interval is $\{x : |x - 34.5| \leq 0.345\} = [34.155, 34.845]$.

35. This set is the intersection of the intervals $[-2, \infty)$ and $(-3, 3)$: $\{x : -2 \leq x < 3\}$. See the picture.

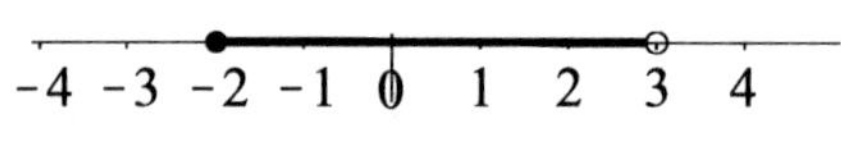

37. This is the intersection of the semi-infinite intervals $\{y : -4 - \sqrt{7} < 2y\}$ and $\{y : 4 - \sqrt{2} \leq y\}$. Since
$$-2 - \sqrt{7}/2 < 4 - \sqrt{2}\,,$$
it is the interval $[4 - \sqrt{2}, \infty)$. See the picture.

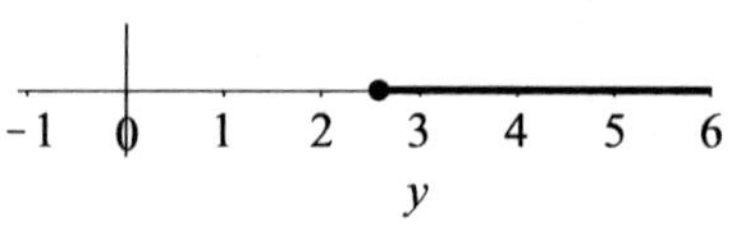

39. The distance from x^2 to 5 must be 4 or more. Therefore, one possibility is $x^2 = 9$ or more and another is $x^2 = 1$ or less. That is, $|x| \geq 3$ or $|x| \leq 1$. See the picture.

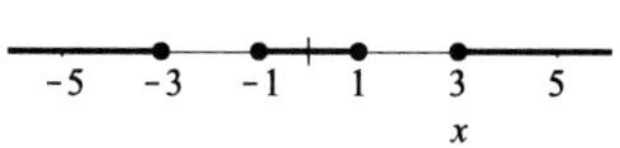

41. This is the intersection of two intervals. One of them is $\{x : -4 \geq x\}$ obtained by subtracting $x+5$ from the left and middle parts of the defining inequalities. The other is $\{x : -3 > x\}$ obtained by subtracting $2x+8$ from the middle and right parts of the defining inequalities. Intersecting yields just one interval: $(-\infty, -4]$. See the picture.

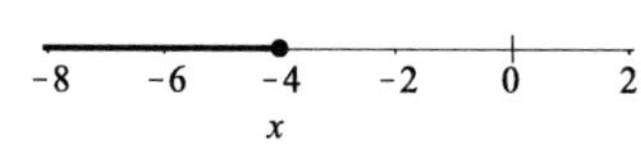

43. x cannot be 0. If $x > 0$, then the inequality is equivalent to $|x+1| > x-1$, which is true for all such x. If $x < 0$, then the inequality is equivalent to $|-x-1| > -x+1$ or $|x+1| > 1-x$ which is true for no such x. See the picture.

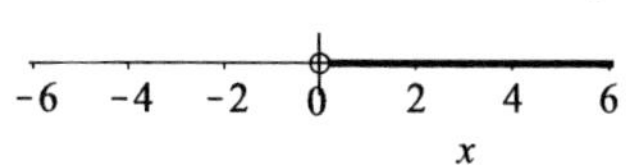

45. $\{x : |x-3| = |x+9|\} = \{-3\}$

47. $\{x : |x-3| < 2 \text{ and } x^2 \leq 8\} = (1,5) \cap [-2\sqrt{2}, 2\sqrt{2}] = (1, 2\sqrt{2}]$

49. $\{s : s < 4 \text{ and } |2s+9| < 4-s\} = \{s : s < 4 \text{ and } s-4 < 2s+9 < 4-s\}$
$= \{s : s < 4 \text{ and } -13 < s < -5-2s\} = \{s : -13 < s < -5/3\}$

51. $\{w : w/(w+1) < 0\} = \{w : -1 < w < 0\} = \{w : |w+1/2| < 1/2\}$

53. When a and b are nonpositive so is $a+b$ and

$$|a+b| = -(a+b) = -a+(-b) = |a| + |b| \leq |a| + |b| .$$

55. If z were the smallest positive number, then $0 < z < 1$. Multiply all three parts by z to conclude that $0 < z^2 < z$ and z^2 would be smaller than z, a contradiction.

57. d. Since $a^2 = 2b^2$, a^2 is an even integer. This implies that a is also even (the square of an odd integer must be odd). Therefore, 2 divides a evenly. That is, $a = 2\alpha$ for some integer α.

e. Substitute for a in the equation $a^2 = 2b^2$ to conclude that $4\alpha^2 = 2b^2$. This implies that $b^2 = 2\alpha^2$.

f. Since the integer b^2 is even, b is also even and 2 divides b with no remainder.

g. From d we conclude that a is even. From f we conclude that b even. This contradicts the fact that a and b are integers with no common factors.

CALCULATOR/COMPUTER EXERCISES

59. Add and subtract 0.0005 to $x = 0.449$. We conclude that y must lie in the open interval $(0.4485, 0.4495)$.

61. Add and subtract $5 \cdot 10^{-4}$ to $x = 0.999 \cdot 10^{-5}$. We conclude that y must lie in the open interval $(-0.49001 \cdot 10^{-3}, 0.50999 \cdot 10^{-3})$.

63. The number $y = 4.001$ has the property that $|x - y| = 0.005$ so it agrees with x to two decimal places.

65. Using *Maple* with 10 significant digits, the product is 0. Using 20 significant digits, the product is 1.036 888 824, and using 30 significant digits, the product is 0.999 999 999 9.

67. The relative errors for $x = 10^n$, $n = 6, 11, 16, 21$ are on the left. The relative errors for $x = -10^n$ are on the right.

x	**Relative Error**
10^6	0.99999
10^{11}	0.090909
10^{16}	0.9999×10^{-8}
10^{21}	1.000000×10^{-11}

x	**Relative Error**
-10^6	1.0001
-10^{11}	0.11111
-10^{16}	0.1000001×10^{-5}
-10^{21}	1.000000×10^{-11}

69. The relative errors for $x = 10^n$, $n = 5, 8, 11, 14$ are on the left. The relative errors for $x = -10^n$ are on the right.

x	**Relative Error**
10^5	0.5
10^8	0.0003334
10^{11}	0.33333×10^{-8}
10^{14}	0.33333×10^{-11}

x	**Relative Error**
-10^5	0.25
-10^8	0.000333
-10^{11}	0.33333×10^{-8}
-10^{14}	0.33333×10^{-11}

1.2 Planar Coordinates and Graphing in the Plone

Problems for Practice

1. The six points are plotted on the right.

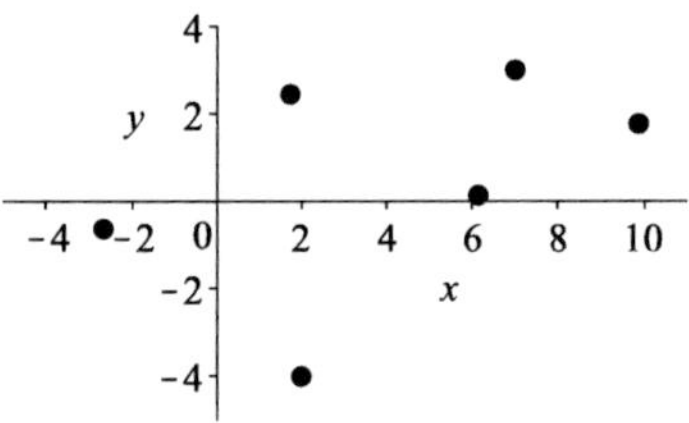

3. This is a straight line. See the plot on the right.

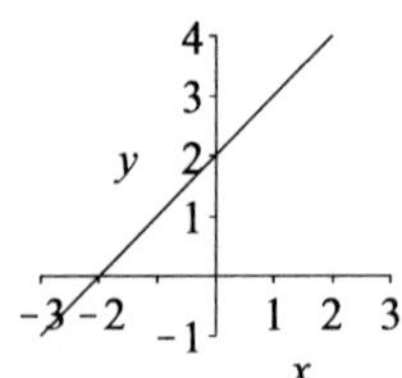

5. The distance from A to B is $|\overline{AB}| = \sqrt{(2+5)^2+(3+6)^2} = 2\sqrt{13}$. The distance from A to C is $|\overline{AC}| = \sqrt{(2+6)^2+(3+6)^2} = 2\sqrt{13}$. The distance from B to C is $|\overline{BC}| = \sqrt{(-4+5)^2+(7+6)^2} = \sqrt{170}$.

7. The center is at $(1,3)$ and the radius is 3. See the plot on the right.

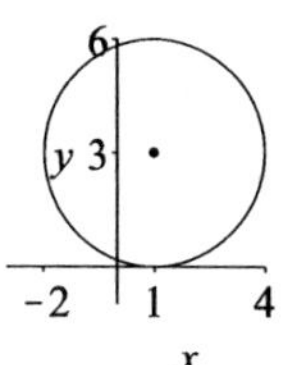

9. The center is at $(0,-5)$ and the radius is $\sqrt{2}$. See the plot on the right.

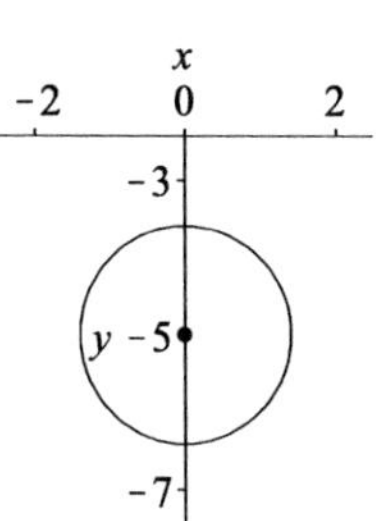

11. Complete the square in y:

$$x^2 + \left(y^2 - y + \tfrac{1}{4}\right) = 0 + \tfrac{1}{4}$$

or $x^2+(y-1/2)^2 = 1/4$. The center is at $(0,1/2)$ and the radius is $1/2$. See the plot on the right.

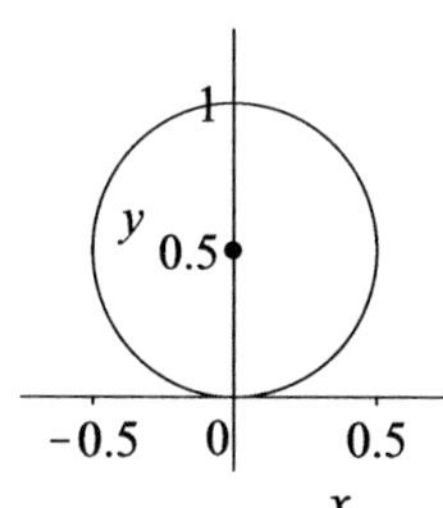

13. Complete the square in x and y:

$$3(x^2+4x+4)+3(y^2-2y+1) = 2+12+3$$

or $(x+2)^2+(y-1)^2 = 17/3$. The center is at $(-2,1)$ and the radius is $\sqrt{17/3}$. See the plot on the right.

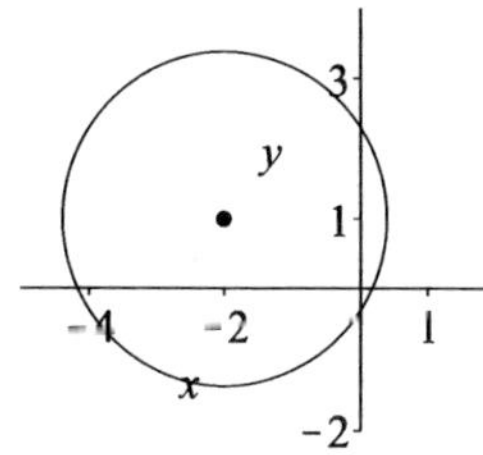

15. Complete the square in x and y:

$$4(x^2-4x+4)+4(y^2+2y+1) = 0+16+4$$

or $(x-2)^2+(y+1)^2 = 5$. The center is at $(2,-1)$ and the radius is $\sqrt{5}$. See the plot on the right.

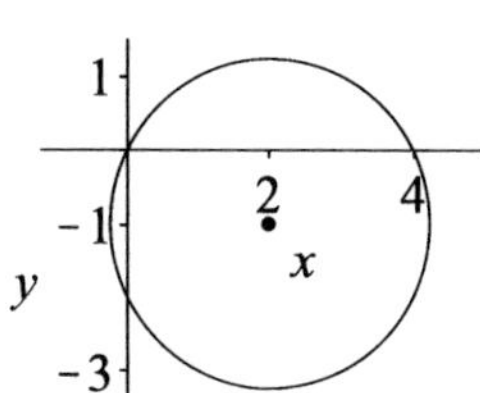

17. Center at $(-3,5)$, radius 6: $(x+3)^2+(y-5)^2 = 36$.

19. Center at $(-4,\pi)$, radius 5: $(x+4)^2+(y-\pi)^2 = 25$.

21. The vertex is at $(0,-3)$ and the axis of symmetry is the line $x=0$ (the y-axis).

23. Since $-B/(2A) = -2/(-2) = 1$, the vertex is at $(1,1)$ and the axis of symmetry is the line $x=1$.

25. Since $y = -x^2 - 6x - 9/2$, $-B/(2A) = -(-6)/(-2) = -3$. The vertex is at $(-3, 9/2)$ and the axis of symmetry is the line $x=-3$.

27. This is an ellipse. Its standard form equation is $\frac{x^2}{(1/2)^2} + \frac{y^2}{1^2} = 1$ and its center is at the origin, (0,0).

29. This is an ellipse. Complete the square in x: $(x^2 + x + 1/4) + 9y^2 = 15/4 + 1/4$ or $(x+1/2)^2 + 9y^2 = 4$. Consequently, its standard form equation is $\frac{(x-(-1/2))^2}{2^2} + \frac{y^2}{(2/3)^2} = 1$, so its center is at the point $(-1/2, 0)$.

31. This is an ellipse. Complete the square in x and y: $(x^2+2x+1)+4(y^2+6y+9) = 12+1+36$ or $(x+1)^2 + 4(y+3)^2 = 49$. Consequently, its standard form equation is $\frac{(x-(-1))^2}{7^2} + \frac{(y-(-3))^2}{(7/2)^2} = 1$, so its center is at the point $(-1,-3)$.

33. This is a hyperbola. Complete the square in y: $2x^2 - 3(y^2 - 2y + 1) = 103 - 3$ or $2x^2 - 3(y-1)^2 = 100$. Consequently, its standard form equation is $\frac{x^2}{(\sqrt{50}\,)^2} - \frac{(y-1)^2}{(10/\sqrt{3}\,)^2} = 1$, so its center is at the point $(0,1)$.

35. The region is all points *between* the vertical lines $x=-3$ and $x=-3$. It is sketched on the right.

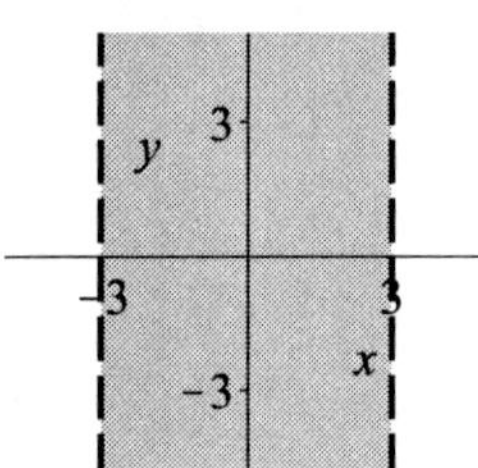

37. This is all points *between* the vertical lines $x=-7$ and $x=7$ and *outside* of the horizontal lines $y=-5$ and $y=-3$. See the picture.

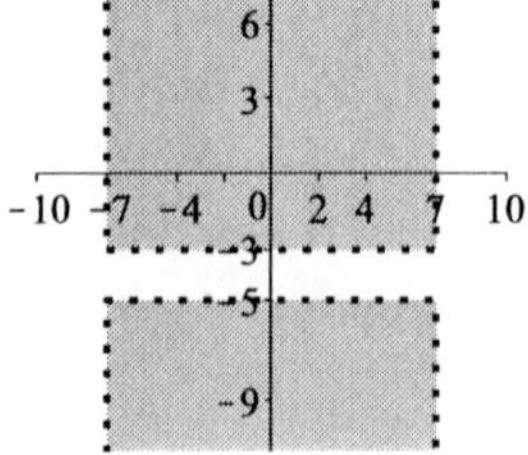

39. This is all points *outside* of the circle centered at the origin having radius 4. See the picture.

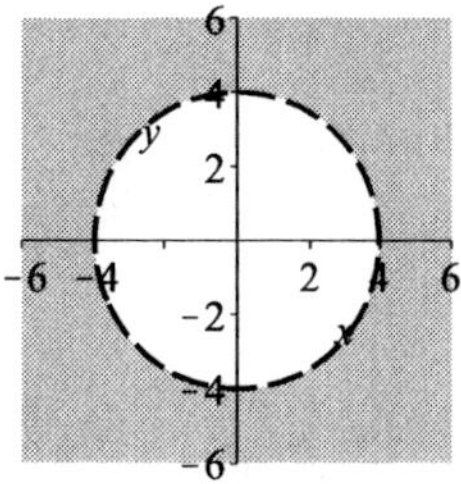

41. This is all points *on and outside* of the circle centered at (2,0) having radius 2. See the picture.

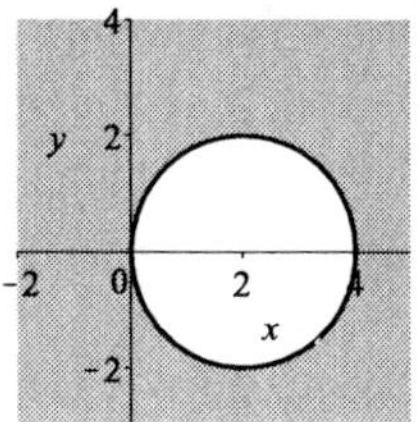

43. This is all points that are *between* the lines $x - y = 1$ and $x - y = -1$ and *on or to the right* of the vertical line $x = 4$. See the picture.

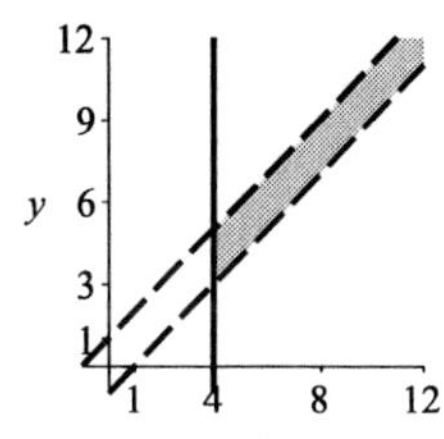

45. This are points that are in the *left half-plane*, *inside* of the circle centered at the origin of radius 3, that also lie *above* the horizontal line $y = -1$. See the picture.

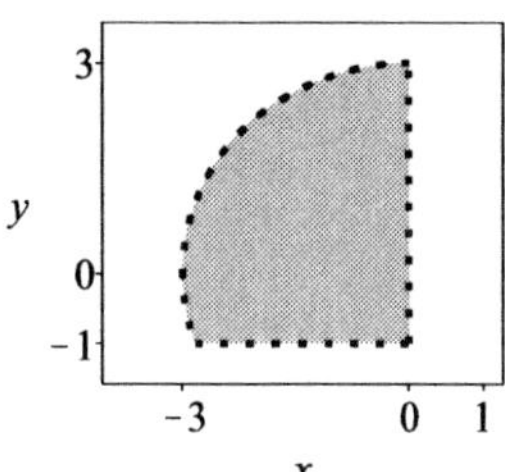

Further Theory and Practice

47. The points equidistant from (2,3) and (8,2) are on the intersection of circles

$$(x-2)^2 + (y-3)^2 = r^2 \quad \text{and} \quad (x-8)^2 + (y-2)^2 = r^2 .$$

This implies that $-4x + 4 - 6y + 9 = -16x + 64 - 4y + 4$ or $12x - 2y = 55$.

Similarly, the points equidistant from (2,3) and (7,9) are on the intersection of circles

$$(x-2)^2 + (y-3)^2 = r^2 \quad \text{and} \quad (x-7)^2 + (y-9)^2 = r^2 .$$

This implies that $-4x+4-6y+9 = -14x+49-18y+81$ or $10x+12y = 117$.

Any point equidistant from the three points must lie on the intersection of these two lines. Solve simultaneously to find that the point is $(447/82, 427/82)$. Verify that the distance from this point to all three points given above is $5\sqrt{4514}/82$.

49. There is just one such point, the origin (0,0). It is one unit from all three.

51. This wants to be the equation of a circle. Complete the square in x and y: $5(x^2 - \frac{2}{5}x + \frac{1}{25}) + 5(y^2 + \frac{3}{5}y + \frac{9}{100}) = 6 + \frac{13}{20}$ or $(x - 1/5)^2 + (y + 3/10)^2 = 133/100$, to see that it is.

53. This can be written in the form $2y^2 - 2x^2 + x - 5y = 7$, which is the equation of a hyperbola.

55. Divide the equation by 2 and complete the square in x: $(x^2 + 9x + 81/4) + y^2 = -k/2 + 81/4$. Let $k = 81/2$ and the equation is $(x + 9/2)^2 + y^2 = 0$ whose graph is the point $(-9/2, 0)$.

57. Write the equation in the form $x^2 + x + y^2 - y = k - 1$ and complete the square in x and y: $(x^2 + x + 1/4) + (y^2 - y + 1/4) = k - 1 + 1/2$. If $k = 1/2$ then $(x + 1/2)^2 + (y - 1/2)^2 = 0$. The graph is the point $(-1/2, 1/2)$.

59. Put the equation of the hyperbola in standard form (complete the square in x): $(x^2 + 6x + 9) - 4y^2 = 7 + 9$ or $\frac{(x+3)^2}{4^2} - \frac{y^2}{2^2} = 1$. The center of the hyperbola is the point $(-3, 0)$ and its vertices are at the points $(-3 \pm 4, 0)$. The values $a = -7$, $b = 1$ produce the widest open vertical strip containing no point of the hyperbola.

61. The four dashed curves in the picture are the points (x, y) where $|x| \cdot |y| = 1$. The points where $|x| \cdot |y| < 1$ are *inside* of these curves. The boundary lines are not included.

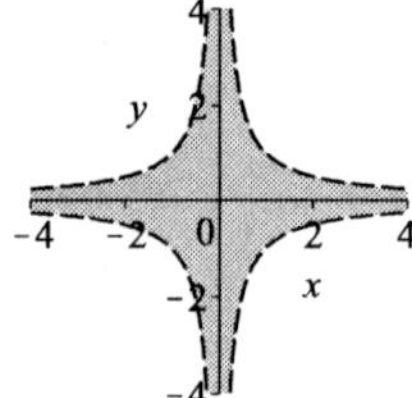

63. This is the region consisting of the points that lie *on and between* the two lines in the picture. The top line is $x + y = 1$ and the bottom line is $x + y = -1$.

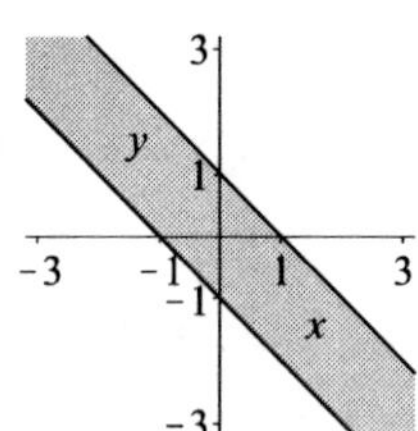

65. This region consists of the points that lie to the *left* of the vertical line $x = 1/3$ and *on or above* the line $y = -2x - 5$. See the picture.

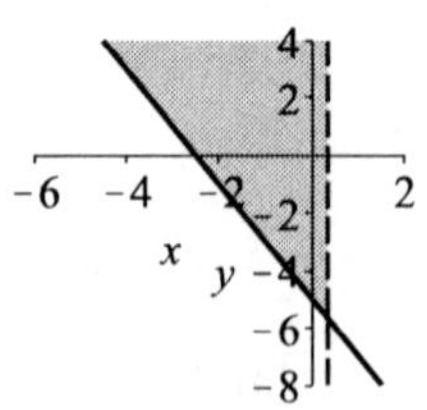

67. These are the points that lie *outside* of the "cup" of the parabola $x = y^2$ and, at the same time, lie between the crossing lines $y = x$ and $y = -x$ as shown in the picture.

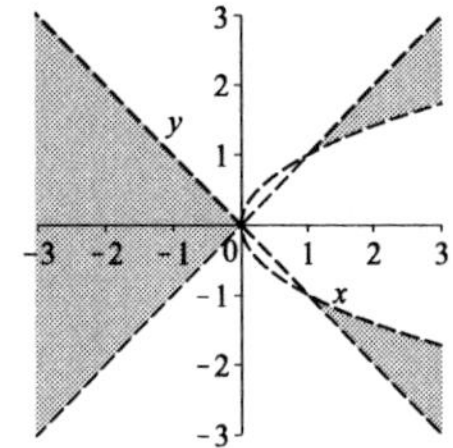

69. These are the points that lie *on* the line $x - 3y = 6$ and are *above* the line $y = 2x$. See the picture.

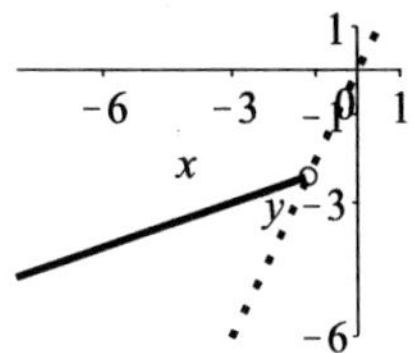

71. The intersection consists of points (x, y) such that $y = 2x$ and $(x - h)^2 + (y - k)^2 = r^2$. Substitute the first equation into the second to see that x must be a solution to $(x - h)^2 + (2x - k)^2 = r^2$, a quadratic in x. There will be either no solutions, exactly one solution, or exactly two solutions.

73. Let the equation of the circle be $(x - h)^2 + (y - k)^2 = r^2$. Substitute the coordinates of the first two points into this equation and set the left sides equal to obtain $-4h + 4 - 8k + 16 = -8h + 16$, which simplifies to $4h - 8k = -4$ or $h - 2k = -1$. Now substitute the coordinates of the second and third points into the equation, equate left sides, and simplify to $18h + 6k = -18$ or $3h + k = -3$. Add twice $3h + k = -3$ to $h - 2k = -1$ to see that $h = -1$ (and $k = 0$). To obtain the circle's radius, let $x = 4$ and $y = 0$ in the equation $(x + 1)^2 + y^2 = r^2$ to get $25 = r^2$ so $r = 5$.

75. Let $P = (x, y)$. Then

$$y = \sqrt{(x - 0)^2 + (y - 1)^2},$$

implying that $y \geq 0$ and $y^2 = x^2 + (y - 1)^2$. That is, $y \geq 0$ and $0 = x^2 + 1 - 2y$. Equivalently, $y = (x^2 + 1)/2$. This is a parabola. See the picture.

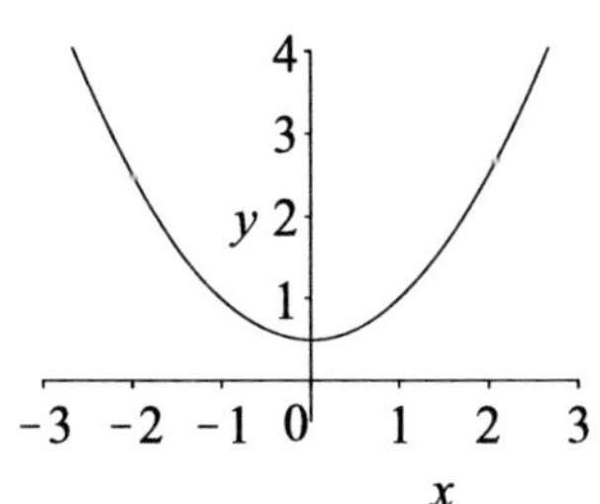

CALCULATOR/COMPUTER EXERCISES

77. These are both parabolas. The picture on the left displays them from a distance. The zoomed versions show that they intersect at the points $(-0.741, 2.181)$ and $(1.149, -1.207)$.

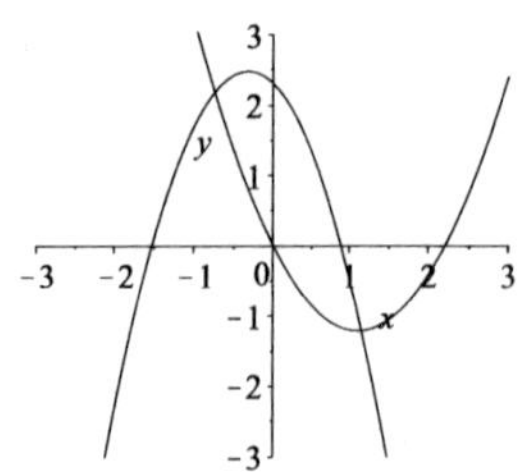

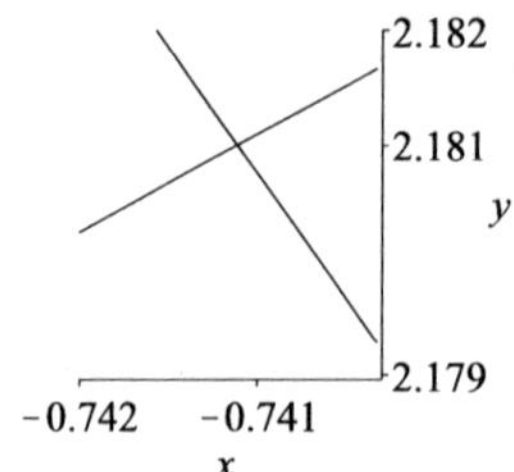

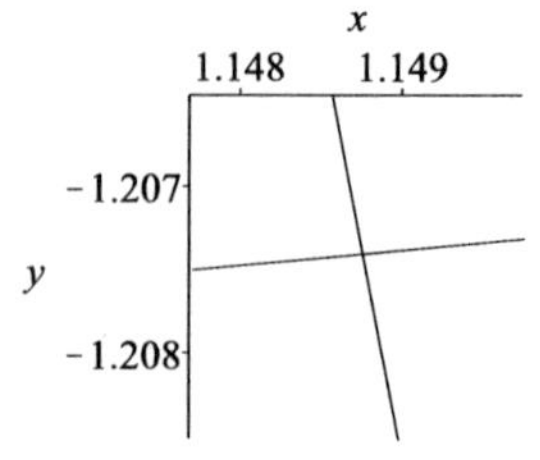

79. The boundary curves are ellipses. The region is all points on and inside the smaller one and on and outside the larger. See the picture.

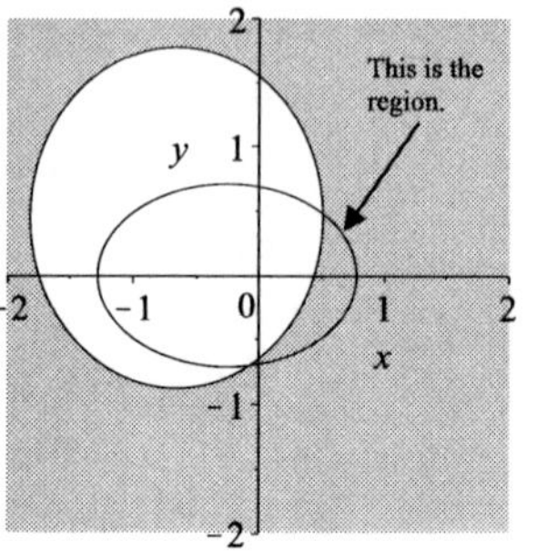

81. The parabola and the two lines are sketched below on the left. On the right is the same parabola and the line $y = 3.8x$.

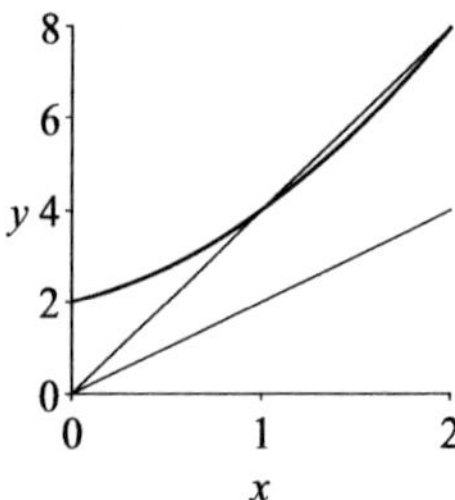

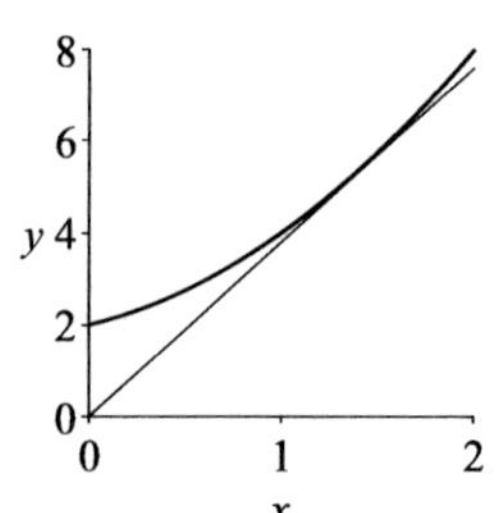

1.3 Lines and Their Slopes

Problems for Practice

1. The line through (0,3) also passes through (4,4). Its slope is $\frac{4-3}{4-0} = \frac{1}{4}$.

   One line through $(-1, -4)$ also passes through $(3, -4)$. Its slope is 0.

   Another line through $(-1, -4)$ goes through $(-3, 0)$, slope: $\frac{0-(-4)}{-3-(-1)} = -2$.

   The line through $(-5, 4)$ also passes through (4,0), slope: $\frac{0-4}{4-(-5)} = -\frac{4}{9}$.

   The line through (3,0) also passes through $(3, -4)$. It is vertical and does not have a slope.

3. The lines are drawn on the right. All of them have slope -3.

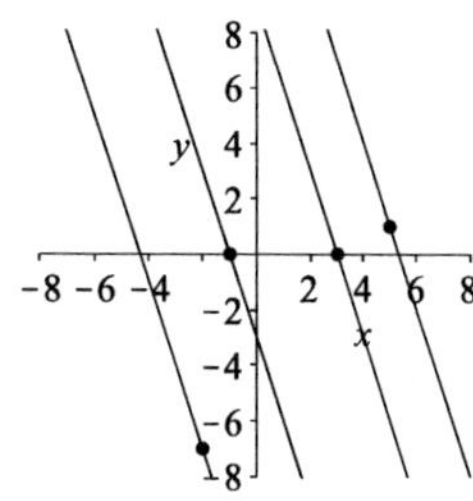

5. $y = 5(x + 3) + 7$

7. $y = -(x + \sqrt{2}\,)$

9. The slope is $(-4 - 7)/(6 - 2) = -11/4$. Using the first point, $y = (-11/4)(x - 2) + 7$.

11. The slope is $(0 - (-4))/(9 - (-7)) = 1/4$. Using the second point, $y = (1/4)(x - 9)$.

13. $y = -4x + 9$

15. $y = \sqrt{2}\,x - \sqrt{3}$

17. Use the point $(-4, 0)$ and the slope 3: $y = 3(x + 4) + 0$ or $y = 3x + 12$.

19. Using the intercept form: $\frac{x}{-2} + \frac{y}{6} = 1$. Therefore, $y = 3x + 6$.

21. The slope is $(10 - 7)/(3 - 2) = 3$. Using the point (2,7), $y = 3(x - 2) + 7$ or $y = 3x + 1$.

23. The slope is $(-9 - 3)/(2 - (-1)) = -4$. Using the point $(-1, 3)$, $y = -4(x + 1) + 3$ or $y = -4x - 1$.

25. $\dfrac{x}{-2} + \dfrac{y}{6} = 1$

27. $\dfrac{x}{-1} + \dfrac{y}{3} = 1$

29. The line has slope $-1/2$ so $y = (-1/2)(x - 1) - 2$ or $y = -x/2 - 3/2$.

31. The line has slope 2 so $y = 2(x - 2) + 1$ or $y = 2x - 3$.

33. The given line has slope $-1/2$ so the line perpendicular to it has slope 2. $y = 2(x - 1) - 2$ or $y = 2x - 4$.

35. The given line has slope $-2/3$ so the line perpendicular to it has slope $3/2$. $y = (3/2)(x - 3) + 4$ or $y = 3x/2 - 1/2$.

37. The x-intercept is $2/3$ and the y-intercept is $1/2$. The slope of the line is $-3/4$.

39. The x-intercept is 3 and the y-intercept is -1. The slope of the line is $1/3$.

41. The line with equation $x/2 = 1$ is vertical, with x-intercept $x = 2$. See the picture.

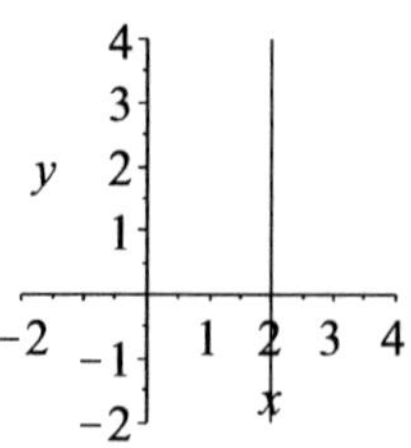

43. This line has slope 3 and it passes through the point $(-1, -2)$. See the picture.

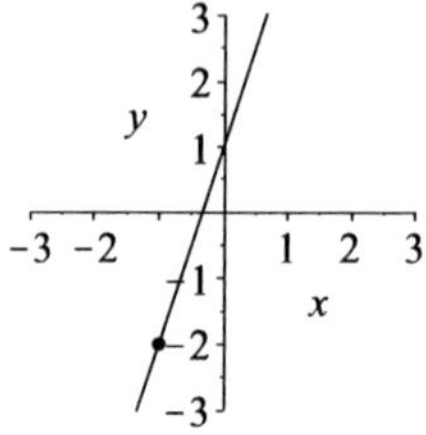

45. This line has intercepts $(-2, 0)$ and (0,4). Its slope is 2. See the picture.

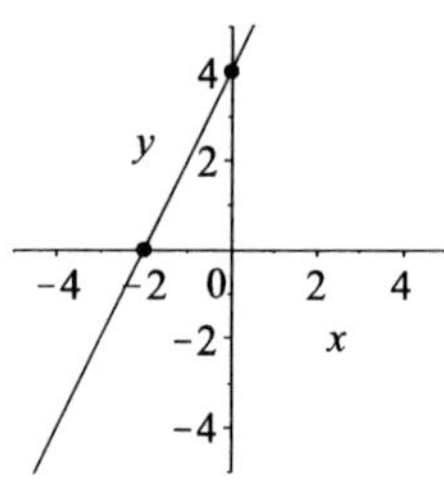

47. This line has intercepts $(3, 0)$ and (0,2). Its slope is $-2/3$. See the picture.

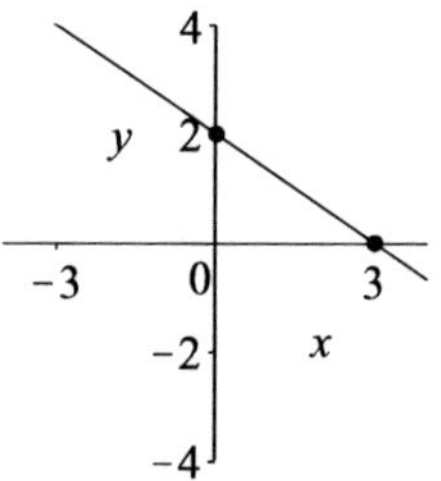

49. The line in question has slope $(7-2)/(4-1) = 5/3$. It is not parallel to the line through $(-4, -3)$ and $(-7, 6)$ because that line has slope $9/(-3) = -3$. Nor is it perpendicular to the line through (7,9) and $(5, -7)$ because that line has slope $-16/(-2) = 8$.

51. The line in the figure has slope $-4/12 = -1/3$. Therefore, the line through $(7, -8)$ that is perpendicular to it has slope 3 and it has the following equation: $y = 3(x - 7) - 8$ or $y = 3x - 29$.

Further Theory and Practice

53. The line in question has the equation $y = 2(x-5)-7$. It will pass through (a, b) when, for example, $a = 6$ and $b = -5$.

55. Since it is perpendicular, the slope of the line in question is the negative reciprocal of $(9-7)/(4-(-2)) = 1/3$ so its equation is $y = -3(x+2)+7$. The point (a, b) will be on this line when, for example, $a = -1$ and $b = 4$.

57. Let (a, b) lie in the first quadrant, on the unit circle ($a^2 + b^2 = 1$), and on the line $y = 2x$. Then $b = 2a$ and $a^2 + 4a^2 = 1$, so $a = \sqrt{5}/5$ and $b = 2\sqrt{5}/5$. Observe that the point $(c, d) = (5a, 5b) = (\sqrt{5}, 2\sqrt{5})$ is 5 units from the origin and also lies on the line $y = 2x$.

59. Subtract twice the second equation from the first to obtain $-5y = -10$. Therefore, $y = 2$ and $x = 7 - 3y = 1$.

61. Add 5 times equation 1 to 2 times equation 2 to obtain $29y = 190 - 16$ or $y = 6$. Substitute this into equation 2: $5x = -8 - 12$, so $x = -4$.

63. The point (a, b) that is nearest to (2,8) must satisfy two equations:

$$3a - 8b = 4 \quad \text{and} \quad \frac{b-8}{a-2} = -\frac{8}{3}.$$

The second equation reflects the fact that the line from (a, b) to $(2, 8)$ must be perpendicular to the line $3x - 8y = 4$. Therefore, $3a - 8b = 4$ and $3b - 24 = -8a + 16$ or $8a + 3b = 40$. Add 3 times the first equation to 8 times the second to obtain $73a = 332$ or $a = 332/73$. Substitute this into either of the two equations to see that $b = 88/73$.

65. If $A = 0$, then the first line is horizontal and the second one is vertical. If $B = 0$, then this is reversed. In either case, the lines are perpendicular. In the case that both A and B are not zero, then the slope of the first line is $-A/B$ and the slope of the second line is B/A. Once more, the lines are perpendicular.

67. The line $y = y_0$ meets the first line at $x_1 = (y_0 - 1)/2$ and it meets the second line at $x_2 = y_0 - 2$. The distance from one intersection point to the other is $d = |x_2 - x_1| = |y_0/2 - 3/2|$. The values of y_0 for which $d = 10^6$ are the solutions to the equation $|y_0 - 3| = 2 \cdot 10^6$. That is, $y_0 = 3 + 2 \cdot 10^6$ and $y_0 = 3 - 2 \cdot 10^6$.

69. The budget line has a Cartesian equation of the form $p_X \cdot x + p_Y \cdot y = C$. Its intercepts are $(C/p_X, 0)$ and $(0, C/p_Y)$ and its slope is $-p_X/p_Y$. The new budget line is parallel to the old one. If $C' > C$, then the new line lies above the old one. Otherwise, it lies below it.

71. Lines perpendicular to ℓ have an equation of the form $-Bx + Ay = C'$. Therefore, the line perpendicular to ℓ that passes through the point (x_0, y_0) has $C' = -Bx_0 + Ay_0$, and intersects ℓ at the point (x_1, y_1) where $x = x_1$ and $y = y_1$ are the solutions to the following two equations.

$$\begin{array}{rcrcl} Ax & + & By & = & C \\ -Bx & + & Ay & = & C' \end{array}$$

Multiply the first equation by A, the second by $-B$, and add to see that

$$\begin{aligned} (A^2 + B^2)x_1 = AC - BC' &= AC - B(-Bx_0 + Ay_0) \\ &= AC + B^2x_0 - ABy_0 \,. \end{aligned}$$

Similarly, $(A^2+B^2)y_1 = AC' + BC = BC + A^2y_0 - ABx_0$. Therefore, the distance from ℓ to (x_0, y_0) is the square root of the following expression.

$$\begin{aligned}(x_1-x_0)^2+(y_1-y_0)^2 &= \left(\frac{AC+B^2x_0-ABy_0-x_0(A^2+B^2)}{A^2+B^2}\right)^2 \\ &\quad + \left(\frac{BC+A^2y_0-ABx_0-y_0(A^2+B^2)}{A^2+B^2}\right)^2 \\ &= \frac{(Ax_0+By_0-C)^2}{A^2+B^2}\end{aligned}$$

The details of the simplification of line 1 to line 2 are left to the reader.

73. Assuming that $s \neq t$, the line ℓ' has slope $(s-t)/(t-s) = -1$. Therefore, ℓ' is perpendicular to the line $y = x$ which has slope 1. The intersection of $y = x$ and ℓ' is the point (a, a) such that $x = a$ and $y = a$ satisfies the equation $y = -(x-s)+t$. Thus $a = -(a-s)+t$ and $2a = s+t$ so $a = (s+t)/2$. The distance squared from (a, a) to (s, t) is

$$(a-s)^2 + (a-t)^2 = ((t-s)/2)^2 + ((s-t)/2)^2 = (s-t)^2/2\,.$$

This is the same as the distance squared from (a, a) to (t, s) (verify).

CALCULATOR/COMPUTER EXERCISES

75. The plot on the left is centered at the point (2,4). The right plot is centered at the same point, but zoomed in. The slope appears to be $m = 4$.

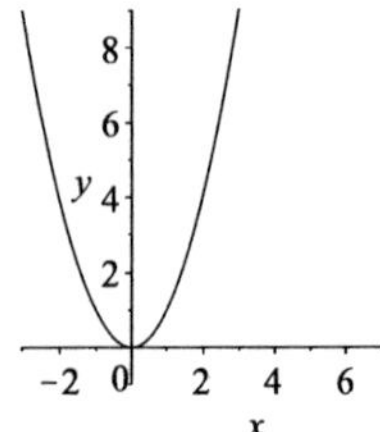

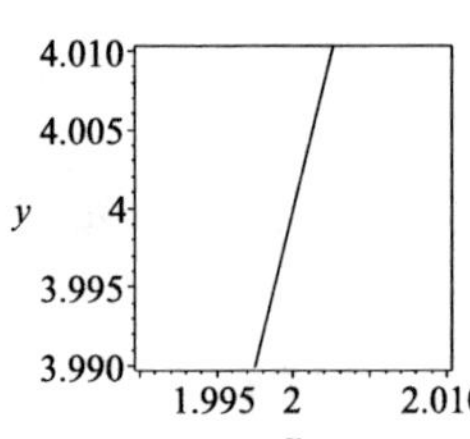

77. The plot on the left is centered at the point (1/2,4/5). The right plot is centered at the same point, but zoomed in. The slope appears to be $m = 1$.

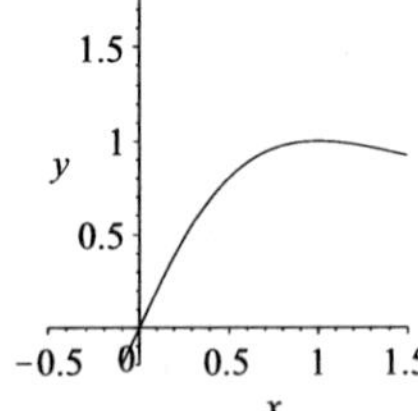

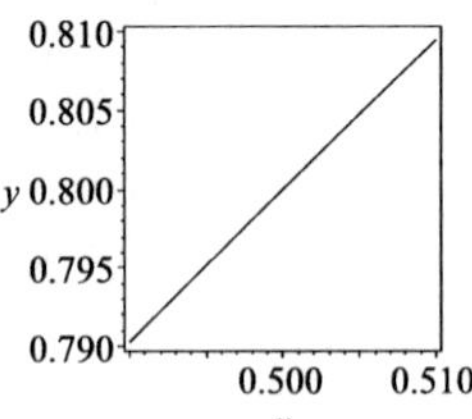

79. The plot on the left is centered at the point (1,1). The right plot is centered at the same point, but zoomed in. The slope appears to be $m = 1/2$.

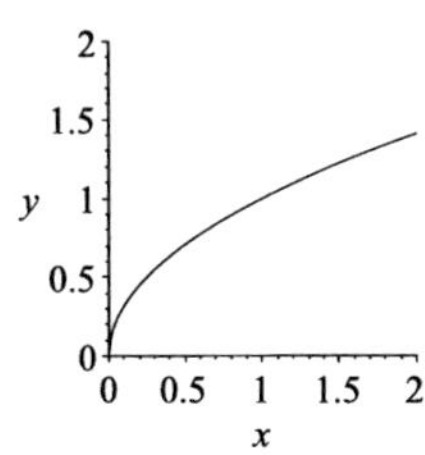

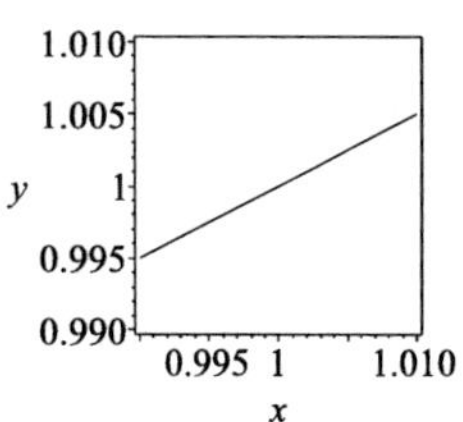

81. The points and the regression line are sketched on the right. The regression line equation is

$$y = 2.309451x + 65.155206\,.$$

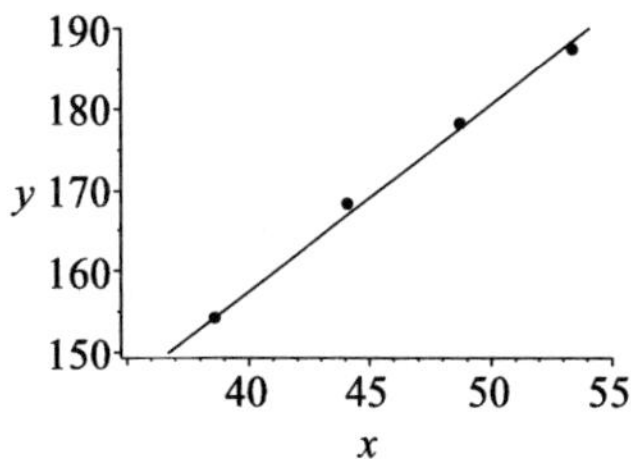

83. The equation of the top half of the circle centered at $(5/4, 0)$ of radius 1 is $y = \sqrt{1-(x-5/4)^2}$. It intersects the curve $y = \sqrt{x}$ when $x = 1-(x-5/4)^2$. That is, $x = 1-x^2+(5/2)x-25/16$, or $x^2-(3/2)x+9/16 = 0$. This simplifies to $(x-3/4)^2 = 0$ and the intersection point is $Q = (3/4, \sqrt{3}/2)$. The equation of the line joining P to Q is $y = \frac{\sqrt{3}/2}{3/4-5/4} \cdot (x-5/4)$ or

$$y = -\sqrt{3}\,(x-5/4)\,.$$

It is sketched on the right along with the half-circle and the line ℓ through Q that is perpendicular to line $\overline{PQ}$. The equation for the line ℓ is

$$y = (\sqrt{3}/3)(x-3/4) + \sqrt{3}/2\,.$$

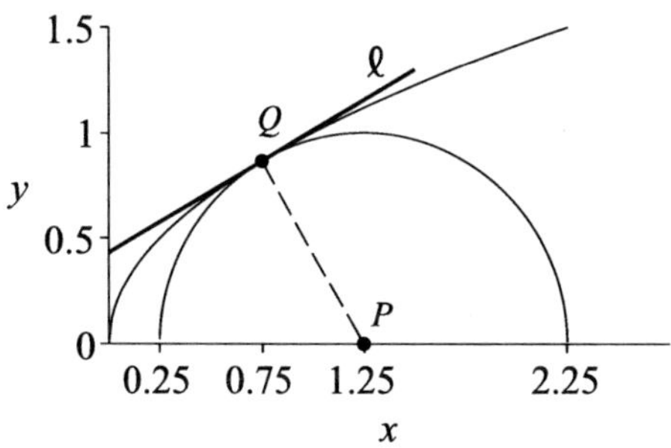

85. The line through P_0 with slope m is $y = m \cdot (x-4) + 16$. The sum of the squares of the errors associated with this line is

$$\begin{aligned} d_1^2 + d_2^2 &= (m(1-4)+16-2)^2 + (m(2-4)+16-6)^2 \\ &= 13m^2 - 124m + 296\,. \end{aligned}$$

The graph of $SQE = d_1^2 + d_2^2$ versus m is displayed on the right. Based upon this picture we estimate the slope of the regression line to be $m_0 = 4.77$. The equation of the regression line $\mathcal{L}$ would then be

$$y = 4.77(x-4) + 16\,.$$

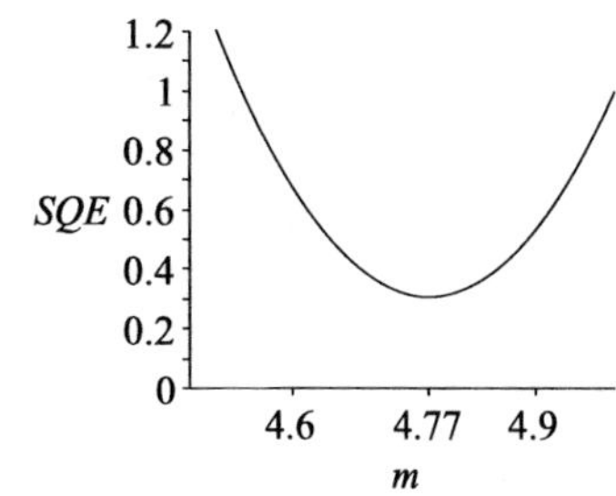

The graph of $ABE = d_1 + d_2$ versus m is displayed on the right. Based upon this picture we estimate the slope of the line minimizing ABE to be $m_1 = 4.66$. The equation of the line is

$$y = 4.66(x - 4) + 16\,.$$

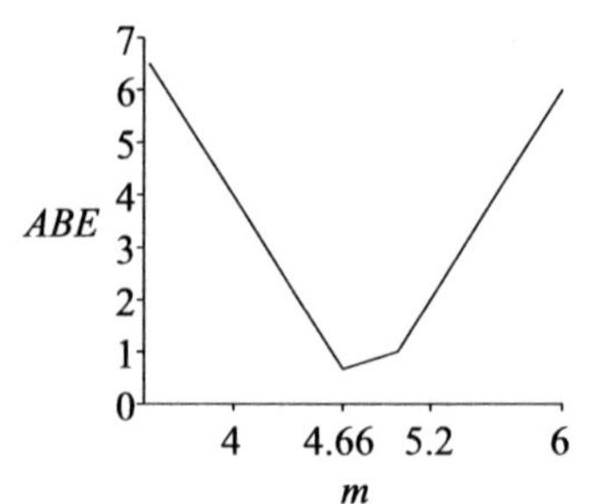

The picture on the right displays the three data points and the two lines. The one minimizing the sum of the squared errors is the solid line $\mathcal{L}$ and the one minimizing the sum of the absolute errors is the dashed line.

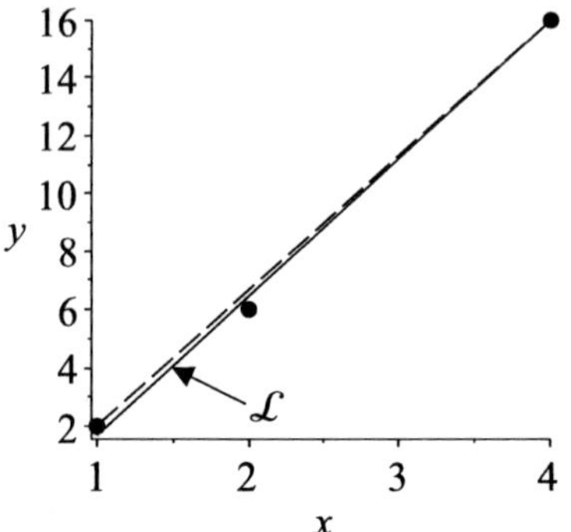

1.4 Functions and Their Graphs

Problems for Practice

1. Division by 0 is not allowed, so the domain is all $x \neq -1$.

3. It must be the case that $x^2 - 2 \geq 0$, so $x^2 \geq 2$ or $|x| \geq \sqrt{2}$. The domain is $(-\infty, -\sqrt{2}] \cup [\sqrt{2}, \infty)$.

5. To avoid division by 0, $x^2 \neq 1$. The domain: $\{x \in \mathbb{R} : x \neq 1 \text{ and } x \neq -1\}$.

7. The quadratic $x^2 - 4x + 5$ has no real roots and it evaluates to a positive number for all x. Therefore, the domain is $\mathbb{R}$.

9. The graph appears on the right.

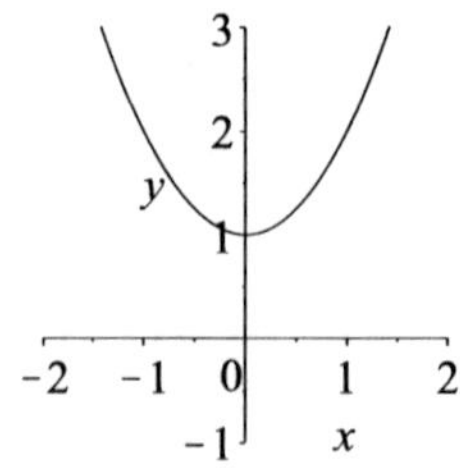

11. The graph appears on the right.

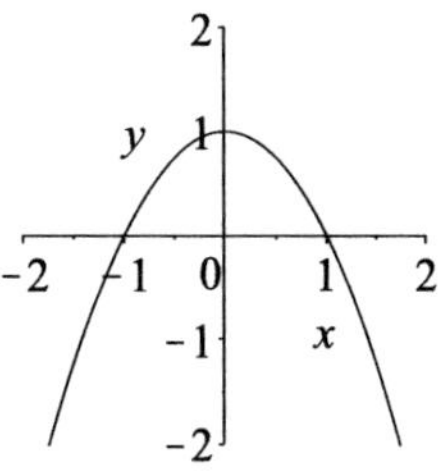

13. The graph appears on the right.

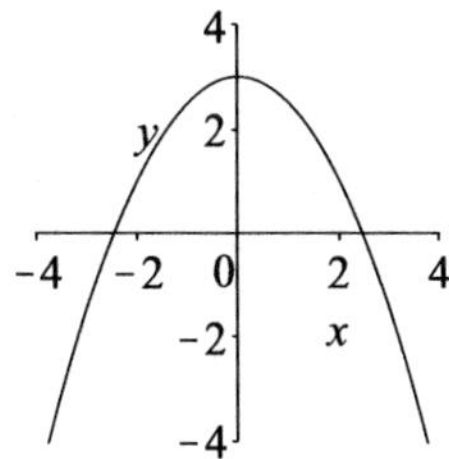

15. The graph appears on the right. The points are plotted for the x values $-2, -1, -0.7, 0.7, 1, 2$.

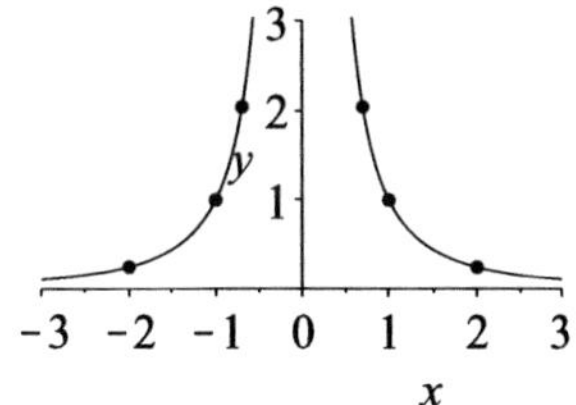

17. The graph appears on the right. The points are plotted for the x values $-2, -1, 0, 1, 2$.

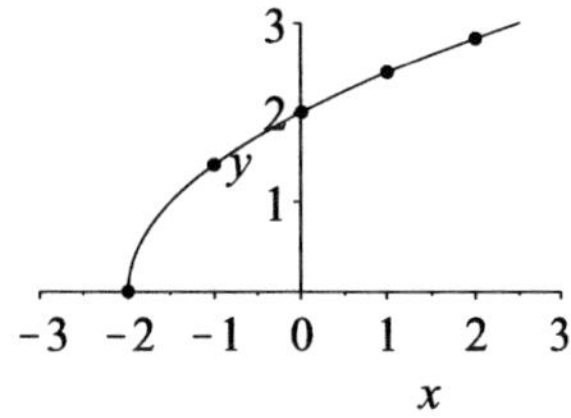

19. The graph appears on the right. The points are plotted for the x values $4.1, 4.5, 5, 6, 7, 8$.

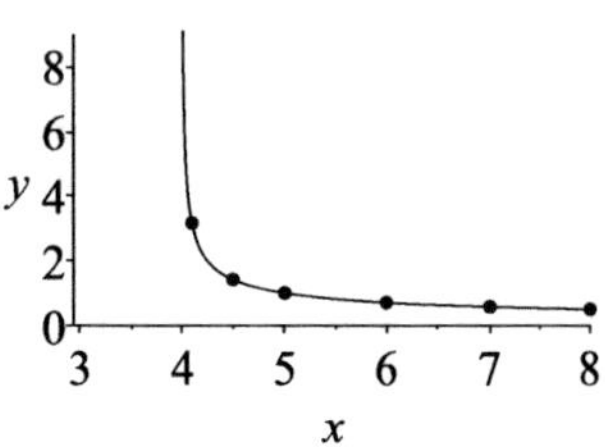

21. The graph appears on the right. The points are plotted for the x values $-0.95, -0.85, -0.5, 0, 1, 2$.

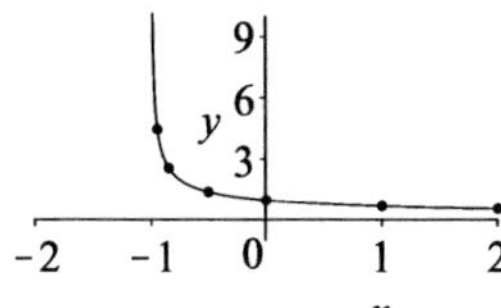

23. The graph appears on the right. The points are plotted for the x values $-1.25, -1, 0, 0.5, 1, 1.5, 2$.

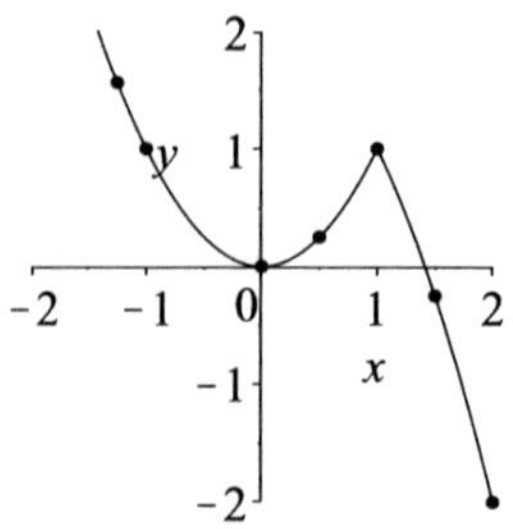

Further Theory and Practice

25. Assuming $s(\alpha) = k\alpha$, since $s(360) = 2\pi r$, we conclude that $k \cdot 360 = 2\pi r$ so $k = \pi r/180$. Consequently, the formula for the arc length function s when the central angle is measured in degrees is $s(\alpha) = \pi r\alpha/180$.

27. Assuming $A(r, \alpha) = kr\alpha$, since $A(r, 360) = \pi r^2$, we may conclude that $k \cdot r \cdot 360 = \pi r^2$ so $k = \pi r/360$. Consequently, the formula for the area function A when the central angle is measured in degrees is $A(r, \alpha) = \pi r^2\alpha/360$.

29. Observe that $x^2 + Ax + B = (x - r)(x - s) = x^2 - (r + s)x + rs$. Since this it true for all x, it follows that $A = -(r + s)$ and $B = rs$. That is, $A(r, s) = -(r + s)$ and $B(r, s) = rs$.

31. From $x = 0$ to $x = 1$, $f(x) = 2x + 1$. From $x = 1$ to $x = 3$, $f(x) = -(3/2)(x - 3)$. From $x = 3$ to $x = 4$, $f(x) = x - 3$. From $x = 4$ to $x = 5$, $f(x) = 1$. This can be expressed in piecewise form as follows.

$$f(x) = \begin{cases} 2x + 1 & \text{if } 0 \le x < 1 \\ -3(x - 3)/2 & \text{if } 1 \le x < 3 \\ x - 3 & \text{if } 3 \le x < 4 \\ 1 & \text{if } 4 \le x \le 5 \end{cases}$$

33. Examination of the tax formulas produces the following piecewise formula for the slope function.

$$m(x) = \begin{cases} 0.10 & \text{if } 0 < x < 8025 \\ 0.15 & \text{if } 8025 < x < 32550 \\ 0.25 & \text{if } 32550 < x < 50000 \end{cases}$$

Since the areas accumulate by multiplying the height by the base, for $x = 0$ to $x = 8025$ the area is $A(x) = 0.10x$. The area at $x = 8025$ is 802.5. From $x = 8025$ to $x = 32550$ the area is $A(x) = 802.5 + 0.15(x - 8025)$ and the area at $x = 32550$ is $A(32550) = 4481.25$. From $x = 32550$ to $x = 50000$ the area function is $A(x) = 4481.25 + 0.25(x - 32550)$. In multicase form

$$A(x) = \begin{cases} 0.10x & \text{if } 0 < x \le 8025 \\ 802.5 + 0.15(x - 8025) & \text{if } 8025 < x \le 32550 \\ 4481.25 + 0.25(x - 32550 & \text{if } 32550 < x \le 50000 \end{cases}$$

Observe that $A(x) = T(x)$.

35. The first few terms are $2, 4, 8, 16, \cdots$. This can be generated by defining $f_1 = 2$ and $f_{n+1} = 2 \cdot f_n$.

37. The first few terms are $1, 2, 6, 24, \cdots$. This can be generated by defining $f_1 = 1$ and $f_{n+1} = (n+1) \cdot f_n$.

39. The first few terms are $1, 3, 6, 10, \cdots$. This can be generated by defining $f_1 = 1$ and $f_{n+1} = f_n + n$. (See Exercise 36.)

41. The function Int can be defined in terms of the floor function as follows.

$$\text{Int}(x) = \begin{cases} \lfloor x \rfloor & \text{if} \quad x \geq 0 \text{ or } x \in \mathbb{Z} \\ \lfloor x+1 \rfloor & \text{if} \quad x < 0 \text{ and } x \notin \mathbb{Z} \end{cases}$$

The graph of the floor function is on the left, the graph of Int is on the right.

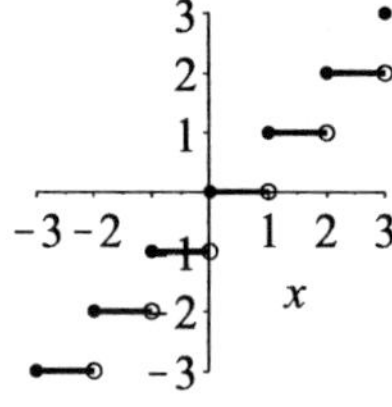

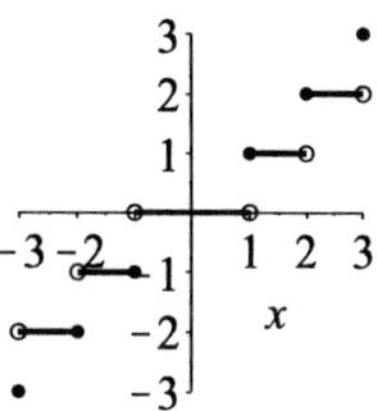

43. A monthly payment of m dollars for n years is a total payment of $m \cdot (12n)$ dollars. If the loan is P dollars, then the interest for n years is $I(P, m, n) = 12mn - P$.

45. For c_n,

$$c_0 = 1, c_1 = c_0c_0 = 1, c_2 = c_0c_1 + c_1c_0 = 2, c_3 = c_0c_2 + c_1c_1 + c_2c_0 = 5,$$
$$c_4 = c_0c_3 + c_1c_2 + c_2c_1 + c_3c_0 = 14, c_5 = c_0c_4 + c_1c_3 + c_2c_2 + c_3c_1 + c_4c_0 = 42,$$
$$c_6 = 2(c_0c_5 + c_1c_4 + c_2c_3) = 132, c_7 = 2(c_0c_6 + c_1c_5 + c_2c_4) + c_3c_3 = 429,$$
$$c_8 = 2(c_0c_7 + c_1c_6 + c_2c_5 + c_3c_4) = 1430$$

The calculations for C_n are easier, and yield the same values. (Verify.)

47. $R(x) = 0 \cdot (x \leq 0) + x \cdot (0 < x \leq 1) + (x > 1)$. See its graph on the right.

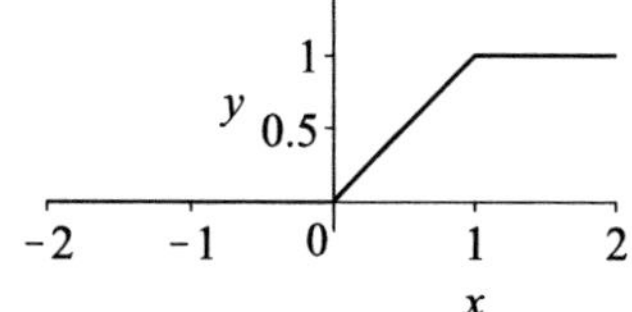

49. $b(x) = 0 \cdot (x \leq 0) + x \cdot (0 < x \leq 1) + (2-x) \cdot (2 < x \leq 2) + 0 \cdot (x > 2)$. See its graph on the right.

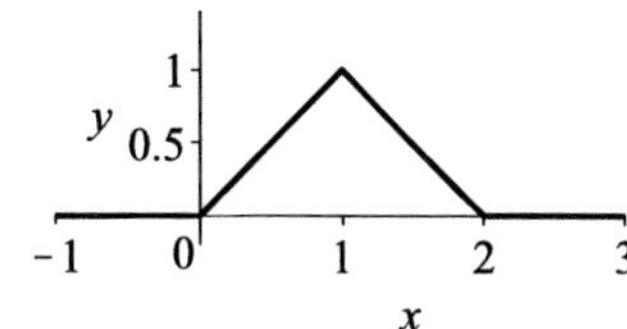

CALCULATOR/COMPUTER EXERCISES

51. The graph of the probability curve $y = P(x)$ is displayed on the right. It indicates that the test should not be used if the disease occurs only rarely in the population, say under 1%. Since $P(0.01) = 0.4975$, if the population is large, then about half of the tests would be false positive.

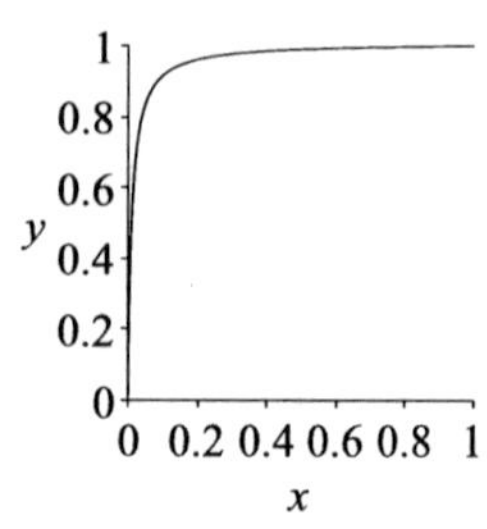

53. (a) The graph of f on the right indicates that it attains a maximum value at or near $x_0 = 4$.

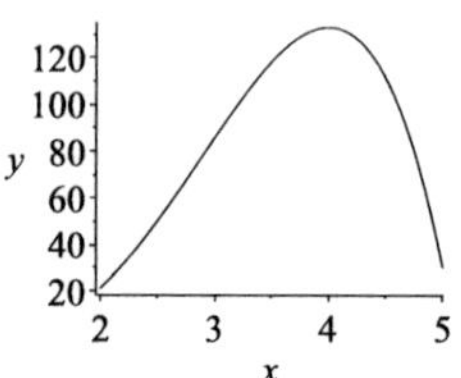

(b) The table below contains the values of $F(4, h)$ and $F(3.99, h)$

h	10^{-3}	10^{-4}	10^{-5}
$F(4, h)$	-7.2×10^{-5}	-7.2×10^{-7}	-7.2×10^{-9}
$F(3.99, h)$	0.00136	0.000143	0.0000141

(c) The graph of $h \mapsto F(3.99, h)$, displayed on the right, indicates that $F(3.99, h) \approx mx$ for $m \approx 1.4$.

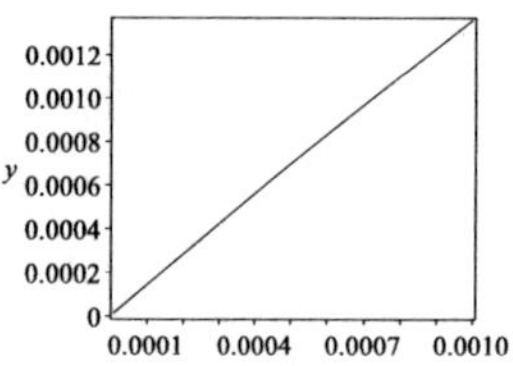

(d) The graph of $h \mapsto F(4, h)$, displayed on the right, indicates that $F(4, h) \approx Ax^2$ for $A \approx -72$.

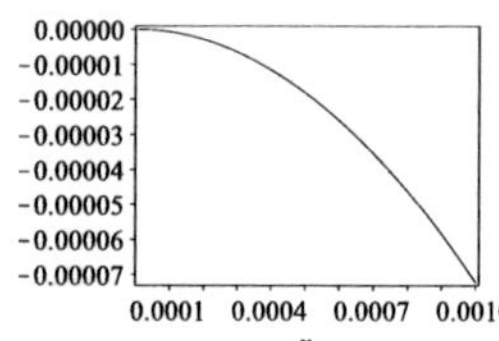

55. (a) Observe that $(x_n - \sqrt{c})^2 = x_n^2 + c - 2x_n\sqrt{c}$ implying that

$$\frac{(x_n - \sqrt{c})^2}{2x_n} = \frac{x_n^2 + c}{2x_n} - \sqrt{c} = x_{n+1} - \sqrt{c}.$$

Since the left side is positive, so it the right side, and $|x_{n+1} - \sqrt{c}| = (x_n - \sqrt{c})^2/(2x_n)$.

(b) x_n approximates $\sqrt{c}$ to k decimal places when $|x_n - \sqrt{c}| < 5 \cdot 10^{-(k+1)}$. Assuming this is the case, and using the identity in part a along with the fact that $x_n > 1$ (why?),

$$|x_{n+1} - \sqrt{c}| = |x_n - \sqrt{c}|^2/(2|x_n|) < \tfrac{1}{2} \cdot 5^2 \cdot 10^{-2(k+1)} < 5 \cdot 10^{-(2k+1)}.$$

(c) Using the recursion formula, $c = 3.75$, and $x_0 = 3/2$, yields the sequence

$$1.5, 2.0, 1.9375, 1.936491936, 1.936491673, 1.936491673, 1.936491673, \cdots$$

It stabilized quickly to 1.936491673, which is correct to nine decimal places.

57. The following table contains the requested data.

n	1	2	3	4	5	6	7	8
q_n	$1/\sqrt{2}$	0.923875	0.980785	0.995180	0.998795	0.999700	0.999925	0.999980
Q_n	$1/\sqrt{2}$	0.653280	0.640728	0.637640	0.636872	0.636680	0.636638	0.636625
p_n	$2\sqrt{2}$	3.06148	3.12145	3.13657	3.14034	3.14130	3.14152	3.14159

1.5 Combining Functions

Problems for Practice

1. $(F + G)(x) = F(x) + G(x) = x^2 + 5 + (x + 1)/(x - 1)$
$= (x^3 - x^2 + 6x - 4)/(x - 1)$

3. $(G \circ H)(x) = G(H(x)) = G(2x - 5) = (2x - 5 + 1)/(2x - 5 - 1)$
$= (x - 2)/(x - 3)$

5. $(H \cdot F - H \circ F)(x) = H(x) \cdot F(x) - H(F(x)) = (2x - 5)(x^2 - 5) - H(x^2 + 5)$
$= (2x - 5)(x^2 - 5) - (2(x^2 - 5) - 5) = 2x^3 - 7x^2 + 10x - 30$

7. $(F \circ G \circ H)(x) = F(G(H(x))) = F(G(2x-5)) = F((2x-5+1)/(2x-5-1))$
$= (x - 2)^2/(x - 3)^2 + 5 = (6x^2 - 34x + 49)/(x + 3)^2$

9. $(G \circ (1/G))(x) = G(1/G(x)) = G((x - 1)/(x + 1))$
$= ((x - 1)/(x + 1) + 1)/((x - 1)/(x + 1) - 1) = -x$

11. $h(x) = (g \circ f)(x)$ where $f(x) = x - 2$ and $g(x) = x^2$.

13. $h(x) = (g \circ f)(x)$ where $f(x) = x^3 + 3x$ and $g(x) = x^4$.

15. Since $h(x) = 3(x + 1)^2 + 1$, $h(x) = (g \circ f)(x)$ where $g(x) = 3x^2 + 1$.

17. Since $h(x) = (x^2 + 1)/((x^2 + 1)^2 + 2)$, $h(x) = (g \circ f)(x)$ where $g(x) = x/(x^2 + 3)$.

19. $(f \circ g)(1/8) = f(g(1/8)) = f(2) = 3$

21. $f^2(11) \cdot g^3(54) = f(11)^2 \cdot g(54)^3 = 27 \cdot (1/54) = 1/2$

23. $x^2 + 4x - 5 = (x + 5)(x - 1)$

25. Both $x = 2$ and $x = -2$ are roots, so $x^2 - 4$ is a factor. Long division yields the factor x^2+2x+2 which is irreducible (complex roots). Therefore, $x^4 + 2x^3 - 2x^2 - 8x - 8 = (x - 2)(x + 2)(x^2 + 2x + 2)$.

27. As s increases, the values $f(s)$ increase steadily from 1 to as large a number as desired. Therefore, f is both one-to-one and onto. It is invertible. Since $t = (f^{-1}(t))^2 + 1$, $t - 1 = (f^{-1}(t))^2$ and, because $f^{-1}(t)$ is not negative, $f^{-1}(t) = \sqrt{t-1}$.

29. As s increases, the values $f(s)$ increase steadily from 0 to 2. Therefore, f is both one-to-one and onto. It is invertible. Since $t = s^2 + s$, $s^2 + s - t = 0$ and $s = (-1 \pm \sqrt{1+4t^2})/2$. Since $s \geq 0$, the plus sign yields the formula for the values of the inverse function: $f^{-1}(t) = (-1 + \sqrt{1+4t^2})/2$.

31. As s increases from -2 to 5, the values $f(s)$ increase steadily from -35 to 98. Therefore, f is both one-to-one and onto. It is invertible. Since $t = (f^{-1}(t))^3 - 27$, $t + 27 = (f^{-1}(t))^3$ and $f^{-1}(t) = (t + 27)^{1/3}$.

33. Examination of the graph of f will show that as s increases from 4, its values $f(s)$ decrease steadily from 16/15 towards 1. Therefore, f is both one-to-one and onto. It is invertible. Since $t = s^2/(s^2-1)$, $t(s^2-1) = s^2$ so $(t-1)s^2 = t$ and $s = \sqrt{t/(t-1)}$. (Take the positive square root because s is not negative.) Therefore, $f^{-1}(t) = \sqrt{t/(t-1)}$.

35. As s increases from 1 to 6, its values $f(s)$ increase steadily from 2 to 3. Therefore, f is both one-to-one and onto. It is invertible. Since $t = \sqrt{f^{-1}(t) + 3}$, $t^2 = f^{-1}(t) + 3$ so $f^{-1}(t) = t^2 - 3$.

37. As s increases from 1 its values $f(s)$ decrease steadily from 1/2 towards 0. f is one-to-one, but it is not onto because 1 is in its range but it is not in its image. f does not have an inverse.

39. All six graphs represent functions but only the graphs in Figure 18 a, e, and f represent functions that are invertible. The functions for Figures b and d are onto, but not one-to-one, and the function for Figure c is one-to-one, but not onto. The graphs of the functions that are the inverses to the functions represented by Figures a, e, and f are displayed below.

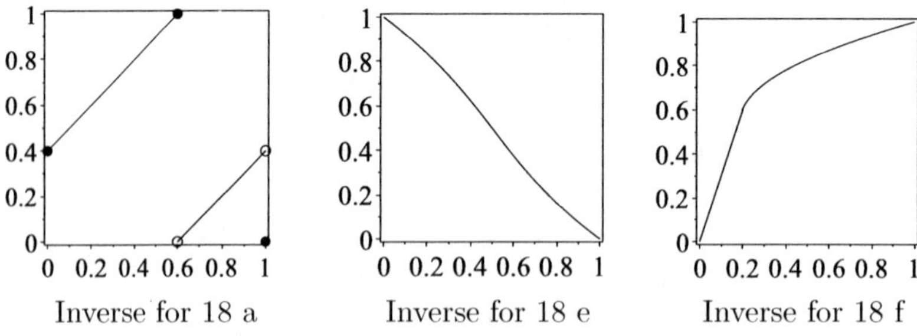

Inverse for 18 a   Inverse for 18 e   Inverse for 18 f

41. $g(x) = f(x - 2)$. This implies that if the the graph of f is translated 2 units to the right, then it will be the graph of g.

43. $g(x) = f(x-1)$. This implies that if the the graph of f is translated 1 unit to the right, then it will be the graph of g.

45. $g(x) = f(x+2) - 12$. This implies that if the graph of f is translated 2 unit to the left and 12 units down, then it will be the graph of g.

47. $g(x) = f(x+3) - 2$. This implies that if the graph of f is translated 3 unit to the left and 2 units down, then it will be the graph of g.

49. Elimitate t to see that the curve is the line with Cartesian equation $y = 3(x-1) - 4$ or $y = 3x - 7$.

51. The curve is a parbola. Eliminating t it has the Cartesian equation $x = 12(y/2)^2 + 1$ or $x = 3y^2 + 1$. This shows that its axis of symmetry is the x-axis, its vertex is (1,0), and it opens to the right.

Further Theory and Practice

53. The degree of the product $p \cdot q$ is the sum of the degrees: $\deg(p \cdot q) = \deg(p) + \deg(q)$. The degree of the composition is the product of the degrees: $\deg(p \circ q) = \deg(p) \cdot \deg(q)$. Because of the previous statement, $\deg(p \circ q) = \deg(q \circ p)$. The degree of $p \pm q$ will never be greater than $\deg(p)$ or $\deg(q)$. It can be less, and will be if, and only if, the degrees are the same that the terms with highest degree cancel one another.

55. $\deg(f) = \deg(p)^2 = n^2$, unless $p(x) = -x + a$, in which case $\deg(f) = 0$.

    If $p(x) = 0$, then $f(x) = p(x + p(x)) = p(x) = 0$ and every root of p is also a root of f. Write p as the product of the linear terms corresponding to each of its roots (real and complex, repeating if necessary): $p(x) = a(x - r_1)(x - r_2) \cdots (x - r_n)$. Then, since each root of p is a root of f, $f(x) = h(x) \cdot (x - r_1)(x - r_2) \cdots (x - r_n) = p(x) \cdot q(x)$, where $q(x) = h(x)/a$. For example, if $p(x) = x^2 - 3x - 4$, then

    $$\begin{aligned} f(x) = p(x + p(x)) &= (x + x^2 - 3x - 4)^2 - 3(x + x^2 - 3x - 4) - 4 \\ &= x^4 - 4x^3 - 7x^2 + 22x + 24 = (x^2 - 3x - 4)(x^2 - x - 6) \end{aligned}$$

    where the factorization of f can be obtained by long division (verify).

57. Let $f(x) = ax + b$ and $g(x) = cx + d$ be affine functions. Then $(g \circ f)(x) = g(ax + b) = c(ax + b) + d = acx + bc + d$, which is also affine.

    The affine function f is invertible if, and only if, $a \neq 0$. Assuming this is the case, $y = af^{-1}(y) + b$ and $f^{-1}(y) = y/a - b/a$.

59. This is a consequence of the observations that if f and g are onto (one-to-one), then $g \circ f$ is also onto (one-to-one). Verification, which is a straightforward application of the definitions, is left to the reader.

61. Since $h(x) = (x+1)^2 + 2$, $h(x) = (g \circ f)(x)$ where $g(x) = x^2 + 2$.

63. Since $h(x) = 2(x^2 - 9) + 18$, $h(x) = (g \circ f)(x)$ where $g(x) = 2x + 18$.

65. Since $h(x) = (x-4)^2 + 2$, $h(x) = (g \circ f)(x)$ where $f(x) = x - 4$.

67. Since $h(x) = ((x^2-1)^{1/3})^3 + 1$, $h(x) = (g \circ f)(x)$ where $f(x) = (x^2-1)^{1/3}$.

69. $f(x) = (x-3)^2$

71. $f(x) = (-x^3+1)/(x^2+1)$

73. If $f(x) = x^p$ and $f \circ f = f \cdot f$, then $(x^p)^p = x^p \cdot x^p$ for all x. Consequently, $x^{(p^2)} = x^{2p}$ and $x^{(p^2-2p)} = 1$ for all x. That is, either $p = 0$ or $p = 2$.

75. Let $\phi = p$. Then $\phi(f(x)) = \phi(x^5) = p(x^5) = \pi = p(x)$ for all x. That is, $\phi \circ f = p$.

    Let $\psi(x) = \pi^{1/5}$ for all x. Then $f(\psi(x)) = f(\pi^{1/5}) = (\pi^{1/5})^5 = \pi = p(x)$. That is, $f \circ \psi = p$.

    Let $\mu(x) = x^{1/5} + \pi$ for all x. Then $\mu(f(x)) = \mu(x^5) = (x^5)^{1/5} + \pi = x + \pi = I(x) + p(x) = (I+p)(x)$. That is, $\mu \circ f = I + p$.

    Let $\lambda(x) = (x+\pi)^{1/5}$ for all x. Then $f(\lambda(x)) = f((x+\pi)^{1/5}) = x + \pi = (I+p)(x)$. That is, $f \circ \lambda = I + p$.

77. The graphs of f and $f \circ f$ are displayed below. The formula for $f \circ f$ can be discerned from the plot on the right.

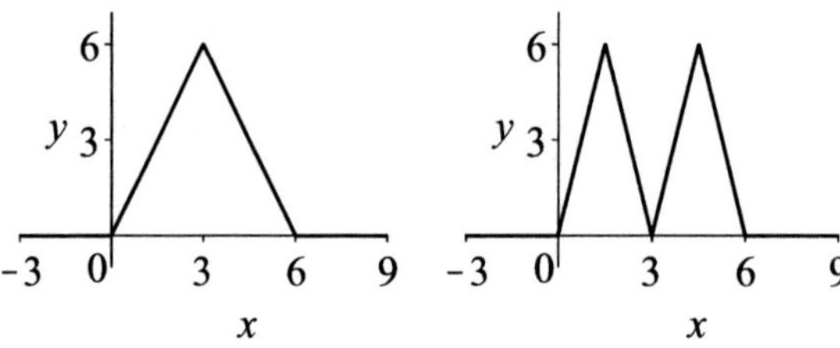

79. Let $G(x) = x^p$ and $F(x) = px$. Then $\log_{10}(x^p) = (\log_{10} \circ G)(x)$ and $p \cdot \log_{10}(x) = (F \circ \log_{10})(x)$.

    Let $H(x) = \log_{10}(p) + x$ for all x. Then $(\log_{10} \circ F)(x) = \log_{10}(px)$ and $(H \circ \log_{10})(x) = H(\log_{10}(x)) = \log_{10}(p) + \log_{10}(x)$.

81. Eliminate t and the curve has the Cartesian equation $x = y^{2/3}$. Note that y can be chosen arbitrarily and $x \geq 0$ for all t. See the picture, which displays the portion of the curve corresponding to $-2 \leq t \leq 2$.

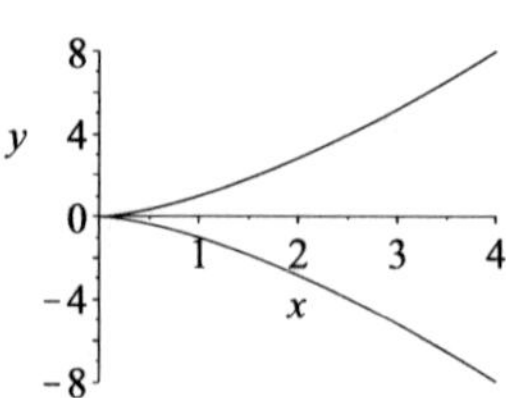

83. Since $\log_{10}(t) = \log_{10}(2) \cdot x$, $t = 10^{\log_{10}(2) \cdot x} = 2^x$. Therefore, the curve is the graph of the Cartesian equation $y = 3 \cdot 2^x$. See the picture, which displays the points corresponding to $0.1 \le t \le 3$.

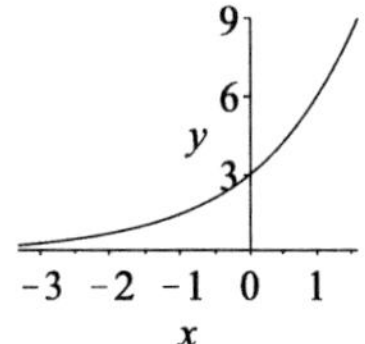

85. f is even when $f(-x) = f(x)$ for all x. Thus f is even if, and only if, $f \circ S = f$.

g is odd when $g(-x) = -g(x)$ for all x. Thus g is odd if, and only if, $g \circ S = S \circ g$.

CALCULATOR/COMPUTER EXERCISES

87. Let $f(x) = \sqrt{x^2 + 2x + 2}$ and plot $y = f(x)$ and $y = f(x - 2) + 1$. See the picture on the right where the translated graph is the dashed curve.

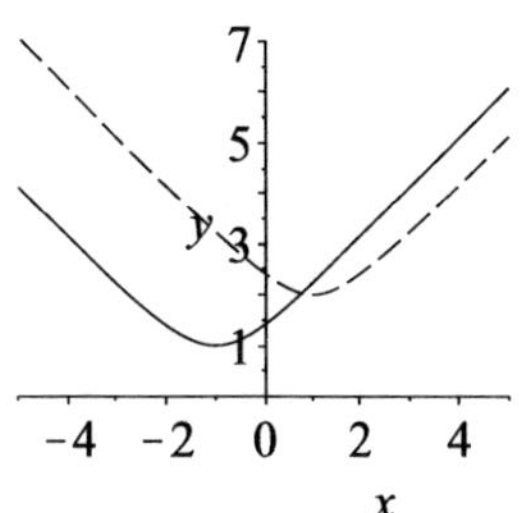

89. Let $f(x) = (x^3+1)/(x^2+1)$ and plot $y = f(x)$ and $y = f(-x)$. See the picture on the right where the reflected graph is the dashed curve.

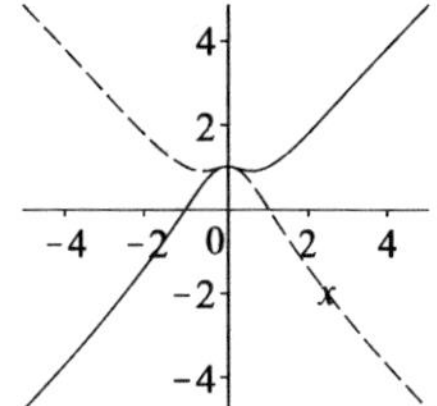

91. The curve and line $y = 18x/25$ are plotted on the right. The line intersects the curve in two points. Using *Maple*'s *fsolve* procedure the t values for the intersection are $t = -0.63068$ and $t = -0.57399$. See the close-up below.

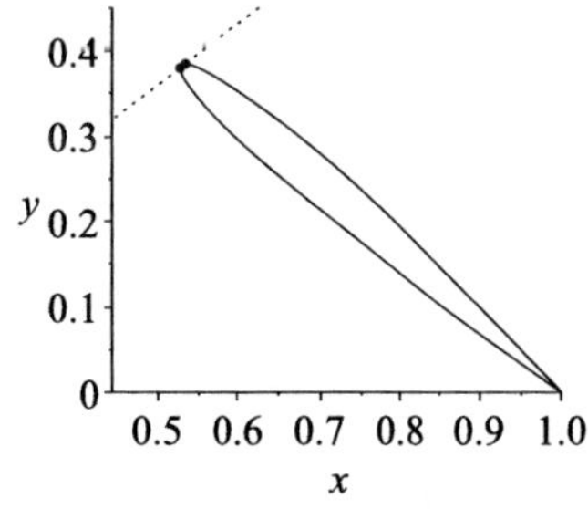

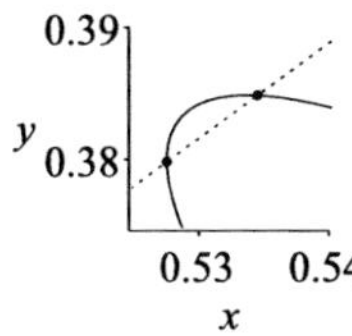

93. The graph of f appears below on the left. Its inverse is drawn on the right. It was obtained by plotting the parametrized curve $y = t$, $x = f(t)$.

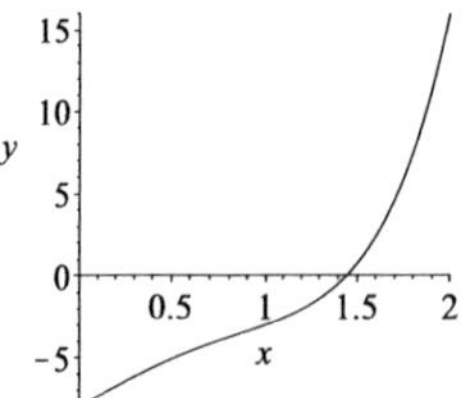

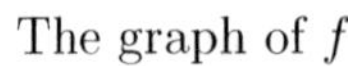
The graph of f.

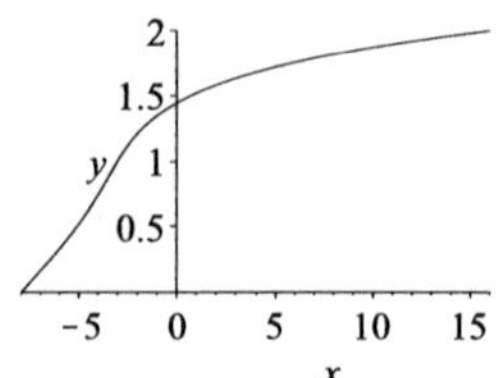

The graph of f^{-1}.

1.6 Trigonometry

Problems for Practice

1. Consult the standard 30-60-90 triangle to see that $\sin(\pi/6) = 1/2$ and $\cos(\pi/6) = \sqrt{3}/2$. Values for the other four functions can be given in terms of these: $\tan(\pi/6) = \sqrt{3}/3$, $\cot(\pi/6) = 3/\sqrt{3} = \sqrt{3}$, $\csc(\pi/6) = 2$, and $\sec(\pi/6) = 2/\sqrt{3} = 2\sqrt{3}/3$.

3. The angle $2\pi/3$ puts the radial line into the second quadrant. With a standard 30-60-90 triangle in the correct position we have $\sin(2\pi/3) = \sqrt{3}/2$ and $\cos(2\pi/3) = -1/2$. Values for the other four functions can be given in terms of these: $\tan(2\pi/3) = -\sqrt{3}$, $\cot(2\pi/3) = -\sqrt{3}/3$, $\csc(2\pi/3) = 2\sqrt{3}/3$, and $\sec(2\pi/3) = -2$.

5. Using a 30-60-90 triangle, $\sin(\pi/3)\sin(\pi/6) = \sqrt{3}/2 \cdot 1/2 = \sqrt{3}/4$.

7. $\cos(\pi/6) + \cos(\pi/3) = \sqrt{3}/2 + 1/2 = (\sqrt{3} + 1)/2$

9. $\tan(\pi/3)/\tan(\pi/6) = \sqrt{3}/(1/\sqrt{3}\,) = 3$

11. $\sin(\pi \cdot \sin(\pi/6)) = \sin(\pi/2) = 1$

13. $\sin(19\pi/2)^{\cos(33\pi)} = \sin(3\pi/2)^{\cos(\pi)} = (-1)^{-1} = -1$

15. If $\sin(\theta) = 1/3$, then $\cos(\theta) = \sqrt{1 - \sin^2(\theta)} = \sqrt{8/9} = 2\sqrt{2}/3$ (the cosine function is positive in the first quadrant).

17. If $\cos(\theta) = 4/5$, then $\sin(\theta) = \sqrt{1 - \cos^2(\theta)} = 3/5$ (the sine function is positive in the first quadrant) and $\sin(2\theta) = 2\sin(\theta)\cos(\theta) = 2 \cdot (3/5) \cdot (4/5) = 24/25$.

19. If $\sin(\theta) = 5/13$, then $\cos(\theta) = \sqrt{1 - \sin^2(\theta)} = 12/13$ (the sine function is positive in the first quadrant) and $\cos^2(\theta/2) = (1 + \cos(\theta))/2 = 25/26$. Therefore, $\cos(\theta/2) = \sqrt{25/26} = 5\sqrt{26}/26$.

21. All six trigonometric functions are positive in the first quadrant.

23. In the third quadrant the tangent and cotangent functions are positive.

25. Since $\sin(2t)$ has period π, this is four complete sine waves.

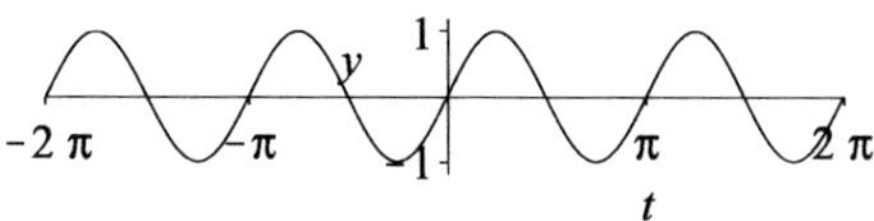

27. Since $\sin(t - \pi/6)$ has period 2π, this is one complete sine wave shifted to the right $\pi/6$ units.

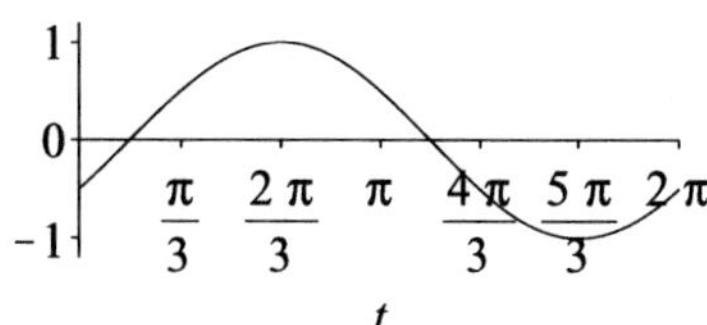

29. The parametrization $\phi_1(t) = \cos(t)$, $\phi_2(t) = \sin(t)$ does the job.

31. The parametrization $\phi_1(t) = \sin(t)$, $\phi_2(t) = \cos(t)$ will handle this. Note that it starts at the right place and, initially, x increases and y decreases.

Further Theory and Practice

33. There is a constant k such that $s(r, \theta) = k \cdot r \cdot \theta$. k can be found by substituting $r = 1$ and $\theta = 2\pi$ to obtain $2\pi = k \cdot 1 \cdot 2\pi$. Therefore, $k = 1$, and $s(r, \theta) = r \cdot \theta$.

35. The second equation, $\sin(2\theta) = 2\sin(\theta)\cos(\theta)$, is an identity, true for all values of θ. The first equation, $\sin(2\theta) = \sin(\theta)\cos(\theta)$, is true only for some values of θ.

    Using the identity, the second equation is equivalent to $\sin(2\theta) = \sin(2\theta)/2$ which is true if, and only if, $\sin(2\theta) = 0$. Thus the second equation is true whenever $2\theta = n\pi$, where n is an integer. That is, $\theta = n\pi/2$ where n can be any integer.

37. Convert to sines and cosines, apply the sum identities, and then divide the top and bottom by $\cos(\theta)\cos(\phi)$.

$$\begin{aligned}\tan(\theta + \phi) &= \frac{\sin(\theta + \phi)}{\cos(\theta + \phi)} = \frac{\sin(\theta)\cos(\phi) + \cos(\theta)\sin(\phi)}{\cos(\theta)\cos(\phi) - \sin(\theta)\sin(\phi)} \\ &= \frac{\tan(\theta) + \tan(\phi)}{1 - \tan(\theta)\tan(\phi)}.\end{aligned}$$

39. Start from the right side, applying the sum identities for cosine. The details are left to the reader (the sine terms cancel).

$$\begin{aligned}\tfrac{1}{2}(\cos(\theta - \phi) + \cos(\theta + \phi)) &= \tfrac{1}{2}(\cos(\theta)\cos(\phi) + \cos(\theta)\cos(\phi)) \\ &= \cos(\theta)\cos(\phi)\end{aligned}$$

41. Start from the right side, applying the sum identity for cosine.

$$\cos\left(\tfrac{\pi}{2} - \theta\right) = \cos\left(\tfrac{\pi}{2}\right)\cos(\theta) + \sin\left(\tfrac{\pi}{2}\right)\sin(\theta) = \sin(\theta)$$

43. Start from the left side, applying the sum identity for cosine.

$$\cos(\theta+\pi)=\cos(\theta)\cos(\pi)-\sin(\theta)\sin(\pi)=-\cos(\theta)$$

45. Begin with the sine and cosine calculation.

$$\begin{aligned}\sin(7\pi/12)&=\sin(\pi/3)\cos(\pi/4)+\cos(\pi/3)\sin(\pi/4)=(\sqrt{3}/2+1/2)\sqrt{2}/2\\&=(\sqrt{6}+\sqrt{2})/4=\tfrac{1}{2}\sqrt{2+\sqrt{3}}\\\cos(7\pi/12)&=\cos(\pi/3)\cos(\pi/4)-\sin(\pi/3)\sin(\pi/4)=(1/2-\sqrt{3}/2)\sqrt{2}/2\\&=(\sqrt{2}-\sqrt{6})/4=-\tfrac{1}{2}\sqrt{2-\sqrt{3}}\end{aligned}$$

In each case, the final answer is obtained by squaring and then simplifying the penultimate expression. The sine and cosine values can be used to find the other four. The final tangent calculation is completed by multiplying the top and bottom by $\sqrt{2+\sqrt{3}}$.

$$\begin{aligned}\tan(7\pi/12)&=\frac{\sin(7\pi/12)}{\cos(7\pi/12)}=\frac{\frac{1}{2}\sqrt{2+\sqrt{3}}}{-\frac{1}{2}\sqrt{2-\sqrt{3}}}=-2-\sqrt{3}\\\cot(7\pi/12)&=\frac{1}{-2-\sqrt{3}}=\frac{-2+\sqrt{3}}{1}=-2+\sqrt{3}\\\csc(7\pi/12)&=\frac{1}{\frac{1}{2}\sqrt{2+\sqrt{3}}}=\frac{2\sqrt{2-\sqrt{3}}}{1}=2\sqrt{2-\sqrt{3}}\\\sec(7\pi/12)&=\frac{1}{-\frac{1}{2}\sqrt{2-\sqrt{3}}}=-\frac{2\sqrt{2+\sqrt{3}}}{1}=-2\sqrt{2+\sqrt{3}}\end{aligned}$$

47. Substitute $\frac{\theta+\phi}{2}$ for θ and $\frac{\theta-\phi}{2}$ for ϕ in the identity in Exercise 40.

49. Substitute $\frac{\theta+\phi}{2}$ for θ and $\frac{\theta-\phi}{2}$ for ϕ in the identity in Exercise 39.

51. Let $C=\sqrt{A^2+B^2}$ and observe that the point $(B/C, A/C)$ lies on the unit circle in the xy plane. Let ϕ be an angle determined by the line from the origin to this point so $B/C=\cos(\phi)$ and $A/C=\sin(\phi)$. Then

$$\begin{aligned}A\cos(\theta)+B\sin(\theta)&=C\cdot\left(\frac{A}{C}\cos(\theta)+\frac{B}{C}\sin(\theta)\right)\\&=C(\sin(\theta)\cos(\phi)+\cos(\theta)\sin(\phi))\\&=C\sin(\theta+\phi)\,.\end{aligned}$$

53. Let d be the distance from the θ vertex to the point where the vertical height line meets the horizontal line. Then $\cot(\theta)=d/h$ and $\cot(\phi)=$

$(\ell+d)/h$. Eliminating d, $\cot(\phi) = (\ell + h\cot(\theta))/h$ or $\cot(\phi) = \ell/h + \cos(\theta)$. Consequently, $\ell/h = \cot(\phi) - \cot(\theta)$ so

$$h = \frac{\ell}{\cot(\phi) - \cot(\theta)} = \frac{\ell}{|\cot(\theta) - \cot(\phi)|}.$$

55. This is a portion of a line. Note that $x/a + y/b = \cos^2(\theta) + \sin^2(\theta) = 1$. Observe that when $\theta = 0$, $(x, y) = (a, 0)$ and when $\theta = \pi/2$, $(x, y) = (0, b)$. The line segment joins the two intercepts.

57. Observe that $(x/a) \cdot (y/b) = 1$ so this parametrization traces out a portion of the curve $xy = ab$.

59. Since 2π is the smallest positive number such that $\sin(x + \sqrt{3} + 2\pi) = \sin(x + \sqrt{3})$ for all x, this function has period 2π.

61. The tangent function has period π. This can be shown using the sum identity:

$$\tan(x + \pi) = \frac{\tan(x) + \tan(\pi)}{1 - \tan(x)\tan(\pi)} = \tan(x)$$

for all x. Examine the graph of the tangent function to see that no smaller positive number has this property.

63. $\tan(2x)$ has period $\pi/2$ and $\sin(3x)$ has period $2\pi/3$. They will both repeat over any interval whose width is a common multiple of $\pi/2$ and $2\pi/3$. The smallest positive number that has this property is $6\pi/3 = 2\pi$.

65. The constant b should be the average of 4 and 51: $b = 27.5$, and the amplitude A will be the common difference between the extremes and the average: $A = 23.5$. We will measure time t in months so the value of ω must be chosen to yield a period of 12. This implies that $12\omega = 2\pi$ so $\omega = \pi/6$. At this point $T(t) = 27.5 + 23.5\sin(\pi t/6 + \phi)$ and it remains to choose a phase angle ϕ so that the maximum value of T is attained on July 1st. This is accomplished by choosing ϕ so that the sine part of the definition attains the value 1 at $t = 6$: $\sin(\pi + \phi) = 1$. So, for example, choose ϕ to solve the equation $\pi + \phi = \pi/2$. That is, $\phi = \pi/2 - \pi = -\pi/2$. See the picture on the right.

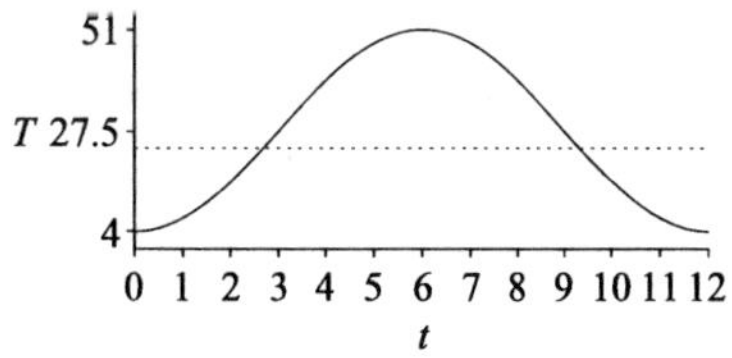

67. (a) The inscribed n-gon splits naturally into n congruent isosceles triangles, each one having a vertex angle $2\pi/n$ and equal sides of length r. Therefore, each triangle has base $2 \cdot r\sin(\pi/n)$. The sum of the n base lengths is $p(n, r)$: $p(n, r) = n \cdot 2r\sin(\pi/n)$.

    (b) The area of each inscribed isosceles triangle is $\frac{1}{2} \cdot (\text{base}) \cdot (\text{height}) = r\sin(\pi/n) \cdot r\cos(\pi/n) = r^2\sin(\pi/n)\cos(\pi/n) = \frac{1}{2}r^2\sin(2\pi/n)$. Since there are n of them, $A(n, r) = \frac{1}{2}r^2 n\sin(2\pi/n)$.

CALCULATOR/COMPUTER EXERCISES

69. The following table contains the initial data, and one more (4-decimal accuracy).

n	1	2	3	4	5	6	7	8
a_n	0.8415	0.9589	0.9816	0.9896	0.9933	0.9954	0.9966	0.9974
n	9	10	11	12	13	14	15	16
a_n	0.9979	0.9983	0.9986	0.9988	0.9990	0.9992	0.9993	0.9993

The next table contains the data for $n = 10^k$, (10-decimal accuracy).

n	10^1	10^2	10^3
a_n	0.9983341665	0.9999833334	0.9999998333
n	10^4	10^5	10^6
a_n	0.9999999983	1.000000000	1.000000000

The tabulated data suggests that the limiting value of a_n as n increases is $\ell = 1$. This is supported by the graph of $y = \sin(x)/x$ displayed on the right.

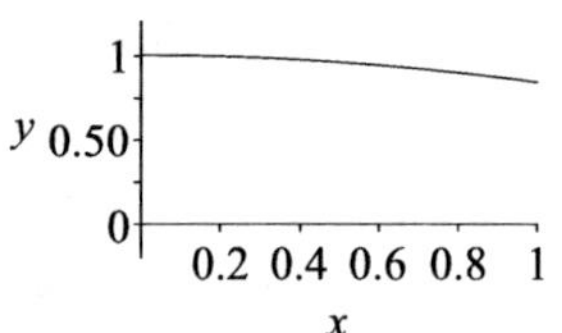

71. The graphs appear below. The zoomed version is on the right.

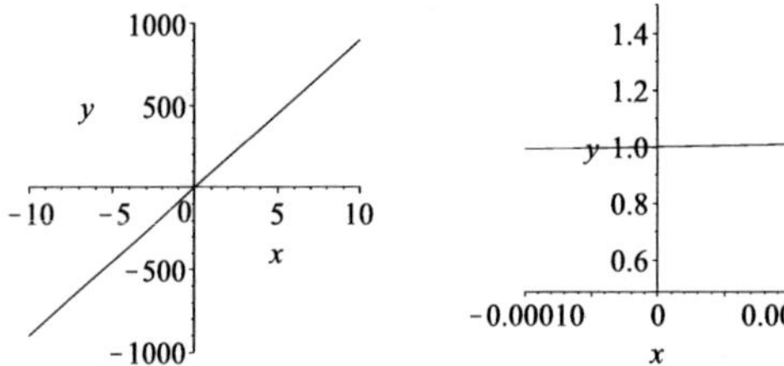

On the left we are too far away from the graph to see the oscillations due to the cosine term. On the right we are so close that the graph appears to be a straight line of slope 0.

The plot on the right gives somewhat of an idea of the behavior of f. The window is 90 times higher than it is wide so the graph of f, which is the solid line, appears to have slope 1. The dashed line is the graph of $y = 90x$ and the effect of the cosine function can be seen, if only barely. If we move closer, then the picture seems to be of two parallel lines. If we move away, even slightly, the two lines appear to coalesce into one.

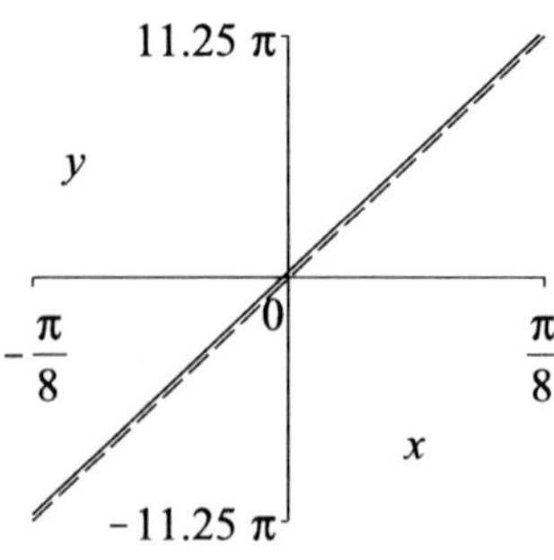

73. Assuming that the normal temperature is recorded at the midpoint of the month we seek a temperature function $T(t) = b + A\sin(\omega t + \phi)$ with period 12 (time is measured in months) having a low of 40 in mid-January and a high of 78 in mid-July. $T(t)$ will oscillate around the average of these two extremes: $b = 59$, with amplitude equal to half their difference: $A = 19$. Choose ω so that the sine wave has period 12. That is, $12\omega = 2\pi$, implying that $\omega = \pi/6$, and $T(x) = 59 + 19\sin(\pi t/6 + \phi)$.

The phase shift ϕ is chosen so that the low is at $t = 0.5$ (mid-January). This is accomplished by choosing ϕ so that $\sin(\pi \cdot 0.5/6 + \phi) = -1$. Let ϕ be the solution to the equation $\pi \cdot 0.5/6 + \phi = 3\pi/2$. That is, $\phi = 3\pi/2 - \pi/12 = 17\pi/12$.

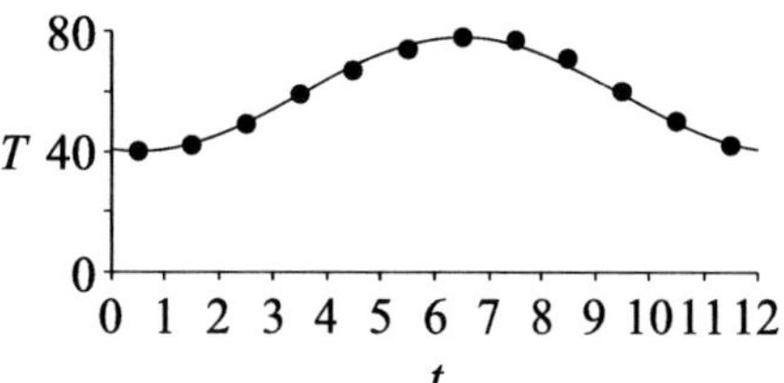

75. Let $\theta = \nu_1 \cdot t$ and $\phi = \nu_2 \cdot t$ in the identity in Exercise 49. This yields

$$\begin{aligned}\cos(\nu_1 \cdot t) + \cos(\nu_2 \cdot t) &= 2\cos((\nu_1 + \nu_2)/2 \cdot t)\cos((\nu_1 - \nu_2)/2 \cdot t) \\ &= A(t)\cos(\omega t).\end{aligned}$$

where $A(t) = 2\cos((\nu_1 - \nu_2)/2 \cdot t)$ and $\omega = (\nu_1 + \nu_2)/2$. If $\nu_1 \approx \nu_2$, then $A(t)$ is a low-frequency cosine wave and ω is about the same as the individual frequencies ν_1 and ν_2. For example, if $\nu_1 = 8$ and $\nu_2 = 6$, then $A(t) = 2\cos(t)$ and $\omega = 7$ yielding $\cos(8t) + \cos(6t) = 2\cos(t)\cos(7t)$.

See the picture on the right which plots the superimposed wave along with the time-varying envelopes $\pm 2\cos(t)$ (the dashed curves). Their frequency is 1.

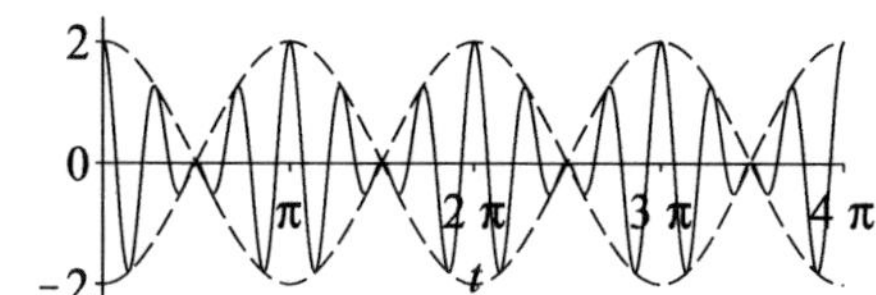

77. Snell's inequality is illustrated on the left with the graph of the expression $\tan(t/3) + 2\sin(t/3) - t$. Cusa's inequality is illustrated on the right with the graph of the expression $t - 3\sin(t)/(2 + \cos(t))$.

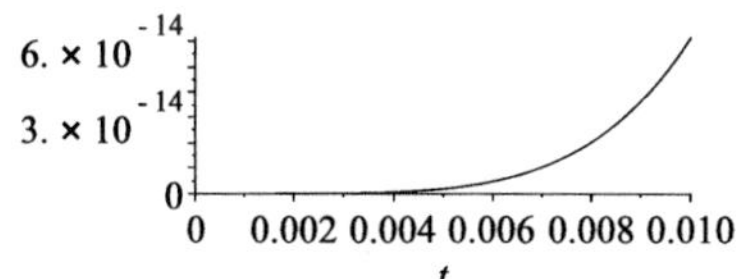

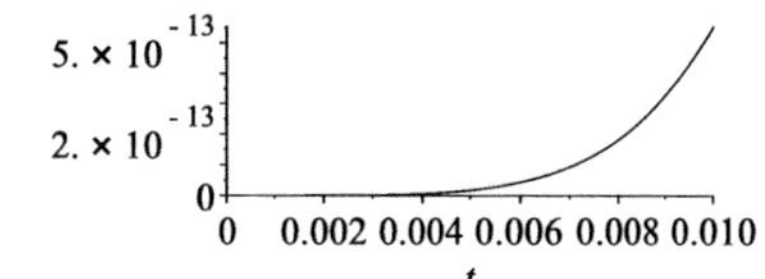

REVIEW EXERCISES FOR CHAPTER 1

1. All x within 6 units of -5: $(-11, 1)$.

3. $x(x-3)$ is 0 when $x = 0$ and $x = 3$. It is negative when $x \in (0, 3)$, so the answer is $[0, 3]$.

5. x is either positive or it is negative. If $x > 0$, then $2x + 1 < 4x$ or $1 < 2x$,

yielding the interval $(1/2, \infty)$. If $x < 0$, then $2x + 1 < -4x$ or $6x < -1$, yielding the interval $(-\infty, -1/6)$. The answer is $(-\infty, -1/6) \cup (1/2, \infty)$.

7. x is either positive or it is negative. If $x > 0$, then $x(x - 1) \leq 6$ or $x^2 - x - 6 \leq 0$. That is, $(x + 2)(x - 3) \leq 0$, implying that $x \leq 3$. This yields the interval $(0, 3]$. If $x < 0$, then $x(x - 1) \geq 6$ or $x^2 - x - 6 \geq 0$. That is, $(x + 2)(x - 3) \geq 0$ is true for all $x \leq -2$ yielding the interval $(-\infty, -2]$. The answer is $(-\infty, -2] \cup (0, 3]$.

9. Use the distance formula: $\sqrt{12^2 + (-5)^2} = \sqrt{144 + 25} = 13$.

11. The midpoint is $((9 + (-3))/2, (-6 + (-1))/2) = (3, -7/2)$.

13. Complete the square: $(x^2 - 2x + 1) + (y^2 + 2y + 1) = 1 + 1$ or $(x - 1)^2 + (y + 1)^2 = 2$. This is a circle centered at $(1, -1)$ having radius $\sqrt{2}$. The axes of symmetry are the lines $x = 1$ and $y = -1$.

15. Complete the square in x and y: $9(x^2+2x+1)+4(y^2-4y+4) = 11+9+16$ or $(x + 1)^2/2^2 + (y - 2)^2/3^2 = 1$. This is an ellipse centered at $(-1, 2)$ having axes of symmetry $x = -1$ and $y = 2$.

17. $y = 3x + b$. When $y = 0, x = 2$, so $0 = 6 + b$ and $b = -6$. The answer is $y = 3x - 6$.

19. The slope is $m = -1/(1) = -1$ so $y = -(x + 1) - 3$ or $y = -x - 4$.

21. The slope of ℓ is $m = -6$, so $y = -6(x + 2) + 1$.

23. The slope of ℓ is $m = (5 - 1)/(0 - (-2)) = 2$, so $y = 2(x + 2) + 1$.

25. $x/4 + y/(-3) = 1$

27. The slope is $m = -1/(1/2) = -2$, so $y = -2(x - 1) + 1$ or $y = -2x + 3$. The intercept form is $x/(3/2) + y/3 = 1$.

29. Add the equations: $3x = 6$, so $x = 2$ and $y = 3$. The lines intersect at $(2, 3)$.

31. Substitute the second equation into the first: $x^2-9 = 3x-5$ or $x^2-3x-4 = 0$. Therefore, $(x + 1)(x - 4) = 0$ and $x = -1$ or $x = 4$. If $x = -1$, then $y = -8$, and if $x = 4$, then $y = 7$. The intersection points are $(-1, -8)$ and $(4, 7)$.

33. The denominator is never 0, so the domain is $\mathbb{R}$.

35. Division by 0 is not allowed, $x \neq 5$. The domain is $(-\infty, 5) \cup (5, \infty)$.

37. To avoid division by 0, the domain is all $x \neq -2$. That is, $(-\infty, -2) \cup (2, \infty)$.

39. First of all, $x \neq \pm 3$, because $x = \pm 3$ makes the denominator 0. In addition, it must be the case that $25 - x^2 \geq 0$. That is, $25 \geq x^2$ or $5 \geq |x|$, implying that $-5 \leq x \leq 5$. The domain is $[-5, -3) \cup (-3, 3) \cup (3, 5]$.

41. The first 5 terms are $1/2$, $2/3$, $3/4$, $4/5$, and $5/6$.

43. These are 4, 8, 14, 24, 42.

45. The first one is $1+2+4=7$. Then 15, 31, 63, and 127.

47. Here are the first 5: 1, 2, 4, 8, $f_5 = 16$.

49. The first 5: 0, 2, 7, 18, $f_5 = 41$.

51. $f^2(\sqrt{x}) = (2\sqrt{x}+3)^2 = 4x + 12\sqrt{x} + 9$

53. $(h \circ g)(x) = h(g(x)) = (1-(1+x^2))/(1+(1+x^2)) = -x^2/(2+x^2)$

55. $(f \circ g^2)(x) = f(g^2(x)) = 2(1+x^2)^2 + 3 = 2x^4 + 4x^2 + 5$

57. $(f \circ g^{1/2})(\sqrt{x}) = f\big(\sqrt{g(\sqrt{x})}\big) = 2\sqrt{1+x} + 3$

59. $$\begin{aligned}\left(f^{-1} \circ \tfrac{1}{f}\right)(x) &= f^{-1}(1/f(x)) = (1/f(x) - 3)/2 = (1/(2x+3) - 3)/2 \\ &= -(3x+4)/(2x+3)\end{aligned}$$

61. As s increases from 1 to 4, $f(s)$ decreases from 1/2 to 1/5, so $T = [1/5, 1/2]$. Since $t = 1/(1+f^{-1}(t))$, $1+f^{-1}(t) = 1/t$ and $f^{-1}(t) = 1/t - 1$.

63. As s increases from 0 to 1, $f(s)$ increases from 2 to 5, so $T = [2,5]$. Since $t = s^2 + 2s + 2$, $s^2 + 2s + 2 - t = 0$ and $s = (-2 \pm \sqrt{4 - 4(2-t)}\,)/2$. That is, $s = -1 \pm \sqrt{t-1}$. Because s is not negative, $s = -1 + \sqrt{t-1}$ and $f^{-1}(t) = -1 + \sqrt{t-1}$.

65. As s increases from -1 towards ∞, $f(s)$ increases to 1, then decreases steadily towards 0. Therefore, f is onto, but it is not one-to-one. Observe, for example, that $f(-1) = f(1) = 1/2$.

67. As s increases from 0 towards ∞, $f(s)$ increases from 0 towards 1, but it never gets there. Therefore, f is one-to-one, but it is not onto.

69. To get the graph of g from the graph of f, translate it 1 unit down and 2 units left.

71. Since $g(x) = f(x-3) - 9$, to get the graph of g from the graph of f, translate it 9 units down and 3 units right.

73. This is a portion of a line. It extends from $(-1/3, 0)$ to $(0, 1/2)$. See the picture.

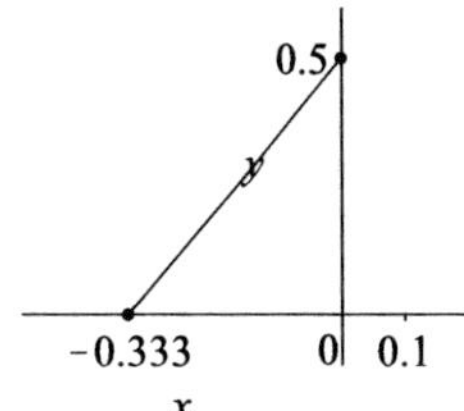

75. This is a portion of a line. It extends from $(1,0)$ to $(0,1)$. See the picture.

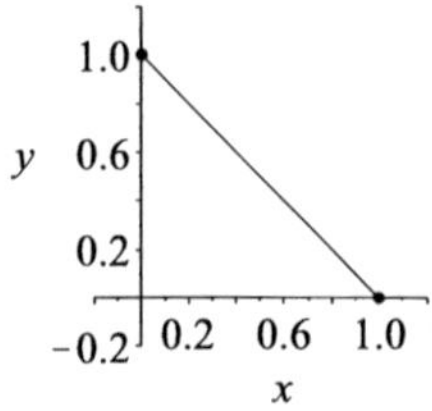

77. The table below contains the values of the six trigonometric functions at the angle A. Note that $a = 3$, $b = 2$, and $c = \sqrt{13}$.

$\sin(A)$	$\cos(A)$	$\tan(A)$	$\cot(A)$	$\csc(A)$	$\sec(A)$
$3/\sqrt{13}$	$2/\sqrt{13}$	$3/2$	$2/3$	$\sqrt{13}/3$	$\sqrt{13}/2$

79. The table below contains the values of the six trigonometric functions at the angle A. Note that $a = 1$, $b = 3$, and $c = \sqrt{10}$.

$\sin(A)$	$\cos(A)$	$\tan(A)$	$\cot(A)$	$\csc(A)$	$\sec(A)$
$1/\sqrt{10}$	$3/\sqrt{10}$	$1/3$	3	$\sqrt{10}$	$\sqrt{10}/3$

81. The table below contains the values of the six trigonometric functions at the angle $\beta = \alpha + \pi/2$, were α is the radian measure of the angle A of Exercise 78. Note that $a = 12$, $b = 5$, $c = 13$, and adding $\pi/2$ does the following to the sine and cosine: $\sin(\alpha + \pi/2) = \cos(\alpha)$ and $\cos(\alpha + \pi/2) = -\sin(\alpha)$. The other values are obtained from these two.

$\sin(\beta)$	$\cos(\beta)$	$\tan(\beta)$	$\cot(\beta)$	$\csc(\beta)$	$\sec(\beta)$
$5/13$	$-12/13$	$-5/12$	$-12/5$	$13/5$	$-13/12$

83. $\tan(\pi/4) - \cot(3\pi/4) = 1 - (-1) = 2$

85. $\csc(\pi/6)\sec(3\pi/4) = 2 \cdot (-\sqrt{2}) = -2\sqrt{2}$

87. $r = \sin(2\alpha) = 2\sin(\alpha)\cos(\alpha) = 2 \cdot (3/4) \cdot \sqrt{1 - 9/16} = 3\sqrt{7}/8$

89. $$r = \tan(2\alpha) = \frac{\sin(2\alpha)}{\cos(2\alpha)} = \frac{2\sin(\alpha)\cos(\alpha)}{\cos^2(\alpha) - \sin^2(\alpha)} = 2 \cdot \frac{\sqrt{1 - \cos^2(\alpha)} \cdot \cos(\alpha)}{2\cos^2(\alpha) - 1}$$

$$= 2 \cdot \frac{\sqrt{1 - 9/25} \cdot (3/5)}{2 \cdot (9/25) - 1} = 2 \cdot \frac{12/25}{-7/25} = -24/7$$

Chapter 2

Limits

2.1 The Concept of Limit

Problems for Practice

1. As $x \to 2$, $x + 3 \to 2 + 3$: $\lim_{x\to 2}(x + 3) = 5$.

3. As $h \to 4$, $3h^2 + 2h + 1 \to 3 \cdot 4^2 + 2 \cdot 4 + 1$: $\lim_{h\to 4}(3h^2 + 2h + 1) = 57$.

5. As $h \to 1$, the numerator approaches -2 and the denominator approaches 2. The quotients approach $-2/2 = -1$: $\lim_{h\to 1} \frac{h-3}{h+1} = -1$.

7. If x is positive, then $g(x) = -4$ so, as $x \to 2$, $g(x) \to -4$: $\lim_{x\to 2} g(x) = -4$.

9. As $x \to 5$, $x - 5 \neq 0$, so $(x^2 - 25)/(x - 5) = (x + 5)(x - 5)/(x - 5) = x + 5$, and $\lim_{x\to 5}((x^2 - 25)/(x - 5)) = \lim_{x\to 5}(x + 5) = 5 + 5 = 10$.

11. As $t \to -7$, $t+7 \neq 0$, so $(t+7)/(t^2-49) = (t+7)/((t-7)(t+7)) = 1/(t-7)$, and $\lim_{t\to -7}((t + 7)/(t^2 - 49)) = \lim_{t\to -7}(1/(t - 7)) = -1/14$.

13. As $x \to -4$, $x^2 + 6x - 8 \to (-4)^2 + 6 \cdot (-4) - 8 = -16$ and $x^2 - 2x - 24 \to (-4)^2 - 2 \cdot (-4) - 24 = 0$ so their quotient, which grows without bound, has no limiting value.

15. When $x < 2$, $f(x) = x^2 - 3x$ so $\lim_{x\to 2^-} f(x) = 2^2 - 3 \cdot 2 = -2 = \ell_L$. When $x > 2$, $f(x) = -x/(x-1)$ so $\lim_{x\to 2^+} f(x) = -2/(2-1) = -2 = \ell_R$. Since $\ell_L = \ell_R = -2$, $\lim_{x\to 2} f(x) = -2$.

17. When $x < 5$, $f(x) = (x^2 - 4)/3$ so $\lim_{x\to 5^-} f(x) = (5^2 - 4)/3 = 7 = \ell_L$. When $x > 5$, $x - 5 \neq 0$ so $f(x) = (x + 2)(x - 5)/((x - 4)(x - 5)) =$

$(x+2)/(x-4)$ and $\lim_{x\to 5^+} f(x) = 7/1 = 7 = \ell_R$. Since $\ell_L = \ell_R = 7$, $\lim_{x\to 5} f(x) = 7$.

19. Since $|f(x)-3| = |x+1-3| = |x-2|$, if $|x-2| < 0.01$, then $|f(x)-3| < 0.01$.

21. Observe that $|f(x)-17| = |5x+2-17| = |5x-15| = 5|x-3|$. Therefore, if $|x-3| < 0.01/5$, then $|f(x)-17| = 5|x-3| < 5\cdot(0.01/5) = 0.01$.

23. Observe that $|f(x)-5| = |5-x^2/4-5| = |x^2/4|$. Therefore, if $|x| < 0.2$, then $|f(x)-5| = |x|^2/4 < (0.2)^2/4 = 0.01$.

25. Observe that if $x < 1$, then $|f(x)-0.999| = |x-0.999|$. Therefore, if $|x-0.999| < 0.001$, then $x < 0.999+0.001 = 1$ so $|f(x)-0.999| = |x-0.999| < 0.001 < 0.01$.

27. Observe that if $x < 2$, then $f(x) = x-2$ and if $x > 2$, then $f(x) = x+2$. (In each case the term in the denominator cancels with a different factor in the numerator.) Therefore, the limit of $f(x)$ as $x \to 2^-$ is 0 and the limit of $f(x)$ as $x \to 2^+$ is 4. f does not have a limit as $x \to 0$.

Further Theory and Practice

29. (a) g has domain all $x \neq 1$, h, f, and k have the same domain, all of $\mathbb{R}$.

    (b) The functions h and f are the same. They have the same domain, range, and values. g and k have the same values as h and f everywhere except at $x = 1$ where g is undefined and k has the value 1. Observe that $h(1) = f(1) = 2$.

    (c) Since all four functions have the same values when $x \neq 1$, all four have a limit at $c = 1$, that limit is 2.

31. (a) h has domain all $x \neq 0$, f, g, and k have the same domain, all of $\mathbb{R}$.

    (b) The functions f and k are the same. They have the same domain, range, and values. h and g have the same values as f and k everywhere except at $x = 0$ where h is undefined and g has the value 1. Observe that $f(0) = k(0) = 0$.

    (c) Since all four functions have the same values when $x \neq 0$, all four have a limit at $c = 0$, that limit is 0.

33. Clearly $\lim_{x\to 0^-} H(x) = 0$ and $\lim_{x\to 0^+} H(x) = 1$. Therefore, $\lim_{x\to 0} H(x)$ does not exist.

35. Let ℓ be a length that measures x feet and y meters. Then $x = 0.3048\,y$ and an error of Δx feet equals an error of $0.3048\,\Delta y$ meters. Therefore, if it is required that $|\Delta x| < 0.001$, then it must be the case that $|0.3048\,\Delta y| < 0.001$. That is, the error in meters can be no more than $0.001/0.3048 = 0.00328$.

37. (a) $\lim_{x\to 0} \lfloor x \rfloor$ does not exist because the left limit is -1 and the right limit is 0.

(b) $\lim_{x\to 1/2}\lfloor x\rfloor = 0$ because $\lfloor x\rfloor = 0$ for all $x \in [0,1)$.

(c) $\lim_{x\to 1} 1/\lfloor x\rfloor$ does not exist because $\lfloor x\rfloor = 0$ for all $x \in [0,1)$ so $1/\lfloor x\rfloor$ is not defined in that interval and there is no limit as $x \to 1^-$.

(d) $\lim_{x\to -1/2} 1/\lfloor x\rfloor = -1$ because $\lfloor x\rfloor = -1$ for all $x \in [-1,0)$ so $1/\lfloor x\rfloor = -1$ for all x sufficiently close to $-1/2$.

39. The graph of $x \mapsto \lfloor x\rfloor$ is displayed on the right. Each break in the graph occurs at an integer value n where $\lim_{x\to n}\lfloor x\rfloor$ does not exist. Observe that both one-sided limits do exist at such points, but they are not the same.

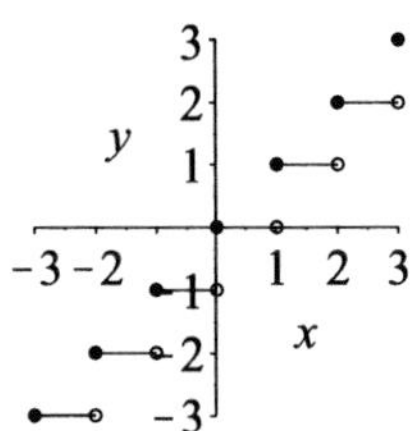

41. $x(t) = t^2$, $y(t) = 0$, the motion is one-dimensional, along the x-axis.

(a) The average velocity for $1 \le t \le 2$: $\dfrac{x(2)-x(1)}{2-1} = \dfrac{4-1}{1} = 3.$

(b) The average velocity for $1 \le t \le 1.5$: $\dfrac{x(1.5)-x(1)}{1.5-1} = \dfrac{2.25-1}{0.5} = 2.5.$

(c) The average velocity for $1 \le t \le 1.1$: $\dfrac{x(1.1)-x(1)}{1.1-1} = \dfrac{1.21-1}{0.1} = 2.1.$

(d) The average velocity for $t = 1$ to $t = 1+h$ is

$$\frac{x(1+h)-x(1)}{h} = \frac{(1+h)^2-1}{h} = \frac{1+2h+h^2-1}{h} = 2+h\,.$$

(e) $\displaystyle\lim_{h\to 0}\frac{x(1+h)-x(1)}{h} = \lim_{h\to 0}(2+h) = 2$

43. $p(t) = t^2 + 2$, the motion is one-dimensional.

(a) Average velocity over the time interval $[2, 2+h]$:

$$\begin{aligned}\overline{v}(h) &= \frac{p(2+h)-p(2)}{h} = \frac{(2+h)^2+2-6}{h} \\ &= \frac{4+4h+h^2+2-6}{h} = \frac{4h+h^2}{h} = 4+h\,.\end{aligned}$$

Therefore, $v_0 = \lim_{h\to 0^+}\overline{v}(h) = \lim_{h\to 0}(4+h) = 4\,.$

(b) If $0 < h < 0.1$, then $4 < 4+h < 4+0.1$. That is, $v_0 < \overline{v}(h) < 0.1$.

(c) If $0 < h < 0.01$, then $4 < 4+h < 4+0.01$. That is, $v_0 < \overline{v}(h) < 0.01$.

(d) If $0 < h < \epsilon$, then $4 < 4+h < 4+\epsilon$. That is, $v_0 < \overline{v}(h) < v_0+\epsilon$.

45. $p(t) = t^2 + 2t$, the motion is one-dimensional.

(a) Average velocity over the time interval $[3, 3+h]$:

$$\begin{aligned}\overline{v}(h) &= \frac{p(3+h)-p(3)}{h} = \frac{(3+h)^2+2(3+h)-15}{h} \\ &= \frac{9+6h+h^2+6+2h-15}{h} = \frac{8h+h^2}{h} = 8+h\,.\end{aligned}$$

Therefore, $v_0 = \lim_{h\to 0^+}\overline{v}(h) = \lim_{h\to 0}(8+h) = 8$.

(b) If $0 < h < 0.1$, then $8 < 8+h < 8+0.1$. That is, $v_0 < \overline{v}(h) < 0.1$.

(c) If $0 < h < 0.01$, then $8 < 8+h < 8+0.01$. That is, $v_0 < \overline{v}(h) < 0.01$.

(d) If $0 < h < \epsilon$, then $8 < 8+h < 8+\epsilon$. That is, $v_0 < \overline{v}(h) < v_0+\epsilon$.

47. $f(x) = x^2$, $c = 3$.

(a) The slope of the line through $(c, f(c))$ and $(c+h, f(c+h))$:

$$\begin{aligned}m(h) &= \frac{f(3+h)-f(3)}{h} = \frac{(3+h)^2-9}{h} \\ &= \frac{9+6h+h^2-9}{h} = \frac{6h+h^2}{h} = 6+h\,.\end{aligned}$$

(b) Therefore, $\lim_{h\to 0} m(h) = \lim_{h\to 0}(6+h) = 6$.

49. $f(x) = x^2+2x$, $c = 1$.

(a) The slope of the line through $(c, f(c))$ and $(c+h, f(c+h))$:

$$\begin{aligned}m(h) &= \frac{f(1+h)-f(1)}{h} = \frac{(1+h)^2+2(1+h)-3}{h} \\ &= \frac{1+2h+h^2+2+2h-3}{h} = \frac{4h+h^2}{h} = 4+h\,.\end{aligned}$$

(b) Therefore, $\lim_{h\to 0} m(h) = \lim_{h\to 0}(4+h) = 4$.

51. If $x > 0$, then $|x| = x$ so $x^{5/3}/|x| = x^{5/3}/x = x^{2/3}$, and $\lim_{x\to 0^+}(x^{5/3}/|x|) = \lim_{x\to 0^+} x^{2/3} = 0$.

If $x < 0$, then $|x| = -x$ so $x^{5/3}/|x| = -x^{5/3}/x = -x^{2/3}$, and $\lim_{x\to 0^-}(x^{5/3}/|x|) = -\lim_{x\to 0^-} x^{2/3} = 0$.

Therefore, $\lim_{x\to 0}(x^{5/3}/|x|) = 0$.

CALCULATOR/COMPUTER EXERCISES

53. Observe that $f(2) = 1$. The plot on the right suggests that $\delta = 0.0095$ has the property that $|x-2| < 0.0095$ implies that $|f(x)-1| < 0.01$.

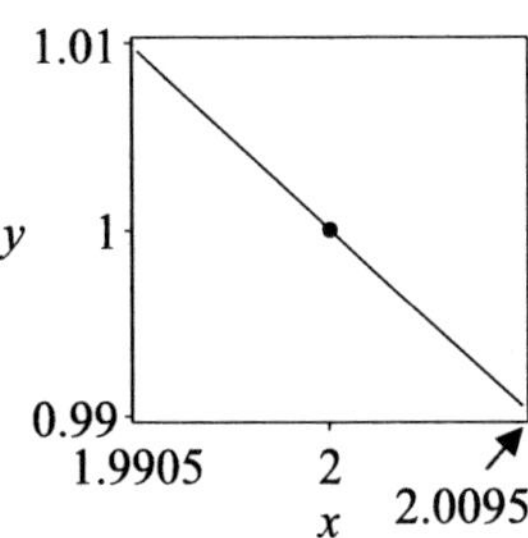

55. Observe that $f(10) = 100$. The plot on the right suggests that $\delta = 0.00046$ has the property that $|x - 10| < 0.00046$ implies that $|f(x) - 100| < 0.01$.

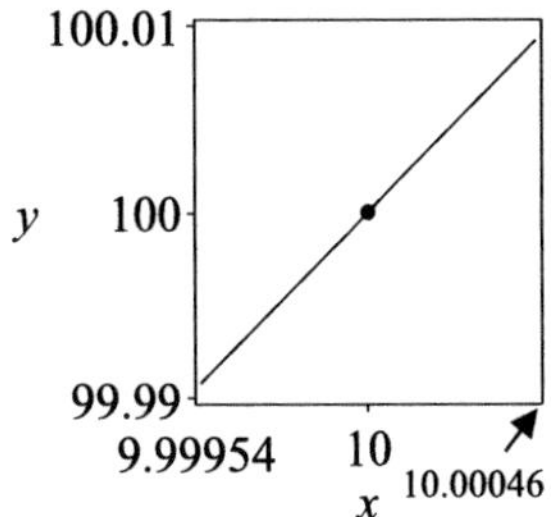

57. Observe that $f(9) = 1/3$. The plot on the right suggests that $\delta = 0.5$ has the property that $|x - 9| < 0.5$ implies that $|f(x) - 1/3| < 0.01$.

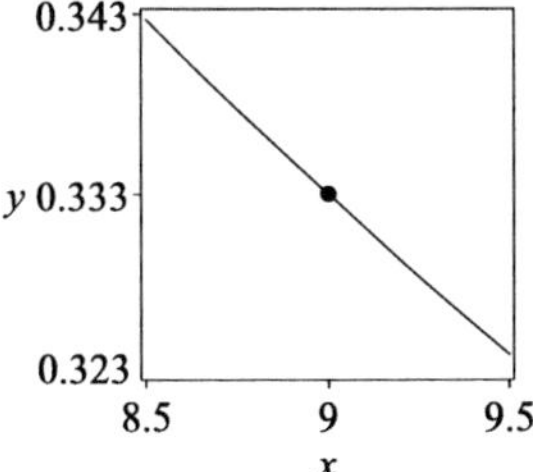

59. Let $T(m) = 1.2|m| + 100m^2$ denote the "tolerance" function. We seek a value of m such that $T(m) < 10^{-3}$. According to *Maple*'s *fsolve* procedure, $T(m) = 10^{-3}$ when $m = 0.0007823299831$. Since $T(m)$ increases with increasing m, $m = 0.000782$ can be used for the accuracy of the settings.

61. $f(x) = \cos(x)/(x - \pi/2)$, $c = \pi/2$

(a) The following table contains the requested values of f.

x	$\pi/2 - 0.01$	$\pi/2 - 0.001$	$\pi/2 - 0.0001$	$\to \pi/2 \leftarrow$	$\pi/2 + 0.0001$	$\pi/2 + 0.001$	$\pi/2 + 0.01$
$f(x)$	-0.999983333	-0.999999833	-0.999999998		-0.999999998	-0.999999833	-0.999983333

(b) The limit appears to be -1.

c, d) The estimate in part b is supported by the graph of f displayed on the left below. It looks like $|f(x) - (-1)| < 0.01$ for all x such that $0 < |x - \pi/2| < 0.24$. More evidence is provided by the plot on the right.

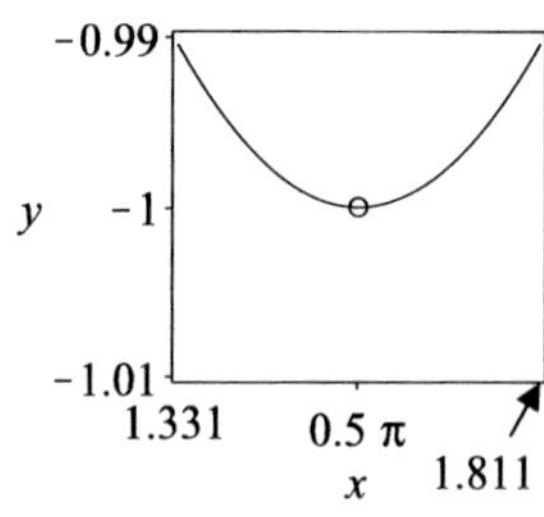

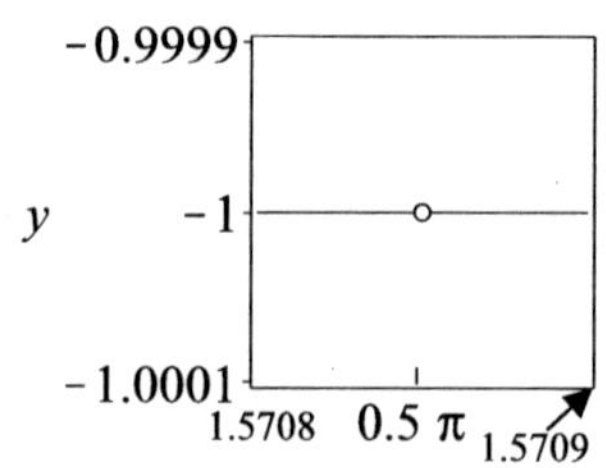

63. $f(x) = (x - \sin(x))/x^3$, $c = 0$

(a) The following table contains the requested values of f.

x	-0.01	-0.001	-0.0001	$\to 0 \leftarrow$	0.0001	0.001	0.01
$f(x)$	0.166665833	0.166666658	0.166666667		0.166666667	0.166666658	0.166665833

(b) The limit appears to be $0.1\overline{6} = 1/6$.

c, d) The estimate in part b is supported by the graph of f displayed on the left below. It looks like $|f(x) - 1/6| < 0.01$ for all x such that $0 < |x| < 1.05$. More evidence is provided by the plot on the right.

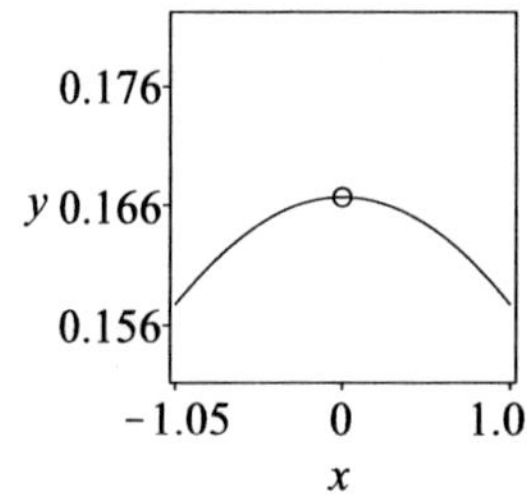

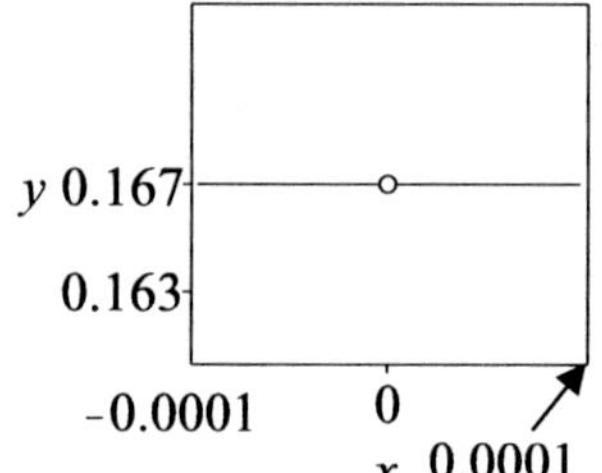

65. $p(t) = t^3 - 8t$

(a) The following table contains several values of $\overline{v}(h) = \frac{p(2+h)-p(2)}{h}$. These values suggest that $v_0 = \lim_{h\to 0^+} \overline{v}(h) = 4$.

h	0.1	0.01	0.001	0.0001	0.00001
$\overline{v}(h)$	4.61000	4.06010	4.00600	4.00060	4.00006

(b) Since $\overline{v}(h) - 4 = 6h + h^2$ (verify), if $0 < h < 1$, then $0 < h^2 < h$ and $0 < \overline{v}(h) - 4 < 7h$. Therefore, if $0 < h < 0.1/7$, then $4 < \overline{v}(h) < 4.1$.

(c) Similarly, if $0 < h < 0.01/7$, then $4 < \overline{v}(h) < 4.01$.

67. $p(t) = 2t - \frac{24}{\pi}\cos\left(\frac{\pi t}{12}\right)$

(a) The following table contains several values of $\overline{v}(h) = \frac{p(2+h)-p(2)}{h}$. These values suggest that $v_0 = \lim_{h\to 0^+} \overline{v}(h) = 3$.

h	0.1	0.01	0.001	0.0001	0.00001
$\overline{v}(h)$	3.022556	3.002266	3.000227	3.000024	3.00002

(b) The graph of $h \mapsto \overline{v}(h) - 3$ displayed on the right indicates that if $0 < h < 0.4$, then $0 < \overline{v}(h) - 3 < 0.1$ so

$$3 < \overline{v}(h) < 3.1\,.$$

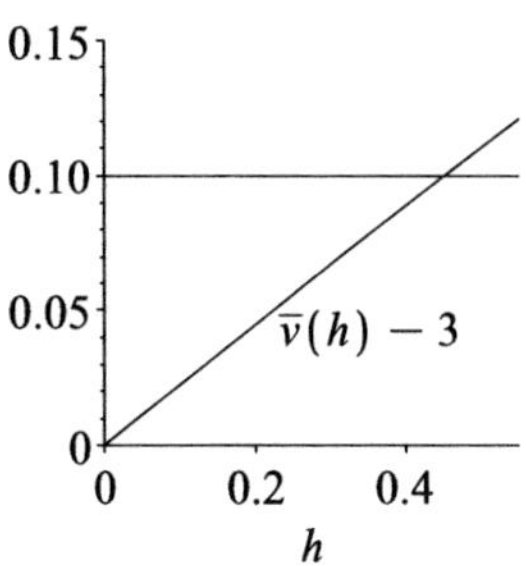

(c) A similar plot will show that if $0 < h < 0.04$, then $0 < \overline{v}(h) - 3 < 0.01$ so $3 < \overline{v}(h) < 3.01$.

69. A spring moves so the amount of extension is $x(t) = \sin(t)$. The average velocity over time interval $[t - 0.0001, t + 0.0001]$ is denoted $\overline{v}(t)$.

(a) An explicit formula for the average velocity is

$$\overline{v}(t) = \frac{\sin(t + 0.0001) - \sin(t - 0.0001)}{0.0002}\,.$$

The graph of $\overline{v}$ for $0 \le t \le 2\pi$ is displayed on the right.

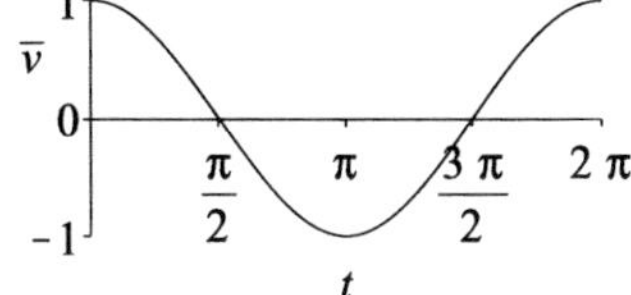

(b) $\overline{v}$ attains its greatest value, 1, at $t = 0$ and $t = 2\pi$. The most negative value of $\overline{v}$ is -1, attained when $t = \pi$.

(c) $\overline{v}(t) = 0$ when $t = \pi/2$ and $t = 3\pi/2$.

(d) The graphs of x and $\overline{v}$ are shown on the right. $\overline{v} < 0$ on the interval $I = (\pi/2, 3\pi/2)$. As t increases through this interval x is decreasing.

$\overline{v}(t) < 0$ means that the velocity is negative. That is, the spring is moving in the negative direction.

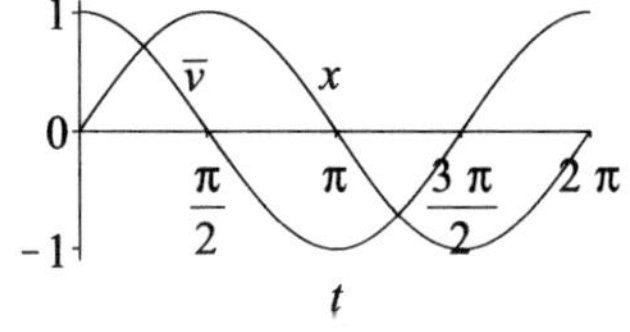

(e) Based upon its graph, $\overline{v}$ appears to approximate $t \mapsto \cos(t)$.

2.2 Limit Theorems

Problems for Practice

1. This is a polynomial. Therefore, $\lim_{x\to 3} 2x = 6$.

3. This is a polynomial. Therefore, $\lim_{x\to 6}(x/3 + 2) = 4$.

5. This is a polynomial. Therefore, $\lim_{x\to 0}(\sqrt{2} - \pi x) = \sqrt{2}$.

7. Since $(x^2 - 4) = (x+2)(x-2)$, $\lim_{x\to -2}\left((x^2-4)/(x-2)\right) = \lim_{x\to -2}(x+2) = 0$.

9. If $x < 5$, then $f(x) = 3x$ so $\lim_{x\to 5^-} f(x) = 15$. If $x > 5$, then $f(x) = 4$ so $\lim_{x\to 5^+} f(x) = 4$. Since the left limit does not equal the right limit, $\lim_{x\to 5} f(x)$ does not exist.

11. If $x < 1$, then $f(x) = 0$ so $\lim_{x\to 1^-} f(x) = 0$. If $x > 1$, then $f(x) = -4$ so $\lim_{x\to 1^+} f(x) = -4$. Since the left limit does not equal the right limit, $\lim_{x\to 1} f(x)$ does not exist.

13. If $x < 4$, then $f(x) = x^2 + 1$ so $\lim_{x\to 4^-} f(x) = 17$. If $x > 4$, then $f(x) = 5x - 3$ so $\lim_{x\to 4^+} f(x) = 17$. Since the left limit equals the right limit, $\lim_{x\to 4} f(x) = 17$.

15. If $x < 4$, then $f(x) = x^3$ so $\lim_{x\to 4^-} f(x) = 48$. If $x > 4$, then $f(x) = 4x^2$ so $\lim_{x\to 4^+} f(x) = 48$. Since the left limit equals the right limit, $\lim_{x\to 4} f(x) = 64$.

17. $\lim_{x\to 4}(2x-6) = \lim_{x\to 4} 2x + \lim_{x\to 4} 6 = 2\lim_{x\to 4} x + 6 = 2\cdot 4 + 6 = 14$

19. $\lim_{x\to 1}(x^2-6) = \lim_{x\to 1} x^2 - \lim_{x\to 1} 6 = (\lim_{x\to 1} x)\cdot(\lim_{x\to 1} x) - 6 = 1\cdot 1 - 6 = -5$

21. $\lim_{x\to 0}\left((x^2+2)/(x+1)\right) = \lim_{x\to 0}(x^2+2)/\lim_{x\to 0}(x+1) = (\lim_{x\to 0} x^2 + \lim_{x\to 0} 2)/(\lim_{x\to 0} x + \lim_{x\to 0} 1) = ((\lim_{x\to 0} x)\cdot(\lim_{x\to 0} x) + 2)/(0+1) = (0+2)/1 = 2$

23. $\lim_{x\to 4}\left((x-5)x/(x+1)\right) = \lim_{x\to 4}\left((x-5)x\right)/\lim_{x\to 4}(x+1)$

$$= \lim_{x\to 4}(x-5)\cdot\lim_{x\to 4} x \Big/ \left(\lim_{x\to 4} x + \lim_{x\to 4} 1\right) = \left(\lim_{x\to 4} x - \lim_{x\to 4}\right)\cdot 4/(4+1)$$
$$= (4-5)\cdot 4/5 = -4/5$$

25. This is a rational function, the limit of the denominator is not 0, so $\lim_{x\to 1}\frac{(x-2)^2}{x+1} = \frac{(1-2)^2}{1+1} = 1/2$.

27. This limit exists because if $x \neq 3$, $(x^2-9)/(x-3) = x+3$. Therefore, $\lim_{x\to 3}\frac{x^2-9}{x-3} = \lim_{x\to 3}(x+3) = 6$.

29. This limit exists because if $x \neq -1$, $(x^2-1)/(x+1) = x-1$. Therefore, $\lim_{x\to -1}\frac{x^2-1}{x+1} = \lim_{x\to -1}(x-1) = -2$.

31. The limit of a quotient is the quotient of the limits:

$$\lim_{x\to\pi}\frac{x-\sqrt[3]{x}}{\sqrt{x}-3} = \frac{\pi-\sqrt[3]{\pi}}{\sqrt{\pi}-3}.$$

Theorem 4: $\lim_{x\to c}\sqrt[n]{x} = \sqrt[n]{c}$, was also used.

33. Since $|\cos(1/x)| \leq 1$ for all $x \neq 0$, $|x^3\cos(1/x)| = |x^3|\cdot|\cos(1/x)| \leq |x^3|$. Consequently, $-|x^3| \leq x^3\cos(1/x) \leq |x^3|$ for all $x \neq 1$. Because $\lim_{x\to 0}|x^3| = 0$, the Pinching Theorem implies that $\lim_{x\to 0} x^3\cos(1/x) = 0$ also.

35. Since $|\sin(\csc(\pi x))| \le 1$ for all x such that $0 < |x-5| < 1$,
$$|(x-5)^2 \sin(\csc(\pi x))| \le |x-5|^2$$
for such x. Therefore, $-|x-5|^2 \le (x-5)^2 \sin(\csc(\pi x)) \le |x-5|^2$ for all x sufficiently close to 5. Since $\lim_{x\to 5} |x-5|^2 = 0$, the Pinching Theorem implies that $\lim_{x\to 5} \big((x-5)^2 \sin(\csc(\pi x))\big) = 0$. Therefore,
$$\lim_{x\to 5} \big(1 + (x-5)^2 \sin(\csc(\pi x))\big) = 1 .$$

37. Because $|\sin(\sec(\pi/x))| \le 1$ whenever $0 < |x-2| < 1$, $|(x-2)^2 \sin(\sec(\pi/2))| \le |x-2|^2$ for such x and $-|x-2|^2 \le (x-2)^2 \sin(\sec(\pi/2)) \le |x-2|^2$. Since $\lim_{x\to 2} |x-2|^2 = 0$, $\lim_{x\to 2} \big((x-2)^2 \sin(\sec(\pi/2))\big) = 0$ by the Pinching Theorem and $\lim_{x\to 2} \big((x+1) + (x-2)^2 \sin(\sec(\pi/2))\big) = 3$.

39. Because $\lim_{x\to 2}(2-|x-2|^3) = 2$ and $\lim_{x\to 2}(2+|x-2|^2) = 2$, the Pinching Theorem implies that $\lim_{x\to 2} f(x) = 2$.

41. Because $\lim_{x\to 0}(-3|x| + 5|x-5|) = 25$ and $\lim_{x\to 0}(6|x| + (x-5)^2) = 25$, the Pinching Theorem can be applied and $\lim_{x\to 0} f(x) = 25$.

43. f is defined on the the closed interval $[1, 2]$. The one-sided limits at the endpoints are both 0: $\lim_{x\to 1^+} f(x) = 0 = \lim_{x\to 2^-} f(x)$.

45. The domain of f is $(-2, 0) \cup (0, 2)$. At these points, $f(x) = \frac{(x-2)(x+2)}{x(x+2)} = \frac{x-2}{x}$. Therefore, $\lim_{x\to -2^+} f(x) = \frac{-4}{-2} = 2$, $\lim_{x\to 0^-} f(x)$ and $\lim_{x\to 0^+} f(x)$ do not exist, and $\lim_{x\to 2^-} f(x) = \frac{0}{8} = 0$.

Further Theory and Practice

47. The limit exists and it is -1. This is because $|x|/x = (-x)/x = -1$ for all $x < 0$.

49. This limit is $7 + 21 = 28$ because $|x| - 3x = -x + 3x = -4x$ for all $x < 0$.

51. This limit is 36 because if $x < 0$, then $(x-|x|)^2 = (x-(-x))^2 = (2x)^2 \to 6^2$ as $x \to -3$.

53. The limit is -8 because if $x > 0$, then $f(x) = 1 - x^2 \to 1 - 9$ as $x \to 3$.

55. This limit does not exist. As $x \to 5^-$, $f(x) \to 6$, and as $x \to 5^+$, $f(x) \to 4$.

57. $\lim_{x\to 0} \frac{\sin(3x)}{x} = 3 \lim_{x\to 0} \frac{\sin(3x)}{3x} = 3 \lim_{h\to 0} \frac{\sin(h)}{h} = 3$. (In the penultimate step we let $3x = h$ and used the fact that $x \to 0$ if, and only if, $h \to 0$.)

59. $\lim_{x\to 0} \frac{\sin(2x)}{\sin(x)} = 2 \lim_{x\to 0} \left(\frac{\sin(2x)}{2x} \cdot \frac{x}{\sin(x)}\right) = 2 \left(\lim_{h\to 0} \frac{\sin(h)}{h}\right) \cdot \left(\lim_{x\to 0} \frac{x}{\sin(x)}\right) = 2 \cdot 1 \cdot 1 = 2$. (Note that for one of the limits we let $2x = h$ and used the fact that $x \to 0$ if, and only if, $h \to 0$.)

61. $\lim_{x\to 0} \frac{1-\cos(x)}{\tan(x)} = \lim_{x\to 0} \left(\frac{1-\cos(x)}{x^2} \cdot \frac{x}{\sin(x)} \cdot x\cos(x)\right) = \frac{1}{2} \cdot 1 \cdot 0 = 0$

63. $\lim_{x\to 0}\frac{\sin(x^\circ)}{x} = \lim_{x\to 0}\frac{\sin\left(x\cdot\frac{\pi}{180}\right)}{x} = \frac{\pi}{180}\lim_{x\to 0}\frac{\sin\left(x\cdot\frac{\pi}{180}\right)}{x\cdot\frac{\pi}{180}} = \frac{\pi}{180}$

65. $\lim_{x\to 0}\frac{\sin(2x)\tan(3x)}{x^2} = 6\lim_{x\to 0}\left(\frac{\sin(2x)}{2x}\cdot\frac{\sin(3x)}{3x}\cdot\frac{1}{\cos(3x)}\right) = 6\cdot 1\cdot 1\cdot 1 = 6$

67. $\lim_{x\to 1}\frac{x^2-1}{\sqrt{x}-1} = \lim_{x\to 1}\frac{(x-1)(x+1)}{\sqrt{x}-1} = \lim_{x\to 1}\frac{(\sqrt{x}-1)(\sqrt{x}+1)(x+1)}{\sqrt{x}-1}$
$= \lim_{x\to 1}(\sqrt{x}+1)(x+1) = 4$

69. $\lim_{h\to 0}\frac{h}{\sqrt{1+2h}-1} = \lim_{h\to 0}\left(\frac{h}{\sqrt{1+2h}-1}\cdot\frac{\sqrt{1+2h}+1}{\sqrt{1+2h}+1}\right)$
$= \lim_{h\to 0}\frac{h(\sqrt{1+2h}+1)}{2h} = \lim_{h\to 0}\frac{\sqrt{1+2h}+1}{2} = 1$

71. $\lim_{h\to 0}\left(\frac{2}{h\sqrt{4+h}} - \frac{1}{h}\right) = \lim_{h\to 0}\frac{2-\sqrt{4+h}}{h\sqrt{4+h}} = \lim_{h\to 0}\left(\frac{2-\sqrt{4+h}}{h\sqrt{4+h}}\cdot\frac{2+\sqrt{4+h}}{2+\sqrt{4+h}}\right)$
$= \lim_{h\to 0}\frac{-h}{h\sqrt{4+h}\,(2+\sqrt{4+h})} = \lim_{h\to 0}\frac{-1}{\sqrt{4+h}\,(2+\sqrt{4+h})} = -1/8$

73. $\lim_{x\to 4}\frac{\sqrt{x}-2}{\sqrt{x+5}-3} = \lim_{x\to 4}\left(\frac{\sqrt{x}-2}{\sqrt{x+5}-3}\cdot\frac{\sqrt{x+5}+3}{\sqrt{x+5}+3}\cdot\frac{\sqrt{x}+2}{\sqrt{x}+2}\right)$
$= \lim_{x\to 4}\frac{(x-4)(\sqrt{x+5}+3)}{(x-4)(\sqrt{x}+2)} = \lim_{x\to 4}\frac{\sqrt{x+5}+3}{\sqrt{x}+2} = \frac{6}{4} = 3/2$

75. When $x<3$, $|x-3| = -(x-3)$ so $\lim_{x\to 3^-}\frac{x-3}{|x-3|} = \lim_{x\to 3^-}\frac{x-3}{-(x-3)} = -1.$

77. $\lim_{x\to 4^-}\frac{x+\sqrt{16-3x}}{x+\sqrt{16-x^2}} = \frac{4+\sqrt{16-12}}{4+\sqrt{16-16}} = \frac{6}{4} = 3/2$

79. $\lim_{x\to 0^+}\frac{\sin(\sqrt{x})}{\sqrt{x}} = \lim_{h\to 0+}\frac{\sin(h)}{h} = 1$. Note that if $h=\sqrt{x}$, then $x\to 0^+$ if, and only if, $h\to 0^+$.

81. Suppose, to the contrary, that there is $\delta_2>0$ such that $|f(x)-\ell_0|<\epsilon_0$ for all values of x satisfying $0<|x-c|<\delta_2$. This leads to a contradiction as follows. Choose an x_0 such that $0<|x_0-c|<\min(\delta_1,\delta_2)$. Then $|f(x_0)-\ell_1|<\epsilon_0$ and $|f(x_0)-\ell_2|<\epsilon_0$ implying that

$$\begin{aligned}|\ell_1-\ell_2| &= |\ell_1 - f(x_0) + f(x_0) - \ell_2| \le |\ell_1 - f(x_0)| + |f(x_0)-\ell_2| \\ &< \epsilon_0 + \epsilon_0 = 2\epsilon_0\,.\end{aligned}$$

But $2\epsilon_0 = |\ell_1-\ell_2|$, and we find that $|\ell_1-\ell_2| < |\ell_1-\ell_2|$, a clear contradiction.

83. The fact that $0\le f(x)\le x^2$ for all x implies that

$$-x^2 \le f(x) \le x^2 \qquad (\star)$$

for all x .

If $x>0$, divide all three parts of $(\star)$ by x to see that $-x\le f(x)/x\le x$, and $\lim_{x\to 0^+} f(x)/x = 0$ by the Pinching Theorem.

If $x<0$, divide all three parts of $(\star)$ by $-x$ to see that $x\le -f(x)/x\le -x$. This implies that $-x\ge f(x)/x\ge x$, and $\lim_{x\to 0^-} f(x)/x = 0$ by the Pinching Theorem once again.

85. Since $-|f(x)| \leq f(x) \leq |f(x)|$ for all x, the Pinching Theorem implies that $\lim_{x\to c} f(x) = 0$.

87. Let $f(x) = x^3/(x^2 - 2.9x + 2)$. The picture on the right indicates that if $2 \leq x \leq 2.006$, then $|f(x) - 40| < 0.1$.

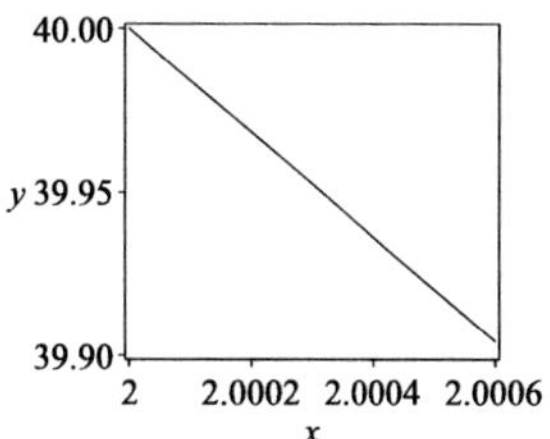

89. Let $f(x) = x^3 - 3x^2 + 2x + 1$. Then $\lim_{x\to 2} f(x) = 1$. According to the picture on the right, $|x-2| \leq 0.045 \implies |f(x)-2| < 0.1$.

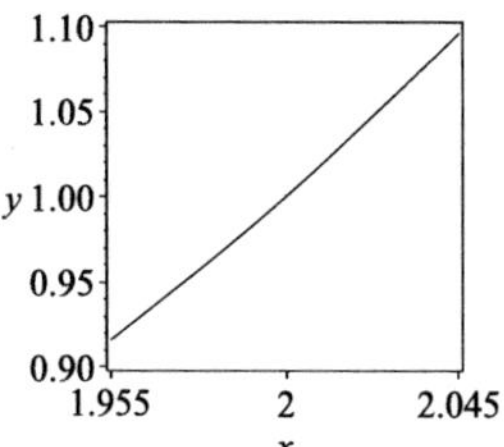

91. The graph of g appears on the right. It appears that $\lim_{x\to 0} g(x) = 0.01746$.

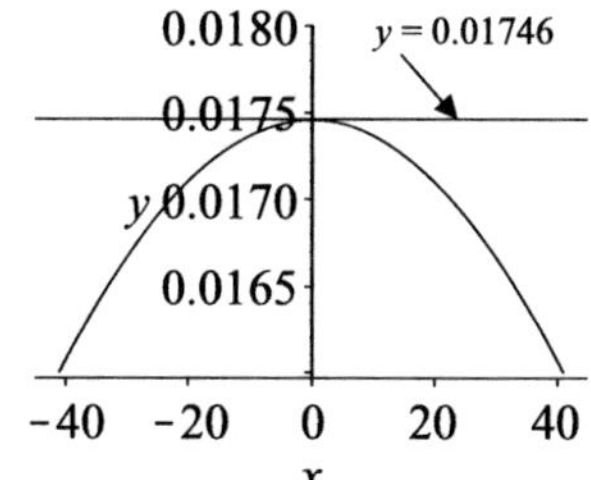

93. The graphs of g, f, and h are shown on the right. According the the Pinching Theorem, $\lim_{x\to 1} f(x) = 2$.

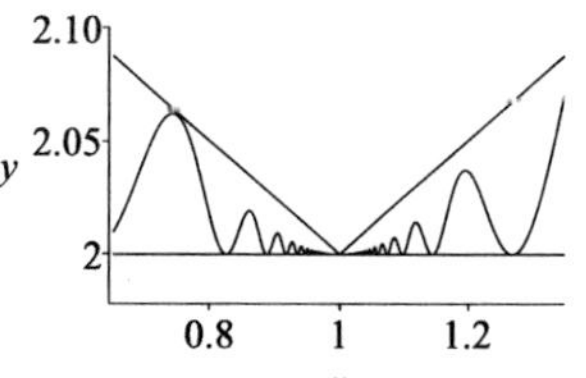

95. The graphs of g, f, and h are shown on the right. According the the Pinching Theorem, $\lim_{x\to 0} f(x) = 0$.

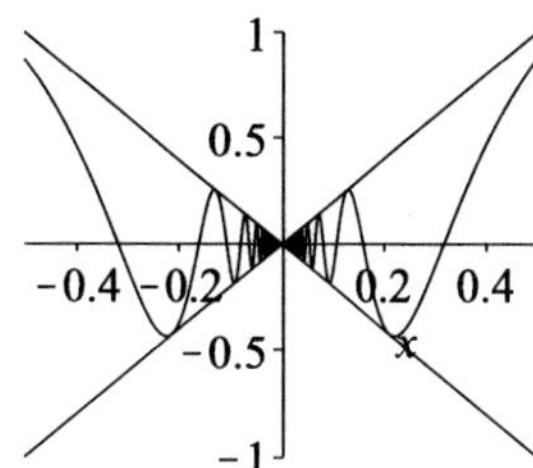

2.3 Continuity

Problems for Practice

1. Near $c = -1$ and $c = 1$, $f(x) = x^2 - 1$, a polynomial, so f is continuous at these two points. f is also continuous at $c = 2$ because $\lim_{x\to 2^-} f(x) = \lim_{x\to 2^+} f(x) = 3 = f(2)$.

3. f is not continuous at $c = -1$ because $\lim_{x\to(-1)^-} f(x) = 0$ and $\lim_{x\to(-1)^+} f(x) = -1$. f is continuous at $c = 1$ because $f(x) = (3x+1)^{1/3}$ for all $x \in (-1,2)$ and $x \mapsto (3x+1)^{1/3}$, as the composition of two continuous functions, is continuous. f is also continuous at $c = 2$. Observe that as $x \to 2^-$, $f(x) \to 8^{1/3} = 2$, and as $x \to 2^+$, $f(x) \to 2 = f(2)$.

5. f is a polynomial. It is continuous at all $x \in \mathbb{R}$.

7. f is a rational function. It is continuous at all points in its domain. That is, $\{x \in \mathbb{R} : x \neq -1\}$.

9. f is a rational function so it is continuous at all points in its domain. That is, all of $\mathbb{R}$.

11. f is a rational function so it is continuous at all points in its domain. Since $f(x) = (x^2 - 5x + 6)/((x+4)(x-3))$, that is $\{x \in \mathbb{R} : x \neq -4, 3\}$.

13. $f(x)$ is a polynomial if $x \neq 3$. Therefore, f is continuous at all $x \neq 3$. Since $\lim_{x\to 3^-} f(x) = 4$ and $\lim_{x\to 3^+} f(x) = 6$, f not continuous at 3. f is continuous on $\{x \in \mathbb{R} : x \neq 3\}$.

15. $f(x)$ is a polynomial if $x \neq -4$. Therefore, f is continuous at all $x \neq -4$. Since $\lim_{x\to(-4)^-} f(x) = 81$ and $\lim_{x\to(-4)^+} f(x) = 81 = f(x)$, f is also continuous at -4. f is continuous on all of $\mathbb{R}$.

17. $f(x)$ is a polynomial if $x \neq 7$. Therefore, f is continuous at all $x \neq 7$. Since $\lim_{x\to 7^-} f(x) = -5$ and $\lim_{x\to 7^+} f(x) = -5 = f(7)$, f is also continuous at 7. f is continuous on all of $\mathbb{R}$.

19. $f(x) = x - 1$ if $x < -1$ and it is a constant if $x > -1$. Therefore, f is continuous at all $x \neq -1$. Since $\lim_{x\to(-1)^-} f(x) = -2$ and $\lim_{x\to(-1)^+} f(x) = 2$, f is not continuous at -1. f is continuous at the points $\{x \in \mathbb{R} : x \neq -1\}$.

21. As the quotient of continuous functions, f is continuous at all $x \neq 0$. However, because $\lim_{x\to 0} f(x) = 1$ and $f(0) = 0$, f is not continuous at 0. f is continuous at all points in $\{x \in \mathbb{R} : x \neq 0\}$.

23. If $x < 3$, then $f(x) = x^2 - 3$, and if $x > 3$, then $f(x) = x + 3$. Therefore, f is continuous at all points $x \neq 3$. Because $\lim_{x\to 3^-} f(x) = 6 = f(3)$ and $\lim_{x\to 3^+} f(x) = 6$, f is also continuous at 3. f is continuous at all points in $\mathbb{R}$.

25. Since $\lim_{x\to 2^-} f(x) = 11 = \lim_{x\to 2^+} f(x)$, define $F(x) = f(x)$ for $x \neq 2$ and $F(2) = 11$ to obtain a continuous extension of f.

27. Since $\lim_{x\to 2^-} f(x) = 4 = \lim_{x\to 2^+} f(x)$, define $F(x) = f(x)$ for $x \neq 2$ and $F(2) = 4$ to obtain a continuous extension of f.

29. Since $f(x)$ is constant if $x \neq 5$, f is continuous for all $x \neq 5$. It is left continuous at $c = 5$ because $\lim_{x\to 5^-} f(x) = 2 = f(5)$. f is not right continuous at $c = 5$ because $\lim_{x\to 5^+} f(x) = 3 \neq f(5)$.

31. Since $f(x)$ is constant if $x > 3$ and $f(x)$ is a polynomial for all $x < 3$, f is continous at all $x \neq 3$. f is not left continuous at 3 because $\lim_{x\to 3^-} f(x) = 4 \neq 2 = f(3)$. However, f is right continuous at 3 because $\lim_{x\to 3^+} f(x) = 2 = f(3)$.

33. The graph of $y = f(x)$ is sketched on the right with dots at the points $(\alpha, f(\alpha))$ and $(\beta, f(\beta))$ corresponding to the extreme values attained at $\alpha = -2$ and $\beta = -3$.

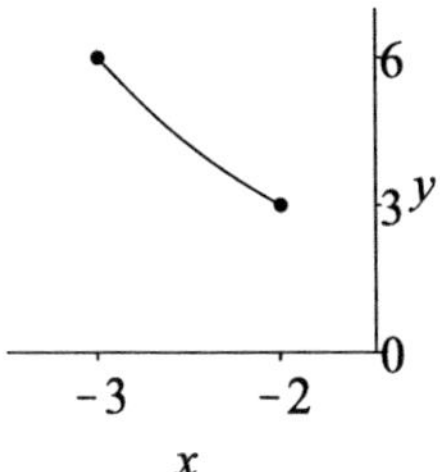

35. The graph of $y = f(x)$ is sketched on the right with dots at the points $(\alpha, f(\alpha))$ and $(\beta, f(\beta))$ corresponding to the extreme values attained at $\alpha = 1$ and $\beta = 2$.

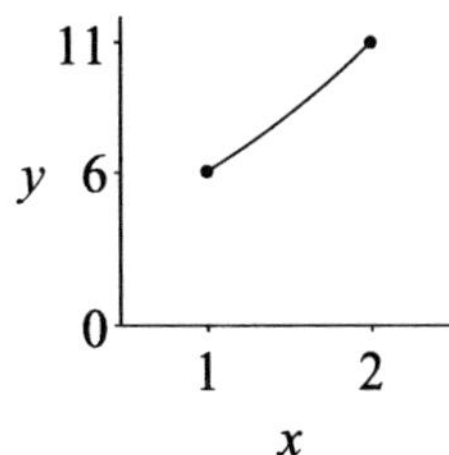

37. The graph of $y = f(x)$ is sketched on the right with dots at the points $(\alpha_1, f(\alpha_1))$, $(\alpha_2, f(\alpha_2))$, and $(\beta, f(\beta))$ corresponding to the extreme values attained at $\alpha_1 = 1$, $\alpha_2 = 2\pi$, and $\beta = \pi$.

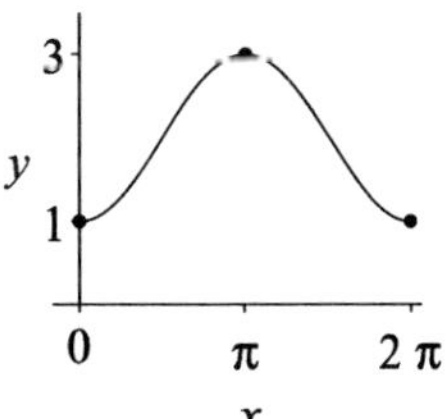

39. The graph of $y = f(x)$ is sketched on the right with dots at the points where f attains its extreme values on the interval I. The extreme values are $\alpha = -1/2$ and $\beta = \sqrt{2}/2$. Given any γ between α and β, there is an $x \in I$ with $f(x) = \gamma$.

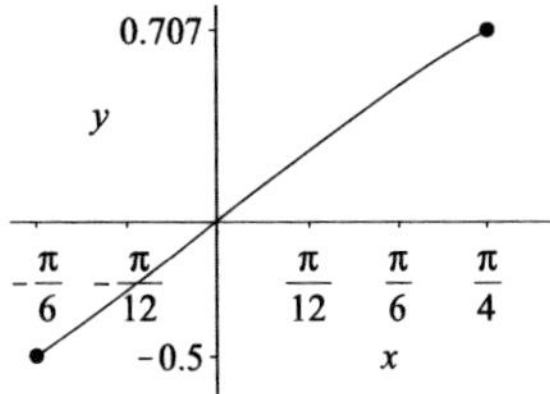

41. The graph of $y = f(x)$ is sketched on the right with dots at the points where f attains its extreme values on the interval I. The extreme values are $\alpha = 1/4$ and $\beta = 49/96 \approx 0.510$. Given any γ between α and β, there is an $x \in I$ with $f(x) = \gamma$.

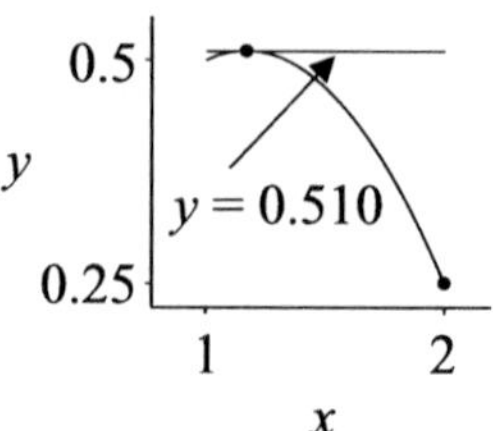

Further Theory and Practice

43. Since $\lim_{x\to 1^-} f(x) = 1$ the line should start at $(1, 1)$. Because $\lim_{x\to 3^+} = -6$, the line should end at $(3, -6)$. Therefore, its slope will be $m = (-6-1)/(3-1) = -7/2$ and its equation is $y = -(7/2)(x-1)+1$ or $y = -7x/2 + 9/2$. Define $F(x) = -7x/2 + 9/2$ for $1 \le x \le 3$.

45. Since $\lim_{x\to 1^-} f(x) = 2/3$ the line should start at $(1, 2/3)$. Because $\lim_{x\to 3^+} = 9$, the line should end at $(3, 9)$. Therefore, its slope will be $m = (9-2/3)/(3-1) = 25/6$ and its equation is $y = (25/6)(x-1)+2/3$ or $y = 25x/6 - 21/6$. Define $F(x) = 25x/6 - 7/2$ for $1 \le x \le 3$.

47. The function f is discontinuous at all integers x such that $\sin(\pi x/2) \neq 0$. That is, all x that are *odd* integers: $x = \pm 1, \pm 3, \pm 5, \ldots$.

49. The function f is discontinuous at each integer x having the property that $2\cos(\pi x/3) \neq 1$. That is, each integer x that is *not* of the form: $x = \pm 1, \pm 7, \pm 13, \ldots$ . .

51. The domain of f is all $x \neq 1$ so it cannot be continuous at $c = 1$. However, f has both a left and a right limit at 1 because if $x \neq 1$, then $f(x) = x + 1$. Consequently, $\lim_{x\to 1^-} f(x) = \lim_{x\to 1^+} f(x) = 2$.

    The domain of g is all $x \in \mathbb{R}$. Since $g(x) = x + 1$ for all $x \neq 1$, it has a left limit and a right limit at $c = 1$ and they are both equal to 2. Because $g(1) = 2$, g is continuous at $c = 1$.

    h is continuous everywhere. At $c = 1$ it has limit 2.

    k has left and right limits at $c = 1$, both are 2. Since $h(1) = 1$, h is not continuous at $c = 1$.

53. Assuming that it is continuous, the new tax function is

$$T(x) = \begin{cases} 0.08x & \text{if} \quad 0 < x \le 8025 \\ 642 + 0.12(x - 8025) & \text{if} \quad 8025 < x \le 32550 \\ 3585 + 0.16(x - 32250) & \text{if} \quad 32250 < x \le 78850 \\ 10993 + 0.20(x - 78850) & \text{if} \quad 78850 < x \le 164550 \\ 28133 + 0.25(x - 164550) & \text{if} \quad 164550 < x \le 357700 \\ 76420 + 0.30(x - 357700) & \text{if} \quad 357700 < x \end{cases}$$

55. (a) $\lim_{x\to 3^-} \lfloor x \rfloor = 2$. To prove it, observe that if $2 < x < 3$, then $\lfloor x \rfloor = 2$.

(b) $\lim_{x\to 2^+}\lfloor 4-2x\rfloor = -1$. To prove it, observe that if $2 < x < 2.5$, then $-4 > -2x > -5$, implying that $0 > 4-2x > -1$ and $\lfloor 4-2x\rfloor = -1$.

(c) $\lim_{x\to 0^-} 1/\lfloor x\rfloor = -1$. To prove it, observe that if $-1 < x < 0$, then $\lfloor x\rfloor = -1$, implying that $1/\lfloor x\rfloor = -1$ also.

(d) $\lim_{x\to 0^-}\lfloor -x\rfloor/x = 0$. To prove it, observe that for all $x \in (-1,0)$, $-x \in (0,1)$ implying that $\lfloor x\rfloor = 0$ and $\lfloor -x\rfloor/x = 0/x = 0$.

57. Since the train traveled 80 km in 1 hour, there had to be some point in its trajectory when it was traveling at more than 60 km/hr. Moreover, as it pulled out of the first station and as it came to a stop in the next station, it had to attain a velocity of less than 60 km/hr. The Intermediate Value Theorem implies that the train must have attained the velocity of 60 km/hr at at least two points between the two stations.

59. Let $f(x) = 10x^3 - 7x^2 + 20x - 14$. Then f is continuous on the closed interval $[0,1]$, $f(0) = -14$, and $f(1) = 9$. The Intermediate Value Theorem implies that there is at least one point $c \in (0,1)$ such that $f(c) = 0$. Since $0 < c < 1$, the Intermediate Value Theorem also implies that there is a point $x \in (0, \pi/2)$ such that $\sin(x_0) = c$. Therefore,

$$10\sin^3(x_0) - 7\sin^2(x_0) + 20\sin(x_0) - 14 = f(c) = 0\,.$$

61. Since $p(-1) = -1$ and $p(0) = 1$, and p is continuous, the Intermediate Value Theorem implies that $p(x) = 0$ for some x in the open interval $(-1,0)$.

63. Evaluate the polynomial p at $x = 1,2,3$ to find that $p(1) = 14$, $p(2) = -10$, and $p(3) = 90$. Since p is continuous, the Intermediate Value Theorem implies that there are numbers x_1 and x_2 such that $1 < x_1 < 2 < x_2 < 3$ and $p(x_1) = p(x_2) = 0$.

65. This statement is false. See the graph sketched on the right.

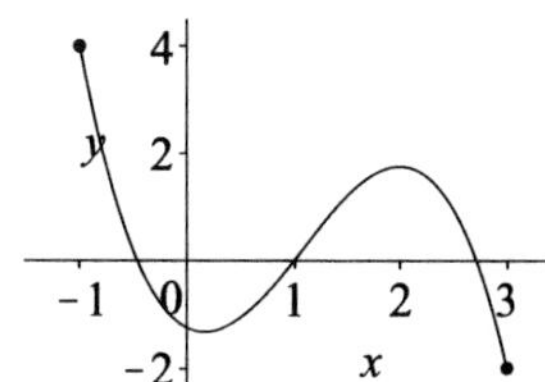

67. This statement is false. See the graph sketched on the right.

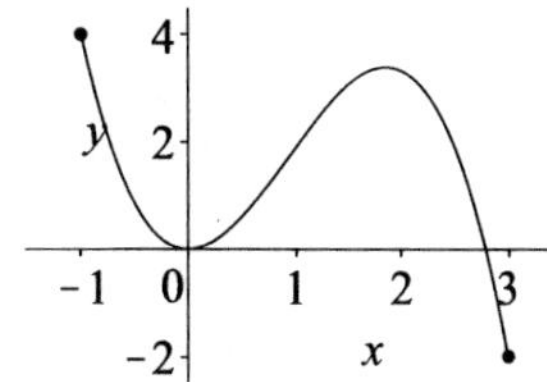

69. This is false. If $f(x) = -1$ for $x < 0$, and $f(x) = 1$ for $x \geq 0$, then $|f|(x) = 1$ for all x. Therefore, $|f|$ is continuous on the closed, bounded interval $[-1, 1]$, but f is not. See the graph of f on the right.

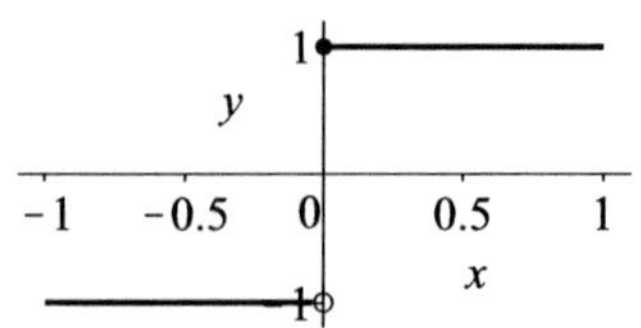

71. This is false. If $f(x) = -1$ for $x < 0$, and $f(x) = 1$ for $x \geq 0$, then $f^2(x) = 1$ for all x. Therefore, f^2 is continuous on the closed, bounded interval $[-1, 1]$, but f is not. See the graph of f on the right.

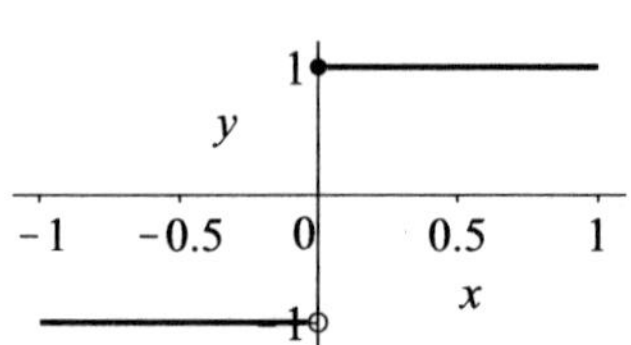

73. This is true. If f were not discontinuous at some point in the interval, then f would attain a maximum value at some point in the interval.

75. $\ell = \lim_{x\to 0} f(x) = 0$ and $L = \lim_{y\to 0} g(y) = 0$ also. Because $g(f(x)) = g(0) = 1$ for all x, $\lim_{x\to 0} g(f(x)) = 1 = g(\ell)$. This shows that the conclusion of Theorem 2 can be true even if its hypotheses are false. (The function f is continuous everywhere, but g is not continuous at $f(0) = 0$.)

77. Observe that

$$g(f(x)) = \begin{cases} 0 & \text{if} \quad x \text{ is irrational} \\ 1 & \text{if} \quad x \text{ is rational} \end{cases}$$

Consequently, if $\lim_{x\to 0} g(f(x)) = \ell$, then $\ell = 0$. However, this is not possible, because every open interval I that contains 0 also contains nonzero rational numbers z and $g(f(z)) = 1$. Thus, for example, if $\epsilon = 1/2$, there is no $\delta > 0$ such that $0 < |x| < \delta$ implies that $|g(f(x))| < \epsilon$.

Comparing this situation to Theorem 2, the function f is continuous at $x = 0$ (why?) and g has a limit at $f(0) = 0$, but $g \circ f$ does not have a limit at $x = 0$.

79. The situation is equivalent to two hikers starting at 8:00 AM, one at the bottom of the path and the other at the top. Since the hikers reach the other end of the path on the same day and at the same time, 6:00 PM, there must have been a time during the day when they met each other on the trail. That is, they were at the same elevation. See the picture on the right. It shows that even if the hikers took *different* paths, they would be at the same elevation sometime during the day.

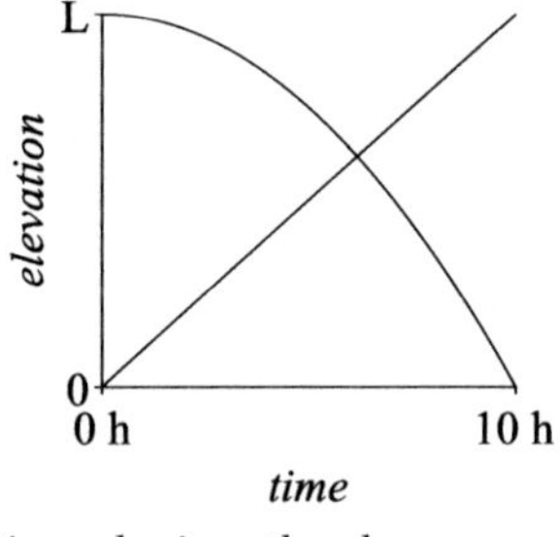

81. Observe that $g(0) = f(1/2) - f(0)$ and $g(1/2) = f(1) - f(1/2)$. Since $f(0) = f(1)$, $g(0) = -g(1/2)$. If $g(0) = 0$, then $g(1/2) = 0$ also, implying that $f(1) = f(1/2)$. That is, $f(1/2) = f(1/2 + 1/2)$, and $c = 1/2$ will satisfy required the condition. If $g(0) \neq 0$, then $g(1/2) \neq 0$, and these two numbers have opposite signs. Therefore, the Intermediate Value Theorem,

applied to g on $[0, 1/2]$, yields a number c in $(0, 1/2)$ such that $g(c) = 0$. That is, $f(c + 1/2) - f(c) = 0$, and $f(c) = f(c + 1/2)$.

CALCULATOR/COMPUTER EXERCISES

83. Let p be a real cubic polynomial. According to Exercise 58, the equation $p(x) = c$ has at least one real solution, x_0. Because the tangent function maps the open interval $I = (-\pi/2, \pi/2)$ *onto* $(-\infty, \infty)$, there is a number $x \in I$ such that $\tan(x) = x_0$, and $p(\tan(x)) = p(x_0) = c$.

    Using *Maple*'s *fsolve* procedure, $x = 1.1601$ is a solution to the equation $\tan^3(x) + 3\tan(x) = 19$.

85. The graph of the function f is plotted on the right. f does not have a continuous extension to $c = 0$ because it has a jump discontinuity at 0. The left limit and the right limit as $x \to 0$ both exist, but they are not the same.

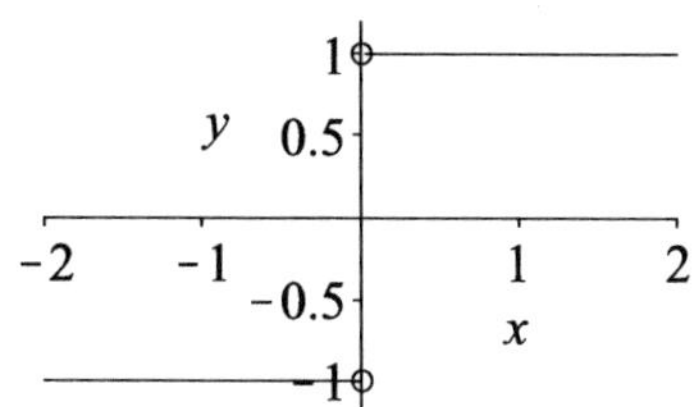

87. The graph of the function f is plotted on the right over the interval $I = [1.5, 2.5]$. f appears to have a continuous extension at $c = 2$ because the left limit and the right limit as $x \to 2$ both exist, and they are the same, -8.

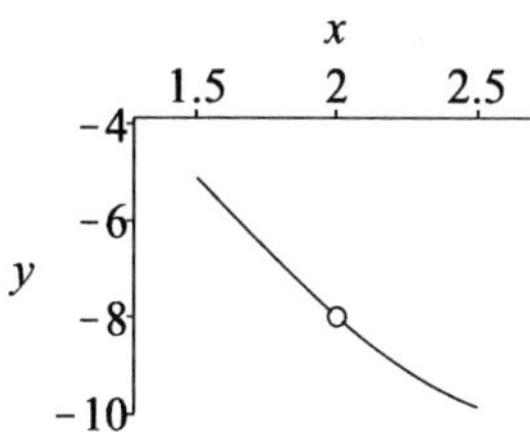

89. The quadratic $x^2 - 6x + 10$ has two complex roots. Since it attains the value 10 at $x = 0$, all of its values must be positive (otherwise it would have a real root by the Intermediate Value Theorem). This implies that $x^4 - 10x^3 + 38x^2 - 64x + 40 \geq 0$ and $p(x) \geq 10$.

    Since $p(x) = (x^2 - 6x + 10)(x - 2)^2 + 10$, $p(2) = 10$. This is the minimum value of p. Moreover, $p(x)$ grows without bound as x grows without bound. Observe, for example, that when $|x| \geq 100$,

$$p(x) = x^4(1 - 10/x + 38/x^2 - 64/x^3 + 50/x^4) \geq x^4/2\,.$$

    Therefore, by the Intermediate Value Theorem, $p(x)$ attains *all* values in the semi-infinite interval $[10, \infty)$. That is, $p(x) = \gamma$ has a solution for all $\gamma \geq 10$.

    Since $p(0) = 50$ and $p(1) = 15$, p attains the value 20 in the interval $(0, 1)$. Because $p(4) = 18$ and $p(5) = 55$, p also attains the value 20 in the interval $(4, 5)$. According to *Maple*'s *fsolve* procedure, $p(x) = 20$ when $x = 0.7266983053$ and $x = 4.113227465$.

91. Using *Maple*'s *Maximize* and *Minimize* procedures we find that on the interval $I = [-1, 2]$ f's continuous extension attains a maximum value of 0.84444 at $x = 0.9369$ and a minimum value of -0.0812 at $x = -1$. See the picture.

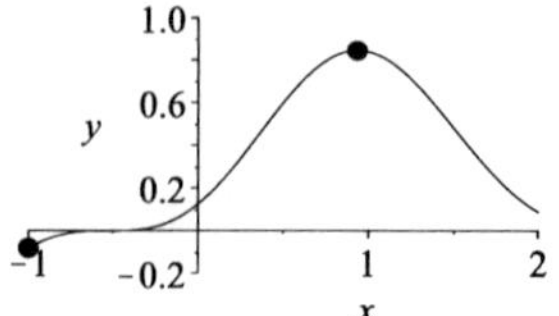

2.4 Infinite Limits and Asymptotes

Problems for Practice

1. $\lim_{x\to 6} \frac{3}{(x-6)^2} = \lim_{x\to 6} \frac{3}{(\text{small }+)} = \infty$

3. $\lim_{x\to+\infty} \frac{x^4+3x-92}{x^5-7x^2+44} = \lim_{x\to+\infty} \frac{1/x+3/x^4-92/x^5}{1-7/x^3+44/x^5} = 0$

5. $\lim_{x\to+\infty} \frac{x+\sqrt{x}}{x-\sqrt{x}} = \lim_{x\to+\infty} \frac{1+1/\sqrt{x}}{1-1/\sqrt{x}} = 1$

7. $\lim_{x\to+\infty} \frac{x+\cos(x)}{x-\sin(x)} = \lim_{x\to+\infty} \frac{1+\cos(x)/x}{1-\sin(x)/x} = 1$

9. $\lim_{x\to 1^+} \frac{1}{\sqrt{x-1}} = \lim_{x\to 1^+} \frac{1}{(\text{small }+)} = \infty$

11. $\lim_{x\to+\infty} \frac{x^2-4x+9}{3x^2-8x+18} = \lim_{x\to+\infty} \frac{1-4/x+9/x^2}{3-8/x+18/x^2} = 1/3$

13. $\lim_{x\to 0^+} \csc^2(x) = \lim_{x\to 0^+} \frac{1}{\sin^2(x)} = \lim_{x\to 0^+} \frac{1}{(\text{small }+)} = \infty$

15. $\lim_{x\to 0} \frac{3}{\sqrt{|x|}} = \lim_{x\to 0} \frac{3}{(\text{small }+)} = \infty$

17. $\lim_{x\to 0^-} \cot(x) = \lim_{x\to 0^-} \frac{1}{\tan(x)} = \lim_{x\to 0^-} \frac{1}{(\text{small }-)} = -\infty$

19. $\lim_{x\to 2^-} \tan(\pi/x) = \lim_{x\to 2^-} \frac{\sin(\pi/x)}{\cos(\pi/x)} = \lim_{x\to 2^-} \frac{+}{(\text{small }-)} = -\infty$

21. There is a vertical asymptote at $x = 7$ and the line $y = 1$ is a horizontal asymptote. See the graph on the right.

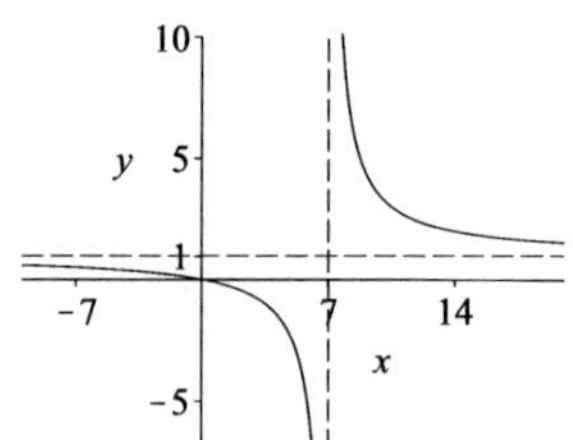

23. The y-axis is a vertical asymptote and the x-axis is a horizontal asymptote. See the graph on the right.

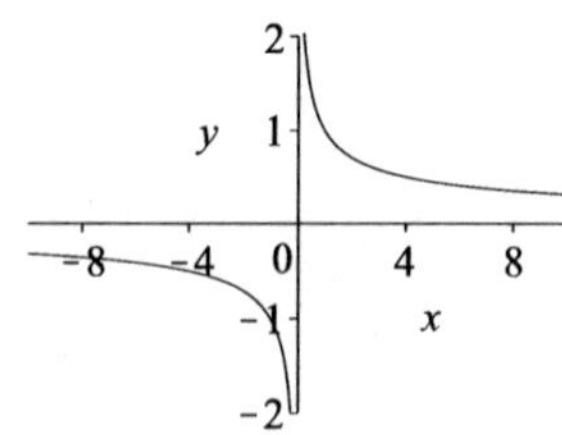

25. There are no vertical asymptotes. The line $y = 1$ is a horizontal asymptote. See the graph on the right.

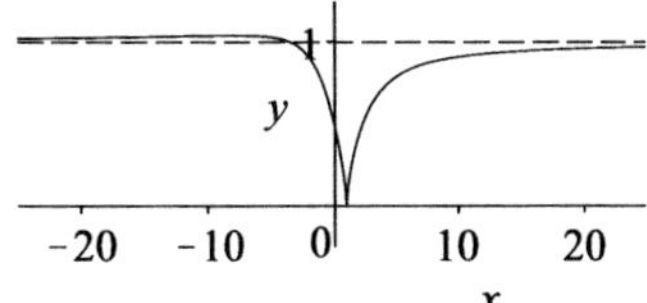

27. There are no vertical asymptotes. The x-axis is a horizontal asymptote. See the graph on the right.

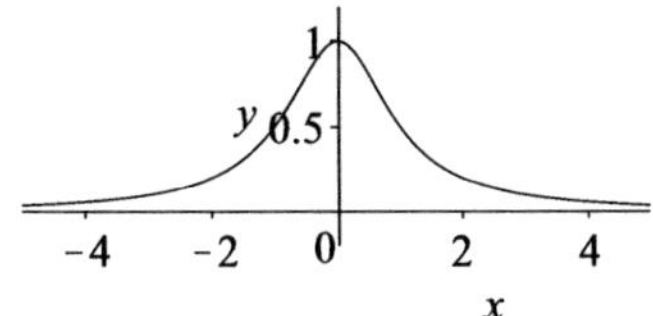

29. The y-axis is a vertical asymptote and the x-axis is a horizontal asymptote. See the graph on the right.

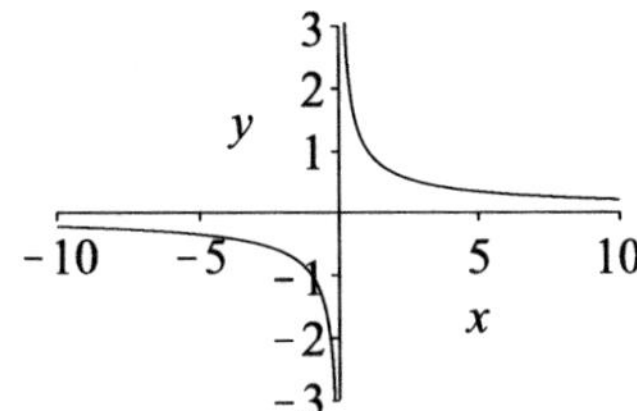

31. There are three vertical asymptotes, one at $x = -4$, another at $x = -1$, and another at $x = 0$, that is, the y-axis. The x-axis is a horizontal asymptote. See the graph on the right.

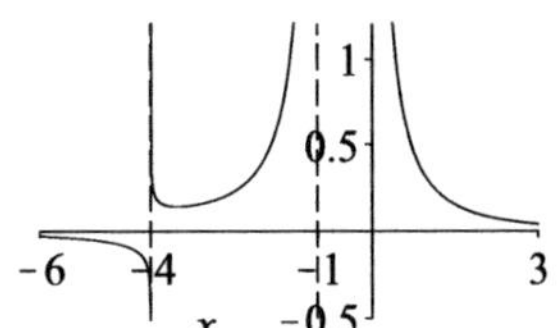

33. The y-axis is a vertical asymptote. There are no horizontal asymptotes. See the graph on the right.

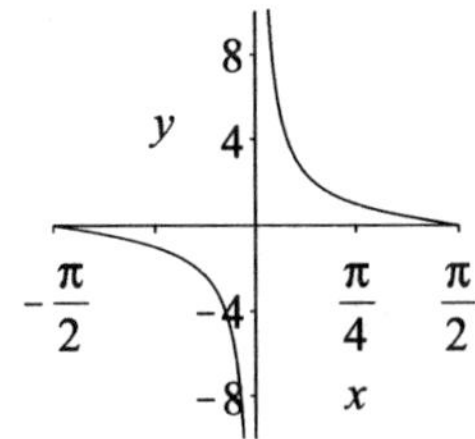

Further Theory and Practice

35. Let f be a function with a domain that contains a set of the form (a, ∞). Then f cannot have two distinct limits as $x \to \infty$. This is verified by assuming to the contrary that there are distinct numbers ℓ and m such that $\lim_{x\to\infty} f(x) = \ell$ and $\lim_{x\to\infty} f(x) = m$. Let $\epsilon = |\ell - m|$. Since $\epsilon > 0$ we can choose x sufficiently large to ensure that $f(x)$ is within $\epsilon/2$ of both ℓ and m. It follows from the Triangle Inequality that

$$\epsilon = |\ell - m| = |\ell - f(x) + f(x) - m| \leq |\ell - f(x)| + |f(x) - m| < \epsilon/2 + \epsilon/2\,.$$

That is, $\epsilon < \epsilon$. This is not possible and we conclude that $\ell = m$.

The argument for uniqueness of the limit of $f(x)$ as $x \to -\infty$ is essentially the same.

37. The Pinching Theorem, $x \to +\infty$. Let f, g, and h be functions with domains that include the interval (a, ∞). Assume further that $g(x) \leq f(x) \leq h(x)$ for all $x > a$. If $\lim_{x\to+\infty} g(x) = \ell$ and $\lim_{x\to+\infty} h(x) = \ell$, then $\lim_{x\to+\infty} f(x) = \ell$. There is a similar Pinching Theorem for the case where $x \to -\infty$. Its proof is exactly like the following proof for $x \to +\infty$ so the details are left to the reader.

    Verification of the Pinching Theorem, $x \to +\infty$. Let $\epsilon > 0$. Choose $N > a$ and large enough so that $|g(x) - \ell| < \epsilon$ and $|h(x) - \ell| < \epsilon$ when $x > N$. Since $g(x) \leq f(x) \leq h(x)$, $g(x) - \ell \leq f(x) - \ell \leq h(x) - \ell$ also. Consequently, if $x > N$, then

$$-\epsilon < -|g(x) - \ell| \leq g(x) - \ell \leq f(x) - \ell \leq h(x) - \ell \leq |h(x) - \ell| < \epsilon,$$

    implying that $|f(x) - \ell| < \epsilon$.

39. Let N be given. Since $\lim_{x\to c} G(x) = +\infty$, there is a $\delta_1 > 0$ such that $G(x) > N + 1 - \ell$ for all x satisfying $0 < |x - c| < \delta_1$. Because $\lim_{x\to c} F(x) = \ell$, there is a $\delta_2 > 0$ such that $F(x) > \ell - 1$ for all x satisfying $0 < |x-c| < \delta_2$. Let $\delta = \min(\delta_1, \delta_2)$. Then if x satisfies $0 < |x-c| < \delta$,

$$F(x) + G(x) > \ell - 1 + N + 1 - \ell = N.$$

    Suppose that $\lim_{x\to c} H(x) = -\infty$. Then $\lim_{x\to c}(-H(x)) = \infty$. Moreover, because $\lim_{x\to c}(-F(x)) = -\ell$, we know that $\lim_{x\to c}(-F(x) - H(x)) = \infty$. Consequently, $\lim_{x\to c}(F(x) + H(x)) = -\infty$.

41. There are several vertical asymptotes, one at every point $x = n\pi$, n a nonzero integer. There are no horizontal asymptotes. See the picture on the right displaying the graph of f and a few plotted points.

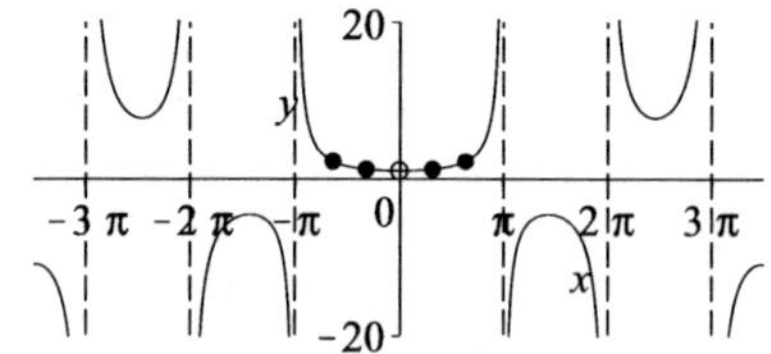

43. There is a vertical asymptote at $x = 0$, i.e. the y-axis. The x-axis is a horizontal asymptote. See the picture on the right displaying the graph of f and several plotted points.

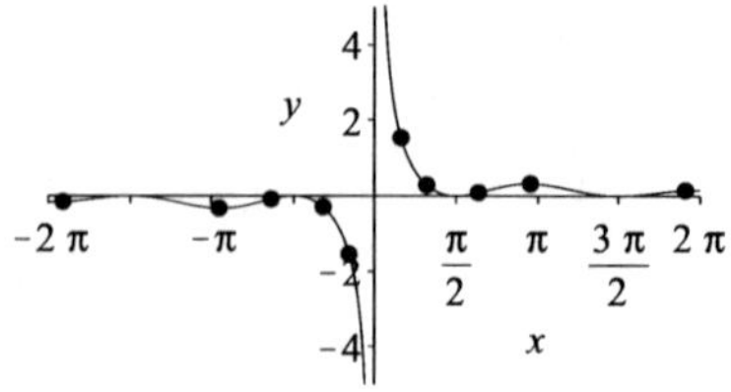

45. There are several vertical asymptotes, one at every point $x = n\pi$, n an integer. There are no horizontal asymptotes. See the picture on the right displaying the graph of f and several plotted points.

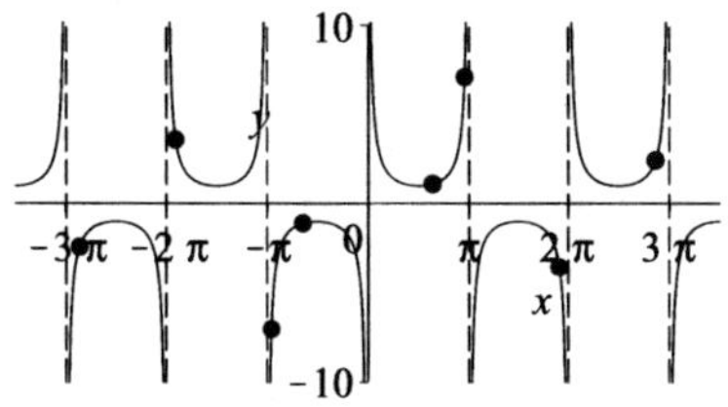

47. In spite of the fact that $f(x)$ is unbounded as $x \to 0$, it also attains the value 0 again and again so it does not have a vertical asymptote at $x = 0$. The graph is $|\sin(x)|$ with an amplitude $1/|x|$ that grows without bound. See the picture on the right displaying the graph of f near the origin

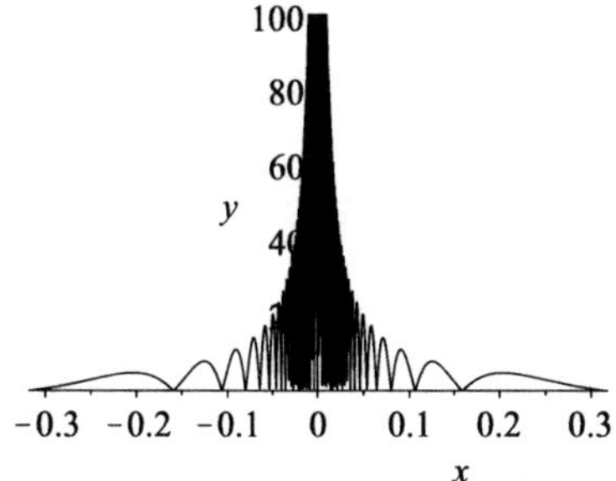

49. $f(10^{80}) = 10^{-20}$ and $g(10^{80}) = 10^{20}$. The f value is 10^{40} times smaller than the g value. In spite of this, $\lim_{x\to\infty} f(x) = \infty$ and $\lim_{x\to\infty} g(x) = 0$. The behavior of f and g on the interval $(0, 10^{80})$ is irrelevant to their limiting behaviors as $x \to \infty$.

51. (Van de Waal's Equation) The domain of the pressure function is $b < V < \infty$. As $V \to \infty$, $P \to 0$. See the graph on the right.

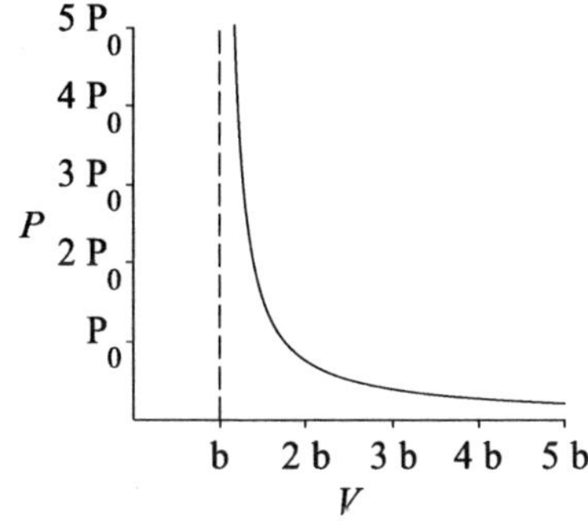

53. No, there can be no vertical asymptote. Since f is continuous and the interval is closed and bounded, f attains a maximum value and a minimum value. This would not be possible if it had a vertical asymptote at some point in the interval.

55. (a) If $m = 0$, then, for example, $\lim_{x\to+\infty}(f(x) - b) = 0$, implying that $\lim_{x\to+\infty} f(x) = b$, and the line $y = b$ is a horizontal asymptote.

(b) Let $\lim_{x\to+\infty}(p(x)/q(x) - (mx + b)) = 0$ where p and q are polynomials and $m \neq 0$. This implies that $\lim_{x\to+\infty} |p(x)/q(x)| = \infty$, so $\deg(p) > \deg(q)$. Moreover, the fact that

$$\lim_{x\to+\infty} \frac{p(x) - (mx + b) \cdot q(x)}{q(x)} = 0$$

implies that the degree of the polynomial $p(x) - mx \cdot q(x) - b \cdot q(x)$ must be *less* the degree of q. This would not be possible if $\deg(p) > \deg(q) + 1$ (why?) and we conclude that $\deg(p) = \deg(q) + 1$.

(c) We say that the graph of a function f is asymptotic to the parabola $y = ax^2 + bx + c$, $a \neq 0$, if either $\lim_{x\to+\infty}(f(x) - (ax^2 + bx + c)) = 0$ or $\lim_{x\to-\infty}(f(x) - (ax^2 + bx + c)) = 0$.

If $f(x) = p(x)/q(x)$ is a quotient of polynomials (with no common factors) this is equivalent to the condition that $\deg(p) = \deg(q) + 2$.

The necessity of this condition can be verified with an argument similar to the one used in part (b).

The sufficiency is established by dividing q into p to obtain a polynomial r such that $\deg(r) < \deg(q)$ and $p(x) = (ax^2+bx+c)\cdot q(x)+r(x)$, where $a \neq 0$. Then $p(x)/q(x) = ax^2 + bx + c + r(x)/q(x)$ and

$$\lim_{x\to\pm\infty}(f(x) - (ax^2 + bx + c)) = \lim_{x\to\pm\infty}(r(x)/q(x)) = 0\,.$$

57. Because $F(x) = x^2-3x+9-27/(x+3)$, $\lim_{x\to\pm\infty}(F(x)-(x^2-3x+9)) = 0$ and $y = x^2 - 3x + 9$ is a parabolic asymptote for F.

$G(x) = x^2 - 5x + 25 - 125/(x+5)$, so $\lim_{x\to\pm\infty}(G(x) - (x^2 - 5x + 25)) = 0$ and $y = x^2 - 5x + 25$ is a parabolic asymptote for G.

Long division will show that

$$H(x) = x^2 + 1 - 1/(x^{1/3} + 1)$$

so

$$H(x) - (x^2 + 1) = -\frac{1}{x^{1/3} + 1} \xrightarrow[x\to\pm\infty]{} 0\,,$$

and $y = x^2 + 1$ is a parabolic asymptote for H. There is also a vertical asymptote at $x = -1$. See the graph on the right.

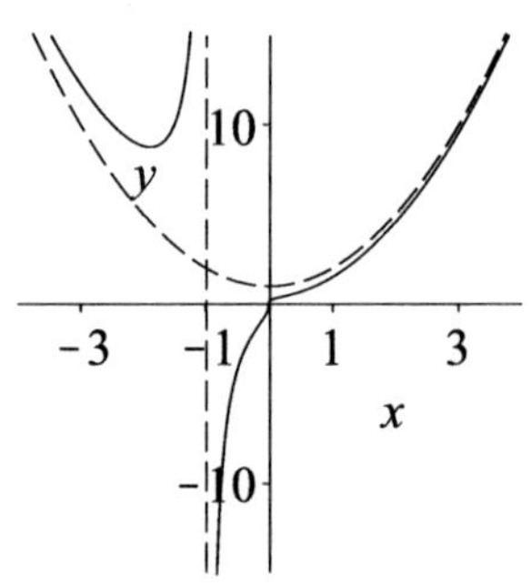

CALCULATOR/COMPUTER EXERCISES

59. Using *Maple*'s *fsolve* procedure we find that $\sqrt{x}/(x + 1)$ is smaller than 0.001 when $x > 999998$.

61. The line $y = 3$ is a horizontal asymptote to the graph of f. See the picture on the left. The picture on the right is in a window where the function and its right asymptote appear to be the same.

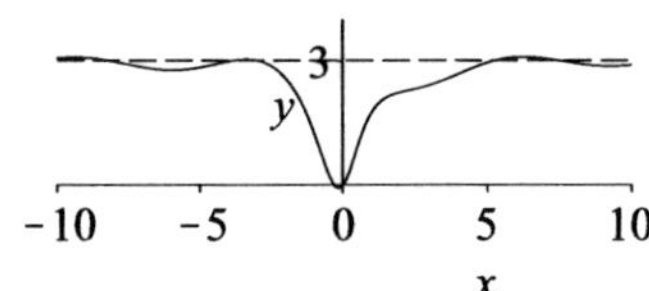

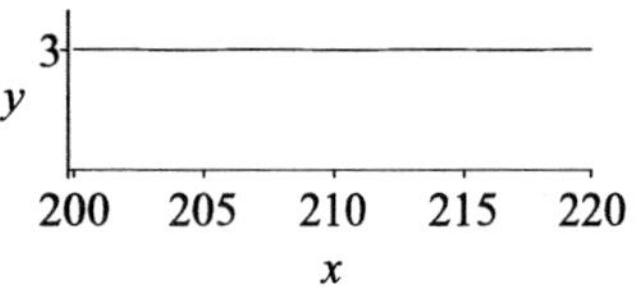

63. The line $y = 1$ is a horizontal asymptote to the graph of f. See the picture on the left. The picture on the right is in a window where the function and its right asymptote appear to be the same.

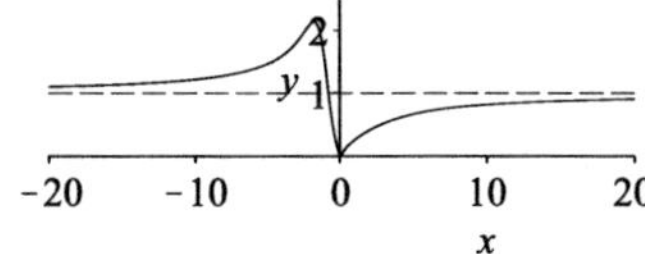

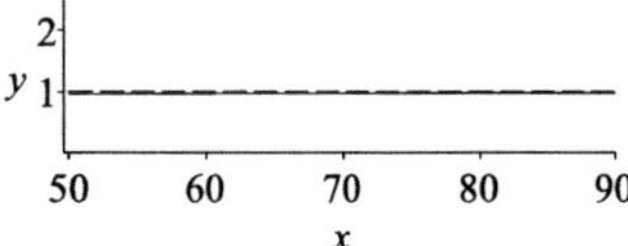

65. The line $x = \pi/2$ is a vertical asymptote. There are no horizontal asymptotes. See the graph on the right.

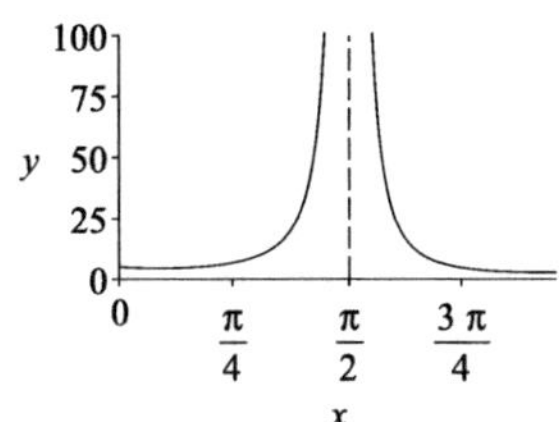

2.5 Limits of Sequences

Problems for Practice

1. Because $0 < a_n < 1/n$, $\lim_{n\to\infty} a_n = 0$.

3. $\lim_{n\to\infty} \frac{3n}{2n+2} = \lim_{n\to\infty} \frac{3}{2+2/n} = 3/2$

5. Because $0 < a_n < 1/n$, $\lim_{n\to\infty} a_n = 0$.

7. Since $a_n = (-1)^n$, $\lim_{n\to\infty} a_n$ does not exist.

9. $\lim_{n\to\infty}((1/3)^n + (1/2)^n) = \lim_{n\to\infty}(1/3)^n + \lim_{n\to\infty}(1/2)^n = 0 + 0 = 0$

11. $\lim_{n\to\infty} \frac{3n^3-5}{4n^2+5} = \lim_{n\to\infty} \frac{3-5/n^3}{4/n+5/n^3} = \infty$ (*diverges* to infinity)

13. $\lim_{n\to\infty} \frac{4+(-n)^3}{n^3+1} = \lim_{n\to\infty} \frac{4/n^3-1}{1+1/n^3} = -1$

15. $\lim_{n\to\infty}(1/100)^{1/n} = 1$ (Theorem 1 f)

17. $\lim_{n\to\infty}(2^{-n} - 6) = \lim_{n\to\infty} 2^{-n} - \lim_{n\to\infty} 6 = -6$

19. $\lim_{n\to\infty} \frac{4n^2+6n+3}{8n^2+3} = \lim_{n\to\infty} \frac{4+6/n+3/n^2}{8+3/n^2} = 4/8 = 1/2$

21. $\lim_{n\to\infty}(3 + 1/n) \cdot (2 - 5/n^2) = \lim_{n\to\infty}(3 + 1/n) \cdot \lim_{n\to\infty}(2 - 5/n^2)$
$= 3 \cdot 2 = 6$

23. $\lim_{n\to\infty}\left(2^{-n}\cdot\frac{n}{n+4}\right) = (\lim_{n\to\infty} 2^{-n})\cdot\left(\lim_{n\to\infty}\frac{1}{1+4/n}\right) = 0\cdot 1 = 0$

25. $\lim_{n\to\infty}\frac{2-(1/2)^n}{4+(1/3)^n} = \frac{\lim_{n\to\infty}(2-(1/2)^n)}{\lim_{n\to\infty}(4+(1/3)^n)} = 2/4 = 1/2$

27. $\lim_{n\to\infty}\frac{2^n-3^n}{3^n+4^n} = \lim_{n\to\infty}\frac{2^n/4^n-3^n/4^n}{3^n/4^n+1} = \lim_{n\to\infty}\frac{(1/2)^n-(3/4)^n}{(3/4)^n+1} = 0$

29. $\lim_{n\to\infty}(2\sin(1/n)+3\cos(1/n)) = \lim_{n\to\infty}2\sin(1/n)+\lim_{n\to\infty}3\cos(1/n)$
$= 0+3 = 3$

31. $\lim_{n\to\infty}\tan\left(\frac{\pi n}{3n+1}\right) = \tan\left(\lim_{n\to\infty}\frac{\pi}{3+1/n}\right) = \tan(\pi/3) = \sqrt{3}$

33. $\lim_{n\to\infty}\frac{n^8}{(2n^4+n^2+1)^2} = \lim_{n\to\infty}\frac{1}{(2n^4+n^2+1)^2/n^8} = \lim_{n\to\infty}\frac{1}{(2+1/n^2+1/n^4)^2}$
$= 1/4$

35. $\lim_{n\to\infty}\tan(\pi\cdot 2^{1/n}) = \tan(\lim_{n\to\infty}(\pi\cdot 2^{1/n}) = \tan(\pi) = 0$

37. $\lim_{n\to\infty}\sin\left(\pi\sin\left(\frac{\pi n}{6n+2}\right)\right) = \sin\left(\pi\sin\left(\lim_{n\to\infty}\frac{\pi}{6+2/n}\right)\right) = \sin(\pi\sin(\pi/6))$
$= \sin(\pi/2) = 1$

39. $-1/n \le \cos(n)/n \le 1/n \implies \lim_{n\to\infty}\cos(n)/n = 0$

41. Since $\lim_{n\to\infty}\frac{2^n}{3\cdot 2^n\pm 1} = \lim_{n\to\infty}\frac{1}{3\pm 2^{-n}} = \frac{1}{3}$,

$$\frac{2^n}{3\cdot 2^n+1} \le \frac{2^n}{3\cdot 2^n+\cos(2^n)} \le \frac{2^n}{3\cdot 2^n-1} \implies \lim_{n\to\infty}\frac{2^n}{3\cdot 2^n+\cos(2^n)} = \frac{1}{3}.$$

43. Since $\lim_{n\to\infty}\frac{\pm n}{5\pm 6n} = \lim_{n\to\infty}\frac{\pm 1}{5/n\pm 6} = \frac{1}{6}$,

$$\frac{n}{5+6n} \le \frac{(-1)^n n}{5+6(-1)^n n} \le \frac{-n}{5-6n} \implies \lim_{n\to\infty}\frac{(-1)^n n}{5+6(-1)^n n} = \frac{1}{6}.$$

45. $1+1/3+1/3^3+1/3^4+\cdots = 1/(1-1/3) = 3/2$

47. $1+1/\sqrt{3}+1/(\sqrt{3})^2+1/(\sqrt{3})^3+\cdots = 1/(1-1/\sqrt{3})$
$= \sqrt{3}/(\sqrt{3}-1) = \sqrt{3}(\sqrt{3}+1)/2 = (3+\sqrt{3})/2$

49. $1+1/10+1/10^2+1/10^3+\cdots = 1/(1-1/10) = 10/(10-1) = 10/9$

Further Theory and Practice

51. $\lim_{n\to\infty}1/(n+7) = 0$. Given $\epsilon > 0$, $|1/(n+7)| < \epsilon$ when $|n+7| > 1/\epsilon$. Choose N to be the smallest positive integer such that $N+7 > 1/\epsilon$. Then if $n > N$, $n+7 > N+7 > 1/\epsilon$ and $|1/(n+7)| = 1/(n+7) < \epsilon$.

53. $\lim_{n\to\infty}(2n+3)/(n+5) = 2$. Since

$$\left|\frac{2n+3}{n+5}-2\right| = \left|\frac{2n+3-(2n+10)}{n+5}\right| = \left|\frac{7}{n+5}\right|,$$

given $\epsilon > 0$, let N be the smallest positive integer such that $N + 5 > 7/\epsilon$. Then if $n > N$, $(n+5)/7 > 1/\epsilon$ and $\left|\frac{2n+3}{n+5} - 2\right| = \frac{7}{n+5} < \epsilon$.

55. $$1/4 + 1/8 + 1/16 + 1/32 + \cdots = 1/4 \cdot (1 + 1/2 + 1/4 + 1/8 + \cdots) = 1/4 \cdot (1/(1 - 1/2)) = 1/(4-2) = 1/2$$

57. $$16/3 + 32/9 + 64/27 + 128/8 + \cdots = 16/3 \cdot (1 + 2/3 + (2/3)^2 + (2/3)^3 + \cdots) = 16/3 \cdot (1/(1 - 2/3)) = 16$$

59. $$123.01 + 23/10^4 + 23/10^6 + 23/10^8 + \cdots$$
$$= \frac{12301}{100} + \frac{23}{10^4} \cdot \left(1 + \left(\frac{1}{10}\right)^2 + \left(\frac{1}{10}\right)^4 + \cdots\right)$$
$$= \frac{12301}{100} + \frac{23}{10^4} \cdot \frac{1}{1-(1/10)^2} = \frac{12301}{100} + \frac{23}{10000-100} = \frac{608911}{4950}$$

61. $\{(-1)^n\}$

63. If $\lim_{n\to\infty} a_n = \ell$, then $\lim_{n\to\infty} a_{n+1} = \ell$ also. Therefore,

$$\lim_{n\to\infty}(a_n - a_{n+1}) = \lim_{n\to\infty} a_n - \lim_{n\to\infty} a_{n+1} = \ell - \ell = 0\,.$$

65. If $\lim_{n\to\infty} a_n = \ell$, then $\lim_{n\to\infty} a_{n+1} = \ell$ also. Therefore, because $\ell \neq 0$,

$$\lim_{n\to\infty} \frac{a_n}{a_{n+1}} = \frac{\lim_{n\to\infty} a_n}{\lim_{n\to\infty} a_{n+1}} = \frac{\ell}{\ell} = 1\,.$$

(a) If $m < n$, then we know that $\lim_{x\to\infty} p(x)/q(x) = 0$. Consequently $\lim_{j\to\infty} p(j)/q(j) = 0$ also.

(b) If $m = n$, then we know that $\lim_{x\to\infty} p(x)/q(x) = p_m/q_n$. Consequently $\lim_{j\to\infty} p(j)/q(j) = p_m/q_n$ also.

(c) If $m > n$, then we know that $\lim_{x\to\infty} p(x)/q(x) = \text{signum}(p_m/q_n)\cdot\infty$. Consequently $\lim_{j\to\infty} p(j)/q(j) = \text{signum}(p_m/q_n)\cdot\infty$ also.

69. Fixing n, $\lim_{m\to\infty} n/(n+m) = 0$, implying that

$$\lim_{n\to\infty}\lim_{m\to\infty} n/(n+m) = 0\,.$$

Fixing m, $\lim_{n\to\infty} n/(n+m) = 1$, implying that

$$\lim_{m\to\infty}\lim_{n\to\infty} n/(n+m) = 1\,.$$

CALCULATOR/COMPUTER EXERCISES

71. When $n = 100$, $n\sin(1/n) = 0.9999833334$.

We know that $\lim_{n\to\infty} n\sin(1/n) = 1$.

73. When $n = 10^6$, $n^{1/n} = 1.000013816$.

    It appears that $\lim_{n\to\infty} n^{1/n} = 1$.

75. When $n = 10^8$, $(\sqrt{n}/\sqrt{n+1}\,)^{\sqrt{n}} = 0.9999500012$.

    It appears that $\lim_{n\to\infty}(\sqrt{n}/\sqrt{n+1}\,)^{\sqrt{n}} = 1$.

77. Using the two graphs of f that are displayed below, it appears that the limiting value of $f(x)$ as $x \to 0^+$ is 1. Since $n^{1/n} = 1/f(1/n)$, and as $n \to \infty$, $1/n \to 0^+$, this tells us that $\lim_{n\to\infty} n^{1/n} = 1$.

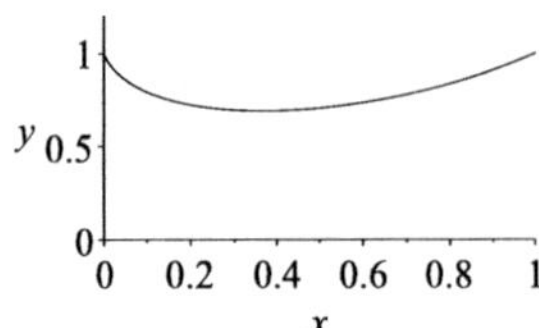

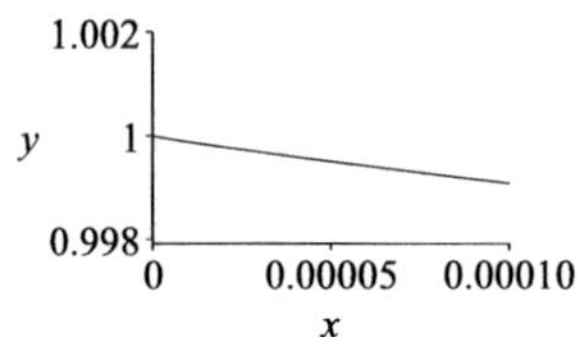

2.6 Exponential Functions and Logarithms

Problems for Practice

1. $\sqrt{2}^{\sqrt{3}}\sqrt{2}^{\sqrt{3}} = \sqrt{2^{\sqrt{3}}}\,\sqrt{2^{\sqrt{3}}} = 2^{\sqrt{3}}$

3. $(1/8)^{-\pi/3} = 8^{\pi/3} = (8^{1/3})^{\pi} = 2^{\pi}$

5. $\left(\sqrt{11}^{\sqrt{2}}\right)^{\sqrt{2}} = (\sqrt{11}\,)^2 = 11$

7. $(3^4 \cdot 9^3/27^2)^{1/2} = (3^{4+6-6})^{1/2} = 3^2 = 9$

9. $\log_5(1/125) = \log_5(5^{-3}) = -3$

11. $e^{\ln(3)} = 3$

13. $\log_{27}(9/3^x) = \log_{27}(3^{2-x}) = (2-x)\log_{27}(3) = (2-x)\log_{27}(27^{1/3}) = (2-x)/3$

15. $\log_2(\log_2(4)) = \log_2(2) = 1$

17. $\log_{4/9}(4^{x/2} \cdot 3^{4x} \cdot 2^{-5x}) = \log_{4/9}(2^x \cdot 9^{2x} \cdot 2^{-5x}) = \log_{4/9}(2^{-4x} \cdot 9^{2x})$
    $= \log_{4/9}((9/4)^{2x}) = -2x$

19. $2^{\log_8(27x^3)} = \left(8^{\log_8(27x^3)}\right)^{1/3} = (27x^3)^{1/3} = 3x$

21. $\exp\left(\sqrt{\ln(3^{\ln(81)})}\right) = \exp\left(\sqrt{\ln(3^4)\ln(3)}\right) = e^{\sqrt{4\ln(3)^2}} = e^{2\ln(3)} = 9$

23. Take the base 2 log of both sides: $\log_2(2^{-x}) = \log_2(2^{3+x})$. This implies that $-x = 3 + x$ and $x = -3/2$.

25. Write the equation in the form $2^{2x} \cdot 3^{2x} = 8 \cdot 6^{-x}$ implying that $6^{2x} = 8 \cdot 6^{-x}$. Therefore, $6^{3x} = 8$, implying that $3x \ln(6) = \ln(8)$ and $x = \ln(8)/(3\ln(6))$. In terms of the base 6 logarithm, $x = \log_6(8)/3 = \log_6(2)$.

27. The graph is sketched on the right. This is an exponential function, shifted up one unit. The line $y = 1$ is a horizontal asymptote.

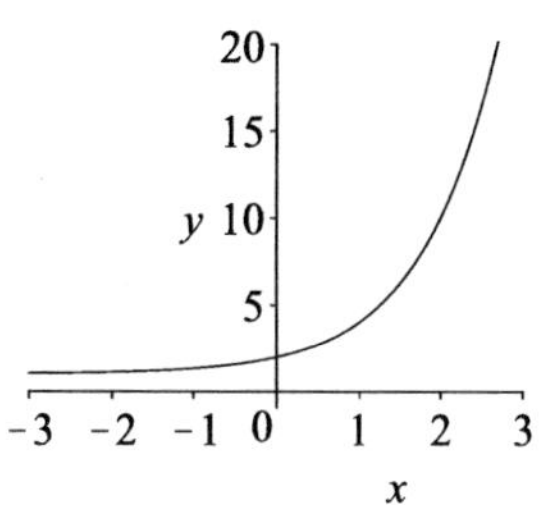

29. The graph is sketched on the right. This is an exponential function flipped over and shifted 3 units up. The line $y = 3$ is a horizontal asymptote.

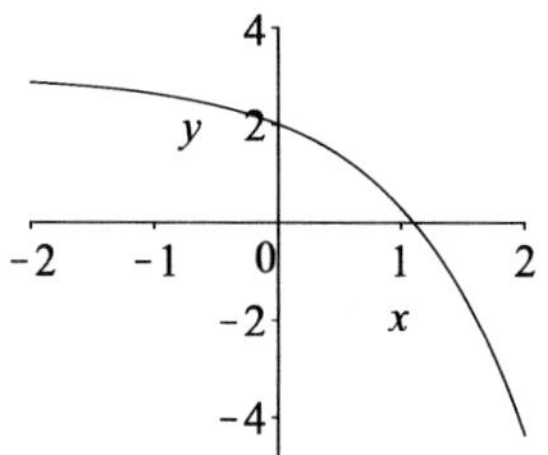

31. The graph is sketched on the right. This is a compressed log function shifted 3 units up. The line $x = 0$ is a vertical asymptote.

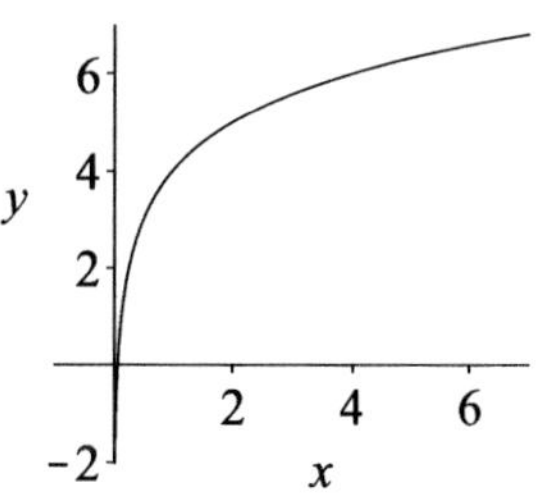

33. $\lim_{n\to\infty}(1 + 1/n)^n = e$

35. $\lim_{n\to\infty}(1 + e/n)^n = e^e$

37. $\lim_{n\to\infty}(1 + 3/(4n))^n = \lim_{n\to\infty}(1 + (3/4)/n)^n = e^{3/4}$

39. $\lim_{n\to\infty}(1 - \pi/n)^n = \lim_{n\to\infty}(1 + (-\pi)/n)^n = e^{-\pi}$

41. $\lim_{n\to\infty}(1 - \pi/(4n))^n = \lim_{n\to\infty}(1 + (-\pi/4)/n)^n = e^{-\pi/4}$

43. $\lim_{x\to e} 2^{\ln(x^3)} = 2^{\ln(e^3)} = 2^3 = 8$

45. $\lim_{x\to e^+}(e/\pi)^{1/(x-e)} = \lim_{y\to\infty}(e/\pi)^y = 0$

47. $\lim_{x\to 1} 5^{\log_3(x)} = 5^{\lim_{x\to 1}\log_3(x)} = 5^0 = 1$

49. $\lim_{x\to -\infty}(1 + 3/x)^x = \lim_{x\to\infty}(1 - 3/x)^{-x} = e^3$

51. $\lim_{x\to\infty} \frac{e^{2x}-e^{-2x}}{e^{2x}+e^{-2x}} = \lim_{x\to\infty} \frac{1-e^{-4x}}{1+e^{-4x}} = 1$

53. As $x \to \infty$, $y \to 1$ so the line $y = 1$ is a right horizontal asymptote.

As $x \to -\infty$, $y \to -1$ and the line $y = -1$ is a left horizontal asymptote.

In addition, the y axis is a vertical asymptote.

55. As $x \to \infty$, $y \to 0$. The x-axis is a right horizontal asymptote.

As $x \to -\infty$, $y \to 4$. The line $y = 4$ is a left horizontal asymptote.

In addition, there is a vertical asymptote at $x = 3$.

57. $P_0 = \$1000$, 6% annual, 12 months (1 year)

Compounded	Annually	Semiannually	Quarterly	Daily	Continuously
Value	$1060	$1060.90	$1061.36	$1061.83	$1061.84

59. $P_0 = \$5000$, 5.25% annual, 36 months (3 years)

Compounded	Annually	Semiannually	Quarterly	Daily	Continuously
Value	$5829.57	$5841.02	$5846.91	$5852.84	$5852.90

61. $P_0 = 5000$, $8000 = P_0 \cdot e^{\lambda \cdot 3}$, $\lambda = \ln(8/5)/3$, $P = P_0 \cdot e^{\lambda \cdot 8} = 17510$

63. $P_0 = 6500$, $8000 = P_0 \cdot e^{\lambda \cdot 2}$, $\lambda = \ln(80/65)/2$, $P = P_0 \cdot e^{\lambda \cdot 1} = 7211$

65. $P_0 = 5$, $4 = P_0 \cdot e^{\lambda \cdot 41}$, $\lambda = \ln(4/5)/41$, $P = P_0 \cdot e^{\lambda \cdot 55} = 3.707$ g

67. $P_0 = 12$, $10 = P_0 \cdot e^{\lambda \cdot 12}$, $\lambda = \ln(10/12)/12$, $P = P_0 \cdot e^{\lambda \cdot (-18)} = 15.77$ g

69. $P_0 = 12$, $8 = P_0 \cdot e^{\lambda \cdot 40}$, $\lambda = \ln(8/12)/40$

$$e^{\lambda \cdot T} = 1/2 \quad \text{when} \quad T = -\ln(2)/\lambda \approx 68.38 \text{ years}$$

71. $P_0 = 14$, $10 = P_0 \cdot e^{\lambda \cdot 100}$, $\lambda = \ln(10/14)/100$

$$e^{\lambda \cdot T} = 1/2 \quad \text{when} \quad T = -\ln(2)/\lambda \approx 206 \text{ years}$$

Further Theory and Practice

73. The graph is sketched on the right. This is the exponential function shifted down 1 unit. Then the portion below the x-axis is rotated above the x-axis. The line $y = 1$ is a left horizontal asymptote.

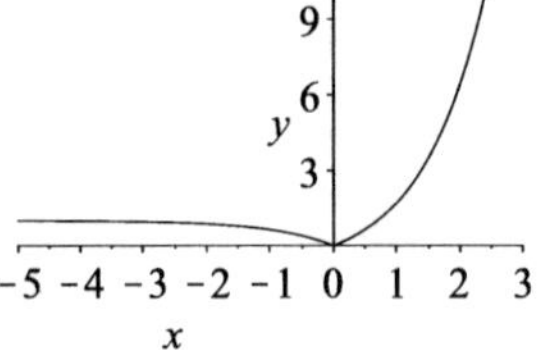

75. The graph is sketched on the right. For $x \geq 2$, this is the natural log function shifted to the right 2 units. That curve is then rotated about the line $x = 2$. There is a vertical asymptote at $x = 2$.

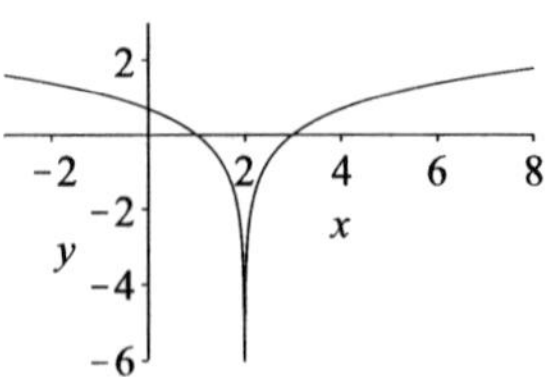

77. $\lim_{n\to\infty} \left(\frac{n+1}{n}\right)^n = \lim_{n\to\infty}(1 + 1/n)^n = e$

79. $\lim_{n\to\infty}\left(1+\frac{3}{n}\right)^{2n} = \lim_{n\to\infty}\left(\left(1+\frac{3}{n}\right)^{n}\right)^{2} = (e^3)^2 = e^6$

81. Since $e^u = \lim_{n\to\infty}\left(1+\frac{u}{n}\right)^n$, give $v > 0$,

$$v = e^{\ln(v)} = \lim_{n\to\infty}\left(1+\frac{\ln(v)}{n}\right)^n .$$

Let $a_n = 1 + \frac{\ln(v)}{n}$.

83. The present value P_0 must satisfy the equation $144000 = P_0 \cdot e^{0.065 \cdot t}$ for $t = 18$. Therefore, $P_0 = 144000 \cdot e^{-0.065 \cdot 18} = 44692.84$.

85. The present value is $P_0 = P_1 + P_2 + P_3$ where P_j must satisfy the equation $1000 = P_j \cdot e^{0.04 \cdot t_j}$ for $t_1 = 2$, $t_2 = 3$, and $t_3 = 4$. Therefore,

$$P_0 = 1000 \cdot (e^{-0.04 \cdot 2} + e^{-0.04 \cdot 3} + e^{-0.04 \cdot 4}) = 2662.18 .$$

87. If P_0 is to double at an annual rate of $r\%$, then it will take T years where $2P_0 = P_0 \cdot e^{r \cdot T/100}$. That is, $r \cdot T/100 = \ln(2)$ and $T = 100\ln(2)/r$ years. This is the "exact" rule. Since $100\ln(2) \approx 69.31$, the integer 69 gives a better approximation.

89. Because $k > 0$, $e^{-kt} \to 0$ as $t \to \infty$. Therefore, $\lim_{t\to+\infty} P(t) = M$. On the other hand, $e^{-kt} \to \infty$ as $t \to -\infty$. This implies that $\lim_{t\to-\infty} P(t) = 0$. Therefore, the line $y = M$ is a right horizontal asymptote, and the line $y = 0$, i.e. the t-axis, is a left horizontal asymptote.

91. Because $\beta > 0$, $e^{-\beta t} \to 0$ as $t \to \infty$. Therefore, $\lim_{t\to+\infty} C(t) = \alpha/\beta$. The line $y = \alpha/\beta$ is a right horizontal asymptote. See the graph on the right.

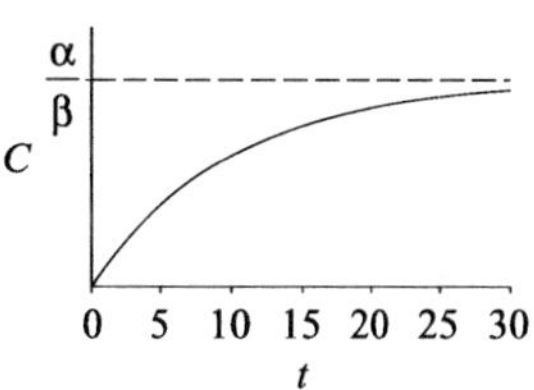

93. Observe that x_n is the point half-way between α and x_{n-1}. Since $x_0 = 1$, $1 \le x_n < \alpha$ for all n, and $\{x_n\}$ is bounded. To see that the sequence is increasing note that $x_n - x_{n-1} = \alpha/2 + x_{n-1}/2 - x_{n-1} = (\alpha - x_{n-1})/2 > 0$ implying that $x_n > x_{n-1}$.

The sequence converges by the Monotone Convergence Theorem. Let ℓ denote its limiting value. Then, because $\lim_{n\to\infty} x_{n-1} = \ell$ also,

$$\ell = \lim_{n\to\infty} x_n = \lim_{n\to\infty} (\alpha + x_{n-1})/2 = (\alpha + \ell)/2 ,$$

implying that $\ell = \alpha$.

95. The percentage $P(t)$ of ^{14}C at the time t years after death decreases exponentially with the same half life as ^{14}C: $P(t) = P_0 \cdot (1/2)^{t/5700}$. For the woolly mammoths on Wrangel Island $t = 1991 + 1700 = 3691$ years so $P(t) = P_0 \cdot (1/2)^{3691/5700} \approx 0.6384 P_0$ and the amount of m_0 was approximately 63.68%.

97. The wood that produced the charcoal was burned t years ago where t satisfies the equation $0.0879 \cdot P_0 = P_0 \cdot (1/2)^{t/5700}$. Therefore, $\ln(0.0879) = (t/5700) \cdot \ln(1/2)$ and $t = 5700 \cdot (\ln(0.0879)/\ln(0.5)) \approx 19996$ years.

CALCULATOR/COMPUTER EXERCISES

99. The graph of $y = \exp(x)$ is on the left, below. On the right is the plot of the same curve with the three secant lines and the three lines with the requested slopes. Indeed, they are indistinguishable. See the table of values of the secant line slopes $m(c)$ and $|\exp(c) - m(c)|$.

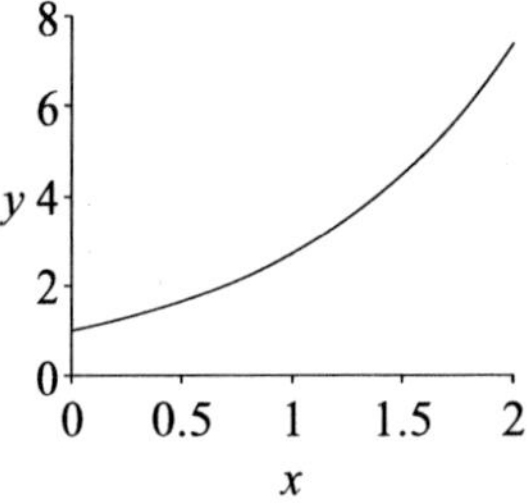

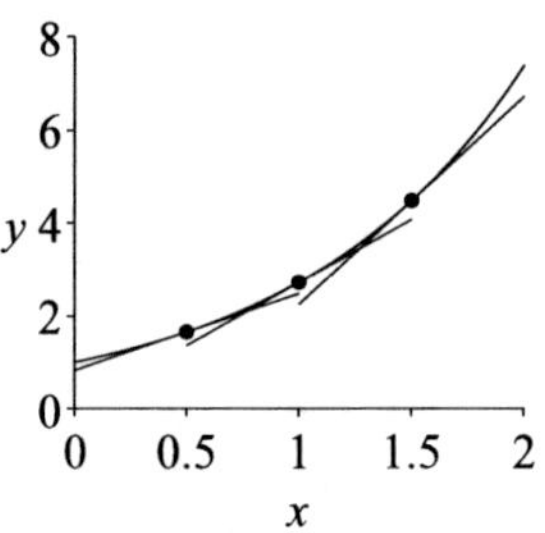

c	$m(c)$	$\lvert\exp(c) - m(c)\rvert$
0.5	1.6487215	2.29×10^{-7}
1.0	2.7182825	6.72×10^{-7}
1.5	4.4816900	9.30×10^{-7}

REVIEW EXERCISES FOR CHAPTER 2

1. Since $\sin(2x) \to 0$ as $x \to \pi$, the limit does not exist.

3. f is continuous at $c = -4$, $\lim_{x\to -4} f(x) = f(-4) = -2$.

5. If $x \neq -6$, then $f(x) = x - 6$ so $\lim_{x\to -6} f(x) = -12$.

7. If $x \neq -5$, $f(x) = 1/(x-5)$, so $\lim_{x\to -5} f(x) = -1/10$.

9. $\lim_{x\to 5^-} f(x) = 6$. Since $f(x) = (x+7)/(x-3)$ when $x > 5$, $\lim_{x\to 5^+} f(x) =$ 6 also, and $\lim_{x\to 5} f(x) = 6$.

11. The tangent function has a vertical asymptote at $x = \pi/2$. Therefore, this function has no limit as $x \to 1/2$.

13. f is continuous at $c = -3$, $\lim_{x\to -3} f(x) = f(-3) = -1$.

15. $\sin(x)/x$ has limit 1 as $x \to 0$, so $\lim_{x\to 0}(x/\sin(x)) = 1$ also.

17. We know that $\lim_{x\to 0}(1/f(x)) = 1/2$. Therefore, $\lim_{x\to 0} f(x) = 2$.

19. Since f is continuous at $c = 64$, $\lim_{x\to 64} f(x) = f(64) = (4+28)/8 = 4$.

21. $\lim_{x\to 0} \frac{1-\cos(x)}{x} = \lim_{x\to 0}\left(x \cdot \frac{1-\cos(x)}{x^2}\right) = \left(\lim_{x\to 0} x\right)\cdot\left(\lim_{x\to 0} \frac{1-\cos(x)}{x^2}\right)$
$= 0 \cdot \frac{1}{2} = 0$

23. $\lim_{x\to 0} \frac{\sin(7x)}{\sin(3x)} = \frac{7}{3}\cdot\lim_{x\to 0}\left(\frac{\sin(7x)}{7x}\cdot\frac{3x}{\sin(3x)}\right) = \frac{7}{3}\cdot\left(\lim_{x\to 0}\frac{\sin(7x)}{7x}\right)\cdot\left(\lim_{x\to 0}\frac{3x}{\sin(3x)}\right)$
$= 7/3$

25. $\lim_{x\to 0} \frac{1-\cos(x)}{\sin(x)} = \lim_{x\to 0}\left(\frac{1-\cos(x)}{x^2}\cdot\frac{x}{\sin(x)}\cdot x\right) = \frac{1}{2}\cdot 1\cdot 0 = 0$

27. $\lim_{x\to 0} \frac{\tan(3x)}{x} = \lim_{x\to 0}\left(\frac{\sin(3x)}{3x}\cdot\frac{3}{\cos(x)}\right) = 1\cdot 3 = 3$

29. $\lim_{x\to 0} \frac{\sin(x)+\cos(x)}{x}$ does not exist. As $x \to 0$, the numerator approaches 1 and the denominator approaches 0.

31. $\lim_{x\to 4} \frac{x^2-16}{\sqrt{x}-2} = \lim_{x\to 4}\frac{(x-4)(x+4)}{\sqrt{x}-2} = \lim_{x\to 4}\frac{(\sqrt{x}-2)(\sqrt{x}+2)(x+4)}{\sqrt{x}-2}$
$= \lim_{x\to 4}(\sqrt{x}+2)(x+4) = 32$

33. $\lim_{x\to 0} \frac{12x}{\sqrt{1+6x}-1} = \lim_{x\to 0}\left(\frac{12x}{\sqrt{1+6x}-1}\cdot\frac{\sqrt{1+6x}+1}{\sqrt{1+6x}+1}\right) = \lim_{x\to 0}\frac{12x(\sqrt{1+6x}+1)}{6x}$
$= \lim_{x\to 0} 2(\sqrt{1+6x}+1) = 4$

35. If $x > 5$, then $(x-5)/|x-5| = 1$. The limit is 1.

37. $\lim_{x\to 4^-}\left(\frac{\sqrt{16-4x}}{\sqrt{16-x^2}}\right) = 2\lim_{x\to 4^-}\sqrt{\frac{4-x}{16-x^2}} = 2\lim_{x\to 4^-}\frac{1}{\sqrt{4+x}} = \frac{2}{\sqrt{8}} = \sqrt{2}/2$

39. $\lim_{x\to 0^-}\frac{\sqrt{x^2}}{x} = \lim_{x\to 0^-}\frac{-x}{x} = -1$

41. $\lim_{x\to -\infty}(9 - e^{2x}) = 9 - 0 = 9$

43. $\lim_{x\to\infty}(\pi + 2^x) = \infty$

45. $\lim_{x\to\infty}(\cos(x) + 2e^{-x}) = \lim_{x\to\infty} \cos(x)$ does not exist

47. $\lim_{x\to\infty} \frac{7x^2+2x+1}{x^3+x+2} = \lim_{x\to\infty} \frac{7/x+2/x^2+1/x^3}{1+1/x^2+2/x^3} = 0$

49. $\lim_{x\to\infty} \frac{3^x+2^x}{3^x+2} = \lim_{x\to\infty} \frac{1+(2/3)^x}{1+2/3^x} = 1$

51. $\lim_{x\to\infty}(1 - 1/x)^x = \lim_{x\to\infty}(1 + (-1)/x)^x = e^{-1}$

53. $\lim_{x\to 0} x\cos^2(1/x) = 0$ by the Pinching Theorem: $|x\cos^2(1/x)| \leq |x|$

55. Note that $\ell = f(c)$. Because

$$|f(x) - f(c)| = |(x+1) - 0| = |x - (-1)| = |x - c|,$$

when $|x - c| < 0.01$, $|f(x) - f(c)| < 0.01$.

57. Note that $\ell = f(c)$ and

$$|f(x) - f(c)| = |2x^2 - 50| = 2|x^2 - 25| = 2 \cdot |x - 5| \cdot |x - c|.$$

Therefore, if $|x - c| < 0.01/(2 \cdot 11)$, then

$$|f(x) - f(c)| < 0.01 \cdot |x - 5|/11 < 0.01.$$

This is because $|x + 5| = |x - c| < 1$ so $|x| < 6$, and $|x - 5| \leq |x| + 5 < 11$.

59. Since $f(x) = 1$ if $x < 0$ and $f(x) = \sin(x)/x$ if $x > 0$, $\lim_{x\to 0} f(x) = 1$. Therefore, f has a continuous extension to F when $F(0) = 1$.

61. Since $f(x) = x - 1$ if $x < -1$, and $f(x) = x^2 + x - 2$ if $x > -1$, $\lim_{x\to(-1)^-} f(x) = -2$ and $\lim_{x\to(-1)^+} f(x) = -2$. Therefore, f has a continuous extension to F when $F(-1) = -2$.

63. This function has a vertical asymptote at $x = -3$. The line $y = 1$ is a horizontal asymptote.

65. This function has no vertical asymptotes. The line $y = 0$ (the x-axis) is a horizontal asymptote.

67. If $x \neq -2$, then $f(x) = 3(x-2)/(x+2)$. Therefore, its graph has a vertical asymptote at $x = -2$, and the line $y = 3$ is a horizontal asymptote.

69. $(1/27)^{-2/3} = 3^2 = 9$

71. $\left(2^4 \cdot 4^3/8^2\right)^{3/2} = \left(2^{4+6-6}\right)^{3/2} = 2^6 = 64$

73. $\log_4(8 \cdot 4^{2x} \cdot 2^{1-2x}) = \log_4(2^{3+4x+1-2x}) = \log_4(2^{4+2x})$

$$= (4 + 2x)\log_4(2) = (4 + 2x) \cdot (1/2) = 2 + x$$

75. $\log_{1/\sqrt{3}}(\log_4(64)) = \log_{1/\sqrt{3}}(3) = -2$

77. Apply $\ln$: $2x\ln(3) = \ln(18) + x\ln(2)$. Therefore, $x = \ln(18)/(\ln(9) - \ln(2))$.

79. At all points x where the floor function is continuous (that is, all non integer values of x), $f(x) = 0$. If x is an integer, then

$$f(x) = \lim_{y \to x^+} \lfloor y \rfloor - \lim_{y \to x^-} \lfloor y \rfloor = x - (x-1) = 1\,.$$

81. $\lim_{n\to\infty} 100^{1/n} = 1$

83. $\lim_{n\to\infty} \left(5^{-n} + 1/7^n\right) = 0 + 0 = 0$

85. $\lim_{n\to\infty} \frac{3n^2-5}{4n^3+5} = \lim_{n\to\infty} \frac{3/n-5/n^3}{4+5/n^3} = 0$

87. Because $A(6) = 75$, K must satisfy the equation $75 = 100(1 - e^{K\cdot 6})$. That is, $e^{6K} = 1/4$, so $6K = -2\ln(2)$ and $K = -\ln(2)/3$. The average score will equal 90 when $90 = 100(1 - e^{Kt})$. That is, $e^{Kt} = 1/10$, so $Kt = -\ln(10)$ and $t = -\ln(10)/K = 3\ln(10)/\ln(2) \approx 9.97$ weeks. Therefore, it will take approximately 3.97 more weeks for the average score to exceed 90.

89. Let A denote the maximum safe level of contamination (grams of ^{226}Ra per cubic centimeter, for example). Then the present contamination level is $5A$, and t years later it will be $C(t) = 5A \cdot (1/2)^{t/1620}$. The area will be unsafe until time T where $C(T) = A$. That is, $(1/2)^{T/1620} = 1/5$. Apply the base 2 logarithm to both sides, and $T = 1620 \cdot \log_2(5) \approx 3761.5$ years.

91. Let $P(t)$ denote the fraction of ^{14}C in the text t years after it was manufactured. Then $P(t) = P_0 \cdot (1/2)^{t/5700}$, where $P_0 = m_0/m$, m being the amount of stable ^{12}C in the text, and m_0 being the amount of ^{14}C in the text at time of manufacture.

   (a) If $t = 1995 + 100 = 2095$, then $P(t) = P_0 \cdot (1/2)^{2095/5700} \approx 0.775P_0$, and the fraction of m_0 would have been 0.775.

   (b) Since $t = 1995 - 180 = 1815$, then $P(t) = P_0 \cdot (1/2)^{1815/5700} \approx 0.802P_0$, and the fraction of m_0 was 0.802.

93. Let $P(t)$ denote the fraction of ^{14}C in the Iceman sample t years after death. Then $P(t) = P_0 \cdot (1/2)^{t/5700}$, where $P_0 = m_0/m$, m being the amount of stable ^{12}C in the sample, and m_0 being the amount of ^{14}C in the sample at the time of death. If $t = 1991 + 3325 = 5316$ years, then $P(t) = P_0 \cdot (1/2)^{5316/5700} = 0.524P_0$. The percentate of m_0 was 52.4%.

95. Let $P(t)$ denote the fraction of ^{14}C in the pigments t years after the paintings were made. Then $P(t) = P_0 \cdot (1/2)^{t/5700}$, where $P_0 = m_0/m$, m being the amount of stable ^{12}C in the pigments, and m_0 being the amount of ^{14}C in the pigments at the time the paintings were made. If one of the samples had 1.837 times the quantity of ^{14}C than the others, then the time difference t satisfies the equation $P(t) = 1.837P_0$. That is, $1.837P_0 = P_0 \cdot (1/2)^{t/5700}$, and $t = -5700 \cdot \log_2(1.837) \approx -5001$ years.

Chapter 3

The Derivative

3.1 Rates of Change and Tangent Lines

Problems for Practice

1. The instantaneous rate of change of a constant function is 0, so $p'(2) = 0$.

3. Using (3.1.6), $p'(c) = 7 \cdot 2\,c^1 = 14c$, so $p'(3) = 42$.

5. Using (3.1.9) and (3.1.6), $p'(c) = 6 + 0 = 6$, so $p'(3) = 6$.

7. Using (3.1.9) and (3.1.6), $p'(c) = 2 \cdot 3\,c^2 - 17 = 6c^2 - 17$, so $p'(2) = 7$.

9. Using (3.1.10) and (3.1.6), $p'(c) = 2\,c - 6 + 0 = 2c - 6$, so $p'(2) = -2$.

11. Using (3.1.13) with $n = -1$, $p'(c) = -1 \cdot c^{-2} = -c^{-2}$, so $p'(1/2) = -4$.

13. $p'(c) = 6$ for all c, so the body is always moving forward.

15. $p'(c) = -3c^2 + 10c$ so $p'(4) = -8$, and the body is moving backward at time $t = 4$.

17. Using (3.1.9), $f'(c) = 10c - 21$, so $f'(3) = 9$.

19. Using (3.1.9), $f'(c) = 6c + 2 \cdot (-1) \cdot c^{-2} = 6c + 2c^{-2}$, so $f'(-2) = -12 - 2/4 = -25/2$.

21. The slope of the tangent line is $m = f'(c)$ with $c = 3$. Since $f'(c) = 2c$, the slope is $m = f'(3) = 6$.

23. The slope of the tangent line is $m = f'(c)$ with $c = -1$. Since $f'(c) = 6c$, the slope is $m = f'(-1) = -6$.

25. The slope of the tangent line is $m = f'(5)$. Since $f'(c) = 4c$, $m = f'(5) = 20$. Moreover, $f(5) = 50$, so the point-slope form of the tangent line equation is $y = 20(x - 5) + 50$.

27. The slope of the tangent line is $m = f'(-2)$. Since $f'(c) = -6c$, $m = f'(-2) = 12$. Moreover, $f(-1) = -7$, so the point-slope form of the tangent line equation is $y = 12(x + 2) - 7$.

29. The slope of the tangent line is $m = f'(2)$. Since $f'(c) = 10c$, $m = f'(2) = 20$. Moreover, $f(2) = 20$, so the point-slope form of the tangent line equation is $y = 20(x-2)+20$. The slope-intercept form of the equation is $y = 20x - 20$.

31. The slope of the tangent line is $m = f'(1)$. Since $f'(c) = 6c + 2$, $m = f'(1) = 8$. Moreover, $f(1) = 5$, so the point-slope form of the tangent line equation is $y = 8(x - 1) + 5$. The slope-intercept form of the equation is $y = 8x - 3$.

33. Because $f'(c) = 4c$, the slope of the tangent line is $m = f'(5) = 20$, and the normal line has slope $-1/m = -1/f'(5) = 1/20$. Moreover, $f(5) = 50$, so the point-slope form of the normal line equation is $y = -(1/20)(x-5)+50$.

35. Because $f'(c) = -6c$, the slope of the tangent line is $m = f'(-2) = 12$. The normal line has slope $-1/m = -1/f'(-2) = 1/12$. Moreover, $f(-2) = -7$, so the point-slope form of the normal line equation is $y = -(1/12)(x + 2) - 7$.

37. Because $f'(c) = 6c$, the slope of the tangent line is $m = f'(-1) = -6$. The normal line has slope $-1/m = -1/f'(-1) = 1/6$. Moreover, $f(-1) = 3$, so the point-slope form of the normal line equation is $y = (1/6)(x+1)+3$. The slope-intercept form of the equation is $y = x/6 + 19/6$.

39. Because $f'(c) = 9c^2 - 4c$, the slope of the tangent line is $m = f'(2) = 28$. The normal line has slope $-1/m = -1/f'(2) = -1/28$. Moreover, $f(2) = 6$, so the point-slope form is $y = (-1/28)(x - 2) + 6$. The slope-intercept form of the equation is $y = -x/28 + 85/14$.

Further Theory and Practice

41. Since the position function is $p(t) = t^3$, the instantaneous velocity at $t = c$ is $p'(c) = 3c^2$. That is, the velocity function is $v(t) = 3t^2$. The instantaneous rate of change of the velocity at time $t = c$ is $v'(c) = 6c$. Therefore, the acceleration at time $t = 1$ is $v'(1) = 6$.

43. Since $C'(x) = 0.1 - 0.001 \cdot 2x = 0.1 - 0.002x$, the marginal cost when $x = 25$ is $C'(25) = 0.1 - 0.002 \cdot 25 = 0.05$ cents per pencil.

45. The position function is $p(t) = t^2 + t$ so $p'(t) = 2t + 1$. Consequently, the instantaneous velocity at $t = c$ is $p'(c) = 2c + 1$ and $p'(2) = 5$.

47. The position function is $p(t) = 3t^2 - 2t$ so $p'(t) = 6t - 2$. Consequently, the instantaneous velocity at $t = c$ is $p'(c) = 6c - 2$ and $p'(3) = 16$.

49. The position function is $p(t) = t + 9/t$ so $p'(t) = 1 + 9\cdot(-1)\cdot t^{-2} = 1 - 9/t^2$. Consequently, the instantaneous velocity at $t = c$ is $p'(c) = 1 - 9/c^2$ and $p'(2) = 1 - 9/4 = -5/4$.

51. Since $f(x) = 3x^2 + 1$, $f'(x) = 6x$, and the slope of the tangent line to the graph of f at the point $(c, f(c)$ is $m = f'(c) = 6c$. When $c = 2$ the slope of the tangent line is 12 and it will be parallel to the line $y = 12x$. Since $f(2) = 13$, this tangent line has the equation $y = 12(x - 2) + 13$.

53. Since $f(x) = x^3 - 15x + 20$, $f'(x) = 3x^2 - 15$, and the slope of the tangent line to the graph of f at the point $(c, f(c)$ is $m = f'(c) = 3c^2 - 15$. Because $3c^2 - 15 = 12$ when $c^2 = 27/3 = 9$, the slope of the tangent line is 12 when $c = 3$ and when $c = -3$. Both of these lines will be parallel to the line $y = 12x$. Since $f(3) = 2$ and $f(-3) = 38$, one tangent line has the equation $y = 12(x - 3) + 2$ and the other tangent line has the equation $y = 12(x + 3) + 38$.

55. The rates of change are positive at points A, B, and F. I would estimate that the rate of change is 20 at point F, 10 at point A, and 3 at point B. The rate of change is 0 at point E and D has the larger negative rate of change so it must be -3 there. That leaves -1 for the rate of change at point C.

57. Let S be the area of the surface of a sphere of radius r. Then $S = f(r) = 4\pi r^2$ and $f'(r) = 4\pi \cdot 2r = 8\pi r$. The rate of growth of the area with respect to the radius when $r = 8$ inches is $f'(8) = 64\pi$ square inches per inch in radius.

59. Since $p'(t) = 2t + 1$, the instantaneous velocity of the body at time t is $2t+1$. The velocity is positive when $2t > -1$ or $t > -1/2$ and it is negative when $t < -1/2$.

61. Since $f'(x) = 9x^2$ the tangent line to the graph of f at the point $(c, f(c)$ has the equation $y = 9c^2 \cdot (x - c) + 3c^3 + 12$ or $y = 9c^2x - 6c^3 + 12$. This line passes through the origin if, and only if, $-6c^3 + 12 = 0$. That is, $c^3 = 2$, so $c = 2^{1/3}$. The equation of this tangent line is $y = 9 \cdot (2^{1/3})^2 x$. The tangent line equation can also be expressed in the form $y = 9 \cdot 2^{2/3} x$ or $y = 18x/2^{1/3}$.

63. Since $f'(x) = 2x$ the tangent line to the graph of f at the point $(c, f(c)$ has the equation $y = 2c \cdot (x - c) + c^2$ or $y = 2cx - c^2$. This line passes through the point $(3, 5)$ if, and only if, c satisfies the equation $5 = 6c - c^2$. That is, $c^2 - 6c + 5 = 0$ or $(c - 1)(c - 5) = 0$. Consequently, either $c = 1$ or $c = 5$.

65. The slope of the line $y = 4 - 9x$ is -9. If the point $(c, f(c))$ on the graph of f has a tangent line with slope -9, then $f'(c) = -9$ so c must be a solution to the equation $3c^2 - 4c - 8 = -9$. That is, $3c^2 - 4c + 1 = 0$ or $(3c - 1)(c - 1) = 0$. Consequently, either $c = 1$ or $c = 1/3$ and the points on the graph of f are $(1, -6)$ and $(1/3, 4/27)$.

67. The tangent line to the graph of f at $(c, f(c))$ has slope $f'(c) = 3c^2 - 8$. At the point $(c, g(c))$ the tangent line to the graph of g has slope $g'(c) = -4/c^2$. These lines are parallel if, and only if, $3c^2 - 8 = -4/c^2$. That is, $3c^4 - 8c^2 + 4 = 0$. The left hand side factors, $(3c^2 - 2)(c^2 - 2) = 0$, and there are four values of c. They are $c = \pm\sqrt{2}$ and $c = \pm\sqrt{3/2} = \pm\sqrt{6}/2$.

69. The slope of the tangent line at $(c, 1/c)$ is $f'(c) = -1/c^2$ so the line L_c has the equation $y = -(x - c)/c^2 + 1/c$ which can also be written in the form $c^2 y = -x + c + c$ or $x + c^2 y = 2c$. Consequently, this line intersects the x- and y-axes at the points $(2c, 0)$ and $(0, 2/c)$ respectively. The right triangle bounded by L_c and the two coordinate axes has base $2c$ and height $2/c$ so its area is $\frac{1}{2} \cdot (2c) \cdot (2/c) = 2$.

71. Since $f'(c) = a \cdot 2c = 2ac$, the equation of the tangent line to the graph of f at the point $(c, f(c))$ is $y = 2ac(x - c) + ac^2 + b$ or $y = 2acx - ac^2 + b$. This line passes through the origin if, and only if, $-ac^2 + b = 0$. That is, $c = \sqrt{b/a}$. Therefore, the graph of f has a tangent line that passes through the origin if, and only if, either b is zero or b is not zero and has the same sign as a. That is, a and b are both positive or they are both negative.

73. Let $f(x) = Ax^2 + Bx + C$ where A, B, and C are constants, $A \neq 0$. The tangent line to the graph of f at the point $P = (c, f(c))$ has the equation $y = (2Ac + B) \cdot (x - c) + f(c)$. Suppose that this line intersects the graph of f at a point $Q = (d, f(d))$, $d \neq c$. Then d is a solution to the equation $f(d) = (2Ac + B)(d - c) + f(c)$, implying that

$$\begin{aligned} 2Ac + B &= \frac{f(d) - f(c)}{d - c} \\ &= \frac{Ad^2 + Bd + C - (Ac^2 + Bc + C)}{d - c} \\ &= \frac{A(d^2 - c^2) + B(d - c)}{d - c} \\ &= A(d + c) + B\,. \end{aligned}$$

But then $2Ac = A(d + c)$ and, since $A \neq 0$, $2c = d + c$, implying that $d = c$. There can be no other intersection point.

75. Since the equations defining T_R and T_L require finite slopes, ℓ_R and ℓ_L are finite. Consequently the graph of T_R is a half-line extending from P into the half-plane to the right of P and the graph of T_L is a half-line extending from P into the half-plane to the left of P. This implies that $0 < \alpha_f(c) < 2\pi$.

    i) There is a corner at P if, and only if, $\ell_R \neq \ell_L$ which, geometrically, is the condition that the graphs of T_R and T_L meet at an angle $\alpha_f(c)$ that is either less than π or greater than π. That is, there is a corner at P if, and only if, $\alpha_f(c) \neq \pi$.

ii) There is a tangent line at P if, and only if, $\ell_R = \ell_L$. In this case the union of the graph of T_R and the graph of T_L is the tangent line at P and $\alpha_f(c) = \pi$.

iii) There is no value for $\alpha_f(c)$ for which P is a cusp. This is because a cusp occurs at P if, and only if, ℓ_R and ℓ_L are both infinite and of opposite sign.

iv) Similarly, there is no value for $\alpha_f(c)$ corresponding to a vertical tangent at P. Vertical tangents occur if, and only if, ℓ_R and ℓ_L are both infinite and of the same sign.

CALCULATOR/COMPUTER EXERCISES

77. (a) Average velocities

i) $\dfrac{H(3) - H(2)}{3 - 2} = -8$

ii) $\dfrac{H(2.1) - H(2)}{2.1 - 2} = 6.4$

iii) $\dfrac{H(2.01) - H(2)}{2.01 - 2} = 7.84$

iv) $\dfrac{H(2.001) - H(2)}{2.001 - 2} = 7.984$

(b) $H'(2) = 8$

79. The following approximations were obtained using *Maple*, 10-digit accuracy, rounded to 4 decimals.

n	0	1	2	3	4
$\dfrac{p(10^{-n}) - p(0)}{10^{-n}}$	0.0794	1.6208	1.7219	1.7310	1.7319

Based upon these values $v(0) \approx 1.73$.

The picture on the right supports the conjectured value. The curve is the graph of p and the straight line has the equation

$$y = 1/2 + 1.73t\,.$$

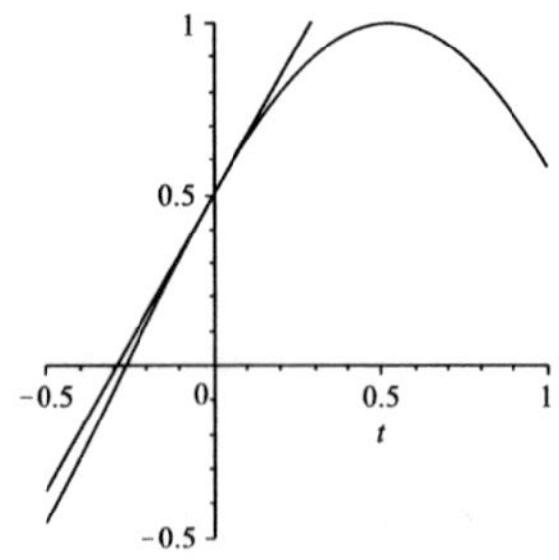

81. The function p in Exercise 80, approximations at $t = 2$.

n	0	1	2	3	4
$\dfrac{p(2+10^{-n})-p(2)}{10^{-n}}$	12.6965	7.7711	7.4261	7.3928	7.3894

Based upon these values $v(2) \approx 7.3894$. We note that $p(2) \approx 7.3891$.

The picture on the right supports the conjectured value. The curve is the graph of p and the straight line has the equation

$$y = 7.3891 + 7.3894(t - 2)\,.$$

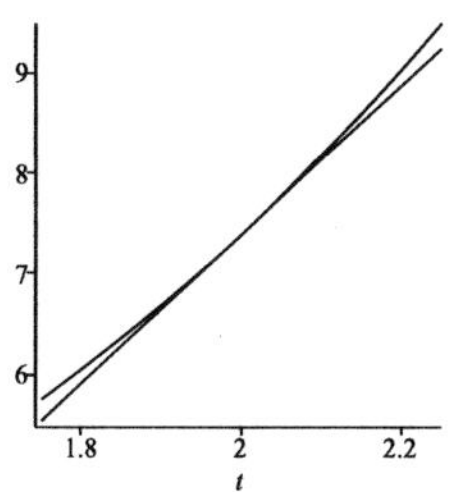

The function p in Exercise 80 with approximations at $t = 3$.

n	0	1	2	3	4
$\dfrac{p(3+10^{-n})-p(3)}{10^{-n}}$	34.5126	21.1241	20.1863	20.0956	20.0866

Based upon these values $v(3) \approx 20.0866$. We note that $p(3) \approx 20.0855$.

The picture on the right supports the conjectured value. The curve is the graph of p and the straight line has the equation

$$y = 20.0855 + 20.0888(t - 3)\,.$$

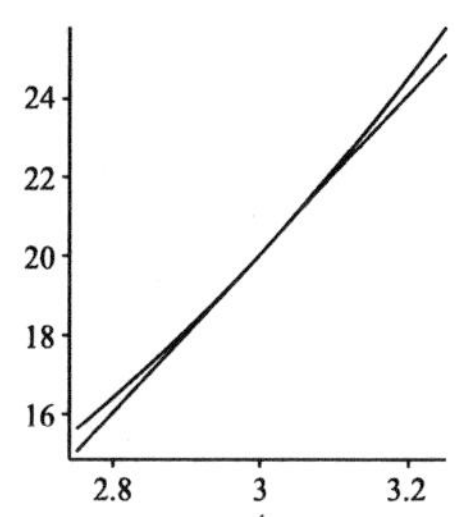

Based upon these examples, it appears that $p(t) = v(t)$.

83. Plots of the graph of the function f, centered at the point $(\pi/4, 1)$, are displayed below. We have zoomed in twice. In each plot the two axes are scaled equally.

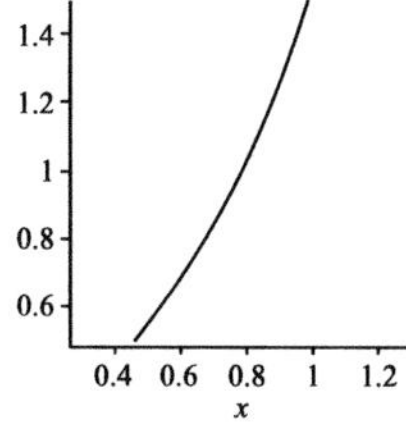

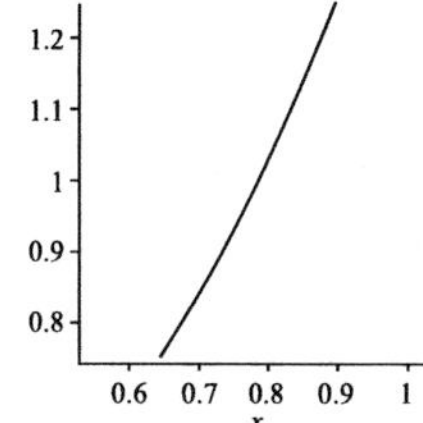

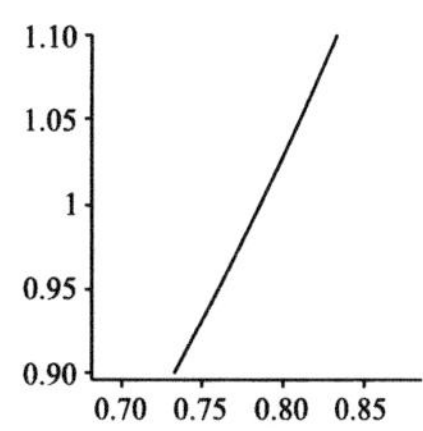

Based upon these pictures it appears that $f'(\pi/4) = 2$.

The following approximations were obtained using *Maple*, 10-digit accuracy, rounded to 4 decimals.

n	1	2	3	4	5
$\dfrac{f(\pi/4 + 10^{-n}) - f(\pi/4)}{10^{-n}}$	2.2305	2.0203	2.0020	2.0002	2.0000

Based upon the plots and the tabulated values $f'(\pi/4) = 2$.

3.2 The Derivative

Problems for Practice

1. Using rule (3.2.12) and the differentiation formulas for sine and cosine, $f'(x) = \cos(x) + \sin(x)$ so $f'(-3) = \cos(-3) + \sin(-3) = \cos(3) - \sin(3)$. The simplified expression for $f'(3)$ is obtained using the fact that the cosine function is even and the sine function is odd.

3. Using rule (3.2.12) and the differentiation formulas, $\frac{dF}{du}(u) = 5 - 6\sin(u)$ so $\frac{dF}{du}(\frac{\pi}{6}) = 5 - 6\sin(\frac{\pi}{6}) = 5 - 6 \cdot \frac{1}{2} = 2$.

5. Using rule (3.2.10) and the differentiation formulas, $\dot{g}(t) = 24t^2 - 8$ so $\dot{g}(5) = 24 \cdot 25 - 8 = 592$.

7. Using rule (3.2.12) and the differentiation formulas for sine and cosine, $f'(x) = 3\cos(x) - 4\sin(x)$ so $f'(c) = 3\cos(c) - 4\sin(c)$.

9. Using rule (3.2.12) and the differentiation formulas, $\frac{dg}{ds}(s) = 9s^2 + \sin(s)$ so $\left.\frac{dg}{ds}\right|_{s=-2} = 36 + \sin(-2) = 36 - \sin(2)$.

11. Using rule (3.2.12), $g'(x) = -\pi + \frac{1}{\pi} \cdot (-1/x^2) = -\pi - 1/(\pi x^2)$.

13. Using rule (3.2.12), extended to three terms, and the differentiation formulas, $g'(x) = 12x^2 + 12x$.

15. $f'(x) = 2x = 6$ when $x = 3$.

17. $f'(x) = 3x^2 = 6$ when $x^2 = 2$. Therefore, $x = \sqrt{2}$ or $x = -\sqrt{2}$.

19. Using the differentiation rules and formulas, $f'(x) = -6x^2$. The picture on the right displays the graph of $y = f'(x)$ (solid line) as well as the graph of $y = f(x)$ (dotted line).

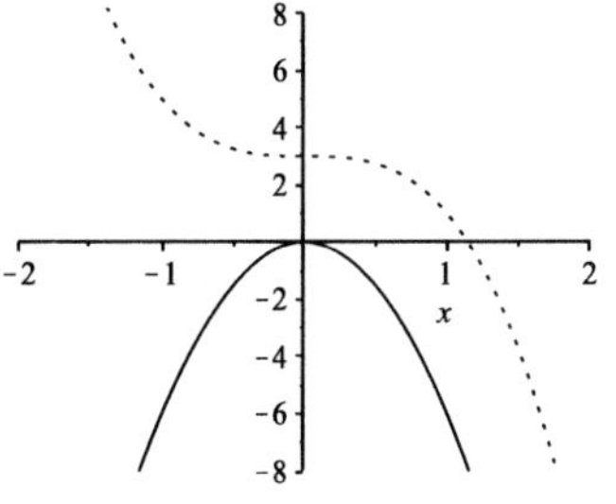

21. Using the differentiation rules and the formula $\frac{d}{dx}(|x|) = \text{signum}(x)$ developed in Example 3, $f'(x) = 2x + \text{signum}(x)$. The picture on the right displays the graph of $y = f'(x)$ (solid line) as well as the graph of $y = f(x)$ (dotted line).

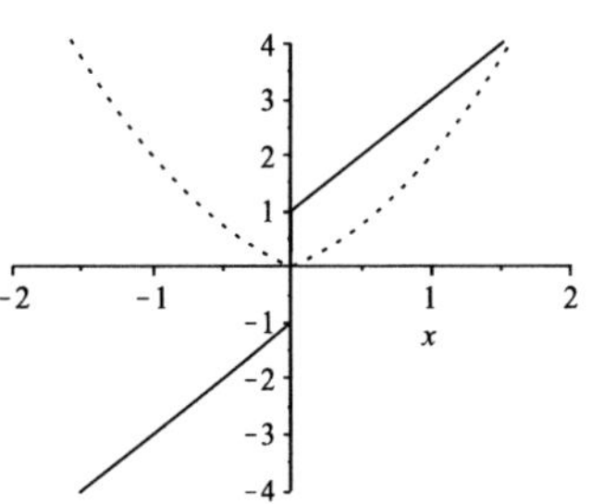

23. $f'(x) = \cos(x)$ so $f'(\pi/3) = \cos(\pi/3) = 1/2$ and the equation of the tangent line to the graph of f at $(\pi/3, \sin(\pi/3))$ is $y = \frac{1}{2}(x - \pi/3) + \sqrt{3}/2$.

25. $f'(x) = -2\sin(x) - 4\cos(x)$ so $f'(3\pi/4) = -2\sin(3\pi/4) - 4\cos(3\pi/4) = -2 \cdot \frac{\sqrt{2}}{2} - 4 \cdot \left(-\frac{\sqrt{2}}{2}\right) = \sqrt{2}$, and the equation of the tangent line to the graph of f at $(3\pi/4, 2\cos(3\pi/4) - 4\sin(3\pi/4)) = (3\pi/4, -3\sqrt{2})$ is $y = \sqrt{2}\,(x - 3\pi/4) - 3\sqrt{2}$.

27. The derivative can be calculated as follows.

$$
\begin{aligned}
f'(x) &= \lim_{\Delta x \to 0} \frac{\sqrt{x+\Delta x} - \sqrt{x}}{\Delta x} = \lim_{\Delta x \to 0} \left(\frac{\sqrt{x+\Delta x} - \sqrt{x}}{\Delta x} \cdot \frac{\sqrt{x+\Delta x} + \sqrt{x}}{\sqrt{x+\Delta x} + \sqrt{x}} \right) \\
&= \lim_{\Delta x \to 0} \frac{x + \Delta x - x}{\Delta x(\sqrt{x+\Delta x} + \sqrt{x})} = \lim_{\Delta x \to 0} \frac{\Delta x}{\Delta x(\sqrt{x+\Delta x} + \sqrt{x})} = \lim_{\Delta x \to 0} \frac{1}{\sqrt{x+\Delta x} + \sqrt{x}} \\
&= \frac{1}{2\sqrt{x}}\,.
\end{aligned}
$$

29. The derivative can be calculated as follows.

$$
\begin{aligned}
f'(x) &= \lim_{\Delta x \to 0} \frac{\sqrt{3(x+\Delta x)+7} - \sqrt{3x+7}}{\Delta x} \\
&= \lim_{\Delta x \to 0} \left(\frac{\sqrt{3(x+\Delta x)+7} - \sqrt{3x+7}}{\Delta x} \cdot \frac{\sqrt{3(x+\Delta x)+7} + \sqrt{3x+7}}{\sqrt{3(x+\Delta x)+7} + \sqrt{3x+7}} \right) \\
&= \lim_{\Delta x \to 0} \frac{3(x+\Delta x) + 7 - (3x+7)}{\Delta x(\sqrt{3(x+\Delta x)+7} + \sqrt{3x+7})} = \lim_{\Delta x \to 0} \frac{3\Delta x}{\Delta x(\sqrt{3(x+\Delta x)+7} + \sqrt{3x+7})} \\
&= \lim_{\Delta x \to 0} \frac{3}{\sqrt{3(x+\Delta x)+7} + \sqrt{3x+7}} = \frac{3}{2\sqrt{3x+7}}\,.
\end{aligned}
$$

Further Theory and Practice

31. Clearly $H' = G$ since the graph of H has a constant (positive) slope. Moreover, $K' = H$ because the slopes of the tangent lines to the graph of K are positive where the function H is positive, the slope is 0 where H is 0, and the slopes are negative where H is negative.

33. As in Example 2, calculate the right limit and the left limit of the quotient that defines the derivative.

    From the right:

$$\lim_{\Delta x \to 0^+} \frac{(0+\Delta x)|0+\Delta x| - 0}{\Delta x} = \lim_{\Delta x \to 0^+} \frac{\Delta x^2}{\Delta x} = \lim_{\Delta x \to 0^+} \Delta x = 0$$

    From the left:

$$\lim_{\Delta x \to 0^-} \frac{(0+\Delta x)|0+\Delta x| - 0}{\Delta x} = \lim_{\Delta x \to 0^-} \frac{\Delta x(-\Delta x)}{\Delta x} = \lim_{\Delta x \to 0^+} (-\Delta x) = 0$$

    The right and left limits are both 0, so $f'(0) = 0$.

35. We first calculate $f'(x)$.

$$\begin{aligned}
f'(x) &= \lim_{\Delta x \to 0} \frac{(x+\Delta x)^{-1/2} - x^{-1/2}}{\Delta x} \\
&= \lim_{\Delta x \to 0} \left(\frac{(x+\Delta x)^{-1/2} - x^{-1/2}}{\Delta x} \cdot \frac{(x+\Delta x)^{-1/2} + x^{-1/2}}{(x+\Delta x)^{-1/2} + x^{-1/2}} \right) \\
&= \lim_{\Delta x \to 0} \frac{(x+\Delta x)^{-1} - x^{-1}}{\Delta x((x+\Delta x)^{-1/2} + x^{-1/2})} = \lim_{\Delta x \to 0} \frac{-\Delta x/((x(x+\Delta x))}{\Delta x((x+\Delta x)^{-1/2} + x^{-1/2})} \\
&= \lim_{\Delta x \to 0} \frac{-\Delta x}{\Delta x((x+\Delta x)^{-1/2} + x^{-1/2})x(x+\Delta x)} \\
&= \lim_{\Delta x \to 0} \frac{-1}{((x+\Delta x)^{-1/2} + x^{-1/2})x(x+\Delta x)} = -\frac{1}{2x^{3/2}}
\end{aligned}$$

    Since $f'(x) = -\frac{1}{2}x^{-3/2}$, $f'(1/16) = -\frac{1}{2} \cdot 16^{3/2} = -32$.

37. Let $p(x) = Ax + B$. Since $p'(x) = A$, the fact that $p'(4) = -5$ implies that $A = -5$ so $p(x) = -5x + B$. Since $p(2) = 6$, $-10 + B = 6$ so $B = 16$ and $p(x) = -5x + 16$.

39. We calculate $f'(c) = \lim_{x \to c} \frac{f(x)-f(c)}{x-c}$.

$$\begin{aligned}
f'(c) &= \lim_{x \to c} \frac{1/(1+x) - 1/(1+c)}{x-c} = \lim_{x \to c} \frac{(c-x)/((1+x)(1+c))}{x-c} \\
&= -\lim_{x \to c} \frac{1}{(1+x)(1+c)} = -\frac{1}{(1+c)^2}
\end{aligned}$$

    Therefore, $f'(x) = -1/(1+x)^2$.

41. We calculate $f'(c) = \lim_{x\to c} \frac{f(x)-f(c)}{x-c}$.

$$f'(c) = \lim_{x\to c} \frac{1/(1+\sqrt{x}) - 1/(1+\sqrt{c})}{x-c} = \lim_{x\to c} \frac{(\sqrt{c}-\sqrt{x})/((1+\sqrt{x})(1+\sqrt{c}))}{x-c}$$
$$= -\lim_{x\to c} \frac{1}{(\sqrt{x}+\sqrt{c})(1+\sqrt{x})(1+\sqrt{c})} = -\frac{1}{2\sqrt{c}(1+\sqrt{c})^2}$$

Therefore, $f'(x) = -\frac{1}{2}x^{-1/2}/(1+\sqrt{x})^2$.

43. f is continuous at $x = 0$, but it is not differentiable there. Observe that its graph has one-sided tangent lines at $(0,0)$, but they do not have the same slope. From the left, where $f(x) = x^2$, the tangent line at $x = 0$ has slope 0. From the right, where $f(x) = x$, the slope is 1.

45. f is continuous at $x = 0$, it is also differentiable there. Its graph has one-sided tangent lines at $(0,0)$, and they have the same slope. From the left, where $f(x) = \sin(x)$, the left tangent line at $x = 0$ has slope $\cos(0) = 1$. For $x \geq 0$, $f(x) = x^2 + x$, and the right tangent line slope is also 1: $\frac{d}{dx}(x^2+x)\Big|_{x=0} = 1$. The line $y = x$ is the tangent line to the graph of f at the point $(0,0)$.

47. $f(x) = \sqrt{x}$, $c = 4$.

49. $f(x) = 1/x$, $c = 5$.

51. Examination of the graph of f confirms the assertions in this Exercise. The following discussion would serve in lieu of a picture.

    $f'(0)$ does not exist (f is not defined for $x < 0$), so we restrict our attention to the open interval $I = (0, \infty)$. Since $0 < a < b$ implies that $\sqrt{a} < \sqrt{b}$, f is an increasing function on I. Moreover, if $0 < a < b$, the fact that $\sqrt{a} < \sqrt{b}$ implies that $2\sqrt{a} < 2\sqrt{b}$ and $\frac{1}{2\sqrt{a}} > \frac{1}{2\sqrt{b}}$. That is, $f'(a) > f'(b)$, because, according to Exercise 27, $f'(x) = \frac{1}{2\sqrt{x}}$.

53. Assume $\alpha \geq 1$. Then

$$f'(0) = \lim_{x\to 0} \frac{x^\alpha - 0^\alpha}{x} = \lim_{x\to 0} \frac{x^\alpha}{x} = \lim_{x\to 0} x^{\alpha-1} = \begin{cases} 1 & \text{if} \quad \alpha = 1 \\ 0 & \text{if} \quad \alpha > 1 \end{cases}$$

If $\alpha < 1$, then 0 is not in the domain of f.

55. $$f'(c) = \lim_{x\to c} \frac{f(x)-f(c)}{x-c} = \lim_{x\to c} \frac{\frac{x}{1+x^2} - \frac{c}{1+c^2}}{x-c} = \lim_{x\to c} \frac{1-xc}{(1+x^2)(1+c^2)} = \frac{1-c^2}{(1+c^2)^2}$$

57. Since $g(x) = f(x+k)$,

$$g'(c-k) = \lim_{h\to 0} \frac{g(c-k+h) - g(c-k)}{h} = \lim_{h\to 0} \frac{f(c+h) - f(c)}{h} = f'(c).$$

59. Observe that

$$E(p) = -\frac{p}{q(p)} \cdot \lim_{\Delta p \to 0} \frac{q(p+\Delta p) - q(p)}{\Delta p} = -p \cdot \frac{q'(p)}{q(p)}\,.$$

61. $P'(t)$ is a quadratic function of $P(t)$ so its graph, when plotted versus $P(t)$, is a parabola. Because $k > 0$, the parabola opens downward. Consequently, $P'(t)$ attains its largest value when $P(t)$ is exactly midway between the two zeros of the quadratic. That is, midway between $P(t) = 0$ and $P(t) = M$.

63. f is continuous at $x = 0$ because when $x \neq 0$,

$$|f(x)| = |x \cdot \sin(1/x)| \leq |x| \underset{x \to 0}{\longrightarrow} 0 = f(0)\,.$$

f is not differentiable at $x = 0$ because the expression

$$\phi(x) = \frac{f(x) - f(0)}{x} = \sin(1/x)$$

does not have a limiting value as $x \to 0$. Note, for example, that if $x_n = 2/(n\pi)$, $n = 1, 2, 3, \ldots$, then $x_n \to 0$ but the values $\phi(x_n)$ continually oscillate between 1 and -1.

65. Because $y = ax + b$ is the tangent line to f at $(c, f(c))$,

$$\begin{aligned}
\lim_{x \to c} \frac{f(x) - (ax+b)}{x - c} &= \lim_{x \to c} \frac{f(x) - f'(c)(x-c) - f(c)}{x - c} \\
&= \lim_{x \to c} \frac{f(x) - f(c) - f'(c)(x-c)}{x - c} \\
&= \lim_{x \to c} \left(\frac{f(x) - f(c)}{x - c} - f'(c) \right) \\
&= f'(c) - f'(c) = 0\,.
\end{aligned}$$

CALCULATOR/COMPUTER EXERCISES

67. $f(x) = x^2 - 2x$ and its tangent line at $(1/2, f(1/2))$.

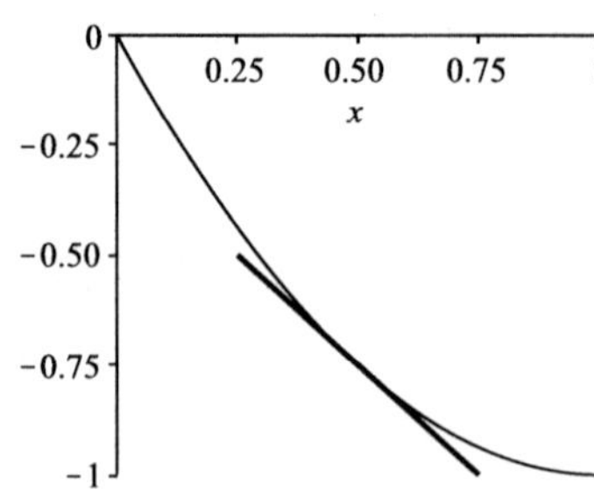

69. The graph of $f(x) = 1/x$ and its tangent line at $(1, f(1))$.

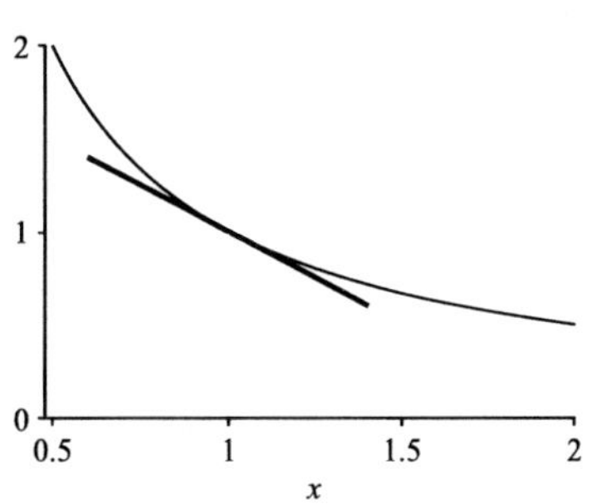

71. Using the estimated slope of the solid secant line between the two closest points on the curve
$$f'(1) \approx 0.5/2 = 0.25\,.$$

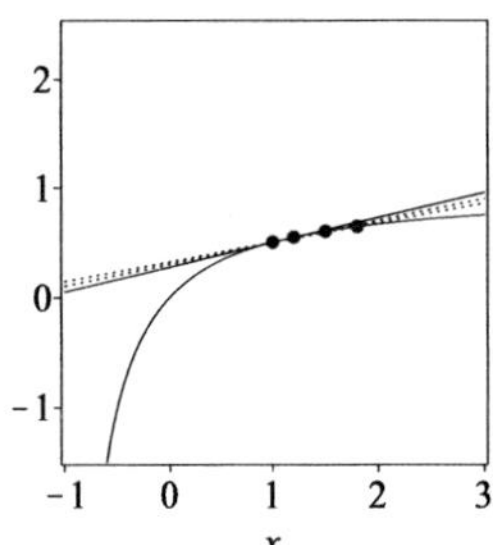

73. Using the estimated slope of the solid secant line between the two closest points on the curve
$$f'(\pi/4) \approx \frac{3\pi/8 - \pi/16}{\pi/2} = \frac{5}{8} = 0.625\,.$$

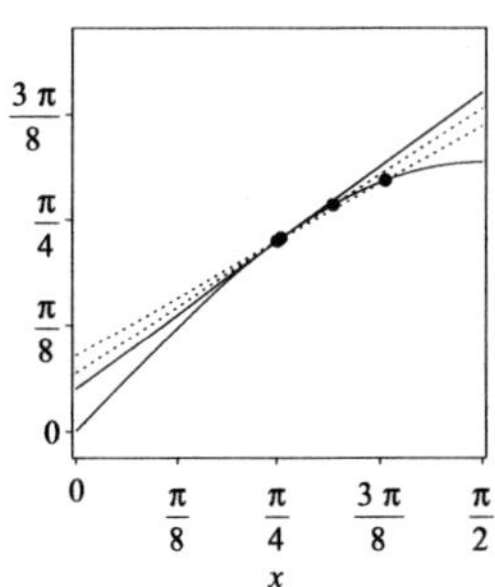

75. Using the picture on the right the estimated slope of the tangent line is $0.003/0.01 = 0.3$.

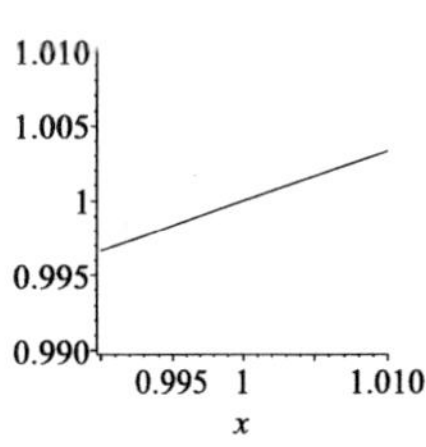

77. Using the picture on the right the estimated slope of the tangent line is $321.5/0.000002 = 3.25 \cdot 10^5$.

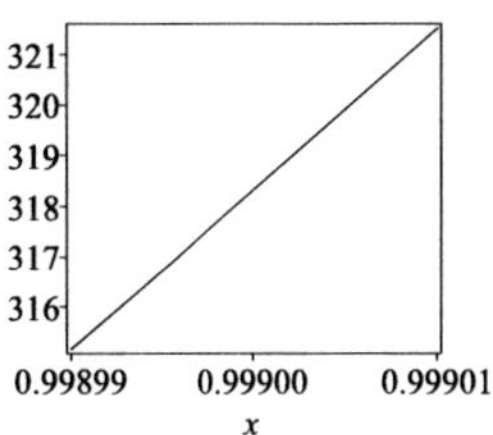

79. The graph of ϕ on the right suggests that ϕ does not have a limiting value as x approaches 1. Therefore, since $f'(1) = \lim_{x\to 1}\phi(x)$, $f'(1)$ does not exist.

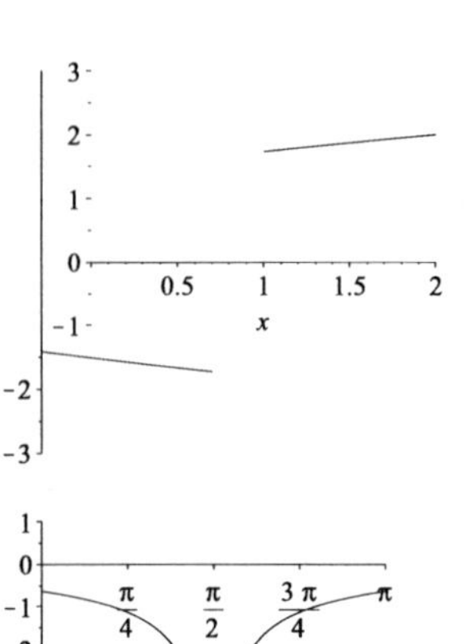

81. The graph of ϕ on the right suggests that ϕ does not have a limiting value as x approaches $\pi/2$. Therefore, since $f'(\pi/2) = \lim_{x\to\pi/2}\phi(x)$, $f'(\pi/2)$ does not exist.

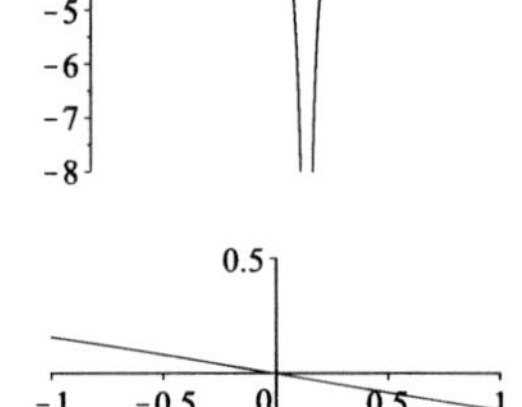

83. The graph of ϕ on the right suggests that ϕ has the limiting value of 0 as x approaches 0. This being the case, f would be differentiable and $f'(0) = \lim_{x\to 0}\phi(x) = 0$.

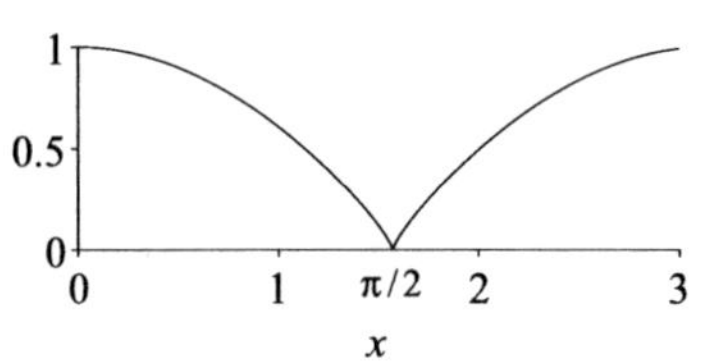

85. The graph on the right suggests that f is not differentiable at the point $c = \pi/2$. Zooming in will confirm that the graph of f has a cusp at $(\pi/2, 0)$.

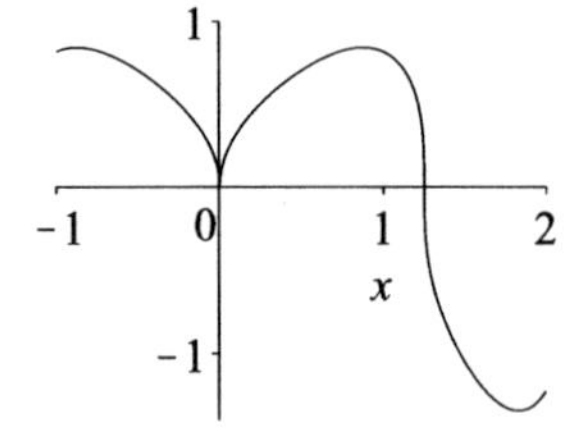

87. The graph on the right suggests that f is not differentiable at two points, $c = 0$ and $c = \sqrt{\pi/2} \approx 1.25$. Zooming in will confirm that the graph of f has a cusp at $(0,0)$ and a vertical tangent at $(\sqrt{\pi/2}, 0)$.

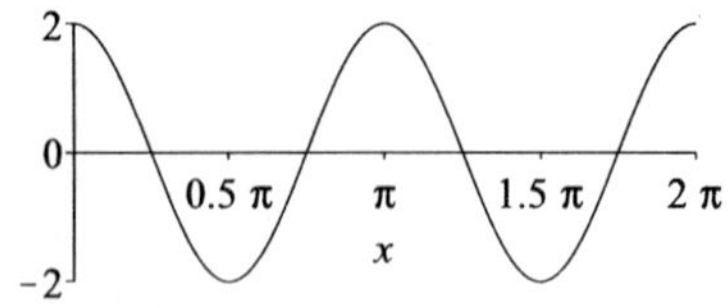

89. This curve is the graph of the approximate derivative for $f(x) = \sin(2x)$, $\Delta x = 10^{-4}$. It appears to be the graph of $y = 2\cos(2x)$ suggesting that $f'(x) = 2\cos(2x)$.

3.3 Rules for Differentiation

1. The derivative of a linear combination.

$$\frac{d}{dx}\Big(5x^3 - 3x^2 + 9\Big) = 5 \cdot (3x^2) - 3 \cdot (2x) + 0 = 15x^2 - 6x$$

3. The derivative of a linear combination.

$$\begin{aligned}\frac{d}{dx}\Big(x^3/\pi + \pi\cos(x) + \sqrt{\pi}\Big) &= 1/\pi \cdot (3x^2) + \pi \cdot (-\sin(x)) + 0\\ &= 3x^2/\pi - \pi\sin(x)\end{aligned}$$

5. The derivative of a linear combination.

$$\begin{aligned}\frac{d}{dx}\Big(\tfrac{1}{5}(4\sin(x) - 3\cos(x)) + 5x\Big) &= \tfrac{1}{5}(4 \cdot \cos(x) - 3 \cdot (-\sin(x)) + 5 \cdot 1\\ &= \tfrac{4}{5}\cos(x) + \tfrac{3}{5}\sin(x) + 5\end{aligned}$$

7. The Product Rule.

$$\begin{aligned}\frac{d}{dx}\Big(x^2(x+3)\Big) &= \frac{d}{dx}\Big(4x\Big) \cdot (x+3) + x^2 \cdot \frac{d}{dx}\Big(x+3\Big)\\ &= (2x) \cdot (x+3) + x^2 \cdot 1 = 3x^2 + 6x\end{aligned}$$

9. The Product Rule.

$$\begin{aligned}\frac{d}{dx}\Big(x^3\cos(x)\Big) &= \frac{d}{dx}\Big(x^3\Big) \cdot \cos(x) + x^3 \cdot \frac{d}{dx}\Big(\cos(x)\Big)\\ &= 3x^2 \cdot \cos(x) + x^3 \cdot (-\sin(x)) = 3x^2\cos(x) - x^3\sin(x)\end{aligned}$$

11. The Product Rule.

$$\begin{aligned}\frac{d}{dx}\Big(\sin(x) \cdot \sin(x)\Big) &= \frac{d}{dx}\Big(\sin(x)\Big) \cdot \sin(x) + \sin(x) \cdot \frac{d}{dx}\Big(\sin(x)\Big)\\ &= \cos(x) \cdot \sin(x) + \sin(x) \cdot \cos(x)\\ &= 2\sin(x)\cos(x)\end{aligned}$$

13. The Product Rule.

$$\begin{aligned}\frac{d}{dx}\Big(\sin(x) \cdot \cos(x)\Big) &= \frac{d}{dx}\Big(\sin(x)\Big) \cdot \cos(x) + \sin(x) \cdot \frac{d}{dx}\Big(\cos(x)\Big)\\ &= \cos(x) \cdot \cos(x) + \sin(x) \cdot (-\sin(x))\\ &= \cos^2(x) - \sin^2(x)\end{aligned}$$

15. The Constant Multiple Rule and the Product Rule.

$$\begin{aligned} 7 \cdot \frac{d}{dx}\Big((x^2 + x) \cdot \sin(x)\Big) &= 7 \cdot \left(\frac{d}{dx}\Big(x^2 + x\Big) \cdot \sin(x) + (x^2 + x) \cdot \frac{d}{dx}\Big(\sin(x)\Big)\right) \\ &= 7 \cdot ((2x + 1)) \cdot \sin(x) + (x^2 + x) \cdot \cos(x)) \\ &= 7(2x + 1)\sin(x) + 7(x^2 + x)\cos(x) \end{aligned}$$

17. The Product Rule.

$$\begin{aligned} \frac{d}{dx}\Big((1/x) \cdot \sin(x)\Big) &= \frac{d}{dx}\Big(1/x\Big) \cdot \sin(x) + (1/x) \cdot \frac{d}{dx}\Big(\sin(x)\Big) \\ &= (-1/x^2) \cdot \sin(x) + (1/x) \cdot \cos(x) \\ &= -\sin(x)/x^2 + \cos(x)/x \end{aligned}$$

19. The Reciprocal Rule.

$$\frac{d}{dx}\Big(1/(x + 1)\Big) = -\tfrac{d}{dx}(x + 1)/(x + 1)^2 = -1/(x + 1)^2$$

21. The Constant Multiple Rule and the Reciprocal Rule.

$$\begin{aligned} 2 \cdot \frac{d}{dx}\Big(1/(3x^2 + 4)\Big) &= 2 \cdot \left(-\tfrac{d}{dx}(3x^2 + 4)/(3x^2 + 4^2)^2\right) \\ &= 2 \cdot \left(-6x/(3x^2 + 4^2)^2\right) = -12x/(3x^2 + 4)^2 \end{aligned}$$

23. The Reciprocal Rule.

$$\begin{aligned} \frac{d}{dx}\Big(1/(x + 3\cos(x))\Big) &= -\tfrac{d}{dx}(x + 3\cos(x))/(x + 3\cos(x))^2 \\ &= -(1 - 3\sin(x))/(x + 3\cos(x))^2 \end{aligned}$$

25. The Constant Multiple Rule and the Reciprocal Rule.

$$\begin{aligned} 9 \cdot \frac{d}{dx}\Big(1/(3\cos(x) + x^3)\Big) &= 9 \cdot \left(-\tfrac{d}{dx}(3\cos(x) + x^3)/(3\cos(x) + x^3)^2\right) \\ &= 9 \cdot \left(-(-3\sin(x) + 3x^2)/(3\cos(x) + x^3)^2\right) \\ &= 27(\sin(x) - x^2)/(3\cos(x) + x^3)^2 \end{aligned}$$

27. The Quotient Rule.

$$\frac{d}{dx}\left(\frac{x}{x + 1}\right) = \frac{(x + 1) \cdot (1) - (x) \cdot (1)}{(x + 1)^2} = \frac{1}{(x + 1)^2}$$

29. The Quotient Rule.

$$\frac{d}{dx}\left(\frac{x-1}{x+1}\right) = \frac{(x+1)\cdot(1)-(x-1)\cdot(1)}{(x+1)^2} = \frac{2}{(x+1)^2}$$

31. The Quotient Rule and the Reciprocal Rule.

$$\frac{d}{dx}\left(\frac{5x-2/x}{x}\right) = \frac{(x)\cdot(5+2/x^2)-(5x-2/x)\cdot(1)}{x^2}$$
$$= \frac{4/x}{x^2} = \frac{4}{x^3}$$

33. The Quotient Rule.

$$\frac{d}{dx}\left(\frac{\sin(x)}{x}\right) = \frac{(x)\cdot(\cos(x))-(\sin(x))\cdot(1)}{(x)^2} = \frac{x\cos(x)-\sin(x)}{x^2}$$

35. The Quotient Rule.

$$\frac{d}{dx}\left(\frac{x^2+7}{\cos(x)}\right) = \frac{(\cos(x))\cdot(2x)-(x^2+7)\cdot(-\sin(x))}{(\cos(x))^2}$$
$$= \frac{2x\cos(x)+(x^2+7)\sin(x)}{\cos^2(x)}$$

37. The Quotient Rule.

$$\frac{d}{dx}\left(\frac{1-\cos(x)}{1+\cos(x)}\right) = \frac{(1+\cos(x))\cdot(\sin(x))-(1-\cos(x))\cdot(-\sin(x))}{(1+\cos(x))^2}$$
$$= \frac{2\sin(x)}{(1+\cos(x))^2}$$

39. The Quotient Rule and the Product Rule.

$$\frac{d}{dx}\left(\frac{x\sin(x)}{x+1}\right) = \frac{(x+1)\cdot\frac{d}{dx}(x\sin(x))-(x\sin(x))\cdot(1)}{(x+1)^2}$$
$$= \frac{(x+1)\cdot(1\cdot\sin(x)+x\cdot\cos(x))-x\sin(x)}{(x+1)^2}$$
$$= \frac{\sin(x)+x(x+1)\cos(x)}{(x+1)^2}$$

41. The Product Rule (twice).

$$\frac{d}{dx}\left(x(x^2+1)(x^3+2)\right) = \frac{d}{dx}\Big(x\cdot(x^2+1)\Big)\cdot(x^3+2)+x(x^2+1)\cdot\frac{d}{dx}(x^3+2)$$
$$= \left(\frac{d}{dx}(x)\cdot(x^2+1)+x\cdot\frac{d}{dx}(x^2+1)\right)\cdot(x^3+2)+x(x^2+1)\cdot(3x^2)$$
$$= \big(1\cdot(x^2+1)+x\cdot 2x\big)\cdot(x^3+2)+3x^3(x^2+1)$$
$$= (3x^2+1)(x^3+2)+3x^3(x^2+1) = 6x^5+4x^3+6x^2+2$$

43. The Product Rule (twice) and $\frac{d}{dx}\sec(x) = \sec(x)\tan(x)$.

$$\begin{aligned}\frac{d}{dx}(\sin(x)\cdot\sin(x)\cdot\sec(x)) &= \frac{d}{dx}\big(\sin(x)\cdot\sin(x)\big)\cdot\sec(x) + \sin^2(x)\cdot\frac{d}{dx}(\sec(x))\\ &= (\cos(x)\cdot\sin(x)+\sin(x)\cdot\cos(x))\cdot\sec(x)+\sin^2(x)\sec(x)\tan(x)\\ &= 2\sin(x)\cos(x)\sec(x)+\sin^2(x)\sec(x)\tan(x)\\ &= 2\sin(x)+\sin^2(x)\sec(x)\tan(x)\end{aligned}$$

45. Since $f'(x) = \frac{(x+1)\cdot 1-(x+3)\cdot 1}{(x+1)^2} = -2/(x+1)^2$, $f'(1) = -1/2$ and the tangent line equation is $y = -(x-1)/2+2$ or $y = -\frac{1}{2}x+\frac{5}{2}$.

47. Since $f'(x) = 6\cdot\big(-2x/(1+x^2)^2\big) = -12x/(1+x^2)^2$, $f'(1) = -12/4 = -3$ and the tangent line equation is $y = -3(x-1)+3$ or $y = -3x+6$.

49. Since $f'(x) = \frac{(x+1)\cdot\cos(x)-\sin(x)\cdot 1}{(x+1)^2} = \frac{(x+1)\cos(x)-\sin(x)}{(x+1)^2}$, $f'(0) = 1$ and the tangent line equation is $y = x$.

51. $\frac{d}{dx}\big(x\sec(x)\big) = 1\cdot\sec(x)+x\cdot\sec(x)\tan(x) = \sec(x)(1+x\tan(x))$

53. $\frac{d}{dx}\big(1/(1+2\sec(x))\big) = -2\sec(x)\tan(x)/(1+2\sec(x))^2$

55. $f'(4) \approx \frac{f(4.1)-f(4)}{4.1-4} = \frac{6.2-5.7}{0.1} = 5.0$

57. $f'(\pi) \approx \frac{f(\pi+0.01)-f(\pi)}{0.01} = \frac{0.2}{0.01} = 20$

Further Theory and Practice

59. $(f\cdot g)'(2) = f'(2)\cdot g(2)+f(2)\cdot g'(2) = 5\cdot 8+3\cdot(-6) = 22$
$(f/g)'(2) = (g(2)\cdot f'(2)-f(2)\cdot g'(2)/g(2)^2 = (8\cdot 5-3\cdot(-6))/64 = 29/32$

61. $(f\cdot g)'(-1) = f'(-1)\cdot g(-1)+f(-1)\cdot g'(-1) = 2\cdot 4+5\cdot(-8) = -32$
$(f^2)'(-1) = (f\cdot f)'(-1) = f'(-1)\cdot f(-1)+f(-1)\cdot f'(-1) = 2\cdot 5+5\cdot 2 = 20$
$(g^2)'(-1) = (g\cdot g)'(-1) = g'(-1)\cdot g(-1)+g(-1)\cdot g'(-1) = -8\cdot 4+4\cdot(-8)$
$= -64$

63. Use the Quotient Rule.

$$\frac{d}{dx}\left(\frac{x+1/x}{1/x}\right) = \frac{(1/x)\cdot(1-1/x^2)-(x+1/x)\cdot(-1/x^2)}{1/x^2} = \frac{2/x}{1/x^2} = 2x$$

Simplify first. $\frac{x+1/x}{1/x} = x^2+1$, so the derivative is $2x$.

65. $\frac{d}{dx}\left(\frac{1}{x^3}\cdot\frac{1}{x^3}\right) = \frac{d}{dx}\left(\frac{1}{x^6}\right) = -\frac{6}{x^7}$ by the Power Rule.

67. Apply the Product Rule, twice.

$$\begin{aligned}\frac{d}{dx}\Big(x\cdot\sin(x)\cos(x)\Big) &= 1\cdot\sin(x)\cos(x)+x\cdot(\cos(x)\cos(x)-\sin(x)\cos(x))\\ &= \sin(x)\cos(x)+x(\cos^2(x)-\sin^2(x))\end{aligned}$$

69. Using the Product Rule and the fact that $(f^2)' = 2f \cdot f'$,

$$\frac{d}{dx}\Big((x^2+1)\cdot(x^2+1)^2\Big) = 2x\cdot(x^2+1)^2 + (x^2+1)\cdot\big(2(x^2+1)\cdot 2x\big)$$
$$= 6x(x^2+1)^2\,.$$

71. Investigate the tangent lines to the graph of $f(x) = 2x/(x+1)$ that pass through the point $P = (2,4)$ shown in the picture. It looks like that there are exactly two such lines.

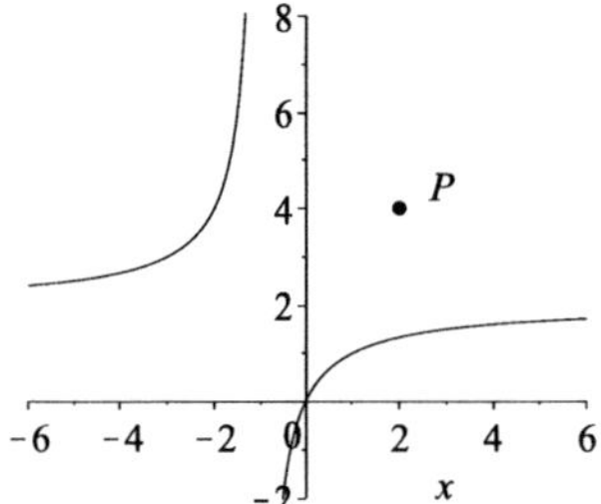

i) Since $f'(x) = ((x+1)\cdot 2 - 2x\cdot 1)/(x+1)^2 = 2/(x+1)^2$, the slope of the tangent line to the graph of f at $(c, f(c))$ is $f'(c) = 2/(c+1)^2$. This is positive for all points c in the domain of f.

ii) The equation of the tangent line to the graph of f at the point $(c, f(c))$ is $y = \frac{2}{(c+1)^2}(x-c) + \frac{2c}{c+1}$, or $y = \frac{2(x-c)+2c(c+1)}{(c+1)^2}$. Equivalently,

$$y = \frac{2x+2c^2}{(c+1)^2}\,.$$

iii) The tangent line through $(c, f(c))$ also passes through the point $P = (2,4)$ if, and only if, c is a solution to the equation

$$4 = \frac{2\cdot 2+2c^2}{(c+1)^2}\,.$$

Since $c \neq -1$, this is equivalent to $4(c+1)^2 = 4+2c^2$ which simplifies to $c^2+4c=0$. There are at exactly two solutions: $c=0$ and $c=-4$.

iv) There are two tangent lines through P. For $c=0$, the tangent line equation is $y=2x$, and for $c=-4$, the tangent line equation is $y = (2x+32)/9$

73. If $f(x) = A\cos(x) + B\sin(x)$, then $f'(x) = -A\sin(x) + B\cos(x)$. The constants A and B must be chosen so that

$$-A\sin(x) + B\cos(x) + A\cos(x) + B\sin(x) = 3\cos(x) - 5\sin(x)$$

for all x. So, for example, if $x=0$, then

$$B + A = 3\,.$$

On the other hand, if $x = \pi/2$, then

$$-A + B = -5\,.$$

Consequently, $2B = -2$, so $B = -1$ and $A = 4$. Observe that substituting these values into the first equation yields an identity.

75. Observe that $(f^2)(x) = f(x)^2 = |x|^2 = x^2$, implying that f^2 is differentiable at all values of x. In particular, $(f^2)'(0) = 0$, while $f'(0)$ does not exist (Example 3, Section 3.2).

77. Let $h = g \cdot (1/g)$. Then $h(x) = 1$ for all x in the domain of $1/g$. Consequently,
$$0 = h'(x) = g'(x) \cdot (1/g(x)) + g(x) \cdot (1/g)'(x)\,.$$
Solve this equation for $(1/g)'(x)$ to obtain
$$(1/g)'(x) = -g'(x)(1/g(x))/g(x) = -g'(x)/g(x)^2\,.$$

79. Let $h > 0$, then
$$\begin{aligned}\frac{1}{2}\Big(D_+f(c,h/2) + D_-f(c,h/2)\Big) &= \frac{1}{2}\left(\frac{f(c+h/2)-f(c)}{h/2} + \frac{f(c)-f(c-h/2)}{h/2}\right)\\ &= \frac{f(c+h/2)-f(c-h/2)}{h} = D_0f(c,h)\,.\end{aligned}$$

81. The elasticity of demand: $E(p) = -q'(p) \cdot \dfrac{p}{q(p)}$, p is price per unit and $q(p)$ is the quantity sold at price p.

(a) Since $q'(p) = \lim_{\Delta p \to 0}(\Delta q/\Delta p)$, when Δp is small we expect that $q'(p) \approx \Delta q/\Delta p$ and, for Δp *sufficiently* small,
$$E(p) \approx -(\Delta q/\Delta p) \cdot \frac{p}{q(p)} = -\frac{\Delta q/q(p)}{\Delta p/p}\,.$$

(b) A 2% increase in the price p of potatoes corresponds to $\Delta p/p = 0.02$. Assuming that $E(p) = 0.31$ and using the approximation in part a), $\Delta q/q(p) \approx -0.31(\Delta p/p) = -0.31 \cdot 0.02 = -0.0062$. This is approximately a 0.62% decrease in potato consumption.

(c) Using $p = 70$, $\Delta p = 10$, $q(p) = 300$, and $\Delta q = -100$ (from the Table), the estimate for $E(p)$ in part a) is $E(70) \approx -\left(\frac{-100/300}{10/70}\right) = 7/3$.

Using $p = 90$, $\Delta p = -10$, $q(p) = 96$, and $\Delta q = 104$, the estimate for $E(p)$ is $E(90) \approx -\left(\frac{104/96}{-10/90}\right) = 39/4$.

The central difference approximation for $E(80)$ can be obtained using the estimate in part a) with $p = 80$, $\Delta p = 20$, $q(p) = 200$, and $\Delta q = -204$. Doing so yields $E(80) \approx -\left(\frac{-204/200}{20/80}\right) = 102/25$.

CALCULATOR/COMPUTER EXERCISES

83. The numbers in the following table were computed in *Maple* using 10 significant digits of which six are exhibited.

h	$D_+f(c,h)$	$D_-f(c,h)$	$D_0f(c,h)$
10^{-1}	0.489434	-0.489434	0.
10^{-2}	0.049343	-0.049343	0.
10^{-3}	0.004934	-0.004934	0.
10^{-4}	0.000493	-0.000493	0.
10^{-5}	0.00005	-0.00005	0.

85. The numbers in the following table were computed in *Maple* using 10 significant digits of which six are exhibited.

h	$D_+f(c,h)$	$D_-f(c,h)$	$D_0f(c,h)$
10^{-1}	0.242799	0.244013	0.243425
10^{-2}	0.243371	0.243492	0.243432
10^{-3}	0.243426	0.243438	0.243433
10^{-4}	0.24344	0.24343	0.24344
10^{-5}	0.2434	0.2434	0.2435

87. $f(x) = \sin(x)$, $f'(\pi/6) = \cos(\pi/6) = \sqrt{3}/2 \approx 0.866025$

$$D_0f(\pi/6, 10^{-3}) = 0.866025$$

89. $f(x) = 1/x$, $f'(c) = -1/c^2 = -\sqrt{2} \approx -1.41421$

$$D_0f(1/\sqrt[4]{2}, 10^{-3}) = -1.41421$$

91. The graph of $x \mapsto D_0f(x, 10^{-5})$ is shown below on the left. The curve seems to be the same as the graph of $f'(x) = -\sin(x)$. The picture on the right displays the graph of f' and f's central difference quotient together. The curves overlap and appear to be the same.

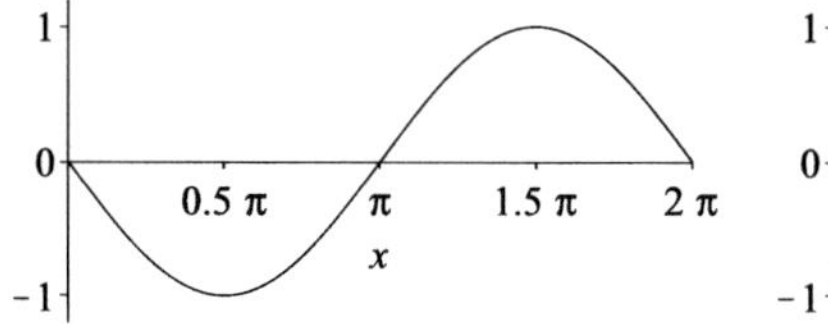

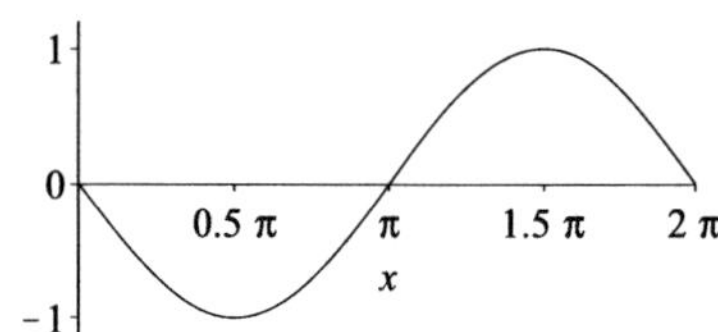

93. $L(a) = 10^5 \cdot (a^{0.5\cdot 10^{-5}} - a^{-0.5\cdot 10^{-5}})$,

$L(2.70) = 0.99322$, $L(2.73) = 1.00345$.

The graph on the right suggests that there is a number a_0 in $[2.70, 2.73]$ such that $L(a_0) = 1$. It looks like a_0 is between 2.718 and 2.72.

After zooming in, we see that, to 3-decimal places, $a_0 = 2.718$.

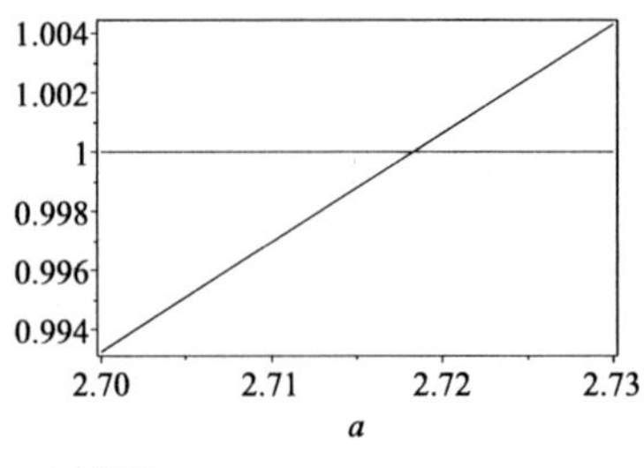

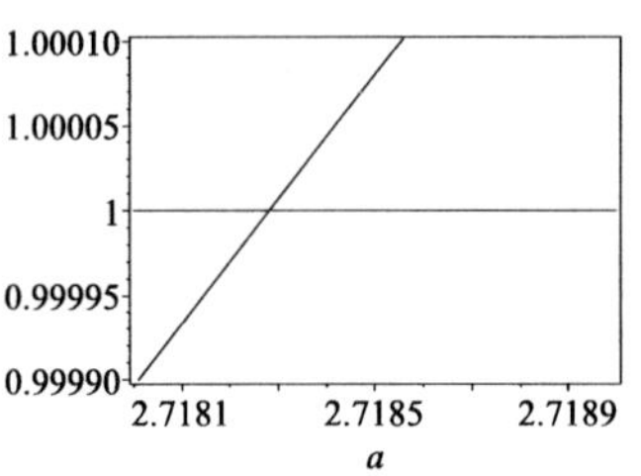

3.4 Differentiation of Some Basic Functions

1. Sum and Power Rule

$$\frac{d}{dx}\left(8x^{10} - 6x^{-5}\right) = 8 \cdot 10x^9 - 6 \cdot (-5)x^{-6} = 80x^9 + 30x^{-6}$$

3. Sum and Power Rule

$$\begin{aligned}\frac{d}{dx}\left(6x^{5/3} - 25x^{3/5}\right) &= 6 \cdot \left(\tfrac{5}{3}x^{2/3}\right) - 25 \cdot \left(\tfrac{3}{5}x^{-2/5}\right)\\ &= 10x^{2/3} - 15x^{-2/5}\end{aligned}$$

5. Sum Rule $\frac{d}{dx}\left(2x + e^x\right) = 2 + e^x$

7. Product Rule $\frac{d}{dx}\left(e^x x^{-3/2}\right) = e^x x^{-3/2} + e^x \cdot \left(-\frac{3}{2}x^{-5/2}\right) = (x - \frac{3}{2})x^{-5/2}e^x$

9. Product Rule

$$\frac{d}{dx}\left(x^{-9}\cot(x)\right) = -9x^{-10}\cot(x) - x^{-9}\csc^2(x) = -9x^{-10}(\cot(x) + x\csc^2(x))$$

11. Product Rule

$$\frac{d}{dx}\left(x^{-1/2}\sec x\right) = -\tfrac{1}{2}x^{-3/2}\sec(x) + x^{-1/2}\sec(x)\tan(x) = -\tfrac{1}{2}x^{-3/2}\sec(x)(1 - 2x\tan(x))$$

13. Sum Rule

$$\frac{d}{dx}\left(\csc(x) + \cot(x)\right) = -\csc\cot(x) - \csc^2(x) = -\csc(x)(\cot(x) + \csc(x))$$

15. Product Rule

$$\frac{d}{dx}(\csc(x)\cot(x)) = -\csc(x)\cot(x)\cot(x) - \csc(x)\csc^2(x) = -\csc(x)(\cot^2(x)+\csc^2(x))$$

17. Product Rule

$$\frac{d}{dx}(\tan(x)\sec(x)) = \sec^2(x)\sec(x)+\tan(x)\sec(x)\tan(x) = \sec(x)(\sec^2(x)+\tan^2(x))$$

19. Product Rule $\frac{d}{dx}(e^x\sin(x)) = e^x\sin(x)+e^x\cos(x) = e^x(\sin(x)+\cos(x))$

21. Reciprocal Rule

$$\frac{d}{dx}(3/(5+7e^x)) = 3\cdot(-7e^x/(5+7e^x)^2) = -21e^x/(5+7e^x)^2$$

23. Trigonometric Identities and Sum Rule

$$\frac{d}{dx}\left(\frac{1+\sec(x)}{\tan(x)}\right) = \frac{d}{dx}\left(\frac{1}{\tan(x)}+\frac{\sec(x)}{\tan(x)}\right) = \frac{d}{dx}(\cot(x)+\csc(x))$$
$$= -\csc^2(x) - \csc(x)\cot(x) = -\csc(x)(\csc(x)+\cot(x))$$

25. Quotient and Power Rule

$$\frac{d}{dx}\left(\frac{x^2}{x+e^x}\right) = \frac{(x+e^x)\cdot 2x - x^2\cdot(1+e^x)}{(x+e^x)^2} = \frac{x(x+2e^x-xe^x)}{(x+e^x)^2}$$

27. Quotient and Power Rule

$$\frac{d}{dx}\left(\frac{x^{1/3}}{1+x^{2/3}}\right) = \frac{(1+x^{2/3})\cdot\frac{1}{3}x^{-2/3} - x^{1/3}\cdot\frac{2}{3}x^{-1/3}}{(1+x^{2/3})^2} = \frac{x^{-2/3}-1}{3(1+x^{2/3})^2}$$

29. $f'(x) = 2x^{-1/3}$ so the slope of the tangent line at P is $f'(8) = 1$ and the equation of the tangent line is $y = (x-8)+12$ or $y = x+4$.

31. $f'(x) = -3x^{-4}\sin(x) + x^{-3}\cos(x)$ so the slope of the tangent line at P is $f'(\pi) = -\pi^{-3}$ and the equation of the tangent line is $y = -\pi^{-3}(x-\pi)$ or $y = -(x-\pi)/\pi^3$.

33. $f'(x) = \sec^2(x)\sec(x) + \tan(x)\sec(x)\tan(x) = \sec^3(x) + \tan^2(x)\sec(x)$ so the slope of the tangent line at P is $f'(\pi/3) = 8 + 6\cdot 3 = 14$ and the equation of the tangent line is $y = 14(x-\pi/3) + 2\sqrt{3}$.

35. $f'(x) = 8\tan(x) + 8x\sec^2(x)$ so the slope of the tangent line at P is $f'(\pi/4) = 8 + 2\pi\cdot 2 = 8+4\pi$ and the equation of the tangent line is $y = (8+4\pi)(x-\pi/4) + 2\pi$ or $y = (8+4\pi)x - \pi^2$.

37. $\frac{d}{dx}(e^x\cdot e^x) = e^x\cdot e^x + e^x\cdot e^x = 2e^{2x}$

39. $\frac{d}{dx}\big(\csc(x)\cdot\csc(x)\big) = -\csc(x)\cot(x)\cdot\csc(x) - \csc(x)\cdot\csc(x)\cot(x)$
$= -2\csc^2(x)\cot(x)$

41. Since $\frac{d}{dx}\big(x^{n+1}/(n+1)\big) = x^n$, and a constant differentiates to 0, the polynomial $p(x) = 7\cdot x^7/7 - 4\cdot x^2/2 + 6\cdot x + C = x^7 - 2x^2 + 6x + C$ will differentiate to this polynomial (C is any constant).

43. Since $\frac{d}{dx}\big(x^{n+1}/(n+1)\big) = x^n$, and a constant differentiates to 0, the polynomial $p(x) = x^9/9 + 6\cdot x^6/6 - 3\cdot x^2/2 + C = x^9/9 + x^6 - 3x^2/2 + C$ will differentiate to this polynomial (C is any constant).

45. Since $1/\sqrt{x} = x^{-1/2}$, an antiderivative can be obtained by adding one to the exponent and dividing by the new exponent. This implies that $x^{1/2}/(1/2) + C = 2\sqrt{x} + C$ differentiates to $1/\sqrt{x}$.

47. Since $\frac{d}{dx}\big(\cot(x)\big) = -\csc^2(x)$, and the derivative of a constant is 0, $\frac{d}{dx}\big(8\cot(x) + C\big) = -8\csc^2(x)$, C is any constant.

49. We will use the fact that if f is differentiable everywhere, then given any constant α, $\frac{d}{dx}\big(f(x-\alpha)\big) = f'(x-\alpha)$. See Exercise 57 in Section 3.2. In particular, if m is a positive integer, then $\frac{d}{dx}\big((x-\alpha)^m\big) = m(x-\alpha)^{m-1}$. By the Product Rule, applied to $p(x) = (x-\alpha)^m q(x)$,

$$\begin{aligned} p'(x) &= m(x-\alpha)^{m-1}q(x) + (x-\alpha)^m q'(x) \\ &= (x-\alpha)^{m-1}(mq(x) + (x-\alpha)q'(x)) \\ &= (x-\alpha)^{m-1}r(x) \end{aligned}$$

where $r(x) = mq(x) + (x-\alpha)q'(x)$. Because $q(\alpha) \neq 0$, and $m > 0$, $r(\alpha) = mq(\alpha) \neq 0$.

51. Since $\tan^2(x) = \sec^2(x) - 1$, and $\frac{d}{dx}\big(\tan(x)\big) = \sec^2(x)$,

$$\frac{d}{dx}\big(\tan(x) - x + C\big) = \sec^2(x) - 1 = \tan^2(x),$$

where C is any constant.

53. Since $f(x) = 2x^6 + x^5$, $f'(x) = 12x^5 + 5x^4$ and $f'(-1) = -7$.

55. Since $f'(x) = (5x^4+4x)(2x^3+7x-2)+(x^5+2x^2-3)(6x^2+7)$, $f'(1) = 63$.

57. Recall that the Product Rule implies that $(g^2)' = 2g\cdot g'$. Therefore, $f'(x) = 2\sec(x)\tan^2(x) - 2\tan(x)\sec^2(x) = 0$. This is obtained more easily using the trigonometric identity $\sec^2(x) - \tan^2(x) = 1$.

59. Let $f(x) = e^x$. Then for $h \neq 0$, $D_0 f(0,h) = \frac{e^{h/2} - e^{-h/2}}{h}$ implying that $\lim_{h\to 0} \frac{e^{h/2}-e^{-h/2}}{h} = f'(0) = e^0 = 1$.

61. If $x > 0$, then $|x| = x$, and $\frac{d}{dx}\big(|x|^p\big) = \frac{d}{dx}\big(x^p\big) = px^{p-1} = p\frac{x^p}{x} = p\frac{|x|^p}{x}$.

If $x < 0$, then $|x| = -x$, and $|x|^p = ((-1)x)^p = (-1)^p x^p$. Therefore, $\frac{d}{dx}\big(|x|^p\big) = (-1)^p \frac{d}{dx}\big(x^p\big) = (-1)^p px^{p-1} = p\frac{(-1)^p x^p}{x} = p\frac{(-x)^p}{x} = p\frac{|x|^p}{x}$.

63. Since $\frac{dy}{dx} = e^x$, the slope of the tangent line to $y = e^x$ at the point $P = (c, e^c)$ is $\left.\frac{dy}{dx}\right|_{x=c} = e^c$. Therefore, the equation the tangent line at P is $y = e^c(x - c) + e^c$ or $e^c x - y = e^c(c - 1)$. The x-intercept of this line is $c - 1$ and its y-intercept is $e^c(1 - c)$. These intercepts are equidistant to the origin if, and only if, $e^c|1 - c| = |c - 1|$. That is, if, and only if, $c = 0$ or $c = 1$.

65. The graph of f has a horizontal tangent line at the point $(c, f(c))$ if, and only if, $f'(c) = 0$. Since $f'(x) = ((x^2 + 1) \cdot px^{p-1} - x^p \cdot 2x)/(x^2 + 1)^2$, this implies that $p(c^2 + 1)c^{p-1} = 2c^{p+1}$. Since $c > 0$, it follows that $p(c^2 + 1) = 2c^2$, so $(p - 2)c^2 = -p$ and $c = \sqrt{p/(2 - p)}$.

67. $\frac{d}{dx}(1/\cos(x)) = -(-\sin(x))/\cos^2(x) = \sec(x)\tan(x)$

69. Let t be time so the units on $(f \cdot g)'(t)$ are (f-units) $\cdot$ (g-units)/(time unit) and the same is true for $f(t) \cdot g'(t)$ and $f'(t) \cdot g(t)$. However, the units for $f'(t) \cdot g'(t)$ are (f-units) $\cdot$ (g-units)/(time unit)^2.

71. Assume $E(p) = -g'(p) \cdot p/q(p)$.

   (a) If $g(p) = a - bp$, then $g'(p) = -b$ and

$$E(p) = -(-b) \cdot p/(a - bp) = -bp/(a - bp)\,.$$

   (b) If $g(p) = a/p^b = ap^{-b}$, then $g'(p) = -abp^{-b-1} = -ab/p^{b+1}$ and

$$E(p) = -(-ab/p^{b+1}) \cdot p/(a/p^b) = (ab/p^b)/(a/p^b) = b\,.$$

73. For the hyperoblic tangent use the Quotient Rule.

$$\tanh'(x) = \frac{d}{dx}\left(\frac{e^x - e^{-x}}{e^x + e^{-x}}\right) = \frac{(e^x + e^{-x})(e^x + e^{-x}) - (e^x - e^{-x})(e^x - e^{-x})}{(e^x + e^{-x})^2}$$

$$= \frac{4}{(e^x + e^{-x})^2} = \left(\frac{2}{e^x + e^{-x}}\right)^2 = \operatorname{sech}^2(x)$$

For the hyperoblic secant use the Reciprocal Rule.

$$\operatorname{sech}'(x) = \frac{d}{dx}\left(\frac{2}{e^x + e^{-x}}\right) = -2\frac{e^x - e^{-x}}{(e^x + e^{-x})^2}$$

$$= -\frac{2}{e^x + e^{-x}} \cdot \frac{e^x - e^{-x}}{e^x + e^{-x}} = -\operatorname{sech}(x)\tanh(x)$$

CALCULATOR/COMPUTER EXERCISES

75. The information in the table on the left is approximate, based upon the graph on the right. The derivative is the dashed curve.

Interval where f increases	Interval where f decreases	Point at which f has a horizontal tangent
$[-0.5, 0.5]$	$[-1.3, -0.5]$	$x = 0.5$
Interval where $f' > 0$	Interval where $f' < 0$	Point(s) where $f' = 0$
(-0.5,0.5)	$(-1.3, -0.5)$	$x = 0.5$

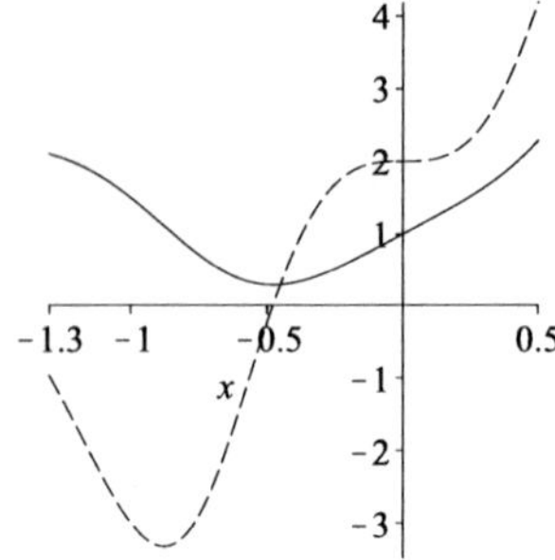

77. The information in the table on the left is approximate, based upon the graph on the right. The derivative is the dashed curve.

Interval where f increases	Interval where f decreases	Point at which f has a horizontal tangent
$[0, 5]$	$[5, 7]$	$x = 0, 5$
Interval where $f' > 0$	Interval where $f' < 0$	Point(s) where $f' = 0$
$(0, 5)$	$(5, 7)$	$x = 0, 5$

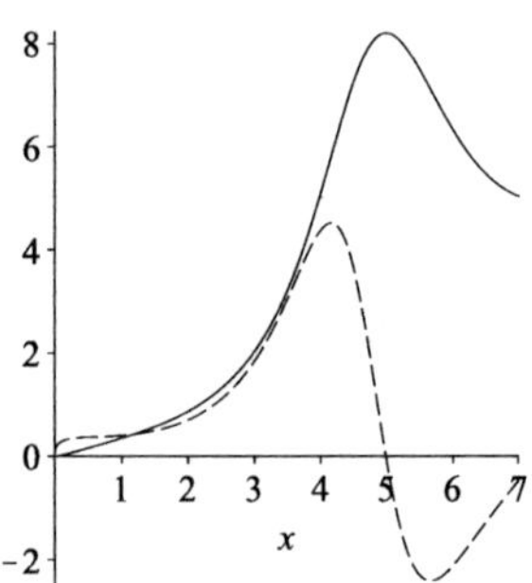

79. The graph of the function

$$k \mapsto D_0 g_k(0, 10^{-4})$$

is plotted on the right. Based upon this picture, $g_k'(0) = k$.

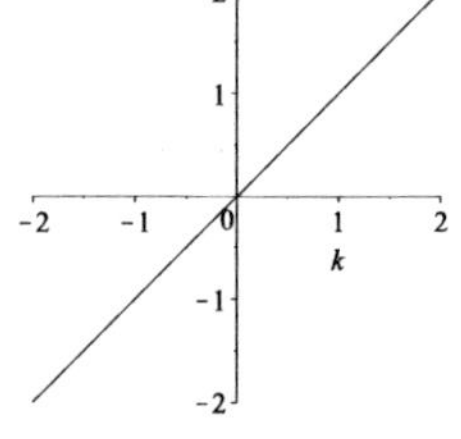

81. The equation $(\tau - \mu^2\tau_0)p - (\tau_0 - \mu^2\tau)p_0 + 2\mu^2\Delta = 0$ defines the *Hugoniot curve*.

(a) Solve the equation for τ: $\tau = \frac{\mu^2\tau_0 p + \tau_0 p_0 - 2\mu^2\Delta}{p + \mu^2 p_0}$, and apply the Quotient Rule to obtain

$$\frac{d\tau}{dp} = \frac{(p + \mu^2 p_0)\cdot \mu^2\tau_0 - (\mu^2\tau_0 p + \tau_0 p_0 - 2\mu^2\Delta)\cdot 1}{(p + \mu^2 p_0)^2}$$
$$= \frac{\mu^4 p_0\tau_0 - p_0\tau_0 + 2\mu^2\Delta}{(p + \mu^2 p_0)^2}.$$

(b) The tangent line equation at $Q = (p_1, \tau_1)$ is

$$\tau = \frac{\mu^4 p_0\tau_0 - p_0\tau_0 + 2\mu^2\Delta}{(p_1 + \mu^2 p_0)^2}\cdot (p - p_1) + \tau_1 .$$

(c) The point (p_0, τ_0) is below the Hugoniot curve if, and only if,

$$\tau_0 < \frac{\mu^2\tau_0 p_0 + \tau_0 p_0 - 2\mu^2\Delta}{p_0 + \mu^2 p_0}.$$

Or, since $p_0 + \mu^2 p_0 > 0$, if, and only if,

$$(p_0 + \mu^2 p_0)\tau_0 < \mu^2\tau_0 p_0 + \tau_0 p_0 - 2\mu^2\Delta.$$

This inequality, in turn, is equivalent to $\Delta < 0$.

(d) The tangent line to a point $Q = (p_1, \tau_1)$ on the Hugoniot curve passes through (p_0, τ_0) when p_0 and τ_0 satisfy the tangent line equation

$$\tau_0 = \frac{\mu^4 p_0\tau_0 - p_0\tau_0 + 2\mu^2\Delta}{(p_1 + \mu^2 p_0)^2} \cdot (p_0 - p_1) + \tau_1.$$

Moreover, τ_1 and p_1 satisfy the Hugoniot equation:

$$\tau_1 = \frac{\mu^2\tau_0 p_1 + \tau_0 p_0 - 2\mu^2\Delta}{p_1 + \mu^2 p_0}.$$

Use this to replace τ_1 in the tangent line equation to obtain

$$\tau_0 = \frac{\mu^4 p_0\tau_0 - p_0\tau_0 + 2\mu^2\Delta}{(p_1 + \mu^2 p_0)^2} \cdot (p_0 - p_1) + \frac{\mu^2\tau_0 p_1 + \tau_0 p_0 - 2\mu^2\Delta}{p_1 + \mu^2 p_0}$$

which can be used to find p_1 in terms of p_0 and τ_0.

(e) With the values given, the last equation is

$$1.7 = \frac{-4.347}{(p_1+0.525)^2} \cdot (2.1 - p_1) + \frac{0.425p_1+4.75}{p_1+0.525}$$

yielding $p_1 = 1.249$ and $p_2 = 4.519$. Substitute these values into the Hugoniot equation above, and the τ-values are $\tau_1 = 2.875$ and $\tau_2 = 1.287$, respectively. The Chapman-Jouquet points are $(1.249, 2.875)$ and $(4.519, 1.287)$. See the picture.

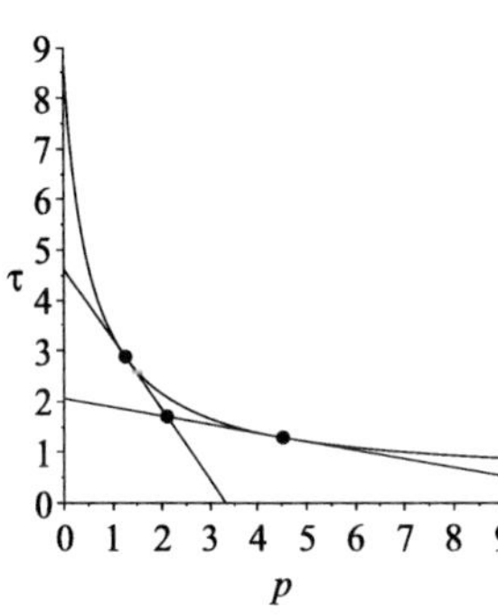

3.5 The Chain Rule

1. Let $u = 3x$, then

$$\frac{d}{dx}\sin(3x) = \frac{d}{dx}\sin(u) = \frac{d}{du}\sin(u) \cdot \frac{du}{dx} = \cos(u) \cdot 3 = 3\cos(3x).$$

3. Let $u = 5x$, then $\frac{d}{dx}e^{5x} = \frac{d}{dx}e^u = \frac{d}{du}e^u \cdot \frac{du}{dx} = e^u \cdot 5 = 5e^{5x}$.

5. Let $u = 1/x$, then

$$\frac{d}{dx}\cos(1/x) = \frac{d}{dx}\cos(u) = \frac{d}{du}\cos(u)\cdot\frac{du}{dx} = -\sin(u)\cdot(-1/x^2) = (1/x^2)\sin(1/x).$$

7. Let $u = x^2 + 3x$, then

$$\frac{d}{dx}\sin(x^2+3x) = \frac{d}{dx}\sin(u) = \frac{d}{du}\sin(u)\cdot\frac{du}{dx} = \cos(u)\cdot(2x+3) = (2x+3)\cos(x^2+3x).$$

9. Let $u = \cos(x)$, then

$$\begin{aligned}\frac{d}{dx}\sec(\cos(x)) = \frac{d}{dx}\sec(u) = \frac{d}{du}\sec(u)\cdot\frac{du}{dx} &= \sec(u)\tan(u)\cdot(-\sin(x)) \\ &= -\sin(x)\sec(\cos(x))\tan(\cos(x)).\end{aligned}$$

11. Let $u = 1 - 7x$, then

$$\frac{d}{dx}\tan(1-7x)) = \frac{d}{dx}\tan(u) = \frac{d}{du}\tan(u)\cdot\frac{du}{dx} = \sec^2(u)\cdot(-7) = -7\sec^2(1-7x).$$

13. Quotient Rule, Chain Rule–Power Rule

$$\begin{aligned}\frac{d}{dx}\left(\frac{x}{\sqrt{1+x}}\right) &= \frac{\sqrt{1+x} - x\cdot\frac{d}{dx}(1+x)^{1/2}}{1+x} = \frac{\sqrt{1+x} - x\cdot\frac{1}{2}(1+x)^{-1/2}}{1+x} \\ &= \frac{1+x-x/2}{(1+x)^{3/2}} = \frac{1+x/2}{(1+x)^{3/2}}\end{aligned}$$

15. Let $u = x^2 + 4$, then

$$\begin{aligned}\frac{d}{dx}\big(\cot(x^2+4) + \csc(x^2+4)\big) &= \frac{d}{dx}\big(\cot(u) + \csc(u)\big) = \frac{d}{du}\big(\cot(u) + \csc(u)\big)\cdot\frac{du}{dx} \\ &= (-\csc^2(u) - \csc(u)\cot(u))\cdot 2x \\ &= -2x\csc(x^2+4)(\csc(x^2+4) + \cot(x^2+4)).\end{aligned}$$

17. Let $u = -1/x^2$, then

$$\frac{d}{dx}\exp(-x^{-2}) = \frac{d}{dx}\exp(u) = \frac{d}{du}\exp(u)\cdot\frac{du}{dx} = \exp(u)\cdot 2x^{-3} = (2/x^3)\exp(-1/x^2).$$

19. The "inside" is $2x + \sin(x)$.

$\frac{d}{dx}\left((2x+\sin(x))^3\right) = 3(2x+\sin(x))^2\cdot\frac{d}{dx}(2x+\sin(x)) = 3(2x+\sin(x))^2(2+\cos(x))$

21. $\frac{d}{dx}\left((4+x^2)^{1/2}\right) = \frac{1}{2}(4+x^2)^{-1/2} \cdot \frac{d}{dx}\left(4+x^2\right) = x/\sqrt{4+x^2}$

23. $\frac{d}{dx}\left((\sin(x))^3\right) = 3(\sin(x))^2 \cdot \frac{d}{dx}(\sin(x)) = 3\sin^2(x)\cos(x)$

25. $12\frac{d}{dx}\left((1+e^x)^{1/2}\right) = 12 \cdot \frac{1}{2}(1+e^x)^{-1/2} \cdot \frac{d}{dx}(1+e^x) = 6e^x/\sqrt{1+e^x}$

27. Since $(f \circ g)(x) = x^2+7x$, $(f \circ g)'(x) = 2x+7$. Since $(g \circ f)(x) = \sqrt{x^4+7x^2}$,

$$(g \circ f)'(x) = \tfrac{1}{2}(x^4+7x^2)^{-1/2} \cdot (4x^3+14x) = (2x^3+7x)/\sqrt{x^4+7x^2}\,.$$

29. Since $(f \circ g)(x) = x^3/(x^3+1)$,

$$(f \circ g)'(x) = ((x^3+1)\cdot 3x^2 - x^3 \cdot 3x^2)/(x^3+1)^2 = 3x^2/(x^3+1)^2\,.$$

Since $(g \circ f)(x) = \big(x/(x+1)\big)^3$,

$$(g \circ f)'(x) = 3\big(x/(x+1)\big)^2 \cdot \left((x+1-x)/(x+1)^2\right) = 3x^2/(x+1)^4\,.$$

31. $\frac{d}{dx}2^x = 2^x \ln(2)$

33. $\frac{d}{dx}3^{2x} = \frac{d}{dx}9^x = 9^x \ln(9) = 3^{2x}\ln(9)$

35. $\frac{d}{dx}\left(8^x - x^8\right) = 8^x \ln(8) - 8x^7$.

37. Let $u = 1/x$, then

$$\begin{aligned}\tfrac{d}{dx}\left((1+e)^{1/x}\right) &= \tfrac{d}{dx}\left((1+e)^u\right) = \tfrac{d}{du}\left((1+e)^u\right)\cdot\frac{du}{dx}\\ &= (1+e)^u\ln(1+e)\cdot(-1/x^2) = -(1/x^2)(1+e)^{1/x}\ln(1+e)\,.\end{aligned}$$

39. Quotient Rule

$$\frac{d}{dx}\left(\frac{3x+2}{5^x}\right) = \frac{5^x\cdot 3 - (3x+2)\cdot 5^x\ln(5)}{5^{2x}} = \frac{3-\ln(5)(3x+2)}{5^x}$$

41. $(g \circ f)'(2) = g'(f(2))\cdot f'(2) = g'(3)\cdot f'(2) = 32$

43. $(g \circ f)'(6) = g'(f(6))\cdot f'(6) = g'(1/2)\cdot f'(6) = 4$

45. Let $H(x) = \sqrt{1+(1+2x)^5}$. If $f(x) = 1+2x$, $g(x) = 1+x^5$, and $k(x) = \sqrt{x}$, then $H = k \circ g \circ f = k \circ (g \circ f)$ and

$$\begin{aligned}H'(x) &= k'\big((g\circ f)(x)\big)\cdot(g\circ f)'(x) = k'(g(f(x))\cdot g'(f(x))\cdot f'(x)\\ &= \tfrac{1}{2}\cdot(1+(1+2x)^5)^{-1/2}\cdot 5(1+2x)^4\cdot 2 = 5(1+2x)^4/\sqrt{1+(1+2x)^5}\,.\end{aligned}$$

47. Let $H(x) = \cos^3(2x))$. If $f(x) = 2x$, $g(x) = \cos(x)$, and $k(x) = x^3$, then $H = k \circ g \circ f = k \circ (g \circ f)$ and

$$\begin{aligned} H'(x) &= k'\big((g \circ f)(x)\big) \cdot (g \circ f)'(x) = k'(g(f(x)) \cdot g'(f(x)) \cdot f'(x) \\ &= 3\cos^2(2x) \cdot (-\sin(2x)) \cdot 2 = -6\cos^2(2x)\sin(2x). \end{aligned}$$

49. Let $H(x) = 2\tan(\sqrt{3x+2})$. If $f(x) = 3x + 2$, $g(x) = \sqrt{x}$, and $k(x) = 2\tan(x)$, then $H = k \circ g \circ f = k \circ (g \circ f)$ and

$$\begin{aligned} H'(x) &= k'\big((g \circ f)(x)\big) \cdot (g \circ f)'(x) = k'(g(f(x)) \cdot g'(f(x)) \cdot f'(x) \\ &= 2\sec^2(\sqrt{3x+2}) \cdot \big(\tfrac{1}{2} \cdot (3x+2)^{-1/2}\big) \cdot 3 \\ &= 3\sec^2(\sqrt{3x+2})/\sqrt{3x+2}. \end{aligned}$$

Further Theory and Practice

51. Observe that $(F \circ G \circ H)'(4) = F'(G(H(4))) \cdot G'(H(4)) \cdot H'(4)$.

    Since $H'(x) = 5/(x+5)^2$, $H'(4) = 5/81$. Since $G'(x) = 1/(2\sqrt{x})$, and $H(4) = 4/9$, $G'(H(4)) = 3/4$. Since $F'(x) = 3$, $F'(G(H(4))) = 3$, and

$$(F \circ G \circ H)'(4) = 3 \cdot (3/4) \cdot (5/81) = 5/36.$$

53. Observe that $(G \circ F \circ F)'(4) = G'(F(F(4))) \cdot F'(F(4)) \cdot F'(4)$.

    Since $F'(x) = 3$ for all x and $G'(x) = 1/(2\sqrt{x})$,

$$(G \circ F \circ F)'(4) = G'(F(F(4))) \cdot 3 \cdot 3 = 9G'(F(13)) = 9G'(40) = 9/(2\sqrt{40}).$$

    Rationalizing the denominator, $9/(2\sqrt{40}) = 9/(4\sqrt{10}) = 9\sqrt{10}/40$.

55. Because x is in meters and $\sin(\omega t)$ is unitless, A must be measured in meters. Since the argument of the sine function is unitless and t in seconds, ω must be measured in units of 1/seconds.

57. The arguments of the exponential and sine function are unitless. Consequently, both k and ω are in units of $(\text{seconds})^{-1}$. The units for A must be centimeters to match the units for x (both the exponential and sine function are unitless). By the Product and Chain Rules

$$x'(t) = -Ake^{-kt}\sin(\omega t) + A\omega e^{-kt}\cos(\omega t).$$

    Both terms on the right side of the equation are in units of centimeters/second matching the units on the left side.

59. Since $m'(t) = 100 \cdot (-0.001213) \cdot e^{-0.001213t} = -0.001213\, m(t)$, $m'(t)/m(t) = -0.001213\ (\text{year})^{-1}$. Both limits are 0.

61. Let A and x denote the area of the triangle and the length of each of its two legs, respectively. Observe that $A = x^2/2$, so $dA/dx = x$ square inches per inch. Since $dx/dt = 2$ inches per minute,

$$dA/dt = dA/dx \cdot dx/dt = 2x \text{ square inches per minute}.$$

When the hypotenuse h is 8 inches, $x = \sqrt{32} = 4\sqrt{2}$ $(2x^2 = h^2)$ so

$$\left.\frac{dA}{dt}\right|_{h=8} = \left.\frac{dA}{dt}\right|_{x=2} = 8\sqrt{2}\ \text{in}^2/\text{min}.$$

63. By the Chain Rule, $g'(t) = 2tf'(t^2)$. Therefore,

$$f'(t^2) - g'(t) = f'(t^2) - 2tf'(t^2) = (1-2t)f'(t^2),$$

and

$$f'(t) - g'(\sqrt{t}\,) = f'(t) - 2\sqrt{t}\,f'(t) = (1-2\sqrt{t}\,)f'(t).$$

65. $$\begin{aligned} f'(x) &= 4\cdot 5\sec^4(\sqrt{7x}\,)\cdot\sec(\sqrt{7x}\,)\tan(\sqrt{7x}\,)\cdot\left(\tfrac{1}{2}(7x)^{-1/2}\cdot 7\right) \\ &= 70\sec^5(\sqrt{7x}\,)\tan(\sqrt{7x}\,)/\sqrt{7x} \end{aligned}$$

67. $$\begin{aligned} f'(x) &= 2\sec^2(\sqrt{x}\,)\cdot 1/(2\sqrt{x}\,) + 6\cdot\tfrac{1}{2}(\tan(x))^{-1/2}\cdot\sec^2(x) \\ &= \sec^2(\sqrt{x}\,)/\sqrt{x} + 3\sec^2(x)/\sqrt{\tan(x)} \end{aligned}$$

69. $$\begin{aligned} f'(x) &= \left(\sqrt{x^2+1}\cdot 2(5x^3+1)\cdot 15x^2 - (5x^3+1)^2\cdot\tfrac{1}{2}\cdot(x^2+1)^{-1/2}\cdot 2x\right)/(x^2+1) \\ &= \left(x(5x^3+1)(30x(x^2+1) - (5x^3+1)\right)/(x^2+2)^{3/2} \\ &= \left(x(5x^3+1)(25x^3+30x-1)\right)/(x^2+2)^{3/2} \end{aligned}$$

71. $$\begin{aligned} f'(x) &= 2\sin(\pi\sin(\pi x^2))\cdot\cos(\pi\sin(\pi x^2))\cdot 2\pi^2 x\cos(\pi x^2) \\ &= 4\pi^2 x\sin(\pi\sin(\pi x^2))\cos(\pi\sin(\pi x^2))\cos(\pi x^2) \end{aligned}$$

73. $$\begin{aligned} f'(x) &= 3\left(\sqrt{2x-1}+\sqrt{2x+1}\,\right)^{-1/2}\cdot\left(\tfrac{1}{2}(2x-1)^{-1/2}\cdot 2 + \tfrac{1}{2}(2x+1)^{-1/2}\cdot 2\right) \\ &= \frac{3(\sqrt{2x-1}+\sqrt{2x+1}\,)}{\sqrt{\sqrt{2x-1}+\sqrt{2x+1}}\,\sqrt{2x-1}\,\sqrt{2x+1}} \\ &= \frac{3\sqrt{\sqrt{2x-1}+\sqrt{2x+1}}}{\sqrt{4x^2-1}} = 3\sqrt{(\sqrt{2x-1}+\sqrt{2x+1}\,)/(4x^2-1)} \end{aligned}$$

75. Both formulas for $f(x)$ evaluate to 1 when $x = 0$ so f is continuous at 0 for any value of the constant c. For differentiability at 0, the right-side tangent line must have the same slope as the left-side tangent line, so c must be chosen so that $3(cx+1)^2\cdot c = 1$ when $x = 0$. Choose $c = 1/3$.

77. (a) The initial population size is $P(0) = P_0$. Since $\lim_{t\to\infty} e^{-kMt} = 0$, $\lim_{t\to\infty} P(t) = M$.

(b) Using the Reciprocal Rule, $P'(t) = \frac{P_0 M^2 (M-P_0) k e^{-kMt}}{(P_0+(M-P_0)e^{-kMt})^2}$.

(c) Referring to the formula for $P'(t)$ in part (b), both exponential terms have a limiting value of 0 as $t \to \infty$. Therefore, $\lim_{t\to\infty} P'(t) = 0$.

79. (a) Both exponential terms have a limiting value of 0 as $t \to \infty$. Therefore, $\lim_{t\to\infty} v(t) = \sqrt{g/\kappa}$.

(b) $v'(t) = \sqrt{\frac{g}{\kappa}}\left(\frac{(1+\exp(-2t\sqrt{g\kappa}))\cdot 2\sqrt{g\kappa}\cdot\exp(-2t\sqrt{g\kappa})-(1-\exp(-2t\sqrt{g\kappa}))(-2\sqrt{g\kappa}\cdot\exp(-2t\sqrt{g\kappa}))}{(1+\exp(-2t\sqrt{g\kappa}))^2}\right)$

$$= \sqrt{\frac{g}{\kappa}}\left(\frac{4\sqrt{g\kappa}\,\exp(-2t\sqrt{g\kappa})}{(1+\exp(-2t\sqrt{g\kappa}))^2}\right) = \frac{4g\,\exp(-2t\sqrt{g\kappa})}{(1+\exp(-2t\sqrt{g\kappa}))^2}$$

(c) Both exponential terms have a limiting value of 0 as $t \to \infty$. Therefore, $\lim_{t\to\infty} v'(t) = 0$.

(d) Compare $g - \kappa v^2(t) = g - g\left(\frac{1-\exp(-2t\sqrt{g\kappa})}{1+\exp(-2t\sqrt{g\kappa})}\right)^2$

$$= g\left(\frac{(1+\exp(-2t\sqrt{g\kappa}))^2-(1-\exp(-2t\sqrt{g\kappa}))^2}{(1+\exp(-2t\sqrt{g\kappa}))^2}\right) = g\left(\frac{4\exp(-2t\sqrt{g\kappa})}{(1+\exp(-2t\sqrt{g\kappa}))^2}\right)$$

to $v'(t)$ in part b).

81. Recall (Exercise 72, Section 3.4) that $\frac{d}{dx}\cosh(x) = \sinh(x)$ and $\frac{d}{dx}\sinh(x) = \cosh(x)$. Therefore, $\frac{d}{dx}\cosh(ax) = a\sinh(ax)$ and $\frac{d}{dx}\sinh(ax) = a\cosh(ax)$.

83. Apply the Chain Rule, $v'(t) = \sqrt{\frac{g}{\kappa}}\,\sqrt{g\kappa}\,\mathrm{sech}^2(t\sqrt{g\kappa}) = g\,\mathrm{sech}^2(t\sqrt{g\kappa})$.

Since $\cosh^2(x) - \sinh^2(x) = 1$ for all x, (verify), $1 - \tanh^2(x) = \mathrm{sech}^2(x)$ implying that $v'(t) = g(1 - \tanh^2(t\sqrt{g\kappa})) = g - \kappa v^2(t)$.

COMPUTER/CALCULATOR EXERCISES

85. According to *Maple*, $g'(x) = (1 + 2x - x^2)/(1 + x^2)^2$,
$f'(x) = (1 + 4x - x^2)/(1 + x^2)^2$, and
$(g\circ f)'(x) = -(x^6 - 2x^5 - 12x^4 + 16x^3 - 28x^2 + 22x + 7)/(x^4 + 3x^2 - 4x + 5)^2$.
Using *Maple*'s *Maximize* procedure the maximum values for g', f' and $(g\circ f)'$ in the interval $[0, 2]$ are attained at 0.26795, 0.38677, and 1.52980, respectively. The graphs of g' (solid line), f' (dashed line), and $(g\circ f)'$ (dotted line) are displayed on the right.

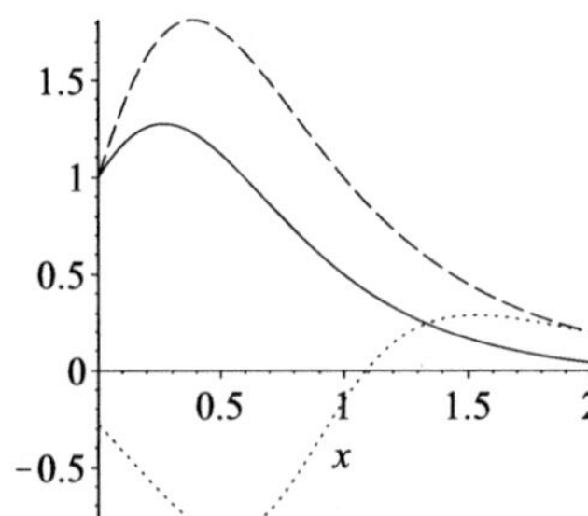

87. Measuring time in billions of years, $0.5 = m(4.51) = e^{-4.51k}$. Consequently, $\ln(0.5) = -4.15k$ and $k = -\ln(0.5)/4.15 = \ln(2)/4.15$. Using

Maple's *fsolve* we find that $m(t_0) = 1/8$ when $t_0 = 13.5$ billion years and $m'(13.5) = -0.01921$ grams per billion years or -1.921×10^{-11} gr/yr.

89. Using the given information,

$$T(t) = 115 + 235e^{-kt}.$$

Applying *fsolve* to $T(25) = 175$ yields $k = 0.05461$.

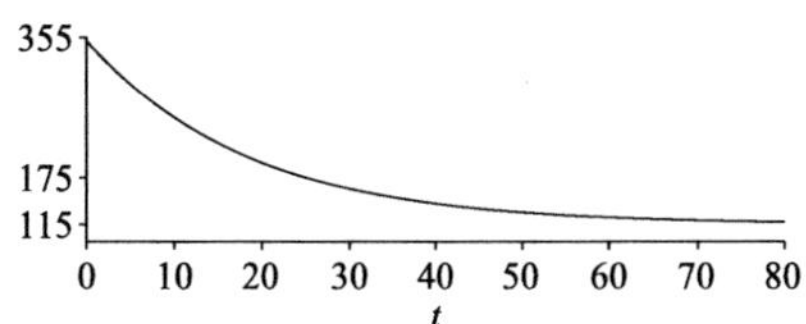

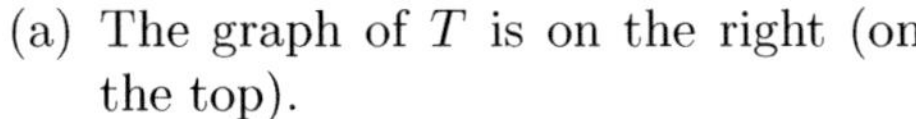

(a) The graph of T is on the right (on the top).
(b) The graph of T' is on the right (on the bottom).
(c) The table displays the rates of change for T at the specified temperatures.

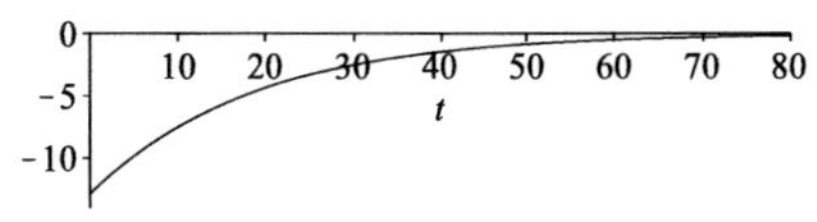

T	175	155	135	125	115.5
T'	-3.276	-2.184	-1.092	-0.546	-0.0273

91. The graph of v and v' are displayed on the right, the velocity v above, and the acceleration v' below.

The velocity curve has a corner at $t = 8$ and the acceleration curve is broken into two separate pieces.

The pictures suggest that v is everywhere continuous and differentiable for all t *except* $t = 8$ seconds. The domain of the acceleration function v' does not include $t = 8$. v' is continuous and differentiable for all other values of t.

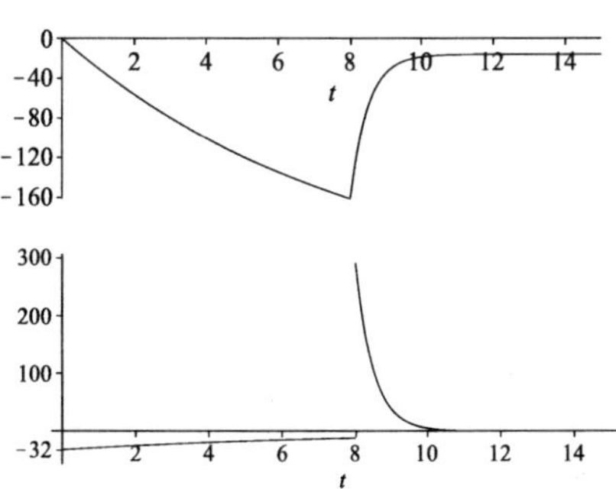

3.6 Derivatives of Inverse Functions

Problems for Practice

For Problems 1–26 use the fact that if $f^{-1}(t) = s$, then $(f^{-1})'(t) = (f'(s))^{-1}$.

1. Since $f^{-1}(4) = 1$, $(f^{-1})'(4) = (f'(1))^{-1} = 1/2$.

3. Since $f^{-1}(4) = -1$, $(f^{-1})'(4) = (f'(-1))^{-1} = 1/3$.

5. Since $f^{-1}(4) = 1$ and $f'(1) = 6$, $(f^{-1})'(4) = (f'(1))^{-1} = 1/6$.

7. Since $f^{-1}(4) = \sqrt{3}$ and $f'(\sqrt{3}) = 5/4$, $(f^{-1})'(4) = (f'(\sqrt{3}))^{-1} = 4/5$.

9. $(f^{-1})'(t) = 1/f'(s) = 1/3$.

11. $(f^{-1})'(t) = 1/f'(s) = -2(s+3)^{3/2} = -2/t^3$, because $\sqrt{s+3} = 1/t$.

13. $(f^{-1})'(t) = 1/f'(s) = 1/(5s^4) = (t-2)^{-4/5}/5$, because $s = (t-2)^{1/5}$.

15. $(f^{-1})'(t) = (f'(s))^{-1} = (2s+2)^{-1}$. Since $s^2 + 2s = t$, $s^2 + 2s - t = 0$ and (quadratic formula) $s = (-2 \pm \sqrt{4+4t}\,)/2 = -1 + \sqrt{1+t}$ (use the plus sign because $s > 1$). Therefore, $2s + 2 = 2\sqrt{1+t}$, implying that $(f^{-1})'(t) = 1/(2\sqrt{1+t}\,)$.

17. $(f^{-1})'(t) = (f'(s))^{-1} = \left(\frac{1}{2}(s^2+9)^{-1/2} \cdot 2s\right)^{-1} = \sqrt{s^2+9}/s$. Since $\sqrt{s^2+9} = t$ and $s = \sqrt{t^2-9}$, $(f^{-1})'(t) = t/\sqrt{t^2-9}$.

19. Because $f(s) = s + 1/s$, $f'(s) = 1 - 1/s^2 = (s^2-1)/s^2$ and $(f^{-1})'(t) = (f'(s))^{-1} = s^2/(s^2-1)$. Moreover, $(s^2+1)/s = t$ so $s^2+1 = st$, $s^2 - ts + 1 = 0$, and (quadratic formula) $s = (t \pm \sqrt{t^2-4}\,)/2 = (t + \sqrt{t^2-4}\,)/2$ because when $t = 5/2$, $s = 2$. Therefore,

$$\begin{aligned}(f^{-1})'(t) &= \frac{(t+\sqrt{t^2-4}\,)^2/4}{(t+\sqrt{t^2-4}\,)^2/4 - 1} = \frac{(t+\sqrt{t^2-4}\,)^2}{(t+\sqrt{t^2-4}\,)^2 - 4} = \frac{t^2 + 2t\sqrt{t^2-4} + t^2 - 4}{t^2 + 2t\sqrt{t^2-4} + t^2} \\ &= \frac{t^2 + t\sqrt{t^2-4} - 2}{t^2 + t\sqrt{t^2-4}} \cdot \frac{t^2 - t\sqrt{t^2-4}}{t^2 - t\sqrt{t^2-4}} = \frac{2t^2 + 2t\sqrt{t^2-4}}{4t^2} = \frac{1}{2} + \frac{\sqrt{t^2-4}}{2t}.\end{aligned}$$

21. $(f^{-1})'(t) = (f'(s))^{-1} = 2\sqrt{s}/\exp(\sqrt{s}\,) = 2\ln(t)/t$, because $\sqrt{s} = \ln(t)$.

23. $(f^{-1})'(t) = (f'(s))^{-1} = (1+s)\ln(2) = 2^t \ln(2)$, because $1 + s = 2^t$.

25. $(f^{-1})'(t) = (f'(s))^{-1} = 1/\left(2^{s^3}\ln(2) \cdot 3s^2\right) = 1/\left(3t\ln(2)(\log_2(t))^{2/3}\right)$, because $s^3 = \log_2(t)$.

27. $\frac{d}{dx}\ln(2x) = 2/(2x) = 1/x$

29. $\frac{d}{dx}(x\ln(x) - x) = 1 \cdot \ln(x) + x \cdot (1/x) - 1 = \ln(x)$

31. $\frac{d}{dx}(\ln(x)/x) = (x \cdot (1/x) - \ln(x) \cdot 1)/x^2 = (1 - \ln(x))/x^2$

33. $\frac{d}{dx}(6\sqrt{\ln(x)}\,) = 6 \cdot \frac{1}{2}(\ln(x))^{-1/2} \cdot (1/x) = 3/\left(x\sqrt{\ln(x)}\,\right)$

35. $\frac{d}{dx}\ln(\sec(x)) = \sec(x)\tan(x)/\sec(x) = \tan(x)$

37. $\frac{d}{dx}(\ln(\ln(x)) = (1/x)/\ln(x) = 1/(x\ln(x))$

39. $\frac{d}{dx}\ln(x + \exp(x)) = (1 + \exp(x))/(x + \exp(x))$

41. $\frac{d}{dx}3^{\log_2(x)} = 3^{\log_2(x)} \cdot \ln(3) \cdot (1/(x \cdot \ln(2)) = \frac{\ln(3)}{\ln(2)} \cdot \frac{3^{\log_2(x)}}{x}$

43. $\frac{d}{dx}\log_{10}(5x+3) = 5/((5x+3)\ln(10))$

45. $\frac{d}{dx}(2^x \log_2(1+4x)) = 2^x \ln(2) \cdot \log_2(1+4x) + 2^x \cdot 4/((1+4x)\ln(2))$
$$= 2^x(\ln(1+4x) + 4/((1+4x)\ln(2)))$$

47. $\frac{d}{dx}(x^3 \log_{10}(6-x)) = 3x^2 \cdot \log_{10}(6-x) + x^3 \cdot (-1/((6-x)\ln(10)))$
$$= x^2(3\log_{10}(6-x) - x/((6-x)\ln(10)))$$

49. $\frac{d}{dx}\log_5(\log_2(x)) = \frac{\frac{d}{dx}\log_2(x)}{\log_2(x)\ln(5)} = \frac{1}{x\ln(2)\log_2(x)\ln(5)} = \frac{1}{x\ln(x)\ln(5)}$

51. Simplify $\ln(f(x))$: $\ln(f(x)) = 4\ln(x^2+x) + 3\ln(x^3+x^2-1) - 2\ln(x^2+2)$

Differentiate: $\frac{f'(x)}{f(x)} = \frac{8x+4}{x^2+x} + \frac{9x^2+6x}{x^2+2} - \frac{4x}{x^2+2}$

Solve for $f'(x)$:

$$\begin{aligned} f'(x) &= f(x) \cdot \left(\frac{8x+4}{x^2+x} + \frac{9x^2+6x}{x^3+x^2-1} - \frac{4x}{x^2+2}\right) \\ &= \frac{(x^2+x)^4(x^3+x^2-1)^3}{(x^2+2)^2} \cdot \left(\frac{8x+4}{x^2+x} + \frac{9x^2+6x}{x^3+x^2-1} - \frac{4x}{x^2+2}\right) \\ &= \frac{x^3(x^2+1)^3(x^3+x^2-1)^2}{(x^2+2)^3} \cdot (13x^6+19x^5+40x^4+50x^3+20x^2-16x-8) \end{aligned}$$

53. Simplify $\ln(f(x))$: $\ln(f(x)) = 3x\ln(x)$. Differentiate: $\frac{f'(x)}{f(x)} = 3\ln(x)+3$, and solve for $f'(x)$: $f'(x) = 3x^{3x}(\ln(x)+1)$.

55. Simplify $\ln(f(x))$: $\ln(f(x)) = x^2\ln(x)$. Differentiate: $\frac{f'(x)}{f(x)} = 2x\ln(x)+x$, and solve for $f'(x)$: $f'(x) = x^{(x^2)}(2x\ln(x)+x) = (2\ln(x)+1)\,x^{(x^2+1)}$.

Further Theory and Practice

57. Chain Rule: $f'(x) = 2x/x^2 = 2/x$

Simplify first: $f(x) = 2\ln(x)$, $f'(x) = 2/x$

59. Chain Rule: $f'(x) = -\frac{1}{2} \cdot x^{-3/2}/x^{-1/2} = -\frac{1}{2} \cdot x^{-1} = -1/(2x)$

Simplify first: $f(x) = -\frac{1}{2}\ln(x)$, $f'(x) = -1/(2x)$

61. $f'(s) = -\frac{1}{2}\cdot(s+3)^{-3/2}$ so the tangent line to the graph of f at $(6, f(6))$ has slope $-1/54$. Therefore, the tangent line to the graph of f^{-1} at $(f(6), 6)$ has slope -54. Its equation is $s = -54(t-1/3)+6$.

63. $f'(s) = 5s^2$ so the tangent line to the graph of f at $(1, f(1))$ has slope 5. Therefore, the tangent line to the graph of f^{-1} at $(f(1), 1)$ has slope $1/5$. Its equation is $s = (t-3)/5+1$.

65. $f(2) = 5$ so $f^{-1}(5) = 2$. Since $f'(2) = 14$, $(f^{-1})'(5) = 1/14$.

67. $f(1) = 3/2$ so $f^{-1}(3/2) = 1$. Since $f'(1) = 1$ (verify), $(f^{-1})'(3/2) = 1$.

69. $f(1) = 0$ so $f^{-1}(0) = 1$. Since $f'(1) = 1$ (verify), $(f^{-1})'(0) = 1$.

71. $f(64) = 9$ so $f^{-1}(9) = 64$. Since $f'(64) = 1/(64\ln(2)) + 1/(64\ln(4)) = (\ln(2)+\ln(4))/(64\ln(2)\ln(4)) = 3/(128\ln(2))$, $(f^{-1})'(9) = 128\ln(2)/3$.

73. $f(1) = 4$ and $f'(1) = 7$ (verify). Therefore, $(f^{-1})(4) = 1/7$.

75. Simplify $\ln(f(x))$: $\ln(f(x)) = \ln(x)\ln(x) = \ln^2(x)$.

    Differentiate: $\frac{f'(x)}{f(x)} = 2\ln(x)/x$, and solve for $f'(x)$: $f'(x) = 2\ln(x)\,x^{\ln(x)-1}$.

77. Simplify $\ln(f(x))$: $\ln(f(x)) = x\ln(\log_2(x))$.

    Differentiate: $\frac{f'(x)}{f(x)} = \ln(\log_2(x)) + x\cdot(1/(x\ln(2))/\log_2(x)) = \ln(\log_2(x)) + 1/\ln(x)$, and solve for $f'(x)$: $f'(x) = \log_2^x(x)\big(\ln(\log_2(x)) + 1/\ln(x)\big)$.

79. Simplify $\ln(f(x))$: $\ln(f(x)) = \ln(x)\ln(\ln(x))$.

    Differentiate: $\frac{f'(x)}{f(x)} = (1/x)\cdot\ln(\ln(x)) + \ln(x)\cdot(1/x)/\ln(x))$, and solve for $f'(x)$: $f'(x) = \ln(x)^{\ln(x)}(\ln(\ln(x)) + 1))/x$.

81. Because f and g are invertible they are both one-to-one and onto. Therefore, their composition $g\circ f$ is also one-to-one and onto, hence invertible. Recall that $(g\circ f)^{-1} = f^{-1}\circ g^{-1}$. Consequently, if $f(s) = t$ and $g(t) = u$, then

$$\frac{d}{du}(g\circ f)^{-1}(u) = \frac{d}{du}f^{-1}(g^{-1}(u)) = (f^{-1})'(t)\cdot(g^{-1})'(u) = (f'(s))^{-1}\cdot(g'(t))^{-1}.$$

    That is, $((g\circ f)^{-1})'(u) = 1/(g'(t)f'(s))$ where $t = f(s)$ and $u = g(t)$.

83. Because the tangent line has slope 1, γ must be chosen so that $(f^{-1})'(\gamma) = 1$. Equivalently, $f'(c) = 1$. Since $f'(s) = \ln(s)$, c must be a solution to the equation $\ln(c) = 1$. Consequently, $c = e$, $\gamma = f(e) = 1$ and the tangent line equation is $s = (t-1) + e$.

85. (a) Because kAt/V is unitless, k is volume/(area·time) so its units are length/time.

    (b) Suppose that $0 \le t_1 < t_2$. Then $-kAt_1/V > -kAt_2/V$ so

$$\exp(-kAt_1/V) > \exp(-kAt_2/V).$$

    If $c(0) < C$, then $c(0) - C < 0$ and $(c(0) - C)\exp(-kAt_1/V) < (c(0) - C)\exp(-kAt_2/V)$, implying that $c(t)$ is increasing.

    If $c(0) > C$, then $c(0) - C > 0$ and $(c(0) - C)\exp(-kAt_1/V) > (c(0) - C)\exp(-kAt_2/V)$, implying that $c(t)$ is decreasing.

    (c) Since the exponential term approaches 0 as $t\to\infty$, $\lim_{t\to\infty} C(t) = C$ and the line $y = C$ is the horizontal asymptote.

    (d) The domain of c is $[0,\infty)$. Since $c(0) > C$, its image is $(C, c(0)]$.

    (e) The domain of c^{-1} is $(C, c(0)]$, its range is $[0,\infty)$.

    To find a formula for $c^{-1}(s)$ solve $s = C + (c(0) - C)\exp(-kAt/V)$ for t to obtain $c^{-1}(s) = -\frac{V}{kA}\ln\left(\frac{s-C}{c(0)-C}\right)$.

(f) $c'(t) = -kA(c(0)-C)e^{-kAt/V}/V$ and, since $c^{-1}(s) = -\frac{V}{kA}\ln\left(\frac{s-C}{c(0)-C}\right) = -\frac{V}{kA}\left(\ln(s-C) - \ln(c(0)-C)\right)$, $(c^{-1})'(s) = -\frac{V/(kA)}{s-C}$.

87. The tangent line to the graph of f^{-1} at the point (q_0, p_0) has the equation $p = m \cdot (q - q_0) + p_0$ where $m = (f^{-1})'(q_0) = 1/f'(p_0)$. Since $f'(p_0) = -E(p_0)f(p_0)/p_0 = -E(p_0)q_0/p_0$, the tangent line to the demand curve is

$$p = -\frac{p_0}{E(p_0)q_0} \cdot (q - q_0) + p_0 \quad \text{or} \quad E(p_0) \cdot \frac{q_0}{p_0} = -\frac{q - q_0}{p - p_0}.$$

CALCULATOR/COMPUTER EXERCISES

89. Using *Maple*'s *fsolve* procedure we find that, to 4 decimal accuracy, $f(0.3239) = 1$ and $f'(0.3239) = 3.267$. Consequently, $(f^{-1})'(1) = 0.3065$.

91. Using *Maple*'s *fsolve* procedure we find that, to 4 decimal accuracy, $f(1.6031) = 15$ and $f'(1.6031) = 37.0224$. Consequently, $(f^{-1})'(15) = 0.0270$.

93. Using *Maple*'s *fsolve* procedure we find that, to 4 decimal accuracy, $f(0.8526) = 2$ and $1/f'(0.8526) = 0.4602$. Consequently, the tangent line equation is

$$x = 0.4602 \cdot (y - 2) + 0.8526\,.$$

The graph of f^{-1} and the tangent line are displayed on the right.

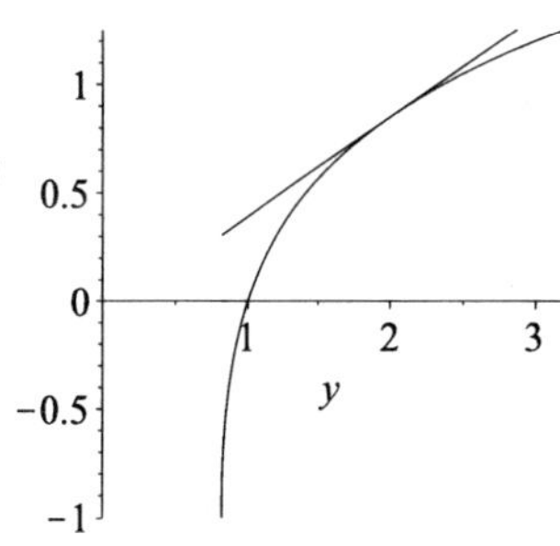

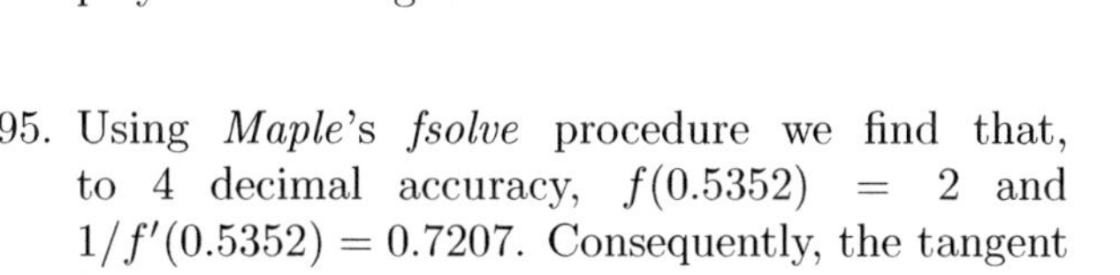

95. Using *Maple*'s *fsolve* procedure we find that, to 4 decimal accuracy, $f(0.5352) = 2$ and $1/f'(0.5352) = 0.7207$. Consequently, the tangent line equation is

$$x = 0.7207 \cdot (y - 2) + 0.5352\,.$$

The graph of f^{-1} and the tangent line are displayed on the right.

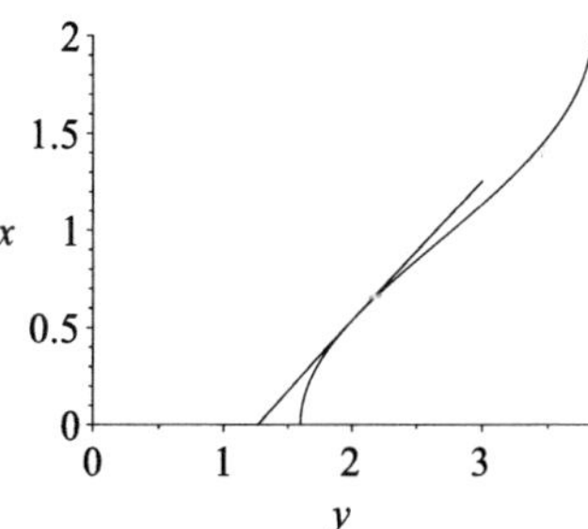

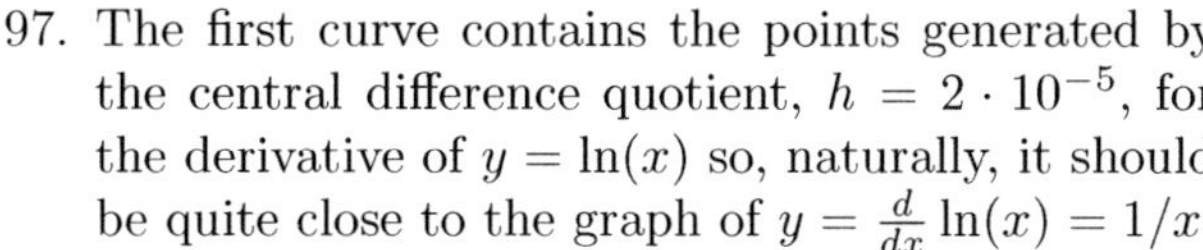

97. The first curve contains the points generated by the central difference quotient, $h = 2 \cdot 10^{-5}$, for the derivative of $y = \ln(x)$ so, naturally, it should be quite close to the graph of $y = \frac{d}{dx}\ln(x) = 1/x$.

In fact, as the picture on the right shows, the two curves are virtually indistinguishable.

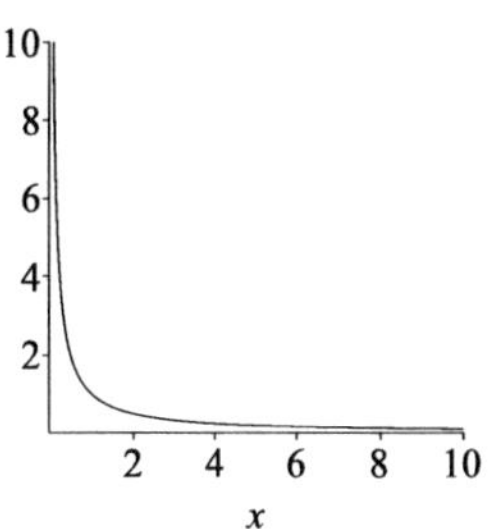

3.7 Higher Derivatives

Problems for Practice

1. $f'(x) = 10x - 6$, $f''(x) = 10$, $f'''(x) = 0$

3. $f'(x) = -1/x^2$, $f''(x) = 2/x^3$, $f'''(x) = -6/x^4$

5. $f'(x) = 1/x$, $f''(x) = -1/x^2$, $f'''(x) = 2/x^3$

7. $f'(x) = 3/\sqrt{6x+5}$, $f''(x) = -9/(6x+5)^{3/2}$, $f'''(x) = 81/(6x+5)^{5/2}$

9. $f'(x) = -3\sin(3x)$, $f''(x) = -9\cos(3x)$, $f'''(x) = 27\sin(3x)$

11. $f'(x) = 2\exp(2x)$, $f''(x) = 4\exp(2x)$, $f'''(x) = 8\exp(2x)$

13. $f'(x) = 1/(x+1)$, $f''(x) = -1/(x+1)^2$, $f'''(x) = 2/(x+1)^3$

15. $f'(x) = (x - 1 - (x+1))/(x-1)^2 = -2/(x-1)^2$, $f''(x) = 4/(x-1)^3$,
$$f'''(x) = -12/(x-1)^4$$

17. $f'(x) = 90x(x^2+1)^{2/3}$,
$$\begin{aligned} f''(x) &= 90(x^2+1)^{2/3} + 90x \cdot \tfrac{2}{3}(x^2+1)^{-1/3} \cdot 2x \\ &= 30(x^2+1)^{-1/3}(3(x^2+1) + 4x^2) = 30(x^2+1)^{-1/3}(7x^2+3)\,, \\ f'''(x) &= -10(x^2+1)^{-4/3} \cdot 2x \cdot (7x^2+3) + 30(x^2+1)^{-1/3} \cdot 14x \\ &= -20x(x^2+1)^{-4/3}(7x^2 + 3 - 21(x^2+1)) \\ &= -20x(x^2+1)^{-4/3}(-14x^2 - 18) = 40x(x^2+1)^{-4/3}(7x^2+9) \end{aligned}$$

19. $f'(x) = \cos(2x) - 2x\sin(2x)$

    $f''(x) = -2\sin(2x) - 2\sin(2x) - 4x\cos(2x) = -4\sin(2x) - 4x\cos(2x)$

    $f''(x) = -8\cos(2x) - 4\cos(2x) + 8x\sin(2x) = -12\cos(2x) + 8x\sin(2x)$

21. $f'(x) = 2^x\ln(2)$, $f''(x) = 2^x\ln^2(2)$, $f'''(x) = 2^x\ln^3(2)$

23. $f'(x) = e^x\sin(x) + e^x\cos(x) = e^x(\sin(x) + \cos(x))$,

    $f''(x) = e^x(\sin(x) + \cos(x)) + e^x(\cos(x) - \sin(x)) = 2e^x\cos(x)$,

    $f'''(x) = 2e^x\cos(x) - 2e^x\sin(x) = 2e^x(\cos(x) - \sin(x))$

25. $f'(x) = 1/(x\ln(3))$, $f''(x) = -1/(x^2\ln(3))$, $f'''(x) = 2/(x^3\ln(3))$

27. $f'(x) = \cos(x)$, $f''(x) = -\sin(x)$, $f'''(x) = -\cos(x)$, $f^{(4)}(x) = \sin(x)$,
$$f^{(5)}(x) = \cos(x)$$

29. Since $\dot{g}(t) = -4(3t^2-2)^{-5} \cdot 6t = -24t(3t^2-2)^{-5}$,
$$\ddot{g}(t) = -24\big((3t^2-2)^{-5} - 5t(3t^2-2)^{-6} \cdot 6t)\big)\,.$$
    Therefore, $\ddot{g}(1) = -24 \cdot (1 - 30) = 696\,.$

31. Four differentiations will kill $-9x^2 + 11$ so

$$\begin{aligned} H'(x) &= -6x^5 + 35x^4 + \cdots & &\Longrightarrow H''(x) = -30x^4 + 140x^3 + \cdots \\ H'''(x) &= -120x^3 + 420x^2 + \cdots & &\Longrightarrow H^{(4)}(x) = -360x^2 + 840x\,. \end{aligned}$$

33. $f'(x) = -4\sin(4x+3)$, $f''(x) = -16\cos(4x+3)$, $f'''(x) = 64\sin(4x+3)$

35. Since $g'(x) = -\sin(\pi\sin(x)) \cdot \pi\cos(x) = -\pi\cos(x)\sin(\pi\sin(x))$,

$$g''(x) = -\pi\big(-\sin(x)\sin(\pi\sin(x)) + \cos(x)\cdot\cos(\pi\sin(x))\cdot\pi\cos(x)\big)\,.$$

Therefore, $g''(\pi/6) = -\pi\big((-1/2)\sin(\pi/2) + \pi\cdot(\sqrt{3}/2)^2\cos(\pi/2)\big) = \pi/2$.

37. Since the velocity is constant, the acceleration is 0, $a(t) = 0$ m/s^2.

39. $a(t) = v'(t) = -16$ m/s^2

41. Since $v(t) = p'(t) = 24$ m/s, $a(t) = 0$ m/s^2.

43. Since $v(t) = p'(t) = 6t + 16$ m/s, $a(t) = 6$ m/s^2.

45. Since $v(t) = p'(t) = 6t^2 + 2t + 6$ m/s, $a(t) = 12t + 2$ m/s^2. Therefore, $a(3) = 38$ m/s^2 and $a(6) = 74$ m/s^2.

47. $v(t) = -32t + 128$ ft/s and $a(t) = -32$ ft/s^2.

49. The maximum height is attained at the instant that the body stops rising, $t = 4$. Therefore, the maximum height is $H(4) = 324$ feet.

51. The object hits the ground when $-16t^2 - 160t + 84 = 0$. That is, when $-4(4t^2+40t-21) = 0$. Since $4t^2+40t-21 = (2t-1)(2t+21)$, the falls for $t = 1/2$ seconds. Since $v(t) = H'(t) = -32t - 160$, it impacts the ground with a velocity of $v(1/2) = -176$ ft/s.

53. The car's acceleration at time t was $a(t) = v'(t) = -6 - 3t$ ft/s^2. It was slowing down when the brakes were applied and continued to slow down until it came to a stop.

Further Theory and Practice

55. Observe that $f(x) = \begin{cases} x^3 & \text{if } x \ge 0 \\ -x^3 & \text{if } x \le 0 \end{cases}$

(a) The left tangent line and the right tangent line to the graph of f at the point $(0,0)$ both have slope 0 so $f'(0)$ exists, and is 0. Moreover, $f'(x) = \begin{cases} 3x^2 & \text{if } x \ge 0 \\ -3x^2 & \text{if } x \le 0 \end{cases}$

(b) Similarly, the left tangent and the right tangent to the graph of f' at $x = 0$ have slope 0, so $f''(0)$ exists, and is 0. That is,

$$f''(x) = \begin{cases} 6x & \text{if } x \ge 0 \\ -6x & \text{if } x \le 0 \end{cases}$$

(c) The two formulas for $f''(x)$ displayed in (b) confirm that $f'''(0)$ does not exist. The second derivative, f'', is differentiable at all $x \neq 0$, but it is not differentiable at 0 because the left tangent line at (0,0) has slope -6 and the right tangent line at (0,0) has slope $+6$.

57. Let $p(x)$ be a polynomial of degree k,

$$p(x) = a_k x^k + a_{k-1} x^{k-1} + \cdots + a_1 x + a_0 \, , a_k \neq 0 \, .$$

Since $\frac{d^n}{dx^n} x^n = n!$ for $n = 1, 2, \ldots$, $\frac{d^k}{dx^k} x^n = 0$ for all $0 \leq n < k$. Therefore, $p^{(k)}(x) = a_k \cdot k!$, a nonzero constant. Moreover, $p^{(k+1)}(x) = 0$ for all x. This verifies the necessity of the two conditions stated in the Exercise.

Conversely, if $p(x)$ is a polynomial such that $p^{(k+1)}(x) = 0$ for all x, then it cannot have degree greater than k and if $p^{(k)}(x_0) \neq 0$ for some x_0 its degree must be at least k.

59. Let $p(x) = ax^3 + bx^2 + cx + d$. Since $p'''(x) = 6a$, $a = 1$ and $p(x) = x^3 + bx^2 + cx + d$. Because $p''(x) = 6x + 2b$, $-6 + 2b = -14$ and $b = -4$. Now we know that $p(x) = x^3 - 4x^2 + cx + d$.

Continuing, since $p'(x) = 3x^2 - 8x + c$ and $p'(2) = -2$, $12 - 16 + c = -2$ so $c = 2$ and $p(x) = x^3 - 4x^2 + 2x + d$. Finally, $p(1) = 1$ implies that $1 - 4 + 2 + d = 1$ and $d = 2$. Conclusion: $p(x) = x^3 - 4x^2 + 2x + 2$.

61. $D^2(f \cdot g)(x) = f''(x)g(x) + 2f'(x)g'(x) + f(x)g''(x)$

$$= 2\cos(x) + 2 \cdot 2x \cdot (-\sin(x)) + x^2(-\cos(x) = (2 - x^2)\cos(x)) - 4x\sin(x)$$

63. $(f \cdot g)''(x) = (2/x^3) \cdot \ln(x) + 2(-1/x^2) \cdot (1/x) + (1/x) \cdot (-1/x^2)$

$$= 2\ln(x)/x^3 - 3/x^3 = (2\ln(x) - 3)/x^3$$

65. $(f \cdot g)^{(2)}(x) = (-1/(4x^{3/2})) \cdot \exp(x) + 2(1/(2x^{1/2}))\exp(x) + \sqrt{x}\,\exp(x)$

$$= x^{-3/2}(x^2 + x - 1/4)\exp(x)$$

67. When $x = c$ the quadradic expression defining the second curve evaluates to $f(c)$ and its derivative, with respect to x, evaluates to $f'(c)$.

69. The Legendre Polynomials

For $n = 0$ the definition states that $P_0(x) = (x^2 - 1)^0 = 1$. Continuing with $n = 1, 2, 3, 4$

$$P_1(x) = \frac{1}{2}\tfrac{d}{dx}(x^2 - 1) = x$$

$$P_2(x) = \frac{1}{8}\frac{d^2}{dx^2}(x^2-1)^2 = \frac{1}{8}\frac{d^2}{dx^2}(x^4 - 2x^2 + 1) = \tfrac{1}{8}(12x^2 - 4) = \tfrac{3}{2}x^2 - \tfrac{1}{2}$$

$$P_3(x) = \frac{1}{48}\frac{d^3}{dx^3}(x^2-1)^3 = \frac{1}{48}\frac{d^3}{dx^3}(x^6 - 3x^4 + 3x^2 - 1) = \tfrac{1}{48}(120x^3 - 72x) = \tfrac{5}{2}x^3 - \tfrac{3}{2}x$$

$$P_4(x) = \frac{1}{384}\frac{d^4}{dx^4}(x^2-1)^4 = \frac{1}{384}\frac{d^4}{dx^4}(x^8 - 4x^6 + 6x^4 - 4x^2 + 1) = \tfrac{1}{384}(1680x^4 - 1440x^2 + 144) = \tfrac{35}{8}x^4 - \tfrac{15}{4}x^2 + \tfrac{3}{8}$$

71. The Hermite Polynomials

For $n = 0$ the definition states that $H_0(x) = 1 \cdot e^{x^2} \cdot e^{-x^2} = 1$. Continuing with $n = 1, 2, 3, 4$

$$H_1(x) = -e^{x^2}\tfrac{d}{dx}e^{-x^2} = -e^{x^2}\cdot(-2xe^{-x^2}) = 2x$$

$$H_2(x) = e^{x^2}\frac{d^2}{dx^2}e^{-x^2} = e^{x^2}\cdot(-2e^{-x^2} + 4x^2e^{-x^2}) = 4x^2 - 2$$

$$H_3(x) = -e^{x^2}\frac{d^3}{dx^3}e^{-x^2} = -e^{x^2}\cdot(4xe^{-x^2} + 8xe^{-x^2} - 8x^3e^{-x^2}) = 8x^3 - 12x$$

$$H_4(x) = e^{x^2}\frac{d^4}{dx^4}e^{-x^2} = e^{x^2}\cdot(12e^{-x^2} - 24x^2e^{-x^2} - 24x^2e^{-x^2} + 16x^4e^{-x^2}) = 16x^4 - 48x^2 + 12$$

73. Let $p(x)$ be a polynomial of degree k:

$$p(x) = a_0 + a_1x + \cdots + a_nx^n + \cdots + a_kx^k\,.$$

Its nth derivative, $p^{(n)}(x)$, is a polynomial of degree $k - n$ with constant term $n!a_n$. Consequently, $p^{(n)}(0) = n!a_n$ and $a_n = p^{(n)}(0)/n!$

75. Let $p(x) = (x - a)^2q(x)$ where q is a polynomial and $q(a) \neq 0$. Then $p(a) = 0$. Moreover, $p'(x) = 2(x - a)q(x) + (x - a)^2q'(x)$, so $p'(a) = 0$, and $p''(x) = 2q(x) + 2(x - a)q'(x) + 2(x - a)q'(x) + (x - a)^2q''(x)$ so $p''(a) = 2q(a) \neq 0$.

Conversely, suppose that p is a polynomial with $p(a) = 0$, $p'(a) = 0$, and $p''(a) \neq 0$. Since a is a root of p, $x - a$ divides $p(x)$ and there is a polynomial $q_1(x)$ such that $p(x) = (x - a)q_1(x)$. Consequently, $p'(x) = q_1(x) + (x - a)q_1'(x)$ and $p'(a) = q_1(a)$. But $p'(a) = 0$, implying that a is also a root of q_1. Let q be the polynomial such that $q_1(x) = (x - a)q(x)$, and $p(x) = (x - a)^2q(x)$.

Applying Leibniz's Rule, $p''(x) = 2q(x) + 2(x - a)q'(x) + (x - a)^2q''(x)$, so $p''(a) = 2q(a)$, implying that $q(a) = p''(a)/2 \neq 0$.

77. If $f(s) = t$ and $f'(s) \neq 0$, then $(f^{-1})'(t) = 1/f'(s)$ and

$$(f^{-1})''(t) = \frac{d}{dt}(1/f'(s)) = \frac{d}{ds}(1/f'(s)) \cdot \frac{ds}{dt} = -\frac{f''(s)}{(f'(s))^2} \cdot (f^{-1})'(t)$$
$$= -\frac{f''(s)}{(f'(s))^3}.$$

Therefore, $(f^{-1})''(\gamma) = -\dfrac{f''(c)}{(f'(c))^3}$.

CALCULATOR/COMPUTER EXERCISES

79. The plot on the right displays the graphs of f, g, h, and k.
    - The solid curve is f.
    - The the dotted line is the linear approximation g.
    - The dashed curve is the quadratic approximation h.
    - The dash-dot curve is the cubic approximation k.

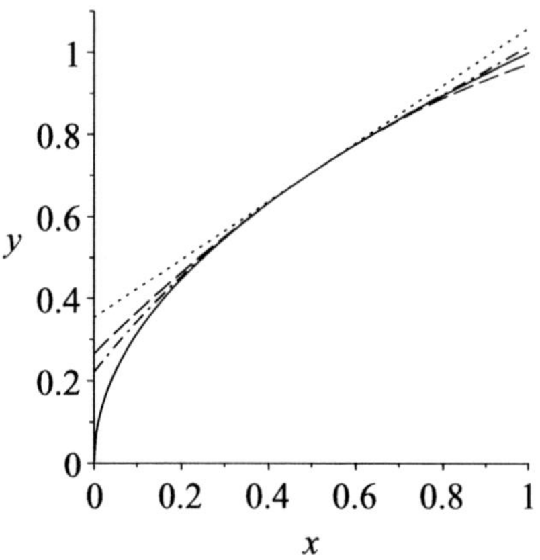

81. The plot on the right displays the graphs of f, g, h, and k.
    - The solid curve is f.
    - The the dotted line is the linear approximation g.
    - The dashed curve is the quadratic approximation h.
    - The dash-dot curve is the cubic approximation k.

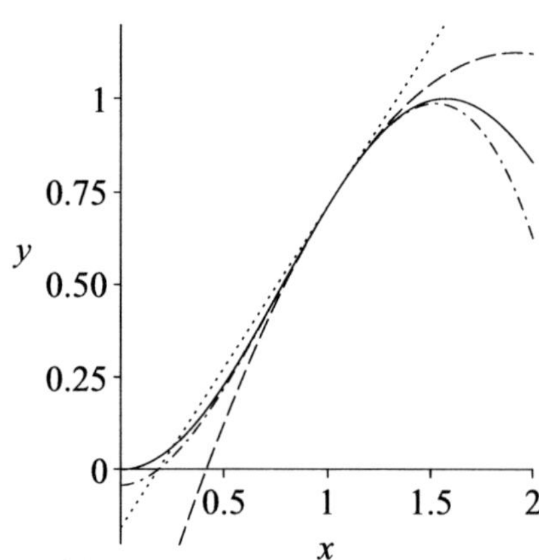

83. The following table displays, to 6 decimals (truncated), the second central difference quotient approximations (scDQ) for $h = 10^{-n}$, $n = 1, 2, 3, 4, 5$. The approxmations were obtained by setting *Maple* to 50 digit accuracy.

n	1	2	3	4	5
scDQ	-0.006083	-0.006134	-0.006134	-0.006134	-0.006134

85. The following table displays, to 6 decimals (truncated), the second central

difference quotient approximations (scDQ) for $h = 10^{-n}$, $n = 1, 2, 3, 4, 5$. The approxmations were obtained by setting *Maple* to 50 digit accuracy.

n	1	2	3	4	5
scDQ	-0.006933	-0.006944	-0.006944	-0.006944	-0.006944

3.8 Implicit Differentiation

Problems for Practice

1. $\frac{d}{dx}\left(4y^{3/2}\right) = 6y^{1/2}\frac{dy}{dx}$

3. $\frac{d}{dx}\left(2x/\sqrt{y}\right) = \frac{\sqrt{y}\cdot 2 - 2x\cdot\frac{d}{dx}(\sqrt{y})}{y} = \frac{2\sqrt{y} - 2x\cdot(1/(2\sqrt{y}))\frac{dy}{dx}}{y} = \frac{2y - x\frac{dy}{dx}}{y^{3/2}}$

5. $\frac{d}{dx}\left(xe^y\right) = e^y + xe^y\frac{dy}{dx}$

7. Differentiate the equation: $\frac{d}{dx}\left(xy^2 + yx^2\right) = \frac{d}{dx}(6)$, and apply the rules of differentiation assuming that y is a function of x.

$$\frac{d}{dx}(xy^2) + \frac{d}{dx}(yx^2) = 0$$
$$y^2 + x\cdot 2y\frac{dy}{dx} + \frac{dy}{dx}\cdot x^2 + y\cdot(2x) = 0$$

At the point (1,2), $4 + 4\frac{dy}{dx} + \frac{dy}{dx} + 4 = 0$, so $\frac{dy}{dx} = -8/5$.

9. Differentiate the equation: $\frac{d}{dx}\left(x^{3/5} + 4y^{3/5}\right) = \frac{d}{dx}(12)$, and apply the rules of differentiation assuming that y is a function of x.

$$\frac{d}{dx}(x^{3/5}) + 4\frac{d}{dx}(y^{3/5}) = 0$$
$$\frac{3}{5}x^{-2/5} + \frac{12}{5}y^{-2/5}\frac{dy}{dx} = 0$$

At the point (32,1), $\frac{3}{20} + \frac{12}{5}\frac{dy}{dx} = 0$, so $\frac{dy}{dx} = -1/16$.

11. Differentiate the equation: $\frac{d}{dx}\left(xy^3 + 3y^2 + x^2\right) = \frac{d}{dx}(21)$, and apply the rules of differentiation assuming that y is a function of x.

$$\frac{d}{dx}(xy^3) + 3\frac{d}{dx}(y^2) + \frac{d}{dx}(x^2) = 0$$
$$y^3 + x\cdot 3y^2\frac{dy}{dx} + 6y\frac{dy}{dx} + 2x = 0$$

At the point (1,2), $8 + 12\frac{dy}{dx} + 12\frac{dy}{dx} + 2 = 0$, so $\frac{dy}{dx} = -5/12$.

13. Differentiate the equation: $\frac{d}{dx}\left(xe^{y-1} + y^2\right) = \frac{d}{dx}(4)$, and apply the rules of differentiation assuming that y is a function of x.

$$\frac{d}{dx}(xe^{y-1}) + \frac{d}{dx}(y^2) = 0$$
$$e^{y-1} + xe^{y-1}\frac{dy}{dx} + 2y\frac{dy}{dx} = 0$$

At the point $(3, 1)$, $1 + 3\frac{dy}{dx} + 2\frac{dy}{dx} = 0$, so $\frac{dy}{dx} = -1/5$.

15. The slope of the tangent line is $\frac{dy}{dx}$ evaluated at P_0.

    To find $\frac{dy}{dx}$, differentiate the equation: $\frac{d}{dx}(xy - 1 = \sqrt{x} + \sqrt{y})$, and apply the rules of differentiation assuming that y is a function of x.

    $$\frac{d}{dx}(xy) - \frac{d}{dx}(1) = \frac{d}{dx}(\sqrt{x}) + \frac{d}{dx}(\sqrt{y})$$
    $$y + x\frac{dy}{dx} = 1/(2\sqrt{x}) + 1/(2\sqrt{y})\frac{dy}{dx}$$

    At the point $P_0 = (4,1)$, $1+4\frac{dy}{dx} = \frac{1}{4}+\frac{1}{2}\frac{dy}{dx}$, so $\frac{dy}{dx} = -3/14$. The equation of the tangent line to the curve at P_0 is $y = -3(x-4)/14+1$.

17. The slope of the tangent line is $\frac{dy}{dx}$ evaluated at P_0.

    To find $\frac{dy}{dx}$, differentiate the equation: $\frac{d}{dx}(\sin^2(x) + \cos^2(y)) = \frac{d}{dx}(5)$, and apply the rules of differentiation assuming that y is a function of x.

    $$\frac{d}{dx}(\sin^2(x)) + \frac{d}{dx}(\cos^2(y)) = 0$$
    $$2\sin(x)\cos(x) - 2\cos(y)\sin(y)\frac{dy}{dx} = 0$$

    At the point $P_0 = (\pi/3, \pi/4)$, $2\cdot\frac{\sqrt{3}}{2}\cdot\frac{1}{2} - 2\cdot\frac{1}{\sqrt{2}}\cdot\frac{1}{\sqrt{2}}\frac{dy}{dx} = 0$, so $\frac{dy}{dx} = \sqrt{3}/2$. The equation of the tangent line at P_0 is $y = \sqrt{3}(x-\pi/3)/2 + \pi/4$.

19. The slope of the tangent line is $\frac{dy}{dx}$ evaluated at P_0.

    To find $\frac{dy}{dx}$, differentiate the equation: $\frac{d}{dx}(x + 3/y + \ln(y)) = \frac{d}{dx}(5)$, and apply the rules of differentiation assuming that y is a function of x.

    $$\frac{d}{dx}(x) + \frac{d}{dx}(3/y) + \frac{d}{dx}(\ln(y)) = 0$$
    $$1 - (3/y^2)\frac{dy}{dx} + (1/y)\frac{dy}{dx} = 0$$

    At the point $P_0 = (2,1)$, $1 - 3\frac{dy}{dx} + \frac{dy}{dx} = 0$, so $\frac{dy}{dx} = 1/2$. The equation of the tangent line at P_0 is $y = (x-2)/2 + 1$ or $y = x/2$.

21. The slope of the tangent line is $\frac{dy}{dx}$ evaluated at P_0.

    To find $\frac{dy}{dx}$, differentiate the equation: $\frac{d}{dx}(e^{xy} = 2y^2 - 1)$, and apply the rules of differentiation assuming that y is a function of x.

    $$\frac{d}{dx}(e^{xy}) = 2\frac{d}{dx}(y^2) - \frac{d}{dx}(1)$$
    $$e^{xy}(y + x\frac{dy}{dx}) = 4y\frac{dy}{dx}$$

    At the point $P_0 = (0,-1)$, $-1 = -4\frac{dy}{dx}$, so $\frac{dy}{dx} = 1/4$. The equation of the tangent line at P_0 is $y = x/4 - 1$.

23. The slope of the tangent line is $\frac{dy}{dx}$ evaluated at P_0.

    To find $\frac{dy}{dx}$, differentiate the equation: $\frac{d}{dx}(\ln(xy-1) + y^2) = \frac{d}{dx}(4)$, and apply the rules of differentiation assuming that y is a function of x.

    $$\frac{d}{dx}(\ln(xy-1)) + \frac{d}{dx}(y^2) = 0$$
    $$(y + x\frac{dy}{dx})/(xy-1) + 2y\frac{dy}{dx} = 0$$

At the point $P_0 = (1,2)$, $2 + \frac{dy}{dx} + 4\frac{dy}{dx} = 0$, so $\frac{dy}{dx} = -2/5$. The equation of the tangent line at P_0 is $y = -2(x-1)/5 + 2$.

25. The slope of the normal line is $-1/(\frac{dy}{dx})$ evaluated at P_0.

To find $\frac{dy}{dx}$, differentiate the equation: $\frac{d}{dx}(xy^4 - x^3y) = \frac{d}{dx}(16)$, and apply the rules of differentiation assuming that y is a function of x.

$$\begin{aligned} \tfrac{d}{dx}(xy^4) - \tfrac{d}{dx}(x^3y) &= 0 \\ y^4 + 4xy^3\tfrac{dy}{dx} - 3x^2y - x^3\tfrac{dy}{dx} &= 0 \end{aligned}$$

At the point $P_0 = (2,2)$, $16 + 64\frac{dy}{dx} - 24 - 8\frac{dy}{dx} = 0$, so $\frac{dy}{dx} = 1/7$ and the normal line has slope -7. The equation of the normal line at P_0 is $y = -7(x-2) + 2$.

27. The slope of the normal line is $-1/(\frac{dy}{dx})$ evaluated at P_0.

To find $\frac{dy}{dx}$, differentiate the equation: $\frac{d}{dx}\big(x - y + \sin(2y)\big) = \frac{d}{dx}(1)$, and apply the rules of differentiation assuming that y is a function of x.

$$\begin{aligned} \tfrac{d}{dx}(x) - \tfrac{d}{dx}(y) + \tfrac{d}{dx}(\sin(2y)) &= 0 \\ 1 - \tfrac{dy}{dx} + 2\cos(2y)\tfrac{dy}{dx} &= 0 \end{aligned}$$

At the point $P_0 = (\pi/4, \pi/4)$, $1 - \frac{dy}{dx} = 0$, so $\frac{dy}{dx} = 1$ and the normal line has slope -1. The equation of the normal line at P_0 is $y = -(x - \pi/4) + \pi/4$ or $y = -x + \pi/2$.

29. Find an equation for $\frac{dy}{dx}$, then another for $\frac{d^2y}{dx^2}$.

The first one is obtained from $\frac{d}{dx}(y^3 + y + x) = \frac{d}{dx}(4)$, which simplifies to Equation (1) below.

$$\begin{aligned} \tfrac{d}{dx}(y^3) + \tfrac{d}{dx}(y) + \tfrac{d}{dx}(x) &= 0 \\ 3y^2\tfrac{dy}{dx} + \tfrac{dy}{dx} + 1 &= 0 \qquad (1) \end{aligned}$$

Differentiate again to obtain Equation (2).

$$\begin{aligned} 3\tfrac{d}{dx}\big(y^2\tfrac{dy}{dx}\big) + \tfrac{d}{dx}\big(\tfrac{dy}{dx}\big) + \tfrac{d}{dx}(1) &= 0 \\ 3\big(2y\tfrac{dy}{dx}\cdot\tfrac{dy}{dx} + y^2\tfrac{d^2y}{dx^2}\big) + \tfrac{d^2y}{dx^2} &= 0 \qquad (2) \end{aligned}$$

Use Equation (1) to determine the value of $\frac{dy}{dx}$ at the point $P_0 = (2,1)$: $3\frac{dy}{dx} + \frac{dy}{dx} + 1 = 0$ implies that $\left.\frac{dy}{dx}\right|_{(2,1)} = -1/4$.

Then substitute $x = 2$, $y = 1$, $\frac{dy}{dx} = -1/4$ into Equation (2)

$$3\big(2\cdot(-\tfrac{1}{4})\cdot(-\tfrac{1}{4}) + \tfrac{d^2y}{dx^2}\big) + \tfrac{d^2y}{dx^2} = 0$$

to see that $\left.\frac{d^2y}{dx^2}\right|_{(2,1)} = -3/32$.

31. Find an equation for $\frac{dy}{dx}$, then another for $\frac{d^2y}{dx^2}$.

The first one is obtained from $\frac{d}{dx}(xy - 6/y) = \frac{d}{dx}(4)$, which simplifies to Equation (1) below.

$$\frac{d}{dx}(xy) - \frac{d}{dx}(6/y) = 0$$
$$y + x\frac{dy}{dx} + (6/y^2)\frac{dy}{dx} = 0 \qquad (1)$$

Differentiate again and simpify to obtain Equation (2).

$$\frac{d}{dx}(y) + \frac{d}{dx}(x\frac{dy}{dx}) + \frac{d}{dx}((6/y^2)\frac{dy}{dx}) = 0$$
$$\frac{dy}{dx} + \frac{dy}{dx} + x\frac{d^2y}{dx^2} + ((-12/y^3)\frac{dy}{dx}\cdot\frac{dy}{dx} + (6/y^2)\frac{d^2y}{dx^2}) = 0$$
$$2\frac{dy}{dx} + (x + 6/y^2)\frac{d^2y}{dx^2} - (12/y^3)(\frac{dy}{dx})^2 = 0 \qquad (2)$$

Use Equation (1) to determine the value of $\frac{dy}{dx}$ at the point $P_0 = (2,3)$: $3 + 2\frac{dy}{dx} + \frac{2}{3}\frac{dy}{dx} = 0$ implies that $\left.\frac{dy}{dx}\right|_{(2,3)} = -9/8$.

Then substitute $x = 2$, $y = 3$, $\frac{dy}{dx} = -9/8$ into Equation (2)

$$-\frac{9}{4} + (2 + \frac{2}{3})\frac{d^2y}{dx^2} - \frac{4}{9}\cdot\frac{81}{64} = 0$$

to see that $\left.\frac{d^2y}{dx^2}\right|_{(2,3)} = 135/128$.

33. Find an equation for $\frac{dy}{dx}$, then another for $\frac{d^2y}{dx^2}$.

The first one is obtained from $\frac{d}{dx}(2e^y) = \frac{d}{dx}(3 - x + y)$, which simplifies to Equation (1) below.

$$\frac{d}{dx}(2e^y) = \frac{d}{dx}(3) - \frac{d}{dx}(x) + \frac{d}{dx}(y)$$
$$2e^y\frac{dy}{dx} = -1 + \frac{dy}{dx} \qquad (1)$$

Differentiate again and simpify to obtain Equation (2).

$$2\frac{d}{dx}(e^y\frac{dy}{dx}) = \frac{d}{dx}((-1) + \frac{dy}{dx})$$
$$2(e^y\frac{dy}{dx}\cdot\frac{dy}{dx} + e^y\frac{d^2y}{dx^2}) = \frac{d^2y}{dx^2}$$
$$2e^y((\frac{dy}{dx})^2 + \frac{d^2y}{dx^2}) = \frac{d^2y}{dx^2} \qquad (2)$$

Use Equation (1) to determine the value of $\frac{dy}{dx}$ at the point $P_0 = (1,0)$: $2\frac{dy}{dx} = -1 + \frac{dy}{dx}$ implies that $\left.\frac{dy}{dx}\right|_{(1,0)} = -1$.

Then substitute $x = 1$, $y = 0$, $\frac{dy}{dx} = -1$ into Equation (2)

$$2((1 + \frac{d^2y}{dx^2}) = \frac{d^2y}{dx^2}$$

to see that $\left.\frac{d^2y}{dx^2}\right|_{(1,0)} = -2$.

35. At $t_0 = 1$, $x_0 = 2$ and $y_0 = 2$. The tangent line passes through $P_0 = (2, 2)$ with slope

$$\left.\frac{dy}{dx}\right|_{(2,2)} = \left.\frac{\phi_2'(t)}{\phi_1'(t)}\right|_{t=1} = \left.\frac{3t^2}{2t}\right|_{t=1} = 3/2\,.$$

Its equation is $y = 3(x-2)/2 + 2$.

37. At $t_0 = 2/3$, $x_0 = 26/27$ and $y_0 = -10/27$. The tangent line passes through $P_0 = (26/27, -10/27)$ with slope

$$\left.\frac{dy}{dx}\right|_{(26/27,-10/27)} = \left.\frac{\phi_2'(t)}{\phi_1'(t)}\right|_{t=2/3} = \left.\frac{3t^2-1}{3t^2+1}\right|_{t=2/3} = 1/7\,.$$

Its equation is $y = (x - 26/27)/7 - 10/27$.

39. At $t_0 = \pi/4$, $x_0 = y_0 = \pi/4 + \sqrt{2}/2$. The tangent line passes through $P_0 = (\pi/4 + \sqrt{2}/2, \pi/4 + \sqrt{2}/2)$ with the following slope.

$$\left.\frac{dy}{dx}\right|_{P_0} = \left.\frac{\phi_2'(t)}{\phi_1'(t)}\right|_{t=\pi/4} = \left.\frac{1+\cos(t)}{1-\sin(t)}\right|_{t=\pi/4} = \frac{1+\sqrt{2}/2}{1-\sqrt{2}/2}$$

The slope simplifies to $3 + 2\sqrt{2}$ (rationalize the denominator and expand) so the equation of the tangent line can be expressed in the form

$$y = (3+2\sqrt{2})(x - \pi/4 - \sqrt{2}/2) + \pi/4 + \sqrt{2}/2\,.$$

41. At $t_0 = 2/3$, $x_0 = 6/13$ and $y_0 = 8/39$. The tangent line passes through $P_0 = (6/13, 8/39)$ with the following slope.

$$\left.\frac{dy}{dx}\right|_{(6/13,8/39)} = \left.\frac{\phi_2'(t)}{\phi_1'(t)}\right|_{t=2/3} = \left.\frac{(1-t^2)/(1+t^2)^2}{t^2(3+t^2)/(1+t^2)^2}\right|_{t=2/3} = 124/45$$

The equation of the tangent line is $y = 124(x - 16/13)/45 + 8/39$ or $y = 124x/45 - 16/15$.

43. At $t_0 = \pi/6$, $x_0 = e^{\sqrt{3}/2}$ and $y_0 = e^{1/2}$. The tangent line passes through $P_0 = (e^{\sqrt{3}/2}, e^{1/2})$ with the following slope.

$$\left.\frac{dy}{dx}\right|_{P_0} = \left.\frac{\phi_2'(t)}{\phi_1'(t)}\right|_{t=\pi/6} = \left.\frac{\cos(t)e^{\sin(t)}}{-\sin(t)e^{\cos(t)}}\right|_{t=\pi/6} = -\sqrt{3}\,e^{(1-\sqrt{3})/2}$$

The equation of the tangent line is $y = -\sqrt{3}\,e^{(1-\sqrt{3})/2}\,(x - e^{\sqrt{3}/2}) + e^{1/2}$.

45. At $t_0 = 1$, $x_0 = 5$ and $y_0 = 5$. The tangent line passes through $P_0 = (5, 5)$ with the following slope.

$$\left.\frac{dy}{dx}\right|_{(5,5)} = \left.\frac{\phi_2'(t)}{\phi_1'(t)}\right|_{t=1} = \left.\frac{3^t\ln(3)}{2^t\ln(2)}\right|_{t=1} = \ln(27)/\ln(4)$$

The equation of the tangent line is $y = \ln(27)(x-5)/\ln(4) + 5$.

Further Theory and Practice

47. The curve crosses the x-axis when $y = 0$. That is, at the points $(1,0)$ and $(-1,0)$. Obtain dy/dx implicitly: $\frac{d}{dx}(x^2 - 4xy + y^2) = \frac{d}{dx}(9)$ implies that $2x - 4y - 4x\frac{dy}{dx} + 2y\frac{dy}{dx} = 0$ or $\frac{dy}{dx} = (x - 2y)/(2x - y)$. If $y = 0$ and $x \neq 0$, then $\frac{dy}{dx} = 1/2$, so the tangent lines are parallel.

49. Obtain dy/dx implicitly: $\frac{d}{dx}(x^2 - xy + y^2) = \frac{d}{dx}(4)$ implies that $2x - y - x\frac{dy}{dx} + 2y\frac{dy}{dx} = 0$ or $\frac{dy}{dx} = (y - 2x)/(2y - x)$. Therefore, the tangent line at the point (a,b) on the ellipse has slope $m_1 = (b - 2a)/(2b - a)$ and the tangent line at the point $(-a,-b)$ on the ellipse has slope $m_2 = (-b + 2a)/(-2b + a)$. Clearly, $m_1 = m_2$.

51. Obtain dy/dx implicitly: $\frac{d}{dx}(x^3 - 3\alpha xy + y^3) = \frac{d}{dx}(0)$ implies that $3x^2 - 3\alpha y - 3\alpha x\frac{dy}{dx} + 3y^2\frac{dy}{dx} = 0$ or $\frac{dy}{dx} = (\alpha y - x^2)/(y^2 - \alpha x)$. Therefore, the tangent line slope at $P = (a,b)$ is $m_1 = (\alpha b - a^2)/(b^2 - \alpha a)$ and the tangent line slope at $Q = (b,a)$ is $m_2 = (\alpha a - b^2)/(a^2 - \alpha b)$. Consequently,

$$m_1 \cdot m_2 = \frac{\alpha b - a^2}{b^2 - \alpha a} \cdot \frac{\alpha a - b^2}{a^2 - \alpha b} = \frac{\alpha b - a^2}{a^2 - \alpha b} \cdot \frac{\alpha a - b^2}{b^2 - \alpha a} = (-1) \cdot (-1) = 1\,.$$

53. We want $\frac{dy}{dx}$ when $x = 2$, $2^y = y^2$, and $x \neq y$. This implies that $y = 4$. Apply implicit differentiation to $\ln(x^y) = \ln(y^x)$, that is, $y\ln(x) = x\ln(y)$: $\frac{d}{dx}(y\ln(x)) = \frac{d}{dx}(x\ln(y)) \Rightarrow \frac{dy}{dx}\ln(x) + y \cdot (1/x) = \ln(y) + x \cdot (1/y)\frac{dy}{dx}$. Therefore, $\frac{dy}{dx} = (\ln(y) - y/x)/(\ln(x) - x/y)$. When $x = 2$ and $y = 4$,

$$\frac{dy}{dx} = \frac{\ln(4) - 2}{\ln(2) - 1/2} = \frac{4(\ln(2) - 1)}{2\ln(2) - 1}\,.$$

55. For (x,y) on the curve, $\frac{d}{dx}(x^{1/2} + y^{1/2}) = 0$, so $\frac{1}{2}x^{-1/2} + \frac{1}{2}y^{-1/2}\frac{dy}{dx} = 0$ and $\frac{dy}{dx} = -(y/x)^{1/2}$. (Note that both x and y must be positive.) Let T be the tangent line at the point (x_0, y_0). Its equation is $y = -\sqrt{y_0/x_0}\,(x - x_0) + y_0$ or $\frac{x}{\sqrt{x_0}} + \frac{y}{\sqrt{y_0}} = \sqrt{x_0} + \sqrt{y_0}$. Consequently, the x- and y-intercepts for the tangent line are $x_0 + \sqrt{x_0 y_0}$ and $y_0 + \sqrt{x_0 y_0}$ respectively. The sum of the intercepts is $x_0 + 2\sqrt{x_0 y_0} + y_0 = (\sqrt{x_0} + \sqrt{y_0}\,)^2 = a^2$.

57. Differentiate the equation with respect to x, assuming y is a function of x, to obtain

$$(-2y^{-3} + 1/y^2) \cdot \frac{dy}{dx} = A \cdot \left(\frac{x^{3/2} \cdot (-bx^{-2})\exp(b/x) - \exp(b/x) \cdot (3/2)x^{1/2}}{x^3} \right).$$

This simplifies to $\displaystyle \frac{y - 2}{y^3} \cdot \frac{dy}{dx} = -A\exp(b/x) \cdot \left(\frac{2b + 3x}{2x^{7/2}} \right)$ so

$$\frac{dy}{dx} = -\tfrac{1}{2}Ax^{-7/2}(2b + 3x)y^3(y - 2)^{-1}\exp(b/x)\,.$$

59. (a) The slope of the demand curve at a point (p, q) is $\frac{dq}{dp}$ which we calculate implicitly. Since $\frac{d}{dp}(p + q + 2p^2q + 3pq^3) = 0$,

$$1 + \frac{dq}{dp} + 4pq + 2p^2\frac{dq}{dp} + 3q^3 + 3p \cdot 3q^2\frac{dq}{dp} = 0\,.$$

Substitute $p = 6$ and $q = 3.454$ to see that $\frac{dp}{dq} = -0.289$.

(b) $E(6) = -q'(6) \cdot 6/3.454 = 0.289 \cdot 6/3.454 = 0.503$

COMPUTER/CALCULATOR EXERCISES

61. According to *Maple*'s *fsolve* procedure, $y_0 = -10.9970$. The slope of the tangent line at $P_0 = (10.0000, -10.9970)$ was approximated by substituting the coordinates of P_0 into the expression obtained using the *implicitdiff* procedure: $\left.\frac{dy}{dx}\right|_{P_0} = -1.0006$.

The approximations were checked by using *implicitplot* to plot the curve in the square $[x_0 - 1, x_0 + 1] \times [y_0 - 1, y_0 + 1]$.

63. According to *Maple*'s *fsolve* procedure, $y_0 = 2.6081$. The slope of the tangent line at $P_0 = (-1.4649, 2.6081)$ was approximated by substituting the coordinates of P_0 into the expression obtained using the *implicitdiff* procedure: $\left.\frac{dy}{dx}\right|_{P_0} = -2.8620$.

The approximations were checked by using *implicitplot* to plot the curve in the square $[x_0 - 1, x_0 + 1] \times [y_0 - 1, y_0 + 1]$.

65. According to *Maple*'s *fsolve* procedure, $y_0 = -10.66574$. The slope of the tangent line at $P_0 = (10.000, -10.66574)$ was approximated by substituting the coordinates of P_0 into the expression obtained using the *implicitdiff* procedure: $\left.\frac{dy}{dx}\right|_{P_0} = -1.00018$. The tangent line equation is $y = -1.00018x - 0.66396$.

The approximations were checked by using *implicitplot* to plot the curve in the square $[x_0 - 1, x_0 + 1] \times [y_0 - 1, y_0 + 1]$.

67. According to *Maple*'s *fsolve* procedure, $y_0 = 4.4599$. The slope of the tangent line at $P_0 = (-2.0125, 4.4599)$ was approximated by substituting the coordinates of P_0 into the expression obtained using the *implicitdiff* procedure: $\left.\frac{dy}{dx}\right|_{P_0} = -3.9162$. The tangent line equation is $y = -3.9162x - 3.4214$.

The approximations were checked by using *implicitplot* to plot the curve in the square $[x_0 - 1, x_0 + 1] \times [y_0 - 1, y_0 + 1]$.

69. According to *Maple*'s *implicitdiff* procedure, the slope of the tangent line at P_0 is $\left.\frac{dy}{dx}\right|_{P_0} = 0.5211$. The tangent line equation is

$$y = 0.5211x - 0.1335.$$

See the plot on the right.

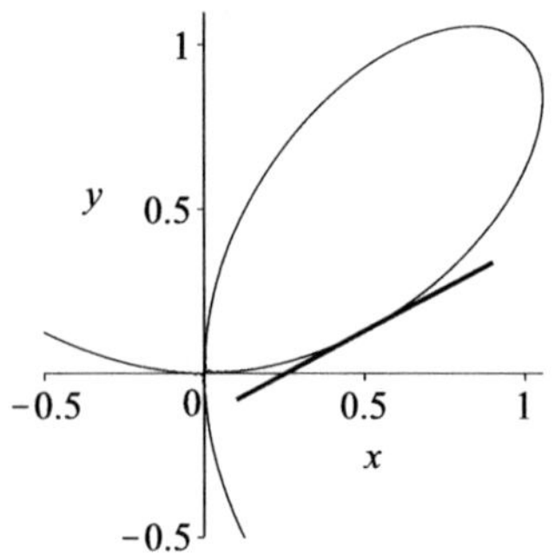

71. According to *Maple*'s *implicitdiff* procedure, the slope of the tangent line at P_0 is $\left.\frac{dy}{dx}\right|_{P_0} = -0.3852$. The tangent line equation is

$$y = -0.3852x + 2.7427.$$

See the plot on the right.

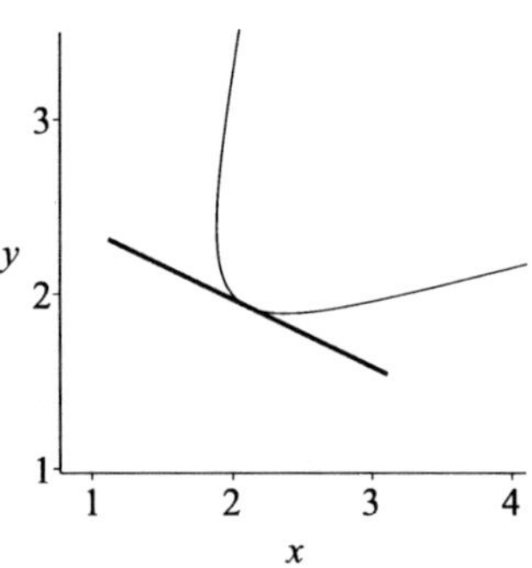

73. Using *Maple*, the slope of the tangent line at P_0 is $\left.\frac{dy}{dx}\right|_{P_0} = -\sqrt{3}/3 = -0.5773$. The tangent line equation is

$$y = -0.5773x + 4.0415.$$

See the plot on the right.

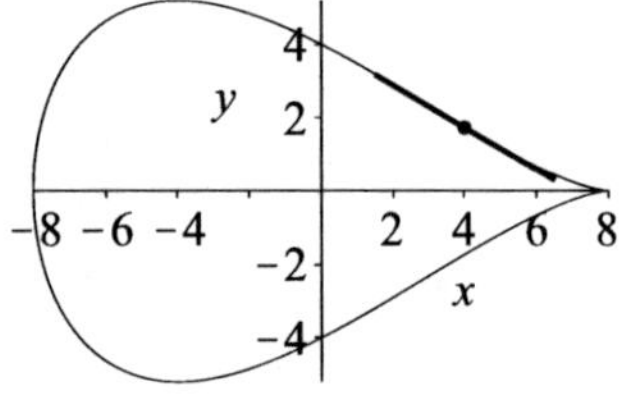

75. Using *Maple*, the slope of the tangent line at P_0 is $\left.\frac{dy}{dx}\right|_{P_0} = 2$. The tangent line equation is

$$y = 2x.$$

See the plot on the right.

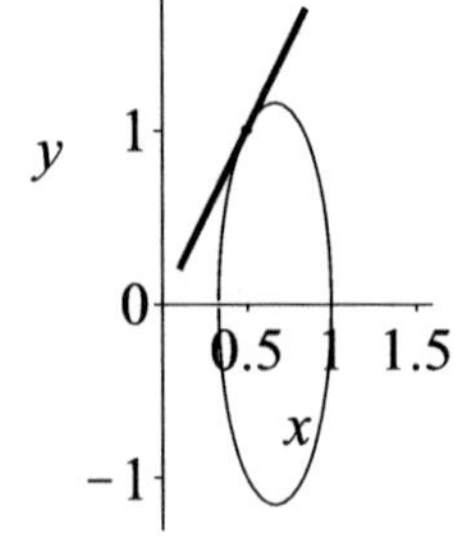

77. Using *Maple*, the slope of the tangent line at P_0 is $\left.\frac{dy}{dx}\right|_{P_0} = -5/12$. The tangent line equation is

$$y = -5x/12 - 1/48.$$

See the plot on the right.

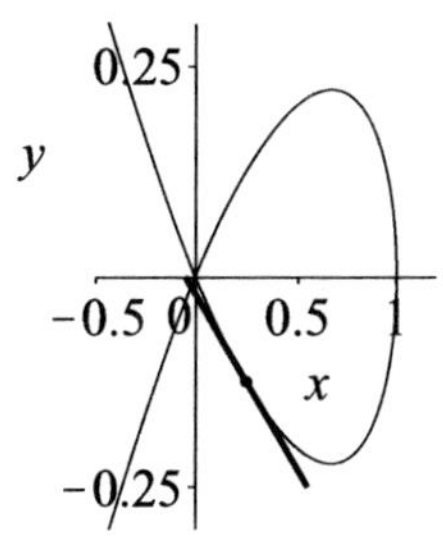

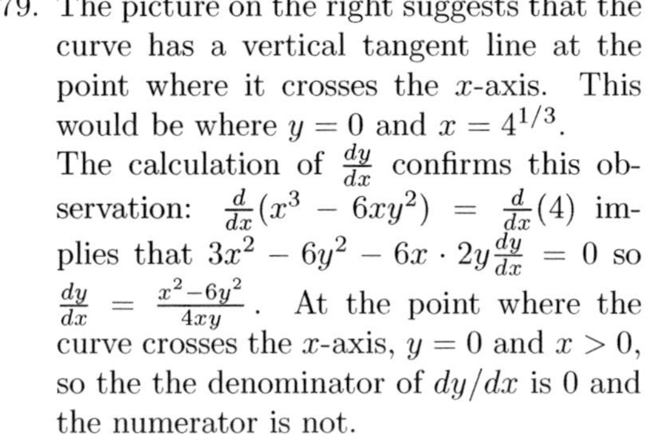

79. The picture on the right suggests that the curve has a vertical tangent line at the point where it crosses the x-axis. This would be where $y = 0$ and $x = 4^{1/3}$. The calculation of $\frac{dy}{dx}$ confirms this observation: $\frac{d}{dx}(x^3 - 6xy^2) = \frac{d}{dx}(4)$ implies that $3x^2 - 6y^2 - 6x \cdot 2y\frac{dy}{dx} = 0$ so $\frac{dy}{dx} = \frac{x^2-6y^2}{4xy}$. At the point where the curve crosses the x-axis, $y = 0$ and $x > 0$, so the the denominator of dy/dx is 0 and the numerator is not.

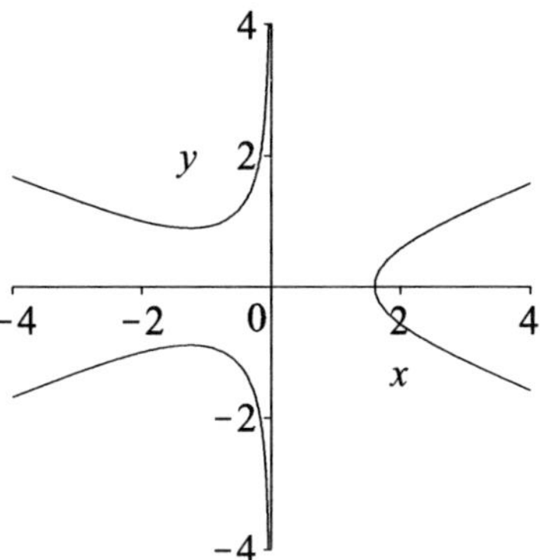

81. The demand curve for $q = q(p)$ is defined by $p + q + 2p^2q + 3pq^3 = 1000$. Using *Maple*'s *fsolve* procedure, when $p = 4$, $q = 4.1521$. Substitute these values into $\frac{dq}{dp} = -\frac{1+4pq+3q^3}{1+2p^2+9pq^2}$ (obtained using *implicitdiff*) to find that $q'(4) = -0.4317$.

For the approximation we ask *fsolve* for $q(4.2)$ and $q(3.8)$ obtaining $q(4.2) = 4.0680$ and $q(3.8) = 4.2408$. Using these values, $q'(4) \approx D_0q(4, 0.4) = -0.4322$.

3.9 Differentials and Approximation of Functions

Problems for Practice

1. With $\Delta x = x - c = -0.1$ and $f'(x) = 1/(2\sqrt{x})$,

$$f(3.9) \approx f(4) + f'(4)\Delta x = 2 + (1/4) \cdot (-0.1) = 2 - 0.025 = 1.975.$$

3. With $\Delta x = x - c = 0.07$ and $f'(x) = -1/(3x^{4/3})$,

$$f(8.07) \approx f(8) + f'(8)\Delta x = 1/2 - (1/48) \cdot (0.07) = 0.4985.$$

5. With $\Delta x = x - c = 0.15$ and $f'(x) = (2/3)x^{-1/3}$,

$$f(8.15) \approx f(8) + f'(8)\Delta x = 4 + (1/3) \cdot (0.15) = 4.05\,.$$

7. With $\Delta x = x - c = -0.2$ and $f'(x) = (14/3)x(1 + 7x^2)^{-2/3}$,

$$f(2.8) \approx f(3) + f'(3)\Delta x = 4 + (7/8) \cdot (-0.2) = 3.825\,.$$

9. With $\Delta x = x - c = 1.06 - \pi/3$ and $f'(x) = -\sin(x)$,

$$f(1.06) \approx f(\pi/3) + f'(\pi/3)\Delta x = 1/2 + (-\sqrt{3}/2) \cdot (1.06 - \pi/3) = 0.4889\,.$$

11. With $\Delta x = x - c = 0.8 - \pi/4$ and $f'(x) = \sec^2(x)$,

$$f(0.8) \approx f(\pi/4) + f'(\pi/4)\Delta x = 1 + (2) \cdot (0.8 - \pi/4) = 1.0292\,.$$

13. With $\Delta x = x - c = 0.5 - \pi/6$ and $f'(x) = \sec(x)\tan(x)$,

$$f(0.5) \approx f(\pi/6) + f'(\pi/6)\Delta x = 2\sqrt{3}/3 + (2/3) \cdot (0.5 - \pi/6) = 1.1390\,.$$

15. With $\Delta x = x - c = 20 - e^3$ and $f'(x) = 1/x$,

$$f(20) \approx f(e^3) + f'(e^3)\Delta x = 3 + (e^{-3}) \cdot (20 - e^3) = 2.9957\,.$$

17. With $\Delta x = x - c = -0.17$ and $f'(x) = e^x$,

$$f(-0.17) \approx f(0) + f'(0)\Delta x = 1 + (1) \cdot (-0.17) = 0.83\,.$$

19. With $\Delta x = x - c = 0.13$ and $f'(x) = 16^x \ln(16)$,

$$f(0.13) \approx f(0) + f'(0)\Delta x = 1 + (4\ln(2)) \cdot (0.13) = 1.3604\,.$$

21. With $\Delta x = x - c = -0.19$ and $f'(x) = \sqrt{2}\,\pi \sin(\pi/(x+3))/(x+3)^2$,

$$f(-0.19) \approx f(1) + f'(1)\Delta x = 1 + (\pi/16) \cdot (-0.19) = 0.9627\,.$$

23. With $\Delta x = x - c = 0.75 - \pi/4$ and $f'(x) = 2\sin(x)\cos(x)$,

$$f(0.75) \approx f(\pi/4) + f'(\pi/4)\Delta x = 1/2 + (1) \cdot (0.75 - \pi/4) = 0.4646\,.$$

25. Use $f(x) = \sqrt{x}$ with $c = 25$, $x = 24$, $\Delta x = x - c = -1$, and $f'(x) = 1/(2\sqrt{x}\,)$,

$$\sqrt{24} \approx f(25) + f'(25)\Delta x = 5 + (1/10) \cdot (-1) = 4.9\,.$$

Since $\sqrt{24} = 4.8990$, the absolute error is $0.00102 = 1.02 \times 10^{-3}$.

27. Use $f(x) = \sqrt{1+\sqrt{x}}$ with $c = 9$, $x = 9.7$, $\Delta x = x - c = 0.7$, and $f'(x) = 1/(4\sqrt{x}\sqrt{1+\sqrt{x}})$,

$$\sqrt{1+\sqrt{9.7}} \approx f(9) + f'(9)\Delta x = 2 + (1/24)\cdot(0.7) = 2.0292\,.$$

Since $\sqrt{1+\sqrt{9.7}} = 2.0284$, the absolute error is $0.000748 = 7.48 \times 10^{-4}$.

29. Use $f(x) = \cos(x)$ with $c = \pi/3$, $x = 55\cdot\pi/180$, $\Delta x = 55\cdot\pi/180 - \pi/3 = -\pi/36$, and $f'(x) = -\sin(x)$,

$$\cos(55°) \approx f(\pi/3) + f'(\pi/3)\Delta x = 1/2 + (-\sqrt{3}/2)\cdot(-\pi/36) = 0.5756\,.$$

Since $\sin(55°) = 0.5738$, the absolute error is $0.002 = 2.0 \times 10^{-3}$.

31. Since $f'(x) = -2(3+2x)^{-2}$, the linearization of f at $c = 0$ is

$$L(x) = 1/3 - 2x/9\,.$$

33. Since $f'(x) = -\frac{1}{2}(x-8)/(4+x)^{5/2}$, the linearization of f at $c = 0$ is

$$L(x) = x/8\,.$$

35. Since $f'(x) = (\sin(x) - 1)/(x + \cos(x))^2$, the linearization of f at $c = 0$ is

$$L(x) = 1 - x\,.$$

37. Since $f'(x) = -\sin(x)$, $f'(\pi/3) = -\sqrt{3}/2$ and the linearization of f at $c = \pi/3$ is $L(x) = 1/2 - \sqrt{3}(x - \pi/3)/2$.

39. Since $f'(x) = (25/9)^x \ln(25/9)$, $f'(1/2) = (5/3)\ln(25/9)$ and the linearization of f at $c = 1/2$ is $L(x) = 5/3 + \frac{10}{3}\ln(5/3)(x - 1/2)$.

41. Since $f'(x) = (x-1)\exp(x-1)/x^2$, $f'(1) = 0$ and the linearization of f at $c = 1$ is $L(x) = 1$.

Further Theory and Practice

43. Since $f'(x) = p\left(\frac{1+ax}{1+bx}\right)^{p-1}\cdot\frac{(1+bx)\cdot a-(1+ax)\cdot b}{(1+bx)^2}$, $f'(0) = p\cdot(a-b)$. Therefore, the linearization of f at $c = 0$ is $L(x) = 1 + p\cdot(a-b)x$.

45. Let $\Delta x = x - c$ and approximate $\Delta(f\cdot g) = (f\cdot g)(x) = (f\cdot g)(c)$ using the linearizations of f and g as follows.

$$\begin{aligned}\Delta(f\cdot g) &= f(x)g(x) - f(c)g(c)\\ &\approx (f(c) + f'(c)\Delta x)(g(c) + g'(c)\Delta x) - f(c)g(c)\\ &= \big(f'(c)g(c) + f(c)g'(c)\big)\Delta x + f'(c)g'(c)(\Delta x)^2\\ &\approx \big(f'(c)g(c) + f(c)g'(c)\big)\Delta x\end{aligned}$$

Voilà... the product rule.

47. $F(x,y) = x + y$, therefore, $y(1.2) \approx y(1) + F(1,2)\Delta x = 2 + 3\cdot 0.2 = 2.6$.

49. $F(x, y) = x^2 - 2y$, therefore,

$$y(1/4) \approx y(0) + F(0, 3)\Delta x = 3 + (-6) \cdot (1/4) = 3/2\,.$$

51. Using implicit differentiation, $\frac{d}{dp}(p + q + 2p^2q + 3pq^3) = 0$. Therefore, $1 + \frac{dq}{dp} + 4pq + 2p^2\frac{dq}{dp} + 3q^3 + 9pq^2\frac{dq}{dp} = 0$. Substitute $p = 6.75$ and $q = 3.248$ to obtain $dq/dp = -0.2613$ for the slope of the demand curve.

Using the differential approximation $dq = \frac{dq}{dp}dp = (-0.2612)dp$,

$$q(6.80) \approx q(6.75) + (-0.2612)(6.80 - 6.75) = 3.235$$

and

$$q(6.60) \approx q(6.75) + (-0.2612)(6.60 - 6.75) = 3.287\,.$$

53. Observe that when $p = 9$, $\frac{9q^2}{190950} + 3q = 8019900$, a quadratic with two solutions (you know the formula). One of them is $q = 381900$ and the point (9,381900) is on the demand curve.

Using implicit differentiation, $\big(q^2 + 2pq\frac{dq}{dp}\big)/190950 + \sqrt{p}\,\frac{dq}{dp} + q/(2\sqrt{p}\,) = 0$. Substitute $p = 9$ and $q = 381900$ to obtain $dq/dp = -21216.\overline{6}$ for the slope of the demand curve.

Using the differential approximation $dq = \frac{dq}{dp}dp = -21216.66\,dp$,

$$q(9.75) \approx q(9) + (-21216.66)(9.75 - 9) = 365987\,.$$

CALCULATOR/COMPUTER EXERCISES

55. According to a calculator, $f(27.09) = 3.03297$.

Approximating with one step, $dx = 0.9$, the approximation is

$$f(27) + f'(27) \cdot 0.9 = 3.03333\,.$$

Using $N = 3$ steps, stepsize $= 0.3$, the approximation is 3.03309.

The absolute error for the one-step approximation is 3.6×10^{-4}. The three-step error is 1.2×10^{-4}.

57. According to a calculator, $f(3) = 1.0986$.

Approximating with one step, $dx = 3 - e$, the approximation is

$$f(e) + f'(e) \cdot (3 - e) = 1.1036\,.$$

Using $N = 2$ steps, stepsize $= (3 - e)/2$, the approximation is 1.1011.

The absolute error for the one step approximation is 5×10^{-3}. The two-step error is 2.5×10^{-3}.

59. The exact solution is correct since $y(x) = 4\exp(x-1) - x - 1$ implies that

$$y'(x) = 4\exp(x-1) - 1 = x + y(x) \text{ and } y(1) = 4 - 1 - 1 = 2\,.$$

The Euler Method approximation to $y(1.2)$ is

$$y_1 = y_0 + F(x_0, y_0) \cdot (0.2) = 2 + (3) \cdot (0.2) = 2.6\,.$$

The Improved Euler Method approximation is

$$z_1 = y_0 + 0.5\big(F(x_0, y_0) + F(x_1, y_1)\big) \cdot (0.2) = 2 + 0.5(3 + 3.8) \cdot (0.2) = 2.68\,.$$

Since $y(1.2) = 2.6856$, the absolute error for the Euler Method approximation is 0.0856. For the Improved Euler Method it is 0.005611.

61. The exact solution is correct since $y(x) = x^2/2 - x/2 + 1/4 + (11/4)\exp(-2x)$ implies that

$$y'(x) = x - 1/2 - (11/2)\exp(-2x) = x^2 - 2y(x) \text{ and } y(0) = 1/4 + 11/4 = 3\,.$$

The Euler Method approximation to $y(0.25)$ is

$$y_1 = y_0 + F(x_0, y_0) \cdot (0.25) = 3 + (-6) \cdot (0.25) = 1.5\,.$$

The Improved Euler Method approximation is

$$z_1 = y_0 + 0.5\big(F(x_0, y_0) + F(x_1, y_1)\big) \cdot (0.2) = 3 + 0.5(-6 - 2.9375) \cdot (0.25) = 1.8828\,.$$

Since $y(1.2) = 1.8554$, the absolute error for the Euler Method approximation is 0.3555. For the Improved Euler Method it is 0.02735.

63. Using *Maple*'s *implictdiff* procedure, $dq/dp = -4 \cdot \dfrac{p\,q^{7/4} + q}{p\,(2p\,q^{3/4} + 1)}\,.$

Its *fsolve* procedure informs us that if $p_0 = 5.10$, then $q_0 = 9611.65$ and if $p_1 = 5$, then $q_1 = 10000$.

Substitute p_0 and q_0 into the formula for dq/dp to find that the slope of the demand curve at (p_0, q_0) is $q'(p_0) = -3769.65$ and the differential approximation to q_1 is $Q_1 = q_0 + q'(p_0) \cdot (p_1 - p_0) = 9988.62$.

The relative error in the approximation is $\left|\dfrac{Q_1 - q_1}{q_1}\right| \times 100 = 0.11\%\,.$

65. Using *Maple*'s *implictdiff* procedure, $dq/dp = -2 \cdot \dfrac{p\,q^{9/5} + 25\,q}{p\,(p\,q^{4/5} + 10)}\,.$

Its *fsolve* procedure informs us that if $p_0 = 1.80$, then $q_0 = 3950655$ and if $p_1 = 2$, then $q_1 = 3200000$.

Substitute p_0 and q_0 into the formula for dq/dp to find that the slope of the demand curve at (p_0, q_0) is $q'(p_0) = -4389810$ and the differential approximation to q_1 is $Q_1 = q_0 + q'(p_0) \cdot (p_1 - p_0) = 3072693$.

The relative error in the approximation is $\left|\dfrac{Q_1 - q_1}{q_1}\right| \times 100 = 3.98\%\,.$

3.10 Other Transcendental Functions

Problems for Practice

1. $\arcsin(1) = \pi/2$

3. $\arcsin(-1) = -\pi/2$

5. $\arccos(1/2) = \pi/3$

7. $\arccos(\sqrt{2}/2) = \pi/4$

9. $\arctan(-\sqrt{3}) = -\pi/3$

11. $\operatorname{arcsec}(\sqrt{2}) = \pi/4$

13. $\operatorname{arccsc}(-\sqrt{2}) = -\pi/4$

15. $\arcsin(\sin(5\pi/4)) = -\pi/4$

17. $\operatorname{arcsec}(\sec(-5\pi/6)) = 5\pi/6$

19. $\frac{d}{dx}\big(\arcsin(1-x)\big) = \frac{1}{\sqrt{1-(1-x)^2}} \cdot (-1) = -\frac{1}{\sqrt{2x-x^2}}$

21. $\frac{d}{dx}\big(x\arcsin(x)\big) = \arcsin(x) + x \cdot \frac{1}{\sqrt{1-x^2}}$

23. $\frac{d}{dx}\big(\log_2(\arccos(x))\big) = \frac{1}{\arccos(x)\ln(2)} \cdot \frac{-1}{\sqrt{1-x^2}}$

25. $\frac{d}{dx}\big(\arctan(e^x)\big) = \frac{e^x}{1+e^{2x}}$

27. $\frac{d}{dx}\big(x \cdot \arctan(x)\big) = \arctan(x) + x \cdot \frac{1}{1+x^2}$

29. $\frac{d}{dx}\big(x^2 \operatorname{arcsec}(x)\big) = 2x\operatorname{arcsec}(x) + x^2 \cdot \frac{1}{|x|\sqrt{x^2-1}} = 2x\operatorname{arcsec}(x) + \frac{|x|}{\sqrt{x^2-1}}$

31. $\frac{d}{dx}\big(\operatorname{arccsc}(x^2)\big) = -\frac{2x}{|x^2|\sqrt{x^4-1}} = -\frac{2}{x\sqrt{x^4-1}}$

33. $\frac{d}{dx}\big(\operatorname{arccot}(\sqrt{x})\big) = -\frac{1}{1+x} \cdot (1/(2\sqrt{x})) = -\frac{1}{2\sqrt{x}(1+x)}$

35. $\frac{d}{dx}\big(\sinh(3x)\big) = 3\cosh(3x)$

37. $\frac{d}{dx}\big(\ln(\sinh(x))\big) = \frac{\cosh(x)}{\sinh(x)} = \coth(x)$

39. $\frac{d}{dx}\big(\cosh(x)/x\big) = \sinh(x)/x + \cosh(x) \cdot (-1/x^2) = (x\sinh(x) - \cosh(x))/x^2$

41. $\dfrac{d}{dx}\big(\ln(\tanh(x)\big) = \dfrac{\operatorname{sech}^2(x)}{\tanh(x)}$

43. $\dfrac{d}{dx}\big(\tanh(\tan(x))\big) = \operatorname{sech}^2(\tan(x))\sec^2(x)$

45. $\dfrac{d}{dx}\big(\operatorname{sech}(\sqrt{x}\,)\big) = -\operatorname{sech}(\sqrt{x}\,)\tanh(\sqrt{x}\,)/(2\sqrt{x}\,)$

47. $\dfrac{d}{dx}\big(\sinh^{-1}(5x)\big) = \dfrac{5}{\sqrt{1+25x^2}}$

49. $\dfrac{d}{dx}\big(\arctan(\sinh(x))\big) = \dfrac{1}{1+\sinh^2(x)}\cdot\cosh(x) = \operatorname{sech}(x)$

51. $\dfrac{d}{dx}\big(\tanh^{-1}(\cos(x))\big) = \dfrac{1}{1-\cos^2(x)}\cdot(-\sin(x)) = -\csc(x)$

53. $\dfrac{d}{dx}\big(\tanh^{-1}(\cosh(x))\big) = \dfrac{1}{1-\cosh^2(x)}\cdot\sinh(x) = -\operatorname{csch}(x)$

Further Theory and Practice

55. Since $f'(x) = 1/\sqrt{1-x^2}$, $f'(1/2) = 2\sqrt{3}/3$ and
$$L(x) = \pi/6 + 2\sqrt{3}\,(x-1/2)/3\,.$$

57. Since $f'(x) = 1/(1+x^2)$, $f'(\sqrt{3}\,) = 1/4$ and
$$L(x) = \pi/3 + (x-\sqrt{3}\,)/4\,.$$

59. If x is in $[-1,1]$, then $\pi/2 - \arccos(x)$ is in $[-\pi/2, \pi/2]$ and
$$\sin(\pi/2-\arccos(x)) = \sin(\pi/2)\cos(\arccos(x)) - \cos(\pi/2)\sin(\arccos(x)) = x\,.$$
Therefore, $\arcsin(x) = \pi/2 - \arccos(x)$.

61. Let x be in the interval $[-1,1]$. Then $\sin(\arcsin(x) = x$ so
$$\cos^2(\arcsin(x)) = 1 - \sin^2(\arcsin(x)) = 1 - x^2\,.$$

63. Using the definitions,
$$\begin{aligned}\cosh(x)\cosh(y) + \sinh(x)\sinh(y) &= \tfrac{1}{4}\big((e^x+e^{-x})(e^y+e^{-y}) + (e^x-e^{-x})(e^y-e^{-y})\big)\\ &= \tfrac{1}{4}\big(2e^{x+y} + 2e^{-(x+y)}\big) = \cosh(x+y)\end{aligned}$$

65. Using the definitions,
$$\begin{aligned}2\cosh^2(x) - 1 &= \tfrac{1}{2}(e^x+e^{-x})^2 - 1 = \tfrac{1}{2}(e^{2x}+2+e^{-2x}) - 1\\ &= \tfrac{1}{2}(e^{2x}+e^{-2x}) = \cosh(2x)\end{aligned}$$

67. Using the definitions,

$$\begin{aligned}\frac{\tanh(x)+\tanh(y)}{1+\tanh(x)\tanh(y)} &= \frac{\frac{e^x-e^{-x}}{e^x+e^{-x}}+\frac{e^y-e^{-y}}{e^y+e^{-y}}}{1+\frac{e^x-e^{-x}}{e^x+e^{-x}}\cdot\frac{e^y-e^{-y}}{e^y+e^{-y}}}\cdot\frac{(e^x+e^{-x})(e^y+e^{-y})}{(e^x+e^{-x})(e^y+e^{-y})}\\ &= \frac{(e^x-e^{-x})(e^y+e^{-y})+(e^x+e^{-x})(e^y-e^{-y})}{(e^x+e^{-x})(e^y+e^{-y})+(e^x-e^{-x})(e^y-e^{-y})}\\ &= \frac{2e^{x+y}-2e^{-(x+y)}}{2e^{x+y}+2e^{-(x+y)}} = \tanh(x+y)\end{aligned}$$

69. If $y(x)=\sinh(\omega x)$, then $y''(x)=\omega^2\sinh(x)$. Consequently, for all x,

$$y''(x)-\omega^2 y(x)=\omega^2\sinh(x)-\omega^2\sinh(x)=0\,.$$

The calculation for $y(x)=\cosh(\omega x)$ is exactly the same.

71. Let f be the function whose graph is the curve traced out by the tip of the boat. Fix a point (x,y) on the curve, and let ℓ be the segment of the tangent line to the curve at (x,y) that extends from (x,y) to the point P on y-axis. The tether lies along ℓ, so its length is a. Apply the Pythagorean Theorem to the right triangle with vertices (x,y), $(0,y)$, and P, to see that $P=(0,y+\sqrt{a^2-x^2}\,)$. Therefore, the slope of ℓ is $-\sqrt{a^2-x^2}/x$ and $f'(x)=-\sqrt{a^2-x^2}/x$. Thus f has the same derivative as the tractrix $T(x)$ defined above. Moreover, $f(a)=T(a)=0$. From this it can be shown (see Chapter 4) that $f(x)=T(x)$ for all $0<x<a$.

73. Given any x, $\arctan(x)$ is in the open interval $(-\pi/2,\pi/2)$ where the secant function is positive. Consequently, $\sec(\arctan(x))=\sqrt{1+x^2}$. Therefore,

$$\sec(gd(u))=\sec(\arctan(\sinh(u))=\sqrt{1+\sinh^2(u)}=\cosh(u)\,.$$

Similar arguments can be used to verify each of the remaining four identities. For example, given any x, because $\arctan(x)$ is in the open interval $(-\pi/2,\pi/2)$, $\sin(\arctan(x))=x/\sqrt{1+x^2}$. Consequently, given any u,

$$\sin(gd(u))=\sin(\arctan(\sinh(u))=\frac{\sinh(u)}{\sqrt{1+\sinh^2(u)}}=\frac{\sinh(u)}{\cosh(u)}=\tanh(u)\,.$$

75. According to the Inverse Function Derivative Rule,

$$\begin{aligned}\frac{d}{d\theta}gd^{-1}(\theta) &= \left.\frac{1}{gd'(u)}\right|_{gd(u)=\theta} = \left.\frac{1}{\operatorname{sech}(u)}\right|_{\arctan(\sinh(u))=\theta}\\ &= \left.\frac{1}{\operatorname{sech}(u)}\right|_{u=\sinh^{-1}(\tan(\theta))} = \frac{1}{\operatorname{sech}(\sinh^{-1}(\tan(\theta)))}\\ &= \cosh(\sinh^{-1}(\tan(\theta)))\,.\end{aligned}$$

However, $\cosh(x) = \sqrt{1 + \sinh^2(x)}$ for any x. Therefore, $\cosh(\sinh^{-1}(x)) = \sqrt{1 + x^2}$ for any x also. In particular, $\cosh(\sinh^{-1}(\tan(\theta))) = \sqrt{1 + \tan^2(\theta)} = \sec(\theta)$, as desired.

77. Let $u = x$ and $v = \sqrt{x} + \sqrt{3}$. Then $v^2 = x + 3 + 2\sqrt{3}\sqrt{x}$ so $v^2 - u - 3 = 2\sqrt{3}\sqrt{x}$ and $(v^2 - u - 3)^2 = 12x$, implying that $(v^2 - u - 3)^2 - 12u = 0$. The polynomial $p(u, v) = (v^2 - u - 3)^2 - 12u$ does the job.

79. Let $u = x$ and $v = \sqrt[3]{x + \sqrt{x}}$. Then $v^3 = u + \sqrt{x}$, implying that $(v^3 - u)^2 = u$, and the polynomial $p(u, v) = (v^3 - u)^2 - u$ does the job.

CALCULATOR/COMPUTER EXERCISES

81. Using the central difference quotient with $h = 10^{-5}$ yields $f'(2.1) \approx 1.7081$. Since $f(2.1) \approx 3.1959$, the tangent line at $(c, f(c)$ has the equation

$$y = 3.1959 + 1.7081(x - 2.1) .$$

See the picture.

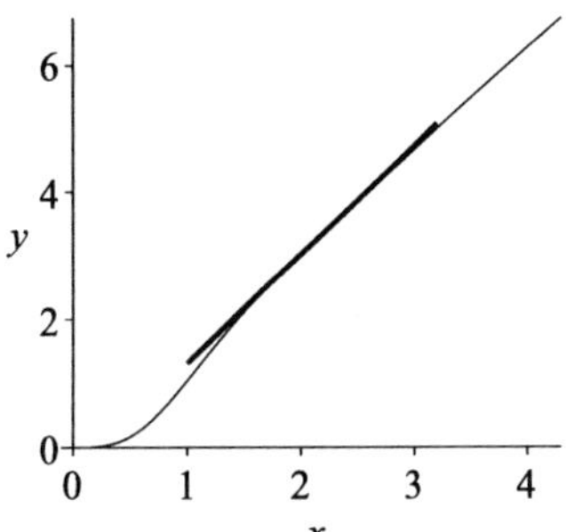

83. Using the central difference quotient with $h = 10^{-5}$ yields $f'(4.5) \approx 0.4721$. Since $f(4.5) \approx 1.9956$, the tangent line at $(c, f(c)$ has the equation

$$y = 1.9956 + 0.4721(x - 4.5) .$$

See the picture.

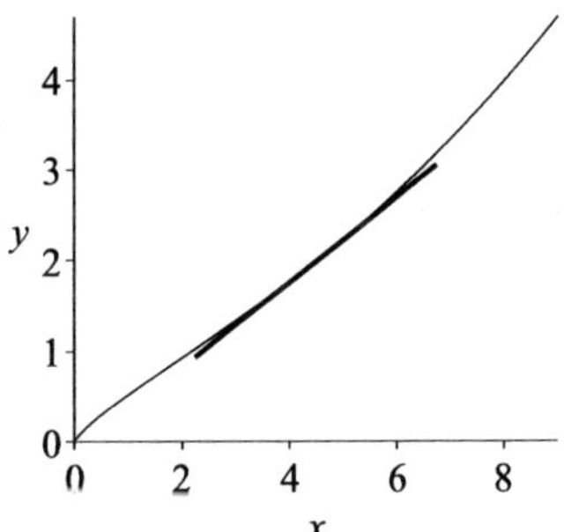

85. Using the central difference quotient with $h = 10^{-5}$ yields $f'(2.5) \approx 0.3482$. Since $f(2.5) \approx 1.0917$, the tangent line at $(c, f(c)$ has the equation

$$y = 1.0917 + 0.3482(x - 2.5) .$$

See the picture.

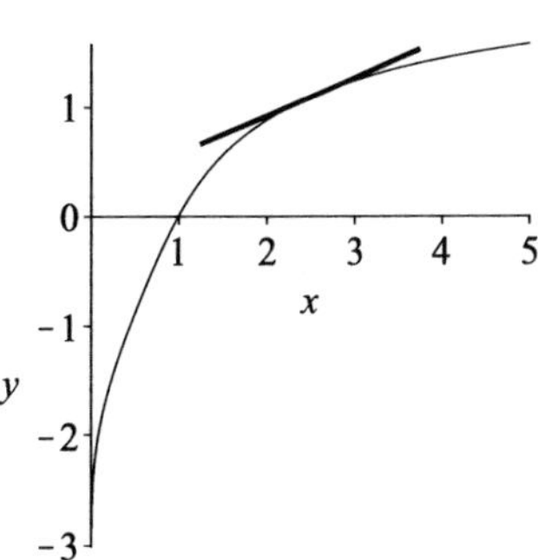

87. The plots appear below, gudermanian on the left and its inverse on the right.

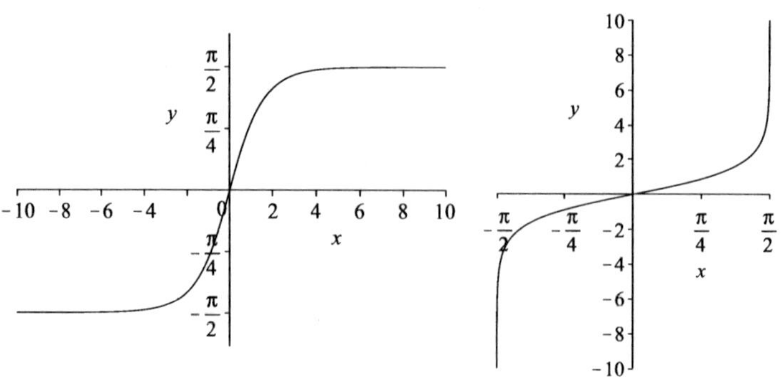

REVIEW EXERCISES FOR CHAPTER 3

1. Since $f'(x) = 3x^2$, $f'(-2) = 12$.
   The tangent line equation is $y = 12x + 16$.

3. Since $f'(x) = -6x^{-3/2}$, $f'(16) = -3/32$.
   The tangent line equation is $y = -3x/32 + 9/2$.

5. Since $f'(x) = 12\sin(\pi x) + 12\pi x\cos(\pi x)$, $f'(1/6) = 6 + \sqrt{3}\,\pi$.
   The tangent line equation is $y = (6 + \sqrt{3}\,\pi)x - \sqrt{3}\,\pi/6$.

7. Since $f'(x) = (2x^2 + 3)/x^2$, $f'(3) = 7/3$.
   The tangent line equation is $y = 7x/3 - 2$.

9. Since $f'(x) = 2/(e\,x) + 1/x^2$, $f'(e) = 3e^{-2}$.
   The tangent line equation is $y = 3x/e^2 - 2/e$.

11. Since $f'(x) = -3\ln 2/2^x$, $f'(0) = -3\ln(2)$.
    The tangent line equation is $y = -3\ln(2)x + 3$.

13. $\frac{d}{dx}\left(\sqrt{x}\,\right) = 1/(2\sqrt{x}\,)$

15. $\frac{d}{dx}\left((3+x)/x^2\right) = (x^2 \cdot 1 - (3+x)\cdot 2x)/x^4 = -(x+6)/x^3$

17. $\frac{d}{dx}\left(x^3 + x^{-3}\right) = 3x^2 - 3x^{-4}$

19. $\frac{d}{dx}\left((x^2 + x - 2)/x\right) = \frac{d}{dx}\left(x + 1 - 2/x\right) = 1 + 2/x^2$

21. $\frac{d}{dx}\left((2 + x^3)^{-3}\right) = -3(2+x^3)^{-4}\cdot(3x^2) = -9x^2(2+x^3)^{-4}$

23. $\frac{d}{dx}\left(\sin(\pi - x)\right) = -\cos(\pi - x) = \cos(x)$

25. $\frac{d}{dx}\left(\cos(2x + \pi/4)\right) = -2\sin(2x + \pi/4)$

27. $\frac{d}{dx}\left(\sec(2x)\right) = 2\sec(2x)\tan(2x)$

29. $\frac{d}{dx}\left(\cot(1/x)\right) = -\csc^2(1/x)\cdot(-1/x^2) = \csc(1/x)/x^2$

31. $\frac{d}{dx}\left(\sin^3(5x^2)\right) = 3\sin^2(5x^2)\cdot\cos(5x^2)\cdot 10x = 30x\sin^2(5x^2)\cos(5x^2)$

33. $\frac{d}{dx}\left(x^3\sin(x^2)\right) = 3x^2\sin(x^2) + x^3\cdot\cos(x^2)\cdot 2x = 3x^2\sin(x^2) + 2x^4\cos(x^2)$

35. $\frac{d}{dx}((1 + \sin(x))/(1 + \cos(x))) = ((1 + \cos(x)) \cdot \cos(x) - (1 + \sin(x)) \cdot (-\sin(x)))/(1 + \cos(x))^2 = (1 + \cos(x) + \sin(x))/(1 + \cos(x))^2$

37. $\frac{d}{dx}\left(\sqrt{1 + \cos(3x^2)}\right) = \frac{1}{2}(1 + \cos(3x^2))^{-1/2} \cdot (-\sin(3x^2) \cdot 6x)$

$$= -3x\sin(3x^2)/\sqrt{1 + \cos(3x^2)}$$

39. $\frac{d}{dx}\big(xe^{3x}\big) = e^{3x} + 3xe^{3x} = (1 + 3x)e^{3x}$

41. $\frac{d}{dx}\big(xe^x/(1+x)\big) = ((1+x)\cdot(e^x+xe^x)-xe^x)/(1+x)^2 = e^x(1+x+x^2)/(1+x)^2$

43. $\frac{d}{dx}\big(2^{-x}\big) = 2^{-x}\ln(2) \cdot (-1) = -2^{-x}\ln(x)$

45. $\frac{d}{dx}\big(1/x + \ln(x)\big) = -1/x^2 + 1/x = (x - 1)/x^2$

47. $\frac{d}{dx}\big(x^3\ln(x)\big) = 3x^2\ln(x) + x^3 \cdot (1/x) = x^2(3\ln(x) + 1)$

49. $\frac{d}{dx}\big(\ln(2\ln(3x))\big) = \frac{d}{dx}\big(\ln(2)+\ln(\ln(3x))\big) = (3/(3x))/\ln(3x) = 1/(x\ln(3x))$

51. $\frac{d}{dx}\big(x^2\log_2(x + 2)\big) = 2x\log_2(x + 2) + x^2 \cdot 1/((x + 2)\ln(2))$

$$= 2x\log_2(x + 2) + x^2/((x + 2)\ln(2))$$

53. $\frac{d}{dx}\big(2^x3^x5^x\big) = \frac{d}{dx}\big(30^x\big) = 30^x\ln(30)$

55. $\frac{d}{dx}\big(\arctan(2/x)\big) = 1/(1 + 4/x^2) \cdot (-2/x^2) = -2/(x^2 + 4)$

57. $\frac{d}{dx}\big(\arcsin(\cos(x))\big) = -\sin(x)/\sqrt{1 - \cos^2(x)} = \sin(x)/|\sin(x)|$

59. $\frac{d}{dx}\big(\cosh(x^2)\big) = \sinh(x^2) \cdot 2x = 2x\sinh(x^2)$

61. $\frac{d}{dx}\big(\sinh^{-1}(e^x)\big) = (1/\sqrt{1 + e^{2x}}\,) \cdot e^x = e^x/\sqrt{1 + e^{2x}}$

63. $v = 60 - 32t$, $a = -32$

65. $v = -20 \cdot 2t/(1 + t^2)^2 = -40t/(1 + t^2)^2$,
$a = -40\big((1 + t^2)^2 - t \cdot 2(1 + t^2) \cdot 2t\big)/(1 + t^2)^2$

$$= 40(3t^2 - 1)/(1 + t^2)^3$$

67. $v = 5e^{-5t}$, $a = -25e^{-5t}$

69. Since $f'(x) = 3x^2 + 3$ and $f(2) = 5$, $(f^{-1})'(5) = 1/f'(2) = 1/15$.

71. Since $f'(x) = ((4 + x^2) \cdot 3x^2 - x^3 \cdot 2x)/(4 + x^2)^2 = x^2(12 + x^2)/(4 + x^2)^2$ and $f(2) = 1$, $(f^{-1})'(1) = 1/f'(2) = 1$.

73. Since $f'(x) = e^x + xe^x$ and $f(1) = e$, $(f^{-1})'(e) = 1/f'(1) = 1/(2e)$.

75. Differentiate: $\frac{d}{dx}(\exp(y) - x + y) = 0$. Simplify, assuming y is a function of x: $\exp(y)\frac{dy}{dx} - 1 + \frac{dy}{dx} = 0$. Solve for $\frac{dy}{dx}$, $\frac{dy}{dx} = 1/(\exp(y) + 1)$.

77. Differentiate: $\frac{d}{dx}(x^2 - y + y^2) = 0$. Simplify, assuming y is a function of x: $2x - \frac{dy}{dx} + 2y\frac{dy}{dx} = 0$. Solve for $\frac{dy}{dx}$, $\frac{dy}{dx} = 2x/(1 - 2y)$.

79. Differentiate: $\frac{d}{dx}(y+1/y) = \frac{d}{dx}(1+\ln(x))$. Simplify, assuming y is a function of x: $(1-1/y^2)\frac{dy}{dx} = 1/x$. Solve for $\frac{dy}{dx}$,
$\frac{dy}{dx} = (1/x)/(1-1/y^2) = y^2/\big(x(y^2-1)\big)$.

81. Differentiate once: $\frac{d}{dx}(y^3+y-x) = 0 \implies 3y^2\frac{dy}{dx} + \frac{dy}{dx} - 1 = 0$.

Solve for $\frac{dy}{dx}$: $\frac{dy}{dx} = 1/(3y^2+1)$, and differentiate again:

$$\frac{d^2y}{dy^2} = (-6y/(3y^2+1)^2)\cdot\frac{dy}{dx} = -6y/(3y^3+1)^3\,.$$

83. Differentiate once: $\frac{d}{dx}(\sin(y)+2\cos(x)) = 0 \implies \cos(y)\frac{dy}{dx} - 2\sin(x) = 0$.

Solve for $\frac{dy}{dx}$: $\frac{dy}{dx} = 2\sin(x)/\cos(y)$, and differentiate again:

$$\begin{aligned}\frac{d^2y}{dy^2} &= 2\cdot\frac{\cos(y)\cdot\cos(x) - \sin(x)\cdot(-\sin(y))\frac{dy}{dx}}{\cos^2(y)} \\ &= 2\cdot\frac{\cos(y)\cos(x) + \sin(x)\sin(y)\cdot\frac{2\sin(x)}{\cos(y)}}{\cos^2(y)} \\ &= 2\cdot\frac{\cos^2(y)\cos(x) + 2\sin^2(x)\sin(y)}{\cos^3(y)} \\ &= 2\sec^2(y)(\cos(y)\cos(x) + 2\sin^2(x)\tan(y))\end{aligned}$$

85. The slope of the tangent line is $\frac{dy}{dx} = \left.\frac{dy/dt}{dx/dt}\right|_{t=2} = \left.\frac{3t^2}{2t-1}\right|_{t=2} = 4$. The equation of the tangent line is $y = 4x-6$.

87. The slope of the tangent line is $\frac{dy}{dx} = \left.\frac{dy/dt}{dx/dt}\right|_{t=\pi/3} = \left.\frac{\sin(t)}{1-\cos(t)}\right|_{t=\pi/3} = \sqrt{3}$. The equation of the tangent line is $y = \sqrt{3}\,x + (3-2\sqrt{3}\,\pi)/6$.

89. The slope of the tangent line is $\frac{dy}{dx} = \left.\frac{dy/dt}{dx/dt}\right|_{t=1} = \left.\frac{1-2/t^2}{1+1/t}\right|_{t=1} = -1/2$. The equation of the tangent line is $y = -x/2 + 7/2$.

91. Since $f'(x) = (7/2)/\sqrt{1+7x}$, $f'(5) = 7/12$ and

$$f(5-0.3) \approx f(5) + (7/12)(-0.3) = 5.825\,.$$

93. Since $f'(x) = (72-4x)/(9-x)^{3/2}$, $f'(5) = 13/2$ and

$$f(5+0.2) \approx f(5) + (13/2)(0.2) = 21.3\,.$$

95. The central difference quotient approximation to $f'(c)$ is

$$D_0 f(c,h) = \frac{f(c+h/2) - f(c-h/2)}{h} = 2.214\,.$$

For the second derivative, the approximation is

$$f''(c) \approx \frac{f(c+h) - 2f(c) + f(c+h)}{h^2} = 4.144\,.$$

97. The central difference quotient approximation to $f'(c)$ is

$$D_0 f(c,h) = \frac{f(c+h/2) - f(c-h/2)}{h} = 27.605\,.$$

For the second derivative, the approximation is

$$f''(c) \approx \frac{f(c+h) - 2f(c) + f(c+h)}{h^2} = 30.5\,.$$

Chapter 4

Applications of the Derivative

4.1 Related Rates

Problems for Practice

1. Since $\frac{dy}{dt} = 3x^2\frac{dx}{dt}$, $\left.\frac{dy}{dt}\right|_{t_0} = 3 \cdot x_0^2 \cdot \left.\frac{dx}{dt}\right|_{t_0} = 3 \cdot 4 \cdot 5 = 60$.

3. Since $\frac{dy}{dt} = -\sin(x)\frac{dx}{dt}$, $\left.\frac{dy}{dt}\right|_{t_0} = -\sin(x_0) \cdot \left.\frac{dx}{dt}\right|_{t_0} = -(1/2) \cdot (-2) = 1$.

5. Since $\frac{dy}{dt} = (1 + \ln(2x))\frac{dx}{dt}$, $\left.\frac{dy}{dt}\right|_{t_0} = (1 + \ln(2x_0))\left.\frac{dx}{dt}\right|_{t_0} = 1 \cdot 3 = 3$.

7. Since $\frac{dy}{dt} = -11\frac{dx}{dt}$, $\left.\frac{dy}{dt}\right|_{t_0} = -11\left.\frac{dx}{dt}\right|_{t_0}$. Therefore, $22 = -11\left.\frac{dx}{dt}\right|_{t_0}$ and $\left.\frac{dx}{dt}\right|_{t_0} = -2$.

9. Since $\frac{dy}{dt} = -18\exp(-3x)\frac{dx}{dt}$, $\left.\frac{dy}{dt}\right|_{t_0} = -18\exp(-3x_0)\left.\frac{dx}{dt}\right|_{t_0}$. Therefore, since $x_0 = 0$ when $y_0 = 6$, $54 = -18\left.\frac{dx}{dt}\right|_{t_0}$ and $\left.\frac{dx}{dt}\right|_{t_0} = -3$.

11. Since $2x\frac{dx}{dt} + 3y^2\frac{dy}{dt} = 0$, $2x_0\left.\frac{dx}{dt}\right|_{P_0} + 3y_0^2\left.\frac{dy}{dt}\right|_{P_0} = 0$. Therefore,
$$\left.\frac{dy}{dt}\right|_{P_0} = -(2x_0/3y_0^2)\left.\frac{dx}{dt}\right|_{P_0} = -4.$$

13. Since $2x\frac{dx}{dt} + x\frac{dy}{dt} + \frac{dx}{dt}y = 0$, $(2x_0 + y_0)\left.\frac{dx}{dt}\right|_{P_0} + x_0\left.\frac{dy}{dt}\right|_{P_0} = 0$. Therefore,
$$-3 \cdot \left.\frac{dx}{dt}\right|_{P_0} - 2 \cdot \left.\frac{dy}{dt}\right|_{P_0} = 0 \text{ and } \left.\frac{dx}{dt}\right|_{P_0} = -(2/3) \cdot \left.\frac{dy}{dt}\right|_{P_0} = -2.$$

15. Since $6x^2\frac{dx}{dt}-x^2\frac{dy}{dt}-2xy\frac{dx}{dt}+3\frac{dy}{dt}=0$, $(6x_0^2-2x_0y_0)\cdot\frac{dx}{dt}\Big|_{P_0}-(x_0^2-3)\cdot\frac{dy}{dt}\Big|_{P_0}=0$. Therefore, $12\frac{dx}{dt}\Big|_{P_0}-\frac{dy}{dt}\Big|_{P_0}=0$ and $\frac{dy}{dt}\Big|_{P_0}=12\frac{dx}{dt}\Big|_{P_0}=-24$.

17. Since $2x\frac{dx}{dt}+3y^2\frac{dy}{dt}=0$, substituting all known values, $-2\cdot 3+12\cdot\frac{dy}{dt}=0$, and $\frac{dy}{dt}\Big|_{P_0}=1/2$. Since $\frac{dx}{dt}\Big|_{P_0}=3$, the slope of the tangent line at P_0 is $(1/2)/3=1/6$, and the equation of the line is $y=(1/6)(x+1)+2$ or $y=x/6+13/6$.

19. Since $6x\frac{dx}{dt}-\frac{dx}{dt}\cdot y-x\frac{dy}{dt}=0$, substituting all known values, $4\cdot\frac{dx}{dt}-1\cdot(-2)=0$, and $\frac{dx}{dt}\Big|_{P_0}=-1/2$. Since $\frac{dy}{dt}\Big|_{P_0}=-2$, the slope of the tangent line at P_0 is $(-2)/(-1/2)=4$, and the equation of the line is $y=4(x-1)+2$ or $y=4x-2$.

21. Since $4x\frac{dx}{dt}-\frac{dx}{dt}\cdot y^2-2xy\frac{dy}{dt}+(1/y)\frac{dy}{dt}=0$, substituting all known values, $8\cdot(-2)-(23/2)\cdot\frac{dy}{dt}=0$, and $\frac{dy}{dt}\Big|_{P_0}=-32/23$. Since $\frac{dx}{dt}\Big|_{P_0}=-2$, the slope of the tangent line at P_0 is $(-32/23)/(-2)=16/23$, and the equation of the line is $y=(16/23)(x-3)+2$ or $y=16x/23-2/23$.

23. Let A, s, and ℓ denote the area, side length, and diagonal length of the square. Then $A=s^2$ and $\ell=\sqrt{2}\,s$, so $\frac{dA}{dt}=2s\frac{ds}{dt}$ and $\frac{d\ell}{dt}=\sqrt{2}\,\frac{ds}{dt}$. Since $\frac{d\ell}{dt}=2$, $\frac{ds}{dt}=2/\sqrt{2}=\sqrt{2}$, and $\frac{dA}{dt}=2\sqrt{2}\,s$. To find $\frac{dA}{dt}\Big|_{\ell=5}$ observe that when $\ell=5$, $s=5/\sqrt{2}$, so $\frac{dA}{dt}\Big|_{\ell=5}=2\sqrt{2}\cdot(5/\sqrt{2})=10$ in^2/min.

25. Let A be the area and s the side length of the triangle. We know that $\frac{ds}{dt}=-3$, and wish to find $\frac{dA}{dt}\Big|_{A=27\sqrt{3}}$. Since $A=\frac{1}{2}\cdot b\cdot h$ where $b=s$, $A=s\cdot h/2$ where h is the height of the triangle.

    Because $h/s=\sin(\pi/3)=\sqrt{3}/2$, $h=\frac{\sqrt{3}}{2}s$ and $A=\frac{\sqrt{3}}{4}s^2$. Consequently, $\frac{dA}{dt}=\frac{\sqrt{3}}{2}s\cdot\frac{ds}{dt}=-\frac{3\sqrt{3}}{2}s$, and when $A=27\sqrt{3}$, $s=2\sqrt{27}=2\cdot 3\sqrt{3}$. From this we may conclude that $\frac{dA}{dt}\Big|_{A=27\sqrt{3}}=-27$ cm^2/s. The area is decreasing at the rate of 27 cm^2/s.

27. Because the radius of the circle is 1, the area of the sector is $A=\frac{1}{2}\theta$ where θ is the central angle. Consequently, $\frac{dA}{dt}=\frac{1}{2}\cdot\frac{d\theta}{dt}=3$.

29. Let θ be the angle of elevation of the loading dock. Then $\sin(\theta)=2/5$ and $h/\ell=2/5$ where h is the height of the container and ℓ is the distance from the container to the bottom of the ramp. Therefore, since $h=0.4\ell$, $\frac{dh}{dt}=0.4\frac{d\ell}{dt}$ and $\frac{dh}{dt}\Big|_{\ell=5}=0.4\cdot\frac{d\ell}{dt}\Big|_{\ell=5}=0.4\cdot 1.2=0.48$ ft/s.

31. Let V and r denote the volume and radius of the balloon. Recall that $V=\frac{4}{3}\pi r^3$, so $\frac{dV}{dt}=4\pi r^2\cdot\frac{dr}{dt}$. Therefore, $\frac{dV}{dt}\Big|_{r=24}=4\pi\cdot 24^2\cdot\frac{dr}{dt}\Big|_{r=24}$. Since $\frac{dV}{dt}=-6$ cm^3/min,

    $$\frac{dr}{dt}\Big|_{r=6}=-6/(4\cdot 24^2\pi)=-1/(384\pi)\approx-8.3\times 10^{-4}\text{ cm/min}.$$

33. Let V denote the volume of the cone as it grows. We wish to find $\left.\frac{dV}{dt}\right|_{h=3}$ where h denotes the cone's height. Recall that if r is the radius of the base of the cone, then $V = \frac{1}{3}\pi r^2 \cdot h \overset{(r=3h/2)}{=} \frac{1}{3}\pi \cdot (3h/2)^2 \cdot h = 3\pi h^3/4$. Consequently, $\frac{dV}{dt} = \frac{9\pi h^2}{4} \cdot \frac{dh}{dt}$, so $\left.\frac{dV}{dt}\right|_{h=3} = \frac{81\pi}{4} \cdot \left.\frac{dh}{dt}\right|_{h=3} = 81\pi/8 \text{ m}^3/\text{min}$.

35. Let c be the distance from the car to the point on the highway that is closest to the policeman, and ℓ the distance from the car to the policeman. Then $\ell^2 = c^2 + 60^2$, so $2c\frac{dc}{dt} = 2\ell\frac{d\ell}{dt}$ and $\frac{dc}{dt} = \frac{\ell}{c} \cdot \frac{d\ell}{dt} = \frac{\ell}{\sqrt{\ell^2 - 60^2}} \cdot \frac{d\ell}{dt}$. Therefore, $\left.\frac{dc}{dt}\right|_{\ell=100} = \frac{100}{\sqrt{100^2-60^2}} \cdot \left.\frac{d\ell}{dt}\right|_{\ell=100} = \frac{100}{80} \cdot 92 = 115$ ft/s. This is approximately 78.4 mph.

37. Let V, S, and r denote the volume, surface area, and radius of the raindrop, respectively. Recall that $V = \frac{4}{3}\pi r^3$ and $S = 4\pi r^2$. Therefore, $\frac{dV}{dt} = 4\pi r^2 \frac{dr}{dt} = S \cdot \frac{dr}{dt}$. If, at the same time, $\left|\frac{dV}{dt}\right| = S$, then $\frac{dr}{dt} = -1$ (unit distance)/(unit time).

39. When $x = 3$, $2y^3 = 16$, so $y = 2$. Since $2x\frac{dx}{dt} + 6y^2\frac{dy}{dt} = 0$ and $\frac{dy}{dt} = y$, when $x = 3$, the quantity $\frac{dx}{dt}$ satisfies the equation $6 \cdot \frac{dx}{dt} + 24 \cdot 2 = 0$, so $\frac{dx}{dt} = -8$.

Further Theory and Practice

41. Let V and H denote the volume and depth of the water in the tank at time t. By similar triangles, the radius r of the circle of water at the top of the the tank is $r = \frac{3}{5}H$ implying that $V = \frac{1}{3}\pi \cdot \left(\frac{3}{5}H\right)^2 \cdot H = \frac{3}{25}\pi H^3$. Therefore, $-120 = \frac{dV}{dt} = \frac{9}{25}\pi H^2 \cdot \frac{dH}{dt}$, and $\frac{dH}{dt} = -\frac{120 \cdot 25}{9\pi \cdot H^2}$.

Since the surface of the water has area πr^2, this area is π when $r = 1$ and $H = 5/3$. Therefore,

$$\left.\frac{dH}{dt}\right|_{r=1,\ H=5/3} = -\left.\left(\frac{120 \cdot 25}{9\pi \cdot H^2}\right)\right|_{H=5/3} = -\frac{120 \cdot 25}{9\pi \cdot (25/9)} = -120/\pi \text{ cm/min}.$$

43. Since $R^{-1} = R_1^{-1} + R_2^{-1}$, $-R^{-2} \cdot \frac{dR}{dt} = -R_1^{-2} \cdot \frac{dR_1}{dt} - R_2^{-2} \cdot \frac{dR_2}{dt}$. If R remains constant, then $\frac{dR}{dt} = 0$, and R_1 and R_2 must satisfy the equation $R_1^{-2} \cdot \frac{dR_1}{dt} + R_2^{-2} \cdot \frac{dR_2}{dt} = 0$. Substituting the known values this implies that $40^{-2} \cdot (0.6) + 20^{-2} \cdot \frac{dR_2}{dt} = 0$ and $\frac{dR_2}{dt} = -(20/40)^2 \cdot (0.6) = -0.15\ \Omega/\text{s}$.

45. Because $y = \sqrt{x}$, $\frac{dy}{dt} = 1/(2\sqrt{x})\frac{dx}{dt}$. The rates of change of the coordinates x and y are equal when $1/(2\sqrt{x}) = 1$. That is, $\sqrt{x} = 1/2$ or $x = 1/4$, and $y = 1/2$. The point is $(1/4, 1/2)$.

47. The volume of the rod is $V = \pi r^2 h$ where r and h are the radius and height of the rod, respectively. Since $\frac{dh}{dt} = 2\frac{dr}{dt}$, $\frac{dV}{dt} = \pi\left(2rh\frac{dr}{dt} + r^2\frac{dh}{dt}\right) = 2\pi r\frac{dr}{dt}(h + r)$. Because $\frac{dV}{dr} = 60\pi$ for all t, when $r = 3$ and $h = 17$, $60\pi = 2\pi \cdot 3 \cdot \frac{dr}{dt} \cdot (17 + 3)$ implying that $\frac{dr}{dt} = 1/2$ cm/min.

49. Assume $t > t_0$. Since $V_0 = \pi r^2 h$, $0 = 2\pi r \frac{dr}{dt} \cdot h + \pi r^2 \frac{dh}{dt}$, and $\frac{dr}{dt} = -\frac{r}{2h} \cdot \frac{dh}{dt}$. Because $h = h_1 t^{-1/2}$, $\frac{dh}{dt} = -\frac{1}{2} h_1 t^{-3/2}$, so the rate equation for r can also be expressed in the following form.

$$r'(t) = -\frac{r(t)}{2h_1 t^{-1/2}} \cdot \frac{-h_1}{2t^{3/2}} = \frac{r(t)}{4t} = \frac{\sqrt{V_0/(\pi h(t))}}{4t} = \frac{1}{4t^{3/4}} \sqrt{\frac{V_0}{\pi h(t)}}$$

51. The volume V of the cone, in terms of its base radius r and height h, is $V = \frac{1}{3}\pi r^2 h$. Therefore, $\frac{dV}{dt} = \frac{1}{3}\pi \cdot \left(2rh\frac{dr}{dt} + r^2 \frac{dh}{dt}\right)$. Since $\frac{dr}{dt} = 6$ and $\frac{dh}{dt} = -4$, $\frac{dV}{dt}\Big|_{r=9,\, h=12} = \frac{1}{3}\pi \cdot \left(18 \cdot 12 \cdot 6 + 81 \cdot (-4)\right) = 324\pi$ cm^3/min.

CALCULATOR/COMPUTER EXERCISES

53. Let r be the radius of the ball and s be the length of one side of the cube. The sum of the two volumes is $24000 = \frac{4}{3}\pi r^3 + s^3$. Consequently, $4\pi r^2 \frac{dr}{dt} + 3s^2 \frac{ds}{dt} = 0$ so $\frac{ds}{dt} = -\frac{4}{3}\pi \cdot \frac{r^2}{s^2} \cdot \frac{dr}{dt}$. The assumption that $\frac{dr}{dt} = -2$ yields $\frac{ds}{dt} = \frac{8}{3}\pi \cdot \frac{r^2}{s^2}$.

    When the diameter of the ball is 26 cm, its radius is 13 and its volume is $\frac{3}{4}\pi \cdot 13^3$. Therefore, the volume of the cube is $24000 - \frac{4}{3}\pi \cdot 13^3$, and its side length is $s = \left(24000 - \frac{4}{3}\pi \cdot 13^3\right)^{1/3}$. This yields the following value for $\frac{ds}{dt}$ at the moment the diameter of the ball is 26 cm.

$$\frac{ds}{dt} = \frac{8}{3}\pi \cdot \frac{13^2}{\left(24000 - \frac{4}{3}\pi \cdot 13^3\right)^{2/3}} \approx 2.349 \text{ cm/sec}$$

4.2 The Mean Value Theorem

Problems for Practice

1. f attains an absolute (and local) minimum value at $c = 2$, $f(2) = 3$. There are no other extreme values.

3. The function f is even, it attains an absolute (and local) maximum value at $c = 0$, $f(0) = 1$. There are no other extreme values.

5. f attains local (and absolute) extreme values of 3 and -3 for $x = c$ when $4c = n\pi/2$, n an odd integer. That is, $c = n\pi/8$, n an odd integer. The extreme values are local maxima when $n = 4k + 1$ and local minima when $n = 4k + 3$ (k an integer).

7. Since $f'(x) = 4x - 24$, $f'(c) = 0$ only when $c = 6$. Because f is differentiable for all x, this is the only candidate for an extreme value.

9. $f'(x) = 4x^3 - 4x = 4x(x^2 - 1) = 4x(x+1)(x-1)$ so $f'(c) = 0$ when $c = -1, 0$, or $+1$. Because f is differentiable for all x, these are the only candidates for an extreme value.

11. Since $f'(x) = (x+5)^3 + 3(x-3)(x+5)^2 = (x+5)^2(x+5+3(x-3))$, $f'(c) = 0$ only when $(c+5)^2(4c-4) = 0$. That is, $c = -5$ or $c = 1$. These are the only candidates for local extreme values.

13. The function f is differentiable for all $x > 0$ and $f'(x) = 1 - 1/x = (x-1)/x$. Therefore, the only candidate for a local extreme value is $c = 1$.

15. f is differentiable for all x and $f'(x) = e^x - 1$. Therefore, the only candidate for a local extreme value is $c = 0$.

17. The graph of f has a vertical asymptote at $x = 0$. Otherwise, f is differentiable and $f'(x) = (xe^x - e^x)/x^2 = (x-1)e^x/x^2$. Therefore, the only candidate for a local extreme value is $c = 1$.

19. f is differentiable for all $x > 0$ and $f'(x) = (x \cdot (1/x) - \ln(x))/x^2 = (1 - \ln(x))/x^2$. Therefore, the only candidate for a local extreme value is $c = e$.

21. f has domain $[0, \infty)$ and is differentiable for all $x > 0$. Since $f'(x) = \frac{3}{2}x^{1/2} - \frac{9}{2}x^{-1/2} = \frac{3}{2}x^{-1/2}(x-3)$, the only candidate for a local extreme value is $c = 3$.

23. f is differentiable at each point in I. Moreover, $f(1/3) = f(3) = 10/3$. Since $f'(x) = 1 - 1/x^2$, the point $c = 1$ satisfies the equation $f'(c) = 0$.

25. f is continuous on I and differentiable at each point in $(0,1)$. Moreover, $f(0) = f(1) = 0$. Since $f'(x) = 1 - \frac{3}{2}x^{1/2}$, the point $c = 4/9$ satisfies the equation $f'(c) = 0$.

27. f is differentiable at each point in I. Moreover, $f(-2) = f(1) = 0$. Since $f'(x) = 3x^2 + 8x + 1$, the points $c_{1,2} = (-8 \pm \sqrt{64 - 12})/6 = -\frac{4}{3} \pm \frac{1}{3}\sqrt{13}$ satisfy the equation $f'(c) = 0$. The point $c_1 = -\frac{4}{3} + \frac{1}{3}\sqrt{13} \approx -0.131$ is in the open interval $(-2, 1)$.

29. f is differentiable at each point in I. Moreover, since $(f(1) - f(0))/(1-0) = 0$, there is a point c in the open interval such that $f'(c) = 0$.

31. f is differentiable on I. Moreover, $(f(32) - f(1))/(32-1) = 1/31$, so there is a point c in the open interval such that $f'(c) = 1/31$.

33. f is differentiable on I. Moreover, $(f(2) - f(-1))/(2-(-1)) = 5$, so there is a point c in the open interval such that $f'(c) = 5$.

35. $F'(x) = 5$ when $F(x) = 5x + C$.

37. $F'(x) = x^2 + \pi$ when $F(x) = \frac{1}{3}x^3 + \pi x + C$.

39. $F'(x) = \cos(x)$ when $F(x) = \sin(x) + C$.

41. $F'(x) = 3x^2$ when $F(x) = x^3 + C$. Since $F(2) = -4$, $8 + C = -4$, and $C = -12$. Therefore, $F(x) = x^3 - 12$ and $F(3) = 15$.

43. $F'(x) = 4/x$ when $F(x) = 4\ln(x) + C$. Since $F(e^2) = 7$, $C = -1$. Therefore, $F(x) = 4\ln(x) - 1$ and $F(e^3) = 11$.

45. $F'(x) = e^x$ when $F(x) = e^x + C$. Since $F(2) = 2 + e^2$, $C = 2$. Therefore, $F(x) = e^x + 2$ and $F(3) = e^3 + 2$.

47. $F'(x) = 2/x^2$ when $F(x) = -2/x + C$. Since $F(-1) = 4$, $C = 2$. Therefore, $F(x) = -2/x + 2$ and $F(4) = 3/2$.

Further Theory and Practice

49. Since $f'(x) = e^x\sin(x) + e^x\cos(x) = e^x(\sin(x) + \cos(x))$, $f'(c) = 0$ when $c = 3\pi/4$.

51. Since $f'(x) = \cos(x) - \sin(x)$, $f'(c) = 0$ when $c = \pi/4$.

53. $f'(x) = -1/(x-1)^2$ and $\frac{f(b)-f(a)}{b-a} = -1/3$. Therefore, $f'(c) = -1/3$ when $(c-1)^2 = 3$. That is, $c = 1 \pm \sqrt{3}$. The answer is $c = 1 + \sqrt{3}$.

55. $f'(x) = 3x^2 + 3$ and $\frac{f(7)-f(1)}{7-1} = 60$. Therefore, $f'(c) = 60$ when $c^2 = 19$. That is, $c = \sqrt{19}$.

57. The graph of f is displayed on the right. $f'(c)$ is never equal to the slope of the line joining the endpoints of of the graph.

The Mean Value Theorem is not contradicted because, although f is continuous on $[4, 7]$, it is not differentiable at all points in $(4, 7)$.

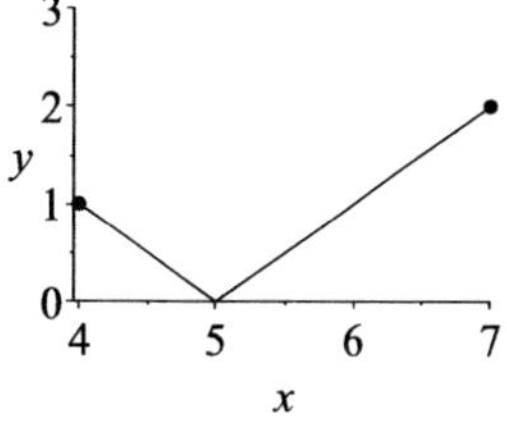

59. The graph of f is displayed on the right. $f'(c)$ is never equal to the slope of the line joining the endpoints of of the graph.

The Mean Value Theorem is not contradicted because, although f is continuous on $[-1, 1]$, it is not differentiable at all points in $(-1, 1)$.

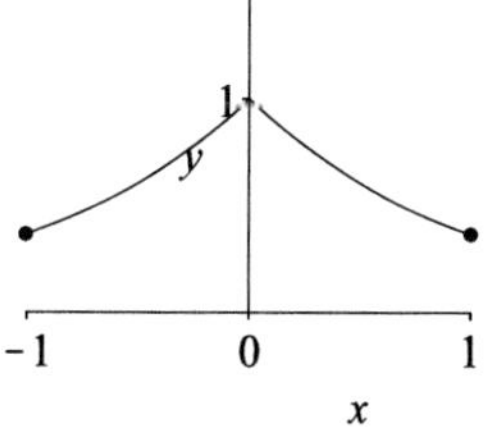

61. $f'(x) = -2\cos(x)\sin(x) + 2\sin(x)\cos(x) = 0$ for all x. Therefore, $f(x) = C$ for some constant C (Theorem 4). Since $f(0) = 1$, $C = 1$, and $f(x) = 1$ for all x. That is, $\cos^2(x) + \sin^2(x) = 1$ for all x.

63. Let $f(x) = 3x^4 - 4x^3 + 6x^2 - 12x + 5$. Suppose there are three points $a < b < c$ on the x-axis where $f(a) = f(b) = f(c) = 0$. By Rolle's Theorem, the derivative of f has at least two real roots, one in (a, b) and another in (b, c). However, $f'(x) = 12x^3 - 12x^2 + 12x - 12 = 12(x^3 - x^2 + x - 1)$

which factors further as $f'(x) = 12(x-1)(x^2+1)$ and, since the roots of x^2+1 are complex, f' has only one real root.

65. Let $f(x) = x^9 + 3x^3 + 2x + 1$. Since $f(0) = 1$ and $f(-1) = -5$, f has a real root in the interval $(-1, 0)$. If f were to have another real root then, by Rolle's Theorem, its derivative $f'(x) = 9x^8 + 9x^2 + 2$ would have at least one real root. However, $f'(x) \geq 2$ for all x, and f' has no real roots.

67. Since $p(x) = x^3 + ax^2 + b$, $p'(x) = 3x^2 + 2ax = x(3x + 2a)$. Therefore, $p'(c) = 0$ when $c = 0$ or $c = -2a/3$, a positive number. Consequently, p cannot have two negative roots for if it did, then p' would have a negative root by Rolle's Theorem.

69. If p were to have two real roots then, by Rolle's Theorem, $p'(x) = 3x^2 + a$ would have at least one real root. However, because $a > 0$, both roots of p' are complex.

71. Assume $1 < p$ and, for $x \geq 0$, let $g(x) = 1 + px$ and $f(x) = (1+x)^p$. Then both f and g are continuous, $g(0) = f(0)$, and for all $0 < x$,

$$g'(x) = p < p(1+x)^{p-1} = f'(x).$$

Consequently, by Exercise 70, $g(x) < f(x)$ for all $0 < x$. That is, $1 + px < (1+x)^p$ when $0 < x$.

73. Let $f(x) = x - \sin(x)$. Then f is differentiable and $f(0) = 0$. If $0 < x$, the Mean Value Theorem implies that there is a point $c \in (0, x)$ such that $f'(c) = (f(x) - f(0))/(x - 0)$. That is, $f(x) = x \cdot f'(c) = x \cdot (1 - \sin(c)) = (+) \cdot (+) > 0$. Therefore, $x > \sin(x)$ when $0 < x$. When $x < 0$, $0 < -x$ and $-x < \sin(-x) = -\sin(x)$. That is, $-x < -\sin(x)$ or, equivalently, $x > \sin(x)$.

75. Let $f(x) = e^{-kx}p(x)$ and observe that $f(a) = f(b) = 0$. Apply Rolle's Theorem to f on the interval $[a, b]$ to obtain a point $c \in (a, b)$ such that $f'(c) = 0$. That is, $e^{-kc}p'(c) - ke^{-kc}p(c) = 0$. The nonzero term e^{-kc} can be cancelled yielding $p'(c) = kp(c)$.

77. Let $f(x) = \sqrt{x}$ and apply the Mean Value Theorem to f on the interval $[x, x+1]$ where $0 < x$. Since $f'(x) = 1/(2\sqrt{x})$, this yields a point c_x in $(x, x+1)$ such that $(f(x+1) - f(x))/(x+1-x) = 1/(2\sqrt{c_x})$. Consequently, for all $x > 0$,

$$0 < \sqrt{x+1} - \sqrt{x} = \frac{1}{2\sqrt{c_x}} < \frac{1}{2\sqrt{x}},$$

and $\lim_{x\to+\infty}(\sqrt{x+1} - \sqrt{x}) = 0$ by the Pinching Theorem.

79. Assume $t < s$ and apply the Mean Value Theorem to f on $[t, s]$ to obtain a point $x \in (t, s)$ such that $(f(s) - f(t))/(s - t) = f'(x)$. That is, $f(s) - f(t) = f'(x)(s-t)$ implying that $|f(s) - f(t)| = |f'(x)| \cdot |s-t| \leq C_1 \cdot |s-t|$. A similar argument will handle the case $s < t$ and, if $s = t$, the inequality is obviously true.

81. Let $F(x) = 2 \cdot \arctan(\sqrt{x})$ and $G(x) = \arcsin\left(\frac{x-1}{x+1}\right)$, $x \geq 0$. Then $F'(x) = 2 \cdot \frac{1}{1+(\sqrt{x})^2} \cdot (1/(2\sqrt{x})) = \frac{1/\sqrt{x}}{1+x} = \frac{1}{\sqrt{x}+x\sqrt{x}}$ and

$$G'(x) = \frac{1}{\sqrt{1-\left(\frac{x-1}{x+1}\right)^2}} \cdot \left(\frac{(x+1)-(x-1)}{(x+1)^2}\right) = \frac{2}{(x+1)\sqrt{(x+1)^2-(x-1)^2}},$$

which simplifies to $G'(x) = \dfrac{2}{(x+1)\sqrt{4x}} = \dfrac{1}{x\sqrt{x}+\sqrt{x}}$. Thus $F'(x) = G'(x)$ for all $x > 0$. This implies, by Theorem 5, that there is a constant C such that $F(x) = G(x) + C$ for all $x \geq 0$. Since $F(1) - G(1) = \pi/2 - 0$, $C = \pi/2$.

83. Observe that $g(a_0) = f(a_0 + (b_0 - a_0)/2) - f(a_0) = f((a_0 + b_0)/2) - f(a_0)$ and $g((a_0 + b_0)/2) = f((a_0 + b_0)/2 + (b_0 - a_0)/2) - f((a_0 + b_0)/2) = f(b_0) - f((a_0 + b_0)/2) = -g(a_0)$. If $g(a_0) = 0$, let $a_1 = a_0$. Otherwise, the Intermediate Value Theorem implies that there is a point $a_1 \in \left(a_0, \frac{a_0+b_0}{2}\right)$ such that $g(a_1) = 0$. In either event, the fact that $g(a_1) = 0$ implies that $f(a_1 + (b_0 - a_0)/2) - f(a_1) = 0$. That is, $f(a_1) = f(b_1)$ where $b_1 = a_1 + (b_0 - a_0)/2$.

This construction can be continued indefinitely to produce the sequence $\{[a_n, b_n]\}_{n=0}^{\infty}$ of closed intervals as described in the text. The sequences $\{a_n\}$ and $\{b_n\}$ are both bounded and monotone, so they converge. Moreover, the fact that $|b_n - a_n| = |b_{n-1} - a_{n-1}|/2$ for all n implies (by the Pinching Theorem) that $\lim_{n\to\infty} a_n = \lim_{n\to\infty} b_n = c$. Since $a_n \leq c < b_n$ for all n, $a < a_1 \leq c$, and $b_1 < b$, $c \in (a, b)$.

Since $f(a_n) = f(b_n)$ for all n, $\lim_{n\to\infty} \frac{f(b_n)-f(a_n)}{b_n-a_n} = 0$. Consequently, $f'(c) = 0$ also. This is because $f'(c)$ cannot be positive (or negative). Suppose, for example, that $f'(c) > 0$. Then there is a $\delta > 0$ such that for all $0 < |x - c| < \delta$, $(f(x) - f(c))/(x - c) > 0$ implying that $f(x) - f(c) > 0$ when $x > c$ and $f(x) < f(c)$ when $x < c$. However, for n sufficiently large, a_n and b_n are both within δ of c. Since $b_n > c$, $f(b_n) > f(c)$, and since $c \geq a_n$, $f(c) \geq f(a_n)$, contradicting the fact that $f(a_n) = f(b_n)$. A similar argument will show that $f'(c) < 0$ is also not possible.

85. Because Φ is continuous, if $\lim_{n\to\infty} x_n = \xi$, then $\Phi(\xi) = \Phi(\lim_{n\to\infty} x_n) = \lim_{n\to\infty} \Phi(x_n) = \lim_{n\to\infty} x_{n+1} = \xi$.

87. By the Mean Value Theorem, there is point z between x_* and x_0 such that $\frac{\Phi(x_0)-\Phi(x_*)}{x_0-x_*} = \Phi'(z)$. Therefore,

$$\Phi(x_0) - \Phi(x_*) = \Phi'(z)(x_0 - x_*),$$

and $|\Phi(x_0) - \Phi(x_*)| = |\Phi'(z)||x_0 - x_*| < K|x_0 - x_*| < |x_0 - x_*| < \delta$. See the picture, the dotted line is the graph of $y = x$.

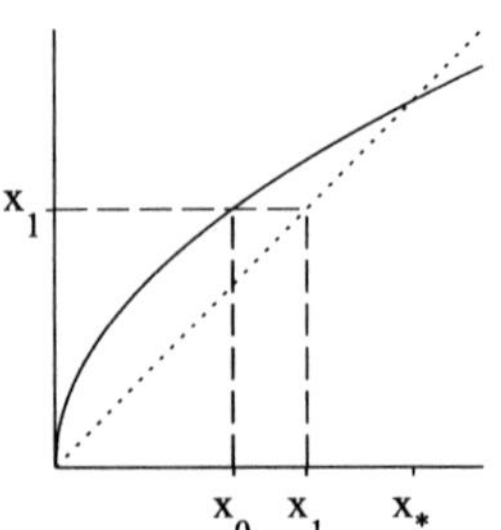

89. Under the assumptions stated in the Exercise, if x_0 is within δ (obtained in Exercise 86) of the equilibrium point x_*, then for every n, $|x_n - x_*| \leq K^n|x_0 - x_*|$ which, because $0 < K < 1$, approaches 0 as $n \to \infty$. By the Pinching Theorem, $\lim_{n\to\infty} x_n = x_*$.

CALCULATOR/COMPUTER EXERCISES

91. The derivative is plotted below on the left. It appears that the function will have extreme values at $x = 1$ and $x \approx -0.6$. This is confirmed by the picture on the right showing the function (solid curve) and derivative (dashed curve) together. There is a local maximum near $x = -0.6$ and a local minimum near $x = 1$.

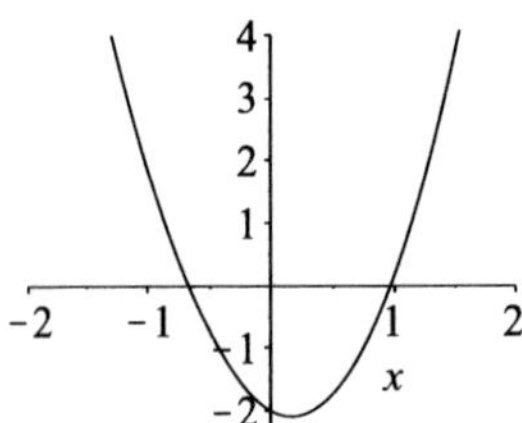

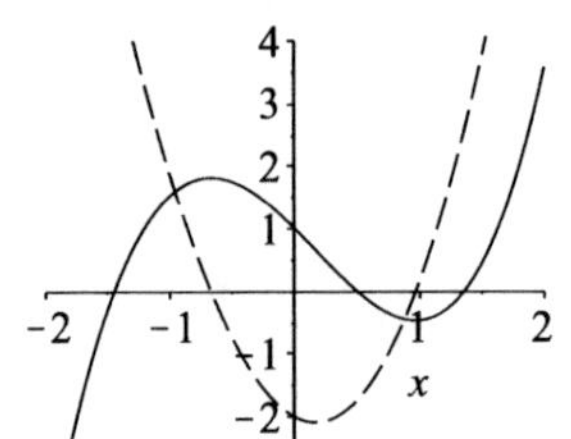

93. The derivative is plotted below on the left. It appears that the function will have extreme values near $x = -1.5$ and $x = -0.2$. This is confirmed by the picture on the right showing the function (solid curve) and derivative (dashed curve) together. There is a local maximum near $x = -1.5$ and a local minimum near $x = -0.2$.

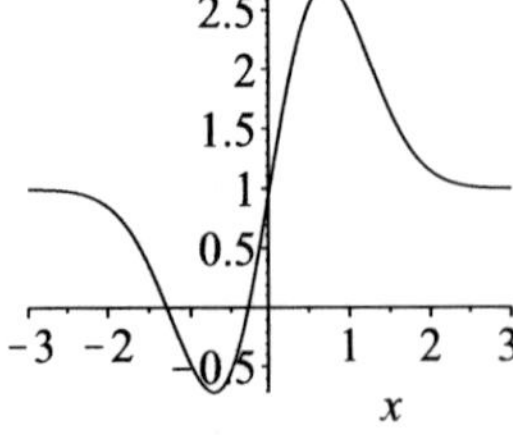

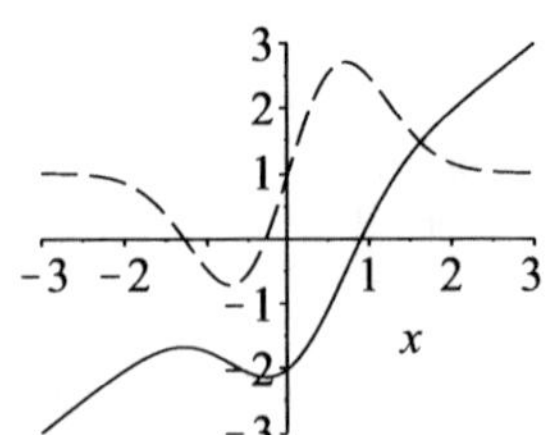

95. The value c guaranteed by the Mean Value Theorem must satisfy $f'(c) = (f(2) - f(0))/(2-0) = 20$. Using *Maple* this is $c \approx 1.312$. See the picture on the right.

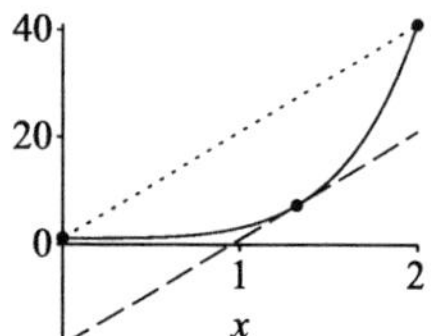

97. The value c guaranteed by the Mean Value Theorem must satisfy $f'(c) = (f(3) - f(-1))/(3-(-1)) = 6$. Using *Maple* this is $c \approx 1.826$. See the picture on the right.

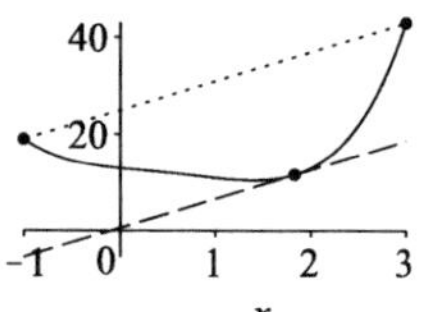

4.3 Maxima and Minima of Functions

Problems for Practice

1. Since $f'(x) = 2x + 1$ there is one critical point, $c = -1/2$. If $x < -1/2$, then $f'(x) < 0$ and f is decreasing on $(-\infty, -1/2)$. If $x > -1/2$, then $f'(x) > 0$ and f is increasing on $(-1/2, \infty)$. $f(c)$ is a local (and absolute) minimum value.

3. Since $f'(x) = 3x^2 + 6x - 45 = 3(x+5)(x-3)$ there are two critical points, $c_1 = -5$ and $c_2 = 3$. If $x < -5$, then $f'(x) > 0$ and f is increasing on $(-\infty, -5)$. If $-5 < x < 3$, then $f'(x) < 0$ and f is decreasing on $(-5, 3)$. If $x > 3$, then $f'(x) > 0$ and f is increasing on $(3, \infty)$. $f(-5)$ is a local maximum value and $f(3)$ is a local minimum value.

5. Since $f'(x) = 12x^3 + 60x^2 + 72x = 12x(x+3)(x+2)$ there are three critical points, $c_1 = -3$, $c_2 = -2$, and $c_3 = 0$. If $x < -3$, then $f'(x) < 0$ and f is decreasing on $(-\infty, -3)$. If $-3 < x < -2$, then $f'(x) > 0$ and f is increasing on $(-3, -2)$. If $-2 < x < 0$, then $f'(x) < 0$ and f is decreasing on $(-2, 0)$. If $x > 0$, then $f'(x) > 0$ and f is increasing on $(0, \infty)$. $f(-3)$ is a local minimum value, $f(-2)$ a local maximum, and $f(0)$ a local minimum value.

7. Since $f'(x) = (x^2 - 25)/(x^2+25)^2 = (x+5)(x-5)/(x^2+25)$ there are two critical points, $c_1 = -5$ and $c_2 = 5$. If $x < -5$, then $f'(x) > 0$ and f is increasing on $(-\infty, -5)$. If $-5 < x < 5$, then $f'(x) < 0$ and f is decreasing on $(-5, 5)$. If $x > 5$, then $f'(x) > 0$ and f is increasing on $(5, \infty)$. $f(-5)$ is a local maximum value and $f(5)$ is a local minimum value.

9. Since $f'(x) = -2/(x-1)^2$, f has no critical points. $f'(x) < 0$ for all $x \neq 1$. f is continuous and decreasing on the interval $(-\infty, 1)$ and continuous and decreasing on the interval $(1, \infty)$. f has no local extreme values.

11. Since $f'(x) = (x+2)(3x+4)$ there are two critical points, $c_1 = -2$ and $c_2 = -4/3$. If $x < -2$, then $f'(x) > 0$ and f is increasing on $(-\infty, -2)$. If $-2 < x < -4/3$, then $f'(x) < 0$ and f is decreasing on $(-2, -4/3)$. If $x > -4/3$, then $f'(x) > 0$ and f is increasing on $(-4/3, \infty)$. $f(-2)$ is a local maximum value and $f(-4/3)$ is a local minimum value.

13. Since $f'(x) = (x+3)(x+1)/(x+2)^2$ there are two critical points, $c_1 = -3$ and $c_2 = -1$, and f is undefined at $x = -2$. If $x < -3$, then $f'(x) > 0$ and f is increasing on $(-\infty, -3)$. If $-3 < x < -2$, then $f'(x) < 0$ and f is decreasing on $(-3, -2)$. If $-2 < x < -1$, then $f'(x) < 0$ and f is decreasing on $(-2, -1)$. If $x > -1$, then $f'(x) > 0$ and f is increasing on $(-1, \infty)$. $f(-3)$ is a local maximum value and $f(-1)$ a local minimum value.

15. Since $f'(x) = 4 - 9/x^2 = (2x+3)(2x-3)/x^2$ there are two critical points, $c_1 = -3/2$ and $c_2 = 3/2$, and f is undefined at $x = 0$. If $x < -3/2$, then $f'(x) > 0$ and f is increasing on $(-\infty, -3/2)$. If $-3/2 < x < 0$, then $f'(x) < 0$ and f is decreasing on $(-3/2, 0)$. If $0 < x < 3/2$, then $f'(x) < 0$ and f is decreasing on $(0, 3/2)$. If $x > 3/2$, then $f'(x) > 0$ and f is increasing on $(3/2, \infty)$. $f(-3/2)$ is a local maximum value and $f(3/2)$ a local minimum value.

17. Since $f'(x) = 3 + 2\cos(x)$ there are no critical points. $f'(x) > 0$ for all x and f is increasing on $(-\infty, \infty)$.

19. Since $f'(x) = e^x - 1$ there is one critical point, $c = 0$. If $x < 0$, then $f'(x) < 0$ and f is decreasing on $(-\infty, 0)$. If $x > 0$, then $f'(x) > 0$ and f is increasing on $(0, \infty)$. $f(0)$ is a local (and absolute) minimum value.

21. Since $f'(x) = -x(x-2)e^{-x}$ there are two critical points, $c_1 = 0$ and $c_2 = 2$. If $x < 0$, then $f'(x) < 0$ and f is decreasing on $(-\infty, 0)$. If $0 < x < 2$, then $f'(x) > 0$ and f is increasing on $(0, 2)$. If $x > 2$, then $f'(x) < 0$ and f is decreasing on $(2, \infty)$. $f(0)$ is a local minimum value and $f(2)$ a local maximum value.

23. Since $f'(x) = 2^x \ln(2) - 1$, there is one critical point, c, where $2^c = 1/\ln(2)$. That is, $c\ln(2) = \ln(1/\ln(2))$, so $c = -\ln(\ln(2))/\ln(2) \approx 0.529$. If $x < c$, then $f'(x) < 0$ and f is decreasing on $(-\infty, c)$. If $x > c$, then $f'(x) > 0$ and f is increasing on (c, ∞). $f(c)$ is a local (and absolute) minimum value.

25. Since $f'(x) = \log_2(x) + x \cdot \frac{1}{x\ln(2)} = \frac{\ln(x)}{\ln(2)} + \frac{1}{\ln(2)}$, there is one critical point, c, where $\ln(c) = -1$. That is, $c = e^{-1}$. If $0 < x < 1/e$, then $f'(x) < 0$ and f is decreasing on $(0, 1/e)$. If $x > 1/e$, then $f'(x) > 0$ and f is increasing on $(1/e, \infty)$. $f(1/e)$ is a local minimum value.

27. Since $f'(x) = (\ln(x) - x \cdot (1/x))/\ln^2(x) = (\ln(x) - 1)/\ln^2(x)$, there is one critical point, $c = e$, and $x = 1$ is not in the domain of f. If $0 < x < 1$, then $f'(x) < 0$ and f is decreasing on $(0, 1)$. If $1 < x < e$, then $f'(x) < 0$

and f is decreasing on $(1, e)$. If $x > e$, then $f'(x) > 0$ and f is increasing on (e, ∞). $f(e)$ is a local minimum value.

29. $f'(x) = (x+1)^{3/2} + \frac{3}{2}x(x+1)^{1/2} = \sqrt{x+1}\cdot(\frac{5}{2}x+1)$. f is differentiable with a critical point at $c = -2/5$ where $f'(c) = 0$. Since $f'(x) < 0$ when $-1 < x < -2/5$ and $f'(x) > 0$ when $-2/5 < x$, f is decreasing to the left of $-2/5$ and increasing to the right. $f(-2/5)$ is a local (and absolute) minimum value for f.

31. $f'(x) = \frac{1}{5}x^{-4/5}(x-12) + x^{1/5} = \frac{(x-12)+5x}{5x^{4/5}} = \frac{6(x-2)}{5x^{4/5}}$. Consequently, f has critical points at $c_1 = 0$, where $f'(c_1)$ does not exist, and at $c_2 = 2$, where $f'(c_2) = 0$. Since $f'(x) < 0$ when $x < 2$ $(x \neq 0)$ and $f'(x) > 0$ when $x > 2$, $f(2)$ is a local minimum value. Because $f'(x) < 0$ near $x = 0$, f is decreasing to the left of 0 and decreasing to the right, so $f(0)$ is not a local extreme value.

33. $f'(x) = 3 - \frac{1}{3}x^{-2/3}$. Consequently, f has critical points at c where $c^{-2/3} = 9$ and at $c_0 = 0$. $f'(0)$ does not exist and $f'(c) = 0$ when $c^{-2/3} = 9$, that is, $c = 9^{-3/2}$ yielding $c_1 = -1/27$ and $c_2 = 1/27$. If $x < -1/27$, then $f'(x) > 0$ (substitute, for example, $x = -1$) and if $-1/27 < x < 1/27$ $(x \neq 0)$, $f'(x) < 0$. Therefore, $f(-1/27)$ is a local maximum value and $f(0)$ is not an extreme value. Since $f'(x) > 0$ when $x > 1/27$, $f(1/27)$ is a local minimum value.

35. $f'(x) = \frac{2}{3}x^{-1/3} - \frac{4}{3}x^{-2/3} = \frac{2}{3}x^{-2/3}\cdot(x^{1/3} - 2)$. Consequently, f has a critical point at $c = 0$ where $f'(0)$ does not exist and at $c = 8$, where $f'(8) = 0$. Since $f'(x) < 0$ when $x < 8$ $(x \neq 0)$, $f(0)$ is not an extreme value. However, since $f'(x) > 0$ when $x > 8$, $f(8)$ is a local minimum value.

37. $$\begin{aligned} f'(x) &= \tfrac{1}{3}(x-1)^{-2/3}(x+5)^{2/3} + \tfrac{2}{3}(x-1)^{1/3}(x+5)^{-1/3} \\ &= \tfrac{1}{3}(x-1)^{-2/3}(x+5)^{-1/3}(x+5+2(x-1)) = \frac{x+1}{(x-1)^{2/3}(x+5)^{1/3}}. \end{aligned}$$

Consequently, f has critical points at $c_1 = -5$ and $c_2 = 1$ where $f'(0)$ does not exist and at $c_0 = -1$ where $f'(-1) = 0$. If $x < -5$, $f'(x) > 0$ and if $-5 < x < -1$, $f'(x) < 0$. Therefore, $f(-5)$ is a local maximum value. If $-1 < x < 1$, $f'(x) > 0$ so and $f(-1)$ is a local minimum value. Since $f'(x) > 0$ when $x > 1$, $f(1)$ is not an extreme value.

39. $$\begin{aligned} f'(x) &= \tfrac{1}{5}(x-4)^{-4/5}(3x+1)^{2/3} + 2(x-4)^{1/5}(3x+1)^{-1/3} \\ &= \tfrac{1}{5}(x-4)^{-4/5}(3x+1)^{-1/3}(3x+1+10(x-4)) = \frac{13(x-3)}{5(x-4)^{4/5}(3x+1)^{1/3}}. \end{aligned}$$

Consequently, f has critical points at $c_1 = 3$ where $f'(3) = 0$ and at $c_2 = -1/3$ and $c_3 = 4$ where f' does not exist. $f'(x)$ changes sign from $(+)$ to $(-)$ at $-1/3$ so $f(-1/3)$ is a local maximum value. $f'(x)$ changes sign from $(-)$ to $(+)$ at 3 so $f(3)$ is a local minimum value. Since $f'(x)$ does not change sign at 4, $f(4)$ is not an extreme value.

41. f is increasing on the intervals where f' is positive. That is, $(-1.5, -1.0)$, $(0.0, 1.5)$, and $(2.0, 2.5)$. f is decreasing on the intervals where f' is negative. That is, $(-1.0, 0.0)$ and $(1.5, 2.0)$.

Further Theory and Practice

43. $f'(x) = 2^{x^2}\ln(2)\cdot 2x - 4x = 2x(2^{x^2}\ln(2) - 2)$. Consequently, f has critical points at $c_0 = 0$ and those c values where $2^{c^2} = 2/\ln(2)$. That is, $c^2\ln(2) = \ln(2/\ln(2))$ or $c_{1,2} = \pm\sqrt{\ln(2/\ln(2))/\ln(2)} \approx \pm 1.236$. The sign of $f'(x)$ changes as x passes through all three of these points, from $(-)$ to $(+)$ to $(-)$ to $(+)$, implying that f has local minima at c_1 and c_2 and a local maximum at 0.

45. Since $x \mapsto |8-3x|$ is not differentiable at $8/3$, f has a critical point at $c_1 = 8/3$. If $x > 8/3$, then $f(x) = 5x - (3x-8) = 2x + 8$ so $f'(x) = 2$ and f has no critical points when $x > 8/3$. If $x < 8/3$, then $f(x) = 5x - (8-3x) = 8x - 8$ so $f'(x) = 8$ and f has no critical points when $x < 8/3$. Since f' does not change sign at $x = 8/3$, $f(8/3)$ is not an extreme value.

47. Neither $x^{1/3}$ nor $|x|$ is differentiable at 0. Neither is f and it has a critical point at $c_0 = 0$. This is not the only critical point. If $x > 0$, then $f(x) = x^{1/3} - x$ so $f'(x) = \frac{1}{3}x^{-2/3} - 1$ and $f'(c) = 0$ when $c^{-2/3} = 3$, that is, $c = 3^{-3/2} = 1/(3\sqrt{3}) = \sqrt{3}/9$. If $x < 0$, $f(x) = x^{1/3} + x$ and $f'(x) = \frac{1}{3}x^{-2/3} + 1 > 0$, so f has no critical points when $x < 0$. Since f is increasing when $x < 0$, and also when $0 < x < \sqrt{3}/9$ (verify), $f(0)$ is not an extreme value. However, $f'(x) < 0$ if $x > \sqrt{3}/9$, (substitute $x = 1$, for example) and $f(\sqrt{3}/9)$ is a local maximum value.

49. $f'(x) = 2x\sqrt{b^2-x^2} + \frac{1}{2}x^2\cdot(-2x/\sqrt{b^2-x^2}\,) = \frac{2x(b^2-x^2)-x^3}{\sqrt{b^2-x^2}} = \frac{x(2b^2-3x^2)}{\sqrt{b^2-x^2}}$.
Therefore, f has a critical point at $c = \sqrt{2/3}\cdot b$. This is a local maximum for f since $f'(x) > 0$ when $x < c$ and $f'(x) < 0$ when $c < x < b$.

51. $f'(x) = \frac{(b-x)\cdot 2x - (x^2+3b^2)\cdot(-1)}{(b-x)^2} = \frac{-x^2+2bx+3b^2}{(b-x)^2} = -\frac{(x+b)(x-3b)}{(b-x)^2}$.

    If $b > 0$, then $c = -b$ is a critical point that is less than b and $f(-b)$ is a local minimum value because $f'(x)$ changes sign from $(-)$ to $(+)$ at $-b$.

    If $b < 0$, then $c = 3b$ is a critical point that is less than b and $f(3b)$ is also a local minimum value because $f'(x)$ changes sign from $(-)$ to $(+)$ at $3b$.

    If $b = 0$, then $f(x) = -x$ which has no critical points.

53. $f'(x) = \frac{(x^2+b^2)\cdot 4x^3 - (x^4+8b^4)\cdot 2x}{(x^2+b^2)^2} = \frac{2x(x^4+2b^2x^2-8b^4)}{(x^2+b^2)^2} = \frac{2x(x^2+4b^2)(x^2-2b^2)}{(x^2+b^2)^2}$.

    If $b > 0$, then $c = \sqrt{2}\,b$ is a critical point that is greater than b and $f(\sqrt{2}\,b)$ is a local minimum value because $f'(x)$ changes sign from $(-)$ to $(+)$ at $\sqrt{2}\,b$.

If $b < 0$, then $c_0 = 0$ and $c_1 = -\sqrt{2}\,b$ are critical points that are greater than b. $f(0)$ is a local maximum value because $f'(x)$ changes sign from $(+)$ to $(-)$ at 0. $f(-\sqrt{2}\,b)$ is a local minimum value because $f'(x)$ changes sign from $(-)$ to $(+)$ at $-\sqrt{2}\,b$.

If $b = 0$, then $f(x) = x^2$ which has no critical points when $0 < x$.

55. $f'(x) = 1/\sqrt{1-x^2} - 1/(2\sqrt{x}) = \frac{2\sqrt{x}-\sqrt{1-x^2}}{2\sqrt{x}\,\sqrt{1-x^2}}$, so f has critical points at those values c such that $2\sqrt{c} = \sqrt{1-c^2}$. That is, $4c = 1 - c^2$, or $c^2 + 4c - 1 = 0$, implying that $c = (-4 \pm \sqrt{20}\,)/2 = -2 \pm \sqrt{5}$. Only $\sqrt{5} - 2$ is in the domain of f. Because f' changes sign: $(-)(+)$, at $\sqrt{5} - 2$, this is a local minimum.

57. Convert to exponentials: $f(x) = \frac{1}{2}(e^x - e^{-x}) + e^x + e^{-x} = \frac{3}{2}e^x + \frac{1}{2}e^{-x}$. Consequently, $f'(x) = \frac{3}{2}e^x - \frac{1}{2}e^{-x}$, and critical points c satisfy the equation $3e^c = e^{-c}$. That is, $e^{2c} = 1/3$ so $2c = \ln(1/3)$ and $c = -\ln(3)/2 \approx -0.549$. This is a local minimum for f because $f'(x)$ changes sign from $(-)$ to $(+)$ at $-\ln(3)/2$.

59. $f'(x) = 2/\sqrt{x^2+1} - 1/x = \frac{2x - \sqrt{x^2+1}}{x\sqrt{x^2+1}}$. Consequently, f has critical points at values c such that $2c = \sqrt{c^2+1}$. This implies that $4c^2 = c^2 + 1$ so $3c^2 = 1$ and $c = \pm\sqrt{1/3} = \pm\sqrt{3}/3$. Since the domain of f is $x \geq 0$, $c = \sqrt{3}/3 \approx 0.577$. This is a local minimum for f because f' changes sign: $(-)(+)$ at $\sqrt{3}/3$.

61. Not necessarily. The fact that $f'(2) > 0$ implies that $f(x) > f(2)$ for all $x > 2$ that are sufficiently close to 2, but 2.000001 may not be close enough.

63. Suppose $f' > 0$. If $a < b$, then f is increasing on an open interval containing $[a, b]$. This implies that $f(a) < f(b)$ so $f(a) \neq f(b)$. If $f' < 0$, then $-f' > 0$ so $-f$ is one-to-one, implying that f is one-to-one also.

65. If f is increasing on an interval I, then $f'(x) > 0$ at all x in the interior of I. Since $\frac{d}{dx}(f^3(x)) = 3f^2(x) \cdot f'(x)$, f^3 has a positive derivative at all points in I where $f(x) \neq 0$. Since there can be at most one point with this property, f^3 must be increasing on I.

67. Let f and g be increasing with the image of f contained in the domain of g. If $a < b$ in the domain of f, then $f(a) < f(b)$ in the domain of g, so $g(f(a)) < g(f(b))$. That is, $g \circ f(a) < g \circ f(b)$, and $g \circ f$ is increasing.

If f and g have positive derivatives. Then $(g \circ f)'(x) = g'(f(x)) \cdot f'(x) > 0$ for all x in the domain of $g \circ f$ so $g \circ f$ is increasing.

69. $f(x) = a(x^2 + (b/a)x) + c = a(x + b/(2a))^2 + c - b^2/(4a)$. If $a > 0$, then $f(x)$ is decreasing on $(-\infty, -b/2a)$ and increasing on $(-b/2a, \infty)$, attaining a minimum value at $-b/2a$. If $a < 0$, then $f(x)$ is increasing on

$(-\infty, -b/2a)$ and decreasing on $(-b/2a, \infty)$, attaining a maximum value at $-b/2a$.

Using calculus, $f'(x) = 2ax + b$ so f has a critical point at $-b/2a$. If $a > 0$, then f' changes sign: $(-)(+)$, at $-b/2a$ and f is decreasing to the left of $-b/2a$ and increasing to the right. If $a < 0$, then f' changes sign: $(+)(-)$, at $-b/2a$ and f is increasing to the left of $-b/2a$ and decreasing to the right.

71. Using Leibniz' derivative formula,

$$f'(x) = (x-b)q(x) + (x-a)q(x) + (x-a)(x-b)q'(x).$$

Therefore, $f'(a)f'(b) = (a-b)q(a) \cdot (b-a)q(b) = (-) \cdot (+) \cdot (+) \cdot (+) < 0$.

73. The marginal product of labor is a good approximation to the rate of change of the output with respect to an increase in the labor input. In general, this function increases at a smaller and smaller rate with increasing intensity of labor. This is illustrated by the tangent lines in the first graph, the incremental changes in the second graph, and the decreasing curve in the third graph corresponding to the decreasing slopes in the first graph.

75. If f has a local maximum at c, then $1/f$ has a local minimum there. The local minima become local maxima.

77. Because $\frac{dT}{dt} < 0$, the white dwarfs are cooling. Because the exponent on T is greater than 1, as T decreases, so does $T^{7/2}$. Therefore, the rate at which the white dwarf's temperature decreases is becoming less negative. That is, it is changing more slowly.

79. If $y > 0$, then $\frac{d}{dy}(ye^y) = e^y + ye^y = (1+y)e^y > 0$.

Differentiate the equation $W(x)\exp(W(x)) = x$ with respect to x to obtain

$$W'(x)\exp(W(x)) + W(x)\exp(W(x)) \cdot W'(x) = 1.$$

This implies that $W'(x) = \exp(-W(x))/(1+W(x))$, which is positive when $W(x) > 0$.

Since $\exp(W(x)) = x/W(x)$, $\exp(-W(x)) = W(x)/x$. Substitute for $\exp(-W(x))$ in the formula for $W'(x)$ in the preceding paragraph to obtain $W'(x) = W(x)/(x(1+W(x))$.

81. (a) g attains a minumum value at some point in $[a, b]$ because it is continuous. The minimum cannot occur at a because $g'(a) = f'(a) - \gamma < 0$ implying that $g(x) < g(a)$ for x to the right of, but near to, a. It cannot occur at b because $g'(b) = f'(b) - \gamma > 0$ implying that $g(x) < g(b)$ for x to the left of, but near to, b.

    (b) At point c where g attains its absolute minimum value, $g'(c) = 0$. That is, $f'(c) - \gamma = 0$ and $f'(c) = \gamma$.

(c) If f increases on (l, c) and decreases on (c, r), then $f' > 0$ on (l, c) and $f' < 0$ on (c, r). By Darboux's Theorem, f' must attain the value 0 on (l, r) and it can only be at c.

CALCULATOR/COMPUTER EXERCISES

83. Using *Maple*, f' has one real root, $c = 2.344225271$, and two complex roots. Evaluating f' to the left and right of c, f is decreasing for $x < c$ and increasing for $x > c$.

85. Using *Maple*'s *fsolve* procedure we find that f has critical points at $c_1 = -1.336311215$, $c_2 = 0.3449001906$, and $c_3 = 2.813363352$. Evaluating f' at intermediate points, we find that f is decreasing on $(-\infty, c_1)$, increasing on (c_1, c_2), decreasing in (c_2, c_3), and increasing on (c_3, ∞).

87. Using *Maple*'s *fsolve* procedure we find that f has critical points at $c_1 = -4.047100313$, $c_2 = -0.4691592058$, and $c_3 = 0.8385322872$. Evaluating f' at intermediate points, we find that f is decreasing on $(-\infty, c_1)$, increasing on (c_1, c_2), decreasing in (c_2, c_3), and increasing on (c_3, ∞).

89. Using *Maple*'s *fsolve* procedure, f has critical points, $c_1 = 0$, $c_2 = 0.3683525321$, and $c_3 = 215.9861099$. Evaluation of f' at intermediate points shows that f is increasing on $(-\infty, c_2)$, decreasing on (c_2, ∞), and increasing on (c_3, ∞).

91. $f'(x) = x^2 - 2\sin(x+1)\cos(x+1)$, which simplifies to $f'(x) = x^2 - \sin(2(x+1))$. f has local extrema near -0.8 and 0.5. Using *Maple*'s *fsolve* procedure we find that the critical points of f are $c_1 = -0.7240683037$ (local max) and $c_2 = 0.4628462915$ (local min).

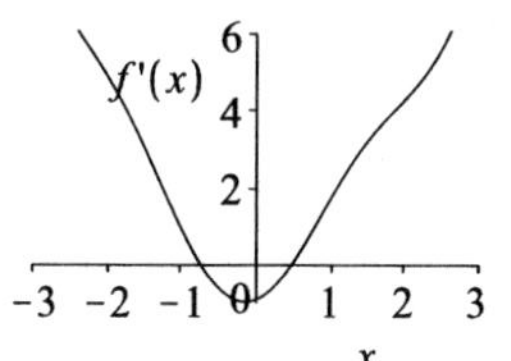

93. $f'(x) = -2xe^{-x^2} - 2(x-2)e^{-(x-2)^2}$
$$-2(x-3)e^{-(x-3)^2}.$$
f has three local extreme values near 0.5, 2, and 3. Using *Maple*'s *fsolve* procedure we find that the critical points are $c_1 = 0.5063812405$ (local max), $c_2 = 2.042890125$ (local min), and $c_3 = 2.956949212$ (local max).

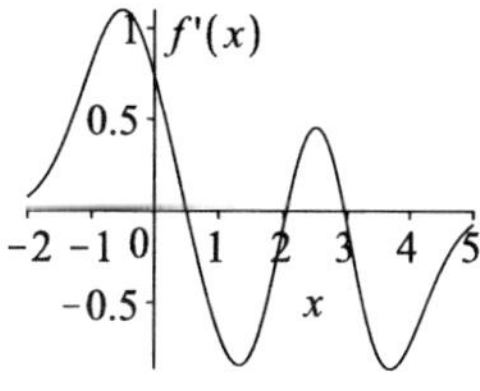

4.4 Applied Maximum-Minimum Problems

Problems for Practice

1. Let A be the area of the rectangle and ℓ and w be its length and width, respectively: $A = \ell \cdot w$. Since $2\ell + 2w = 100$, $\ell = 50 - w$ and we must maximize $A = w(50 - w) = 50w - w^2$ for $0 < w < 50$. Since $dA/dw = 50 - 2w$, the only critical point is $c = 25$. This is an absolute maximum

for A since $dA/dw > 0$ when $w < c$ and $dA/dw < 0$ for $w > c$. The rectangle with maximum area is a square with $w = \ell = 25$ m. The area is $25^2 = 625$ m^2.

3. The square of the distance from (x, y) to the origin is $s = x^2 + y^2$. Since $xy^2 = 1$, $y = 1/\sqrt{x}$ and $s = x^2 + 1/x$. We wish to minimize s for $x \in (0, \infty)$. Since $\frac{ds}{dx} = 2x - 1/x^2 = (2x^3 - 1)/x^2$, the only critical point is $c = (1/2)^{1/3} = 2^{-1/3}$ where s attains an absolute minimum value (note that $ds/dx < 0$ if $x < (1/2)^{1/3}$ and $ds/dx > 0$ if $x > (1/2)^{1/3}$). The point in the first quadrant closest to the origin is $(2^{-1/3}, 2^{1/6})$.

5. Let x denote the length of the side of each square so, when the box is formed, it has volume $f(x) = (20-2x)^2 \cdot x = 4 \cdot (x^3 - 20x^2 + 100x)$. We wish to maximize this function for $0 < x < 10$. Since $f'(x) = 4(x-10)(3x-10)$, it attains an absolute maximum value at $x = 10/3$. (Check how the sign of f' changes.) The maximum volume is $f(10/3) = 16000/27 \approx 592.6$ cubic inches.

7. The intercepts a and b satisfy $x/a + y/b = 1$ for all (x, y) on the line. In particular, $1/a + 2/b = 1$, so $a = 1/(1 - 2/b) = b/(b - 2)$. The slope of the line is $m = -b/a$.

   We will minimize the function $f(b) = \frac{b}{b-2} \cdot b = \frac{b^2}{b-2}$ for $2 < b < \infty$ and use the result to obtain the slope. Since $f'(b) = \frac{b(b-4)}{(b-2)^2}$, f has an absolute minimum value of 8 at $c = 4$ (note that $f' < 0$ if $2 < b < 4$ and $f' > 0$ if $b > 4$). When $b = 4$, $a = 4/2 = 2$, and $m = -4/2 = -2$. This is the slope that minimizes the product ab.

9. We will maximize $f(x) = x - x^3$ for $x > 0$. Since $f'(x) = 1 - 3x^2$, there is one critical point, $c = \sqrt{1/3} = \sqrt{3}/3$, where f has an absolute maximum value because $f' > 0$ when $0 < x < c$ and $f' < 0$ when $x > c$.

11. Suppose the page is w inches wide and h inches high. Then the area of the page is $A = w \cdot h$. This is to be minimized for a type area of $80 = (w-2) \cdot (h-2)$. Therefore, $h = 80/(w-2) + 2$ and $A(w) = 80w/(w-2) + 2w$. This is to be minimized for $2 < w$. Since $A'(w) = 2(w^2 - 4w - 76)/(w-2)^2$ there is one critical point $c = 2 + 4\sqrt{5} \approx 10.94$ where A attains an absolute minimum value. This is because $A' < 0$ when $2 < w < c$ and $A' > 0$ when $c < w$. The width should be $2 + 4\sqrt{5}$ inches and the height should also be $2 + 4\sqrt{5}$ inches. The total page area is $A = 84 + 16\sqrt{5} \approx 119.78$ in^2.

13. Let (x, y) be the point on the graph that is the upper right corner of the rectangle. Then $0 < x < 2$ and the area of the rectangle is $A(x) = 2x \cdot (4 - x^2) = 8x - 2x^3$. This is to be maximized for $x \in (0, 2)$. Since $A'(x) = 8 - 6x^2$, there is one critical point, $c = \sqrt{4/3} = 2\sqrt{3}/3$, where A attains an absolute maximum value. This is because $A' < 0$ if $0 < x < c$ and $A' > 0$ if $c < r < 2$. Thus the base of the rectangle should be $4\sqrt{3}/3$ and its height should be $4 - 12/9 = 24/9 = 8/3$.

15. Let x be the length of one side of the base and y be the height of the planter. Suppose the cost of the concrete is 1 dollar per square meter and the brick is 12 dollars per square meter. Then the total cost is $C = x^2 + 12 \cdot (4 \cdot x \cdot y)$. The planter is to have volume 3 so $x^2 y = 3$ and $y = 3/x^2$. Therefore, $C(x) = x^2 + 48x \cdot 3/x^2 = x^2 + 144/x$. This is to be minimized for $x > 0$. Since $C'(x) = 2x - 144/x^2 = \frac{2(x^3-72)}{x^2}$, there is one critical point, $c = 72^{1/3}$. C attains an absolute minimum value at c because $C' < 0$ if $0 < x < c$ and $C' > 0$ if $c < x$. The dimensions that minimize cost are $x = 72^{1/3} = 2 \cdot 9^{1/3}$ m and $y = 3/72^{2/3} = \frac{3}{72}x = \frac{1}{24}x$ m.

17. Let x be the width and y the height of the rectangle. The perimeter of the pentagonal shape is $3x + 2y$. The triangle has side length x and area $\frac{1}{2}x \cdot x \sin(\pi/3) = \sqrt{3}\,x^2/4$. The area enclosed by the pentagonal shape is $\sqrt{3}\,x^2/4 + xy = 9801$ so $y = (9801 - \sqrt{3}\,x^2/4)/x$. We wish to minimize the perimeter $f(x) = 3x + 2(9801 - \sqrt{3}\,x^2/4)/x$. This is $f(x) = k \cdot (x + b/x)$ where $k = 3 - \sqrt{3}/2$ and $b = 7128 + 1188\sqrt{3}$. Since $f'(x) = k \cdot (1 - b/x^2) = k \cdot \frac{x^2 - b}{x^2}$, f has one critical point, $c = \sqrt{b}$, where f attains an absolute minimum value. (Check the sign of f'). The base length that will minimize the area is $b = \sqrt{7128 + 1188\sqrt{3}} = 6\sqrt{33(6 + \sqrt{3})}$ m.

19. Let r be the radius of the cylinder and h its height. Its surface area is $\frac{1}{2} \cdot 4\pi r^2 + \pi r^2 + 2\pi r \cdot h = 5\pi$. Therefore, $h = (5\pi - 3\pi r^2)/(2\pi r)$ and we wish to maximize $f(r) = \pi r^2 \cdot (5\pi - 3\pi r^2)/(2\pi r) + \frac{1}{2} \cdot \frac{4}{3}\pi r^3$ for $r > 0$. This simplifies to $f(r) = \frac{5\pi}{2} \cdot (r - r^3/3)$ so $f'(r) = \frac{5\pi}{2} \cdot (1 - r^2)$ and f has one critical point, $c = 1$, where f attains an absolute maximum value. (Check the sign of f'). Thus the radius of the cylinder of maximum volume is $r = 1$, its height is $h = 1$ also, and its volume is $5\pi/3$ in^2.

21. If the price per unit is p, then for x units sold, $p = 1602 - 8x$, implying that the profit $P(x) = p \cdot x - C(x)$ is $P(x) = (1602 - 8x) \cdot x - (1000 + 2x)$. This simplifies to $P(x) = 1600x - 8x^2 - 1000$ which must be maximized for $x > 0$. Since $P'(x) = 1600 - 16x$, there is one critical point $c = 100$. This is an absolute maximum for P as can be verified by checking the sign of P'.

23. If the price per unit is p, then for x units sold, $x = 51 - p/8$, implying that $p = 408 - 8x$. Therefore, the profit $P(x) = p \cdot x - C(x)$ is $P(x) = (408 - 8x) \cdot x - (1200 + 8x)$. This simplifies to $P(x) = 400x - 8x^2 - 1200$ which must be maximized for $x > 0$. Since $P'(x) = 400 - 16x$, there is one critical point $c = 25$. This is an absolute maximum for P as can be verified by checking the sign of P'.

25. Since $f'(x) = 2x + 1$, there is one critical point, $c = -1/2$. This is an absolute minimum for f (check the sign of f'). Thus the minimum value attained on the interval is $f(-1/2) = -1/4$. The maximum value must be attained at one of the endpoints. Since $f(-1) = 0$ and $f(1) = 2$ the maximum value is $f(1) = 2$.

27. Since $f'(x) = 3x^2 - 3x = 3x(x-1)$, there are two critical points, $c_1 = 0$ and $c_2 = 1$. Both are in the interval. Checking values of f, end points: $f(-1) = -1/2$, $f(3) = 31/2$, and critical points: $f(0) = 2$, $f(1) = 3/2$. Therefore, $-1/2$ is the minimum value attained and $31/2$ is the maximum value attained on the interval.

29. Since $f'(x) = (x+2)(3x+2)$, there are two critical points, $c_1 = -2$ and $c_2 = -2/3$. Both are in the interval. Checking values of f, end points: $f(-3) = -3$, $f(1) = 9$, and critical points: $f(-2) = 0$ and $f(-2/3) = -32/27$. Therefore, -3 is the minimum value attained and 9 is the maximum value attained on the interval.

31. Since $f'(x) = \frac{4x^2-9}{x^2}$, there are two critical points, $c_1 = -3/2$ and $c_2 = 3/2$. c_2 is in the interval. Checking values of f, end points: $f(1) = 13$, $f(2) = 25/2$, and critical point: $f(3/2) = 12$. Therefore, 12 is the minimum value attained and 13 is the maximum value attained on the interval.

33. Since $f'(x) = e^x - 1$, there is one critical point, $c = 0$. It is in the interval. Checking values of f at the end points: $f(-1) = e^{-1} + 1 \approx 1.37$, $f(1) = e - 1 \approx 1.72$, and at the critical point: $f(0) = 1$. Therefore, 1 is the minimum value attained and $e - 1$ is the maximum value attained on the interval.

Further Theory and Practice

35. The number of cars passing through the tunnel per hour, as a function of cars per km, is $f(x) = x \cdot s = x \cdot (\alpha - \beta x)$. Since $f'(x) = \alpha - 2\beta x$, f is maximized when $x = \alpha/(2\beta)$. The actual number of cars would be the integer closest to $\alpha/(2\beta)$.

37. The profit P satisfies the equation $P(x) = R(x) - C(x)$. If x_0 is the production level that maximizes profit, then $P'(x_0) = 0$, which is equivalent to $R'(x_0) = C'(x_0)$.

39. The average cost is minimized when $\overline{C}\,'(x) = \frac{xC'(x)-C(x)}{x^2} = 0$. That is, at the value x_0 where $C'(x_0) = C(x_0)/x_0$ (the marginal cost equals the average cost).

41. Since

$$f'(\theta) = -\frac{a(\cos^2(\theta) - \sin^2(\theta)) + 2b\cos(\theta)\sin(\theta)}{(a\sin(\theta)\cos(\theta) - b\cos^2(\theta))^2} = -\frac{a\cos(2\theta) + b\sin(2\theta)}{(a\sin(\theta)\cos(\theta) - b\cos^2(\theta))^2},$$

f has a critical point at the angle θ_0 where $a\cos(2\theta_0) = -b\sin(2\theta_0)$. That is, $\tan(2\theta_0) = -a/b$. Consequently, $\frac{2\tan(\theta_0)}{1-\tan^2(\theta_0)} = -a/b$ (Chapter 1.6, Exercise 37), so $(a/b)\tan^2(\theta_0) - 2\tan(\theta_0) - a/b = 0$. Solving for $\tan(\theta_0)$, $\tan(\theta_0) = \left(2 \pm \sqrt{4 + 4a^2/b^2}\,\right)/(2a/b)$ and $\tan(\theta_0) = \frac{b+\sqrt{a^2+b^2}}{a}$.

43. Let x the the length of wire that is shaped into a circle. Then $L - x$ is the length shaped into a square. The radius of the circle is $r = x/(2\pi)$

and the length of the side of the square is $\ell = (L - x)/4$. Therefore, we wish to minimuze $A(x) = \pi \cdot (x/(2\pi))^2 + (L - x)^2/16$. This simplifies to $A(x) = \frac{x^2}{4\pi} + \frac{1}{16}(x - L)^2$, to be minimized for $0 \le x \le L$. Since $A'(x) = x/(2\pi) + (x - L)/8 = (1/(2\pi) + 1/8)x - L/8$, there is one critical point, $c = \frac{L/8}{1/(2\pi)+1/8} = \pi L/(4 + \pi)$, which is an absolute minimum for A (verify). This is the length of wire that should be shaped into a circle. The length of wire shaped into a square is $L(1 - \pi/(4+\pi)) = 4L/(4+\pi)$.

45. Let x be the length of wire shaped into a closed semicircle of radius r. Then $L - x$ is the length shaped into a circle of radius ρ. The circumference of the semicircle is $2r + \pi r$. This must equal x, so $r = x/(2 + \pi)$. The circumference of the circle is $2\pi\rho$ which must equal $L - x$. Therefore, $\rho = (L-x)/(2\pi)$ and the total area enclosed by the semicircle and the circle is $A(x) = \frac{1}{2} \cdot \pi(x/(2 + \pi))^2 + \pi(L - x)^2/(4\pi^2)$. This simplifies to $A(x) = \pi x^2/(2(2 + \pi)^2) + (x - L)^2/(4\pi)$ which is to be maximized and minimized for $0 \le x \le L$. Since $A'(x) = \pi x/(2 + \pi)^2 + (x - L)/(2\pi) = \left(\frac{\pi}{(2+\pi)^2} + \frac{1}{2\pi}\right)x - L/(2\pi)$, there is one critical point, $c = \frac{L/(2\pi)}{\frac{\pi}{(2+\pi)^2}+\frac{1}{2\pi}} = \frac{(2+\pi)^2 L}{2\pi^2+(2+\pi)^2}$, which is an absolute minimum for A (verify). $A(c) = \frac{\pi L^2}{6\pi^2+8\pi+8}$.

The maximum value for A will occur at one of the two endpoints. Since

$$A(0) = L^2/(4\pi) = 0.0796L^2 \quad \text{and} \quad A(L) = \pi L^2/(2(2 + \pi)^2) \approx 0.0594L^2,$$

the maximum area is obtained by forming a circle.

47. The drag δ as a function of speed v is $\delta(v) = av^2 + b/v^2$. This is to be minimized for $v > 0$. Since $\delta'(v) = 2av - 2b/v^3 = 2 \cdot \frac{av^4-b}{v^3}$, there is one critical point, $c = (b/a)^{1/4}$. This is an absolute minimum for δ.

49. Since $U(r) = A(br^{-12} - r^{-6})$, $U'(r) = A(-12br^{-13} + 6r^{-7})$. This can also be expressed as $U'(r) = 6Ar^{-13}(r^6 - 2b)$, implying that U has a critical point at $r_0 = (2b)^{1/6}$. If $0 < r < r_0$, then $U' < 0$, and if $r > r_0$, then $U' > 0$, so r_0 is an absolute minimum for U.

51. Let x and y denote the lengths of the horizontal and vertical portions of the fence, respectively. Then $x + 2y = 100$ and we wish to maximize $A = xy + C$ where C is the area of the portion of the garden with the curved boundary. Let θ be the central angle of the arc bounding the garden. The radial lines from the center of the circle to the two points on the circle are the legs of an isosceles triangle with equal sides 80, and base x. Draw a picture to see that $\sin(\theta/2) = (x/2)/80$, so $\theta = 2\arcsin(x/160)$. This yields the following formula for the area C (the area of the sector minus the area of the isosceles triangle):

$$C(x) = 160 \arcsin(x/160) - (x/2) \cdot \sqrt{80^2 - x^2/4}.$$

The area of the garden, as a function of x, is $A(x) = x \cdot (100 - x)/2 + C(x)$. This is to be maximized for $0 < x < 100$.

53. Since $\Omega'(\theta) = \dfrac{b}{R^4}\csc^2(\theta) - \dfrac{b}{r^4}\cdot\dfrac{\cos(\theta)}{\sin^2(\theta)} = \dfrac{b}{\sin^2(\theta)}\cdot\left(\dfrac{1}{R^4} - \dfrac{\cos(\theta)}{r^4}\right)$, Ω is maximized when $\cos(\theta) = (r/R)^4$.

55. Let x be the length of the extension of the initial 100 m side $(0 \le x < 50)$. Let y be the length of the side perpendicular to the extension. Then the area of the rectangle is $A = (100 + x)\cdot y$. Since $200 + 2x + 2y = 300$, $y = 50 - x$, and the area, as a function of x, is $A(x) = (100 + x)(50 - x)$. This is to be maximized. Since $A'(x) = -2x - 50$, A is decreasing for $x \in [0, 50)$ so A is maximum for $x = 0$, $A(0) = 5000$ m^2.

57. Let r and h denote the radius of the cylinder and its height, respectively. The amount of metal used is proportional to the sum of the areas of the two squares and the area of the sides: $A = 2\cdot(2r)^2 + 2\pi r h$. Since $V = \pi r^2 h$, $h = V/(\pi r^2)$ and $A(r) = 8r^2 + 2\pi r\cdot V/(\pi r^2)$. Thus we must minimize $A(r) = 8r^2 + \frac{2V}{r}$ for $0 < r < \infty$. Since $A'(r) = 16r - \frac{2V}{r^2} = 2\cdot\frac{8r^3 - V}{r^2}$, A has one critical point $c = (V/8)^{1/3}$ where A attains an absolute minimum value (verify). Thus $r = (V/8)^{1/3}$. Since $h = V/(\pi r^2)$,

$$\frac{h}{r} = \frac{V}{\pi r^3} = \frac{V}{\pi\cdot V/8} = \frac{8}{\pi}.$$

59. This expression, as a function of x, can be minimized by minimizing the function $f(x) = \frac{\ell^2 + 18x^2}{12x + \ell}$ for $x > 0$. Since $f'(x) = \frac{12(3x+\ell)(6x-\ell)}{(12x+\ell)^2}$, there is one critical point, $\ell/6$, where f has an absolute minimum value (verify).

61. We wish to find the maximum value of $f(\alpha) = \frac{\tan(\alpha)}{2+\tan^2(\alpha)}$ for $0 < \alpha < \pi/2$. But $\tan(\alpha)$ attains all positive real values for $\alpha \in (0, \pi/2)$, so f has the same maximum value as $g(x) = \frac{x}{2+x^2}$ for $x > 0$. We maximize g instead. Since $g'(x) = \frac{x^2-2}{(x^2+2)^2}$, there is one critical point, $c = \sqrt{2}$, where g attains the maximum value of $g(\sqrt{2}) = \sqrt{2}/4$. Thus $\tan(\beta_0) = \sqrt{2}/4$, implying that $\sin(\beta_0) = 1/3$ (verify).

CALCULATOR/COMPUTER EXERCISES

63. The plot on the right is the graph of f and its derivative (dashed line). Using *Maple*'s *fsolve* procedure the critical points for the low point and (possible) high point are $c_1 = -0.515$ and $c_2 = 1.839$. Evaluating f at these points and the two endpoints yields the absolute minimum $f(c_1) = -0.223$ and the absolute maximum $f(-4) = 1.060$.

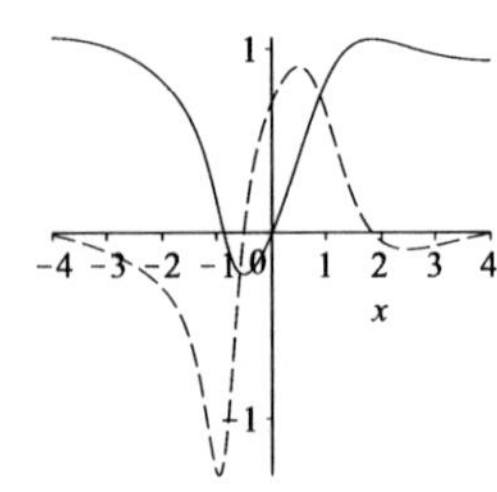

65. The hotter source is at $(0, 0)$ and the other source at $(10, 0)$. The heat received at point $(x, 0)$ is $H(x) = 2/x^2 + 1/(10 - x)^2$. The plot on the right is the graph of H. Using *Maple*'s *fsolve* procedure the critical point for the low point is $x_0 \approx 5.575$. This is the coolest point between the two sources.

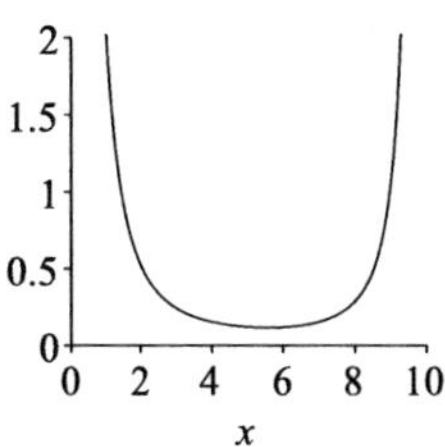

67. According to *Maple*, R is maximum for $\alpha = \arctan\left(\frac{v}{\sqrt{v^2+2gh}}\right)$.

69. Y/x is a continuous function of x on $[0, \infty)$ and has limit 0 as $x \to \infty$. This implies that it attains an absolute maximum value. Using *Maple* Y/x has two critical points in $[0, \infty)$, $c_1 = 0.1316$ and $c_2 = 0.5128$. The values of Y/x at these points are 0.9445 and 1.0288 respectively. Since the value of Y/x at $a = 0$ is 1, its absolute maximum value is 1.0288.

71. In $n = 16$ steps, $c \approx 0.644$.

4.5 Concavity

Problems for Practice

1. $f'(x) = \frac{\sqrt{x}-2}{\sqrt{x}}$, $f''(x) = \frac{1}{x^{3/2}}$. There are no inflection points, f is concave up on $(0, \infty)$. There is one critical point, $c = 4$, which (by the Second Derivative Test) is local minimum for f.

3. $f'(x) = \frac{x-3}{2x^{3/2}}$, $f''(x) = -\frac{x-9}{4x^{5/2}}$. There is an inflection point at $x_0 = 9$. f is concave up on $(0, 9)$ and concave down on $(9, \infty)$. There is one critical point, $c = 3$, which (by the Second Derivative Test) is local minimum for f.

5. $f'(x) = \frac{2(x^3-27)}{x^2}$, $f''(x) = \frac{2(x^3+54)}{x^3}$. There is an inflection point at $x_0 = -54^{1/3} = -3 \cdot 2^{1/3}$. f is concave up on $(-\infty, x_0)$, concave down on $(x_0, 0)$, and concave up $(0, \infty)$. There is one critical point, $c = 3$, which (by the Second Derivative Test) is local minimum for f.

7. $f'(x) = 3(x + 7)(x - 1)$, $f''(x) = 6(x + 3)$. There is an inflection point at $x_0 = -3$. f is concave down on $(-\infty, -3)$ and concave up on $(-3, \infty)$. There are two critical points, -7 and 1. By the Second Derivative Test, -7 is a local maximum and 1 is a local minimum for f.

9. $f'(x) = 6(x+1)(x-2)$, $f''(x) = 6(2x-1)$. There is an inflection point at $x_0 = 1/2$. f is concave down on $(-\infty, 1/2)$ and concave up on $(1/2, \infty)$. There are two critical points, -1 and 2. By the Second Derivative Test, f has a local maximum at -1 and a local minimum at 2.

11. $f'(x) = 2x(2x+3)(x-3)$, $f''(x) = 6(2x^2 - 2x - 3)$. There are inflection points at $x_1 = (1-\sqrt{7})/2$ and $x_2 = (1+\sqrt{7})/2$. f is concave up on $(-\infty, x_1)$, concave down on (x_1, x_2), and concave up on (x_2, ∞). There are three critical points, $-3/2$, 0, and 3. By the Second Derivative Test, f has a local minimum at $-3/2$, a local maximum at 0, and a local minimum at 3.

13. $f'(x) = 15(x+1)^2(x-1)^2$, $f''(x) = 60x(x+1)(x-1)$. There are inflection points at -1, 0, and 1. f is concave down on $(-\infty, -1)$, concave up on $(-1, 0)$, concave down on $(0, 1)$, and concave up on $(1, \infty)$. There are two critical points, -1 and 1. f does not attain a local extreme value at either one because $f' \geq 0$ for all x so f is never decreasing.

15. $f'(x) = -\frac{x^2-3}{(x^3+3)^2}$, $f''(x) = \frac{2x(x-3)(x+3)}{(x^2+3)^3}$. There are inflection points at -3, 0, and 3. f is concave down on $(-\infty, -3)$, concave up on $(-3, 0)$, concave down on $(0, 3)$, and concave up on $(3, \infty)$. There are two critical points, $-\sqrt{3}$ and $\sqrt{3}$. By the Second Derivative Test, f has a local minimum value at $-\sqrt{3}$ and a local maximum value at $\sqrt{3}$.

17. $f'(x) = \frac{4x}{(x^2+1)^2}$, $f''(x) = -\frac{4(3x^2-1)}{(x^2+1)^3}$. There are inflection points at $-\sqrt{3}/3$ and $\sqrt{3}/3$. f is concave down on $(-\infty, -\sqrt{3}/3)$, concave up on $(-\sqrt{3}/3, \sqrt{3}/3)$, and concave down on $(\sqrt{3}/3, \infty)$. There is one critical point, 0, where f has a local minimum value by the Second Derivative Test.

19. $f'(x) = -\sin(x) + 1$, $f''(x) = -\cos(x)$. There are inflection points at $\pi/2$ and $3\pi/2$. f is concave down on $(0, \pi/2)$, concave up on $(\pi/2, 3\pi/2)$, and concave down on $(3\pi/2, 2\pi)$. There is one critical point, $\pi/2$. However, f does not have a local extreme value there because f' is never negative.

21. $f'(x) = \frac{1}{2}\sec^2(x/2)$, $f''(x) = \frac{1}{2}\sec^2(x/2)\tan(x/2)$. There is an inflection point at 0. f is concave down on $(-\pi, 0)$ and concave up on $(0, \pi)$. There are no critical points.

23. $f'(x) = 2\sin(x)\cos(x) = \sin(2x)$, $f''(x) = 2\cos(2x)$. There are two inflection points, $\pi/4$ and $3\pi/4$. f is concave up on $(0, \pi/4)$, concave down on $(\pi/4, 3\pi/4)$, and concave up on $(3\pi/4, \pi)$. There is a critical point at $\pi/2$ where, by the Second Derivative Test, f attains a local maximum value.

25. $f'(x) = e^x - e^{-x}$, $f''(x) = e^x + e^{-x}$. There are no inflection points. f is concave up on $(-\infty, \infty)$. There is a critical point at 0 where, by the Second Derivative Test, f attains a local minimum value.

27. $f'(x) = \frac{x-1}{x}$, $f''(x) = 1/x^2$. There are no inflection points. f is concave up on $(0, \infty)$. There is a critical point at 1 where, by the Second Derivative Test, f attains a local minimum value.

29. $f'(x) = \frac{2(x^2-4)}{x}$, $f''(x) = \frac{2(x^2+4)}{x^2}$. There are no inflection points. f is concave up on $(0, \infty)$. There is a critical point at 2 where, by the Second

Derivative Test, f attains a local minimum value.

31. $f'(x) = \frac{4x-5}{3x^{2/3}}$, $f''(x) = \frac{2(2x+5)}{9x^{5/3}}$. There are inflection points at $-5/2$ and 0. f is concave up on $(-\infty, -5/2)$, concave down on $(-5/2, 0)$, and concave up on $(0, \infty)$. There are critical points at 0 and $5/4$. By the Second Derivative Test, f attains a local minimum value at $5/4$. At 0 the tangent line is vertical so f does not attain an extreme value.

33. $f'(x) = 1 + (1/2)^x \ln(1/2) = 1 - \ln(2)/2^x$, $f''(x) = \ln(2)^2/2^x$. There are no inflection points. f is concave up on $(-\infty, \infty)$. There is a critical point at c where $2^c = \ln(2)$. That is, $c\ln(2) = \ln(\ln(2))$, or $c = \ln(\ln(2))/\ln(2)$. By the Second Derivative Test, f has a local minimum value at c.

35. $f'(x) = \frac{1}{\sqrt{1-x^2}} - 2x$, $f''(x) = \frac{x}{(1-x^2)^{3/2}} - 2$. There are possible inflection points at those values x where $x = 2(1-x^2)^{3/2}$. We solve $x^2 = 4(1-x^2)^3$. Let $x^2 = y$. Then $y = 4(1-y)^3$ when $y = 1/2$ (verify) so $x^2 = 1/2$ and there is one inflection point: $\sqrt{2}/2$, (you should verify that $-\sqrt{2}/2$ is not a zero of f''). f has an inflection point at $\sqrt{2}/2$ because it is concave down on $(-1, \sqrt{2}/2)$ and concave up on $(\sqrt{2}/2, 1)$ (check $f''(0)$ and $\lim_{x\to 1^-} f''(x)$). As for critical points, they will occur only at those x values where $2x = 1/\sqrt{1-x^2}$ or $4x^2(1-x^2) = 1$. Thus $4x^4 - 4x^2 + 1 = 0$ so $x^2 = (4 \pm \sqrt{0}\,)/8 = 1/2$. This yields only one critical point: $c = \sqrt{2}/2$, (you should verify that $-\sqrt{2}/2$ is not a zero of f'). Since f is concave down to the left of c and concave up to the right, f does not have an extreme value at c.

Further Theory and Practice

37. $f'(x) = x(x-1)$ and $f''(x) = 2x - 1$. The critical points are 0 and 1. Since $f''(0) < 0$ and $f''(1) > 0$, f has a local maximum at 0 and a local minimum at 1.

39. $f'(x) = x^2 - 1$ and $f''(x) = 2x$. The critical points are -1 and 1. Since $f''(-1) < 0$ and $f''(1) > 0$, f has a local maximum at -1 and a local minimum at $x = 1$.

41. $f'(x) = (x^2-4)^5 = (x+2)^5(x-2)^5$ and $f''(x) = 10x(x^2-4)^4$. The critical points are -2 and 2. Since f' changes sign from $(+)$ to $(-)$ at -2, f has a local maximum at -2. It has a local minimum at 2 because f' changes sign from $(-)$ to $(+)$ at 2.

43. $f''(x) = x(x+1)$ changes sign at both -1 and 0. f has an inflection point at each one.

45. f'' is never 0, f has no inflection points.

47. $f''(x) = (x+5)^3(x-5)^3$, it changes sign at both -5 and 5. f has an inflection point at -5 and 5.

49. Since $y''(x) = \frac{16a^3(3x^2-4a^2)}{(4a^2+x^2)^3}$, the Witch of Agnesi has inflection points at

$x_{1,2} = \mp 2a/\sqrt{3} = \mp\frac{2\sqrt{3}}{3} \cdot a$. The curve is concave down on the open interval $(-\frac{2\sqrt{3}}{3} \cdot a, \frac{2\sqrt{3}}{3} \cdot a)$, and concave up on the two complementary intervals.

51. Let $g = f^2$, then $g'(x) = 2f(x)f'(x)$ so
$$g''(x) = 2(f'(x))^2 + 2f(x)f''(x) > 0 \quad \text{for all } x\,.$$
Therefore, f^2 is concave up on $(-\infty, \infty)$.

53. Observe that $m'(t) = -k(m(t)^2 - 2c\,m(t)) = -k((m(t) - c)^2 - c^2)$ so when $0 < m(t) < 2c$, $m'(t)$ is positive and increasing for $0 < m(t) < c$ and positive and decreasing for $c < m(t) < 2c$. Thus the graph of m is rising and concave up until $m(t) = c$, at which point it inflects to become concave down, but still rising.

55. When the equations $x = x(t)$, $y = y(t)$ define a curve that is the graph of a differentiable function $y = y(x)$, we know that $\frac{dy}{dx} = \frac{dy/dt}{dx/dt} = \frac{\dot{y}}{\dot{x}}$. Therefore, $\frac{d^2y}{dx^2} = \frac{d}{dx}\left(\frac{dy}{dx}\right) = \frac{d}{dx}\left(\frac{\dot{y}}{\dot{x}}\right)$.

57. Using the formulas in Exercises 55 and 56, $\frac{dy}{dx} = \frac{\sin(t)}{1-\cos(t)}$ and $\frac{d^2y}{dx^2} = -\frac{1}{a(\cos(t)-1)^2}$. Therefore, the arch of the cycloid is concave down, rising from 0 at the left endpoint to a maximum at the midpoint of the interval, and then falling back to 0 at the right endpoint. (The only critical point is at $t = \pi$ and $\frac{dy}{dx}$ changes sign from (+) to (−) at that point.)

59. Using the formulas discussed in Exercises 55 and 56, $\frac{dy}{dx} = -\frac{\sin(2t)}{1+\tan^2(t)}$ and $\frac{d^2y}{dx^2} = -2\cos(t)^4(4\cos^2(t) - 3)$. The graph has inflection points at $t_{1,2} = \mp\pi/6$ (indicated on the graph). It is concave up when $t < -\pi/6$ and $t > \pi/6$ and concave down otherwise. There is a local max at $t = 0$. The plot on the right is for $\arctan(-4) \le t \le \arctan(4)$. The y-axis is a horizontal asymptote.

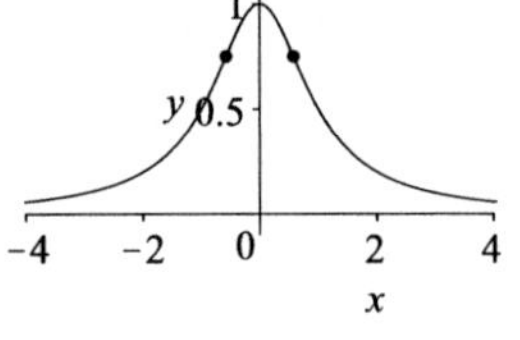

61. Since $f'(x) = -\sin(x)$ and $f''(x) = -\cos(x)$, the curvature is $\frac{|\cos(x)|}{(1+\sin^2(x))^{3/2}}$.

63. Since $f'(x) = -1/x^2$ and $f''(x) = 2/x^3$, the curvature is $\frac{2}{|x|^3(1+1/x^4)^{3/2}}$. This simplifies to $\frac{2|x^3|}{(4x+1)^{3/2}}$.

65. Since $f'(x) = x^2$ and $f''(x) = 2x$, the curvature function, for $x > 0$, simplifies to $\kappa(x) = \frac{2x}{(x^4+1)^{3/2}}$. Differentiate: $\kappa'(x) = -\frac{2(5x^4-1)}{(x^4+1)^{5/2}}$, implying that κ has one positive critical point, $c = 1/5^{1/4}$. The First Derivative Test shows this to be an absolute maximum for κ.

67. Since $f'(x) = e^x$ and $f''(x) = e^x$, the curvature function simplifies to $\kappa(x) = \frac{e^x}{(1+e^{2x})^{3/2}}$. Differentiate: $\kappa'(x) = \frac{e^x(2e^{2x}-1)}{(1+e^{2x})^{5/2}}$, implying that κ

has one critical point at the value c where $e^{2c} = 1/2$. That is, $c = \ln(1/2)/2 = -\ln(2)/2$. The First Derivative Test shows this to be an absolute maximum for κ.

69. Since $(g \circ f)'' = \left((g' \circ f) \cdot f'\right)' = (g'' \circ f) \cdot f' \cdot f' + (g' \circ f) \cdot f''$,

$$(g \circ f)''(c) = g''(f(c)) \cdot (f'(c))^2 + g'(f(c) \cdot f''(c)\,.$$

Therefore, if f and g are concave up and $g' > 0$, then $g \circ f$ is also concave up.

71. Since $R - C$ is maximized at q_0, we expect that $R''(q_0) - C''(q_0)$ will be negative. That is, $R''(q_0) < C''(q_0)$, making graph of C'' the curve that starts lower and ends higher.

CALCULATOR/COMPUTER EXERCISES

73. Start with the sketch of the graph on the right. Using *Maple*'s *fsolve* procedure we find that there is a critical point at $c \approx 1.278$. This is the local maximum. There are two points where the graph changes concavity: -2.399 and 2.399. The concavity changes from up to down to up.

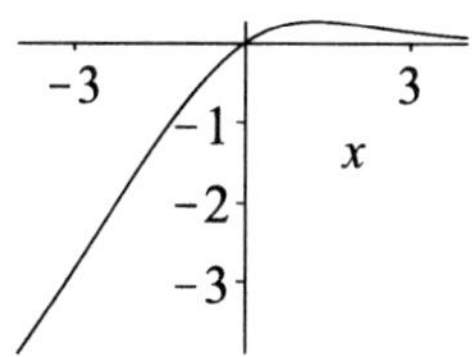

75. Start with the sketchs of the graph below. The left one is from far away, the right one is close up. Using *Maple*'s *fsolve* procedure we find that the critical points are at -0.8078 and 6.1417. The first is a local minimum, the second a local maximum. The inflection points are at -0.3437, 0.5329, and 5.0177. The concavity changes from up to down to up, then back to down.

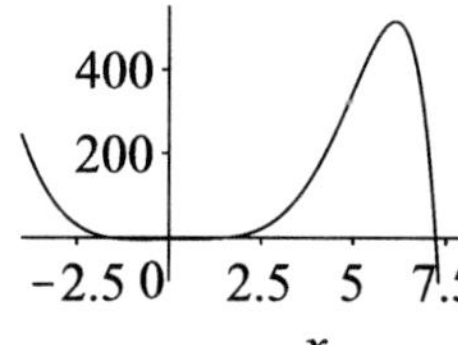

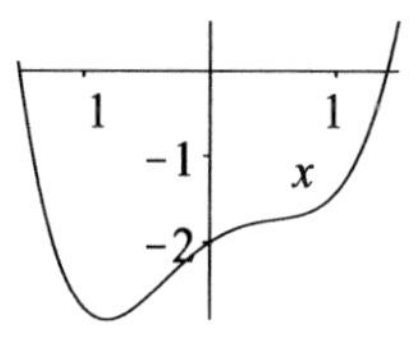

77. Start with the sketch of the graph on the right. Using *Maple*'s *fsolve* procedure we find that the two critical points, a local max and a local min, are at -1.4440 and 0.5604, respectively. There are three inflection points where the concavity changes from up to down, then to up, then back to down. Their approximate locations are -1.8252, -0.7587, and 1.2263, respectively.

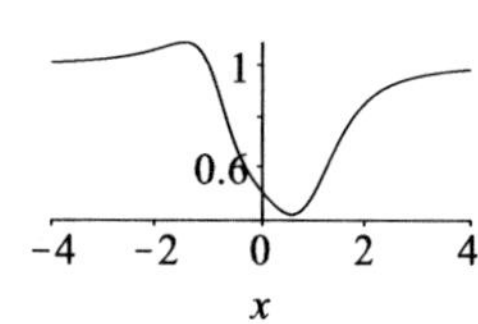

79. Start with the sketch of the graph on the right. Using *Maple*'s *fsolve* procedure we find that there is one critical point at .7087. This is a local maximum. There are two inflection points where the concavity changes from up to down, then back to up. Their approximate locations are -0.4030 and 1.4414.

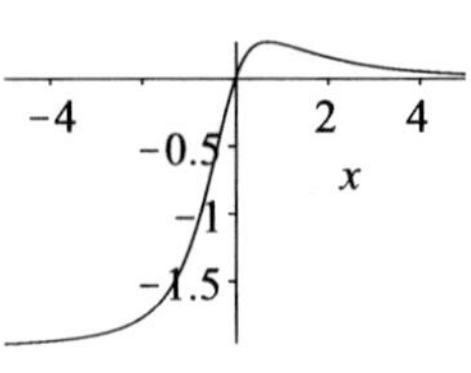

4.6 Graphing Functions

Problems for Practice

1. Since f is a polynomial there are no asymptotes. $f'(x) = 3(x+1)(x-3)$ and $f''(x) = 6(x-1)$ implying that $(-1, 12)$ is a local maximum, $(3, -20)$ a local minimum, and $(1, -4)$ an inflection point. f is increasing on $(\infty, -1)$, decreasing in $(-1, 3)$, and increasing on $(3, \infty)$. The concavity changes from down to up at $(1, -4)$.

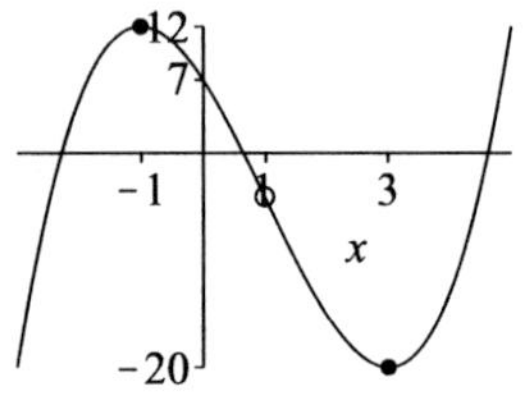

3. f is a polynomial so no asymptotes. $f'(x) = 3x(x-2)$ and $f''(x) = 6(x-1)$ implying that $(0, 4)$ is a local maximum, $(2, 0)$ is a local minimum, and $(1, 2)$ is an inflection point. f is increasing on $(-\infty, 0)$ decreasing on $(0, 2)$ and increasing on $(2, \infty)$. The concavity changes from down to up at $(1, 2)$.

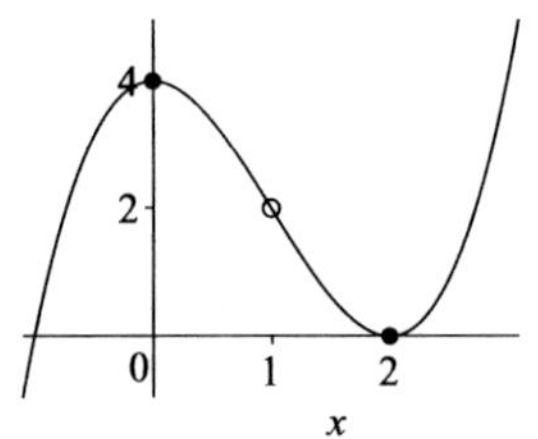

5. $\lim_{|x|\to\infty} f(x) = 0$, the y-axis is a horizontal asymptote. Since $f'(x) = -\frac{x^2-4}{(x^2+4)^2}$ and $f''(x) = \frac{2x(x^2-12)}{(x^2+4)^3}$, $(-2, -\frac{1}{4})$ is a global min, $(2, \frac{1}{4})$ a global max, and $(-2\sqrt{3}, -\sqrt{3}/8)$, $(0, 0)$, and $(2\sqrt{3}, \sqrt{3}/8)$ are inflection points. f is decreasing on the interval $(-\infty, -2)$, increasing on $(-2, 2)$, and decreasing on $(2, \infty)$. The concavity changes from down to up at the first inflection point, back to down at $(0, 0)$, then to up at the third inflection point.

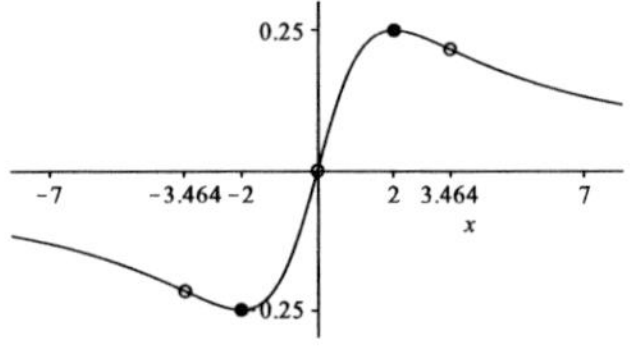

7. The points $x = \pm 2$ are not in the domain of f. The lines $x = \pm 2$ are vertical asymptotes. Since $\lim_{|x|\to\infty} f(x) = 1$, the line $y = 1$ is a horizontal asymptote. $f'(x) = -\frac{16x}{(x^2-4)^2}$ and $f''(x) = \frac{16(3x^2+4)}{(x^2-4)^3}$ implying that there is a local maximum at $(0, -1)$ and no local minima or inflection points. f is increasing on $(-\infty, -2)$ and $(-2, 0)$ and is de-

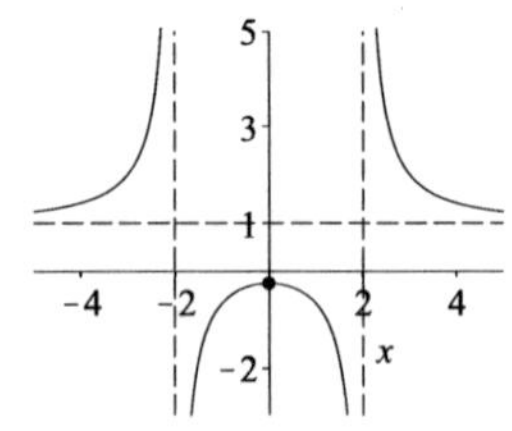

creasing on $(0, 2)$ and on $(2, \infty)$. The concavity changes from up to down to up as x increases through -2, 0, and 2.

9. The domain of f is $[0, \infty)$, there are no asymptotes. $f'(x) = \frac{4x^{3/2}-1}{2x^{1/2}}$ and $f''(x) = \frac{8x^{3/2}+1}{4x^{3/2}}$ implying that there is a global minimum at $((1/4)^{2/3}, f((1/4)^{2/3}))$ and no local maxima. There are no inflection points. f is decreasing on $(0, (1/4)^{2/3})$ and increasing on $((1/4)^{2/3}, \infty)$.

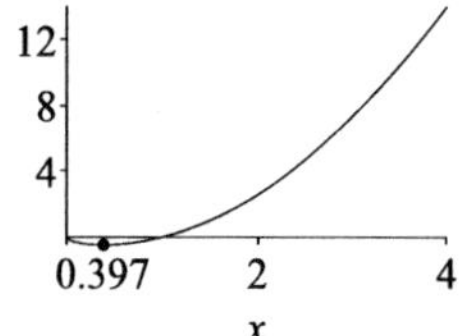

11. There are no asymptotes. $f'(x) = \cos(x) - 1$ and $f''(x) = -\sin(x)$ so there are horizontal tangents at every point $(2n\pi, -2n\pi)$, n an integer. These are not local extrema since $f' \leq 0$ implying that f never increases. There is an inflection point at every point $(n\pi, -n\pi)$. f is concave down on the intervals $(2n\pi, (2n+1)\pi)$ and concave up on the intervals $((2n-1)\pi, 2n\pi)$, n an integer.

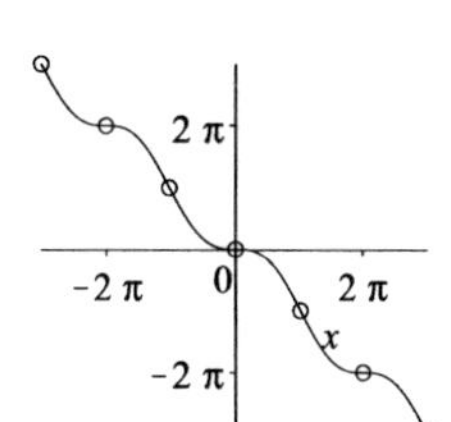

13. 0 is not in the domain of f, the y-axis is a vertical asymptote. Since $\lim_{|x|\to\infty} f(x) = 0$, the x-axis is a horizontal asymptote. $f'(x) = -\frac{2x^{1/3}+3}{3x^2}$ and $f''(x) = \frac{2(5x^{1/3}+9)}{9x^3}$ so there is a local maximum at $(-27/8, 4/27)$ and an inflection point at $(-729/125, 100/729)$. f is concave up on the interval $(-\infty, -729/125)$, concave down on $(-729/125, 0)$, and concave up on $(0, \infty)$.

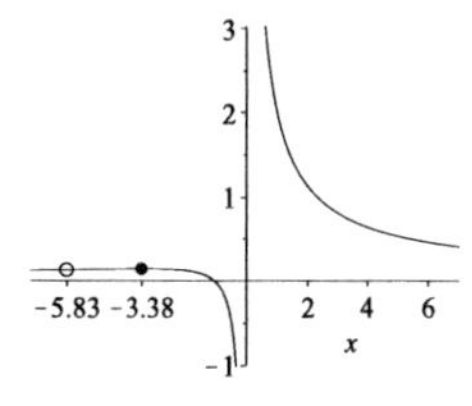

15. -3 is not in the domain of f, the line $x = -3$ is a vertical asymptote. Since $\lim_{|x|\to\infty} f(x) = 0$, the x-axis is a horizontal asymptote. $f'(x) = -\frac{2x-3}{3x^{2/3}(x+3)^2}$ and $f''(x) = \frac{2(5x^2-15x-9)}{9x^{5/3}(x+3)^3}$ so there is a local maximum at $(3/2, f(3/2))$ and a vertical tangent at $(0, 0)$. There are also inflection points at $(x_1, f(x_1))$ and $(x_2, f(x_2))$ where $x_{1,2} = (15 \mp \sqrt{15^2 + 180}\,)/10 = 3/2 \mp 9\sqrt{5}/10$. f is concave up on the interval $(-\infty, -3)$, concave down on $(-3, x_1)$, concave up on $(x_1, 0)$, concave down on $(0, x_2)$, and concave up on (x_2, ∞). Therefore, the point $(0, 0)$ is also an inflection point.

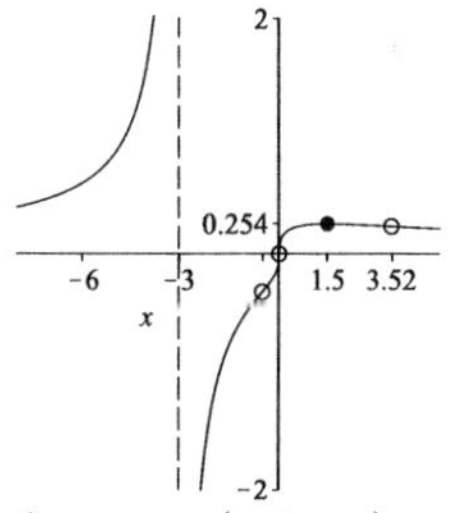

17. f is defined everywhere and differentiable except at -1. $f'(x) = \frac{6x+5}{5(x+1)^{4/5}}$ and $f''(x) = \frac{2(3x+5)}{25(x+1)^{9/5}}$ so there is a local minimum at $(-5/6, f(-5/6))$, an inflection point at $(-5/3, f(-5/3))$, and a vertical tangent at $(-1, 0)$. The concavity changes from down to up at $(-1, 0)$ so this is also an inflection point. f decreases on $(-\infty, -5/6)$ and increases on

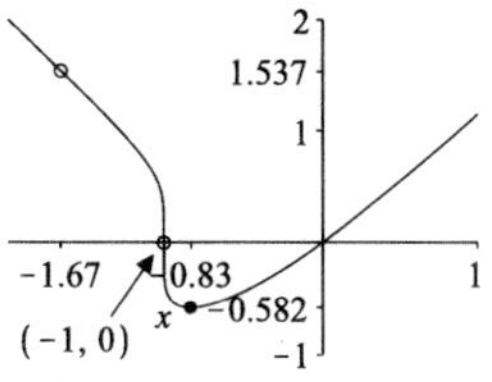

$(-5/6, \infty)$ so f attains a global minimum at $-5/6$. f is concave up on $(-\infty, -5/3)$, concave down on $(-5/3, -1)$, and concave up on $(-1, \infty)$.

19. There are no asymptotes. $f'(x) = \frac{2(2x-1)}{3x^{2/3}}$ and $f''(x) = \frac{4(x+1)}{9x^{5/3}}$, implying that f is decreasing on $(-\infty, 1/2)$ and increasing on $(1/2, \infty)$, $(1/2, f(1/2))$ is a global minimum. There is an inflection point at $(-1, 3)$ where the concavity changes from up to down. f is concave down until $x = 0$ where the concavity changes to up and $(0, 0)$ is also an inflection point. f is concave up on $(0, \infty)$.

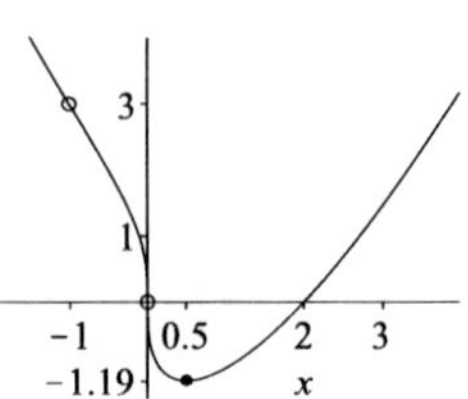

21. The domain of f is $(0, \infty)$. Since $\lim_{x\to\infty} f(x) = 0$, the x-axis is a horizontal asymptote. $f'(x) = -\frac{x-1}{2\sqrt{x}(x+1)^2}$ and $f''(x) = \frac{3x^2-6x-1}{4x^{3/2}(x+1)^3}$, implying that f is increasing on $(0, 1)$ and decreasing on $(1, \infty)$. f has a global maximum at $(1, 1/2)$. There is an inflection point at $(x_0, f(x_0))$ where $x_0 = 1 + \frac{2}{3}\sqrt{3} \approx 2.155$ (verify). f is concave down on $(0, x_0)$ and concave up on (x_0, ∞).

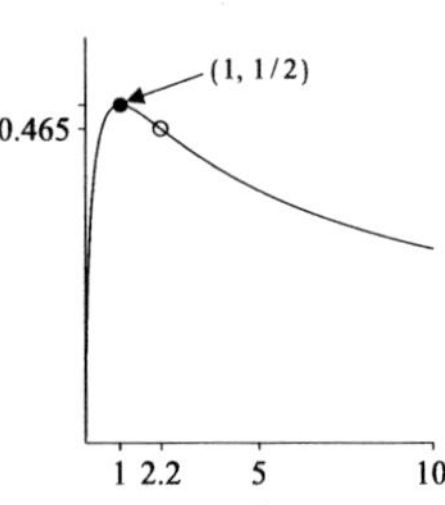

23. 0 is not in the domain of f, the y-axis is a vertical asymptote. $f'(x) = \frac{2(x^3-8)}{x^2}$ and $f''(x) = \frac{2(x^3+16)}{x^3}$, implying that f is decreasing on $(-\infty, 0)$ and on $(0, 2)$. f is increasing on $(2, \infty)$ so f has a local minimum at $(2, 12)$. There is an inflection point at $(-16^{1/3}, 0)$, f is concave up on $(-\infty, -16^{1/3})$, concave down on $(-16^{1/3}, 0)$ and concave up on $(0, \infty)$.

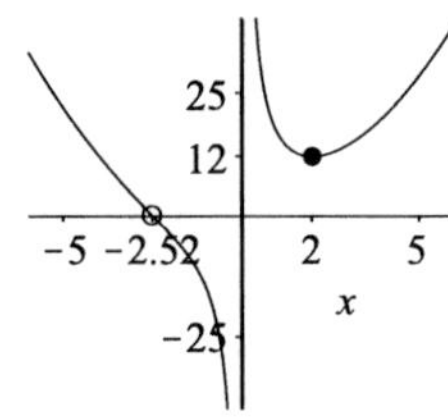

Further Theory and Practice

25. 4 is not in the domain of f, $x = 4$ is a vertical asymptote and, since $\lim_{|x|\to\infty} f(x) = 0$, the x-axis is a horizontal asymptote. $f'(x) = -\frac{2(x+2)}{3x^{2/3}(x-4)^2}$ and $f''(x) = \frac{2(5x^2+20x-16)}{9x^{5/3}(x-4)^3}$ so f increases on $(-\infty, -2)$, decreases on $(-2, 4)$, and decreases on $(4, \infty)$ implying that $(-2, f(-2))$ is a local maximum and $(0, 0)$ has a vertical tangent. There are inflection points at $(c_1, f(c_1))$ and $(c_2, f(c_2))$ where $c_{1,2} = -2 \mp \frac{6}{5}\sqrt{5}$ (verify). f is concave up on $(-\infty, c_1)$, concave down on $(c_1, 0)$, concave up on $(0, c_2)$, and concave down on $(c_2, 4)$ and on $(4, \infty)$. This makes $(0, 0)$ an inflection point also.

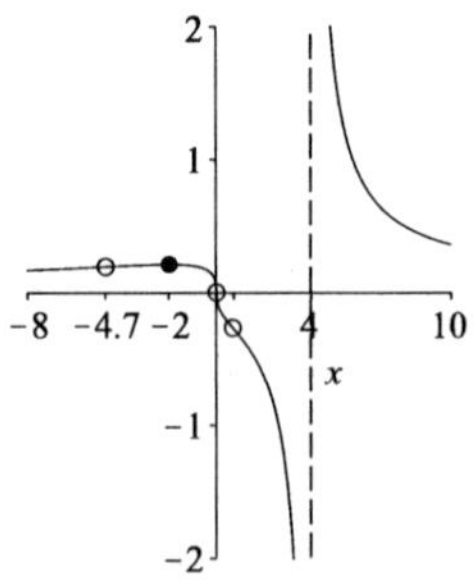

27. The domain of f is $[-1, 1]$. There are no asymptotes. $f'(x) = \frac{x(3x^2-2)}{\sqrt{1-x^2}}$ and $f''(x) = \frac{6x^4-9x^2+2}{(1-x^2)^{3/2}}$ so f increases on $(-1, -\sqrt{2/3}\,)$, decreases on $(-\sqrt{2/3}\,, 0)$, increases on $(0, \sqrt{2/3}\,)$, and then decreases on $(\sqrt{2/3}\,, 1)$ implying that

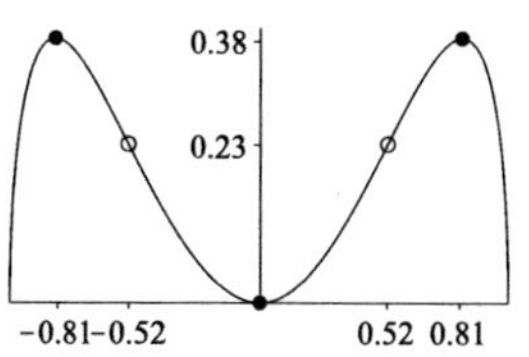

$(-\sqrt{2/3}\,, 2\sqrt{3}/9)$ and $(\sqrt{2/3}\,, 2\sqrt{3}/9)$ are global maxima and $(-1, 0)$, $(0, 0)$, and $(1, 0)$ are global minima. There are inflection points at $(x_1, f(x_1))$ and $(x_2, f(x_2))$ where $x_{1,2} = \mp\frac{1}{6}\sqrt{27 - 3\sqrt{33}}$ (verify). f is concave down on $(-1, x_1)$, concave up on (x_1, x_2), and concave down on $(x_2, 1)$.

29. f is not defined at 0, $x = 0$ is a vertical asymptote. Since $\lim_{x\to-\infty} f(x) = -1$ and $\lim_{x\to+\infty} f(x) = 1$, $y = -1$ is a left horizontal asymptote and $y = 1$ is a right horizontal asymptote. $f'(x) = -\frac{2(2x+9)}{x^2\sqrt{x^2+8x+18}}$ and $f''(x) = -\frac{2(x+6)(4x^2+27x+54)}{x^3(x^2+8x+18)^{3/2}}$ so f increases on $(-\infty, -9/2)$, decreases on $(-9/2, 0)$, and decreases on $(0, \infty)$. This implies that f has

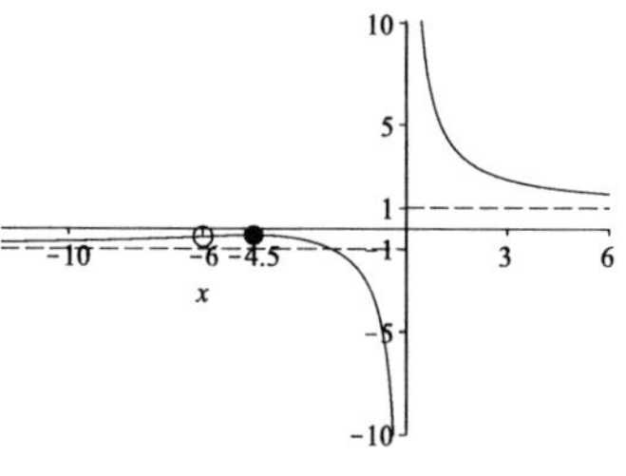

a local maximum at $(-9/2, -1/3)$. There are no local minima. There is an inflection point at $(-6, -\sqrt{6}/6)$. f is concave up on $(-\infty, -6)$, concave down on $(-6, 0)$, and concave up on $(0, \infty)$.

31. f is defined for all x but is not differentiable at -2. There are no asymptotes. If $x > -2$, then $f'(x) = \frac{3x-1}{2\sqrt{x+2}}$ and $f''(x) = \frac{3x+13}{4(x+2)^{3/2}}$, so f decreases on $(-2, 1/3)$ and increases on $(1/3, \infty)$. f is concave up and has a local minimum at $(1/3, -14\sqrt{21}/9)$. When $x < -2$,

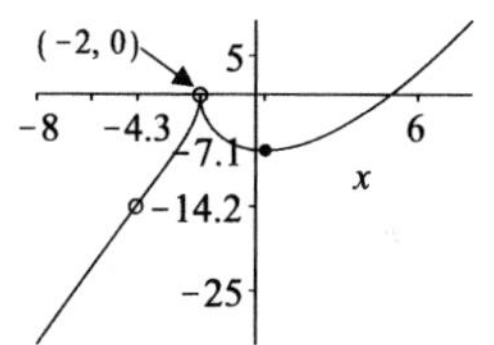

$f'(x) = -\frac{3x-1}{\sqrt{-x-2}}$ and $f''(x) = \frac{3x+13}{4(-x-2)^{3/2}}$. f increases with an inflection point at $(-13/3, -28\sqrt{21}/9)$ where it changes concavity from down to up. At $(-2, 0)$, both one-sided tangents are vertical producing a cusp. Since f is continuous and its concavity changes from down to up, $(2, 0)$, which is a local maximum, can be regarded as another point of inflection.

33. f is not defined at 0, the y-axis is a vertical asymptote. There are no horizontal asymptotes. $f'(x) = \frac{2x^3-1}{x^2}$ and $f''(x) = \frac{2(x^3+1)}{x^3}$ so f decreases on $(-\infty, 0)$ and on $(0, 1/2^{1/3})$ and increases on $(1/2^{1/3}, \infty)$ implying that $(1/2^{1/3}, f(1/2^{1/3}))$ is a local minimum. There are no local maxima. There is an inflection point at $(-1, 1)$. f is concave up on $(-\infty, -1)$, concave down on $(-1, 0)$, and concave up on $(0, \infty)$.

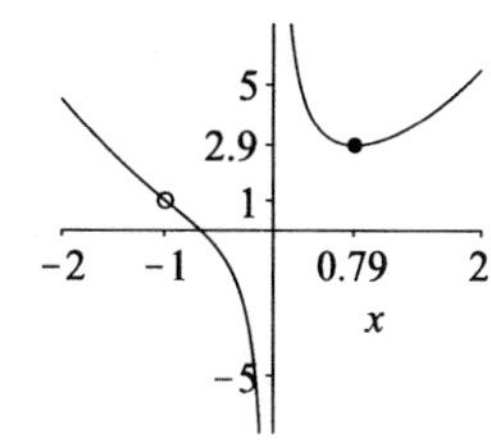

35. f is periodic with period $p = 2\pi$, so we consider the graph on $[-\pi, \pi]$. f is not defined at $\pm\pi/2$. There is a vertical asymptote at $\pi/2$ but $\lim_{x\to-\pi/2} f(x) = 0$ (verify). There are no horizontal asymptotes. The derivatives simplify to $f'(x) = \frac{1}{1-\sin(x)}$ and $f''(x) = \frac{\cos(x)}{(1-\sin(x))^2}$ so $f' \geq 0$ and f increases on $(-\pi, -\pi/2)$, on $(-\pi/2, \pi/2)$, and on $(\pi/2, \pi)$, implying that there are no local maxima or minima. There are also no inflection points. f is concave down on $(-\pi, -\pi/2)$, concave up on $(-\pi/2, \pi/2)$, and concave down on $(\pi/2, \pi)$. This repeats over every adjoining interval of width 2π.

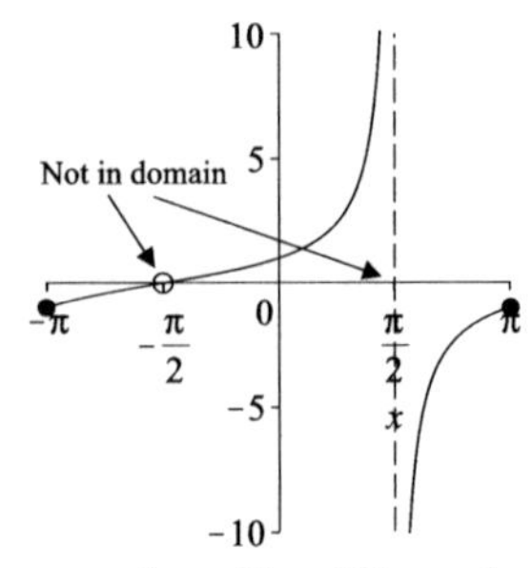

37. There are no asymptotes. When $x > 0$, $f'(x) = (x + 1)(3x + 1)$ and $f''(x) = 6x + 4$ so f increases and is concave up. When $x < 0$, $f'(x) = -(x + 1)(3x + 1)$ and $f''(x) = -(6x + 4)$ so f is decreasing on $(-\infty, -1)$, increasing on $(-1, -1/3)$ and decreasing on $(-1/3, 0)$, implying that $(-1, 0)$ and $(0, 0)$ are global minima and $(-1/3, 4/27)$ is a local maximium. There is an inflection point at $(-2/3, 2/27)$ where f changes from concave up to concave down. This also makes $(0, 0)$ a point of inflection.

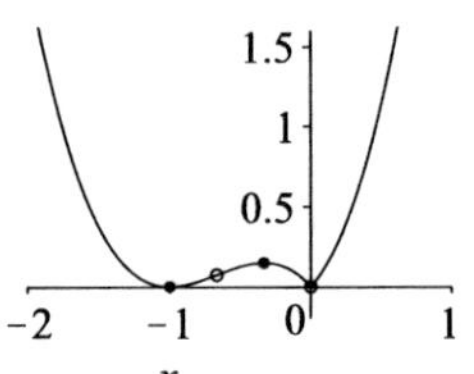

39. f is not defined at ± 1, $x = \pm 1$ are vertical asymptotes. Since $\lim_{|x|\to\infty} f(x) = 0$, the x-axis is a horizontal asymptote. $f'(x) = -\frac{13x^2+28x+13}{(x^2-1)^2}$ and $f''(x) = \frac{2(x+2)(13x^2+16x+7)}{(x^2-1)^3}$ so f has critical points at $c_{1,2} = -\frac{14}{13} \mp \frac{3}{13}\sqrt{3}$ (verify). f decreases on $(-\infty, c_1)$ and increases on $(c_1, -1)$ so $(c_1, f(c_1))$ is a local minimum. f increases on $(-1, c_2)$ and decreases on $(c_2, 1)$ so $(c_2, f(c_2))$ is a local maximum. f also decreasing on $(1, \infty)$. f has an inflection point at $(-2, -4)$. f is concave down on $(-\infty, -2)$, concave up on $(-2, -1)$, concave down in $(-1, 1)$, and concave up on $(1, \infty)$.

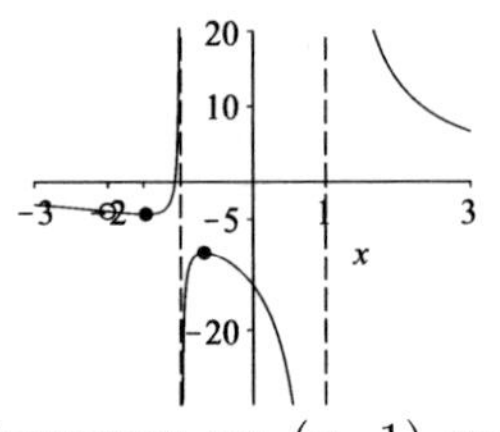

41. f is not defined at 0, $x = 0$ is a vertical asymptote. Since $f(x) = x/2 - 2/x$, the line $y = x/2$ is a skew asymptote. $f'(x) = \frac{x^2+4}{2x^2}$ and $f''(x) = -\frac{4}{x^3}$ so f is increasing on $(-\infty, 0)$ and on $(0, \infty)$. There are no inflection points.

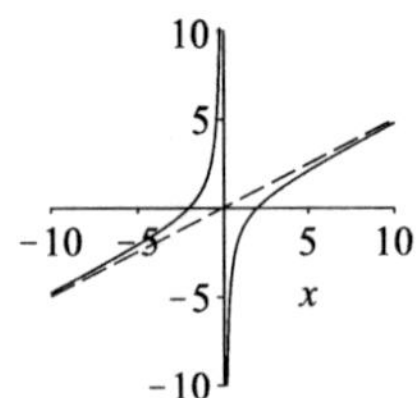

43. By the Mean Value Theorem, given $x > 0$, $(f'(x) - f'(0))/x \geq 1$, implying that $f'(x) \geq f'(0) + x$. Consequently, there is an $x_0 > 0$ such that $f'(x) \geq 1$ for all $x \geq x_0$. Using the Mean Value Theorem once more, given $x > x_0$, $(f(x) - f(x_0))/(x - x_0) \geq 1$. This implies that $f(x) \geq f(x_0) + x - x_0$ and $\lim_{x\to\infty} f(x) = \infty$. A similar argument will show that

$\lim_{x\to-\infty} f(x) = \infty$.

CALCULATOR/COMPUTER EXERCISES

45. f is a polynomial, no asymptotes. f has a global minimum at $A \approx (-4.399, -125.33)$ and inflection points at $B \approx (-2.943, -76.223)$ and $C \approx (-0.0566, -4.054)$. f is concave up on $(-\infty, B)$, concave down on (B, C) and concave up on (C, ∞).

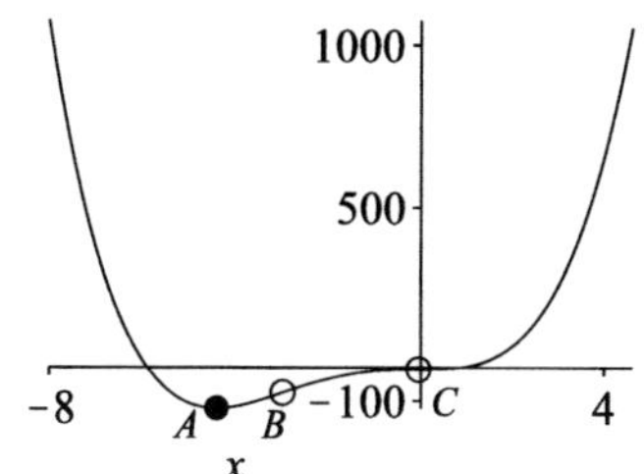

47. f is a polynomial, no asymptotes. f has local maxima at $A \approx (-1.601, 0.148)$ and $E \approx (0.399, 7.316)$, local minima at $C = (-1.184, -0.176)$ and $G \approx (1.586, -0.646)$, and inflects at the points: $B \approx (-1.412, -0.00259)$, $D = (-0.3145, 3.781)$, and $F = (1.126, 2.694)$. f is concave down on $(-\infty, B)$, concave up on (B, D), concave down on (D, F) and concave up on (F, ∞).

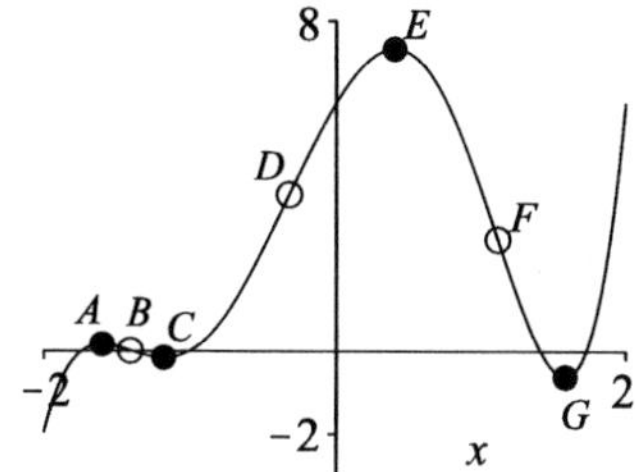

49. The domain of f is $[-1, 1]$. f has a global minimum at $A \approx (-0.3837, -0.2071)$, a global maximum at $C \approx (0.9239, 1.207)$ and an inflection point at $B \approx (0.5120, 0.7020)$. f is concave up on $(-1, B)$ and concave down on $(B, 1)$.

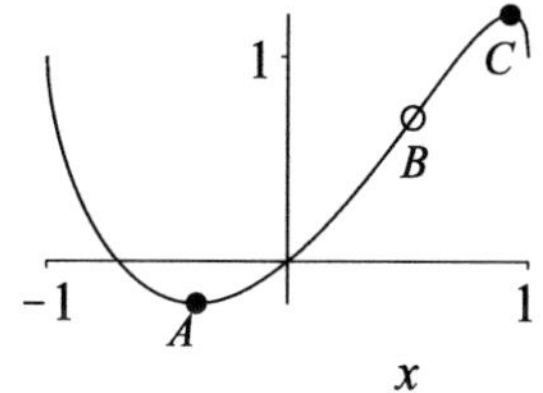

51. The domain of f is $[-2, 2]$. f has a global maximum at $C = (0, 2)$, and global minima at the endpoints. There are four inflection points, $A \approx (-1.732, 0.25)$, $B \approx (-0.583, 1.428)$, $D \approx (0.583, 1.428)$, and $E \approx (1.732, 0.25)$. f is concave down on $(-1, A)$ and concave up on (A, B), concave down on (B, D), concave up on (D, E), and concave down on $(E, 2)$.

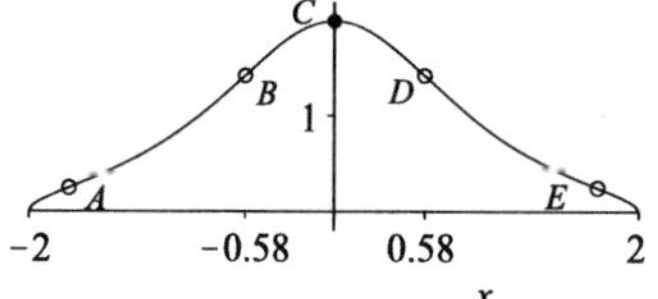

53. f is defined everywhere and has no asymptotes. It has a local maximum at the point $A \approx (-2.225, 10.59)$ and a local minimum at $C \approx (2.440, -4.113)$. There is an inflection point at $B \approx (0.0960, 3.269)$. f increases on $(-\infty, A)$, decreases on (A, C), and increases on (C, ∞). The graph is con-

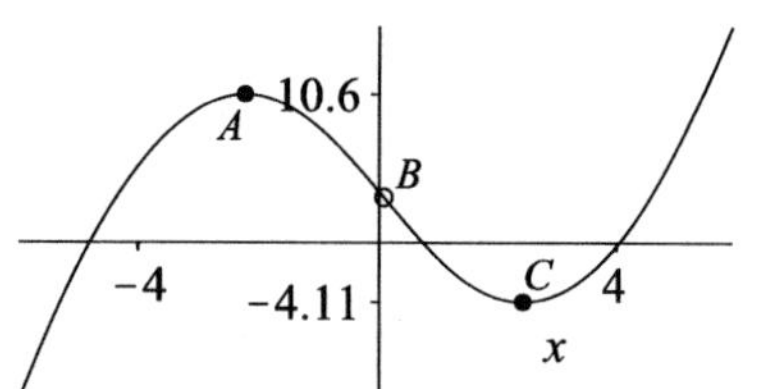

cave down on $(-\infty, B)$ and concave up on (B, ∞).

55. The four plots appear below, f on top and g on the bottom.

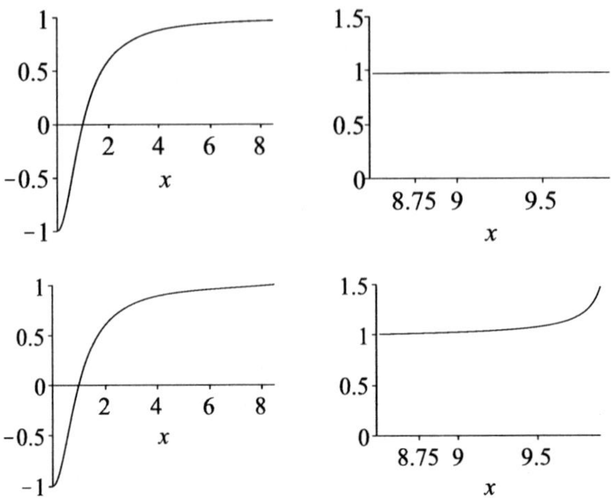

The similarity of the graphs on $[0, 8.5]$ is due to the fact that

$$f(x) = \frac{x^2 - 1}{x^2 + 1} \quad \text{and} \quad g(x) \approx \frac{x^2 - 0.9900}{x^2 + 1} \cdot \frac{x^2 - 101.01}{x^2 - 100}\,.$$

On most of the second interval the graphs are still similar but, because f has the horizontal asymptote $y = 1$, and g has the vertical asymptote $x = 10$, the points on their graphs separate as $(x, g(x))$ begins to move upward at a faster rate.

4.7 L'Hôpital's Rule

Problems for Practice

1. $\displaystyle \lim_{x\to 0} \frac{1 - e^x}{x} = \lim_{x\to 0} \frac{\frac{d}{dx}(1 - e^x)}{\frac{d}{dx}x} = \lim_{x\to 0} \frac{-e^x}{1} = -1$

3. $\displaystyle \lim_{x\to 5} \frac{\ln(x/5)}{x - 5} = \lim_{x\to 5} \frac{\frac{d}{dx}\ln(x/5)}{\frac{d}{dx}(x - 5)} = \lim_{x\to 5} \frac{1/x}{1} = 1/5$

5. $$\lim_{x\to \pi/2} \frac{\ln(\sin(x))}{(\pi - 2x)^2} = \lim_{x\to \pi/2} \frac{\frac{d}{dx}\ln(\sin(x))}{\frac{d}{dx}(\pi - 2x)^2} = \lim_{x\to \pi/2} \frac{\cot(x)}{-4(\pi - 2x)}$$
$$= \lim_{x\to \pi/2} \frac{\frac{d}{dx}\cot(x)}{\frac{d}{dx}\big(-4(\pi - 2x)\big)} = \lim_{x\to \pi/2} \frac{-\csc^2(x)}{8} = -1/8$$

7. $$\lim_{x\to -1} \frac{\cos(x+1)-1}{x^3+x^2-x-1} = \lim_{x\to -1} \frac{\frac{d}{dx}(\cos(x+1)-1)}{\frac{d}{dx}(x^3+x^2-x-1)} = \lim_{x\to -1} \frac{-\sin(x+1)}{3x^2+2x-1}$$
$$= \lim_{x\to -1} \frac{\frac{d}{dx}(-\sin(x+1))}{\frac{d}{dx}(3x^2+2x-1)} = \lim_{x\to -1} \frac{-\cos(x+1)}{6x+2} = 1/4$$

9. $$\lim_{x\to 0} \frac{e^x - e^{-x}}{x^2} = \lim_{x\to 0} \frac{\frac{d}{dx}(e^x - e^{-x})}{\frac{d}{dx}x^2} = \lim_{x\to 0} \frac{e^x + e^{-x}}{2x} \text{ does not exist}$$

11. $$\lim_{x\to 1} \frac{\ln(x)}{x-\sqrt{x}} = \lim_{x\to 1} \frac{\frac{d}{dx}\ln(x)}{\frac{d}{dx}(x-\sqrt{x})} = \lim_{x\to 1} \frac{1/x}{1-1/(2\sqrt{x}\,)} = 2$$

13. $$\lim_{x\to\infty} \frac{\sin(3/x)}{\sin(9/x)} = \lim_{x\to\infty} \frac{\frac{d}{dx}\sin(3/x)}{\frac{d}{dx}\sin(9/x)} = \lim_{x\to\infty} \frac{\cos(3/x)\cdot(-3/x^2)}{\cos(9/x)\cdot(-9/x^2)} = 1/3$$

15. $$\lim_{x\to -\infty} \frac{\ln(1+1/x)}{\sin(1/x)} = \lim_{x\to -\infty} \frac{\frac{d}{dx}\ln(1+1/x)}{\frac{d}{dx}\sin(1/x)} = \lim_{x\to -\infty} \frac{(1+1/x)^{-1}\cdot(-1/x^2)}{\cos(1/x)\cdot(-1/x^2)}$$
$$= 1$$

17. $$\lim_{x\to\infty} \frac{x^2}{e^{3x}} = \lim_{x\to\infty} \frac{\frac{d}{dx}x^2}{\frac{d}{dx}e^{3x}} = \lim_{x\to\infty} \frac{2x}{3e^{3x}} = \lim_{x\to\infty} \frac{\frac{d}{dx}(2x)}{\frac{d}{dx}(3e^{3x})} = \lim_{x\to\infty} \frac{2}{9e^{3x}} = 0$$

19. $$\lim_{x\to 0} \frac{\sin^2(x)}{x^2} = \lim_{x\to 0} \left(\frac{\sin(x)}{x}\right)^2 = 1$$

21. $$\lim_{x\to 1} \frac{(x-1)^2}{\cos^2(\pi x/2)} = \lim_{x\to 1} \frac{\frac{d}{dx}(x-1)^2}{\frac{d}{dx}\cos^2(\pi x/2)} = \lim_{x\to 1} \frac{2(x-1)}{2\cos(\pi x/2)\cdot(-\sin(\pi x/2)\cdot\pi/2)} =$$
$$\lim_{x\to 1} \frac{2(x-1)}{-\pi\sin(\pi x)/2} = -\tfrac{4}{\pi}\cdot\lim_{x\to 1} \frac{x-1}{\sin(\pi x)} = -\tfrac{4}{\pi}\cdot\lim_{x\to 1} \frac{\frac{d}{dx}(x-1)}{\frac{d}{dx}\sin(\pi x)}$$
$$= -\tfrac{4}{\pi}\cdot\lim_{x\to 1} \frac{1}{\pi\cos(\pi x)} = 4/\pi^2$$

23. $$\lim_{x\to 0} \frac{\sin(2x)-2x}{x^3} = \lim_{x\to 0} \frac{\frac{d}{dx}(\sin(2x)-2x)}{\frac{d}{dx}x^3} = \lim_{x\to 0} \frac{2\cos(2x)-2}{3x^2}$$
$$= \tfrac{2}{3}\cdot\lim_{x\to 0} \frac{\frac{d}{dx}(\cos(2x)-1)}{\frac{d}{dx}x^2} = -\tfrac{4}{3}\cdot\lim_{x\to 0} \frac{\sin(2x)}{2x} = -\frac{4}{3}$$

25. $$\lim_{x\to\infty} x\cdot e^{-2x} = \lim_{x\to\infty} \frac{x}{e^{2x}} = \lim_{x\to\infty} \frac{\frac{d}{dx}x}{\frac{d}{dx}e^{2x}} = \lim_{x\to\infty} \frac{1}{2e^{2x}} = 0$$

27. $$\lim_{x\to\infty} e^{-x}\cdot\ln(x) = \lim_{x\to\infty} \frac{\ln(x)}{e^x} = \lim_{x\to\infty} \frac{\frac{d}{dx}\ln(x)}{\frac{d}{dx}e^x} = \lim_{x\to\infty} \frac{1/x}{e^x} = 0$$

29. $$\lim_{x\to 1^+} (x-1)^{-1}\ln(x) = \lim_{x\to 1^+} \frac{\ln(x)}{x-1} = \lim_{x\to 1^+} \frac{\frac{d}{dx}\ln(x)}{\frac{d}{dx}(x-1)} = \lim_{x\to 1^+} \frac{1/x}{1} = 1$$

31. $\lim_{x\to -2}(x+2)\tan(\pi x/4) = \lim_{x\to -2}\dfrac{x+2}{\cot(\pi x/4)} = \lim_{x\to -2}\dfrac{1}{-\pi\csc^2(\pi x/4)/4} = -\dfrac{4}{\pi}$

33. Let $u = x^{\sqrt{x}}$. Since $\ln(u) = \ln\left(x^{\sqrt{x}}\right) = \sqrt{x}\ln(x) = \ln(x)/x^{-1/2}$, which, as $x \to 0^+$, is of the form $-\infty/\infty$, l'Hôpital's Rule yields

$$\lim_{x\to 0^+}\ln(u) = \lim_{x\to 0^+}\frac{\ln(x)}{x^{-1/2}} = \lim_{x\to 0^+}\frac{1/x}{-x^{-3/2}/2} = \lim_{x\to 0^+}\frac{\sqrt{x}}{-2} = 0\,.$$

Therefore, $\lim_{x\to 0^+} x^{\sqrt{x}} = e^0 = 1$.

35. Let $u = \left(\frac{x-5}{x}\right)^x$. Since $\ln(u) = x\ln(1-5/x) = \frac{\ln(1-5/x)}{x^{-1}}$, which, as $x \to \infty$, is of the form $0/0$, l'Hôpital's Rule yields

$$\lim_{x\to\infty}\ln(u) = \lim_{x\to\infty}\frac{(1-5/x)^{-1}\cdot(5/x^2)}{-1/x^2} = -5\lim_{x\to\infty}(1-5/x)^{-1} = -5\,.$$

Therefore, $\lim_{x\to\infty}\left(\frac{x-5}{x}\right)^x = e^{-5}$.

37. Let $u = x^{1/x}$. Since $\ln(u) = (1/x)\ln(x) = \frac{\ln(x)}{x}$, which, as $x \to \infty$, is of the form ∞/∞, l'Hôpital's Rule yields

$$\lim_{x\to\infty}\ln(u) = \lim_{x\to\infty}\frac{1/x}{1} = 0\,.$$

Therefore, $\lim_{x\to\infty}(x\cdot\ln(x))^{-1/x} = e^0 = 1$.

39. Let $u = x^{1/\sqrt{x}}$. Since $\ln(u) = \ln\left(x^{1/\sqrt{x}}\right) = x^{-1/2}\ln(x) = \ln(x)/x^{1/2}$, which, as $x \to \infty$, is of the form ∞/∞, l'Hôpital's Rule yields

$$\lim_{x\to\infty}\ln(u) = \lim_{x\to\infty}\frac{\ln(x)}{x^{1/2}} = \lim_{x\to\infty}\frac{1/x}{\frac{1}{2}x^{-1/2}} = 2\lim_{x\to\infty}\frac{1}{\sqrt{x}} = 0\,.$$

Therefore, $\lim_{x\to\infty} x^{1/\sqrt{x}} = e^0 = 1$.

41. $\lim_{x\to\infty}\left(\frac{x^3}{\sqrt{x^2+5}} - \frac{x^3}{\sqrt{x^2+3}}\right) = \lim_{x\to\infty}\frac{x^3(\sqrt{x^2+3}-\sqrt{x^2+5})}{\sqrt{x^2+5}\,\sqrt{x^2+3}}$

$= \lim_{x\to\infty}\left(\frac{x^3(\sqrt{x^2+3}-\sqrt{x^2+5})}{\sqrt{x^2+5}\,\sqrt{x^2+3}}\cdot\frac{\sqrt{x^2+3}+\sqrt{x^2+5}}{\sqrt{x^2+3}+\sqrt{x^2+5}}\right)$

$= \lim_{x\to\infty}\frac{-2x^3}{\sqrt{x^2+5}\,\sqrt{x^2+3}\,(\sqrt{x^2+3}+\sqrt{x^2+5})}$

$= \lim_{x\to\infty}\frac{-2}{\sqrt{1+5/x}\,\sqrt{1+3/x}\,(\sqrt{1+3/x}+\sqrt{1+5/x})} = -1$

43. $\lim_{x\to 0}\left(\frac{x}{1-\cos(x)} - \frac{2}{x}\right) = \lim_{x\to 0}\frac{x^2-2(1-\cos(x))}{x(1-\cos(x))}$ (0/0, apply l'Hôpital)

$= \lim_{x\to 0}\frac{2x-2\sin(x)}{1-\cos(x)+x\sin(x)} = \lim_{x\to 0}\frac{2-2\cos(x)}{2\sin(x)+x\cos(x)}$ (again)

$= \lim_{x\to 0}\frac{2\sin(x)}{3\cos(x)-x\sin(x)} = 0$

45. $\lim_{x\to 0}\left(\frac{1}{\sin(x)} - \frac{1}{\sin(2x)}\right) = \lim_{x\to 0}\frac{\sin(2x)-\sin(x)}{\sin(x)\sin(2x)}$ (0/0, apply l'Hôpital)

$= \lim_{x\to 0}\frac{2\cos(2x)-\cos(x)}{\cos(x)\sin(2x)+2\sin(x)\cos(2x)}$ Does not exist

47. $\lim_{x\to\infty}\left(\sqrt{4x-5}-2\sqrt{x}\right)=\lim_{x\to\infty}\left(\frac{\sqrt{4x-5}-2\sqrt{x}}{1}\cdot\frac{\sqrt{4x-5}+2\sqrt{x}}{\sqrt{4x-5}+2\sqrt{x}}\right)$
$=\lim_{x\to\infty}\frac{-5}{\sqrt{4x-5}+2\sqrt{x}}=0$

49. $\lim_{x\to\infty}(x+1)\ln\left(\frac{x}{x+1}\right)=\lim_{x\to\infty}\frac{\ln\left(\frac{x}{x+1}\right)}{(x+1)^{-1}}$ (0/0, l'Hôpital)
$=\lim_{x\to\infty}\frac{1/x-1/(x+1)}{-(x+1)^{-2}}=\lim_{x\to\infty}\frac{x+1}{-x}=-1$

51. $\lim_{x\to\infty}\left(\sqrt{e^{2x}-e^{x}}-e^{x}\right)=\lim_{x\to\infty}\left(\frac{\sqrt{e^{2x}-e^{x}}-e^{x}}{1}\cdot\frac{\sqrt{e^{2x}-e^{x}}+e^{x}}{\sqrt{e^{2x}-e^{x}}+e^{x}}\right)$
$=\lim_{x\to\infty}\frac{-e^{x}}{\sqrt{e^{2x}-e^{x}}+e^{x}}=\lim_{x\to\infty}\frac{-1}{\sqrt{1-1/e^{x}}+1}=-1/2$

53. $\lim_{x\to 0}\frac{\arctan(x)}{x}=\lim_{x\to 0}\frac{1/(1+x^2)}{1}=1$ (0/0, l'Hôpital)

55. $\lim_{x\to 0}\frac{\sinh(x)}{\exp(x)-1}=\lim_{x\to 0}\frac{\cosh(x)}{\exp(x)}=1$ (0/0, l'Hôpital)

57. $\lim_{x\to 0}\frac{\cosh(x)-1}{\sinh^2(x)}=\lim_{x\to 0}\frac{\sinh(x)}{2\sinh(x)\cosh(x)}$ (0/0, l'Hôpital)
$=\lim_{x\to 0}\frac{1}{2\cosh(x)}=1/2$

59. $\lim_{x\to 0}\frac{\arcsin(x)-x}{\arctan(x)-x}=\lim_{x\to 0}\frac{1/\sqrt{1-x^2}-1}{1/(1+x^2)-1}$ (0/0, l'Hôpital)
$=\lim_{x\to 0}\frac{-x(1-x^2)^{-3/2}}{2x(1+x^2)^{-2}}=\lim_{x\to 0}\frac{-(1-x^2)^{-3/2}}{2(1+x^2)^{-2}}=-1/2$

61. Let $u=(\cos(2x))^{1/x^2}$. Since $\ln(u)=(1/x^2)\ln(\cos(2x))=\frac{\ln(\cos(2x))}{x^2}$, which, as $x\to 0$, is of the form $0/0$, l'Hôpital's Rule (twice) yields

$$\lim_{x\to 0}\ln(u)=\lim_{x\to 0}\frac{-2\tan(2x)}{2x}=\lim_{x\to 0}\frac{-4\sec^2(2x)}{2}=-2\,.$$

Therefore, $\lim_{x\to 0}(\cos(2x))^{1/x^2}=e^{-2}$.

63. Let $u=\left(\frac{5+x}{2+x}\right)^{x}$. Since $\ln(u)=x\ln\left(\frac{5+x}{2+x}\right)=\frac{\ln\left(\frac{5+x}{2+x}\right)}{x^{-1}}$, which, as $x\to\infty$, is of the form $0/0$, l'Hôpital's Rule yields

$$\lim_{x\to\infty}\ln(u)=\lim_{x\to\infty}\frac{(5+x)^{-1}-(2+x)^{-1}}{-x^{-2}}=\lim_{x\to\infty}\frac{3x^2}{(5+x)(2+x)}=3\,.$$

Therefore, $\lim_{x\to\infty}\left(\frac{5+x}{2+x}\right)^{x}=e^{3}$.

Further Theory and Practice

65. Let $u=(1+h)^{1/h}$. Since $\ln(u)=(1/h)\ln(1+h)=\frac{\ln(1+h)}{h}$, which, as $h\to 0^{+}$, is of the form $0/0$, l'Hôpital's Rule yields

$$\lim_{h\to 0^{+}}\ln(u)=\lim_{h\to 0^{+}}\frac{(1+h)^{-1}}{1}=1\,.$$

Therefore, $\lim_{h\to 0^{+}}(1+h)^{1/h}=e$.

Given x, let $u = (1 + hx)^{1/h}$. Since $\ln(u) = (1/h)\ln(1 + hx) = \frac{\ln(1+hx)}{h}$, which, as $h \to 0^+$, is of the form $0/0$, l'Hôpital's Rule yields

$$\lim_{h\to 0^+} \ln(u) = \lim_{h\to 0^+} \frac{(1 + hx)^{-1} \cdot x}{1} = x\,.$$

Therefore, $\lim_{h\to 0^+} (1 + hx)^{1/h} = e^x$.

67. The limit, as $x \to 0$, is of the form $0/0$ so, by l'Hôpital's Rule, three times,

$$\lim_{x\to 0} \frac{a\sin(x)-\sin(ax)}{\tan(bx)-b\tan(x)} = \lim_{x\to 0} \frac{a\cos(x)-a\cos(ax))}{b\sec^2(bx)-b\sec^2(x))}$$

$$= \lim_{x\to 0} \frac{-a\sin(x)+a^2\sin(ax))}{2b^2\sec^2(bx)\tan(bx)-2b\sec^2(x)\tan(x)}$$

$$= \lim_{x\to 0} \frac{-a\cos(x)+a^3\cos(ax))}{4b^3\sec^2(bx)\tan^2(bx)+2b^3\sec^4(bx)-4b\sec^2(x)\tan^2(x)-2b\sec^4(x)} = \frac{a^3-a}{2(b^3-b)}\,.$$

69. This does not require l'Hôpital's Rule: $\lim_{x\to(\pi/2)^-} (x - \pi/2)\cot(x) = 0$.

71. l'Hôpital, $0/0$: $\lim_{x\to 0^+} \frac{\ln(1+x)}{\ln(1+3x)} = \lim_{x\to 0^+} \frac{(1+x)^{-1}}{3(1+3x)^{-1}} = 1/3$.

73. l'Hôpital, ∞/∞: $\lim_{x\to 0^+} \frac{2\ln(x)}{\ln(2x)} = \lim_{x\to 0^+} \frac{2/x}{1/x} = 2$.

75. Let $u = x^{-1/\ln(x)}$. Since $\ln(u) = (-1/\ln(x))\ln(x) = -1$, $\lim_{x\to 0^+} \ln(u) = \lim_{x\to 0^+} (-1) = -1$.

Therefore, $\lim_{x\to 0^+} x^{-1/\ln(x)} = e^{-1}$.

77. The graph is on the right. Using l'Hôpital's Rule,

$$\lim_{x\to\infty} f(x) = \lim_{x\to\infty} \frac{x}{e^x} = \lim_{x\to\infty} \frac{1}{e^x} = 0\,,$$

so the x-axis is a right horizontal asymptote.

79. The graph is on the right. Since $\lim_{x\to -\infty} f(x) = 0$, the x-axis is a left horizontal asymptotote. Using l'Hôpital's Rule,

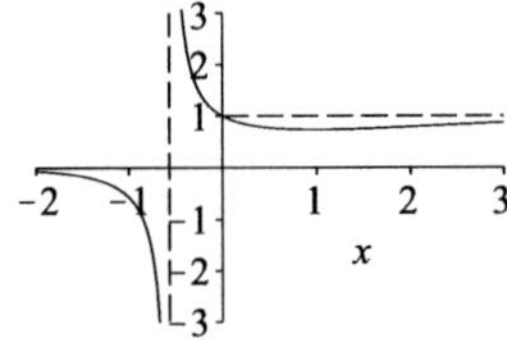

$$\lim_{x\to\infty} f(x) = \lim_{x\to\infty} \frac{e^x}{e^x+1} = \lim_{x\to\infty} \frac{e^x}{e^x} = 1\,,$$

and the line $y = 1$ is a right horizontal asymptote.

81. The graph is on the right. Using l'Hôpital's Rule,

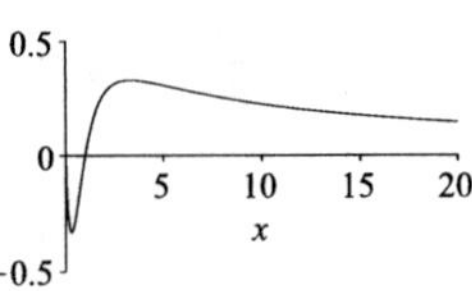

$$\lim_{x\to\infty} f(x) = \lim_{x\to\infty} \frac{\ln(x)+1}{2x} = \lim_{x\to\infty} \frac{1/x}{2} = 0\,,$$

and the x-axis is a right horizontal asymptote.

83. Since $r^x = e^{x\ln(r)} = e^{ax}$ where $a = \ln(r) > 0$, this is equivalent to showing that

$$\lim_{x\to\infty} \frac{x^\alpha}{e^{ax}} = 0 \quad \text{when } a > 0\,.$$

If $\alpha \le 1$ and $x > 1$, then $x^\alpha \le x$, so $0 \le \frac{x^\alpha}{e^{ax}} \le \frac{x}{e^{ax}}$. Since $\lim_{x\to\infty} \frac{x}{e^{ax}} = \lim_{x\to\infty} \frac{1}{ae^{ax}} = 0$ by l'Hôpital's Rule, the Pinching Theorem implies that $\lim_{x\to\infty} \frac{x^\alpha}{e^{ax}} = 0$.

If $\alpha > 1$, then l'Hôpital's Rule also applies to $\lim_{x\to\infty} \frac{x^\alpha}{e^{ax}}$. Consequently, after n applications, $0 < \alpha - n \le 1$,

$$\lim_{x\to\infty} \frac{x^\alpha}{e^{ax}} = \frac{\alpha\cdot(\alpha-1)\cdot(\alpha-2)\cdots(\alpha-n+1)}{a^n} \lim_{x\to\infty} \frac{x^{\alpha-n}}{e^{ax}} = 0\,.$$

The fact that $\lim_{x\to\infty} \frac{x^\alpha}{r^x} = 0$ implies that for any polynomial $p(x)$, $\lim_{x\to\infty} \frac{p(x)}{r^x} = 0$. Therefore,

$$\lim_{x\to\infty} \frac{a\cdot r^x + p(x)}{b\cdot r^x + q(x)} = \lim_{x\to\infty} \frac{a + p(x)/r^x}{b + q(x)/r^x} = \frac{a}{b}\,.$$

85. If $\alpha \le 0$, then for $x > e$, $\ln(x)^\alpha \le 1$ and $0 \le \frac{\ln(x)^\alpha}{x^\beta} \le \frac{1}{x^\beta}$. Since $\lim_{x\to\infty} \frac{1}{x^\beta} = 0$, the Pinching Theorem implies that $\lim_{x\to\infty} \frac{\ln(x)^\alpha}{x^\beta} = 0$.

If $\alpha > 0$ then, as $x \to \infty$, $\frac{\ln(x)^\alpha}{x^\beta}$ is indeterminate of the form ∞/∞ so, applying l'Hôpital's Rule,

$$\lim_{x\to\infty} \frac{\ln(x)^\alpha}{x^\beta} = \lim_{x\to\infty} \frac{\alpha\ln(x)^{\alpha-1}\cdot(1/x)}{\beta x^{\beta-1}} = \frac{\alpha}{\beta}\cdot \lim_{x\to\infty} \frac{\ln(x)^{\alpha-1}}{x^\beta}\,.$$

Repeat this n times, where $-1 < \alpha - n \le 0$, to obtain

$$\lim_{x\to\infty} \frac{\ln(x)^\alpha}{x^\beta} = \left(\frac{\alpha}{\beta}\right)^{n-1} \cdot \lim_{x\to\infty} \frac{\ln(x)^{\alpha-n}}{x^\beta} = 0\,.$$

87. The equation of the line is $y = \frac{f(b)-f(a)}{g(b)-g(a)} \cdot (x - g(a)) + f(a)$. If $P = (x, y)$ is a point in the plane, then the vertical distance from P to this line is

$$\frac{f(b)-f(a)}{g(b)-g(a)} \cdot (x - g(a)) + f(a) - y\,.$$

When $P = (g(t), f(t))$, the vertical distance from P to the line is

$$\frac{f(b)-f(a)}{g(b)-g(a)} \cdot (g(t) - g(a)) + f(a) - f(t) = \frac{r(t)}{g(b)-g(a)}\,.$$

Consequently, $r(t)$ is the scaled vertical distance between the parametrized curve $t \mapsto (g(t), f(t))$, $a \le t \le b$, and the line passing through its endpoints $(g(a), f(a))$ and $(g(b), f(b))$. Points $\xi \in (a, b)$ where $r'(\xi) = 0$ are values of the parameter t where $r(t)$ attains local maximum and local minimum values.

CALCULATOR/COMPUTER EXERCISES

89. The graph on the right suggests that the limit is 1/2. The table below provides more evidence.

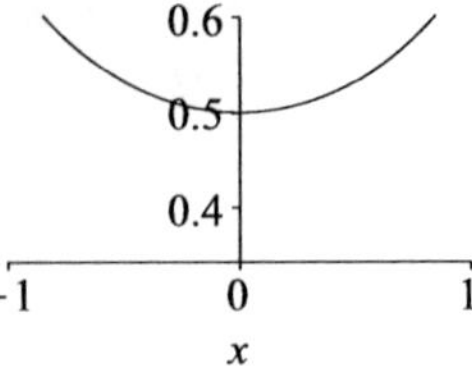

n	1	2	3	4
$f(10^{-n})$	0.50123	0.50001	0.50000	0.50000

91. The graph on the right suggests that the limit is 1. The table below provides more evidence.

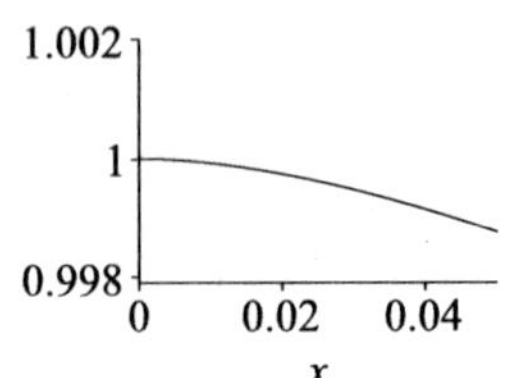

n	1	2	3	4
$f(10^{-n})$	0.99617	0.99992	0.99999	0.99999

93. The picture on the right shows the graphs of f/g (solid line) and f'/g' (dashed line) near $c = 0$. Both have limit 0.5 as $x \to 0$.

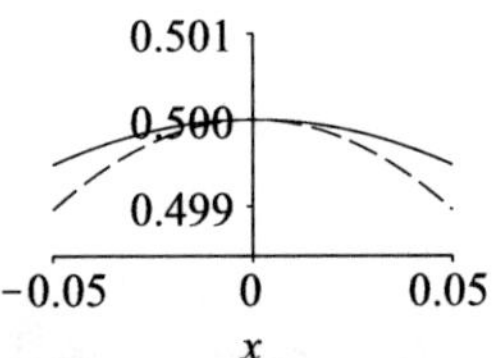

95. The picture on the right shows the graphs of f/g (solid line) and f'/g' (dashed line) near $c = \pi$. Both have limit -0.5 as $x \to 0$.

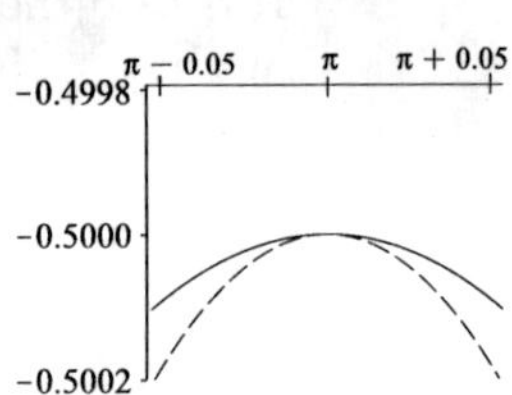

4.8 The Newton-Raphson Method

Problems for Practice

1. $\Phi(x) = x - \frac{x^2-6}{2x}$, $x_1 = 2.0, x_2 = 2.5, x_3 = 2.45$

3. $\Phi(x) = x - \frac{x^2-5x+2}{2x-5}$, $x_1 = 4.0, x_2 = 4.6667, x_3 = 4.5641$

5. $\Phi(x) = x - \frac{x^5-24}{5x^4}$, $x_1 = 2.0, x_2 = 1.9, x_3 = 1.8883$

7. $f(x) = x^3 - 4$, $\Phi(x) = x - \frac{x^3-4}{3x^2}$, $x_1 = 2.0, x_2 = 1.6667, x_3 = 1.5911$

9. $f(x) = x^2 - 6$, $\Phi(x) = x - \frac{x^2-6}{2x}$, $x_1 = 2.5, x_2 = 2.45, x_3 = 2.4495$

11. $f(x) = x^2 - 10$, $\Phi(x) = x - \frac{x^2-10}{2x}$, $x_1 = 2.0, x_2 = 3.5, x_3 = 3.1786$

Further Theory and Practice

13. One iteration because the tangent line and the graph of the function are one and the same. It does not depend upon the initial estimate.

15. (a) Since $\frac{f(x)f''(x)}{f'(x)^2} = -\tan^2(x)$, $\left|\frac{f(x)f''(x)}{f'(x)^2}\right| < 1$ when $|x| < \pi/4$.

    (b) Since $\frac{f(x)f''(x)}{f'(x)^2} = \frac{2(x^2-x)}{(2x-1)^2}$, $\lim_{x\to(1/2)^+} \frac{f(x)f''(x)}{f'(x)^2} = \infty$, implying that $\left|\frac{f(x)f''(x)}{f'(x)^2}\right|$ is not less than 1 on $(1/2, 2)$.

    (c) Since $\frac{f(x)f''(x)}{f'(x)^2} = \frac{2(x^3+8)}{3x^3}$, $\lim_{x\to(-1)^-} \frac{f(x)f''(x)}{f'(x)^2} = -14/3$, implying that $\left|\frac{f(x)f''(x)}{f'(x)^2}\right|$ is not less than 1 on $(-3, -1)$.

    (d) Let $F(x) = \frac{f(x)f''(x)}{f'(x)^2} = \frac{1}{6}\cdot(3 - 12x^{-2})$. Since $F'(x) = 4/x^3$, F increases from $F(1.5) = -0.3\overline{8}$ to $F(2.5) = 0.18$ on $(1.5, 2.5)$. Consequently, $\left|\frac{f(x)f''(x)}{f'(x)^2}\right| < 1$ on this interval.

17. When $f(x) = x^2 - c$, $\Phi(x) = x - \frac{x^2-c}{2x} = x/2 + c/(2x) = \frac{1}{2}\left(x + \frac{c}{x}\right)$.

    If $x_j \approx \sqrt{c}$, then $c/x_j \approx \sqrt{c}$ also. If one approximation is too small, then the other is too large. Therefore, it is natural to assume that the average of the two approximations will be better than either one.

19. Suppose that f is always concave up. Then $f'' > 0$ and f' is an increasing function. Since f' has no root, f' must also be positive. Thus f is increasing and concave up implying that it has only one root, x_*. Consider an initial guess, x_1, that is to the right of x_*. Then, since f is concave up, the graph of f lies above its tangent line at $(x_1, f(x_1))$, implying that $x_* < x_2 < x_1$. Similarly, $x_* < x_3 < x_2$, and, in general, $x_* < x_{j+1} < x_j$ for all $j = 1, 2, 3, \ldots$. By the Monotone Convergence Theorem the sequence $\{x_j\}$ converges and its limit must be x_* because Φ has only one equilibrium point. If the initial guess, x_1, lies to the left of x_*, then x_2 will be to the right (why?), and from there the Newton-Raphson sequence decreases monotonically to x_*.

    If f is always concave down, apply the above argument to $-f$, which is always concave up.

21. This follows from Exercise 89 (Section 4.2). All that is required is that x_1 be chosen in the open interval centered at x_* where Φ is continuous and $|\Phi'(x)| < 1$.

23. Assume f'' exists and is continuous. Let x_* be a root of f with $f'(x_*) \neq 0$. Then, because $f(x_*) = 0$, $\Phi'(x_*) = \frac{f(x_*)f''(x_*)}{f'(x_*)^2} = 0$. Since Φ' is continuous, there is an open interval I centered at x_* such that $|\Phi'(x)| < 1$ when $x \in I$. Applying Exercise 21, the Newton-Raphson Method will converge to x_* provided x_1 is chosen sufficiently close to x_*.

CALCULATOR/COMPUTER EXERCISES

25. Starting with $x_1 = 1.0$, $x_2 = 1.1$, $x_3 = 1.089073$, $x_4 = 1.088916$, and $x_5 = 1.088916$ with $|x_5 - x_4| \approx 3.2 \times 10^{-8}$.

27. Starting with $x_1 = 2.0$, $x_2 = 1.845360$, $x_3 = 1.802926$, $x_4 = 1.800013$, $x_5 = 1.800000$, and $x_6 = 1.800000$ with $|x_6 - x_5| \approx 0$.

29. Starting with $x_1 = 1.0$, $x_5 = x_6 = 0.9955987720$. This approximates the larger of two positive real roots. (The other one is approximately 0.8690473896.)

31. Starting with $x_1 = 0.6$, $x_5 = x_6 = 0.62902$ with $|x_6 - x_5| \approx 5.1 \times 10^{-9}$.

33. The picture on the right shows the graph of p in the given viewing window. It appears that $c = 1$ is a multiple root. Starting with $x_1 = 1.5$, the smallest value of N for which $|x_N - 1| < 5 \times 10^{-4}$ is $N = 12$, $x_{12} = 1.000340615$.

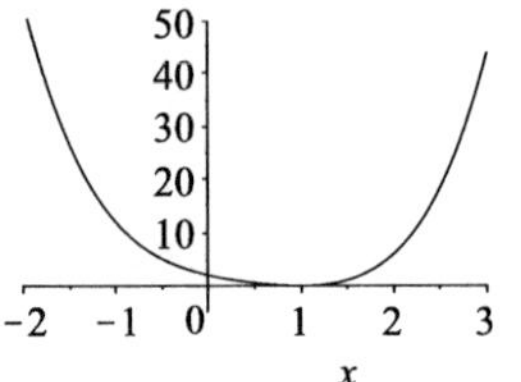

35. The picture on the right shows the graph of p in the given viewing window. It appears that $c = 0.5$ is a multiple root. Starting with $x_1 = 1.0$, the smallest value of N for which $|x_N - c| < 5 \times 10^{-4}$ is $N = 11$, $x_{11} = 0.5004395571$.

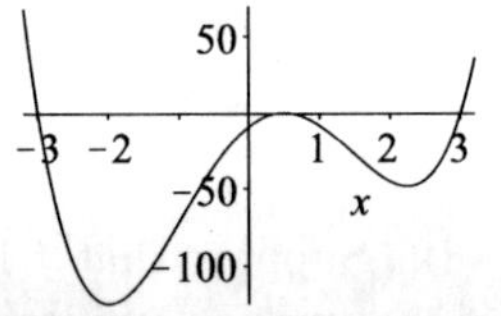

37. Exercise 33. $p(x) = x^4 - 2x^3 + 3x^2 - 4x + 2$, $c = 1$, $x_1 = 1.5$, Modified Newton-Raphson.

    The smallest value of M for which $|x_M - c| < 5 \times 10^{-4}$ is $M = 4$, $x_4 = 1.000001290$. Using the unmodifed Newton-Raphson Method, $N = 11$.

    Exercise 34. $p(x) = x^4 + x^3 - 6x^2 - 4x + 8$, $c = -2$, $x_1 = -1.5$, Modified Newton-Raphson.

    The smallest value of M for which $|x_M - c| < 5 \times 10^{-4}$ is $M = 4$, $x_4 = -2.000002455$. Using the unmodifed Newton-Raphson Method, $N = 11$.

    Exercise 35. $p(x) = 4x^4 - 4x^3 - 35x^2 + 36x - 9$, $c = 1$, $x_1 = 1.5$, Modified Newton-Raphson.

    The smallest value of M for which $|x_M - c| < 5 \times 10^{-4}$ is $M = 3$, $x_3 = 0.4999410600$. Using the unmodifed Newton-Raphson Method, $N = 11$.

    Exercise 36. $p(x) = x^4 - 2x^3 + 3x^2 - 4x + 2$, $c = 1$, $x_1 = 1.5$, Modified Newton-Raphson.

    The smallest value of M fo0r which $|x_M - c| < 5 \times 10^{-4}$ is $M = 4$, $x_4 = 1.999994637$. Using the unmodifed Newton-Raphson Method, $N = 10$.

39. Define $f(x) = 2\pi\sqrt{\frac{\ell}{g}} \cdot \left(1 + \frac{x^2}{16} + \frac{11x^4}{3072}\right) - T$. Using a plot of f we start with $x_1 = 1.0$ as a first estimate of a root and obtain $x_3 = x_4 = 1.080$. To four significant digits, $\phi = 1.08$.

41. (a) As in Example 5 of this section, $P = 10000$, $r = 0.0675$, $n = 20$, and $V = 9125$. Let

$$f(x) = 10000 \cdot \left((1+x)^{-20} + \frac{0.0675}{2} \cdot \frac{1-(1+x)^{-20}}{\sqrt{1+x}-1}\right) - 9125\,,$$

and $\Phi(x) = x - \frac{f(x)}{f'(x)}$. Beginning with $x_1 = 0.07$, $x_2 = 0.07711$, $x_3 = 0.7753$, and $x_4 = 0.7753$. Stopping here we estimate the effective yield to be 7.75%.

(b) Now $P = 10000$, $r = 0.065$, $n = 10$, and $V = 10575$. Let

$$f(x) = 10000 \cdot \left((1+x)^{-10} + \frac{0.065}{2} \cdot \frac{1-(1+x)^{-10}}{\sqrt{1+x}-1}\right) - 10575\,,$$

and $\Phi(x) = x - \frac{f(x)}{f'(x)}$. Beginning with $x_1 = 0.06$, $x_2 = 0.0582$, $x_3 = x_4 = 0.0582$. Stopping here we estimate the effective yield to be 5.82%.

43. With $P = 10000$, $r = 0.05$, $n = 20$, and $V = 8681$,

$$f(x) = 10000 \cdot \left((1+x)^{-20} + \frac{0.05}{2} \cdot \frac{1-(1+x)^{-20}}{\sqrt{1+x}-1}\right) - 8681\,,$$

and $\Phi(x) = x - \frac{f(x)}{f'(x)}$. Beginning with $x_1 = 0.06$, $x_2 = 0.0624$, $x_3 = x_4 = 0.0625$. Stopping here the estimated effective yield is 6.25%.

With $P = 10000$, $r = 0.07$, $n = 20$, and $V = 10674$,

$$f(x) = 10000 \cdot \left((1+x)^{-20} + \frac{0.07}{2} \cdot \frac{1-(1+x)^{-20}}{\sqrt{1+x}-1}\right) - 10674\,,$$

and $\Phi(x) = x - \frac{f(x)}{f'(x)}$. Beginning with $x_1 = 0.06$, $x_2 = 0.0648$, $x_3 = x_4 = 0.0650$. Stopping here the estimated effective yield is 6.50%.

The second investment is the better of the two.

45. With $P = 10000$, $r = 0.08$, $n = 8$, and $V = 10679$,

$$f(x) = 10000 \cdot \left((1+x)^{-8} + \frac{0.08}{2} \cdot \frac{1-(1+x)^{-8}}{\sqrt{1+x}-1}\right) - 10679\,,$$

and $\Phi(x) = x - \frac{f(x)}{f'(x)}$. Beginning with $x_1 = 0.08$, $x_2 = 0.069$, $x_3 = x_4 = 0.0700$. Stopping here the estimated effective yield is 7%.

With $P = 10000$, $r = 0.09$, $n = 20$, and $V = 11052$,

$$f(x) = 10000 \cdot \left((1+x)^{-8} + \frac{0.09}{2} \cdot \frac{1-(1+x)^{-8}}{\sqrt{1+x}-1} \right) - 11052\,,$$

and $\Phi(x) = x - \frac{f(x)}{f'(x)}$. Beginning with $x_1 = 0.09$, $x_2 = 0.0727$, $x_3 = x_4 = 0.0737$. Stopping here the estimated effective yield is 7.37%.

The second investment is the better of the two.

47. If $x_3 = x_1$, then x_1 is a solution to the equation $x = F(F(x))$, where $F(x) = x - \frac{x+\sin(x)}{1+\cos(x)}$. Let $g(x) = F(F(x)) - x$ and apply the Newton-Raphson algorithm to g, starting with $z_1 = 1.8$, to obtain $z_2 = 1.803074298$, $z_3 = 1.803005957$, and $z_4 = z_5 = 1.803005922$. The approximate value of the cycling point is $x_1 = 1.803005922$.

49. The graph of Y/x and its first derivative on the right provides the starting point for a Newton-Raphson search to find the second positive root of the derivative. With $x_1 = 0.5$, $x_2 = 0.5128299121$, and $x_3 = x_4 = 0.5128209155$. Since Y/x evaluates to 1 at 0 and to 1.028759831 at 0.5128209155, this as the approximate absolute maximum value of Y/x for $x \geq 0$.

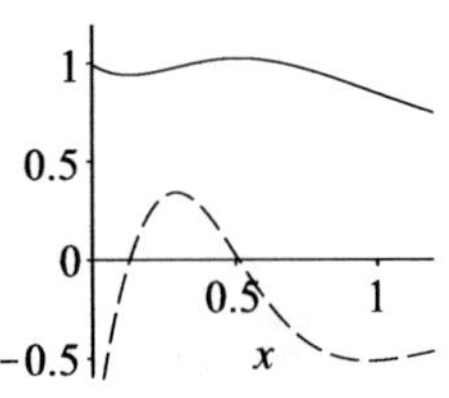

4.9 Antidifferentiation and Applications

Problems for Practice

1. $\int (x^2 - 5x)\,dx = \frac{1}{3}x^3 - \frac{5}{2}x^2 + C$

3. $\int e^x\,dx = e^x + C$

5. $\int \sqrt{x+2}\,dx = \int (x+2)^{1/2}\,dx = \frac{2}{3}(x+2)^{3/2} + C$

7. $\int \left(\frac{x^2 + x^{-3}}{x^4} \right) dx = \int (x^{-2} + x^{-7})\,dx = -x^{-1} - \frac{1}{6}x^{-6} + C$

9. $\int (x+1)^2\,dx = \frac{1}{3}(x+1)^3 + C$ (verify)

11. $\int (\exp(e \cdot x)\,dx = \frac{1}{e}\exp(e \cdot x) + C$ (verify)

13. $\int (x^{-7/3} - 4x^{-2/3})\,dx = -\frac{3}{4}x^{-4/3} - 12x^{1/3} + C$

15. $\int (3\cos(4x) + 2x)\,dx = \frac{3}{4}\sin(4x) + x^2 + C$

17. $\int \sec^2(8x)\,dx = \frac{1}{8}\tan(8x) + C$

19. $\int (3x-2)^3\,dx = \frac{1}{3} \cdot \frac{1}{4}(3x-2)^4 + C = \frac{1}{12}(3x-2)^4 + C$

21. Integrating, $F(x) = \int 6x^2\,dx = 2x^3 + C$. $F(1) = 3$ implies that $C = 1$, so $F(0) = 2 \cdot 0^3 + 1 = 1$.

23. Integrating, $F(x) = \int \cos(x)\,dx = \sin(x) + C$. $F(\pi/2) = -1$ implies that $C = -2$, so $F(\pi/6) = \sin(\pi/6) - 2 = 1/2 - 2 = -3/2$.

25. Integrating, $F(x) = \int 6e^{2x}\,dx = 3e^{2x} + C$. $F(0) = -1$ implies that $C = -4$, so $F(1/2) = 3e - 4$.

27. Let $p(t)$, $v(t)$, and $a(t)$ denote the sprinter's position, velocity, and acceleration at time t seconds after she begins to run. Then $p(0) = v(0) = 0$, and $v'(t) = a(t) = 4$. Consequently, $v(t) = \int a(t)\,dt = 4t + C$, and, since $v(0) = 0$, $C = 0$, so $v(t) = 4t$. To find how fast she is running at the 20 m mark, we need $p(t)$. We calculate $p(t) = \int v(t)\,dt = 2t^2 + D$. Since $p(0) = 0$, $D = 0$, and $p(t) = 2t^2$. The time T that it takes her to run 20 m must satisfy the equation $p(T) = 20$ so $T^2 = 10$, and $T = \sqrt{10} \approx 3.167$ seconds. At this time the she has a speed of $v(\sqrt{10}) = 4\sqrt{10} \approx 12.649\ \frac{\text{meters}}{\text{sec}}$.

29. Let $h(t)$, $v(t)$, and $a(t)$ denote the baseball's height, velocity, and acceleration at time t seconds after it is thrown. Then $h(0) = 0$, $v(0) = 100$ ft/s, and $v'(t) = a(t) = -32.16$ ft/s^2. Consequently, $v(t) = \int a(t)\,dt = -32.16t + C$, and, since $v(0) = 100$, $C = 100$, so $v(t) = -32.16t + 100$. To find how high the ball goes, we need $h(t)$. We calculate $h(t) = \int v(t)\,dt = -16.08t^2 + 100t + D$. Since $h(0) = 0$, $D = 0$, and $h(t) = -16.08t^2 + 100t$. The ball reaches its maximum height at time T when $v(T) = 0$. That is, $-32.16t + 100 = 0$ so $T = 100/32.16 \approx 3.109$ seconds. The maximum height is $h(T) = -16.08 \cdot T^2 + 100T \approx 155.5$ feet.

The ball is at the height of 25 feet when $h(t) = 25$. That is, $-16.08t^2 + 100t = 25$. Using the quadratic formula, we find $t_1 \approx 0.2609$ and $t_2 \approx 5.958$. The second time is when the ball is on the way down, so its velocity will be $v(t_2) \approx -91.61$ ft/s^2.

31. Let $-\alpha$ denote the constant rate of acceleration ($\alpha > 0$). Then $a(t) = -\alpha$. Let $v(t)$ and $p(t)$ denote the velocity and position of the car t hours after it has started to decelerate. Then $v(0) = 50$ and we take $p(0) = 0$. Calculating, $v(t) = \int a(t)\,dt = -\alpha t + C$, where $C = 50$ because $v(0) = 50$. Consequently, $p(t) = \int v(t)\,dt = -\frac{1}{2}\alpha t^2 + 50t + D$, where $D = 0$ because $p(0) = 0$.

From the information given, at the time T that the car stops, $v(T) = 0$, and $p(T) = 1200/5280 = 5/22$. Therefore, T and α satisfy the equations

$$-\alpha T + 50 = 0 \quad \text{and} \quad -\tfrac{1}{2}\alpha T^2 + 50T = 5/22\,.$$

Therefore, $T = 50/\alpha$ and $-\frac{1}{2} \cdot 50^2/\alpha + 50^2/\alpha = 5/22$, implying that $\alpha = \frac{50^2}{2} \cdot \frac{22}{5} = 5500\,\text{miles/hr}^2$ ($= 121/54$ ft/s^2).

33. Let $p(t)$, $v(t)$, and $a(t)$ be the position, velocity, and acceleration of the ball t seconds after it is dropped. Then $p(0) = 361$, $v(0) = 0$, and $a(t) =$

-32.16 ft/s^2. We calculate $v(t) = \int a(t)\,dt = -32.16t + C$ where $C = 0$ because $v(0) = 0$, and $p(t) = \int v(t)\,dt = -16.08t^2 + D$, where $D = 361$ because $p(0) = 361$. The ball strikes the ground at the time T that $p(T) = 0$. That is $-16.08T^2 + 361 = 0$, so $T = \sqrt{361/16.06} \approx 4.738$ seconds.

Further Theory and Practice

35. $\int \sin(x)\cos(x)\,dx = \int \frac{1}{2}\sin(2x)\,dx = -\frac{1}{4}\cos(2x) + C$

37. $\int \tan^2(x)\,dx = \int(\sec^2(x) - 1)\,dx = \tan(x) - x + C$

39. $\int \cos^2(x)\,dx = \int \frac{1}{2}(1 + \cos(2x))\,dx = \frac{1}{2}\big(x + \frac{1}{2}\sin(2x)\big) + C$

41. $\int \dfrac{\sin(x)}{\cos^2(x)}\,dx = \int \sec(x)\tan(x)\,dx = \sec(x) + C$

43. $\int 2^{x\ln(2)}\,dx = \int e^{x\ln^2(2)}\,dx = \frac{1}{\ln^2(2)}e^{x\ln^2(x)} + C = \frac{1}{\ln^2(2)}2^{x\ln(2)} + C$

45. Since $F'(x) = f(x)$, F is increasing on $(-\infty, -3)$, decreasing on $(-3, 2)$, and increasing on $(2, \infty)$. Therefore, F has a local maximum at -3 and a local minimum at 2.

47. Assuming that $\int g(x)\,dx = G(x) + C$, $G'(x) = g(x)$. By the Chain Rule,

$$\frac{d}{dx}G(f(x)) = G(f(x))f'(x) = g(f(x))f'(x)\,,$$

implying that $\int g(f(x))f'(x)\,dx = G(f(x)) + C$.

49. $\int 2x\cos(x^2 + 7)\,dx = \sin(x^2 + 7) + C$

51. $\int x\exp(x^2)\,dx = \frac{1}{2}\int 2x\exp(x^2)\,dx = \frac{1}{2}\exp(x^2) + C$

53. Because $F'(x) = f(x) = G'(x)$ for all $x \in I$, we know that there is a constant C such that $F(x) = G(x) + C$ on I. Substitute x_0 and $F(x_0) = G(x_0) + C$ implying that $C = 0$ and $F(x) = G(x)$ for every x in I.

This is false for $f(x) = 1/x$ on $I = (-\infty, 0) \cup (0, \infty)$. Let $F(x) = \ln(|x|)$ and $G(x) = \begin{cases} \ln(x) & \text{if } x > 0 \\ \ln(|x|) + 1 & \text{if } x < 0 \end{cases}$. Then F and G are antiderivatives for f on I and $F(1) = G(1) = 0$ but $F(x) \neq G(x)$ on $(-\infty, 0)$.

55. Let $u = \sqrt{x}$, by the Chain Rule,

$$\frac{d}{dx}\mathrm{erf}(\sqrt{x}\,) = \frac{d}{du}\mathrm{erf}(u)\frac{du}{dx} = \frac{2\exp(-u^2)}{\sqrt{\pi}}\cdot\frac{1}{2\sqrt{x}} = \frac{\exp(x)}{\sqrt{\pi x}}\,.$$

57. Let $x(t)$ and $y(t)$ be the horizontal and vertical distances from the point of origin to the arrow at time t. Then $x(0) = y(0) = 0$, $x'(0) = 190\cos(60°) = 95$ ft/s and $y'(0) = 190\sin(60°) = 95\sqrt{3}$ ft/s. Since $x''(t) = 0$, $x'(t) = C$ where $C = 95$ and $x(t) = \int x'(t)\,dt = 95t + D$ where $D = 0$ since $x(0) = 0$.

As for the vertical component, $y''(t) = -32.16$ ft/s^2, so $y'(t) = -32.16t + y'(0) = -32.16t + 95\sqrt{3}$ and $y(t) = \int y'(t)\,dt = -16.08t^2 + 95\sqrt{3}\,t + y(0) = -16.08t^2 + 95\sqrt{3}\,t$.

The arrow reaches its highest point at time T where $y'(T) = 0$. That is $-32.16T + 95\sqrt{3} = 0$, or $T = 95\sqrt{3}/32.16$. The maximum height is $y(T) \approx 420.9$ feet.

The horizontal component of the velocity is constant, $x'(t) = 95$ for all t.

59. The initial speed of the train is $30\frac{\text{mi}}{\text{hr}} = 30 \cdot \frac{5280 \text{ ft}}{60 \text{ min}} = 2640$ ft/min. Let $p(t)$, $v(t)$, and $a(t)$ denote the position, velocity, and acceleration of the train t minutes after the engineer applies the first set of brakes. Then $p(0) = 0$, $v(0) = 2640$, and $a(t) = -5280/5 = -1056$ ft/min^2. Therefore, for $0 < t < 15/60$, $v(t) = -1056t + 2640$ and $p(t) = \int v(t)\,dt = -528t^2 + 2640t + p(0) = -528t^2 + 2640t$. The train has gone $p(15/60) = 627$ feet and is moving at a speed of $v(15/60) = 2376$ ft/min.

    Subsequently, $a(t) = -5280 \cdot 0.3 = -1584$ ft/min^2 and, resetting the clock at $t = 0$ and measuring $p(t)$ from the current train position, $v(t) = \int a(t)\,dt = -1584t + C$, where $C = 2376$ because $v(0) = 2376$. Therefore, $p(t) = \int v(t)\,dt = -792t^2 + 2376t + D$ where $D = 0$ because $p(0) = 0$. Ignoring the barrier, the train will come to a stop in T minutes where $v(T) = 0$. That is, $-1584T + 2376 = 0$, or $T = 3/2$ minutes. It will have traveled $p(3/2) = 1782$ feet for a total stopping distance of $627 + 1782 = 2409$ feet. This is 709 feet beyond the barrier. The train will hit the barrier when $p(t) = 1700 - 627 = 1073$ feet. That is, $-792t^2 + 2376t = 1073$. Using the quadratic formula, $t \approx 0.5538$ minutes, and the train's velocity will be $v(0.5538) \approx 1498.7$ ft/min $= 17$ mi/hr.

61. Sound travels at the approximate speed of 968 ft/sec. Assuming that the sound traveled for T seconds, the well would be $968T$ feet deep. To find T we observe that the stone fell for $6 - T$ seconds. Let $p(t)$ be the distance the stone has traveled in t seconds after it is dropped. Then $p''(t) = 32.18$ ft/s^2, $p'(t) = 32.18t$, and $p(t) = 16.08t^2$. Since $p(6 - T) = 968T$, we find T by solving the equation $16.08(6 - T)^2 = 968T$. Using the quadratic formula, $T \approx 0.5021$ seconds, and the well is $968T \approx 486$ feet deep.

63. Let $p(t)$ be the height of the rocket t seconds after its fuel is spent. Then $p(0) = 800$ feet, $p'(0) = 100$ ft/s, and $p''(t) = -32.16$ ft/s^2. Consequently, $p'(t) = \int p''(t)\,dt = -32.16t + 100$, and $p(t) = -16.08t^2 + 100t + 800$. The rocket reaches its maximum height at time T when $p'(T) = 0$. That is, $T = 100/32.16 \approx 3.109$ seconds. The maximum height is $p(T) \approx 955.5$ feet.

    The rocket strikes the ground when $p(t) = 0$. Using the quadratic formula, this is at time $t_* \approx 10.82$ seconds. Its velocity is $p'(t_*) \approx -247.9$ ft/s.

65. Let $A(t)$ and $B(t)$ be the distance, in miles, traveled by car A and car

B respectively, t hours after the cars begin to accelerate. Then $A(0) = B(0) = 0$, $A'(0) = B'(0) = 90$, and $A''(t) = 6\frac{\text{mi}}{\text{min}^2} = \frac{6}{(1/60)^2}\frac{\text{mi}}{\text{hr}^2} = 21600$ mi/hr^2, $B''(t) = \frac{9}{(1/60)^2} = 32400$ mi/hr^2. Integrating two times and using the initial values, $A(t) = \frac{1}{2}\cdot 21600t^2 + 90t$ and $B(t) = \frac{1}{2}\cdot 32400t^2 + 90t$. The advantage of car B over car A at time t is $D(t) = B(t) - A(t) = 5400t^2$. This is 6 miles at time T satisfying $5400T^2 = 6$ or $T = \sqrt{1/900} = 1/30$ hours (2 minutes).

67. Since $v(t) = -kg(1 - e^{-t/k})$, the height, $y(t)$, of the object t time units after it is dropped, is $y(t) = \int v(t)\,dt = -kg\big(t + ke^{-kt}\big) + C$. Since $y(0) = H$, C must satisfy the equation $y(0) = H$. That is, $-k^2g + C = H$, and $C = H + k^2g$. This yields $y(t) = -kgt - k^2ge^{-kt} + H + k^2g$, or $y(t) = H - kgt + k^2g(1 - e^{-t/k})$.

CALCULATOR/COMPUTER EXERCISES

69. Since $m(t) = \int -0.1213 \cdot e^{-0.0001213t}\,dt = \frac{0.12132}{0.0001213} \cdot e^{-0.0001213t} + C = 1000e^{-0.0001213t} + C$, and $m(0) = 1000$, the constant if integration $C = 0$ and $m(t) = 1000e^{-0.0001213t}$. The equation $m(t) = 800$ is satisfied when $1000e^{-0.0001213t} = 800$ so $-0.0001213t = \ln(800/1000)$. That is, $t = \ln(5/4)/0.0001213 \approx 1839.6$ years.

71. $T(t) = \int T'(t)\,dt = \frac{0.2}{0.02}e^{-0.02t} + C = 10e^{-0.02t} + C$. Since $T(0) = 50$, the constant of integration $C = 40$ and $T(t) = 10e^{-0.02t} + 40$. Consequently, $T_\infty = \lim_{t\to\infty} T(t) = 40°$C.

The object has cooled 4° at time τ_1 satisfying $T(\tau_1) = 46$. That is, $10e^{-0.02\tau_1} + 40 = 46$ or $e^{-0.02\tau_1} = 6/10$, $\tau_1 = \ln(5/3)/0.02 \approx 25.54$ seconds.

The object has cooled 2° more at time t_2 satisfying $T(t_2) = 44$. That is, $10e^{-0.02t_2} + 40 = 44$ or $e^{-0.02t_2} = 4/10$, $t_2 = \ln(5/2)/0.02 \approx 45.81$ seconds. The additional time is $\tau_2 = t_2 - \tau_1 = 20.27$ seconds.

The object has cooled one more degree at time t_3 satisfying $T(t_3) = 43$. That is, $10e^{-0.02t_3} + 40 = 43$ or $e^{-0.02t_3} = 3/10$, $t_3 = \ln(10/3)/0.02 \approx 60.20$ seconds. The additional time is $\tau_3 = t_3 - t_2 = 14.39$ seconds.

73. Using the integral formula in Exercise 54,

$$v(t) = \frac{39.2}{4.427} \cdot \big(1 + e^{4.427t}\big)^{-1} + C.$$

Since $v(0) = 0$, $C = -\frac{39.2}{2\cdot 4.427}$ and

$$v(t) = \frac{39.2}{4.427} \cdot \big(1 + e^{4.427t}\big)^{-1} - \frac{39.2}{8.854}.$$

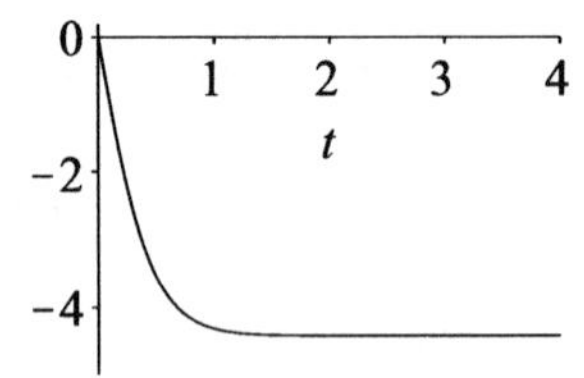

See the plot on the right. Terminal velocity is $v_\infty = -\frac{39.2}{8.854} \approx -4.427$. $v(t) = 0.999v_\infty$ at time t satisfying the equation $v(t) = 0.999v_\infty$. Applying the Newton-Raphson Method to the function $f(t) = v(t) - 0.999v_\infty$ with $t_1 = 1.5$ we find that $t_2 = 1.639500078$, $t_3 = 1.704995715$, $t_4 =$

1.716524773, $t_5 = 1.716828903$, and $t_6 = t_7 = 1.716829107$. To three decimals, $v(1.717) = 0.999v_\infty$.

75. The graph of $\frac{d}{dx}(\mathrm{Si}(x)) = \sin(x)/x$ appears on the right. $\mathrm{Si}(x)$ has a positive derivative on the intervals $(-3\pi, -2\pi)$, $(-\pi, \pi)$, and $(2\pi, 3\pi)$, so Si is increasing on these open intervals and decreasing on the complementary open intervals. There are inflection points on the graph of Si at the points where $\sin(x)/x$ attains its extreme values. These are half way between its zeros (verify). Consequently, Si changes conavity from up to down at the midpoint of each interval in which it is increasing, and from down to up at the midpoint of each interval in which it is decreasing.

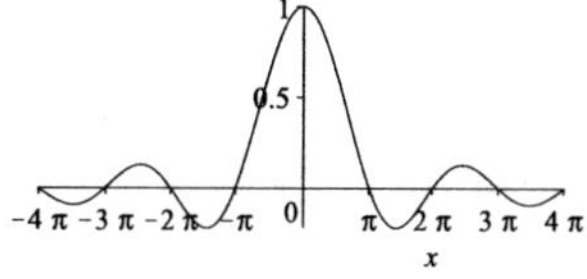

The graph of $\mathrm{Si}(x))$ appears on the right (solid curve). Its derivative, $\sin(x)/x$, is the dashed curve. The description of the relationship between the two graphs in the previous paragraph is confirmed by what we see in the picture.

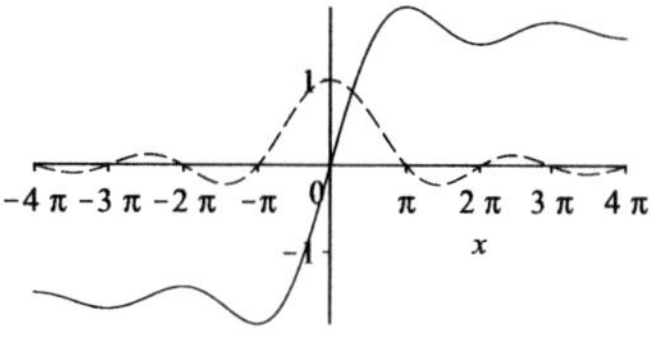

REVIEW EXERCISES FOR CHAPTER 4

1. Differentiate the equation with respect to t, $\frac{dx}{dt} + 2y\frac{dy}{dt} = 3x^2\frac{dx}{dt}$. Substitute the given data: $2 + 10\frac{dy}{dt}\Big|_{P_0} = 27 \cdot 2$. Therefore, $\frac{dy}{dt}\Big|_{P_0} = 26/5$.

3. Differentiate the equation with respect to t, $\left(x\frac{dy}{dt} - y\frac{dx}{dt}\right)/x^2 - (2/y^2)\frac{dy}{dt} = 0$. Substitute the given data: $\left(\frac{1}{4}\frac{dy}{dt}\Big|_{P_0} + \frac{3}{2}\right)/(1/16) - 8\frac{dy}{dt}\Big|_{P_0} = 0$. Therefore, $\frac{1}{4}\frac{dy}{dt}\Big|_{P_0} + \frac{3}{2} - \frac{1}{2}\frac{dy}{dt}\Big|_{P_0} = 0$, and $\frac{dy}{dt}\Big|_{P_0} = 6$.

5. Let S denote the surface area of the snowball. Then $S = 4\pi r^2$, and $\frac{dS}{dt} = 8\pi r\frac{dr}{dt}$. When $r = 4$, $-3 = 32\pi\frac{dr}{dt}\Big|_{r=4}$, so $\frac{dr}{dt}\Big|_{r=4} = -\frac{3}{32\pi}$ cm/min.

7. Differentiate the volume equation with respect to t, $0 = \frac{1}{3}\pi\left(2rh\frac{dr}{dt} + r^2\frac{dh}{dt}\right)$. When $r = 3/2$, $h = \frac{8\pi}{\pi\cdot(3/4)} = 32/3$. Substitute this data into the equation relating the derivatives to obtain $2 \cdot \frac{3}{2} \cdot \frac{32}{3} \cdot \frac{3}{160} + \frac{9}{4}\frac{dh}{dt}\Big|_{r=3/2}$, implying that $\frac{dh}{dt}\Big|_{r=3/2} = -4/15$ m/min.

9. Since $y = 1/x$, $\frac{dy}{dt} = -(1/x^2)\frac{dx}{dt}$. At $t = 7$, $x = 1/6$, $y = 6$, and $\frac{dx}{dt} = 3$. Substitute into the derivative equation to obtain $\frac{dy}{dt}\Big|_{t=7} = -36 \cdot 3 = -108$. Therefore, $\frac{d}{dt}(x + y)\Big|_{t=7} = \frac{dx}{dt}\Big|_{t=7} + \frac{dy}{dt}\Big|_{t=7} = 3 - 108 = -105$.

11. Let x and ℓ denote the distance from the pulley to the dock and the length of the rope, respectively. Then $x^2 + 6^2 = \ell^2$ and $2x\frac{dx}{dt} = 2\ell\frac{d\ell}{dt}$. When $x = 8$, $\ell = \sqrt{64 + 36} = 10$ so, at that instant, $8 \cdot \frac{dx}{dt}\Big|_{x=8} = 10 \cdot \left(-\frac{2}{5}\right)$, so $\frac{dx}{dt}\Big|_{x=8} = -1/2$ ft/s.

13. Since $(f(4) - f(1))/(4 - 1) = 21$ (verify), we must solve the equation $f'(c) = 21$. That is, $3c^2 = 21$. The answer is $c = \sqrt{7}$.

15. Since $(f(-27) - f(-3))/(-27 - (-3)) = -1/81$ (verify), we must solve the equation $f'(c) = -1/81$. That is, $-\frac{1}{c^2} = -\frac{1}{81}$. The answer is $c = -9$.

17. Since $f'(x) = 1 - 1/x^2$, we must solve the equation $1 - 1/c^2 = 0$. The answer is $c = -1$.

19. Since $f'(x) = \frac{1}{1-e} + \frac{1}{x}$, we must solve the equation $\frac{1}{1-e} + \frac{1}{c} = 0$. The answer is $c = e - 1$.

21. $F(x) = \int (-x^{-2} + x^{-1})\,dx = x^{-1} + \ln(|x|) + C$. Since $F(1) = 7$, $C = 6$ and $F(x) = 1/x + \ln(|x|) + 6$. Therefore, $F(1/2) = 2 + \ln(1/2) + 6 = 8 - \ln(2)$.

23. $F(x) = \int 4/x\,dx = 4\ln(|x|) + C$. Since $F(e^2) = 7$, $C = -1$ and $F(x) = 4\ln(|x|) - 1$. Therefore, $F(e^3) = 11$.

25. Since $f'(x) = 3(x - 2)(x + 2)$, the critical points are $c_{1,2} = -2, 2$. Since f increases on $(-\infty, -2)$, decreases on $(-2, 2)$, and increases on $(2, \infty)$, f has a local maximum at -2 and a local minimum at 2.

27. The domain of f is $(0, \infty)$. $f'(x) = 3 \cdot \frac{x^3-16}{x}$, so there is one critical point, $c = 16^{1/3}$. Since f decreases on $(0, c)$ and increases on (c, ∞), f has a global minimum at c.

29. The domain of f does no include 0. $f'(x) = 3 \cdot \frac{(x^2+4)(x^2-4)}{x^2}$, so there are two critical point, $c_{1,2} = \mp 2$. Since f increases on $(-\infty, -2)$ and decreases on $(-2, 0)$, f has a local maximum at -2. Since f decreases on $(0, 2)$ and increases on $(2, \infty)$, it has a local minimum at 2.

31. The domain of f is $(0, \infty)$. $f'(x) = \frac{\ln(x)(2-\ln(x))}{x^2}$, so there are two critical points. One is $c_1 = 1$, and the other is c_2, where $\ln(c_2) = 2$. That is, $c_2 = e^2$. Since f decreases on $(0, 1)$, increases on $(1, e^2)$, and decreases on (e^2, ∞), it has a local minimum at 1 and a local maximum at e^2.

33. The domain of f is $(0, \infty)$. $f'(x) = -\frac{x-5}{2\sqrt{x}(x+5)^2}$, so there is a critical point at $c = 5$. Since f increases on $(0, 5)$ and decreases on $(5, \infty)$, it has a local maximum at 5.

35. The domain of f is $(-1, \infty)$. $f'(x) = \frac{(x-1)(x-2)}{x+1}$, so there are two critical points, $c_{1,2} = 1, 2$. f increases on $(-1, 1)$, decreases on $(1, 2)$, and increases on $(2, \infty)$, so it has a local maximum at 1 and a local minimum at 2.

37. Since $f'(x) = 3x^2(2x - 1)(x - 1)^2$, the critical points are 0, 1/2, and 1. f decreases on $(-\infty, 0)$ and on $(0, 1/2)$ and increases on $(1/2, 1)$ and on $(1, \infty)$, so it has has a global minimum at 1/2.

39. $f'(x) = 3x^2 - 12 = 3(x - 2)(x + 2)$. The critical points are -2 and 2. Since $f(-3) = 11$, $f(5) = 67$, $f(-2) = 18$, and $f(2) = -14$ the minimum value is $f(2) = -14$ and the maximum value is $f(5) = 67$.

41. $f'(x) = -e^{-x}(x-1)$. There is one critical point, 1. Since $f(-1) = -e \approx -2.72$, $f(2) = 2e^{-2} \approx 0.271$, and $f(1) = e^{-1} \approx 0.368$, the minimum value is $f(-1) = -e$ and the maximum value is $f(1) = e^{-1}$.

43. Let the width of the rectangle be $2x$ and its height be y. We wish to maximize the area $A = 2xy + \frac{1}{2}\pi x^2$. Since the perimeter of the window is $5 = 2x + 2y + \pi x$, $y = (5 - (2+\pi)x)/2$ and, in terms of x, we maximize the area function

$$f(x) = 2x(5 - (2+\pi)x)/2 + \pi x^2/2 = 5x - (2+\pi/2)x^2 .$$

Since $f'(x) = 5 - (4+\pi)x$, f has a critical point at $c = 5/(4+\pi)$ which, because $f''(x) < 0$, is an absolute maximum for f. The base length $2x = 10/(4+\pi)$ maximizes the area.

45. Observe that

$$A'(\theta) = \tfrac{3}{2}b^2\big(-\sqrt{3}\csc(\theta)\cot(\theta) - \csc^2(\theta)\big) = -\tfrac{3}{2}b^2\csc(\theta)\big(\sqrt{3}\cot(\theta) - \csc(\theta)\big) .$$

This is 0 for acute angle θ if, and only if, $\frac{\csc(\theta)}{\cot(\theta)} = \sqrt{3}$. That is,

$$\sqrt{3} = \frac{1/\sin(\theta)}{\cos(\theta)/\sin(\theta)} = \sec(\theta) .$$

Since $\lim_{\theta\to 0^+}\cot(\theta) = \infty$, $A'(\theta)$ changes sign from negative to positive as θ passes through the angle $\theta_0 = \arccos(1/\sqrt{3}\,)$ and θ_0 is a global minimum for A on $(0, \pi/2)$.

47. Points on the two curves are of the form $(x, \sqrt{x})$ and (x, x^2), $0 \le x \le 1$. They can also be described as (y^2, y) and $(\sqrt{y}, y)$, $0 \le y \le 1$. For a given y value, the horizontal distance between two points is $h(y) = \sqrt{y} - y^2$. We will maximize h for $0 \le y \le 1$. Since

$$h'(y) = \frac{1}{2\sqrt{y}} - 2y = \frac{1 - 4y^{3/2}}{2\sqrt{y}} ,$$

there is a critical point at the value c such that $c^{3/2} = 1/4$. That is, $c = (1/4)^{2/3}$. A sign check on h' shows that this is an absolute maximum for h on $(0, 1)$. The longest horizontal line segment has length

$$h((1/4)^{2/3}) = (1/4)^{1/3} - (1/4)^{4/3} \approx 0.4725 .$$

49. Let x, y, and z denote the length, width, and depth of the box. Then $y = 2z$ and $144 = 2xy + 2xz + 2yz = 4xz + 2xz + 4z^2 = 6xz + 4z^2$. We wish to minimize $V = xyz$. Replacing y with $2z$, $V = 2xz^2$, and, because $x = (144 - 4z^2)/(6z) = 24/z - 2z/3$,

$$V(z) = 2\left(\frac{24}{z} - \frac{2z}{3}\right)\cdot z^2 = 48z - \tfrac{4}{3}z^3 .$$

Since $V'(z) = 48 - 4z^2$, the only positive critical point is $c = \sqrt{12} = 2\sqrt{3}$. Because $V''(z) < 0$ for $z > 0$, V attains an absolute maximum value when $z = 2\sqrt{3}$. The maximum volume is $V(2\sqrt{3}) = 96\sqrt{3} - \frac{4}{3}(2\sqrt{3})^3 = 96\sqrt{3} - 32\sqrt{3} = 64\sqrt{3} \approx 110.85 \text{ in}^3$.

51. Since $f''(x) = 12(x^2 - 1)$, f is concave up on $(-\infty, -1)$, concave down on $(-1, 1)$, and concave up on $(1, \infty)$. There are two inflection points, $(-1, -9)$ and $(1, 5)$.

53. Since $f''(x) = -\dfrac{4e^x e^6(e^{2x} - e^6)}{(2e^6 + e^{2x})^{5/2}}$, f is concave up on $(-\infty, 3)$ and concave down on $(3, \infty)$. There is one inflection point, $(3, \sqrt{3}/3)$.

55. The domain of f is $(0, \infty)$. Since $f''(x) = -\dfrac{x-6}{x^3}$, f is concave up on $(0, 6)$ and concave down on $(6, \infty)$. There is one inflection point, $(6, 1/2 + \ln(6))$.

57. The domain of f is $(0, \infty)$. Since $f''(x) = \dfrac{6\ln(x) - 5}{x^4}$, f is concave down on $(0, e^{5/6})$ and concave up on $(e^{5/6}, \infty)$. There is one inflection point, $(e^{5/6}, f(e^{5/6}))$.

59. Since $f''(x) = \dfrac{36(x+3)}{25x^{6/5}}$, f is concave down on $(-\infty, -3)$, concave up on $(-3, 0)$, and concave up on $(0, \infty)$. There is one inflection point, $(-3, f(-3))$.

61. $\displaystyle \lim_{x\to\infty} \frac{2e^x + x}{e^x + 2x} = \lim_{x\to\infty} \frac{2e^x + 1}{e^x + 2} = \lim_{x\to\infty} \frac{2e^x}{e^x} = 2$

63. $\displaystyle \lim_{x\to\pi/8} \frac{\cos(12x)}{\cos(4x)} = \lim_{x\to\pi/8} \frac{-12\sin(12x)}{-4\sin(4x)} = \frac{12}{-4} = -3$

65. $\displaystyle \lim_{x\to\pi/2} \frac{(2x - \pi)}{\cot(x)} = \lim_{x\to\pi/2} \frac{2}{(-\csc^2(x))} = -2$

67. $\displaystyle \lim_{x\to\infty} \frac{x^3}{e^{x/2}} = \lim_{x\to\infty} \frac{3x^2}{(1/2)e^{x/2}} = \lim_{x\to\infty} \frac{6x}{(1/4)e^{x/2}} = \lim_{x\to\infty} \frac{6}{(1/8)e^{x/2}} = 0$

69. $$\lim_{x\to\infty} \left(\sqrt{x^2 + 8x} - x\right) = \lim_{x\to\infty} \left(\frac{\sqrt{x^2 + 8x} - x}{1} \cdot \frac{\sqrt{x^2 + 8x} + x}{\sqrt{x^2 + 8x} + x}\right)$$
$$= \lim_{x\to\infty} \frac{8x}{\sqrt{x^2 + 8x} + x}$$

This is l'Hôpital, ∞/∞. However, repeated application of the Rule leads to cycling expressions that remain of the form ∞/∞. The limit, 4, is found easily by dividing the numerator and denominator by x and simplifying.

71. Let $u = \left(\frac{x-2}{x+1}\right)^x$. Then $\ln(u) = x\ln\left(\frac{x-2}{x+1}\right) = \frac{\ln\left(\frac{x-2}{x+1}\right)}{x^{-1}}$ which, as $x \to \infty$, is

indeterminate of the form 0/0. Applying l'Hôpital's Rule,

$$\lim_{x\to\infty} \ln(u) = \lim_{x\to\infty} \frac{(x-2)^{-1} - (x+1)^{-1}}{-x^{-2}} = \lim_{x\to\infty} \frac{3x^2}{-(x-2)(x+1)} = -3\,.$$

Since $\lim_{x\to\infty} \ln(u) = -3$, $\lim_{x\to\infty} \left(\frac{x-2}{x+1}\right)^x = e^{-3}$.

73. $\displaystyle \lim_{x\to 0}\left(\frac{1}{x} - \frac{1}{e^x-1}\right) = \lim_{x\to 0} \frac{e^x-1-x}{x(e^x-1)} = \lim_{x\to 0} \frac{e^x-1}{e^x+xe^x-1} = \lim_{x\to 0} \frac{e^x}{2e^x+xe^x} = \frac{1}{2}$

75. $\Phi(x) = x - \dfrac{2+\ln(x)-x}{1/x-1}$. With $x_1 = 4.0$,

$$x_2 = \Phi(x_1) = 3.181725815 \quad \text{and} \quad x_3 = \Phi(x_2) = 3.146284845\,.$$

77. $\Phi(x) = x - \dfrac{\sqrt{x}+1/\sqrt{x}-3}{1/(2\sqrt{x})-1/(2x^{3/2})}$. With $x_1 = 8.0$,

$$x_2 = \Phi(x_1) = 6.823500280 \quad \text{and} \quad x_3 = \Phi(x_2) = 6.854079461\,.$$

79. $\int \sec(x)\tan(x)\,dx = \sec(x) + C$

81. $\int (3x+1)^2\,dx = \frac{1}{3}\int 3(3x+1)^2\,dx = \frac{1}{3}\cdot\frac{1}{3}(3x+1)^3 + C = \frac{1}{9}(3x+1)^3 + C$

83. $\int x(x-1)(x+1)\,dx = \int (x^3-x)\,dx = \frac{1}{4}x^4 - \frac{1}{2}x^2 + C$

85. $\int (x^{4/3} - 4x^{-1/3})\,dx = \frac{3}{7}x^{7/3} - 6x^{2/3} + C$

87. $\int \sec^2(3x)\,dx = \frac{1}{3}\tan(3x) + C$

89. $\int (x^{7/2} + 2x^3)\,dx = \frac{2}{9}x^{9/2} + \frac{1}{2}x^4 + C$

Chapter 5

The Integral

5.1 Introduction to Integration—The Area Problem

Problems for Practice

1. $\sum_{j=1}^{4} 3j = 3 + 6 + 9 + 12 = 30$

3. $\sum_{j=2}^{5}(-2j^2) = -8 - 18 - 32 - 50 = -108$

5. $\sum_{n=2}^{5} \frac{2n}{n-1} = 4 + 3 + \frac{8}{3} + \frac{5}{2} = \frac{73}{6}$

7. $\sum_{m=3}^{6}(2m^2 - 3m) = 9 + 20 + 35 + 54 = 118$

9. $\sum_{j=1}^{3} j \cdot \sin^2(j\pi/6) = \frac{1}{4} + \frac{3}{2} + 3 = \frac{19}{4}$

11. $2 + 3 + 4 + 5 + 6 = \sum_{j=2}^{6} j$

13. $9 + 13 + 17 + 21 + 25 + 29 = \sum_{j=0}^{5}(9 + 4j)$

15. $\frac{1}{4} + \frac{1}{5} + \frac{1}{6} + \frac{1}{7} + \frac{1}{8} = \sum_{j=4}^{8} \frac{1}{j}$

17. $\sum_{j=1}^{12} j^2 = \frac{12 \cdot 13 \cdot 25}{6} = 650$

19. $\sum_{j=1}^{12}(j^2 + 2j) = \sum_{j=1}^{12} j^2 + 2\sum_{j=1}^{12} j = \frac{12 \cdot 13 \cdot 25}{6} + 2 \cdot \frac{12 \cdot 13}{2} = 806$

21. $\sum_{j=12}^{24}(3j - 2) = 3\sum_{j=12}^{24} j - 2\sum_{j=12}^{24} 1$

$= 3\left(\sum_{j=1}^{24} j - \sum_{j=1}^{11} j\right) - 2 \cdot 13 = 3\left(\frac{24 \cdot 25}{2} - \frac{11 \cdot 12}{2}\right) - 26 = 676$

23. $\displaystyle\sum_{j=0}^{19} \exp(j) = \sum_{j=0}^{19} e^j \overset{(5.1.3)}{=} \frac{e^{20} - 1}{e - 1}$

25. $\sum_{j=2}^{6} \ln(j) = \ln(2 \cdot 3 \cdot 4 \cdot 5 \cdot 6) = \ln(6!) = \ln(720)$

27. $\Delta x = \frac{5-3}{2} = 1$ and $x_j = 3 + j \cdot 1 = 3 + j$. Consequently,

$$I \approx \Delta x \sum_{j=1}^{2} f(3 + j) = f(4) + f(5) = 23\,.$$

29. $\Delta x = \frac{1-1/2}{2} = 1/4$ and $x_j = 1/2 + j/4$. Consequently,

$$I \approx \Delta x \sum_{j=1}^{2} f(1/2 + j/4) = \frac{1}{4}(f(3/4) + f(1))$$

$$= -\frac{1}{4}(\log_2(3/4) + \log_2(1)) = -\log_2(3/4)/4\,.$$

Since $\log_2(1/4) = -2$, the approximation can also be expressed in the form $(2 - \log_2(3))/4$.

31. $\Delta x = \frac{2\pi - \pi/2}{3} = \pi/2$ and $x_j = \pi/2 + j \cdot (\pi/2)$. Consequently,

$$I \approx \Delta x \sum_{j=1}^{3} f(\pi/2 + j \cdot (\pi/2)) = \frac{\pi}{2}(f(\pi) + f(3\pi/2) + f(2\pi)) = 3\pi\,.$$

33. $\Delta x = \frac{-2-(-4)}{4} = 1/2$ and $x_j = -4 + j/2$. Consequently,

$$I \approx \Delta x \sum_{j=1}^{4} f(-4 + j/2) = \frac{1}{2}(f(-7/2) + f(-3) + f(-5/2) + f(-2)) = 197/60\,.$$

35. $\Delta x = \frac{\pi-(-\pi)}{4} = \pi/2$ and $x_j = -\pi + j \cdot (\pi/2)$. Consequently,

$$I \approx \Delta x \sum_{j=1}^{4} f(-\pi + j \cdot (\pi/2)) = \frac{\pi}{2}(f(-\pi/2) + f(0) + f(\pi/2) + f(\pi)) = \pi^2/2\,.$$

37. $\Delta x = \frac{5/2-(-1)}{7} = 1/2$ and $x_j = -1 + j/2$. Consequently,

$$I \approx \Delta x \sum_{j=1}^{7} f(-1 + j/2) = \frac{1}{2}(f(-1/2) + f(0) + \cdots + f(2) + f(5/2)) = 14\,.$$

Further Theory and Practice

39. $\Delta x = \frac{\pi/3-0}{2} = \pi/6$ and $x_j = j \cdot (\pi/6)$. Consequently,

$$L = \Delta x \sum_{j=1}^{2} f((j-1)\cdot(\pi/6)) = \frac{\pi}{6}(f(0) + f(\pi/6)) = \pi/12 \approx 0.2168$$

$$R = \Delta x \sum_{j=1}^{2} f(j\cdot(\pi/6)) = \frac{\pi}{6}(f(\pi/6) + f(\pi/3)) = (\pi/12)\cdot(1+\sqrt{3}\,) \approx 0.7155$$

$$\frac{L+R}{2} = \frac{\pi}{24}(2+\sqrt{3}\,) \approx 0.4886 \quad \text{and} \quad A = 0.5\,.$$

41. $\Delta x = \frac{2-0}{2} = 1$ and $x_j = j$. Consequently,

$$L = \Delta x \sum_{j=1}^{2} f(j-1) = f(0) + f(1) = 5/2 \approx 2.5$$

$$R = \Delta x \sum_{j=1}^{2} f(j) = f(1) + f(2) = 19/6 \approx 3.167$$

$$\frac{L+R}{2} = 17/6 \approx 2.833 \quad \text{and} \quad I = 4 - \ln(3) \approx 2.901\,.$$

43. $\Delta x = \frac{2-1}{3} = 1/3$ and $x_j = 1 + j\cdot(1/3)$. Consequently,

$$L = \Delta x \sum_{j=1}^{3} f(1+(j-1)\cdot(1/3)) = \frac{1}{3}(f(1) + f(4/3) + f(5/3)) = 431/1200 \approx 0.3592$$

$$R = \Delta x \sum_{j=1}^{3} f(1+j\cdot(1/3)) = \frac{1}{3}(f(4/3) + f(5/3) + f(2)) = 731/1200 \approx 0.6062$$

$$\frac{L+R}{2} = 581/1200 \approx 0.4842 \quad \text{and} \quad A = 1/2 = 0.5\,.$$

45. The area is $A = \lim_{N\to\infty} R(N)$, N a positive integer, where $R(N)$ is the right sum obtained from a uniform partition of $[0,2]$ into N subintervals, each having width $\Delta x = (2-0)/N = 2/N$. The jth partition point is $x_j = 2j/N$ and $R(N) = \frac{2}{N}\sum_{j=1}^{N} f(2j/N) = \frac{2}{N}\sum_{j=1}^{N}(4-(2j/N)^2)$. The limiting value of $R(N)$, as $N\to\infty$, can be determined by manipulating the summation formula as follows.

$$\begin{aligned} R(N) &= \frac{2}{N}\left(4\sum_{j=1}^{N} 1 - \frac{4}{N^2}\sum_{j=1}^{N} j^2\right) = \frac{8}{N}\cdot N - \frac{8}{N^3}\cdot\frac{N(N+1)(2N+1)}{6} \\ &= 8 - \frac{4}{3}\cdot\frac{N(N+1)(2N+1)}{N^3} \xrightarrow[N\to\infty]{} 8 - \frac{8}{3} = \frac{16}{3} \end{aligned}$$

47. If $e^p = e \cdot e^2 \cdot e^3 \cdots e^{100}$, then, by taking the natural logarithm of both sides, $p = 1 + 2 + \cdots + 100 = \frac{100 \cdot 101}{2} = 5050$.

If $\ln(q) = \sum_{n=1}^{100} \ln(n)$, then, by applying the exponential function to both sides, $q = e^{\ln(1)} \cdot e^{\ln(2)} \cdot e^{\ln(3)} \cdots e^{\ln(100)} = 1 \cdot 2 \cdot 3 \cdots 100 = 100!$.

49. $S = \sum_{j=1}^{N} (2j)^2 = 4 \sum_{j=1}^{N} j^2 = 4 \cdot \frac{N(N+1)(2N+1)}{6}$. Therefore,

$$\begin{aligned}\sum_{j=1}^{N} (2j-1)^2 = \sum_{j=1}^{2N} j^2 - S &= \frac{2N(2N+1)(4N+1)}{6} - 4 \cdot \frac{N(N+1)(2N+1)}{6} \\ &= \frac{N}{3} \cdot \Big((2N+1)(4N+1) - 2(N+1)(2N+1) \Big) \\ &= \frac{N}{3} \cdot (4N^2 - 1) = \frac{N(2N-1)(2N+1)}{3}.\end{aligned}$$

51. If $M = 1$, then $\sum_{j=1}^{N} (a_j - a_{j-1}) = a_N - a_0 = a_N - a_{M-1}$. Otherwise,

$$\begin{aligned}\sum_{j=M}^{N} (a_j - a_{j-1}) = \sum_{j=1}^{N} (a_j - a_{j-1}) - \sum_{j=1}^{M-1} (a_j - a_{j-1}) &= a_N - a_0 - (a_{M-1} - a_0) \\ &= a_N - a_{M-1}.\end{aligned}$$

53. The bottom row is $2S = \underbrace{(N+1) + (N+1) + \cdots + (N+1)}_{N \text{ times}}$. From this if follows that $2S = N(N+1)$ and $S = N(N+1)/2$.

55. In terms of S, the sum on the left is $3S - 3 \cdot \frac{N(N+1)}{2} + N$, and the sum on the right telescopes to N^3. Consequently, $3S - 3 \cdot \frac{N(N+1)}{2} + N = N^3$, implying that

$$\begin{aligned}S = \frac{1}{3} \cdot \left(N^3 - N + \frac{3N(N+1)}{2} \right) &= \frac{1}{3} \cdot \frac{2N(N^2-1) + 3N(N+1)}{2} \\ &= \frac{N(N+1)(2(N-1)+3)}{6} = \frac{N(N+1)(2N+1)}{6}.\end{aligned}$$

57. $F(x) = x^3/3$ is an antiderivative of f. The point c satisfies the equation $(F(b) - F(0))/(b - 0) = F'(c)$. That is, $(b^3/3)/b = c^2$, so $c = b/\sqrt{3}$. Using one rectangle with base $\Delta x = b$ and height $f(c) = b^2/3$, the area approximation is $f(c)\Delta x = (b^2/3) \cdot b = b^3/3$.

59. To obtain the area under the curve and above the interval $[0, b]$ let $R(N)$ be the approximation obtained using a uniform partition of $[0, b]$ with N subintervals, each of width $\Delta x = b/N$. Then the jth partition point is $x_j = j \cdot (b/N)$ so $R(N) = (b/N) \sum_{j=1}^{N} (jb/N)^3$. Using the summation

formula in Exercise 58 yields the following simpification of $R(N)$.

$$R(N) = \frac{b}{N}\sum_{j=1}^{N}\left(\frac{jb}{N}\right)^3 = \frac{b^4}{N^4}\sum_{j=1}^{N} j^3 = \frac{b^4}{N^4}\cdot\frac{N^2(N+1)^2}{4}$$
$$= \frac{b^4}{4}\cdot\frac{N^2(N+1)^2}{N^4} \xrightarrow[N\to\infty]{} \frac{b^4}{4}.$$

The area from $x = 0$ to $x = b$, $b^4/4$, is the sum of the area from $x = 0$ to $x = a$, $a^4/4$, and the area A from $x = a$ to $x = b$: $b^4/4 = a^4/4 + A$. Therefore, $A = b^4/4 - a^4/4 = (b^4 - a^4)/4$.

61. According to Exercise 56, $A(b) = b^3/3$. Consequently, $A'(b) = b^2 = f(b)$.

63. (a) Using the formula for $S_N(k)$ and $P_k(N) = a_k N^k + \cdots + a_1 N + a_0$,

$$\lim_{N\to\infty}\frac{S_N(k)}{N^{k+1}} = \frac{1}{k+1} + \frac{a_k}{N} + \cdots + \frac{a_1}{N^k} + \frac{a_0}{N^{k+1}} \xrightarrow[N\to\infty]{} \frac{1}{k+1}.$$

(b) let $R(N)$ be the approximation obtained using a uniform partition of $[0, b]$ with N subintervals, each of width $\Delta x = b/N$. Then the jth partition point is $x_j = j\cdot(b/N)$ so $R(N) = (b/N)\sum_{j=1}^{N}(jb/N)^k$. Using the limit in part a yields the following simpification of $R(N)$.

$$R(N) = \frac{b}{N}\sum_{j=1}^{N}\left(\frac{jb}{N}\right)^k = \frac{b^{k+1}}{N^{k+1}}\sum_{j=1}^{N} j^k = \frac{b^{k+1}}{N^{k+1}}\cdot S_N(k)$$
$$= b^{k+1}\cdot\frac{S_N(k)}{N^{k+1}} \xrightarrow[N\to\infty]{} \frac{b^{k+1}}{k+1}.$$

(c) The area from $x = 0$ to $x = b$, $b^{k+1}/(k+1)$, is the sum of the area from $x = 0$ to $x = a$, $a^{k+1}/(k+1)$, and the area A from $x = a$ to $x = b$: $b^{k+1}/(k+1) = a^{k+1}/(k+1) + A$. Therefore,

$$A = \frac{b^{k+1}}{k+1} - \frac{a^{k+1}}{k+1} = \frac{b^{k+1} - a^{k+1}}{k+1}.$$

65. The two areas sum to the area of the rectangle with base a and height b. Therefore the area under the graph of $x = f^{-1}(y)$ and the interval $[0, b]$ is $ab - A$.

67. Let $R(N)$ be the approximation obtained using a uniform partition of $[0, b]$ with N subintervals, each of width $\Delta x = b/N$. The kth partition point is $x_k = k\cdot(b/N)$, so $R(N) = (b/N)\sum_{k=1}^{N}\cos(kb/N)$. Using the summation formula developed in Exercise 66, with $t = b/N$, yields the

following simpification of $R(N)$.

$$R(N) = \frac{b}{N}\sum_{k=1}^{N}\cos(kb/N) = \frac{b}{N}\cdot\frac{\sin(b/2)\cos\left(\frac{N+1}{N}\cdot(b/2)\right)}{\sin((b/2)/N)}$$

$$= 2\sin(b/2)\cos\left(\tfrac{N+1}{N}\cdot(b/2)\right)\cdot\frac{(b/2)/N}{\sin((b/2)/N)} \xrightarrow[N\to\infty]{} \sin(b)\,.$$

In evaluating the last limit, we made use of the trigonometric identity $2\sin(b/2)\cos(b/2) = \sin(b)$, and the limit formula $\lim_{x\to 0}\dfrac{x}{\sin(x)} = 1$.

CALCULATOR/COMPUTER EXERCISES

69. The following table contains approximations $R(25\cdot k)$ for $k = 1, 2, 3$.

k	1	2	3
$R(25\cdot k)$	0.16	0.16	0.16

Maple approximates the area as 0.1666666667.

71. The following table contains approximations $R(25\cdot k)$ for $k = 1, 2, 3$.

k	1	2	3
$R(25\cdot k)$	1.57	1.57	1.57

Maple approximates the area as 1.570796327.

73. Using *Maple*'s *fsolve* procedure (and a plot of the function), we find that $a = -1.841405660$ and $b = 1.146193221$. The following table contains approximations $R(25\cdot k)$ for $k = 1, 2, 3$.

k	1	2	3
$R(25\cdot k)$	1.94	1.94	1.94

Maple approximates the area as 1.949090927.

75. Using *Maple*'s *fsolve* procedure (and a plot of the function), we find that $a = 0.1010258483$ and $b = 1.679630610$. The following table contains approximations $R(25\cdot k)$ for $k = 1, 2, 3$.

k	1	2	3
$R(25\cdot k)$	0.30	0.30	0.30

Maple approximates the area as 0.3072939161.

5.2 The Riemann Integral

Problems for Practice

1. (a) Since $\Delta x = 4$, $\mathcal{R}(f, \mathcal{S}) = 4(f(4) + f(8) + f(12)) = 11/6$.

(b) $\int_2^{14} 1/x\,dx = \ln(x)\Big|_2^{14} = \ln(14) - \ln(2) = \ln(7)$

3. (a) Since $\Delta x = 2$, $\mathcal{R}(f, \mathcal{S}) = 2(f(-1) + f(1) + f(3) + f(5)) = 72$.

   (b) $\int_{-2}^{6} x^2\,dx = x^3/3\Big|_{-2}^{6} = 72 - (-8/3) = 224/3$

5. (a) Since $\Delta x = 2$, $\mathcal{R}(f, \mathcal{S}) = 2(f(1) + f(3) + f(5)) = 378$.

   (b) $\int_1^7 (x^3 + 4x)\,dx = (x^4/4 + 2x^2)\Big|_1^7 = 2793/4 - 9/4 = 696$

7. (a) Since $\Delta x = 3$, $\mathcal{R}(f, \mathcal{S}) = 3(f(-7) + f(-4) + f(-1) + f(2)) = 42$.

   (b) The function $F(x) = \begin{cases} x^2/2 & \text{if} \quad x \geq 0 \\ -x^2/2 & \text{if} \quad x < 0 \end{cases}$ is an antiderivative for $|x|$. Therefore, $\int_{-7}^{5} |x|\,dx = F(x)\Big|_{-7}^{5} = 25/2 - (-49/2) = 37$.

9. (a) Since $\Delta x = 2$, $\mathcal{R}(f, \mathcal{S}) = 2\big(f(-2) + f(0) + f(2)\big) = 2(e^{-2} + 1 + e^2)$.

   (b) $\int_{-4}^{2} e^x\,dx = e^x\Big|_{-4}^{2} = e^2 - e^{-4}$.

11. (a) Since $\Delta x = \frac{2\pi}{3}$, $\mathcal{R}(f, \mathcal{S}) = \frac{2\pi}{3}\big(f(-\frac{\pi}{3}) + f(\frac{\pi}{3}) + f(\pi) + f(\frac{5\pi}{3})\big) = \frac{\pi}{3}$.

    (b) $\int_{-\pi}^{5\pi/3} \cos(x)\,dx = \sin(x)\Big|_{-\pi}^{5\pi/3} = -\sqrt{3}/2 - 0 = -\sqrt{3}/2$.

13. $\Delta x = 5/4$ and the two subintervals of the second order uniform partition are $I_1 = [-1/2, 3/4]$ and $I_2 = [3/4, 2]$.

    The minimum values of f on the two intervals are $f(0) = 0$ and $f(3/4) = 9/16$, yielding the lower Riemann sum $\mathcal{R}(f, \mathcal{L}_2) = \frac{5}{4}\big(0 + \frac{9}{16}\big) = \frac{45}{64}$.

    The maximum values of f on the two intervals are $f(3/4) = 9/16$ and $f(2) = 4$, yielding the upper Riemann sum $\mathcal{R}(f, \mathcal{U}_2) = \frac{5}{4}\big(\frac{9}{16} + 4\big) = \frac{365}{64}$.

15. $\Delta x = 2\pi/3$ and the two subintervals of the second order uniform partition are $I_1 = [0, 2\pi/3]$ and $I_2 = [2\pi/3, 4\pi/3]$.

    The minimum values of f on the two intervals are $f(2\pi/3) = -1/2$ and $f(\pi) = -1$, yielding the lower Riemann sum $\mathcal{R}(f, \mathcal{L}_2) = \frac{2\pi}{3}\big(-\frac{1}{2} - 1\big) = -\pi$.

    The maximum values of f on the two intervals are $f(0) = 1$ and $f(2\pi/3) = -1/2$, yielding the upper Riemann sum $\mathcal{R}(f, \mathcal{U}_2) = \frac{2\pi}{3}\big(1 - \frac{1}{2}\big) = \frac{\pi}{3}$.

17. The integrand is sketched on the right with bounding lines for the area $A = 2\sqrt{2}$. We conclude that:

$$\int_1^3 \sqrt{2}\,dx = 2\sqrt{2}.$$

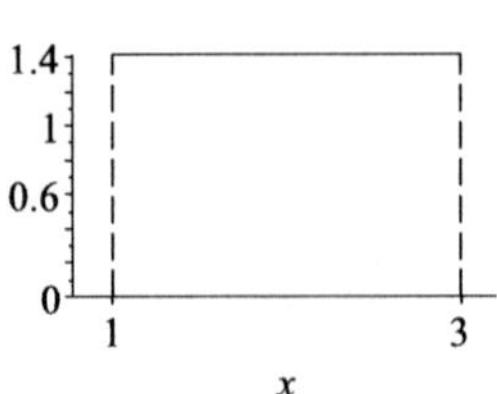

19. The integrand is sketched on the right with bounding lines for the area $A = 5$. We conclude that:

$$\int_{-1}^{3} |x|\,dx = 5\,.$$

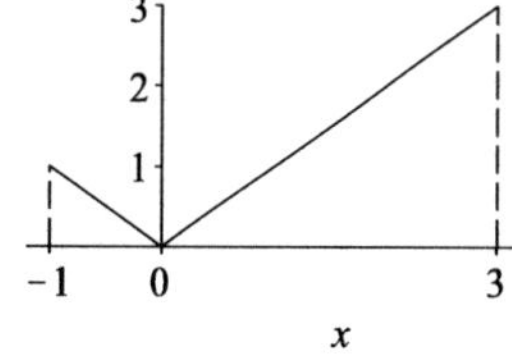

21. The integrand is sketched on the right with bounding lines for the area $A = 9/2$. We conclude that:

$$\int_{-1}^{2} (1+x)\,dx = 9/2\,.$$

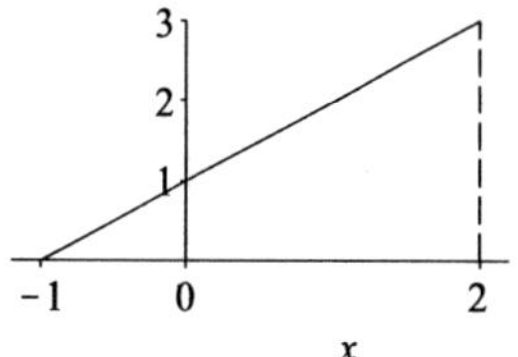

23. The integrand is sketched on the right with bounding lines for the area $A = 13/2$. We conclude that:

$$\int_{-2}^{3} |x-1|\,dx = 13/2\,.$$

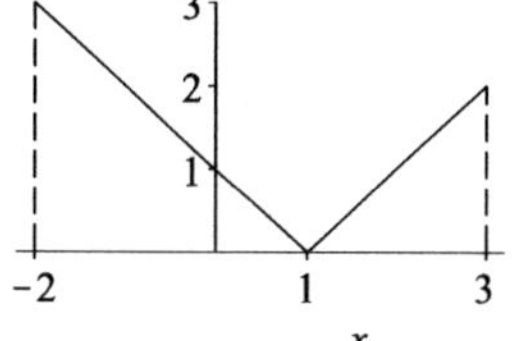

25. $\int_1^2 (6x^2 - 2x)\,dx = (2x^3 - x^2)\Big|_1^2 = 12 - 1 = 11$

27. $\int_1^4 \sqrt{x}\,dx = \left(\frac{2}{3}x^{3/2}\right)\Big|_1^4 = \frac{16}{3} - \frac{2}{3} = \frac{14}{3}$

29. $\int_0^{\pi/4} \sec^2(x)\,dx = \tan(x)\Big|_0^{\pi/4} = 1 - 0 = 1$

31. $\int_0^{\pi/3} \sec(x)\tan(x)\,dx = \sec(x)\Big|_0^{\pi/3} = 2 - 1 = 1$

33. $\int_8^{27} x^{1/3}\,dx = \left(\frac{3}{4}x^{4/3}\right)\Big|_8^{27} = \frac{243}{4} - 12 = \frac{195}{4}$

35. $\int_0^1 e^x\,dx = e^x\Big|_0^1 = e - 1$

37. $\int_{-e}^{-1} (1/x)\,dx = \ln(|x|)\Big|_{-e}^{-1} = \ln(1) - \ln(e) = -1$

39. $\int_0^3 2^x\,dx = \left(\frac{2^x}{\ln(2)}\right)\Big|_0^3 = \frac{8}{\ln(2)} - \frac{1}{\ln(2)} = \frac{7}{\ln(2)}$

41. $\int_{\pi/4}^{\pi/3} \csc^2(x)\,dx = (-\cot(x))\Big|_{\pi/4}^{\pi/3} = -\sqrt{3}/3 - (-1) = 1 - \sqrt{3}/3$

43. $\int_{-1}^{2} (x^2+5x+6)/(x+2)\,dx = \int_{-1}^{2} (x+3)\,dx = \left(\frac{x^2}{2} + 3x\right)\Big|_{-1}^{2} = 8 - (-\frac{5}{2}) = \frac{21}{2}$

Further Theory and Practice

45. $\Delta x = 1$, there is only one subinterval, I.

    The minimum value of f on I is $f(0) = 1$ and the maximum value is $f(1) = e$, yielding the lower Riemann sum $\mathcal{R}(f, \mathcal{L}_1) = 1$ and the upper Riemann sum $\mathcal{R}(f, \mathcal{U}_1) = e$.

The midpoint value is $f(1/2) = \sqrt{e}$ so for $\mathcal{S}_1 = \{1/2\}$, $\mathcal{R}(f, \mathcal{S}_1) = \sqrt{e}$.

47. $\Delta x = 1$, there are two subintervals, $I_1 = [1, 2]$ and $I_2 = [2, 3]$, and f is decreasing on I.

    Because f is decreasing, $\mathcal{L}_2 = \{2, 3\}$ and $\mathcal{U}_2 = \{1, 2\}$. The midpoint selection is $\mathcal{S}_2 = \{3/2, 5/2\}$. Consequently,

$$\mathcal{R}(f, \mathcal{L}_2) = f(2) + f(3) = \frac{1}{3} + \frac{1}{4} = \frac{7}{12}$$
$$\mathcal{R}(f, \mathcal{S}_2) = f(\tfrac{3}{2}) + f(\tfrac{5}{2}) = \frac{2}{5} + \frac{2}{7} = \frac{24}{35}$$
$$\mathcal{R}(f, \mathcal{U}_2) = f(1) + f(2) = \frac{1}{2} + \frac{1}{3} = \frac{5}{6}.$$

49. $\Delta x = 1$, there are three subintervals, $I_1 = [0, 1]$, $I_2 = [1, 2]$, and $I_3 = [2, 3]$, and f is increasing on I.

    Because f is increasing, $\mathcal{L}_3 = \{0, 1, 2\}$ and $\mathcal{U}_3 = \{1, 2, 3\}$. The midpoint selection is $\mathcal{S}_3 = \{1/2, 3/2, 5/2\}$. Consequently,

$$\mathcal{R}(f, \mathcal{L}_3) = f(0) + f(1) + f(2) = \ln(2) + \ln(3) = \ln(6) \approx 1.7918$$
$$\mathcal{R}(f, \mathcal{S}_3) = f(\tfrac{1}{2}) + f(\tfrac{3}{2}) + f(\tfrac{5}{2}) = \ln(\tfrac{3}{2}) + \ln(\tfrac{5}{2}) + \ln(\tfrac{7}{2}) = \ln(\tfrac{105}{8}) \approx 2.5746$$
$$\mathcal{R}(f, \mathcal{U}_3) = f(1) + f(2) + f(3) = \ln(2) + \ln(3) + \ln(4) = \ln(24) \approx 3.1780.$$

51. $\int_0^{\pi/3} \sin(3x)\,dx = \big(-\frac{\cos(3x)}{3}\big)\big|_0^{\pi/3} = \frac{1}{3} - (-\frac{1}{3}) = \frac{2}{3}$

53. $\int_0^{\pi/8} \sec^2(2x)\,dx = \big(\frac{\tan(2x)}{2}\big)\big|_0^{\pi/8} = \frac{1}{2}$

55. $\int_0^1 e^{-x}\,dx = \big(-e^{-x}\big)\big|_0^1 = -e^{-1} - (-1) = 1 - 1/e$

57. $\Delta x = 1$, there are three subintervals, $I_1 = [0, 1]$, $I_2 = [1, 2]$, and $I_3 = [2, 3]$. It is easy to verify that f is increasing on I_1, decreasing on I_2, and increasing on I_3. Therefore, $\mathcal{L}_3 = \{0, 2, 2\}$ and $\mathcal{U}_3 = \{1, 1, 3\}$, and

$$\mathcal{R}(f, \mathcal{L}_3) = f(0) + f(2) + f(2) = 11$$
$$\int_0^3 (2x^3 - 9x^2 + 12x + 1)\,dx = (x^4/2 - 3x^3 + 6x^2 + x)\big|_0^3 = 33/2$$
$$\mathcal{R}(f, \mathcal{U}_3) = f(1) + f(1) + f(3) = 22.$$

59. The points s_1 and s_2 should be chosen to satisfy the Mean Value Theorem for the integrand's *antiderivative* $F(x) = x^4$ on the intervals of the partition, $I_1 = [1, 3]$ and $I_2 = [3, 5]$, respectively. This yields the selection of points $\mathcal{S}_2 = \{10^{1/3}, 68^{1/3}\}$ (verify). Since $\Delta x = 2$, the Riemann sum for this selection is

$$\mathcal{R}(f, \mathcal{S}_3) = 2(f(10^{1/3}) + f(68^{1/3})) = 624.$$

    This is the same as the integral of f over the interval $[1, 5]$ (verify).

61. The points s_1 and s_2 should be chosen to satisfy the Mean Value Theorem for the integrand's *antiderivative* $F(x) = \ln(x)$ on the intervals of the partition, $I_1 = [1, 3/2]$ and $I_2 = [3/2, 2]$, respectively. This yields the selection $\mathcal{S}_2 = \{1/(2\ln(3/2)), 1/(2\ln(4/3))\}$ (verify). Since $\Delta x = 1/2$, the Riemann sum for this selection is

$$\mathcal{R}(f, \mathcal{S}_2) = \frac{1}{2}(f(1/(2\ln(3/2))) + f(1/(2\ln(4/3))) = \ln(2)\,.$$

This is the same as the integral of f over the interval $[1, 2]$ (verify).

63. If (x, y_C) lies on the circle and (x, y_E) lies on the ellipse, as in the figure, then

$$y_E = b\sqrt{1 - \frac{x^2}{a^2}} = \frac{b}{a}\sqrt{a^2 - x^2} = \frac{b}{a} \cdot y_C\,.$$

Equivalently, if f_E and f_C are the functions defined by the top half of the ellipse and the circle, respectively, then $f_E(x) = \frac{b}{a} \cdot f_C(x)$ for all $x \in [0, a]$. Therefore, given any selection $\mathcal{S}_N$ for the order N uniform partition of $[0, a]$, $\mathcal{R}(f_E, \mathcal{S}_N) = \frac{b}{a} \cdot \mathcal{R}(f_C, \mathcal{S}_N)$, implying that $\lim_{N\to\infty} \mathcal{R}(f_E, \mathcal{S}_N) = \frac{b}{a} \cdot \lim_{N\to\infty} \mathcal{R}(f_C, \mathcal{S}_N)$, and the area of the portion of the ellipse in the first quadrant is b/a times the area of the quarter circle. By symmetry, this extends to the area of the entire ellipse and the entire circle. That is, Ellipse Area $= \frac{b}{a} \cdot \pi a^2 = \pi ab$.

65. By Theorem 3,

$$\int_a^b (m(x - c) + h)\, dx = (mx^2/2 + (h - mc)x)\big|_a^b = \frac{m}{2} \cdot (b^2 - a^2) + (h - mc)(b - a)$$
$$= h(b - a) + m \cdot \frac{b^2 - a^2}{2} - m \cdot \frac{b + a}{2} \cdot (b - a) = h(b - a)\,.$$

67. The area of a trapezoid is half the sum of its bases times its height. Applying this to the region in question, $A_T = \frac{1}{2} \cdot (f(a) + f(b)) \cdot (b - a)$. This is the right side of the desired inequality. The validity of the inequality follows from the fact that the integral on the left side evaluates to the area under the graph of f.

69. Suppose, for example, that $\ell_j < u_j$. The Mean Value Theorem, applied to f on $[\ell_j, u_j]$, yields a point c_j such that $(f(u_j) - f(\ell_j))/(u_j - \ell_j) = f'(c_j)$. Consequently, $f(u_j) - f(\ell_j) = f'(c_j)(u_j - \ell_j)$, implying that

$$|f(u_j) - f(\ell_j)| \le M_1 \cdot |u_j - \ell_j| \le M_1 \cdot \Delta x\,.$$

If $u_j < \ell_j$ we would obtain $|f(\ell_j) - f(u_j)| \le M_1 \cdot \Delta x$. Since $f(u_j) \ge f(\ell_j)$, either conclusion implies that $f(u_j) - f(\ell_j) \le M_1 \cdot \Delta x$, and this inequality is also valid if $u_j = \ell_j$.

Multiply both sides of the jth inequality above by Δx and sum from $j = 1$ to $j = N$ to obtain $\Delta x \sum_{j=1}^N f(u_j) - \Delta x \sum_{j=1}^N f(\ell_j) \le M_1 \cdot (\Delta x)^2 \cdot N$.

That is,

$$\mathcal{R}(f,\mathcal{U}_N) - \mathcal{R}(f,\mathcal{L}_N) \le M_1 \cdot \left(\frac{b-a}{N}\right)^2 \cdot N = M_1 \frac{(b-a)^2}{N}.$$

CALCULATOR/COMPUTER EXERCISES

71. Using *Maple*, $N = 100$, $\mathcal{R}(f,\mathcal{S}_{100}) = 0.3749999999$. *Maple* approximates the integral's value as 0.3750000001.

73. Using *Maple*, $N = 60$, $\mathcal{R}(f,\mathcal{S}_{60}) = 3.141951551$. *Maple* approximates the integral's value as 3.141592654.

75. The function is plotted on the right. Since it is increasing, the lower sum uses the left endpoints and the upper sum uses the right endpoints.

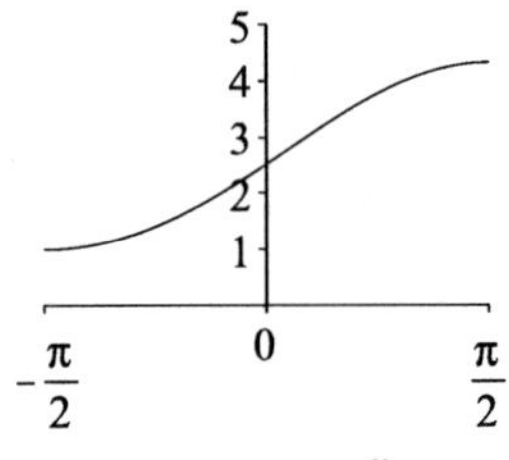

$$\mathcal{R}(f,\mathcal{L}_{50}) = 8.035809924$$
$$\mathcal{R}(f,\mathcal{U}_{50}) = 8.244835710$$

77. The function is plotted on the right. Since it is increasing, the lower sum uses the left endpoints and the upper sum uses the right endpoints.

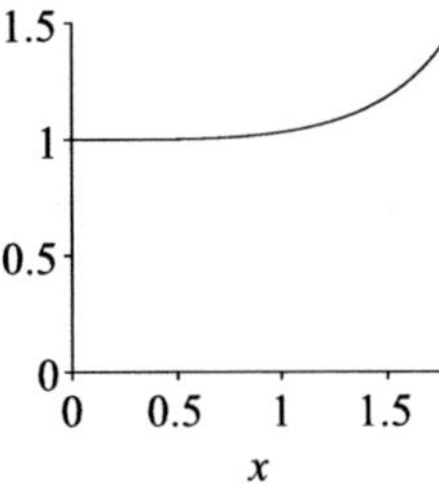

$$\mathcal{R}(f,\mathcal{L}_{50}) = 1.893551751$$
$$\mathcal{R}(f,\mathcal{U}_{50}) = 1.908235239$$

5.3 Properties of the Integral

Problems for Practice

1. $\int_1^7 f(x)\,dx = \int_1^3 f(x)\,dx + \int_3^7 f(x)\,dx = 4$

3. $\int_{-7}^3 (4f(x) - 9g(x))\,dx = 4\int_{-7}^3 f(x)\,dx - 9\int_{-7}^3 g(x)\,dx = 4\cdot(-7) - 9\cdot(-4) = 8$

5. $\int_9^{-2} f(x)\,dx = \int_9^7 f(x)\,dx + \int_7^{-2} f(x)\,dx = -(-4) + 6 = 10$

7. $\int_9^4 (6f(x) - 7g(x))\,dx = 6\int_9^4 f(x)\,dx - 7\int_9^4 g(x)\,dx = 6\cdot(5) - 7\cdot(15) = -75$. Therefore, $\int_4^9 (6f(x) - 7g(x))\,dx = 75$.

9. Since $\frac{-3}{4}\int_5^{-3} f(x)\,dx = 7$, $\int_5^{-3} f(x)\,dx = -\frac{28}{3}$. Consequently,

$$\begin{aligned}\int_5^{-3} (6f(x)+1)\,dx &= 6\int_5^{-3} f(x)\,dx + \int_5^{-3} 1\,dx \\ &= 6\cdot\left(-\frac{28}{3}\right) + 1\cdot(-3-5) = -64\,.\end{aligned}$$

11. $\int_4^1 3f(x)\,dx = 9\int_4^1 f(x)/3\,dx = 9\cdot(-2) = -18$

13. $\int_2^1 (f(x) - 3g(x) + 5)\,dx = \int_2^1 f(x)\,dx - 3\int_2^1 g(x)\,dx + \int_2^1 5\,dx = 5\cdot(1-2)$
$= -5$

15. The average value of f over the interval $[1,5]$ is

$$f_{\text{ave}} = \frac{1}{5-1}\int_1^5 (1+x)\,dx = 4\,.$$

This value is attained by f at $c = 3$. See the picture. It illustrates the fact that the area of the region under the graph of f and over the interval $[1,5]$ equals the area of the rectangle having base $[1,5]$ and height 4. That is, $\int_1^5 (1+x)\,dx = f_{\text{ave}}\cdot(5-1)$.

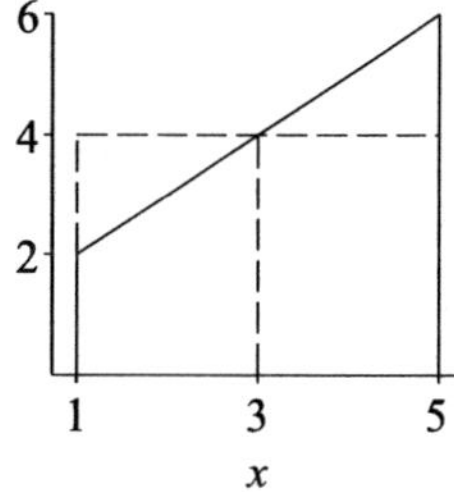

17. The average value of f over the interval $[1,4]$ is

$$f_{\text{ave}} = \frac{1}{4-1}\int_1^4 \sqrt{x}\,dx = 14/9 \approx 1.56\,.$$

This value is attained by f at $c = 196/81 \approx 2.42$. See the picture. It illustrates the fact that the area of the region under the graph of f and over the interval $[1,4]$ equals the area of the rectangle having base $[1,4]$ and height 14/9. That is, $\int_1^4 \sqrt{x}\,dx = f_{\text{ave}}\cdot(4-1)$.

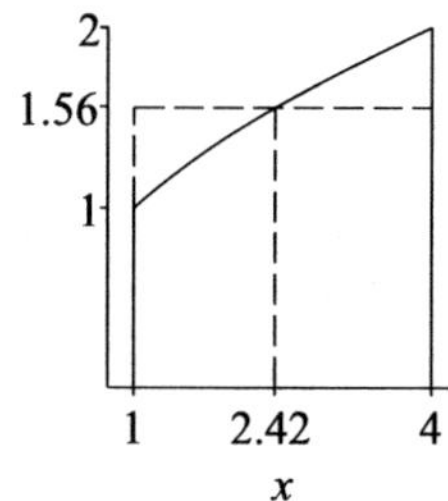

19. The average value of f over the interval $[0,3]$ is

$$f_{\text{ave}} = \frac{1}{3-0}\int_0^3 (12x^2+5)\,dx = 41\,.$$

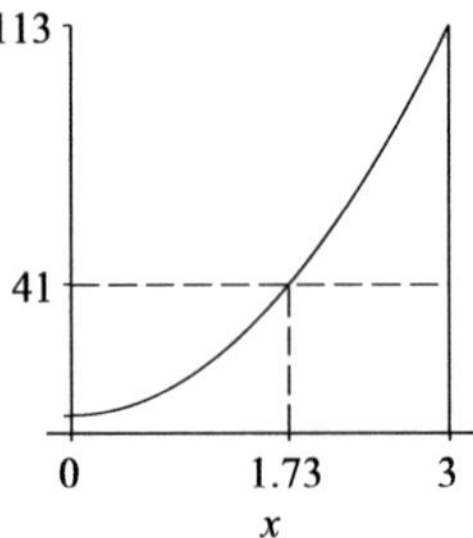

This value is attained by f at $c = \sqrt{3} \approx 1.73$. See the picture. It illustrates the fact that the area of the region under the graph of f and over the interval $[0,3]$ equals the area of the rectangle having base $[0,3]$ and height 41. That is, $\int_0^3 (12x^2+5)\,dx = f_{\text{ave}} \cdot (3-0)$.

21. The average value of f over the interval $[-4,2]$ is

$$f_{\text{ave}} = \frac{1}{2-(-4)}\int_{-4}^2 (8-x^3)\,dx = \frac{1}{6}\cdot\left(8x - \frac{x^4}{4}\right)\Bigg|_{-4}^2 = \frac{12-(-96)}{6} = 18\,.$$

23. The average value of f over the interval $[1,4]$ is

$$f_{\text{ave}} = \frac{1}{4-1}\int_1^4 (\sqrt{x}-1/\sqrt{x})\,dx = \frac{1}{3}\cdot\left(\frac{2}{3}x^{3/2} - 2\sqrt{x}\right)\Bigg|_1^4 = \frac{4/3-(-4/3)}{3} = \frac{8}{9}\,.$$

25. The average value of f over the interval $[0,\pi/2]$ is

$$f_{\text{ave}} = \frac{1}{\pi/2-0}\int_0^{\pi/2} \cos(x)\,dx = \frac{1}{\pi/2}\cdot\big(\sin(x)\big)\Big|_0^{\pi/2} = \frac{1-0}{\pi/2} = \frac{2}{\pi}\,.$$

27. The average value of f over the interval $[0,\pi/4]$ is

$$f_{\text{ave}} = \frac{1}{\pi/4-0}\int_0^{\pi/4} \sec^2(x)\,dx = \frac{1}{\pi/4}\cdot\big(\tan(x)\big)\Big|_0^{\pi/4} = \frac{1-0}{\pi/4} = \frac{4}{\pi}\,.$$

29. The integrand f increases on $[0,3]$ so its minimum value is $m = f(0) = 3$ and its maximum value is $M = f(3) = 6$. Therefore,

$$3\cdot(3-0) \le \int_0^3 \sqrt{9+x^3}\,dx \le 6\cdot(3-0) \implies 9 \le \int_0^3 \sqrt{9+x^3}\,dx \le 18\,.$$

31. The integrand f increases on $[0,4]$ so its minimum value is $m = f(0) = 1/5$ and its maximum value is $M = f(4) = 1/3$. Therefore,

$$\frac{1}{5}\cdot(4-0) \le \int_0^4 (25-x^2)^{-1/2}\,dx \le \frac{1}{3}\cdot(4-0) \implies \frac{4}{5} \le \int_0^4 (25-x^2)^{-1/2}\,dx \le \frac{4}{3}\,.$$

Further Theory and Practice

33. Let $I = \int_a^b f(x)\,dx$ and $J = \int_a^b g(x)\,dx$. Then the two integral equations are equivalent to $I - 3J = 3$ and $-6J + 9I = 6$, or $3I - 2J = 2$. Multiply the first equation by -3 and add to the third equation to obtain $7J = -7$, so $J = -1$ and $I = 0$. That is, $\int_a^b f(x)\,dx = 0$ and $\int_a^b g(x)\,dx = -1$.

35. Let $I = \int_a^b f(x)\,dx$ and $J = \int_a^b g(x)\,dx$. Then the two integral equations are equivalent to $I + 2J = -7$ and $J - I = 4$. Add these equations to obtain $3J = -3$, so $J = -1$ and $I = J - 4 = -5$. That is, $\int_a^b f(x)\,dx = -5$ and $\int_a^b g(x)\,dx = -1$.

37. $$\int_1^3 \frac{3x-5}{x}\,dx = \int_1^3 (3 - 5/x)\,dx = 3\int_1^3 1\,dx - 5\int_1^3 1/x\,dx$$
$$= 3\cdot(3-1) - 5\cdot\big(\ln(x)\big)\Big|_1^3 = 6 - 5\ln(3)$$

39. $$\int_{-1}^0 (6x^2 + e^x)\,dx = 6\int_{-1}^0 x^2\,dx + \int_{-1}^0 e^x\,dx$$
$$= 6\cdot\left(\frac{x^3}{3}\right)\Big|_{-1}^0 + \left(e^x\right)\Big|_{-1}^0 = 6\cdot\left(\frac{0-(-1)}{3}\right) + 1 - e^{-1} = 3 - 1/e$$

41. $$\int_0^1 \sqrt{x}\,(1-x))\,dx = \int_0^1 x^{1/2}\,dx - \int_0^1 x^{3/2}\,dx$$
$$= \left(\frac{2}{3}x^{3/2}\right)\Big|_0^1 - \left(\frac{2}{5}x^{5/2}\right)\Big|_0^1 = \frac{2}{3}\cdot(1-0) - \frac{2}{5}\cdot(1-0) = \frac{4}{15}$$

43. $$\int_0^1 3e^x(2 - e^{-x})\,dx = 6\int_0^1 e^x\,dx - 3\int_0^1 1\,dx$$
$$= 6\cdot\left(e^x\right)\Big|_0^1 - 3\cdot(1-0) = 6\cdot(e-1) - 3 = 6e - 9$$

45. The left side integrates to $\left(\dfrac{x^{p+q+1}}{p+q+1}\right)\Big|_0^1 = \dfrac{1}{p+q+1}$ and the product on the right side evaluates to $\dfrac{1}{p+1}\cdot\dfrac{1}{q+1}$. These two expressions are equal if, and only if, $(p+1)(q+1) = p+q+1$. That is, $pq+p+q+1 = p+q+1$, or $pq = 0$. Since $p > 0$ and $q > 0$, the equality is never true.

47. Let $f(x) = 1 + x$, $g(x) = x$, and $a = 1$, $b = 2$. Then the left side evaluates to $1 + \ln(2)$, and the right side evaluates to $5/3$ (verify).

49. Let f denote the integrand. Since $f'(x) = -\dfrac{2(x-2)}{(x^2-4x+5)^2}$, f has only one critical point in the interval of integration, $c = 2$. Checking endpoints and c, $f(1) = 1/2$, $f(4) = 1/5$, and $f(2) = 1$. Therefore, the minimum and maximum values of f on $[1,4]$ are $m = 1/5$ and $M = 1$. Consequently,

$$\frac{1}{5}\cdot(4-1) \le \int_1^4 f(x)\,dx \le 1\cdot(4-1) \implies \frac{3}{5} \le \int_1^4 \frac{1}{x^2-4x+5}\,dx \le 3\,.$$

51. Let f denote the integrand. Since $f'(x) = -e^{-x/2}(x-2)/2$, f has only one critical point in the interval of integration, $c = 2$. Checking endpoints and

c, $f(1) = e^{-1/2} \approx 0.61$, $f(3) = 3e^{-3/2} \approx 0.67$, and $f(2) = 2e^{-1} \approx 0.74$. Therefore, the minimum and maximum values of f on $[1, 3]$ are $m = e^{-1/2}$ and $M = 2e^{-1}$. Consequently,

$$e^{-1/2} \cdot (3-1) \leq \int_1^3 f(x)\,dx \leq 2e^{-1} \cdot (3-1) \implies 2e^{-1/2} \leq \int_1^3 xe^{-x/2}\,dx \leq 4e^{-1}.$$

53. $F(x) = \int_1^x (3t^2+1)\,dt = \left(t^3+t\right)\Big|_1^x = x^3 + x - 2$

55. $F(x) = \int_x^1 (2t+1/t^2)\,dt = -\int_1^x (2t - 1/t)\,dt = -\left(t^2 - 1/t\right)\Big|_1^x = -x^2 + 1/x$

57. $F(x) = \int_1^{x^2} \frac{1}{\sqrt{t}}\,dt = \left(2\sqrt{t}\right)\Big|_1^{x^2} = 2x - 2$

59. Since $|f(x)| = -(\pi x - 3)$ if $x \leq 3/\pi$ and $|f(x)| = \pi x - 3$ otherwise,

$$\begin{aligned}
\int_0^1 |\pi x - 3|\,dx &= -\int_0^{3/\pi} (\pi x - 3)\,dx + \int_{3/\pi}^1 (\pi x - 3)\,dx \\
&= -\left(\pi x^2/2 - 3x\right)\Big|_0^{3/\pi} + \left(\pi x^2/2 - 3x\right)\Big|_{3/\pi}^1 \\
&= \pi/2 - 3 - 2(9/(2\pi) - 9/\pi) = 9/\pi + \pi/2 - 3 \approx 1.436\,.
\end{aligned}$$

This is larger than $|\int_0^1 (\pi x - 3)\,dx| = \left|\left(\pi x^2/2 - 3x\right)\Big|_0^1\right| = 3 - \pi/2 \approx 1.430$.

61. Since $|f(x)| = 1 - 9x^2$ for $x \in [-1/3, 1/3]$ and $|f(x)| = 9x^2 - 1$ otherwise,

$$\begin{aligned}
\int_{-2}^1 |f(x)|\,dx &= \int_{-2}^{-1/3} (9x^2-1)\,dx + \int_{-1/3}^{1/3} (1-9x^2)\,dx + \int_{1/3}^1 (9x^2-1)\,dx \\
&= \frac{200}{9} + \frac{4}{9} + \frac{20}{9} = \frac{224}{9} \approx 24.889\,.
\end{aligned}$$

This is larger than $|\int_{-2}^1 (1-9x^2)\,dx| = 24$.

63. Calculate

$$\begin{aligned}
\int_{-1}^1 P_2(x) \cdot Q(x)\,dx &= \frac{1}{2}\int_{-1}^1 (3x^2-1)(Bx+C)\,dx \\
&= \frac{1}{2} \cdot \left(3B\int_{-1}^1 x^3\,dx + 3C\int_{-1}^1 x^2\,dx - B\int_{-1}^1 x\,dx - C\int_{-1}^1 1\,dx\right) \\
&= \frac{1}{2} \cdot \left(3B \cdot \left(\frac{x^4}{4}\right)\Big|_{-1}^1 + 3C\left(\frac{x^3}{3}\right)\Big|_{-1}^1 - B\left(\frac{x^2}{2}\right)\Big|_{-1}^1 - C \cdot 2\right) \\
&= \frac{1}{2} \cdot (2C - 2C) = 0\,.
\end{aligned}$$

CALCULATOR/COMPUTER EXERCISES

65. Using *Maple*'s *Maximize* procedure the minimum value of f on $[0, 2]$ is $3/17$ and the maximum value is approximately 1.42031924533774 so

$$\frac{6}{17} \le \int_0^2 \frac{1+x}{1+x^4}\,dx \le 2.84063849067548$$

67. Using *Maple*'s *Maximize* procedure the minimum value of f on $[0, \pi/2]$ is 1 and the maximum value is approximately 3.33673426933339 so

$$\pi/2 \le \int_0^{\pi/2} (3\sin(x^2) + \cos(x))\,dx \le 5.24132993444392$$

69. The average value of f over the interval $[0, \pi/2]$ is

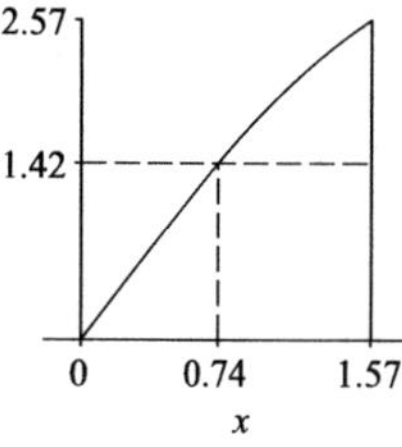

$$f_{\text{ave}} = \frac{1}{\pi/2}\int_0^{\pi/2} (x + \sin(x))\,dx = \frac{1 + \pi^2/8}{\pi/2} \approx 1.42\,.$$

Using *Maple*'s *fsolve* procedure, this value is attained by f at $c \approx 0.7444503032$. The picture illustrates the fact that the area of the region under the graph of f and over the interval $[0, \pi/2]$ equals the area of the rectangle having base $[0, \pi/2]$ and height f_{ave}: $\int_0^{\pi/2} (x + \sin(x))\,dx = f_{\text{ave}} \cdot (\pi/2 - 0)$.

71. The average value of f over the interval $[2, 6]$ is

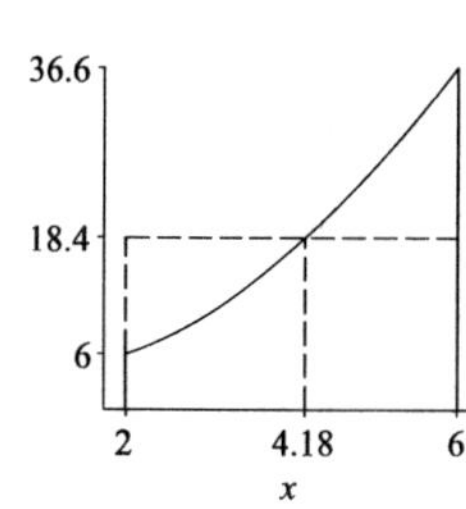

$$f_{\text{ave}} = \frac{1}{6-2}\int_2^6 (x^2 + 4/x)\,dx = \frac{52}{3} + \ln(3) \approx 18.4\,.$$

Using *Maple*'s *fsolve* procedure, this value is attained by f at $c \approx 4.180320710$. The picture illustrates the fact that the area of the region under the graph of f and over the interval $[2, 6]$ equals the area of the rectangle having base $[2, 6]$ and height f_{ave}: $\int_2^6 (x^2 + 4/x)\,dx = f_{\text{ave}} \cdot (6 - 2)\,.$

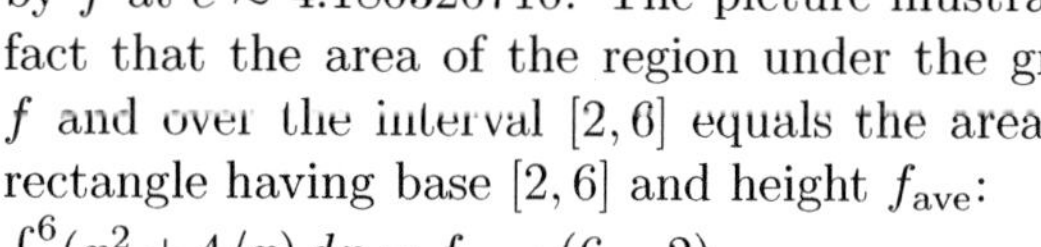

5.4 The Fundamental Theorem of Calculus

Problems for Practice

1. $\int_0^2 e^{3x}\,dx = \frac{1}{3}e^{3x}\big|_0^2 = \frac{1}{3}(e^6 - 1)$

3. $\int_{\pi/6}^{\pi/4} 12\cos(2x)\,dx = 6\sin(2x)\big|_{\pi/6}^{\pi/4} = 6(1 - \sqrt{3}/2) = 6 - 3\sqrt{3}$

5. $\int_{-\pi/3}^{\pi/2} 4\sin(2x)\,dx = -2\cos(2x)\big|_{-\pi/3}^{\pi/2} = -2(-1 - (-1/2)) = 1$

7. $\int_{\pi/2}^{\pi} \sec(x/3)\tan(x/3)\,dx = 3\sec(x/3)\Big|_{\pi/2}^{\pi} = 3(2 - 2/\sqrt{3}) = 6 - 2\sqrt{3}$

9. $\int_0^{\pi/4} \sin^2(2x)\,dx = \int_0^{\pi/4} \left(\frac{1-\cos(4x)}{2}\right)dx = \left(\frac{x - \sin(4x)/4}{2}\right)\Big|_0^{\pi/4} = \frac{\pi}{8}$

11. $F(x) = \int_{-1}^{x} (3t^2+1)\,dx = (t^3+t)\big|_{-1}^{x} = x^3 + x - (-1-1) = x^3 + x + 2$

13. $F(x) = \int_8^x 4t^{1/3}\,dx = 3t^{4/3}\big|_8^x = 3x^{4/3} - 48$

15. $F(x) = \int_0^x (e^t + e^{-t})/2\,dx = ((e^t - e^{-t})/2)\big|_0^x = (e^x - e^{-x})/2$

17. $F(x) = \int_{-\pi/4}^{x} \sec(t)\tan(t)\,dx = \sec(t)\big|_{-\pi/4}^{x} = \sec(x) - \sqrt{2}$

19. $F(x) = \int_1^x 15(t^{1/2}+t^{3/2})\,dx = \left(15\left(\frac{2}{3}t^{3/2}+\frac{2}{5}t^{5/2}\right)\right)\Big|_1^x = 10x^{3/2}+6x^{5/2}-16$

21. $\frac{d}{dx}\int_{-2}^{x}(2t+1)(t+2)\,dt = (2x+1)(x+2)$

23. $\frac{d}{dx}\int_{-1}^{x}(t^3+2t^2+3t+6)/(t+2)\,dt = (x^3+2x^2+3x+6)/(x+2) = x^2+3$

25. $\frac{d}{dx}\int_0^x \tan(t^2)\,dt = \tan(x^2)$

27. $\frac{d}{dx}\int_{-1}^{x} t\sqrt{1+t}\,dt = x\sqrt{1+x}$

29. $\frac{d}{dx}\int_x^{\pi/4}\cos(4t)\,dt = -\frac{d}{dx}\int_{\pi/4}^{x}\cos(4t)\,dt = -\cos(4x)$

31. $\frac{d}{dx}\int_x^{\pi/4}\cot(t)\,dt = -\frac{d}{dx}\int_{\pi/4}^{x}\cot(t)\,dt = -\cot(x)$

33. $\frac{d}{dx}\int_x^5 \sqrt{2-\sin^2(t)}\,dt = -\frac{d}{dx}\int_5^x \sqrt{2-\sin^2(t)}\,dt = -\sqrt{2-\sin^2(x)}$

35. Let $u = 2x$. Then

$$\begin{aligned}
\frac{d}{dx}\int_x^{2x}\sin(2t)\,dt &= \frac{d}{dx}\int_0^u \sin(2t)\,dt - \frac{d}{dx}\int_0^x \sin(2t)\,dt \\
&= \left(\frac{d}{du}\int_0^u \sin(2t)\,dt\right)\cdot\frac{du}{dx} - \sin(2x) \\
&= \sin(2u)\cdot 2 - \sin(2x) = 2\sin(4x) - \sin(2x)\,.
\end{aligned}$$

37. Let $u = x^2$. Then

$$\begin{aligned}
\frac{d}{dx}\int_x^{x^2}\cot(t)\,dt &= \frac{d}{dx}\int_0^u \cot(t)\,dt - \frac{d}{dx}\int_0^x \cot(t)\,dt \\
&= \left(\frac{d}{du}\int_0^u \cot(t)\,dt\right)\cdot\frac{du}{dx} - \cot(x) \\
&= \cot(u)\cdot 2x - \cot(x) = 2x\cot(x^2) - \cot(x)\,.
\end{aligned}$$

39. Let $u = \exp(-x)$ and $v = \exp(x)$. Then

$$\begin{aligned}\frac{d}{dx}\int_{\exp(-x)}^{\exp(x)} \ln(t)\,dt &= \frac{d}{dx}\int_0^v \ln(t)\,dt - \frac{d}{dx}\int_0^u \ln(t)\,dt \\ &= \left(\frac{d}{dv}\int_0^v \ln(t)\,dt\right)\cdot\frac{dv}{dx} - \left(\frac{d}{du}\int_0^u \ln(t)\,dt\right)\cdot\frac{du}{dx} \\ &= \ln(v)\cdot\exp(x) - \ln(u)\cdot(-\exp(-x)) = x(e^x - e^{-x})\,.\end{aligned}$$

41. Let $u = 2x$ and $v = \sqrt{x}$. Then

$$\begin{aligned}\frac{d}{dx}\int_{2x}^{\sqrt{x}} \sqrt{2t^2-1}\,dt &= \frac{d}{dx}\int_0^v \sqrt{2t^2-1}\,dt - \frac{d}{dx}\int_0^u \sqrt{2t^2-1}\,dt \\ &= \left(\frac{d}{dv}\int_0^v \sqrt{2t^2-1}\,dt\right)\cdot\frac{dv}{dx} - \left(\frac{d}{du}\int_0^u \sqrt{2t^2-1}\,dt\right)\cdot\frac{du}{dx} \\ &= \sqrt{2x-1}\cdot 1/(2\sqrt{x}) - \sqrt{2\cdot 4x^2-1}\cdot 2 = \frac{\sqrt{2x-1}}{2\sqrt{x}} - 2\sqrt{8x^2-1}\,.\end{aligned}$$

Further Theory and Practice

43. Since $F'(x) = x(x-1)$ and $F''(x) = 2x-1$, F is increasing on $(-\infty, 0)$, decreasing on $(0,1)$, and increasing on $(1,\infty)$. There is one inflection point, at $c = 1/2$. F is concave down on $(-\infty, 1/2)$ and concave up on $(1/2,\infty)$.

45. The domain of F is $(0,\infty)$. Since $F'(x) = x\ln(x)$ and $F''(x) = \ln(x)+1$, F is decreasing on $(0,1)$ and increasing on $(1,\infty)$. There is one inflection point, at $c = 1/e$. F is concave down on $(0, 1/e)$ and concave up on $(1/e,\infty)$.

47. Since $F'(x) = f(x) - f(-x)\cdot(-1) = f(x) + f(-x)$, and f is odd, $F'(x) = f(x) - f(x) = 0$ for all $x \in (-a,a)$. Therefore, there is a constant C such that $F(x) = C$ for all $x \in (-a,a)$. Because $F(0) = 0$, $C = 0$, and $F(x) = 0$ for all $x \in (-a,a)$.

49. Since $v(t) = p'(t)$, $\int_a^b v(t)\,dt = \int_a^b p'(t)\,dt = p(b) - p(a)$.

51. The point P is on the graph of F because $F(e) = \int_1^e 1/t\,dt = \ln(t)\big|_1^e = 1$. Since $F'(e) = 1/e$, the tangent line to the graph of F at P has the equation $y = (1/e)(x-e)+1$ or $y = x/e$.

53. The point P is on the graph of F because $F(\pi/6) = \int_0^{\pi/6} \cos(t)\,dt = \sin(t)\big|_0^{\pi/6} = 1/2$. Since $F'(\pi/6) = \cos(\pi/6) = \sqrt{3}/2$, the tangent line to the graph of F at P has the equation $y = (\sqrt{3}/2)(x - \pi/6) + 1/2$ or $y = \frac{1}{2}\sqrt{3}\,x + \frac{1}{2} - \frac{1}{12}\sqrt{3}\,\pi$.

55. Observe that $F(x) = \begin{cases} 2x & \text{if } 0 \le x < 1 \\ 2 & \text{if } 1 \le x < 2 \\ 4-x & \text{if } 2 \le x \le 3 \end{cases}$. Since $f(x) = F'(x)$ at all points where f is continuous, f can be defined as follows.

$$f(x) = \begin{cases} 2 & \text{if } 0 \le x \le 1 \\ 0 & \text{if } 1 < x \le 2 \\ -1 & \text{if } 2 < x \le 3 \end{cases}$$

57. The integrand is not defined at 0. Therefore, the integral, over the interval $[-2, 2]$, is also not defined.

59. Let $u = g(h(x))$. Then

$$\begin{aligned} \frac{d}{dx}\int_a^{g(h(x))} f(t)\,dt &= \frac{d}{dx}\int_a^u f(t)\,dt = \frac{d}{du}\int_a^u f(t)\,dt \cdot \frac{du}{dx} \\ &= f(u)\cdot g'(h(x))h'(x) = f(g(h(x))g'(h(x))h'(x)\,. \end{aligned}$$

61. The average rate of change of f over the interval $[a, b]$ is $\dfrac{f(b)-f(a)}{b-a}$. The average value of f' over $[a, b]$, calculated using the Fundamental Theorem of Calculus, is $f_{\text{ave}} = \dfrac{2}{b-a}\displaystyle\int_a^b f'(t)\,dt = \dfrac{f(b)-f(a)}{b-a}$, which is exactly the same.

63. Observe that, if $a \le x \le b$,

$$\begin{aligned} g(x) - f(x) &= \int_a^x g'(t)\,dt + g(a) - \left(\int_a^x f'(t)\,dt + f(a)\right) \\ &= \int_a^x (g'(t) - f'(t))\,dt + (g(a) - f(a)) \ge 0\,. \end{aligned}$$

65. Using the Product Rule and the Fundamental Theorem of Calculus,

$$\begin{aligned} F'(x) &= -2x\exp(-x^2)\int_0^x \exp(t^2)\,dt + \exp(-x^2)\cdot\exp(x^2) \\ &= -2x\exp(-x^2)\int_0^x \exp(t^2)\,dt + 1 = 1 - 2xF(x)\,. \end{aligned}$$

67. Using the Fundamental Theorem of Calculus, $C'(x) = \dfrac{x^4}{\sqrt{1+x^2}}$. Therefore, C is increasing. Moreover, because $C''(x) = \dfrac{x^3(4+3x^2)}{(1+x^2)^{3/2}}$, C is concave down on $(-\infty, 0)$ and concave up on $(0, \infty)$.

69. By the Fundamental Theorem of Calculus, when $x \in I$, $F'(x) = f(x)$ and $G'(x) = f(x)$. Therefore, $F'(x) - G'(x) = 0$ on I, implying that there is a constant C such that $F(x) - G(x) = C$. That is, $F(x) = G(x) + C$.

By Theorem 3 in Section 5.3,

$$F(x) = \int_a^x f(t)\,dt = \int_a^b f(t)\,dt + \int_b^x f(t)\,dt = G(x) + C\,,$$

where $C = \int_a^b f(t)\,dt$.

71. The function A_f for $h = 1/2$ is

$$A_f(x) = \frac{1}{2\cdot(1/2)}\int_{x-1/2}^{x+1/2} f(t)\,dt = F(x+1/2) - F(x-1/2)\,,$$

where F is defined as follows.

$$F(x) = \begin{cases} 0 & \text{if } x < 0 \\ x^2/2 & \text{if } 0 \le x \end{cases}$$

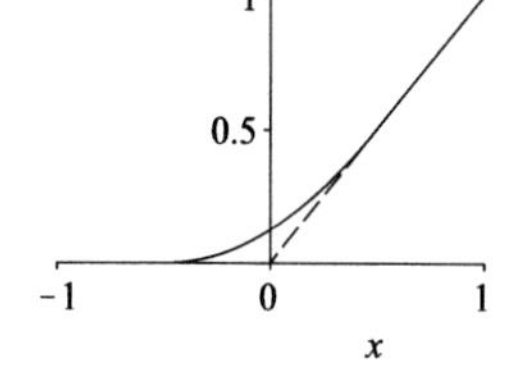

See the picture, f is the dashed curve and A_f is the solid curve.

73. If $0 \le x < 1$, then $f(x) = 0$. Therefore, $F(x) = 0$. If $1 \le x < 2$, then $f(x) = 1$. Therefore, $F(x) = \int_0^x f(t)\,dt = \int_0^1 f(t)\,dt + \int_1^x f(t)\,dt = 0 + \int_1^x 1\,dx = x - 1$. For x near to 1, the values of F are given by the following formulas.

$$F(x) = \begin{cases} 0 & \text{if } 0 \le x < 1 \\ x - 1 & \text{if } 1 \le x < 2 \end{cases}$$

Consequently, at $c = 1$, the graph of F has a left tangent line with slope 0 and a right tangent line with slope 1. F is not differentiable at $c = 1$.

CALCULATOR/COMPUTER EXERCISES

75. The plot on the right displays the graph of F' (solid line) and the graph of F'' (dashed line).

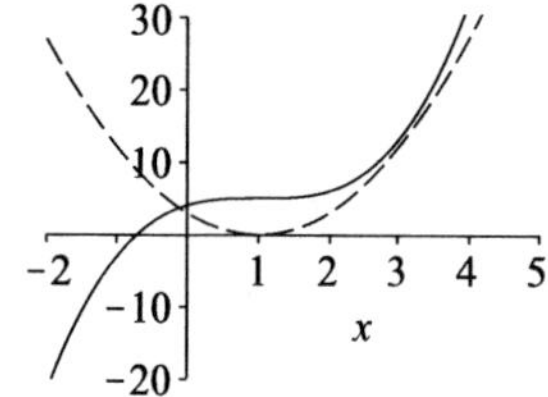

Since $F'' \ge 0$, F is concave up on $(-\infty, \infty)$. Moreover, F is decreasing on $(-\infty, c)$ and increasing on (c, ∞), where $c \approx -0.7099759467$.

77. The plot on the right displays the graph of F' (solid line) and the graph of F'' (dashed line).

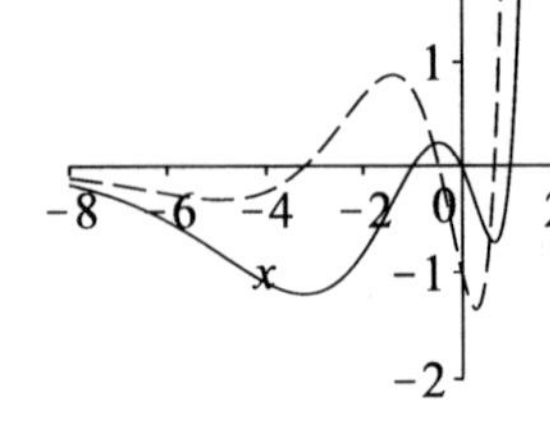

F is decreasing on $(-\infty, -1)$, increasing on $(-1, 0)$, decreasing on $(0, 1)$ and increasing on $(1, \infty)$.

Using *Maple*'s *fsolve* procedure, F is concave down on $(-\infty, -3.2143)$, concave up on $(-3.2143, -0.4608)$, concave down on $(-0.4608, 0.6751)$, and concave up on $(0.6751, \infty)$.

79. The plot on the right displays the graph of $y = N(T)$. Observe that $N(37) \approx 359$. Consequently, the temperature remains below 37°C for approximately 359 days and reaches at least 37°C on about 6 days of the year.

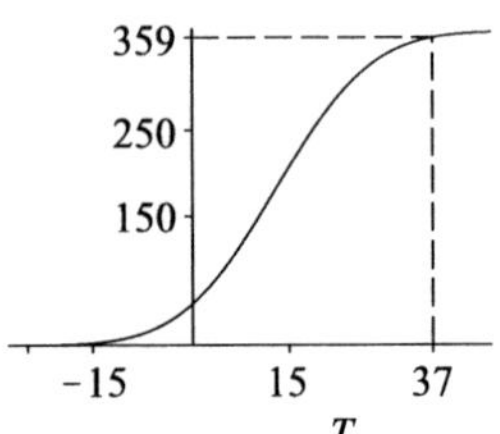

81. The Cornu spiral is displayed on the right.

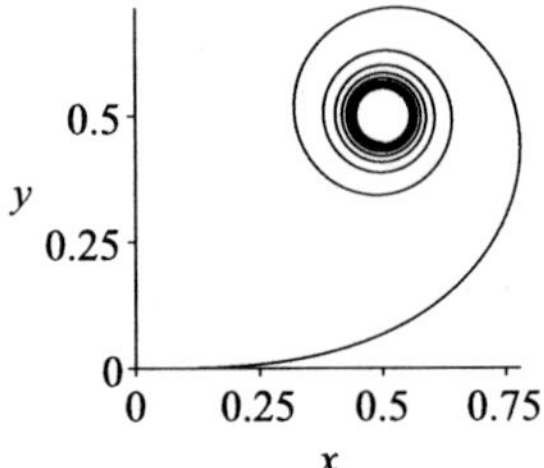

83. For $f(t) = \sqrt{9 + t^4}$ and $F(x) = \int_{0.25}^{x} f(t)\, dt$,

$$\frac{F(2 + 0.001/2) - F(2 - 0.001/2)}{0.001} \approx 5.00000\,,$$

and $F'(2) = f(2) = 5$.

85. For $f(t) = \sqrt{t}\, e^{t-4}$ and $F(x) = \int_{0.25}^{x} f(t)\, dt$,

$$\frac{F(4 + 0.001/2) - F(4 - 0.001/2)}{0.001} \approx 2.000000000\,,$$

and $F'(4) = f(4) = 2$.

5.5 A Calculus Approach to the Logarithm and Exponential Function

Problems for Practice

1. Given that $\ln(u) = 4$, $\int_1^{1/u} 1/t\,dt = \ln(1/u) = -\ln(u) = -4$.

3. Given that $\ln(u) = 4$, $\int_1^{u/e} 1/t\,dt = \ln(u/e) = \ln(u) - \ln(e) = 4 - 1 = 3$.

5. Given that $\ln(u) = 4$, $\int_3^{3u} 1/t\,dt = \ln(3u) - \ln(3) = \ln(3) + \ln(u) - \ln(3) = 4$.

7. Since $\ln(u) = 4$ and $\ln(v) = 12$, $\int_1^{v/u} 1/t\,dt = \ln(v/u) = \ln(v) - \ln(u) = 8$.

9. Since $\ln(v) = 12$, $\int_u^{uv} 1/t\,dt = \ln(uv) - \ln(u) = \ln(uv/u) = \ln(v) = 12$.

11. Since $\ln(v) = 12$, $\int_1^{2v} 1/t\,dt = \ln(2v) = \ln(2) + \ln(v) = \ln(2) + 12$.

13. $\dfrac{d}{dx}\ln(4x) = \dfrac{d}{dx}(\ln(4) + \ln(x)) = 1/x$

15. $\dfrac{d}{dx}\ln(2x+3) = 2/(2x+3)$

17. $\dfrac{d}{dx}\ln(\sqrt{1-x}) = \dfrac{d}{dx}\left(\frac{1}{2}\ln(1-x)\right) = \dfrac{-1}{2(1-x)}$

19. $\dfrac{d}{dx}\left(\dfrac{1}{4+\ln(x)}\right) = \dfrac{-1/x}{(\ln(4)+\ln(x))^2}$

21. $\dfrac{d}{dx}\ln^2(x) = 2\ln(x)\cdot(1/x) = 2\ln(x)/x$

23. $\dfrac{d}{dx}(x2^x) = 2^x + x2^x\ln(2) = 2^x(1 + x\ln(2))$

25. $\dfrac{d}{dx}2^{\ln(x)} = 2^{\ln(x)}\ln(2)\cdot(1/x) = 2^{\ln(x)}\ln(2)/x$

27. $\dfrac{d}{dx}\log_{10}(5/x) = \dfrac{d}{dx}\left(\dfrac{\ln(5/x)}{\ln(10)}\right) = \dfrac{d}{dx}\left(\dfrac{\ln(5)-\ln(x)}{\ln(10)}\right) = \dfrac{-1}{x\ln(10)}$

29. $\dfrac{d}{dx}\log_2(3x) = \dfrac{d}{dx}\left(\dfrac{\ln(3x)}{\ln(2)}\right) = \dfrac{d}{dx}\left(\dfrac{\ln(3)+\ln(x)}{\ln(2)}\right) = \dfrac{1}{x\ln(2)}$

31. $\int_0^1 2^x\,dx = \left.\dfrac{2^x}{\ln(2)}\right|_0^1 = (2-1)/\ln(2) = 1/\ln(2)$

33. $$\int_0^1 \frac{1+10^x}{10^x}\,dx = \int_0^1 ((1/10)^x + 1)\,dx = \left.\left(\frac{(1/10)^x}{\ln(1/10)} + x\right)\right|_0^1$$
$$= \frac{1/10 - 1}{-\ln(10)} + 1 = 1 + 9/(10\ln(10))$$

Further Theory and Practice

35. Let $f(x) = 2^x - x$. Then $f'(x) = 2^x\ln(2) - 1$ and $f''(x) = 2^x\ln^2(2)$. There is one critical point, c, satisfying the equation $2^c\ln(2) = 1$. That

is, $2^c = 1/\ln(2)$, so $c = -\ln(\ln(2))/\ln(2)$. Since $f''(c) > 0$, f has a local minimum at c.

37. Let $f(x) = x10^{-x}$. Then $f'(x) = 10^{-x}(1 - x\ln(10))$ and the second derivative simplifies to $f''(x) = \ln(10)10^{-x}(-2 + x\ln(10))$. There is one critical point, c, satisfying the equation $c\ln(10) = 1$. That is, $c = 1/\ln(10)$. Since $f''(c) < 0$, f has a local maximum at c.

39. Let $f(x) = x\log_{10}(x)$. The domain of f is $(0, \infty)$. Since the derivative is $f'(x) = (\ln(x)+1)/\ln(10)$, and $f''(x) = 1/(x\ln(10))$. There is one critical point, c, satisfying the equation $\ln(c) = -1$. That is, $c = e^{-1}$. Since $f''(c) > 0$, f has a local minimum at $1/e$.

41. Let $f(x) = \ln^2(x)$. The domain of f is $(0, \infty)$. Since the derivative is $f'(x) = 2\ln(x)/x$, and $f''(x) = 2(1 - \ln(x))/x^2$. There is one critical point, c, satisfying the equation $\ln(c) = 0$. That is, $c = 1$. Since $f''(c) > 0$, f has a local minimum at 1.

43. $F(x) = \int_a^{\ln(x)} 1/t\,dt$, so $F'(x) = \dfrac{1}{\ln(x)} \cdot (1/x) = \dfrac{1}{x\ln(x)}$, and the second derivative is $F''(x) = -\dfrac{\ln(x)+1}{x^2\ln^2(x)}$

45. $F(x) = \int_a^{1/x} \ln(t)\,dt$, so $F'(x) = \ln(1/x)\cdot(-1/x^2) = \dfrac{\ln(x)}{x^2}$, and the second derivative is $F''(x) = \dfrac{1 - 2\ln(x)}{x^3}$

47. $F(x) = \int_a^{\ln(x)} \ln(t)\,dt$, so $F'(x) = \ln(\ln(x)) \cdot (1/x) = \ln(\ln(x))/x$, and the second derivative is $F''(x) = \dfrac{1 - \ln(\ln(x))\ln(x)}{x^2\ln(x)}$

49. Calculating derivatives yields

$$\begin{aligned}
y &= A\cos(\ln(x)) + B\sin(\ln(x)) \\
y' &= -A \cdot \frac{\sin(\ln(x))}{x} + B \cdot \frac{\cos(\ln(x))}{x} \\
y'' &= -A \cdot \frac{x\cos(\ln(x))/x - \sin(\ln(x))}{x^2} + B \cdot \frac{-x\sin(\ln(x))/x - \cos(\ln(x))}{x^2}.
\end{aligned}$$

Therefore,

$$\begin{aligned}
x^2y'' + xy' + y &= -A(\cos(\ln(x)) - \sin(\ln(x))) + B(-\sin(\ln(x)) - \cos(\ln(x))) \\
&\quad - A\sin(\ln(x)) + B\cos(\ln(x)) + A\cos(\ln(x)) + B\sin(\ln(x)) \\
&= 0.
\end{aligned}$$

51. This is not valid because $\dfrac{d}{dx}\ln(|x|) = 1/x$, not $1/|x|$.

53. Let $F(x) = A \cdot x\ln(x) + B \cdot x$. Then $F'(x) = A\ln(x) + A + B$, so $F'(x) = \ln(x)$ when $A = 1$ and $B = -1$. That is, $F(x) = x\ln(x) - x$ is an antiderivative for $\ln(x)$: $\int \ln(x)\,dx = x\ln(x) - x + C$.

55. i) Differentiating: $\dfrac{d}{dx}f(xy) = \dfrac{d}{dx}(f(x) + f(y))$. The Chain Rule must be used on the left side, implying that $f'(xy) \cdot y = f'(x)$.

    ii) Replace x with t and y with $1/t$ in the last equation to obtain the identity $f'(1) \cdot (1/t) = f'(t)$. This shows that f is an antiderivative for the function $t \mapsto f'(1)/t$.

    iii) Continuing part ii, observe that $t \mapsto f'(1)\ln(t)$ is also an antiderivative for $t \mapsto f'(1)/t$. Consequently, there is a constant C such that $f(t) = f'(1) \cdot \ln(t) + C$. Substituting $t = 1$, $f(1) = C$.

    iv) Substitute $x = y = 1$ into the given property of f to obtain $f(1) = f(1) + f(1)$, implying that $f(1) = 0$ so $f(t) = f'(1) \cdot \ln(t)$ for all $t > 0$.

57. We wish to show that, for $a > 0$, $a^{x+y} = a^x \cdot a^y$, using

$$(1)\ a^x = \exp(x \cdot \ln(a)) \quad \text{and} \quad (2)\ \exp(s+t) = \exp(s) \cdot \exp(t)\,.$$

Proceed as follows, assuming that $a > 0$.

$$\begin{aligned} a^{x+y} &\overset{(1)}{=} \exp((x+y) \cdot \ln(a)) = \exp(x \cdot \ln(a) + y \cdot \ln(a)) \\ &\overset{(2)}{=} \exp(x \cdot \ln(a)) \cdot \exp(y \cdot \ln(a)) \overset{(1)}{=} a^x \cdot a^y\,. \end{aligned}$$

59. We wish to show that, for $a > 0$, $(a^x)^y = a^{x \cdot y}$, using

$$(1)\ a^x = \exp(x \cdot \ln(a)) \quad \text{and} \quad (2)\ (\exp(s))^t = \exp(st)\,.$$

Proceed as follows, assuming that $a > 0$.

$$\begin{aligned} (a^x)^y &\overset{(1)}{=} \exp(y \cdot \ln(a^x)) = \exp(xy \cdot \ln(a)) \\ &\overset{(2)}{=} (\exp(\ln(a)))^{x \cdot y} = a^{x \cdot y}\,. \end{aligned}$$

61. We wish to show that, for $x, y, a > 0$ and $a \neq 1$, $\log_a(xy) = \log_a(x) + \log_a(y)$, using

$$(1)\ \log_a(x) = \frac{\ln(x)}{\ln(a)} \quad \text{and} \quad (2)\ \ln(x \cdot y) = \ln(x) + \ln(y)\,.$$

Proceed as follows, assuming that $x, y, a > 0$ and $a \neq 1$.

$$\log_a(xy) \overset{(1)}{=} \frac{\ln(xy)}{\ln(a)} \overset{(2)}{=} \frac{\ln(x) + \ln(y)}{\ln(a)} = \frac{\ln(x)}{\ln(a)} + \frac{\ln(y)}{\ln(a)} \overset{(1)}{=} \log_a(x) + \log_a(y)\,.$$

63. We wish to show that, for $x, y, a > 0$ and $a \neq 1$, $\log_a\left(\frac{x}{y}\right) = \log_a(x) - \log_a(y)$, using

$$(1)\ \log_a(x) = \frac{\ln(x)}{\ln(a)} \quad \text{and} \quad (2)\ \ln\left(\frac{x}{y}\right) = \ln(x) - \ln(y)\,.$$

Proceed as follows, assuming that $x, y, a > 0$ and $a \neq 1$.

$$\log_a\left(\frac{x}{y}\right) \stackrel{(1)}{=} \frac{\ln(x/y)}{\ln(a)} \stackrel{(2)}{=} \frac{\ln(x) - \ln(y)}{\ln(a)} = \frac{\ln(x)}{\ln(a)} - \frac{\ln(y)}{\ln(a)} \stackrel{(1)}{=} \log_a(x) - \log_a(y)\,.$$

65. Differentiate, $V'(x) = e^x + xe^x = e^x(1+x)$, and we see that V is increasing on $(-1, \infty)$ so V is a one-to-one function. Since $V(-1) = -1/e$ and $\lim_{x\to\infty} V(x) = \infty$ (verify), $V : (-1, \infty) \to (-1/e, \infty)$ is one-to-one and onto, therefore invertible. $W : (-1/e, \infty) \to (-1, \infty)$ denotes its inverse. Given $y \in (-1/e, \infty)$, $W(y) = x$ if, and only if, $V(x) = y$. That is, $W(y) = x$ iff $y = x\exp(x)$. Equivalently, $W(y)\exp(W(y)) = y$ for all $y > -1/e$.

67. Because $W(0)\exp(W(0)) = 0$, $W(0) = 0$, and the limit is indeterminate of the form $0/0$. Using Exercise 66,

$$\lim_{y\to 0} \frac{W(y)}{y} = \lim_{y\to 0} W'(y) = \lim_{y\to 0} \frac{1}{y + \exp(W(y))} = \frac{1}{\exp(W(0))} = 1\,.$$

69. The function $f(x) = x\ln(x)$ has the derivative $f'(x) = \ln(x) + 1$, which is positive for $x > 1/e$. Therefore, f is increasing on $(1/e, \infty)$. Since $f(1) = 0$ and $\lim_{x\to\infty} x\ln(x) = \infty$, $f : (1, \infty) \to (0, \infty)$ is one-to-one and onto. This guarantees that, given $y > 0$, there is a unique $x > 1$ such that $x\ln(x) = y$.

Suppose that $x\ln(x) = y$. Then $\ln(x) = y/x$ so $x = e^{y/x}$ and, multiplying both sides by y/x, $y = (y/x)\exp(y/x)$. Thus $z = y/x$ is a solution to the equation $z\exp(z) = y$. But $W(y)$ is the unique solution to this equation implying that $y/x = W(y)$, and $x = y/W(y)$.

71. The function $f(x) = x + e^x$ has the derivative $f'(x) = 1 + e^x$, which is positive for all x. Therefore, f is increasing on $(-\infty, \infty)$. Since $\lim_{x\to-\infty} f(x) = -\infty$ and $\lim_{x\to\infty} f(x) = \infty$, $f : (-\infty, \infty) \to (-\infty, \infty)$ is one-to-one and onto. This guarantees that, given any real number u, there is a unique x such that $x + e^x = u$.

Substitute $x = u - W(e^u)$ into $x + e^x = u$ to obtain

$$u - W(e^u) + e^u \cdot e^{-W(e^u)} = u\,.$$

Cancel u on each side and rearrange terms to obtain the equivalent equation

$$W(e^u)\exp(W(e^u)) = e^u\,.$$

Since this is true for all u, $x = u - W(e^u)$ provides the solution to $x + e^x = u$ for all u.

73. (a) Let $f(x) = x + 1/x$. Then $f'(x) = 1 - 1/x^2$ so, if $x > 0$, the only critical point for f is $c = 1$. Since $f''(x) = 2/x^3$, $f'' > 0$ on $(0, \infty)$ and $f(1) = 2$ is an absolute minimum value for f.

(b) This follows from part a. Given $t > 0$ and $c > 1$, let $x = c^t$. Then $1/x = c^{-t}$ and, because $x > 1$, $x + 1/x > 2$. That is, $c^t + c^{-t} > 2$. From this we obtain $c^t - 1 > 1 - c^{-t}$ by replacing 2 with $1 + 1$ and rearranging.

(c) Since $(c^t - 1) - (1 - c^{-t}) > 0$ when $c > 1$ and $t > 0$,

$$\int_0^a ((c^t - 1) - (1 - c^{-t}))\, dt > 0\,.$$

This implies that $\int_0^a (c^t - 1)\, dt > \int_0^a (1 - c^{-t})\, dt$. Therefore, if $0 < b \leq a$, then $\int_0^a (c^t - 1)\, dt > \int_0^b (1 - c^{-t})\, dt$. Consequently, if $\int_0^a (c^t - 1)\, dt = \int_0^b (1 - c^{-t})\, dt$, then it must be the case that $b > a$.

(d) Consider the assertion in part c with $c = e$. Since $\int_0^a (e^t - 1)\, dt = e^a - 1 - a$ and $\int_0^b (1 - e^{-t})\, dt = b - e^b + 1$, the assumption that the integrals are equal is equivalent to the assumption that $e^a - 1 - a = b + e^{-b} - 1$. That is, $e^a - e^{-b} = a + b$. Therefore, if this is true, then it must be the case that $b > a$.

CALCULATOR/COMPUTER EXERCISES

75. Let $f(x) = \sqrt{x - 1}/\log_3(x)$. Then $f'(x) = \dfrac{\ln(3)(x\ln(x) - 2x + 2)}{2x\sqrt{x - 1}\,\ln^2(x)}$. Using *Maple*'s *solve* procedure we find one critical point,

$$c = \exp(W(-2e^{-2}) + 2) \approx 4.921553634\,.$$

(W denotes the Lambert W function.) The *fsolve* procedure agrees. Thus $f(c) \approx 1.365172715$ is an absolute minimum for f (First Derivative Test).

77. Let $f(x) = x + \log_2(1 + x + x^2)$. Then $f'(x) = \dfrac{\ln(2) \cdot x^2 + (\ln(2) + 2) \cdot x + \ln(2) + 1}{\ln(2) \cdot (1 + x + x^2)}$ and, using the quadratic formula, we find two critical points:

$$c_{1,2} = \frac{-(\ln(2) + 2) \mp \sqrt{4 - 3\ln^2(2)}}{2\ln(2)}\,.$$

These are $c_1 \approx -3.096544676$ and $c_2 \approx -0.7888454060$. Evaluating f' in the complementary intervals reveals that f is increasing on $(-\infty, c_1)$, decreasing on (c_1, c_2), and increasing on (c_2, ∞). Therefore, f attains a local maximum at c_1 and a local minimum at c_2.

5.6 Integration by Substitution

Problems for Practice

1. With $u = 3x$ and $du = 3\,dx$,

$$\int \sin(3x)\,dx = \frac{1}{3}\int \sin(3x)\cdot 3\,dx = \frac{1}{3}\int \sin(u)\,du$$
$$= -\frac{1}{3}\cos(u) + C = -\frac{1}{3}\cos(3x) + C\,.$$

3. With $u = x^8 + 1$ and $du = 8x^7\,dx$,

$$\int 64(x^8+1)^{-5}x^7\,dx = 8\int (x^8+1)^{-5}\cdot 8x^7\,dx = 8\int u^{-5}\,du$$
$$= -2u^{-4} + C = -2(x^8+1)^{-4} + C\,.$$

5. With $u = x^3 - 5$ and $du = 3x^2\,dx$,

$$\int 30x^2(x^3-5)^{3/2}\,dx = 10\int (x^3-5)^{3/2}\cdot 3x^2\,dx = 10\int u^{3/2}\,du$$
$$= 4u^{5/2} + C = 4(x^3-5)^{5/2} + C\,.$$

7. With $u = \cos(s)$ and $du = -\sin(s)\,ds$,

$$\int \sin(s)\cos^4(s)\,ds = -\int \cos^4(s)\cdot(-\sin(s)\,ds) = -\int u^4\,du$$
$$= -\frac{1}{5}u^5 + C = -\frac{1}{5}\cos^5(s) + C\,.$$

9. With $u = \pi/t$ and $du = -\pi/t^2\,dt$,

$$\int \sin(\pi/t)/t^2\,dt = -\frac{1}{\pi}\int \sin(\pi/t)\cdot(-\pi/t^2\,dt) = -\frac{1}{\pi}\int \sin(u)\,du$$
$$= \frac{1}{\pi}\cos(u) + C = \frac{1}{\pi}\cos(\pi/t) + C\,.$$

11. With $u = x^2 + 1$ and $du = 2x\,dx$,

$$\int 16x(x^2+1)^7\,dx = 8\int (x^2+1)^7\cdot 2x\,dx = 8\int u^7\,du$$
$$= u^8 + C = (x^2+1)^8 + C\,.$$

13. With $u = 1 + x^2$ and $du = 2x\,dx$,

$$\int \frac{x}{\sqrt{1+x^2}}\,dx = \frac{1}{2}\int \frac{1}{\sqrt{1+x^2}}\cdot 2x\,dx = \frac{1}{2}\int u^{-1/2}\,du$$
$$= \sqrt{u} + C = \sqrt{1+x^2} + C\,.$$

15. With $u = \cos(x)$ and $du = -\sin(x)\,dx$,

$$\int \frac{\sin(x)}{\cos^2(x)}\,dx = -\int \frac{1}{\cos^2(x)}\cdot(-\sin(x)\,dx) = -\int u^{-2}\,du$$
$$= \frac{1}{u} + C = \frac{1}{\cos(x)} + C = \sec(x) + C\,.$$

17. With $u = \sin(2x)$ and $du = 2\cos(2x)\,dx$,

$$\int 24\sin^5(2x)\cos(2x)\,dx = 12\int \sin^5(2x)\cdot 2\cos(2x)\,dx = 12\int u^5\,du$$
$$= 2u^6 + C = 2\sin^6(2x) + C\,.$$

19. With $u = 1 + x^2$ and $du = 2x\,dx$,

$$\int \frac{24x}{\sqrt{1+x^2}}\,dx = 12\int \frac{1}{\sqrt{1+x^2}}\cdot 2x\,dx = 12\int u^{-1/2}\,du$$
$$= 24u^{1/2} + C = 24\sqrt{1+x^2} + C\,.$$

21. With $u = \sqrt{x}$ and $du = \frac{1}{2}x^{-1/2}\,dx$,

$$\int \frac{\sin(\sqrt{x})}{\sqrt{x}}\,dx = 2\int \sin(\sqrt{x})\cdot\left(\frac{1}{2}\cdot x^{-1/2}\,dx\right) = 2\int \sin(u)\,du$$
$$= -2\cos(u) + C = -2\cos(\sqrt{x}) + C\,.$$

23. Let $u = t^2 - t$ so $du = (2t-1)\,dt$. Since $t = 1 \Longrightarrow u = 0$ and $t = 2 \Longrightarrow u = 2$, the definite integral can be evaluated as follows.

$$\int_1^2 (t^2 - t)^5(2t-1)\,dt = \int_0^2 u^5\,du = \frac{1}{6}u^6\Big|_0^2 = \frac{64}{6} = \frac{32}{3}$$

25. Let $u = \frac{1}{2}x^2 + 1$ so $du = x\,dx$. Since $x = 0 \Longrightarrow u = 1$ and $x = 4 \Longrightarrow u = 9$, the definite integral can be evaluated as follows.

$$\int_0^4 x\sqrt{\tfrac{1}{2}x^2 + 1}\,dx = \int_0^9 \sqrt{u}\,du = \frac{2}{3}u^{3/2}\Big|_0^9 = \frac{2}{3}(27-1) = \frac{52}{3}$$

27. Let $u = 4x^2 - \pi$ so $du = 8x\,dx$ and $x\,dx = \frac{1}{8}\,du$. Since $x = \sqrt{\pi/4} \Longrightarrow u = 0$ and $x = \sqrt{3\pi/8} \Longrightarrow u = \pi/2$, the definite integral can be evaluated as follows.

$$\int_{\sqrt{\pi/4}}^{\sqrt{3\pi/8}} 24x\cos(4x^2 - \pi)\,dx = 24\int_0^{\pi/2} \cos(u)\cdot\frac{1}{8}\,du = 3\sin(u)\Big|_0^{\pi/2} = 3$$

29. Let $u = x^3 - 1$ so $du = 3x^2\,dx$ and $x^2\,dx = \frac{1}{3}\,du$. Since $x = -1 \Longrightarrow u = -2$ and $x = 0 \Longrightarrow u = -1$, the definite integral can be evaluated as follows.

$$\int_{-1}^{0} 24\frac{x^2}{(x^3-1)^5}\,dx = 24\int_{-2}^{-1} \frac{1}{u^5}\cdot\frac{1}{3}\,du = -\frac{8}{4}u^{-4}\Big|_{-2}^{-1} = -2\left(1 - \frac{1}{16}\right) = -\frac{15}{8}$$

31. Let $u = \sqrt{x}$ so $du = \frac{1}{2\sqrt{x}}\,dx$ and $\frac{1}{\sqrt{x}}\,dx = 2\,du$. Since $x = \pi^2/4 \Longrightarrow u = \pi/2$ and $x = \pi^2 \Longrightarrow u = \pi$, the definite integral can be evaluated as follows.

$$\int_{\pi^2/4}^{\pi^2} \frac{\cos(\sqrt{x})}{\sqrt{x}}\,dx = \int_{\pi/2}^{\pi} \cos(u)\cdot 2\,du = 2\sin(u)\Big|_{\pi/2}^{\pi} = 2\,(0-1) = -2$$

33. Let $u = \sec(\theta)$ so $du = \sec(\theta)\tan(\theta)\,dx$. Since $\theta = 0 \Longrightarrow u = 1$ and $\theta = \pi/3 \Longrightarrow u = 2$, the definite integral can be evaluated as follows.

$$\int_0^{\pi/3} \sec^2(\theta)\cdot\sec(\theta)\tan(\theta)\,d\theta = \int_1^2 u^2\,du = \frac{1}{3}u^3\Big|_1^2 = \frac{1}{3}\,(8-1) = \frac{7}{3}$$

35. Let $u = 1 + 2\tan(x)$ so $du = 2\sec^2(x)\,dx$, $\sec^2(x)\,dx = \frac{1}{2}\,du$, and $\tan(x) = (u-1)/2$. Since $x = 0 \Longrightarrow u = 1$ and $x = \pi/4 \Longrightarrow u = 3$, the definite integral can be evaluated as follows.

$$\begin{aligned}\int_0^{\pi/4} \frac{24\tan(x)\sec^2(x)}{(1+2\tan(x))^2}\,dx &= 24\int_1^3 \frac{(u-1)/2}{u^2}\cdot\frac{1}{2}\,du = 6\int_1^3 (u^{-1} - u^{-2})\,du \\ &= 6\left(\ln(u) + \frac{1}{u}\right)\Big|_1^3 = 6\left(\ln(3) + \frac{1}{3} - 1\right) \\ &= 6\ln(3) - 4\end{aligned}$$

37. $\int_0^{\pi/8} \tan(2x)\,dx = \frac{1}{2}\ln(|\sec(2x)|)\Big|_0^{\pi/8} = \frac{1}{2}\ln(\sec(\pi/4)) = \frac{1}{2}\ln(\sqrt{2})$
$= \frac{1}{4}\ln(2)$

39. $\int_0^{\pi} \sec(x/4)\,dx = 4\ln(|\sec(x/4) + \tan(x/4)|)\Big|_0^{\pi} = 4\ln(\sqrt{2}+1)$

41. Let $u = \pi x^2$ so $du = 2\pi x\,dx$ and $x\,dx = \frac{1}{2\pi}\,du$. Since $x = 0 \Longrightarrow u = 0$ and $x = 1/2 \Longrightarrow u = \pi/4$, the integral evaluates as follows.

$$\int_0^{1/2} x\tan(\pi x^2)\,dx = \int_0^{\pi/4} \tan(u)\cdot\left(\frac{1}{2\pi}\,du\right) = \left(\frac{1}{2\pi}\ln(|\sec(u)|)\right)\Big|_0^{\pi/4}$$
$$= \frac{1}{2\pi}\ln(\sqrt{2}\,) = \frac{1}{4\pi}\ln(2)$$

43. Let $u = \pi\sin(x)/4$ so $du = \pi\cos(x)/4\,dx$ and $\cos(x)\,dx = \frac{4}{\pi}\,du$. Since $x = 0 \Longrightarrow u = 0$ and $x = \pi/2 \Longrightarrow u = \pi/4$, the integral evaluates as follows.

$$\int_0^{\pi/2} \cos(x)\sec(\pi\sin(x)/4)\,dx = \int_0^{\pi/4} \sec(u)\cdot\left(\frac{4}{\pi}\,du\right)$$
$$= \left(\frac{4}{\pi}\ln(|\sec(u)+\tan(u)|)\right)\Big|_0^{\pi/4}$$
$$= \frac{4}{\pi}\ln(\sqrt{2}+1)$$

45. Factor 4 out of the first term in the denominator, then apply Forumula 23 with $a = 3$ and $b = 3/4$.

$$\int \frac{1}{(4t+3)(t+3)}\,dt = \frac{1}{4}\int \frac{1}{(t+3/4)(t+3)}\,dt = \frac{1}{4}\cdot\frac{1}{9/4}\ln\left(\left|\frac{t+3/4}{t+3}\right|\right) + C$$
$$= \frac{1}{9}\ln\left(\left|\frac{4t+3}{t+3}\right|\right) + C$$

The two constants are not the same, here is why.

$$\ln\left(\left|\frac{t+3/4}{t+3}\right|\right) = \ln\left(\left|\frac{4t+3}{4t+12}\right|\right) = \ln\left(\left|\frac{4t+3}{t+3}\cdot\frac{1}{4}\right|\right) = \ln\left(\left|\frac{4t+3}{t+3}\right|\right) - \ln(4)$$

47. The numerator is a perfect square, factor it, complete the square in the denominator, and make the substitution $u = t+3$, $du = dt$. Then use Formula 35 with $a = 4$.

$$\int \frac{9+6t+t^2}{25+6t+t^2}\,dt = \int \frac{(t+3)^2}{(t+3)^2+16}\,dt = \int \frac{u^2}{16+u^2}\,du$$
$$= u - 4\arctan\left(\frac{u}{4}\right) + C$$
$$= t - 4\arctan\left(\frac{t+3}{4}\right) + C$$

The two constants are not the same.

49. Make the substitution $u = \exp(t)$, $du = \exp(t)\,dt$. Then $\exp(2t) = u^2$ and $dt = \exp(-t)\,du = \frac{1}{u}\,du$. Then use Formula 39 with $a = 2$.

$$\begin{aligned}\int \frac{1}{4+\exp(2t)}\,dt &= \int \frac{1}{4+u^2}\cdot\left(\frac{1}{u}\,du\right) = \frac{1}{8}\ln\left(\frac{u^2}{4+u^2}\right)+C\\ &= \frac{1}{8}\ln\left(\frac{\exp(2t)}{4+\exp(2t)}\right)+C\end{aligned}$$

The two constants are the same.

51. Make the substitution $u = \sqrt{t}$, $du = \frac{1}{2\sqrt{t}}\,dt$, so $dt = 2u\,du$. Then use Formula 7 with $a = 2$ and $b = 1$.

$$\begin{aligned}\int \frac{1}{2+\sqrt{t}}\,dt &= \int \frac{1}{2+u}\cdot 2u\,du = 2\cdot\left(2+u-2\ln(|2+u|)\right)+C\\ &= 2\sqrt{t}-4\ln\left(|2+\sqrt{t}\,|\right)+C\end{aligned}$$

The two constants are not the same.

Further Theory and Practice

53. Substitute $u = 2x+3$ and $du = 2\,dx$, so $x = (u-3)/2$ and $dx = \frac{1}{2}\,du$.

$$\begin{aligned}\int x\cdot(2x+3)^{1/2}\,dx &= \int \frac{u-3}{2}\cdot u^{1/2}\cdot\left(\frac{1}{2}\,du\right) = \frac{1}{4}\int(u^{3/2}-3u^{1/2})\,du\\ &= \frac{1}{4}\left(\frac{2}{5}u^{5/2}-2u^{3/2}\right)+C = \frac{1}{10}u^{3/2}\,(u-5)+C\\ &= \frac{1}{5}(2x+3)^{3/2}(x-1)+C\end{aligned}$$

55. Substitute $u = x+3$ and $du = dx$, so $x = u-3$ and $dx = du$.

$$\begin{aligned}&\int \frac{x}{\sqrt{x+3}}\,dx = \int \frac{u-3}{u^{1/2}}\,du = \int(u^{1/2}-3u^{-1/2})\,du\\ &= \frac{2}{3}u^{3/2}-6u^{1/2}+C = \frac{2}{3}u^{1/2}(u-9) = \frac{2}{3}(x-6)\sqrt{x+3}+C\end{aligned}$$

57. Substitute $u = 6+x^2$ and $du = 2x\,dx$, so $x^2 = u-6$ and $x\,dx = \frac{1}{2}du$.

$$\begin{aligned}&\int x^4\cdot(6+x^2)^{-1/2}\cdot x\,dx = \int(u-6)^2\cdot u^{-1/2}\cdot\frac{1}{2}\,du\\ &= \frac{1}{2}\int(u^2-12u+36)\cdot u^{-1/2}\,du = \frac{1}{2}\int(u^{3/2}-12u^{1/2}+36u^{-1/2})\,du\\ &= \frac{1}{2}\left(\frac{2}{5}u^{5/2}-8u^{3/2}+72u^{1/2}\right)+C = \frac{1}{5}u^{1/2}\left(u^2-20u+180\right)+C\\ &= \frac{1}{5}(x^4-8x^2+96)\sqrt{6+x^2}+C\end{aligned}$$

59. Substitute $u = x - 5$ and $du = dx$, so $x = u + 5$ and $dx = du$.

$$\int (x+2)\sqrt{x-5}\,dx = \int (u+7)\cdot u^{1/2}\,du = \int (u^{3/2} + 7u^{1/2})\,du$$
$$= \frac{2}{5}u^{5/2} + \frac{14}{3}u^{3/2} + C = \frac{2}{15}u^{3/2}\,(3u+35) + C$$
$$= \frac{2}{15}(3x+20)(x-5)^{3/2} + C$$

61. Substitute $u = \frac{1}{2}\sin(x)$ and $du = \frac{1}{2}\cos(x)\,dx$, so $\sin(x) = 2u$ and $\cos(x)\,dx = 2\,du$.

$$\int \frac{\cos(x)}{\sqrt{4-\sin^2(x)}}\,dx = \int \frac{1}{\sqrt{4-4u^2}}\cdot(2\,du) = \int \frac{1}{\sqrt{1-u^2}}\,du$$
$$= \arcsin(u) + C = \arcsin\left(\frac{1}{2}\sin(x)\right) + C$$

63. Combine the log terms. Then substitute $u = \ln(x)$, $du = \frac{1}{x}\,dx$, and proceed as shown below.

$$\int \frac{\ln(x^{11}) - \ln(x^7)}{x}\,dx = \int \frac{\ln(x^4)}{x}\,dx = 4\int \frac{\ln(x)}{x}\,dx$$
$$= 4\int u\,du = 2u^2 + C = 2\ln^2(x) + C$$

65. Simpify the integrand. Then substitute $u = \ln(x)$, $du = \frac{1}{x}\,dx$, and proceed as shown below.

$$\int \left(\frac{2}{x\ln(x^2)} - \frac{2}{x\ln^2(x)}\right)dx = \int \left(\frac{1}{x\ln(x)} - \frac{2}{x\ln^2(x)}\right)dx = \int \left(\frac{1}{u} - \frac{2}{u^2}\right)du$$
$$= \ln(|u|) + \frac{2}{u} + C = \ln(|\ln(x)|) + \frac{2}{\ln(x)} + C$$

67. Let $u = \arctan(x)$ so $du = \frac{1}{1+x^2}\,dx$. The integral evaluates as follows.

$$\int \frac{\arctan(x)}{1+x^2}\,dx = \int u\,du = \frac{1}{2}u^2 + C = \frac{1}{2}\arctan^2(x) + C$$

69. Let $u = \exp(x)$ so $du = \exp(x)\,dx$ and $\exp(2x) = u^2$. The integral evaluates as follows.

$$\int \frac{\exp(x)}{1+\exp(2x)}\,dx = \int \frac{1}{1+u^2}\,du = \arctan(u) + C = \arctan(\exp(x)) + C$$

71. Let $u = \exp(x)$ so $du = \exp(x)\,dx$ and $\exp(2x) = u^2$. The integral evaluates with another substitution indicated below.

$$\int \frac{\exp(x)}{1 + 2\exp(x) + \exp(2x)}\,dx = \int \frac{1}{u^2 + 2u + 1}\,du = \int \frac{1}{(u+1)^2}\,du$$
$$\overset{(u+1=v)}{=} \int \frac{1}{v^2}\,dv = -\frac{1}{v} + C = -\frac{1}{u+1} + C$$
$$= -\frac{1}{\exp(x) + 1} + C$$

73. Split the integral in two. Substitute $u = 4 - x^2$, $du = -2x\,dx$, and $x^2 = 4 - u$ into the first part and $v = \frac{1}{2}x$, $dv = \frac{1}{2}\,dx$, and $x = 2v$ into the second. Only one constant of integration should be used.

$$\int \frac{6x^3 + 3x + 1}{\sqrt{4 - x^2}}\,dx = \int \frac{6x^2 + 3}{\sqrt{4 - x^2}} \cdot (x\,dx) + \int \frac{1}{\sqrt{4 - x^2}}\,dx$$
$$= \int \frac{6(4 - u) + 3}{\sqrt{u}} \cdot \left(-\frac{1}{2}\,du\right) + \int \frac{1}{\sqrt{4 - 4v^2}} \cdot 2\,dv$$
$$= -\frac{1}{2}\int (27u^{-1/2} - 6u^{1/2})\,du + \int \frac{1}{\sqrt{1 - v^2}}\,dv$$
$$= -27u^{1/2} + 2u^{3/2} + \arcsin(v) + C = u^{1/2}(2u - 27) + \arcsin(v) + C$$
$$= \arcsin(x/2) - (2x^2 + 19)\sqrt{4 - x^2} + C$$

75. Substitute $u = 1 - \exp(2x)$ and $du = -2\exp(2x)\,dx$. The integral evaluates as follows.

$$\int \frac{\exp(2x)}{\sqrt{1 - \exp(2x)}}\,dx = \int \frac{1}{\sqrt{u}} \cdot \left(-\frac{1}{2}\,du\right) = -\sqrt{u} + C = -\sqrt{1 - \exp(2x)} + C$$

77. Replace $\tan(t)$ with $\sin(t)/\cos(t)$. Then substitute $u = \cos(t)$ and $du = -\sin(t)\,dt$. The du integral evaluates using Formula 74 with $a = 4$ and $b = 1$.

$$\int \frac{\sin(t)}{\cos(t)\sqrt{4 + \cos(t)}}\,dt = -\int \frac{1}{u\sqrt{4 + u}}\,du = -\frac{1}{2}\ln\left(\frac{\sqrt{4 + u} - 2}{\sqrt{4 + u} + 2}\right) + C$$
$$= -\frac{1}{2}\ln\left(\frac{\sqrt{4 + \cos(t)} - 2}{\sqrt{4 + \cos(t)} + 2}\right) + C$$

79. Make the substitution $x = t^2$, $dx = 2t\,dt$ to obtain $\int \frac{12}{t\sqrt{4+t^4}}\,dt = \int \frac{12}{2x\sqrt{4+x^2}}\,dx$.

The dx integral evaluates using Formula 95 with $a = 2$.

$$\int \frac{12}{t\sqrt{4+t^4}}\,dt = 6\int \frac{1}{x\sqrt{4+x^2}}\,dx = -3\ln\left(\left|\frac{2+\sqrt{4+x^2}}{x}\right|\right) + C$$
$$= -3\ln\left(\frac{2+\sqrt{4+t^4}}{t^2}\right) + C$$

81. Substitute $u = \sec(t)$, $du = \sec(t)\tan(t)\,dt$. The du integral integral evaluates using Formula 35 with $a = 1$.

$$\int \frac{\sec^3(t)\tan(t)}{1+\sec^2(t)}\,dt = \int \frac{u^2}{1+u^2}\,du = u - \arctan(u) + C$$
$$= \sec(t) - \arctan(\sec(t)) + C$$

83. Direct evaluation: $\int_{-1}^{1} 3t^2\,dt = t^3\Big|_{-1}^{1} = 1 - (-1) = 2$. Evaluation via substitution: $u = t^2$, $du = 2t\,dt$, $dt = \frac{1}{2\sqrt{u}}\,du$ for the right half: $[0, 1]$, and $u = t^2$, $du = 2t\,dt$, $dt = -\frac{1}{2\sqrt{u}}\,du$ for the left half: $[-1, 0]$.

$$\int_{-1}^{1} 3t^2\,dt = \int_{0}^{1} 3t^2\,dt + \int_{-1}^{0} 3t^2\,dt$$
$$= \int_{0}^{1} 3u\cdot\left(\frac{1}{2\sqrt{u}}\,du\right) + \int_{1}^{0} 3u\cdot\left(-\frac{1}{2\sqrt{u}}\,du\right)$$
$$= \int_{0}^{1} 3u\cdot\left(\frac{1}{\sqrt{u}}\,du\right) = 2u^{3/2}\Big|_{0}^{1} = 2$$

85. Split the integral into two pieces and substitute $x = -t$, $dx = -dt$ on the interval $[-a, 0]$.

$$\int_{-a}^{a} f(x)\,dx = \int_{a}^{0} f(x)\,dx + \int_{0}^{a} f(x)\,dx$$
$$= \int_{a}^{0} f(-t)\cdot(-1)\,dt + \int_{0}^{a} f(x)\,dx$$
$$= \int_{0}^{a} f(t)\,dt + \int_{0}^{a} f(x)\,dx = 2\int_{0}^{a} f(x)\,dx$$

87. Substitute $x = b - t$, $dx = -dt$. Since $x = a \implies t = b - a$ and $x = b \implies t = 0$, the integral transforms as follows.

$$\int_{a}^{b} f(x)\,dx = \int_{b-a}^{0} f(b-t)\cdot(-1)\,dt \overset{(t=x)}{=} \int_{0}^{b-a} f(b-x)\,dx$$

89. Substitute $x = a + (b - a)t$, $dx = (b - a)\,dt$. Since $x = a \Longrightarrow t = 0$ and $x = b \Longrightarrow t = 1$, the integral transforms as follows.

$$\begin{aligned}\int_a^b f(x)\,dx &= \int_0^1 f(a + (b-a)t)\cdot(b-a)\,dt \\ &\overset{(t=x)}{=} (b-a)\int_0^1 f(a + (b-a)x)\,dx\end{aligned}$$

91. Using the identity the integral evaluates as follows.

$$\begin{aligned}\int \sin(\alpha x)\cos(\beta x)\,dx &= \frac{1}{2}\int \big(\sin((\alpha+\beta)x) + \sin((\alpha-\beta)x)\big)\,dx \\ &= -\frac{1}{2}\left(\frac{\cos((\alpha+\beta)x)}{\alpha+\beta} + \frac{\cos((\alpha-\beta)x)}{\alpha-\beta}\right) + C\end{aligned}$$

93. Put the right side of the first equation over a common denominator to obtain the left side. The right side integrates in two pieces, make the substitution $u = t^2 + 1$, $du = 2t\,dt$ in the first one and $v = t + 1$, $dv = dt$ in the second. Only one constant should be used.

$$\begin{aligned}\int \left(\frac{2t}{t^2+1} + \frac{1}{(t+1)^2}\right) dt &= \int \frac{1}{u}\,du + \int \frac{1}{v^2}\,dv = \ln(|u|) - \frac{1}{v} + C \\ &= \ln(t^2+1) - \frac{1}{t+1} + C\end{aligned}$$

CALCULATOR/COMPUTER EXERCISES

95. The substitution $u = \tan(t^3 + t)\,dt$, $du = (3t^2 + 1)\sec^2(t^3 + t)\,dt$ converts the left integrand into the right one. Equality of the definite integrals requires that $\tan(b^3 + b) = 1$ or $b^3 + b = \pi/4$. Using *Maple*'s *fsolve* procedure, $b = 0.5850969677$.

97. The substitution $u = x + e^x$, $du = (1 + e^x)\,dx$ converts the left integrand into the right one. Equality of the definite integrals requires that $x + e^x = 9$. Using *Maple*, $b = 1.952651453$. The exact value can be expressed in terms of the Lambert W function: $b = 9 - W(e^9)$.

5.7 More on the Calculation of Area

Problems for Practice

1. Since $\cos(x) \geq 0$ on $[\pi/4, \pi/2]$ and $\cos(x) \leq 0$ on $[\pi/2, 2\pi/3]$, the area is sum of two integrals.

$$\int_{\pi/4}^{\pi/2} 2\cos(x)\,dx + \left(-\int_{\pi/2}^{2\pi/3} 2\cos(x)\,dx\right) = \left(2\sin(x)\Big|_{\pi/4}^{\pi/2}\right) + \left(2\sin(x)\Big|_{2\pi/3}^{\pi/2}\right)$$
$$= 2(1-\sqrt{2}/2) + 2(1-\sqrt{3}/2) = 4-\sqrt{2}-\sqrt{3}$$

3. Since $f(x) = 2(x+2)(x-2)$, $f(x) \geq 0$ on $[-3,-2]$ and on $[2,5]$ and $f(x) \leq 0$ on $[-2,2]$. The area is sum of three integrals.

$$\int_{-3}^{-2} f(x)\,dx = \left(\frac{2}{3}x^3 - 8x\right)\Big|_{-3}^{-2} = \frac{32}{3} - 6$$
$$-\int_{-2}^{2} f(x)\,dx = -\left(\frac{2}{3}x^3 - 8x\right)\Big|_{-2}^{2} = -\left(-\frac{32}{3} - \frac{32}{3}\right)$$
$$\int_{2}^{5} f(x)\,dx = \left(\frac{2}{3}x^3 - 8x\right)\Big|_{2}^{5} = \frac{130}{3} - \left(-\frac{32}{3}\right)$$

Therefore, the area is $4 \cdot \frac{32}{3} + \frac{130}{3} - 6 = 80$.

5. Since $f(x) \leq 0$ on $[-2,-1]$ and $f(x) \geq 0$ on $[-1,1]$, the area is sum of two integrals.

$$-\int_{-2}^{-1} f(x)\,dx + \int_{-1}^{1} f(x)\,dx = \left(\frac{1}{4}x^4 + \frac{1}{2}x^2 + 2x\right)\Big|_{-1}^{-2} + \left(\frac{1}{4}x^4 + \frac{1}{2}x^2 + 2x\right)\Big|_{-1}^{1}$$
$$= 2 - \left(-\frac{5}{4}\right) + \frac{11}{4} - \left(-\frac{5}{4}\right) = \frac{29}{4}$$

7. Since $f(x) \leq 0$ on $[-1/2, 0]$ and $f(x) \geq 0$ on $[0,1]$, the area is sum of two integrals. Using the substitution $u = 1 - x^2$, $F(x) = -\frac{4}{3}(1-x^2)^3$ is an antiderivative for f.

$$-\int_{-1/2}^{0} f(x)\,dx + \int_{0}^{1} f(x)\,dx = \left(-\frac{4}{3}(1-x^2)^3\Big|_{0}^{-1/2}\right) + \left(-\frac{4}{3}(1-x^2)^3\Big|_{0}^{1}\right)$$
$$= -\frac{9}{16} - \left(-\frac{4}{3}\right) + \left(0 - \left(-\frac{4}{3}\right)\right) = \frac{101}{48}$$

9. Since $f(x) \leq 0$ on $[-2,0]$ and $[2,3]$ and $f(x) \geq 0$ on $[0,2]$, the area is sum of three integrals. Using the substitution $u = 4 - x^2$, $F(x) = -\frac{1}{8}(4-x^2)^4$

is an antiderivative for f.

$$-\int_{-2}^{0} f(x)\,dx = \left(-\frac{1}{8}(4-x^2)^4\right)\Big|_{0}^{-2} = 0-(-32) = 32$$
$$\int_{0}^{2} f(x)\,dx = \left(-\frac{1}{8}(4-x^2)^4\right)\Big|_{0}^{2} = 0-(-32) = 32$$
$$-\int_{2}^{3} f(x)\,dx = \left(-\frac{1}{8}(4-x^2)^4\right)\Big|_{3}^{2} = 0-\left(-\frac{625}{8}\right) = \frac{625}{8}$$

Therefore, the area is $64 + \frac{625}{8} = \frac{1137}{8}$.

11. Since $f(x) = -3(x+4)(x-1)$, $f(x) \geq 0$ on $[-2,1]$ and $f(x) \leq 0$ on $[1,2]$. The area is sum of two integrals. Note that $F(x) = 12x - 9x^2/2 - x^3$ is an antiderivative for the integrand.

$$\int_{-2}^{1} f(x)\,dx + \left(-\int_{1}^{2} f(x)\,dx\right) = F(x)\Big|_{-2}^{1} + F(x)\Big|_{2}^{1} = \frac{81}{2} + \frac{17}{2} = 49$$

13. Inspection of its graph reveals that $f(x) \leq 0$ on $[-3,1]$, where $F(x) = -x^3/3$ is an antiderivative, and $f(x) \leq 0$ on $[1,3/2]$, where $G(x) = x^2 - 3x$ is an antiderivative. G is also an antiderivative for f on $[3/2,4]$ where $f(x) \geq 0$. The area calculation requires three integrals.

$$-\int_{-3}^{1} f(x)\,dx + \left(-\int_{1}^{3/2} f(x)\,dx\right) + \int_{3/2}^{4} f(x)\,dx = F(x)\Big|_{1}^{-3} + G(x)\Big|_{3/2}^{1} + G(x)\Big|_{3/2}^{4}$$
$$= \frac{28}{3} + \frac{1}{4} + \frac{25}{3} = \frac{95}{6}$$

15. Observe that $f(x) \geq 0$ on the interval $[0,\pi/2]$ and $f(x) \leq 0$ on $[\pi/2,\pi]$. However, three integrals are required for the area because $F(x) = -\cos(x)$ is an antiderivative on $[0,\pi/4]$ and $G(x) = \sin(x)$ is an antiderivative on $[\pi/4,\pi]$.

$$\int_{0}^{\pi/4} f(x)\,dx + \int_{\pi/4}^{\pi/2} f(x)\,dx + \left(-\int_{\pi/2}^{\pi} f(x)\,dx\right) = F(x)\Big|_{0}^{\pi/4} + G(x)\Big|_{\pi/4}^{\pi/2} + G(x)\Big|_{\pi}^{\pi/2}$$
$$= \left(-\frac{\sqrt{2}}{2} - (-1)\right) + \left(1 - \frac{\sqrt{2}}{2}\right) + (1-0) = 3 - \sqrt{2}$$

17. Sketch the graph of f to see that $f(x) \geq 0$ on $[-2,2]$ where $F(x) = 4x - x^3/3$ is an antiderivative. F is also an antiderivative for f on $[2,3]$ where $f(x) \leq 0$. $f(x) \leq 0$ on the remainder of the interval of integration as well, where the function $G(x) = x^3/3 - 4x^2 + 10x$ is an antiderivative.

Therefore, three integrals are required to obtain the area.

$$\int_{-2}^{2} f(x)\,dx + \left(-\int_{2}^{3} f(x)\,dx\right) + \left(-\int_{3}^{6} f(x)\,dx\right) = F(x)\Big|_{-2}^{2} + F(x)\Big|_{3}^{2} + G(x)\Big|_{6}^{3}$$
$$= \frac{32}{3} + \frac{7}{3} + 15 = 28$$

19. The graphs of f and g meet when $x^2 - 1 = 8$, that is, $x^2 = 9$ or $x = \pm 3$. The area between the graphs is $\int_{-3}^{3}(g(x) - f(x))\,dx = \int_{-3}^{3}(8 - (x^2 - 1))\,dx = \int_{-3}^{3}(9 - x^2)\,dx = (9x - x^3/3)\big|_{-3}^{3} = 27 - 9 - (-27 + 9) = 18 + 18 = 36$.

21. The graphs of f and g intersect when $x^2 - x - 2 = 0$, that is, $(x-2)(x+1) = 0$ or $x = -1$, $x = 2$. Since $f(0) < g(0)$, the area between the graphs is $\int_{-1}^{2}(g(x) - f(x))\,dx = \int_{-1}^{2}(-x^2 + x + 2)\,dx = 9/2$.

23. The graphs of f and g intersect when $x^2 + 2x - 8 = 0$, that is, $(x+4)(x-2) = 0$ or $x = -4$, $x = 2$. Since $f(0) > g(0)$, the area between the graphs is $\int_{-4}^{2}(f(x) - g(x))\,dx = \int_{-4}^{2}(-x^2 - 2x + 8)\,dx = 36$.

25. The graphs of f and g meet when $x^2 - 4 = 0$, that is, $x = \pm 2$. Since $f(0) > g(0)$, the area between the two graphs is $\int_{-2}^{2}(f(x) - g(x))\,dx = \int_{-2}^{2}(4 - x^2)\,dx = 32/3$.

27. The graphs of f and g intersect when $x^3 - 3x^2 + 2x = 0$, that is,

$$x(x-1)(x-2) = 0 \quad \text{or} \quad x = 0, x = 1, x = 2\,.$$

Since $\int_{0}^{1}(f(x) - g(x))\,dx = 1/4$ and $\int_{1}^{2}(f(x) - g(x))\,dx = -1/4$, the area between the two graphs is $1/4 + 1/4 = 1/2$.

29. The graphs of f and g meet when $x^4 - 5x^2 + 4 = 0$ or $(x^2 - 1)(x^2 - 4) = 0$. That is, $x = \pm 1$, $x = \pm 2$. Since $\int_{-2}^{-1}(f(x) - g(x))\,dx = -22/15$, $\int_{-1}^{1}(f(x) - g(x))\,dx = 76/15$, $\int_{1}^{2}(f(x) - g(x))\,dx = -22/15$, the area between the two graphs is $22/15 + 76/15 + 22/15 = 8$.

31. The graphs of f and g meet when $\tan(x) = -\sqrt{3}$, that is, $x = -\pi/3$ and $x = 2\pi/3$. Since $\int_{-\pi}^{-\pi/3}(f(x) - g(x))\,dx = 3$ and $\int_{-\pi/3}^{2\pi/3}(f(x) - g(x))\,dx = -4$, the area between the two graphs is $3 + 4 = 7$.

33. The graphs of f and g meet when $2\sin(x)\cos(x) = \cos(x)$. That is, $\cos(x) = 0$ or $\sin(x) = 1/2$ so $x = \pi/6$ and $x = \pi/2$. Since $\int_{0}^{\pi/6}(f(x) - g(x))\,dx = -1/4$ and $\int_{\pi/6}^{\pi/2}(f(x) - g(x))\,dx = 1/4$, the area between the two graphs is $1/4 + 1/4 = 1/2$.

35. The curves intersect when $2y^2 = 4$. That is, $y = \pm\sqrt{2}$. Since

$$\int_{-\sqrt{2}}^{\sqrt{2}} (y^2 - (-y^2 + 4))\,dy = \int_{-\sqrt{2}}^{\sqrt{2}} (2y^2 - 4)\,dy = \left(\frac{2}{3}y^3 - 4y\right)\Big|_{-\sqrt{2}}^{\sqrt{2}}$$
$$= \left(\frac{4}{3}\sqrt{2} - 4\sqrt{2}\right) - \left(-\frac{4}{3}\sqrt{2} + 4\sqrt{2}\right) = \left(\frac{8}{3} - 8\right)\sqrt{2} = -\frac{16}{3}\sqrt{2},$$

the area between the two curves is $\frac{16}{3}\sqrt{2}$.

37. The curves meet when $\cos(y) = \sin(y)$. That is, $y = \pi/4$ and $y = 5\pi/4$. Since $\int_{\pi/4}^{5\pi/4} (\cos(y) - \sin(y))\,dy = -2\sqrt{2}$, the area between the two curves is $2\sqrt{2}$.

39. The curves meet when $\ln(x) = \frac{x-1}{e-1}$. By inspection, $x = 1$ and $x = e$ (graph the two curves). The y-coordinates of the intersection points are $y = 0$ and $y = 1$. The integral is easier to evaluate when the curves are described as $x = e^y$ and $x = (e-1)y + 1$. Since

$$\int_0^1 (e^y - ((e-1)y + 1))\,dy = \left(e^y - \frac{1}{2}(e-1)y^2 - y\right)\Big|_0^1$$
$$= \left(e - \frac{1}{2}(e-1) - 1\right) - 1 = \frac{1}{2}e - \frac{3}{2},$$

the area between the two curves is $(3 - e)/2$.

Further Theory and Practice

41. The graph of f lies above the graph of g over the interval. Therefore, the area is

$$\int_0^1 \left(x(1 + x^2) - \frac{1}{2}x\right)\,dx = \left(\frac{1}{4}x^2 + \frac{1}{4}x^4\right)\Big|_0^1 = \frac{1}{2}.$$

43. Over the given range of x, the graphs of f and g intersect when $x = 0$, $x = \pi$, and $2\sin(x) = 2\sin(x)\cos(x)$. That is, $\cos(x) = 1$ or $x = 0$ (again). Since

$$\int_0^{\pi} (2\sin(x) - \sin(2x)))\,dx = \left(-2\cos(x) - \frac{1}{2}\cos(2x)\right)\Big|_0^{\pi}$$
$$= \left(2 - \frac{1}{2}\right) - \left(-2 - \frac{1}{2}\right) = 4,$$

the area between the curves is 4.

45. The curves intersect when $2y^2 = 8$ or $y = \pm 2$. Since $\int_{-2}^{2} (y^2 + 6 - (-y^2 + 14))\,dy = -64/3$, the area between the curves is $64/3$.

47. The curves intersect at $(0, 0)$ and $(1, 1)$. Since $\int_0^1 (y^2 - y^3)\,dy = 1/12$, the area between the curves is $1/12$.

49. For $0 \le x \le 4$, $y = \sqrt{x}$ is equivalent to $x = y^2$ and $0 \le y \le 2$. Therefore, the area below $y = \sqrt{x}$ and above $[0, 4]$ on the x-axis can also be regarded as the area of the region that lies between the line $x = 4$ and the curve $x = y^2$ and to the right of the interval $[0, 2]$ on the y-axis: $\int_0^2 (4 - y^2)\, dy$.

51. For $0 \le x \le 1$, $y = \arcsin(x)$ is equivalent to $x = \sin(y)$ and $0 \le y \le \pi/2$. Therefore, the area below the arcsin curve and above $[0, 1]$ on the x-axis can also be regarded as the area of the region that lies between the line $x = 1$ and the curve $x = \sin(y)$ and to the right of the interval $[0, \pi/2]$ on the y-axis: $\int_0^{\pi/2} (1 - \sin(y))\, dy$.

53. The integral $\int_0^4 (\sqrt{x} - x/2)\, dx$ represents the area of the region shown on the right. The same area can be obtained with the following integral.

$$\int_0^2 (2y - y^2)\, dy$$

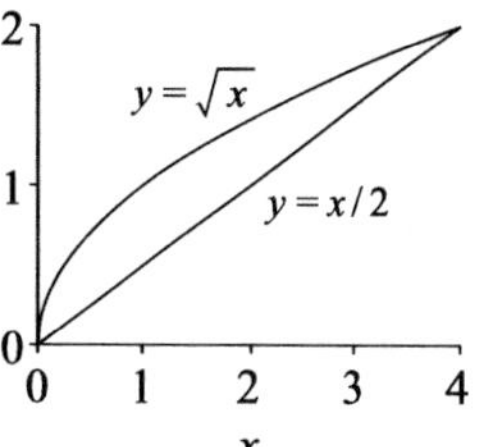

55. The integral $\int_{-3}^3 (3 - |x|)\, dx$ represents the area of the region shown on the right. The same area can be obtained with the following integral.

$$\int_0^3 (y - (-y))\, dy$$

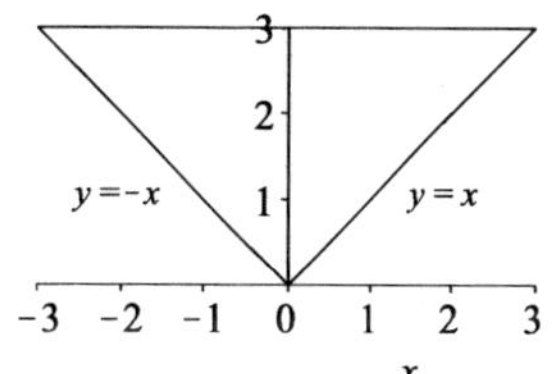

57. The triangle is displayed on the right. Its area can be obtained as the following sum of integrals.

$$\int_1^2 (x - 1)\, dx + \int_2^3 (3 - x)\, dx$$

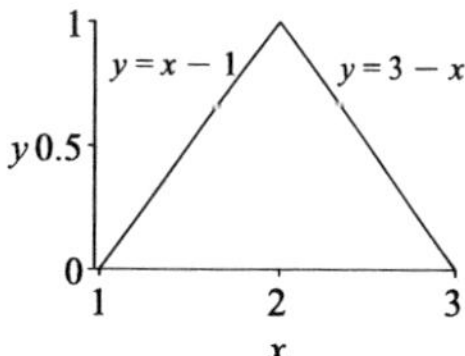

59. The region is displayed on the right. Its area can be obtained using the following sum.

$$\int_{-1}^0 (2 - x^2 - (-x))\, dx + \int_0^1 (2 - x^2 - x)\, dx\,.$$

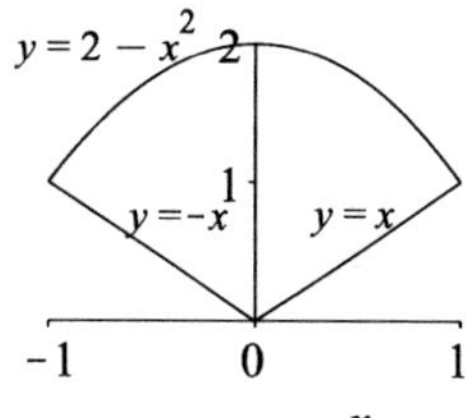

61. The triangle is displayed on the right. Its area can be obtained as the following dy-integral.

$$\int_0^1 (3 - y - (y + 1))\, dy$$

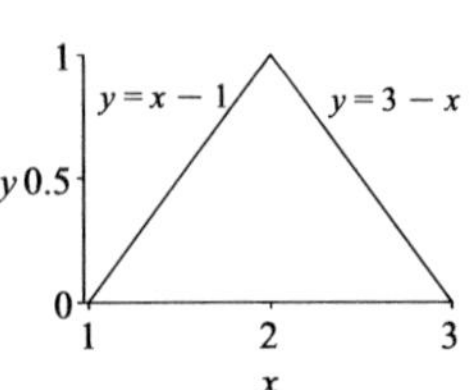

63. The region is displayed on the right. Its area can be obtained using the following sum of dy-integrals.

$$\int_0^1 (y - (-y))\, dy + \int_1^2 (\sqrt{2 - y} - (-\sqrt{2 - y}))\, dx\,.$$

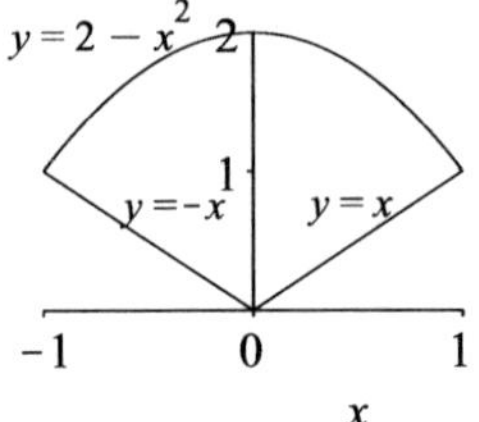

65. The region whose area is represented by the two dx-integrals is displayed on the right. The area can also be obtained using the dy-integral $\int_0^{\sqrt{2}} (4 - y^2 - y^2)\, dy$.

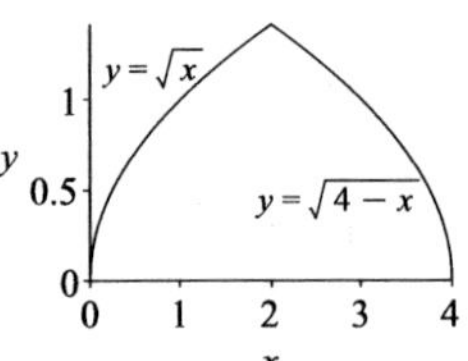

67. The region whose area is represented by the two dx-integrals is displayed on the right. The area can also be obtained using the dy-integral $\int_0^{\pi/4} (\cos(y) - \sin(y))\, dy$.

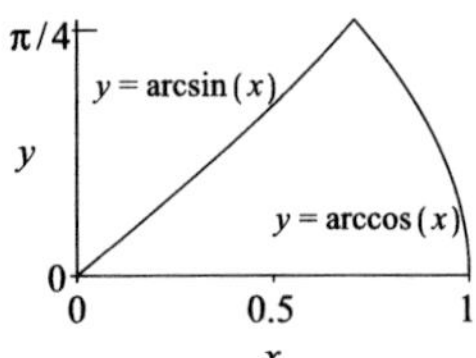

69. The region whose area is represented by the two dx-integrals is displayed on the right. The area can also be obtained using the dy-integral $\int_0^2 (\sqrt{(2 - y)/2} - (-(y - 2)^2/4))\, dy$.

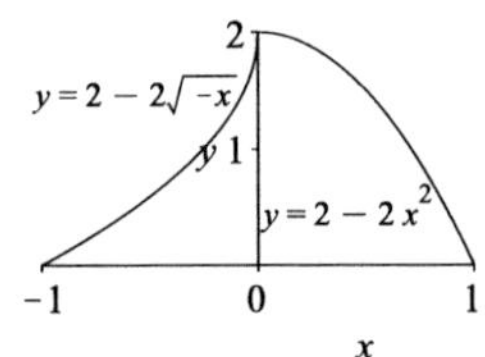

71. Substitute $y = (x - 1)/4$ into the second equation and simplify to obtain

$$25x^2 - 30x - 11 = 0\,.$$

The quadratic formula yields the x-coordinates of the points of intersection of the two curves: $a = (3 - 2\sqrt{5})/5$ and $b = (3 + 2\sqrt{5})/5$. See the picture on the right.

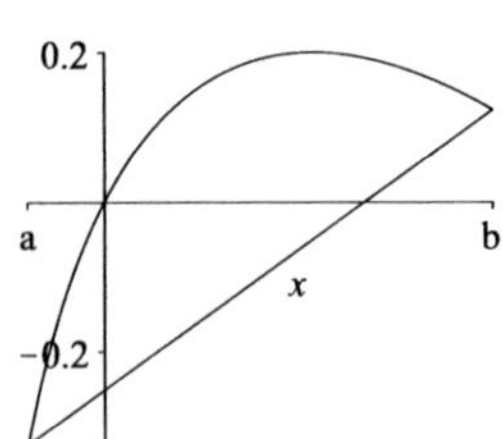

Using the quadratic formula, solve the second equation for y:

$$y = \frac{-2x - 3 \pm \sqrt{20x+9}}{2},$$

and the area is $\displaystyle\int_a^b \left(\frac{-2x - 3 + \sqrt{20x+9}}{2} - \frac{x-1}{4} \right) dx = \frac{\sqrt{5}}{6}.$

CALCULATOR/COMPUTER EXERCISES

73. The x coordinates of the points of intersection are $a = -2$ and $b = 1.275682204$. The area between the curves for $a \le x \le b$ is

$$\int_a^b (6 - x^4 - (x^3 + x))\, dx \approx 17.10264920\,.$$

See the picture.

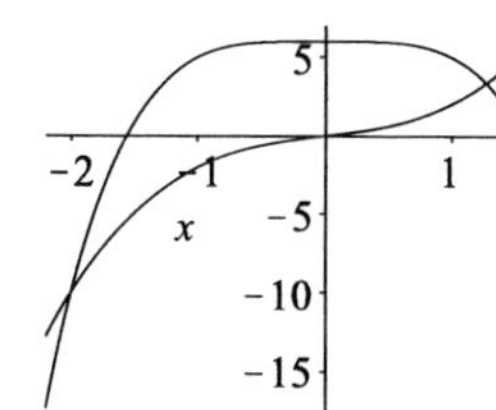

75. The x coordinates of the points of intersection are $a = 1$ and $b = 3.808878806$. The area between the curves for $a \le x \le b$ is

$$\int_a^b (\ln(x)/x - (x-1)/8)\, dx \approx 0.4011197670\,.$$

See the picture.

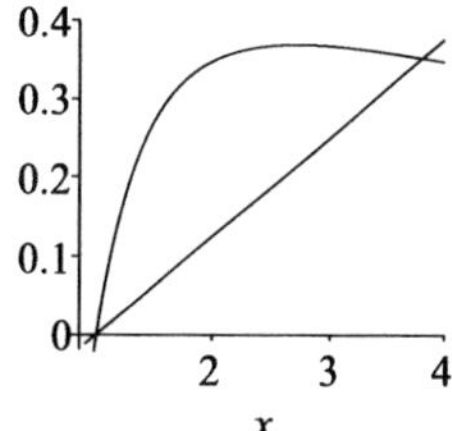

77. The x coordinates of the points of intersection are $a = -1$ and $b = 0.5436890127$. The area between the curves for $a \le x \le b$ is

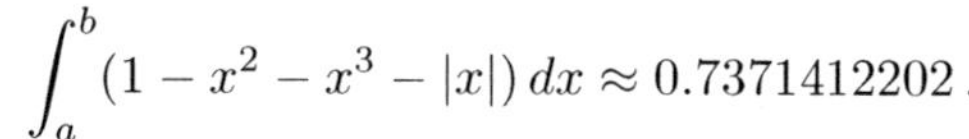

$$\int_a^b (1 - x^2 - x^3 - |x|)\, dx \approx 0.7371412202\,.$$

See the picture.

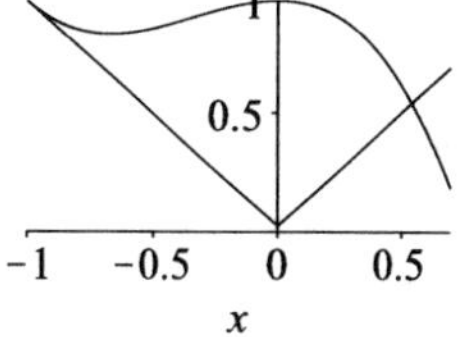

79. The x coordinates of the points of intersection are $a = -0.5767698076$ and $b = 0.9492450330$. The area between the curves for $a \le x \le b$ is

$$\int_a^b (x + 2 - \sec^2(x))\, dx \approx 1.289515941\,.$$

See the picture.

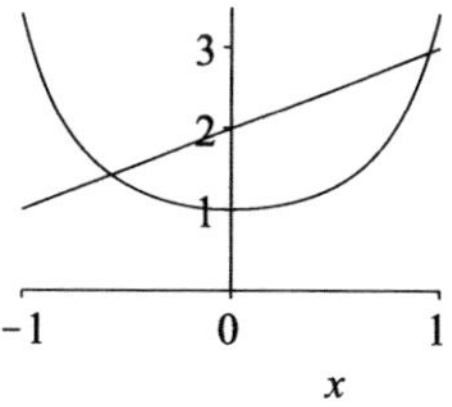

5.8 Numerical Techniques of Integration

Problems for Practice

1. The partition is $\{1, 2, 3\}$, $\Delta x = 1$.

$$\begin{aligned}\mathcal{M}_2 &= 1 \cdot (f(3/2) + f(5/2)) = \frac{9}{4} + \frac{25}{4} = \frac{17}{2}\\ \mathcal{T}_2 &= \frac{1}{2} \cdot (f(1) + 2 \cdot f(2) + f(3)) = \frac{1}{2}(1 + 8 + 9) = 9\\ \mathcal{S}_2 &= \frac{1}{3} \cdot (f(1) + 4f(2) + f(3)) = \frac{1}{3}(1 + 16 + 9) = \frac{26}{3}\end{aligned}$$

3. The partition is $\{-1, 0, 1\}$, $\Delta x = 1$.

$$\begin{aligned}\mathcal{M}_2 &= 1 \cdot (f(-1/2) + f(1/2)) = \frac{\sqrt{7}}{2} + \frac{\sqrt{11}}{2} = \frac{\sqrt{7} + \sqrt{11}}{2}\\ \mathcal{T}_2 &= \frac{1}{2} \cdot (f(-1) + 2 \cdot f(0) + f(1)) = \frac{1}{2}(\sqrt{2} + 2\sqrt{2} + 2) = 1 + \frac{3\sqrt{2}}{2}\\ \mathcal{S}_2 &= \frac{1}{3} \cdot (f(-1) + 4f(0) + f(1)) = \frac{1}{3}(\sqrt{2} + 4\sqrt{2} + 2) = \frac{2}{3} + \frac{5\sqrt{2}}{3}\end{aligned}$$

5. The partition is $\{1, 2, 3\}$, $\Delta x = 1$.

$$\begin{aligned}\mathcal{M}_2 &= 1 \cdot (f(3/2) + f(5/2)) = \ln(3/2) + \ln(5/2)\\ \mathcal{T}_2 &= \frac{1}{2} \cdot (f(1) + 2 \cdot f(2) + f(3)) = \frac{1}{2}(0 + 2\ln(2) + \ln(3)) = \ln(2) + \frac{1}{2}\ln(3)\\ \mathcal{S}_2 &= \frac{1}{3} \cdot (f(1) + 4f(2) + f(3)) = \frac{1}{3}(0 + 4\ln(2) + \ln(3)) = \frac{4}{3}\ln(2) + \frac{1}{3}\ln(3)\end{aligned}$$

7. The partition is $\{\pi/4, \pi/2, 3\pi/4, \pi, 5\pi/4\}$, $\Delta x = \pi/4$. The calculation of $\mathcal{M}_4$ takes advantage of the fact that $\cos(\pi \pm \theta) = -\cos(\theta)$.

$$\begin{aligned}\mathcal{M}_4 &= \frac{\pi}{4} \cdot (f(3\pi/8) + f(5\pi/8) + f(7\pi/8) + f(9\pi/8))\\ &= \frac{\pi}{4} \cdot (\cos(3\pi/8) - \cos(3\pi/8) - \cos(\pi/8) - \cos(\pi/8)) = -\frac{\pi}{2}\cos(\pi/8)\\ \mathcal{T}_4 &= \frac{\pi}{8} \cdot (f(\pi/4) + 2 \cdot (f(\pi/2) + f(3\pi/4) + f(\pi)) + f(5\pi/4))\\ &= \frac{\pi}{8}\left(\frac{\sqrt{2}}{2} + 2 \cdot \left(0 - \frac{\sqrt{2}}{2} - 1\right) - \frac{\sqrt{2}}{2}\right) = -\frac{2 + \sqrt{2}}{8}\pi\\ \mathcal{S}_4 &= \frac{\pi}{12} \cdot (f(\pi/4) + 4f(\pi/2) + 2f(3\pi/4) + 4f(\pi) + f(5\pi/4))\\ &= \frac{\pi}{12}\left(\frac{\sqrt{2}}{2} + 4 \cdot 0 + 2 \cdot \left(-\frac{\sqrt{2}}{2}\right) + 4 \cdot (-1) - \frac{\sqrt{2}}{2}\right) = -\frac{4 + \sqrt{2}}{12}\pi\end{aligned}$$

9. The partition is $\{-1/2, 1/2, 3/2, 5/2, 7/2\}$, $\Delta x = 1$.

$$\begin{aligned}\mathcal{M}_4 &= 1 \cdot (f(0) + f(1) + f(2) + f(3)) \\ &= 0 + 1 + 8 + 27 = 36 \\ \mathcal{T}_4 &= \frac{1}{2} \cdot (f(-1/2) + 2 \cdot (f(1/2) + f(3/2) + f(5/2)) + f(7/2)) \\ &= \frac{1}{2}\left(-\frac{1}{8} + 2 \cdot \left(\frac{1}{8} + \frac{27}{8} + \frac{125}{8}\right) + \frac{343}{8}\right) = \frac{81}{2} \\ \mathcal{S}_4 &= \frac{1}{3} \cdot (f(-1/2) + 4f(1/2) + 2f(3/2) + 4f(5/2) + f(7/2)) \\ &= \frac{1}{3}\left(-\frac{1}{8} + 4 \cdot \frac{1}{8} + 2 \cdot \frac{27}{8} + 4 \cdot \frac{125}{8} + \frac{343}{8}\right) = \frac{75}{2}\end{aligned}$$

11. According to (5.8.3), if $|f''(x)| \le C$ for $x \in I = [a, b]$, then

$$\left|\int_a^b f(x)\,dx - \mathcal{M}_N\right| \le \frac{C}{24} \cdot \frac{(b-a)^3}{N^2} = E_N\,.$$

For $f(x) = x + 1/x$, $f''(x) = 2/x^3$ and, for the interval $I = [1/2, 7/2]$, $C = 16$. If $N = 10$, then $E_{10} = 9/50 = 0.18$. Since the integral A satisfies the inequality $\mathcal{M}_{10} - E_{10} \le A \le \mathcal{M}_{10} + E_{10}$, we expect that $7.75202 \le A \le 8.11202$. These are the values for α and β.

As a check, $A = \int_{1/2}^{7/2} (1 + 1/x)\,dx = 6 + \ln(7) = 7.94591\ldots$.

13. According to (5.8.3), if $|f''(x)| \le C$ for $x \in I = [a, b]$, then

$$\left|\int_a^b f(x)\,dx - \mathcal{M}_N\right| \le \frac{C}{24} \cdot \frac{(b-a)^3}{N^2} = E_N\,.$$

For $f(x) = (2 + x)^{-1/3}$, $f''(x) = 4/(9(2 + x)^{7/3})$ and, for the interval $I = [-1, 3]$, $C = 4/9$. If $N = 10$, then $E_{10} = 8/675 = 0.01185$. Since the integral A satisfies the inequality $\mathcal{M}_{10} - E_{10} \le A \le \mathcal{M}_{10} + E_{10}$, we expect that $2.87223 \le A \le 2.89594$. These are the values for α and β.

As a check, $A = \int_{-1}^{3} (2 + x)^{-1/3}\,dx = -\frac{3}{2} + \frac{3}{2} \cdot 5^{2/3} = 2.88602\ldots$.

15. According to (5.8.6), if $|f''(x)| \le C$ for $x \in I = [a, b]$, then

$$\left|\int_a^b f(x)\,dx - \mathcal{T}_N\right| \le \frac{C}{12} \cdot \frac{(b-a)^3}{N^2} = E_N\,.$$

For $f(x) = x^3$, $f''(x) = 6x$ and, for the interval $I = [-5, 1]$, $C = 30$. If $N = 10$, then $E_{10} = 27/5 = 5.4$. Since the integral A satisfies the inequality $\mathcal{T}_{10} - E_{10} \le A \le \mathcal{T}_{10} + E_{10}$, we expect that $-163.56 \le A \le -152.76$. These are the values for α and β.

As a check, $A = \int_{-5}^{1} x^3\,dx = -156$.

17. According to (5.8.6), if $|f''(x)| \leq C$ for $x \in I = [a, b]$, then

$$\left| \int_a^b f(x)\,dx - \mathcal{T}_N \right| \leq \frac{C}{12} \cdot \frac{(b-a)^3}{N^2} = E_N\,.$$

For $f(x) = x\exp(x^2)$, $f''(x) = 2x\exp(x^2)(3+2x^2)$ and, for the interval $I = [3/2, 2]$, $C = 44\exp(4)$. If $N = 10$, then $E_{10} = 11\exp(4)/2400 = 0.25024$. Since the integral A satisfies the inequality $\mathcal{T}_{10} - E_{10} \leq A \leq \mathcal{T}_{10} + E_{10}$, we expect that $22.39635 \leq A \leq 22.89684$. These are the values for α and β.

As a check, $A = \int_{3/2}^{2} x\exp(x^2)\,dx = \frac{1}{2} \cdot (\exp(4) - \exp(9/4)) = 22.55520\ldots$.

19. According to (5.8.8), if $|f^{(4)}(x)| \leq C$ for $x \in I = [a, b]$, then

$$\left| \int_a^b f(x)\,dx - \mathcal{S}_N \right| \leq \frac{C}{180} \cdot \frac{(b-a)^5}{N^4} = E_N\,.$$

For $f(x) = x^5$, $f^{(4)}(x) = 120x$ and, for the interval $I = [1, 4]$, $C = 480$. If $N = 10$, then $E_{10} = 81/1250 = 0.0648$. Since the integral A satisfies the inequality $\mathcal{S}_{10} - E_{10} \leq A \leq \mathcal{S}_{10} + E_{10}$, we expect that $682.4757 \leq A \leq 682.6053$. These are the values for α and β.

As a check, $A = \int_1^4 x^5\,dx = 1365/2 = 682.5$.

21. According to (5.8.8), if $|f^{(4)}(x)| \leq C$ for $x \in I = [a, b]$, then

$$\left| \int_a^b f(x)\,dx - \mathcal{S}_N \right| \leq \frac{C}{180} \cdot \frac{(b-a)^5}{N^4} = E_N\,.$$

For $f(x) = \sqrt{x}$, $f^{(4)}(x) = -\frac{15}{16} \cdot x^{-7/2}$ and, for the interval $I = [1, 9]$, $C = 15/16$. If $N = 10$, then $E_{10} = 32/1875 = 0.017066$. Since the integral A satisfies the inequality $\mathcal{S}_{10} - E_{10} \leq A \leq \mathcal{S}_{10} + E_{10}$, we expect that $17.31568 \leq A \leq 17.34981$. These are the values for α and β.

As a check, $A = \int_1^9 \sqrt{x}\,dx = 52/3 = 17.\overline{3}$.

23. Using the partition $\{0, 20, 40, 60, 80, 100\}$ and $\Delta x = 20$, apply the Trapezoidal Rule to the function L having values $0, 5, 20, 30, 55, 100$ at the partition points to obtain

$$\int_0^{100} L(x)\,dx \approx \frac{20}{2}(0 + 2 \cdot (5 + 20 + 30 + 55) + 100) = 3200\,.$$

The approximate coefficient of inequality is $1 - 3200/5000 = 9/25 = 0.36$.

25. Using the partition $\{0, 10, 20, \ldots, 100\}$ and $\Delta x = 10$, apply the Trapezoidal Rule to the function L having values $0, 4, 8, 14, 22, 32, 42, 56, 70, 82, 100$

at the partition points to obtain

$$\int_0^{100} L(x)\,dx \approx \frac{10}{2}(0+2\cdot(4+8+14+\cdots+82)+100) = 3800\,.$$

The approximate coefficient of inequality is $1-2600/5000 = 6/25 = 0.24$.

27. Using the partition $\{0,10,20,\ldots,100\}$ and $\Delta x = 10$, apply Simpson's Rule to the function L having values $0,4,8,14,22,32,42,56,70,82,100$ at the partition points to obtain

$$\int_0^{100} L(x)\,dx \approx \frac{10}{3}(0+4\cdot4+2\cdot8+4\cdot14+\cdots+4\cdot82+100) = 11360/3\,.$$

The approximate coefficient of inequality is $1-11360/(3\cdot5000) = 91/375 \approx 0.2426$.

29. Use the partition $\{0,1,2,3,4,5,6,7,8\}$ of the interval $[0,8]$, $\Delta t = 1$. The Simpson's Rule estimate for the integral is

$$\int_0^{8} c(t)\,dt \approx \frac{1}{3}\cdot(0+4(1.9)+2(5.8)+4(9.4)+2(10.4)+4(9.1)+2(5.9)+4(2.1)+0) \approx 44.73\,.$$

Therefore, r is approximately $5/44.73\ L/s$ or $300/44.73 = 6.706\ L/min$.

Further Theory and Practice

31. For the Trapezoid rule, if $|f''(x)| \le C$ for $x \in [a,b]$, then

$$\left|\int_a^b f(x)\,dx - \mathcal{T}_N\right| \le \frac{C}{12}\cdot\frac{(b-a)^3}{N^2} = ET_N\,.$$

When $f(x) = \exp(-x/2)$ and the interval is $[1,9]$, $f''(x) = \exp(-x/2)/4$ so $C = \exp(-1/2)/4$, and $ET_N = \frac{32}{3}\cdot\frac{\exp(-1/2)}{N^2}$. $ET_N = 10^{-4}$ when $N = 254.35$ so $N = 255$ will guarantee an error bound of 10^{-4}.

For Simpson's Rule, if $|f^{(4)}(x)| \le C$ for $x \in [a,b]$, then

$$\left|\int_a^b f(x)\,dx - \mathcal{S}_N\right| \le \frac{C}{180}\cdot\frac{(b-a)^5}{N^4} = ES_N\,.$$

Since $f^{(4)}(x) = \exp(-x/2)/16$, $C = \exp(-1/2)/16$, and the error bound is $ES_N = \frac{512}{45}\cdot\frac{\exp(-1/2)}{N^4}$. $ES_N = 10^{-4}$ when $N = 16.21$ so $N = 18$ (N must be even) will guarantee an error bound of 10^{-4}.

33. According to (5.8.3), if $|f''(x)| \le C$ for $x \in [a,b]$, then

$$\left|\int_a^b f(x)\,dx - \mathcal{M}_N\right| \le \frac{C}{24}\cdot\frac{(b-a)^3}{N^2} = E_N\,.$$

For $f(x)=\sqrt{6+x}$, $f''(x)=-\frac{1}{4(6+x)^{3/2}}$. When $x\in[-2,2]$, the maximum value of $|f''(x)|$ is $C=1/32$. Consequently, $E_N=\frac{1}{12N^2}$, and $E_N=10^{-3}$ when $N\approx 9.13$. Therefore, if $N=10$, then the Midpoint Rule approximation is accurate to within 10^{-3}.

35. According to (5.8.3), if $|f''(x)|\le C$ for $x\in[a,b]$, then

$$\left|\int_a^b f(x)\,dx-\mathcal{M}_N\right|\le\frac{C}{24}\cdot\frac{(b-a)^3}{N^2}=E_N\,.$$

For $f(x)=\sin(\pi x)$, $f''(x)=-\pi^2\sin(\pi x)$. When $x\in[-2,2]$, the maximum value of $|f''(x)|$ is $C=\pi^2$. Consequently, $E_N=\frac{8\pi^2}{3N^2}$, and $E_N=10^{-3}$ when $N\approx 162.23$. Therefore, if $N=163$, then the Midpoint Rule approximation is accurate to within 10^{-3}.

37. According to (5.8.8), if $|f^{(4)}(x)|\le C$ for $x\in[a,b]$, then

$$\left|\int_a^b f(x)\,dx-\mathcal{S}_N\right|\le\frac{C}{180}\cdot\frac{(b-a)^5}{N^4}=E_N\,.$$

For $f(x)=\sqrt{6+x}$, $f^{(4)}(x)=-\frac{15}{16(6+x)^{7/2}}$. When $x\in[-2,2]$, the maximum value of $|f^{(4)}(x)|$ is $C=15/2048$. Consequently, $E_N=\frac{1}{24N^4}$, and $E_N=10^{-3}$ when $N\approx 2.54$. Therefore, if $N=4$, then the Simpson Rule approximation is accurate to within 10^{-3}.

39. According to (5.8.8), if $|f^{(4)}(x)|\le C$ for $x\in[a,b]$, then

$$\left|\int_a^b f(x)\,dx-\mathcal{S}_N\right|\le\frac{C}{180}\cdot\frac{(b-a)^5}{N^4}=E_N\,.$$

For $f(x)=\sin(\pi x)$, $f^{(4)}(x)=\pi^2\sin(\pi x)$. When $x\in[-2,2]$, the maximum value of $|f^{(4)}(x)|$ is $C=\pi^4$. Consequently, $E_N=\frac{256\pi^4}{45N^4}$, and $E_N=10^{-3}$ when $N\approx 27.28$. Therefore, if $N=28$, then the Simpson Rule approximation is accurate to within 10^{-3}.

41. The average value of x^3 over $[\gamma-h,\gamma+h]$ $(h>0)$ is

$$\frac{1}{2h}\int_{\gamma-h}^{\gamma+h}x^3\,dx=\frac{1}{2h}\cdot\left(\frac{1}{4}x^4\right)\Bigg|_{\gamma-h}^{\gamma+h}=\frac{(\gamma+h)^4-(\gamma-h)^4}{8h}$$

$$=\frac{\gamma^4+4\gamma^3h+6\gamma^3h^2+4\gamma h^3+h^4-(\gamma^4-4\gamma^3h+6\gamma^3h^2-4\gamma h^3+h^4)}{8h}$$

$$=\frac{8\gamma^3h+8\gamma h^3}{8h}=\gamma\,(\gamma^2+h^2)\,.$$

Moreover,

$$\frac{1}{6}\Big((\gamma-h)^3+4\gamma^3+(\gamma+h)^3\Big)=\frac{1}{6}(\gamma^3-\gamma^2h+3\gamma h^2-h^3)+\frac{2}{3}\gamma^3$$
$$+\frac{1}{6}(\gamma^3+\gamma^2h+3\gamma h^2+h^3)$$
$$=\gamma^3+\gamma h^2=\gamma(\gamma^2+h^2)\,.$$

Combining these two facts, given any γ and any $h>0$,

$$\int_{\gamma-h}^{\gamma+h} x^3\,dx = 2h\cdot\frac{1}{6}\cdot\Big((\gamma-h)^3+4\gamma^3+(\gamma+h)^3\Big)\,.$$

In particular, if $\{x_0, x_1, \ldots, x_N\}$ is a uniform partition of an interval $[a,b]$, with N an even integer and $\Delta x = (b-a)/N$, then for every odd integer j, $1 \le j \le N-1$,

$$\int_{x_{j-1}}^{x_{j+1}} x^3\,dx = 2\Delta x\cdot\frac{1}{6}\cdot(x_{j-1}^3+4x_j^3+x_{j+1}^3)$$
$$=\frac{\Delta x}{3}\cdot(x_{j-1}^3+4x_j^3+x_{j+1}^3)\,.$$

Summing these integrals for $j=1,3,\ldots,N-1$,

$$\int_a^b x^3\,dx = \frac{\Delta x}{3}\cdot(x_0^3+4x_1^3+2x_2^3+4x_3^3+\cdots+2x_{N-2}^3+4x_{N-1}^3+x_N^3)\,,$$

and the Nth order Simpson Rule approximation is exact when applied to x^3. This extends to $x\mapsto Dx^3$ for any D and, since the approximation is also exact for $x\mapsto Ax^2+Bx+C$, it will be exact for an arbitrary polynomial of degree 3: $P(x)=Ax^2+Bx+C+Dx^3$, $(D\neq 0)$.

43. With $N=2$ the partition is $\{-1,0,1\}$ and $\Delta x = 1$.

$$\mathcal{M}_2 = f(-1/2)+f(1/2) = \frac{\sqrt{2}}{2}+\frac{\sqrt{2}}{2}=\sqrt{2}$$
$$\mathcal{S}_2 = \frac{1}{3}(f(-1)+4f(0)+f(1)) = \frac{1}{3}(1+0+1)=\frac{2}{3}$$

Since $A=\int_{-1}^{1}\sqrt{|x|}\,dx = \int_{-1}^{0}(-x)^{1/2}\,dx+\int_0^1 x^{1/2}\,dx = 2\int_0^1 x^{1/2}\,dx = 4/3$, the absolute errors are $\left|\frac{4}{3}-\sqrt{2}\right| = 0.08088\ldots$ for $\mathcal{M}_2$ and $\left|\frac{4}{3}-\frac{2}{3}\right| = 0.66666\ldots$ for $\mathcal{S}_2$.

With $N = 4$ the partition is $\{-1, -1/2, 0, 1/2, 1\}$ and $\Delta x = 1/2$.

$$\begin{aligned} \mathcal{M}_4 &= \frac{1}{2}(f(-3/4) + f(-1/4) + f(1/4) + f(3/4)) = \frac{1}{2}\left(\frac{\sqrt{3}}{2} + \frac{1}{2} + \frac{1}{2} + \frac{\sqrt{3}}{2}\right) \\ &= \frac{\sqrt{3}+1}{2} \\ \mathcal{S}_4 &= \frac{1}{6}(f(-1) + 4f(-1/2) + 2f(0) + 4f(1/2) + f(1)) = \frac{1}{6}\left(1 + 2\sqrt{2} + 0 + 2\sqrt{2} + 1\right) \\ &= \frac{2\sqrt{2}+1}{3}. \end{aligned}$$

The absolute errors are $\left|\frac{4}{3} - \frac{\sqrt{3}+1}{2}\right| = 0.03269\ldots$ for $\mathcal{M}_4$ and $\left|\frac{4}{3} - \frac{2\sqrt{2}+1}{3}\right| = 0.05719\ldots$ for $\mathcal{S}_4$.

45. Apply the Mean Value Theorem to f''' on the interval with endpoints ξ and η to obtain a point ζ between them such that $\left|\frac{f'''(\xi)-f'''(\eta)}{\xi-\eta}\right| = |f^{(4)}(\zeta)|$. This implies that if C is a constant such that $C \geq |f^{(4)}(x)|$ for all $x \in [a, b]$, then $|f'''(\xi) - f'''(\eta)| \leq C(b - a)$. Therefore,

$$\left|\frac{(b-a)^4}{180N^4}(f'''(\xi) - f'''(\eta))\right| \leq \frac{C}{180} \cdot \frac{(b-a)^5}{N^4}.$$

47. The trapezoidal approximation for data points (a, α) and (b, β), $a < b$ is $\frac{1}{2}(\alpha + \beta)(b - a)$. Applying this to the data in the table,

$$\begin{aligned} \int_0^{100} L(x) \approx \frac{1}{2}\Big(&(0+3)(16) + (3+8)(12) + (8+24)(23) + (24+46)(24) \\ &+(46+69)(13) + (69+88)(9) + (88+100)(3)\Big) = 3034. \end{aligned}$$

The approximate coefficient of inequality is $1 - 3034/5000 = 983/2500 = 0.3932$.

49. The x and y data for the pool profile is summarized in the following table.

x	0	2	4	6	8	10	12	14	16
y	0.77	1.5	2.0	2.4	2.8	3.1	3.4	3.6	3.9

The Midpoint estimate of the area of the side of the pool is

$$\mathcal{M}_4 = 4(1.5 + 2.4 + 3.1 + 3.6) = 42.4,$$

the pool volume is approximately $42.4 \cdot 6 = 254.4$ m^3.

The Trapezoidal estimate of the area of the side of the pool is

$$\mathcal{T}_8 = \frac{2}{2}(0.77 + 2(1.5 + 2.0 + 2.4 + 2.8 + 3.1 + 3.4 + 3.6) + 3.9) = 42.27,$$

the pool volume is approximately $42.27 \cdot 6 = 253.62$ m^3.

The Simpson Rule estimate of the area of the side of the pool is

$$\mathcal{S}_8 = \frac{2}{3}(0.77{+}4(1.5){+}2(2.0){+}4(2.4){+}2(2.8){+}4(3.1){+}2(3.4){+}4(3.6){+}3.9) = 42.31\overline{3}\,,$$

the pool volume is approximately $42.31\overline{3} \cdot 6 = 253.88$ m^3.

51. $f(x) = x^4$ and the partitions are $\{0, 1, 2, 3, 4\}$ and $\{0, 2, 4\}$.

$$\mathcal{S}_4 = \frac{1}{3}(f(0) + 4f(1) + 2f(2) + 4f(3) + f(4)) = 616/3 = 205.\overline{3}$$
$$\mathcal{M}_2 = 2(f(1) + f(3)) = 164$$
$$\mathcal{T}_2 = \frac{2}{2}(f(0) + 2f(2) + f(4)) = 288$$

53. $f(x) = 1/x$ and the partitions are $\{1, 2, 3, 4, 5, 6, 7\}$ and $\{1, 3, 5, 7\}$.

$$\mathcal{S}_6 = \frac{1}{3}(f(1) + 4f(2) + 2f(3) + 4f(4) + 2f(5) + 4f(6) + f(7)) = \frac{617}{315} = 1.9587\ldots$$
$$\mathcal{M}_3 = 2(f(2) + f(4) + f(6)) = \frac{11}{6} = 1.8\overline{3}$$
$$\mathcal{T}_3 = \frac{2}{2}(f(1) + 2f(3) + 2f(5) + f(7)) = \frac{232}{105} = 2.2095\ldots$$

55. Recall that if $p'(t) = v(t)$ is the velocity of the car, then $\int_a^b v(t)\,dt = p(b) - p(a)$ is its displacement for $a \le t \le b$. Measuring time in hours ($\Delta t = 1/60$), the Simpson Rule estimate for $\int_0^{6/60} v(t)\,dt$ is

$$\frac{1/60}{3}\Big(90 + 4(80) + 2(75) + 4(80) + 2(80) + 4(70) + 60\Big) = 7.\overline{6}\ \text{km}\,.$$

CALCULATOR/COMPUTER EXERCISES

57. The graph of $y = f''(x)$ is on the right. $C = 0.022$ has the property that $C \ge |f''(x)|$ for $x \in [1.25, 4]$. Consequently, an absolute error bound for the Midpoint Approximation using N subintervals is $E_N = \frac{0.019}{N^2}$.

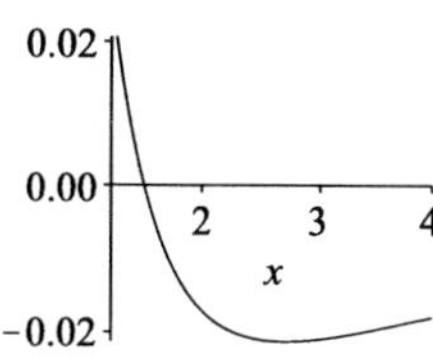

$E_N \approx 5 \cdot 10^{-4}$ when $N = 6.17\ldots$ implying that $\mathcal{M}_7 = 3.876\ldots$ is a three decimal approximation to integral.

59. The graph of $y = f''(x)$ is on the right. $C = 0.25$ has the property that $C \geq |f''(x)|$ for $x \in [1,5]$. Consequently, an absolute error bound for the Midpoint Approximation using N subintervals is $E_N = \frac{0.67}{N^2}$.

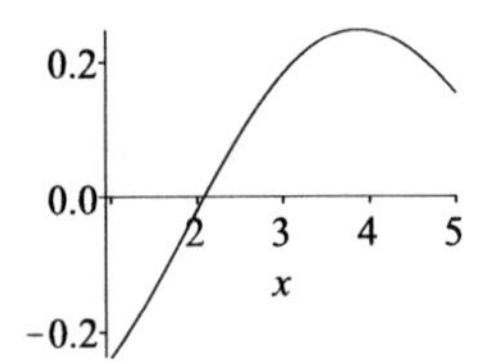

$E_N \approx 5 \cdot 10^{-4}$ when $N = 36.51\ldots$ implying that $\mathcal{M}_{37} = 0.603\ldots$ is a three decimal approximation to integral.

61. Using *Maple* we define a function $S(N)$ that outputs the Nth Simpson Rule approximation for the function f on the given interval. We find that $S(10) = 9.037913573$, $S(20) = 9.037285340$, and $S(30) = 9.037284330$. Rounded to four decimal places our estimate is 9.0373.

63. Using *Maple* we define a function $S(N)$ that outputs the Nth Simpson Rule approximation for the function f on the given interval. We find that $S(10) = 13.98241795$, $S(20) = 13.98286919$, $S(30) = 13.98289100$, and $S(40) = 13.98289468$. Rounded to four decimal places our estimate is 13.9829.

65. Calculate $A = \int_1^4 \sqrt{x}\,dx = 14/3$ and

$$\mathcal{S}_6 = \frac{1}{2} + \frac{1}{3}\left(\sqrt{6} + \sqrt{10} + \sqrt{14} + \sqrt{2} + \sqrt{3}\right).$$

The absolute error is $\epsilon = |A - \mathcal{S}_6| \approx 0.0001036141$ and, using *Maple*'s *fsolve* procedure, $\epsilon = (b-a)^5 \cdot |f^{(4)}(c)|/(180 \cdot 6^4)$ when $c = 1.898305267$.

67. Calculate $A = \int_1^8 (1+x)^{1/3}\,dx = \frac{27}{4} \cdot 9^{1/3} - \frac{3}{2} \cdot 2^{1/3}$ and

$$\mathcal{S}_4 = \frac{7}{12}\left(2^{1/3} + 9^{1/3} + 4^{2/3}(15^{1/3} + 29^{1/3}) + 11^{1/3} \cdot 2^{2/3}\right).$$

The absolute error is $\epsilon = |A - \mathcal{S}_4| \approx 0.001850859$ and, using *Maple*'s *fsolve* procedure, $\epsilon = (b-a)^5 \cdot |f^{(4)}(c)|/(180 \cdot 4^4)$ when $c = 3.210462718$.

69. The $N = 6$ Simpson's Rule approximation simplifies to

$$\mathcal{S}_6 = \frac{1}{9}\sqrt{\frac{2}{\pi}}\left(2 + e^{-1/2} + 4e^{-2/9} + 2e^{-1/18}\right) \approx 0.6827586136.$$

71. Using the *PolynomialInterpolation* procedure (*Maple*) we find that

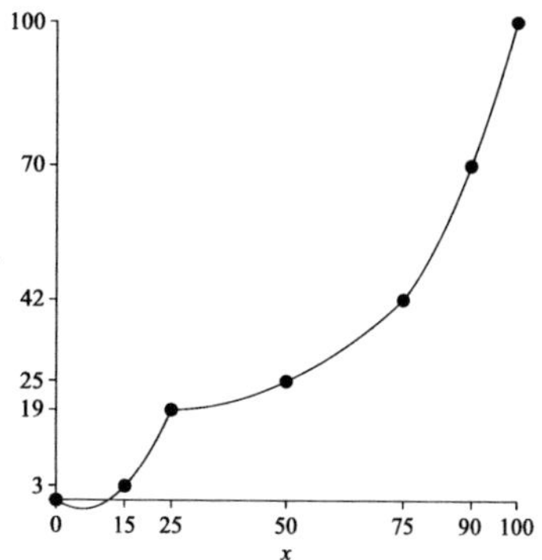

$$p_1(x) = \frac{7}{125}x^2 - \frac{16}{25}x\,,$$
$$p_2(x) = \frac{11}{1250}x^2 - \frac{21}{50}x + 24\,,$$
$$p_3(x) = \frac{17}{375}x^2 - \frac{421}{75}x + 208\,.$$

See the picture. Calculate:

$$\int_0^{25} p_1(x)\,dx = 275/3\,, \quad \int_{25}^{75} p_2(x)\,dx = 4025/3\,, \quad \int_{75}^{100} p_3(x)\,dx = 2985/18\,,$$

and $\int_0^{100} L(x)\,dx \approx 55625/18 \approx 3090.27\ldots$.

73. (a) $\mathcal{S}_{20} \approx 364.9962047$. The exact value should be 365 (the number of days in a year).

(b) $F(36) \approx 361.6260813$ using the $N = 20$ Simpson's Rule approximation. Therefore, there are 3 or 4 heat alerts a year.

(c) With $\mu = 12$ and a temperature variation $-21 \leq T \leq 41$, $F(36) \approx 360.2884420$ using the $N = 20$ Simpson's Rule approximation. Therefore there are, on average, about 1.4 more heat alerts a year. This is roughly a 40% increase in heat alerts per year.

REVIEW EXERCISES FOR CHAPTER 5

1. $1 + 4 + 9 + \cdots + 25^2 = \sum_{j=1}^{25} j^2$

3. $1/2 + 1/3 + 1/4 + \cdots + 1/100 = \sum_{j=2}^{100}(1/j)$

5. Since the selection is $\mathcal{S} = \{3, 4, 5\}$ and $\Delta x = 1$,

$$\mathcal{R}(f, \mathcal{S}) = 3^3 + 4^3 + 5^3 = 216\,.$$

7. Since the selection is $\mathcal{S} = \{2, 3, 4\}$ and $\Delta x = 1$,

$$\mathcal{R}(f, \mathcal{S}) = \frac{1}{3} + \frac{1}{2} + \frac{3}{5} = \frac{43}{30}\,.$$

9. Since $\Delta x = \pi/3$,

$$\mathcal{R}(f, \mathcal{S}) = \frac{\pi}{3}\big(\sin(\pi/6) + \sin(\pi/2)\big) = \frac{\pi}{3}\left(\frac{1}{2} + 1\right) = \frac{\pi}{2}\,.$$

11. Since $\Delta x = 1/2$,

$$\mathcal{R}(f,\mathcal{S}) = \frac{1}{2}\left(f(5/4) + f(7/4) + f(9/4) + f(11/4)\right) = \frac{1}{2}(25 + 49 + 81 + 121) = 138\,.$$

13. $\int_1^3 6x^2\,dx = (2x^3)\Big|_1^3 = 54 - 2 = 52$

15. $\int_{-2}^{-1} 1/x\,dx = \ln(|x|)\Big|_{-2}^{-1} = 0 - \ln(2) = -\ln(2)$

17. $\int_1^4 (2x - 1/\sqrt{x})\,dx = \left(x^2 - 2\sqrt{x}\right)\Big|_1^4 = 12 - (-1) = 13$

19. $\int_{\pi/6}^{\pi/3} (4\sin(x) - \cos(x))\,dx = \left(-4\cos(x) - \sin(x)\right)\Big|_{\pi/6}^{\pi/3}$

$$= -2 - \tfrac{\sqrt{3}}{2} - (-2\sqrt{3} - \tfrac{1}{2}) = \tfrac{3}{2}(\sqrt{3} - 1)$$

21. $\int_{16}^{81} 1/x^{1/4}\,dx = \left(\tfrac{4}{3}x^{3/4}\right)\Big|_{16}^{81} = 36 - \tfrac{32}{3} = \tfrac{76}{3}$

23. $\int_1^2 (24/x^2)\,dx = \left(-\tfrac{24}{x}\right)\Big|_1^2 = -12 - (-24) = 12$

25. $\int_0^2 3^x\,dx = \left(\tfrac{3^x}{\ln(3)}\right)\Big|_0^2 = \tfrac{9}{\ln(3)} - \tfrac{1}{\ln(3)} = 8/\ln(3)$

27. $\int_{\pi/4}^{\pi/2} \csc(x)\cot\,dx = \left(-\csc(x)\right)\Big|_{\pi/4}^{\pi/2} = -1 - (-\sqrt{2}) = \sqrt{2} - 1$

29. The average value is $\frac{1}{8-1}\int_1^8 1/x^{2/3}\,dx = \frac{3}{7}$ and $f(c) = \frac{3}{7}$ when $c = \frac{7}{9}\sqrt{21}$.

31. The average value is $\frac{1}{4-1}\int_1^4 \sqrt{x}\,dx = \frac{14}{9}$ and $f(c) = \frac{14}{9}$ when $c = \frac{196}{81}$.

33. Since $F'(x) = \frac{x+3}{x+1}$, $F'(3) = \frac{3}{2}$.

35. Since $F'(x) = \frac{x^3+20}{x+2}$, $F'(2) = \frac{28}{4} = 7$.

37. Since $F'(x) = -\sin^2(\pi/x)$, $F'(4) = -1/2$.

39. Since $F'(x) = \ln(1+3x)\cdot 3$, $F'(1) = 3\ln(4) = 6\ln(2)$.

41. Since $F'(x) = -\left((6x)^2 + 6x + 2\right)^{2/3}\cdot 6$, $F'(1/3) = -6(8)^{2/3} = -24$.

43. Since $F'(x) = \sqrt{5+x^2} - \sqrt{5+x^2}\cdot(-1) = 2\sqrt{5+x^2}$, $F'(2) = 6$.

45. $\int_1^{\alpha\beta}(1/t)\,dt = \int_1^{\alpha}(1/t)\,dt + \int_{\alpha}^{\alpha\beta}(1/t)\,dt \overset{(t=\alpha u)}{=} 3 + \int_1^{\beta}(1/u)\,du = 3 + 6 = 9$

47. $\int_{\alpha}^{\beta}(1/t)\,dt = \int_{\alpha}^{1}(1/t)\,dt + \int_1^{\beta}(1/t)\,dt = -3 + 6 = 3$

49. $\int_1^{\alpha^4}(1/t)\,dt \overset{(t=u^4)}{=} \int_1^{\alpha}(1/u^4)\cdot(4u^3)\,du = 4\int_1^{\alpha}(1/u)\,du = 12$

51. Substitute $u = \sin(x)$, $du = \cos(x)\,dx$ to obtain

$$\int \frac{\cos(x)}{\sin^3(x)}\,dx = \int \frac{1}{u^3}\,du = -\frac{1}{2u^2} + C = -\frac{1}{2}\csc^2(x) + C\,.$$

53. Substitute $u = 2 + x^4$, $du = 4x^3\,dx$ to obtain

$$6\int 4x^3(2+x^4)^{-1/2}\,dx = 6\int u^{-1/2}\,du = 12u^{1/2} + C = 12\sqrt{2+x^4} + C\,.$$

55. Substitute $u = 3 + 2\cos(x)$, $du = -2\sin(x)\,dx$ to obtain

$$-\frac{1}{2}\int \frac{-2\sin(x)}{3+2\cos(x)}\,dx = -\frac{1}{2}\int \frac{1}{u}\,du = -\frac{1}{2}\ln(|u|)+C = -\frac{1}{2}\ln(3+2\cos(x))+C\,.$$

57. Substitute $u = \sqrt{x} + 1$, $du = \frac{1}{2\sqrt{x}}\,dx$ to obtain

$$2\int \frac{(\sqrt{x}+1)^{1/3}}{2\sqrt{x}}\,dx = 2\int u^{1/3}\,du = \frac{3}{2}u^{4/3} + C = \frac{3}{2}(\sqrt{x}+1)^{4/3} + C\,.$$

59. Substitute $u = \ln(x)$, $du = \frac{1}{x}\,dx$ to obtain

$$2\int \frac{1}{x\ln^2(x)}\,dx = 2\int \frac{1}{u^2}\,du = -2\cdot\frac{1}{u} + C = -\frac{2}{\ln(x)} + C\,.$$

61. Substitute $u = 4x + 1$, $du = 4\,dx$, and $x = (u-1)/4$ to obtain

$$3\int 4x\sqrt{4x+1}\,dx = 3\int \frac{u-1}{4}\cdot\sqrt{u}\,du = \frac{3}{4}\int (u^{3/2} - u^{1/2})\,du = \frac{3}{4}\left(\frac{2}{5}u^{5/2} - \frac{2}{3}u^{3/2}\right) + C$$
$$= \frac{1}{10}u^{3/2}\,(3u-5) + C = \frac{1}{10}(4x+1)^{3/2}\,(12x-2) + C = \frac{1}{5}(4x+1)^{3/2}\,(6x-1) + C\,.$$

63. Substitute $u = x + 2$, $du = dx$, and $x = u - 2$ to obtain

$$\int \frac{x}{\sqrt{x+2}}\,dx = \int \frac{u-2}{u^{1/2}}\,du = \int (u^{1/2} - 2u^{-1/2})\,du = \frac{2}{3}u^{3/2} - 4u^{1/2} + C$$
$$= \frac{2}{3}u^{1/2}\,(u-6) + C = \frac{2}{3}(x-4)\sqrt{x+2} + C\,.$$

65. Substitute $u = \arctan(x)$, $du = \frac{1}{1+x^2}\,dx$ to obtain

$$\int \frac{\arctan^2(x)}{1+x^2}\,dx = \int u^2\,du = \frac{1}{3}u^3 = \frac{1}{3}\arctan^3(x) + C\,.$$

67. Substitute $u = \arcsin(x)$, $du = \frac{1}{\sqrt{1-x^2}}\,dx$ to obtain

$$\int \frac{\arcsin(x)}{\sqrt{1-x^2}}\,dx = \int u\,du = \frac{1}{2}u^2 + C = \frac{1}{2}\arcsin^2(x) + C\,.$$

69. Split the integral in two and in the second one substitute $u = 1 - x^2$, $du = -2x\,dx$ to obtain

$$\int \frac{1-2x}{\sqrt{1-x^2}}\,dx = \int \frac{1}{\sqrt{1-x^2}}\,dx + \int \frac{-2x}{\sqrt{1-x^2}}\,dx = \arcsin(x) + \int \frac{1}{\sqrt{u}}\,du$$
$$= \arcsin(x) + 2\sqrt{u} + C = \arcsin(x) + 2\sqrt{1-x^2} + C\,.$$

71. $\int_0^{\pi/9} 24\tan(3x)\,dx = 8\ln(|\sec(3x)|)\big|_0^{\pi/9} = 8\ln(2)$

73. $\int_0^{\pi/2} \sec(x/3)\,dx = 3\ln(|\sec(x) + \tan(x)|)\big|_0^{\pi/2} = 3\ln(\sqrt{3}\,) = \frac{3}{2}\ln(3)$

75. Substitute $u = 9 + x^4$, $du = 4x^3\,du$.

$$\frac{1}{4}\int_0^2 4x^3\sqrt{9+4x^2}\,dx = \frac{1}{4}\int_9^{25} \sqrt{u}\,du = \left(\frac{1}{6}u^{3/2}\right)\bigg|_9^{25} = \frac{125}{6} - \frac{27}{6} = \frac{49}{3}$$

77. The graphs intersect when $4 - x^2 = 3x$. That is, $x^2 + 3x - 4 = 0$, or $(x+4)(x-1) = 0$, so $x = -4$ and $x = 1$. Since $f(0) > g(0)$, the area is $\int_{-4}^{1}(f(x) - g(x))\,dx = 125/6$.

79. The graphs intersect when $-x^4 + x^2 + 16 = 2x^4 - 2x^2 - 20$. This simplifies to $x^4 - x^2 - 12 = 0$, or $(x^2-4)(x^2+3) = 0$, so $x = \pm 2$. Since $f(0) > g(0)$, the area is $\int_{-2}^{2}(f(x) - g(x))\,dx = 608/5$.

81. The graphs intersect when $x^3 - 2x^2 - 3x = 0$. That is, $x(x+1)(x-3) = 0$, so $x = -1$, $x = 0$, and $x = 3$. Since f is positive on $[-1, 0]$ and negative on $[0, 3]$, the area is $\int_{-1}^{0} f(x)\,dx - \int_0^3 f(x)\,dx = 71/6$.

83. The graphs intersect when $x^3 - x^2 - 9x + 9 = 0$. One solution is $x = 1$ and long division yields two more: $x = -3$ and $x = 3$. Since $\int_{-3}^{1}(f(x) - g(x)\,dx = 128/3$ and $\int_1^3(f(x) - g(x)\,dx = -20/3$, the area is $\frac{128}{3} + \frac{20}{3} = \frac{148}{3}$.

85. The partition is $\{-1, 1, 3\}$ and $\Delta x = 2$.

$$\mathcal{M}_2 = 2\big(f(0) + f(2)\big) = 2(8 + 0) = 16$$
$$\mathcal{T}_2 = \frac{2}{2}\big(f(-1) + 2f(1) + f(3)\big) = 9 + 14 - 19 = 4$$
$$\mathcal{S}_2 = \frac{2}{3}\big(f(-1) + 4f(1) + f(3)\big) = \frac{2}{3}(9 + 28 - 19) = 12$$

87. The partition is $\{0, 1/2, 1, 3/2, 2\}$ and $\Delta x = 1/2$.

$$\mathcal{M}_4 = \frac{1}{2}\big(f(1/4) + f(3/4) + f(5/4) + f(7/4)\big) = \frac{57}{64}$$
$$\mathcal{T}_4 = \frac{1}{4}\big(f(0) + 2f(1/2) + 2f(1) + 2f(3/2) + f(2)\big) = \frac{101}{105}$$
$$\mathcal{S}_4 = \frac{1}{6}\big(f(0) + 4f(1/2) + 2f(1) + 4f(3/2) + f(2)\big) = \frac{292}{315}$$

89. For the Midpoint Rule Approximation $\mathcal{M}_N$, N equal subintervals,

$$\left|\int_a^b f(x)\,dx - \mathcal{M}_N\right| \le \frac{C}{24}\cdot\frac{(b-a)^3}{N^2} = E_N\,,$$

provided that $C \ge |f''(x)|$ for all $x \in [a,b]$. Since $f''(x) = -\frac{1}{4(5+x)^{3/2}}$, $C = |f''(-1)| \approx 0.032$, and the error bound is $E_{10} \approx 0.0225$. Consequently, $37.3183\ldots \le \int_{-1}^{11} f(x)\,dx \le 37.3633\ldots$. These are the values for α and β.

91. For the Midpoint Rule Approximation $\mathcal{M}_N$, N equal subintervals,

$$\left|\int_a^b f(x)\,dx - \mathcal{M}_N\right| \le \frac{C}{24}\cdot\frac{(b-a)^3}{N^2} = E_N\,,$$

provided that $C \ge |f''(x)|$ for all $x \in [a,b]$. Since $f''(x) = -\frac{2}{9(1+x)^{5/3}}$, $C = |f''(0)| = 2/9$, and the error bound is $E_{10} \approx 0.03176$. Consequently, $11.2232\ldots \le \int_0^7 f(x)\,dx \le 11.2867\ldots$. These are the values for α and β.

93. The following table contains the xy data.

x	0	0.85	1.70	2.55	3.40	4.25	5.10
y	0	8.67	5.79	7.02	15.22	20.46	0

Since $\Delta x = 0.85$, the Trapezoidal and Simpson's Rule area estimates are

$$\mathcal{T}_6 = \frac{0.85}{2}\big(0 + 2(8.67) + 2(5.79) + 2(7.02) + 2(15.22) + 2(20.46) + 0\big) = 48.59 \text{ km}^2$$
$$\mathcal{S}_6 = \frac{0.85}{3}\big(0 + 4(8.67) + 2(5.79) + 4(7.02) + 2(15.22) + 4(20.46) + 0\big) = 52.88 \text{ km}^2\,.$$

95. Using time intervals of 0.5 seconds, the Simpson's Rule estimate for $\int_5^{10} v(t)\,dt$ is

$$\mathcal{S}_{10} = \frac{0.5}{3}\big(10.66 + 4(10.69) + 2(10.86) + 4(10.98) + \cdots + 2(10.64) + 4(10.58) + 10.43\big)$$
$$= 53.87 \text{ m}\,.$$

97. First make a slight simplification. Then substitute $u = 2x+1$, $du = 2\,dx$, and $2x+3 = u+2$, and use Formula 20 with $a = 2$ and $b = 1$.

$$\int \frac{2}{(2x+1)(x+3/2)^2}\,dx = 4\int \frac{2}{(2x+1)(2x+3)^2}\,dx = 4\int \frac{1}{u(u+2)^2}\,du$$
$$= -\left(\ln\left(\left|\frac{u+2}{u}\right|\right) + \frac{u}{u+2}\right) + C = -\left(\ln\left(\left|\frac{2x+3}{2x+1}\right|\right) + \frac{2x+1}{2x+3}\right) + C$$

The antiderivative can be simplified in two ways. First of all, the minus sign can be absorbed into the log term by inverting the argument. Secondly, the fact that $\frac{2x+1}{2x+3} = 1 - \frac{2}{2x+3}$ means that the second term can be replaced with $-\frac{2}{2x+3}$. Doing this yields the following integral formula.

$$\int \frac{2}{(2x+1)(x+3/2)^2}\,dx = \frac{2}{2x+3} + \ln\left(\left|\frac{2x+1}{2x+3}\right|\right) + C$$

99. Substitute $u = \exp(t)$, $du = \exp(t)\,dt$, $dt = \frac{1}{u}\,du$, $\exp(3t) = u^3$, and $\exp(2t) = u^2$. Then use Formula 48 with $a = 2$.

$$\int \frac{\exp(3t)}{4-\exp(2t)}\,dt = \int \frac{u^3}{4-u^2}\cdot\frac{1}{u}\,du = \int \frac{u^2}{4-u^2}\,du = -u + \ln\left(\left|\frac{2+u}{2-u}\right|\right) + C$$
$$= \ln\left(\left|\frac{\exp(t)+2}{\exp(t)-2}\right|\right) - \exp(t) + C$$

Chapter 6

Techniques of Integration

6.1 Integration by Parts

Problems for Practice

1. Let $u = x$ and $dv = e^x\,dx$. Then $du = dx$ and $v = e^x$ so

$$\int xe^x\,dx = xe^x - \int e^x\,dx = xe^x - e^x + C = (x-1)e^x + C\,.$$

3. Let $u = 2x+5$ and $dv = e^{x/3}\,dx$. Then $du = 2\,dx$ and $v = 3e^{x/3}$ so

$$\begin{aligned}\int (2x+5)e^{x/3}\,dx &= 3(2x+5)e^{x/3} - \int 6e^{x/3}\,dx \\ &= 3(2x+5)e^{x/3} - 18e^{x/3}\,dx + C = 3(2x-1)e^{x/3} + C\,.\end{aligned}$$

5. Let $u = 4x+2$ and $dv = \sin(2x)\,dx$. Then $du = 4\,dx$ and $v = -\frac{1}{2}\cos(2x)$ so

$$\begin{aligned}\int (4x+2)\sin(2x)\,dx &= -\frac{1}{2}(4x+2)\cos(2x) + 4\int \frac{1}{2}\cos(2x)\,dx \\ &= -(2x+1)\cos(2x) + \sin(2x) + C\,.\end{aligned}$$

7. Let $u = \ln(4x)$ and $dv = x\,dx$. Then $du = \frac{1}{x}\,dx$ and $v = \frac{1}{2}x^2$ so

$$\begin{aligned}\int x\ln(4x)\,dx &= \frac{1}{2}x^2\ln(4x) - \frac{1}{2}\int x\,dx \\ &= \frac{1}{2}x^2\ln(4x) - \frac{1}{4}x^2 + C = \frac{1}{4}x^2(2\ln(4x) - 1) + C\,.\end{aligned}$$

9. Let $u = \ln(x)$ and $dv = 9x^2\,dx$. Then $du = \frac{1}{x}\,dx$ and $v = 3x^3$ so

$$\begin{aligned}\int 9x^2 \ln(x)\,dx &= 3x^3 \ln(x) - 3\int x^2\,dx \\ &= 3x^3\ln(x) - x^3 + C = x^3(3\ln(x) - 1) + C\,.\end{aligned}$$

11. Let $u = x$ and $dv = e^x\,dx$. Then $du = dx$ and $v = e^x$ so

$$\begin{aligned}\int_0^1 xe^x\,dx &= xe^x\Big|_0^1 - \int_0^1 e^x\,dx = xe^x\Big|_0^1 - e^x\Big|_0^1 \\ &= e - (e-1) = 1\,.\end{aligned}$$

13. Let $u = x$ and $dv = \cos(x)\,dx$. Then $du = dx$ and $v = \sin(x)$ so

$$\begin{aligned}\int_0^\pi x\cos(x)\,dx &= x\sin(x)\Big|_0^\pi - \int_0^\pi \sin(x)\,dx = x\sin(x)\Big|_0^\pi + \cos(x)\Big|_0^\pi \\ &= 0 + (-1-1) = -2\,.\end{aligned}$$

15. Let $u = 4x$ and $dv = \sin(x/3)\,dx$. Then $du = 4\,dx$ and $v = -3\cos(x/3)$ so

$$\begin{aligned}\int_0^{\pi/2} 4x\sin(x/3)\,dx &= -12x\cos(x/3)\Big|_0^{\pi/2} + 12\int_0^{\pi/2} \cos(x/3)\,dx \\ &= -12x\cos(x/3)\Big|_0^{\pi/2} + 36\sin(x/3)\Big|_0^{\pi/2} = -3\pi\sqrt{3} + 18 = 18 - 3\sqrt{3}\pi\,.\end{aligned}$$

17. Let $u = \ln(x)$ and $dv = x\,dx$. Then $du = \frac{1}{x}\,dx$ and $v = \frac{1}{2}x^2$ so

$$\begin{aligned}\int_1^e x\ln(x)\,dx &= \frac{1}{2}x^2\ln(x)\Big|_1^e - \frac{1}{2}\int_1^e x\,dx = \frac{1}{2}x^2\ln(x)\Big|_1^e - \frac{1}{4}x^2\Big|_1^e \\ &= \frac{1}{2}e^2 - \left(\frac{1}{4}e^2 - \frac{1}{4}\right) = \frac{1}{4}(e^2+1)\,.\end{aligned}$$

19. Let $u = \ln(x)$ and $dv = x^{-1/2}\,dx$. Then $du = \frac{1}{x}\,dx$ and $v = 2x^{1/2}$ so

$$\begin{aligned}\int_1^4 \frac{\ln(x)}{\sqrt{x}}\,dx &= 2x^{1/2}\ln(x)\Big|_1^4 - 2\int_1^4 x^{-1/2}\,dx = 2x^{1/2}\ln(x)\Big|_1^4 - 4x^{1/2}\Big|_1^4 \\ &= 4\ln(4) - (8-4) = 8\ln(2) - 4\,.\end{aligned}$$

21. Let $u = x^2$ and $dv = e^x\,dx$. Then $du = 2x\,dx$ and $v = e^x$ so

$$\int x^2e^x\,dx = x^2e^x - \int 2xe^x\,dx\,.$$

Continue with $u = 2x$, $dv = e^x\,dx$, and $du = 2\,dx$, $v = e^x$, to obtain

$$\int x^2 e^x\,dx = x^2e^x - \left(2xe^x - \int 2e^x\,dx\right)$$
$$= x^2e^x - 2xe^x + 2e^x + C = (x^2 - 2x + 2)e^x + C\,.$$

23. Let $u = x^2$ and $dv = \cos(x)\,dx$. Then $du = 2x\,dx$ and $v = \sin(x)$ so

$$\int x^2\cos(x)\,dx = x^2\sin(x) - \int 2x\sin(x)\,dx\,.$$

Continue with $u = 2x$, $dv = \sin(x)\,dx$, and $du = 2\,dx$, $v = -\cos(x)$, to obtain

$$\int x^2\cos(x)\,dx = x^2\sin(x) - \left(-2x\cos(x) + \int 2\cos(x)\,dx\right)$$
$$= x^2\sin(x) + 2x\cos(x) - 2\sin(x) + C = (x^2 - 2)\sin(x) + 2x\cos(x) + C\,.$$

25. Let $u = 4x^2$ and $dv = \cos(2x)\,dx$. Then $du = 8x\,dx$ and $v = \frac{1}{2}\sin(2x)$ so

$$\int 4x^2\cos(2x)\,dx = 2x^2\sin(2x) - \int 4x\sin(2x)\,dx\,.$$

Continue with $u = 4x$, $dv = \sin(2x)\,dx$, and $du = 4\,dx$, $v = -\frac{1}{2}\cos(2x)$, to obtain

$$\int 4x^2\cos(2x)\,dx = 2x^2\sin(2x) - \left(-2x\cos(2x) + \int 2\cos(2x)\,dx\right)$$
$$= 2x^2\sin(2x) + 2x\cos(2x) - \sin(2x) = (2x^2 - 1)\sin(2x) + 2x\cos(2x) + C\,.$$

27. Let $u = x^3$ and $dv = e^x\,dx$. Then $du = 3x^2\,dx$ and $v = e^x$ so

$$\int x^3e^x\,dx = x^3e^x - \int 3x^2e^x\,dx\,.$$

Continue with $u = 3x^2$, $dv = e^x\,dx$, and $du = 6x\,dx$, $v = e^x$, to obtain

$$\int x^3e^x\,dx = x^3e^x - \left(3x^2e^x - \int 6xe^x\,dx\right)$$
$$= x^3e^x - 3x^2e^x + \int 6xe^x\,dx\,.$$

Finish with $u = 6x$, $dv = e^x\,dx$, and $du = 6\,dx$, $v = e^x$ to obtain

$$\int x^3e^x\,dx = x^3e^x - 3x^2e^x + \left(6xe^x - \int 6e^x\,dx\right)$$
$$= x^3e^x - 3x^2e^x + 6xe^x - 6e^x + C$$
$$= (x^3 - 3x^2 + 6x - 6)e^x + C$$

29. Let $u = \ln^2(x)$ and $dv = 16x^3\,dx$. Then $du = 2\ln(x)\cdot\frac{1}{x}\,dx$ and $v = 4x^4$ so

$$\int 16x^3\ln^2(x)\,dx = 4x^4\ln^2(x) - \int 8x^3\ln(x)\,dx\,.$$

Continue with $u = \ln(x)$, $dv = 8x^3\,dx$, and $du = \frac{1}{x}\,dx$, $v = 2x^4$, to obtain

$$\begin{aligned}\int 16x^3\ln^2(x)\,dx &= 4x^4\ln^2(x) - \left(2x^4\ln(x) - \int 2x^3\,dx\right)\\ &= 4x^4\ln^2(x) - 2x^4\ln(x) + \frac{1}{2}x^4 + C = 2x^4\ln(x) - \frac{1}{4}x^4 + C\\ &= x^4\left(4\ln^2(x) - 2\ln(x) + \frac{1}{2}\right) + C\,.\end{aligned}$$

31. Let $u = \arctan(x)$ and $dv = 2\,dx$. Then $du = \frac{1}{1+x^2}\,dx$ and $v = 2x$ so

$$\int_0^1 2\arctan(x)\,dx = 2x\arctan(x)\Big|_0^1 - \int_0^1 \frac{2x}{1+x^2}\,dx\,.$$

Continue with the substitution $u = 1+x^2$, $dv = 2x\,dx$ to obtain

$$\begin{aligned}\int_0^1 2\arctan(x)\,dx &= 2x\arctan(x)\Big|_0^1 - \int_1^2 \frac{1}{u}\,du\\ &= 2x\arctan(x)\Big|_0^1 - \ln(u)\Big|_1^2\\ &= 2\cdot\frac{\pi}{4} - \ln(2) = \frac{\pi}{2} - \ln(2)\,.\end{aligned}$$

33. Let $u = \arcsin(x/2)$ and $dv = dx$. Then $du = \frac{1}{\sqrt{1-x^2/4}}\cdot\frac{1}{2}\,dx$ and $v = x$ so

$$\int_0^{\sqrt{3}} \arcsin(x/2)\,dx = x\arcsin(x/2)\Big|_0^{\sqrt{3}} - \int_0^{\sqrt{3}} \frac{x}{\sqrt{4-x^2}}\,dx\,.$$

Continue with the substitution $u = 4-x^2$, $dv = -2x\,dx$ to obtain

$$\begin{aligned}\int_0^{\sqrt{3}} \arcsin(x/2)\,dx &= x\arcsin(x/2)\Big|_0^{\sqrt{3}} - \int_4^1 \frac{1}{\sqrt{u}}\cdot\left(-\frac{1}{2}\right)\,du\\ &= x\arcsin(x/2)\Big|_0^{\sqrt{3}} + \sqrt{u}\Big|_4^1\\ &= \sqrt{3}\cdot\frac{\pi}{3} + (1-2) = \frac{1}{3}\pi\sqrt{3} - 1\,.\end{aligned}$$

35. Let $u = \arccos(x)$ and $dv = 2\,dx$. Then $du = -\frac{1}{\sqrt{1-x^2}}\,dx$ and $v = 2x$ so

$$\int_{1/2}^{1} 2\arccos(x)\,dx = 2x\,\arccos(x)\Big|_{1/2}^{1} + \int_{1/2}^{1} \frac{2x}{\sqrt{1-x^2}}\,dx\,.$$

Continue with the substitution $u = 1 - x^2$, $du = -2x\,dx$ to obtain

$$\begin{aligned}\int_{1/2}^{1} 2\arccos(x)\,dx &= 2x\ \arccos(x)\Big|_{1/2}^{1} - \int_{3/4}^{0} \frac{1}{\sqrt{u}}\,du\\ &= 2x\ \arccos(x)\Big|_{1/2}^{1} - 2\sqrt{u}\Big|_{3/4}^{0}\\ &= 0 - \frac{\pi}{3} - (0 - \sqrt{3}\,) = \sqrt{3} - \frac{\pi}{3}\,.\end{aligned}$$

37. Using the reduction formula, $a = 1/2$,

$$\begin{aligned}\int_0^1 x^2 e^{x/2}\,dx &= 2x^2 e^{x/2}\Big|_0^1 - 4\int_0^1 x e^{x/2}\,dx\\ &= 2x^2 e^{x/2}\Big|_0^1 - 4\left(2xe^{x/2}\Big|_0^1 - 2\int_0^1 e^{x/2}\,dx\right)\\ &= 2x^2 e^{x/2}\Big|_0^1 - 8xe^{x/2}\Big|_0^1 + 16e^{x/2}\Big|_0^1\\ &= (2e^{1/2} - 0) - (8e^{1/2} - 0) + (16e^{1/2} - 16)\\ &= 10\sqrt{e} - 16\,.\end{aligned}$$

39. Using the reduction formula, $a = 1$,

$$\begin{aligned}\int_0^1 x^3 e^x\,dx &= x^3 e^x\Big|_0^1 - 3\int_0^1 x^2 e^x\,dx\\ &= x^3 e^x\Big|_0^1 - 3\left(x^2 e^x\Big|_0^1 - 2\int_0^1 xe^x\,dx\right)\\ &= x^3 e^x\Big|_0^1 - 3x^2 e^x\Big|_0^1 + 6\left(xe^x\Big|_0^1 - \int_0^1 e^x\,dx\right)\\ &= x^3 e^x\Big|_0^1 - 3x^2 e^x\Big|_0^1 + 6xe^x\Big|_0^1 - 6e^x\Big|_0^1\\ &= e - 3e + 6e - (6e - 6) = 6 - 2e\,.\end{aligned}$$

Further Theory and Practice

41. Simplify then let $u = \ln(x)$, $dv = \frac{1}{2}\,dx$, and $du = \frac{1}{x}\,dx$, $v = \frac{1}{2}x$.

$$\begin{aligned}\int \ln(\sqrt{x}\,)\,dx &= \int \frac{1}{2}\ln(x)\,dx = \frac{1}{2}x\ln(x) - \int \frac{1}{2}\,dx \\ &= \frac{1}{2}x\ln(x) - \frac{1}{2}x + C = \frac{1}{2}x\big(\ln(x) - 1\big) + C\end{aligned}$$

43. Simplify then let $u = \ln(x)$, $dv = -x^3\,dx$, and $du = \frac{1}{x}\,dx$, $v = -\frac{1}{4}x^4$.

$$\begin{aligned}\int x^3\ln(1/x)\,dx &= \int -x^3\ln(x)\,dx = -\frac{1}{4}x^4\ln(x) + \int \frac{1}{4}x^3\,dx \\ &= -\frac{1}{4}x^4\ln(x) + \frac{1}{16}x^4 + C = \frac{1}{16}x^4(1 - 4\ln(x)) + C\end{aligned}$$

45. Let $u = x$, $dv = 2^{-x}\,dx$, and $du = dx$, $v = -\frac{2^{-x}}{\ln(2)}$, and proceed as follows.

$$\begin{aligned}\int x\,2^{-x}\,dx &= -\frac{x\,2^{-x}}{\ln(2)} + \frac{1}{\ln(2)}\int 2^{-x}\,dx \\ &= -\frac{x\,2^{-x}}{\ln(2)} - \frac{2^{-x}}{\ln^2(2)} + C = -\frac{2^{-x}}{\ln^2(2)}(x\ln(2) + 1) + C\end{aligned}$$

47. Let $u = \ln^2(x)$, $dv = x^{-2}\,dx$, and $du = 2\ln(x)\cdot\frac{1}{x}\,dx$, $v = -x^{-1}$, and proceed as follows. In the second integral let $u = \ln(x)$, $dv = 2x^{-2}\,dx$, and $du = \frac{1}{x}\,dx$, $v = -2x^{-1}$.

$$\begin{aligned}\int x^{-2}\ln^2(x)\,dx &= -\frac{\ln^2(x)}{x} + \int 2x^{-2}\ln(x)\,dx \\ &= -\frac{\ln^2(x)}{x} - \frac{2\ln(x)}{x} + \int 2x^{-2}\,dx \\ &= -\frac{\ln^2(x)}{x} - \frac{2\ln(x)}{x} - \frac{2}{x} + C \\ &= -\frac{\ln^2(x) + 2\ln(x) + 2}{x} + C\,.\end{aligned}$$

49. Let $u = x$, $dv = \sec(x)\tan(x)\,dx$, and $du = dx$, $v = \sec(x)$, and proceed as follows.

$$\begin{aligned}\int x\sec(x)\tan(x)\,dx &= x\sec(x) - \int \sec(x)\,dx \\ &= x\sec(x) - \ln(|\sec(x) + \tan(x)|) + C\end{aligned}$$

51. Let $u = x$, $dv = \cos(x + \pi/4)\,dx$, and $du = dx$, $v = \sin(x + \pi/4)$, and proceed as follows.

$$\begin{aligned}\int x\cos(x + \pi/4)\,dx &= x\sin(x + \pi/4) - \int \sin(x + \pi/4)\,dx \\ &= x\sin(x + \pi/4) + \cos(x + \pi/4) + C\end{aligned}$$

53. Let $u = \cos(\ln(x))$, $dv = 2\,dx$, and $du = -\sin(\ln(x)) \cdot \frac{1}{x}\,dx$, $v = 2x$, and proceed as follows.

$$\int 2\cos(\ln(x))\,dx = 2x\cos(\ln(x)) + \int 2\sin(\ln(x))\,dx$$

Now, in the sine integral, let $u = \sin(\ln(x))$, $dv = 2\,dx$, and $du = \cos(\ln(x))\,dx$, $v = 2x$, to obtain the equation

$$\int 2\cos(\ln(x))\,dx = 2x\cos(\ln(x)) + 2x\sin(\ln(x)) - \int 2\cos(\ln(x))\,dx\,,$$

which implies that $\int 2\cos(\ln(x))\,dx = x\big(\cos(\ln(x)) + \sin(\ln(x))\big) + C$.

55. Let $u = x$, $dv = (x+3)^{-1/2}\,dx$, and $du = dx$, $v = 2(x+3)^{1/2}$, to obtain

$$\int_{-2}^{1} x(x+3)^{-1/2}\,dx = 2x(x+3)^{1/2}\Big|_{-2}^{1} - \int_{-2}^{1} 2(x+3)^{1/2}\,dx\,.$$

The integral that remains can be evaluated by substituting $u = x + 3$. Anticipating the outcome,

$$\begin{aligned}\int_{-2}^{1} x(x+3)^{-1/2}\,dx &= 2x(x+3)^{1/2}\Big|_{-2}^{1} - \frac{4}{3}(x+3)^{3/2}\Big|_{-2}^{1} \\ &= (4-(-4)) - \left(\frac{32}{3} - \frac{4}{3}\right) = -\frac{4}{3}\,.\end{aligned}$$

57. Make the substitution $y = \sqrt{x}$, $dy = \frac{1}{2\sqrt{x}}\,dx$, and $dx = 2y\,dy$ to obtain

$$\int \sin(\sqrt{x}\,)\,dx = \int 2y\sin(y)\,dy\,.$$

Now integrate by parts: $u = 2y$, $dv = \sin(y)\,dy$ and $du = 2\,dy$, $v = -\cos(y)$.

$$\begin{aligned}\int \sin(\sqrt{x}\,)\,dx &= -2y\cos(y) + \int 2\cos(y)\,dy \\ &= -2y\cos(y) + 2\sin(y) + C \\ &= -2\sqrt{x}\cos(\sqrt{x}\,) + 2\sin(\sqrt{x}\,) + C\end{aligned}$$

59. Make the substitution $y = x^2$, $dy = 2x\,dx$ to obtain

$$\int x^2 \cdot 2x\exp(x^2)\,dx = \int y\exp(y)\,dy\,.$$

Now integrate by parts: $u = y$, $dv = \exp(y)\,dy$ and $du = dy$, $v = \exp(y)$.

$$\begin{aligned}\int x^3 \exp(x^2)\,dx &= y\exp(y) - \int \exp(y)\,dy \\ &= y\exp(y) - \exp(y) + C \\ &= x^2\exp(x^2) - \exp(x^2) + C \\ &= \exp(x^2)(x^2-1) + C\end{aligned}$$

61. Expand the integrand and apply integration by parts to the term $2xe^x$.

$$\begin{aligned}\int_0^1 (e^x - x)^2\,dx &= \int_0^1 (e^{2x} + x^2 - 2xe^x)\,dx \\ &= \left(\frac{1}{2}e^{2x} + \frac{1}{3}x^3\right)\Big|_0^1 - \int_0^1 2xe^x\,dx\end{aligned}$$

Now integrate by parts: $u = 2x$, $dv = e^x\,dx$ and $du = 2\,dx$, $v = e^x$.

$$\begin{aligned}\int_0^1 (e^x - x)^2\,dx &= \left(\frac{1}{2}e^{2x} + \frac{1}{3}x^3\right)\Big|_0^1 - \left(2xe^x\Big|_0^1 - \int_0^1 2e^x\,dx\right) \\ &= \left(\frac{1}{2}e^{2x} + \frac{1}{3}x^3\right)\Big|_0^1 - \left(2xe^x\Big|_0^1 - 2e^x\Big|_0^1\right) \\ &= \left(\frac{1}{2}e^2 + \frac{1}{3} - \frac{1}{2}\right) - (2e - (2e-2)) = \frac{1}{2}e^2 - \frac{13}{6}\end{aligned}$$

63. If $p = -1$ the integral evaluates via substitution to $\frac{1}{2}\ln^2(x) + C$ (verify). Otherwise, use integration by parts, $u = \ln(x)$, $dv = x^p\,dx$ and $du = \frac{1}{x}\,dx$, $v = \frac{1}{p+1}x^{p+1}$.

$$\begin{aligned}\int x^p \ln(x)\,dx &= \frac{1}{p+1}x^{p+1}\ln(x) - \int \frac{1}{p+1}x^p\,dx \\ &= \frac{x^{p+1}\ln(x)}{p+1} - \frac{x^{p+1}}{(p+1)^2} + C \\ &= \frac{x^{p+1}}{p+1}\left(\ln(x) - \frac{1}{p+1}\right) + C\end{aligned}$$

65. Let $u = \arctan(x/\sqrt{3}\,)$, $dv = 2x\,dx$ and $du = \frac{1}{1+x^2/3}\cdot\frac{1}{\sqrt{3}}\,dx$, $v = x^2$. The

first term on the right hand side evaluates to $\pi/6$.

$$\begin{aligned}\int_0^1 2x\arctan(x/\sqrt{3})\,dx &= x^2\arctan(x/\sqrt{3})\Big|_0^1 - \frac{3}{\sqrt{3}}\int_0^1 \frac{x^2/3}{1+x^2/3}\,dx\\ &= \frac{\pi}{6} - \frac{3}{\sqrt{3}}\int_0^1 \frac{1+x^2/3-1}{1+x^2/3}\,dx\\ &= \frac{\pi}{6} - \left(\frac{3}{\sqrt{3}}\int_0^1 dx - 3\int_0^1 \frac{1}{1+x^2/3}\cdot\frac{1}{\sqrt{3}}\,dx\right)\\ &= \frac{\pi}{6} - \frac{3}{\sqrt{3}}x\Big|_0^1 + 3\arctan(x/\sqrt{3})\Big|_0^1\\ &= \frac{\pi}{6} - \sqrt{3} + 3\cdot\frac{\pi}{6} = \frac{2\pi}{3} - \sqrt{3}\end{aligned}$$

67. Let $u = \ln(1+x^2)$, $dv = dx$ and $du = \frac{2x}{1+x^2}\,dx$, $v = x$. The first term on the right hand side evaluates to $\ln(2)$.

$$\begin{aligned}\int_0^1 \ln(1+x^2)\,dx &= x\ln(1+x^2)\Big|_0^1 - 2\int_0^1 \frac{x^2}{1+x^2}\,dx\\ &= \ln(2) - 2\int_0^1 \left(\frac{1+x^2-1}{1+x^2}\right)dx\\ &= \ln(2) - 2\int_0^1 \left(1 - \frac{1}{1+x^2}\right)dx\\ &= \ln(2) - 2(x - \arctan(x))\Big|_0^1\\ &= \ln(2) - 2\left(1 - \frac{\pi}{4}\right) = \frac{\pi}{2} + \ln(2) - 2\end{aligned}$$

69. Let $u = \ln(2+x)$, $dv = 9x^2\,dx$ and $du = \frac{1}{2+x}\,dx$, $v = 3x^3$. The first term on the right hand side evaluates to 0. Use long division to convert $x^3/(2+x)$ to an integrable expression.

$$\begin{aligned}\int_{-1}^0 9x^2\ln(2+x)\,dx &= 3x^3\ln(2+x)\Big|_{-1}^0 - 3\int_{-1}^0 \frac{x^3}{2+x}\,dx\\ &= -3\int_{-1}^0 \left(x^2 - 2x + 4 - \frac{8}{2+x}\right)dx\\ &= -3\left(\frac{1}{3}x^3 - x^2 + 4x - 8\ln(2+x)\right)\Big|_{-1}^0\\ &= 24\ln(2) - (1+3+12) = 24\ln(2) - 16\end{aligned}$$

71. Let $u = \ln^n(x)$, $dv = dx$ and $du = n\ln^{n-1}(x)\cdot\frac{1}{x}\,dx$, $v = x$.

$$\int \ln^n(x)\,dx = x\ln^n(x) - \int n\ln^{n-1}(x)\,dx$$

73. Use the fact that $\ln(x^2-1) = \ln(x-1) + \ln(x+1)$ to obtain

$$\begin{aligned}\int \ln(x^2-1)\,dx &= \int \ln(x-1)\,dx + \int \ln(x+1)\,dx \\ &= (x-1)\ln(x-1) - x + (x+1)\ln(x+1) - x + C \\ &= x\ln(x^2-1) - 2x + \ln\left(\frac{x+1}{x-1}\right) + C\end{aligned}$$

75. The graph of $y = \pi x/2$ lies above the graph of $y = \arcsin(x)$ for $0 < x < 1$ so the area is given by the following integral. Split it into two parts and use integration by parts on the second part: $u = \arcsin(x)$, $dv = dx$ and $du = \frac{1}{\sqrt{1-x^2}}\,dx$, $v = x$. Then use substitution: $u = 1-x^2$, $du = -2x\,dx$.

$$\begin{aligned}\int_0^1 (\pi x/2 - \arcsin(x))\,dx &= \int_0^1 \pi x/2\,dx - \int_0^1 \arcsin(x)\,dx \\ &= \frac{\pi x^2}{4}\bigg|_0^1 - x\arcsin(x)\bigg|_0^1 + \int_0^1 \frac{x}{\sqrt{1-x^2}}\,dx \\ &= \frac{\pi x^2}{4}\bigg|_0^1 - x\arcsin(x)\bigg|_0^1 - \sqrt{1-x^2}\bigg|_0^1 \\ &= \frac{\pi}{4} - \frac{\pi}{2} - (0-1) = 1 - \frac{\pi}{4}\end{aligned}$$

77. Split the integral into two parts and use integration by parts on the second part: $u = x$, $dv = \cos(\pi x/2)\,dx$ and $du = dx$, $v = \frac{2}{\pi}\sin(\pi x/2)$.

$$\begin{aligned}\int_{-1}^1 (\cos(\pi x/2) - x\cos(\pi x/2))\,dx &= \int_{-1}^1 \cos(\pi x/2)\,dx - \int_{-1}^1 x\cos(\pi x/2)\,dx \\ &= \frac{2}{\pi}\sin(\pi x/2)\bigg|_{-1}^1 - \left(\frac{2}{\pi}x\sin(\pi x/2)\bigg|_{-1}^1 - \frac{2}{\pi}\int_{-1}^1 \sin(\pi x/2)\,dx\right) \\ &= \frac{2}{\pi}\sin(\pi x/2)\bigg|_{-1}^1 - \frac{2}{\pi}x\sin(\pi x/2)\bigg|_{-1}^1 - \frac{4}{\pi^2}\cos(\pi x/2)\bigg|_{-1}^1 \\ &= \left(\frac{2}{\pi} - \left(-\frac{2}{\pi}\right)\right) - \left(\frac{2}{\pi} - \frac{2}{\pi}\right) - 0 = \frac{4}{\pi}\end{aligned}$$

79. Use integration by parts $u = \ln(1+x^4)$, $dv = dx$ and $du = \frac{4x^3}{1+x^4}\,dx$, $v = x$.

$$\begin{aligned}\int_0^1 (\ln(1+x^4)\,dx &= x\ln(1+x^4)\Big|_0^1 - 4\int_0^1 \frac{x^4}{1+x^4}\,dx \\ &= \ln(2) - 4\int_0^1 \frac{1+x^4-1}{1+x^4}\,dx \\ &= \ln(2) - 4\int_0^1 \left(1 - \frac{1}{1+x^4}\right)\,dx \\ &= \ln(2) - 4 + 4\alpha\end{aligned}$$

81. Use integration by parts: $u = \frac{1}{(x^2+a^2)^n}$, $dv = dx$ and $du = \frac{-2nx}{(x^2+a^2)^{n+1}}\,dx$, $v = x$ on the integral $\int \frac{1}{(x^2+a^2)^n}\,dx$ to obtain the following equation.

$$\int \frac{1}{(x^2+a^2)^n}\,dx = \frac{x}{(x^2+a^2)^n} + 2n\int \frac{x^2}{(x^2+a^2)^{n+1}}\,dx$$

Now manipulate the integral on the right hand side.

$$\begin{aligned}\int \frac{1}{(x^2+a^2)^n}\,dx &= \frac{x}{(x^2+a^2)^n} + 2n\int \frac{x^2+a^2-a^2}{(x^2+a^2)^{n+1}}\,dx \\ &= \frac{x}{(x^2+a^2)^n} + 2n\int \left(\frac{1}{(x^2+a^2)^n} - \frac{a^2}{(x^2+a^2)^{n+1}}\right)\,dx\end{aligned}$$

Finally, split the integral on the right hand side into two,

$$\int \frac{1}{(x^2+a^2)^n}\,dx = \frac{x}{(x^2+a^2)^n} + 2n\int \frac{1}{(x^2+a^2)^n}\,dx - 2na^2\int \frac{1}{(x^2+a^2)^{n+1}}\,dx\,,$$

and solve for $\int \frac{1}{(x^2+a^2)^{n+1}}\,dx$ to obtain the desired reduction formula.

83. The easiest way to obtain formula (6.1.6) is via the identity

$$\sin(ax)\sin(bx) = \frac{1}{2}\big(\cos((a-b)x) - \cos((a+b)x)\big)\,.$$

It implies that

$$\begin{aligned}\int \sin(ax)\sin(bx) &= \frac{1}{2}\int \big(\cos((a-b)x) - \cos((a+b)x)\big)\,dx \\ &= \frac{1}{2}\left(\frac{\sin((a-b)x)}{a-b} - \frac{\sin((a+b)x)}{a+b}\right) + C\,.\end{aligned}$$

This can also be obtained using integration by parts, twice. But the final form of the antiderivative requires the use of trigonometric identities similar to the one stated above.

85. Use integration by parts twice. First $u = e^{ax}$, $dv = \cos(bx)\,dx$, and $du = ae^{ax}\,dx$, $v = \frac{1}{b}\sin(bx)$.

$$\int e^{ax}\cos(bx)\,dx = \frac{1}{b}e^{ax}\sin(bx) - \frac{a}{b}\int e^{ax}\sin(bx)\,dx$$

Then do it again, with $u = e^{ax}$, $dv = \sin(bx)\,dx$, and $du = ae^{ax}\,dx$, $v = -\frac{1}{b}\cos(bx)$ to obtain the following equation.

$$\int e^{ax}\cos(bx)\,dx = \frac{1}{b}e^{ax}\sin(bx) - \frac{a}{b}\left(-\frac{1}{b}e^{ax}\cos(bx) + \frac{a}{b}\int e^{ax}\cos(bx)\,dx\right)$$

Solve for the integral it two steps. Add the constant of integration at the end.

$$\left(1 + \frac{a^2}{b^2}\right)\int e^{ax}\cos(bx)\,dx = \frac{1}{b}e^{ax}\sin(bx) + \frac{a}{b^2}e^{ax}\cos(bx)$$

$$\int e^{ax}\cos(bx)\,dx = \frac{e^{ax}\big(a\cos(bx) + b\sin(bx)\big)}{a^2 + b^2} + C$$

87. (a) Let $u = \text{FresnelS}(x)$, $dv = dx$, and $du = \sin(\pi x^2/2)\,dx$, $v = x$. Then

$$\int \text{FresnelS}(x)\,dx = x\,\text{FresnelS}(x) - \int x\sin(\pi x^2/2)\,dx$$

Now make the substitution $u = \pi x^2/2$, $du = \pi x\,dx$ to obtain

$$\begin{aligned}\int \text{FresnelS}(x)\,dx &= x\,\text{FresnelS}(x) - \int \sin(u)\cdot\frac{1}{\pi}\,du \\ &= x\,\text{FresnelS}(x) + \frac{1}{\pi}\cos(u) + C \\ &= x\,\text{FresnelS}(x) + \frac{1}{\pi}\cos(\pi x^2/2) + C\,.\end{aligned}$$

(b) Let $u = \text{FresnelC}(x)$, $dv = dx$, and $du = \cos(\pi x^2/2)\,dx$, $v = x$. Then

$$\int \text{FresnelC}(x)\,dx = x\,\text{FresnelC}(x) - \int x\cos(\pi x^2/2)\,dx$$

Now make the substitution $u = \pi x^2/2$, $du = \pi x\,dx$ to obtain

$$\begin{aligned}\int \text{FresnelC}(x)\,dx &= x\,\text{FresnelC}(x) - \int \cos(u)\cdot\frac{1}{\pi}\,du \\ &= x\,\text{FresnelC}(x) - \frac{1}{\pi}\sin(u) + C \\ &= x\,\text{FresnelC}(x) - \frac{1}{\pi}\sin(\pi x^2/2) + C\,.\end{aligned}$$

(c) Let $u = \text{FresnelS}(x)$, $dv = 2x\,dx$, and $du = \sin(\pi x^2/2)\,dx$, $v = x^2$. Then

$$\int 2x\,\text{FresnelS}(x)\,dx = x^2\,\text{FresnelS}(x) - \int x^2 \sin(\pi x^2/2)\,dx$$

Now apply integration by parts, $u = x$, $dv = x\sin(\pi x^2/2)\,dx$, and $du = dx$, $v = -\frac{1}{\pi}\cos(\pi x^2/2)$ to obtain

$$\begin{aligned}\int 2x\,\text{FresnelS}(x)\,dx &= x^2\,\text{FresnelS}(x) - \left(-\frac{x}{\pi}\cos(\pi x^2/2) + \frac{1}{\pi}\int \cos(\pi x^2/2)\,dx\right)\\ &= x^2\,\text{FresnelS}(x) + \frac{x}{\pi}\cos(\pi x^2/2) - \frac{1}{\pi}\,\text{FresnelC}(x) + C\,.\end{aligned}$$

(d) Let $u = \text{FresnelC}(x)$, $dv = 2x\,dx$, and $du = \cos(\pi x^2/2)\,dx$, $v = x^2$. Then

$$\int 2x\,\text{FresnelC}(x)\,dx = x^2\,\text{FresnelC}(x) - \int x^2 \cos(\pi x^2/2)\,dx$$

Now apply integration by parts, $u = x$, $dv = x\cos(\pi x^2/2)\,dx$, and $du = dx$, $v = \frac{1}{\pi}\sin(\pi x^2/2)$ to obtain

$$\begin{aligned}\int 2x\,\text{FresnelC}(x)\,dx &= x^2\,\text{FresnelC}(x) - \left(\frac{x}{\pi}\sin(\pi x^2/2) - \frac{1}{\pi}\int \sin(\pi x^2/2)\,dx\right)\\ &= x^2\,\text{FresnelC}(x) - \frac{x}{\pi}\sin(\pi x^2/2) + \frac{1}{\pi}\,\text{FresnelS}(x) + C\,.\end{aligned}$$

89. (a) Let $u = \text{Si}(x)$, $dv = dx$ and $du = \frac{\sin(x)}{x}\,dx$, $v = x$. Then

$$\int \text{Si}(x)\,dx = x\,\text{Si}(x) - \int \sin(x)\,dx = x\,\text{Si}(x) + \cos(x) + C\,.$$

(b) Let $u = x$, $dv = \text{Si}(x)\,dx$ and $du = dx$, $v = x\,\text{Si}(x) + \cos(x)$. Then

$$\int x\,\text{Si}(x)\,dx = x^2\text{Si}(x) + x\cos(x) - \int (x\,\text{Si}(x) + \cos(x))\,dx$$

implying that

$$2\int x\,\text{Si}(x)\,dx = x^2\text{Si}(x) + x\cos(x) - \sin(x)$$

and

$$\int x\,\text{Si}(x)\,dx = \frac{1}{2}\big(x^2\text{Si}(x) + x\cos(x) - \sin(x)\big) + C\,.$$

CALCULATOR/COMPUTER EXERCISES

91. Using *Maple*'s *fsolve* procedure the curves intersect at $a = -0.6715530943$ and $b = 1.075063145$. The area between them is

$$\int_a^b (1 + 2x - xe^x)\, dx = 1.377424425\ldots.$$

93. The graphs of $f(x) = \arctan(x) - 1$ and $g(x) = \ln(x)/x$ are displayed on the right. Using *Maple*'s *fsolve* procedure the points of intersection are at $a = 0.7670950937$ and $b = 4.259674557$. The area of the region between them is

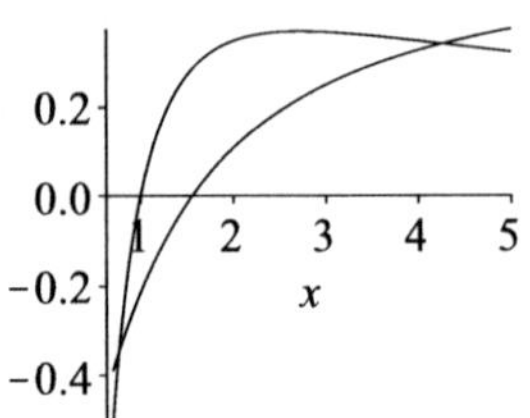

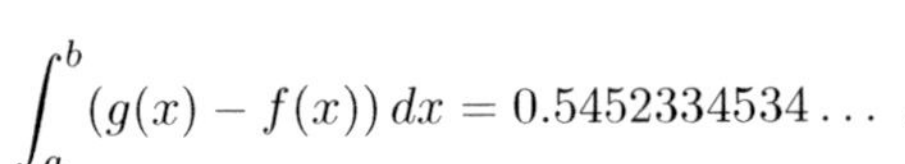

$$\int_a^b (g(x) - f(x))\, dx = 0.5452334534\ldots.$$

95. The nth coefficient simplifies to $a_n = (-4)^n/n^2$ and $a_0 = 2\pi^2/3$. The 3rd and 5th Fourier cosine-polynomial approximations of f are

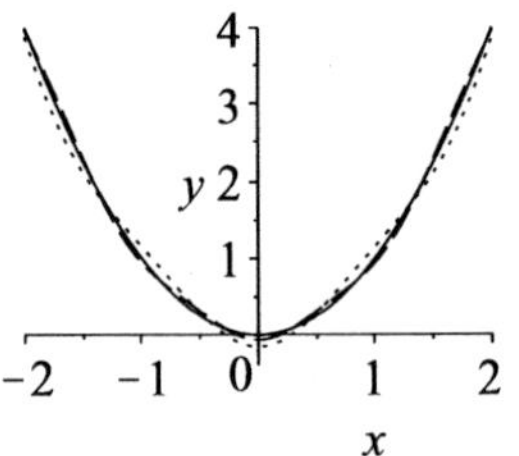

$$\frac{\pi^2}{3} - 4\cos(x) + \cos(2x) - \frac{4}{9}\cos(3x)$$

and

$$\frac{\pi^2}{3} - 4\cos(x) + \cos(2x) - \frac{4}{9}\cos(3x) + \frac{1}{4}\cos(4x) - \frac{4}{25}\cos(5x).$$

They are plotted on the right along with the graph of f.

6.2 Powers and Products of Trigonometric Functions

Problems for Practice

1. $\int \sin^2(x/2)\, dx = \int \frac{1}{2}(1 - \cos(x))\, dx = \frac{1}{2}(x - \sin(x)) + C$

3. $\int \cos^2(2x)\, dx = \int \frac{1}{2}(1 + \cos(4x))\, dx = \frac{1}{2}(x + \frac{1}{4}\sin(4x)) + C$

5. $\int \tan^2(x/2)\, dx = \int (\sec^2(x/2) - 1)\, dx = 2\tan(x/2) - x + C$

7. The sine function to an odd power, split off one sine and proceed as follows,

then substitute $u = \cos(2x)$, $du = -2\sin(2x)\,dx$.

$$\begin{aligned}\int 6\sin^3(2x)\,dx &= 6\int \sin^2(2x)\sin(2x)\,dx = 6\int (1-\cos^2(2x))\sin(2x)\,dx \\ &= -3\int (1-u^2)\,du = -3u + u^3 + C \\ &= \cos(2x)(\cos^2(2x) - 3) + C\end{aligned}$$

9. The sine function to an odd power, split off one sine and proceed as follows, then substitute $u = \cos(x/3)$, $du = -\frac{1}{3}\sin(x/3)\,dx$.

$$\begin{aligned}\int \sin^3(x/3)\,dx &= \int \sin^2(x/3)\sin(x/3)\,dx = \int (1-\cos^2(x/3))\sin(x/3)\,dx \\ &= -3\int (1-u^2)\,du = -3u + u^3 + C \\ &= \cos(x/3)(\cos^2(x/3) - 3) + C\end{aligned}$$

11. The sine function to an odd power, split off one sine and proceed as follows, then substitute $u = \cos(x/3)$, $du = -\frac{1}{3}\sin(x/3)\,dx$.

$$\begin{aligned}\int 3\sin^5(x/3)\,dx &= 3\int \sin^4(x/3)\sin(x/3)\,dx = 3\int (1-\cos^2(x/3))^2\sin(x/3)\,dx \\ &= -9\int (1-u^2)^2\,du = -9\int (1-2u^2+u^4)\,du \\ &= -9\left(u - \frac{2}{3}u^3 + \frac{1}{5}u^5\right) + C \\ &= -9\cos(x/3) + 6\cos^3(x/3) - \frac{9}{5}\cos^5(x/3) + C\end{aligned}$$

13. The cosine function to an odd power, split off one cosine and proceed as follows, then substitute $u = \sin(x/3)$, $du = \frac{1}{3}\cos(x/3)\,dx$.

$$\begin{aligned}\int 5\sin^2(x/3)\cos^3(x/3)\,dx &= 5\int \sin^2(x/3)\cos^2(x/3)\cos(x/3)\,dx \\ &= 5\int \sin^2(x/3)(1-\sin^2(x/3))\cos(x/3)\,dx \\ &= 15\int u^2(1-u^2)\,du = 15\int (u^2 - u^4)\,du \\ &= 15\left(\frac{1}{3}u^3 - \frac{1}{5}u^5\right) + C \\ &= 5\sin^3(x/3) - 3\sin^5(x/3) + C\end{aligned}$$

15. The cosine function to an odd power, split off one cosine and proceed as follows, then substitute $u = \sin(x)$, $du = \cos(x)\,dx$.

$$\begin{aligned}\int \frac{\cos^3(x)}{\sin(x)}\,dx &= \int \frac{\cos^2(x)}{\sin(x)} \cdot \cos(x)\,dx = \int \frac{1-\sin^2(x)}{\sin(x)} \cdot \cos(x)\,dx \\ &= \int \frac{1-u^2}{u}\,du = \int \left(\frac{1}{u} - u\right)\,du \\ &= \ln(|u|) - \frac{1}{2}u^2 + C = \ln(|\sin(x)|) - \frac{1}{2}\sin^2(x) + C\end{aligned}$$

17. The sine function to an odd power, split off one sine and proceed as follows, then substitute $u = \cos(x)$, $du = -\sin(x)\,dx$.

$$\begin{aligned}\int_0^\pi \sin^3(x)\cos^4(x)\,dx &= \int_0^\pi \sin^2(x)\cos^4(x)\sin(x)\,dx \\ &= \int_0^\pi (1-\cos^2(x))\cos^4(x)\sin(x)\,dx \\ &= -\int_1^{-1} (1-u^2)u^4\,du = \int_{-1}^1 (u^4 - u^6)\,du \\ &= \left(\frac{1}{5}u^5 - \frac{1}{7}u^7\right)\Big|_{-1}^1 = \left(\frac{1}{5} - \frac{1}{7}\right) - \left(-\frac{1}{5} + \frac{1}{7}\right) \\ &= \frac{2}{5} - \frac{2}{7} = \frac{4}{35}\end{aligned}$$

19. The sine function to an odd power, split off one sine and proceed as follows, then substitute $u = \cos(x)$, $du = -\sin(x)\,dx$.

$$\begin{aligned}\int_0^\pi \sin^3(x)\cos^6(x)\,dx &= \int_0^\pi \sin^2(x)\cos^6(x)\sin(x)\,dx \\ &= \int_0^\pi (1-\cos^2(x))\cos^6(x)\sin(x)\,dx \\ &= -\int_1^{-1} (1-u^2)u^6\,du = \int_{-1}^1 (u^6 - u^8)\,du \\ &= \left(\frac{1}{7}u^7 - \frac{1}{9}u^9\right)\Big|_{-1}^1 = \frac{2}{7} - \frac{2}{9} = \frac{4}{63}\end{aligned}$$

21. The cosine function to an odd power, split off one cosine and proceed as

follows, then substitute $u = \sin(x)$, $du = \cos(x)\,dx$.

$$\begin{aligned}
\int_0^{\pi/2} 15\sin^2(x)\cos^5(x)\,dx &= 15\int_0^{\pi/2} \sin^2(x)\cos^4(x)\cos(x)\,dx \\
&= 15\int_0^{\pi/2} \sin^2(x)(1-\sin^2(x))^2\cos(x)\,dx \\
&= 15\int_0^1 u^2(1-u^2)^2\,du = 15\int_0^1 u^2(1-2u^2+u^4)\,du \\
&= 15\int_0^1 (u^2-2u^4+u^6)\,du = 15\left(\frac{1}{3}u^3 - \frac{2}{5}u^5 + \frac{1}{7}u^7\right)\Big|_0^1 \\
&= 5 - 6 + \frac{15}{7} = \frac{8}{7}
\end{aligned}$$

23. Use formula (6.2.12),

$$\begin{aligned}
\int \cos^4(u)\,du &= \frac{1}{4}\sin(u)\cos^3(u) + \frac{3}{4}\int \cos^2(u)\,du \\
&= \frac{1}{4}\sin(u)\cos^3(u) + \frac{3}{4}\left(\frac{1}{2}\sin(u)\cos(u) + \frac{1}{2}\int du\right) \\
&= \frac{1}{4}\sin(u)\cos^3(u) + \frac{3}{8}\sin(u)\cos(u) + \frac{3}{8}u + C\,.
\end{aligned}$$

25. Using Exercise 23,

$$\int_0^{2\pi} \cos^4(x)\,dx = \left(\frac{1}{4}\sin(x)\cos^3(x) + \frac{3}{8}\sin(x)\cos(x) + \frac{3}{8}x\right)\Big|_0^{2\pi} = \frac{3\pi}{4}\,.$$

27. Using Exercise 24,

$$\begin{aligned}
\int_{\pi/2}^{\pi} \sin^4(x)\,dx &= \left(-\frac{1}{4}\sin^3(x)\cos(x) - \frac{3}{8}\cos(x)\sin(x) + \frac{3}{8}x\right)\Big|_{\pi/2}^{\pi} \\
&= \frac{3\pi}{8} - \frac{3\pi}{16} = \frac{3\pi}{16}
\end{aligned}$$

29. Using Exercise 23, start by replacing $\sin^2(t)$ with $1-\cos^2(t)$, then split the integral into two parts and use the half-angle identity on the first part,

$$\begin{aligned}
\int_0^{\pi/4} \cos^2(t)\sin^2(t)\,dt &= \int_0^{\pi/4} \cos^2(t)(1-\cos^2(t))\,dt \\
&= \int_0^{\pi/4} \cos^2(t)\,dt - \int_0^{\pi/4} \cos^4(t)\,dt \\
= \frac{1}{2}(t + \sin(2t)/2)\Big|_0^{\pi/4} &- \left(\frac{1}{4}\sin(t)\cos^3(t) + \frac{3}{8}\sin(t)\cos(t) + \frac{3}{8}t\right)\Big|_0^{\pi/4} \\
= \frac{\pi}{8} + \frac{1}{4} - \left(\frac{1}{16} + \frac{3}{16} + \frac{3\pi}{32}\right) &= \frac{\pi}{32}\,.
\end{aligned}$$

31. First substitute $u = x/2$, $du = \frac{1}{2}\,dx$. Then use the reduction formula for cosine, followed by Exercise 23.

$$\begin{aligned}\int_0^{\pi} \cos^6(x/2)\,dx &= 2\int_0^{\pi/2} \cos^6(u)\,du \\ &= 2\left(\frac{1}{6}\sin(u)\cos^5(u)\Big|_0^{\pi/2} + \frac{5}{6}\int_0^{\pi/2} \cos^4(u)\,du\right) \\ &= \frac{5}{3}\left(\frac{1}{4}\cos^3(u)\sin(u) + \frac{3}{8}\cos(u)\sin(u) + \frac{3}{8}u\right)\Big|_0^{\pi/2} = \frac{5\pi}{16}\end{aligned}$$

33. First replace $\cos^2(x)$ with $1-\sin^2(x)$ and split the integral into two parts. Use the reduction formula for the sine on the second part, followed by Exercise 24 applied to what remains.

$$\begin{aligned}\int_{\pi/6}^{\pi/3} \cos^2(x)\sin^4(x)\,dx &= \int_{\pi/6}^{\pi/3} (1-\sin^2(x))\sin^4(x)\,dx \\ &= \int_{\pi/6}^{\pi/3} \sin^4(x)\,dx - \int_{\pi/6}^{\pi/3} \sin^6(x)\,dx \\ &= \int_{\pi/6}^{\pi/3} \sin^4(x)\,dx - \left(-\frac{1}{6}\sin^5(x)\cos(x)\Big|_{\pi/6}^{\pi/3} + \frac{5}{6}\int_{\pi/6}^{\pi/3} \sin^4(x)\,dx\right) \\ &= \frac{1}{6}\left(-\frac{1}{4}\sin^3(x)\cos(x) - \frac{3}{8}\cos(x)\sin(x) + \frac{3}{8}x\right)\Big|_{\pi/6}^{\pi/3} - \left(-\frac{\sqrt{3}}{48}\right) \\ &= \left(\frac{\pi}{96} - \frac{\sqrt{3}}{192}\right) + \frac{\sqrt{3}}{48} = \frac{\pi}{96} + \frac{\sqrt{3}}{64}\end{aligned}$$

35. Using formula (6.2.14),

$$\begin{aligned}\int \tan^4(x)\,dx &= \frac{1}{3}\tan^3(x) - \int \tan^2(x)\,dx \\ &= \frac{1}{3}\tan^3(x) - \left(\tan(x) - \int dx\right) \\ &= \frac{1}{3}\tan^3(x) - \tan(x) + x + C\,.\end{aligned}$$

37. Substitute $u = \sec(x)$, $du = \sec(x)\tan(x)\,dx$.

$$\int \sec^2(x)\sec(x)\tan(x)\,dx = \int u^2\,du = \frac{1}{3}u^3 + C = \frac{1}{3}\sec^3(x) + C$$

39. Using formula (6.2.13),

$$\int_0^{\pi/4} \sec^4(x)\,dx = \frac{1}{3}\sec^2(x)\tan(x)\Big|_0^{\pi/4} + \frac{2}{3}\int_0^{\pi/4} \sec^2(x)\,dx$$
$$= \frac{2}{3} + \left(\frac{2}{3}\tan(x)\right)\Big|_0^{\pi/4} = \frac{4}{3}.$$

41. Using formula (6.2.13),

$$\int_0^{\pi/3} 10\sec^6(x)\,dx = 10\left(\frac{1}{5}\sec^4(x)\tan(x)\Big|_0^{\pi/3} + \frac{4}{5}\int_0^{\pi/3}\sec^4(x)\,dx\right)$$
$$= 32\sqrt{3} + 8\left(\frac{1}{3}\sec^2(x)\tan(x)\Big|_0^{\pi/3} + \frac{2}{3}\int_0^{\pi/3}\sec^2(x)\,dx\right)$$
$$= 32\sqrt{3} + \frac{32}{3}\sqrt{3} + \left(\frac{16}{3}\tan(x)\right)\Big|_0^{\pi/3}$$
$$= 32\sqrt{3} + \frac{32}{3}\sqrt{3} + \frac{16}{3}\sqrt{3} = 48\sqrt{3}.$$

43. Convert to secants and use formula (6.2.13).

$$\int_0^{\pi/4} \sec^4(x)\tan^2(x)\,dx = \int_0^{\pi/4}\sec^4(x)(\sec^2(x)-1)\,dx$$
$$= \int_0^{\pi/4}(\sec^6(x)-\sec^4(x))\,dx = \int_0^{\pi/4}\sec^6(x)\,dx - \int_0^{\pi/4}\sec^4(x)\,dx$$
$$= \frac{1}{5}\sec^4(x)\tan(x)\Big|_0^{\pi/4} + \frac{4}{5}\int_0^{\pi/4}\sec^4(x)\,dx - \int_0^{\pi/4}\sec^4(x)\,dx$$
$$= \frac{4}{5} - \frac{1}{5}\left(\frac{1}{3}\sec^2(x)\tan(x)\Big|_0^{\pi/4} + \frac{2}{3}\int_0^{\pi/4}\sec^2(x)\,dx\right)$$
$$= \frac{4}{5} - \frac{1}{5}\cdot\frac{2}{3} - \left(\frac{2}{15}\tan(x)\right)\Big|_0^{\pi/4} = \frac{4}{5} - \frac{2}{15} - \frac{2}{15} = \frac{8}{15}$$

45. Convert to secants and use formula (6.2.13).

$$\int_0^{\pi/3} 2\sec(x)\tan^2(x)\,dx = \int_0^{\pi/3} 2\sec(x)(\sec^2(x)-1)\,dx$$
$$= \int_0^{\pi/3} 2(\sec^3(x)-\sec(x))\,dx = 2\int_0^{\pi/3}\sec^3(x)\,dx - 2\int_0^{\pi/3}\sec(x)\,dx$$
$$= 2\left(\frac{1}{2}\sec(x)\tan(x)\Big|_0^{\pi/3} + \frac{1}{2}\int_0^{\pi/3}\sec(x)\,dx\right) - 2\int_0^{\pi/3}\sec(x)\,dx$$
$$= 2\sqrt{3} - \Big(\ln(|\sec(x)+\tan(x)|)\Big)\Big|_0^{\pi/3} = 2\sqrt{3} - \ln(2+\sqrt{3})$$

47. Convert to sines and cosines, then let $u = \cos(x)$, $du = -\sin(x)\,dx$.

$$\int 8\tan(x)\sec^4(x)\,dx = 8\int \frac{\sin(x)}{\cos(x)} \cdot \frac{1}{\cos^4(x)}\,dx = -8\int (\cos(x))^{-5} \cdot (-\sin(x))\,dx$$
$$= -8\int u^{-5}\,du = 2u^{-4} + C = 2\sec^4(x) + C$$

49. Convert to sines and cosines, then split off one sine, convert the remaining sines to cosines, and let $u = \cos(x)$, $du = -\sin(x)\,dx$.

$$\int 3\tan^3(x)\sec(x)\,dx = 3\int \frac{\sin^3(x)}{\cos^3(x)} \cdot \frac{1}{\cos(x)}\,dx = 3\int \frac{\sin^2(x)}{\cos^4(x)} \cdot \sin(x)\,dx$$
$$= 3\int \frac{1-\cos^2(x)}{\cos^4(x)} \cdot \sin(x)\,dx = -3\int \frac{1-u^2}{u^4}\,du = -3\int (u^{-4} - u^{-2})\,du$$
$$= u^{-3} - 3u^{-1} + C = \sec^3(x) - 3\sec(x) + C$$

51. Convert to sines and cosines and make the substitution $u = \sin(x/3)$, $du = \frac{1}{3}\cos(x/3)\,dx$.

$$\int \cot(x/3)\csc^3(x/3)\,dx = \int \frac{\cos(x/3)}{\sin(x/3)} \cdot \frac{1}{\sin^3(x/3)}\,dx = 3\int u^{-4}\,du$$
$$= -u^{-3} + C = -\csc^3(x/3) + C$$

53. Convert to sines and cosines, split off one cosine, convert the remaining cosines into sines, and make the substitution $u = \sin(x/3)$, $du = \frac{1}{3}\cos(x/3)\,dx$.

$$\int \cot^3(x/3)\csc(x/3)\,dx = \int \frac{\cos^3(x/3)}{\sin^3(x/3)} \cdot \frac{1}{\sin(x/3)}\,dx = \int \frac{\cos^2(x/3)}{\sin^4(x/3)} \cdot \cos(x/3)\,dx$$
$$= \int \frac{1-\sin^2(x/3)}{\sin^4(x/3)} \cdot \cos(x/3)\,dx = 3\int \frac{1-u^2}{u^4}\,du = 3\int (u^{-4} - u^{-2})\,du$$
$$= -u^{-3} + 3u^{-1} + C = -\csc^3(x/3) + 3\csc(x/3) + C$$

55. Convert to sines and cosines, cancel two cosines, and make the substitution $u = \sin(x)$, $du = \cos(x)\,dx$.

$$\int 6\cos^3(x)\tan^2(x)\,dx = \int 6\cos^3(x) \cdot \frac{\sin^2(x)}{\cos^2(x)}\,dx = 6\int \cos(x)\sin^2(x)\,dx$$
$$= 6\int u^2\,du = 2u^3 + C = 2\sin^3(x) + C$$

57. Multiply top and bottom by $1+\sin(x)$. Simplify into two parts and finish by observing what the substitution $u = \cos(x)$, $du = -\sin(x)\,dx$ will do to the second part of the integrand.

$$\begin{aligned}\int \frac{1}{1-\sin(x)}\,dx = \int \frac{1+\sin(x)}{1-\sin^2(x)}\,dx &= \int \frac{1+\sin(x)}{\cos^2(x)}\,dx \\ &= \int (\sec^2(x) + (\cos(x))^{-2} \cdot \sin(x))\,dx \\ &= \tan(x) + \cos(x)^{-1} + C = \tan(x) + \sec(x) + C\end{aligned}$$

59. Multiply top and bottom by $1 + \cos(x)$. Simplify into two parts and finish by observing what the substitution $u = \sin(x)$, $du = \cos(x)\,dx$ will do to the first part of the integrand and converting the second part to cotangents, then to cosecants in order to integrate it.

$$\begin{aligned}\int \frac{\cos(x)}{1-\cos(x)}\,dx = \int \frac{\cos(x)(1+\cos(x))}{1-\cos^2(x)}\,dx &= \int \frac{\cos(x)+\cos^2(x)}{\sin^2(x)}\,dx \\ &= \int ((\sin(x)^{-2}\cos(x) + \cot^2(x))\,dx \\ = -\sin(x)^{-1} + \int (\csc^2(x) - 1)\,dx &= -\csc(x) - \cot(x) - x + C\end{aligned}$$

61. Multiply top and bottom by $\sec(x) + 1$, then convert the denominator to $\tan^2(x)$ and split the integrand into two parts, converting the first part to sines and cosines and the second part to cotangent. The first part simplifies to $\csc(x)$ and the second part is $\cot(x)$.

$$\begin{aligned}\int \frac{\tan(x)}{\sec(x)-1}\,dx &= \int \frac{\tan(x)(\sec(x)+1)}{\sec^2(x)-1}\,dx = \int \frac{\tan(x)(\sec(x)+1)}{\tan^2(x)}\,dx \\ &= \int \frac{\sec(x)+1}{\tan(x)}\,dx = \int \left(\frac{1}{\cos(x)} \cdot \frac{\cos(x)}{\sin(x)} + \cot(x)\right)\,dx \\ &= \int (\csc(x) + \cot(x))\,dx = -\ln(|\csc(x) + \cot(x)|) + \ln(|\sin(x)|) + C\end{aligned}$$

63. Use the half-angle identity $\cos^2(x) = \frac{1}{2}(1 + \cos(2x))$, three times, then $\cos^2(2x) = \frac{1}{2}(1 + \cos(4x))$. In the last integral use the outcome for the

substitution $u = \sin(2x)$, $du = 2\cos(2x)\,dx$.

$$\begin{aligned}\int \cos^6(x)\,dx &= \int \frac{1}{2^3}(1+\cos(2x))^3\,dx = \frac{1}{8}\int (1+3\cos(2x)+3\cos^2(2x)+\cos^3(2x))\,dx \\ &= \frac{1}{8}\int (1+3\cos(2x)+\tfrac{3}{2}(1+\cos(4x)))\,dx + \frac{1}{8}\int \cos^2(2x)\cdot\cos(2x)\,dx \\ &= \frac{1}{8}\left(\frac{5}{2}x+\frac{3}{2}\sin(2x)+\frac{3}{8}\sin(4x)\right)+\frac{1}{8}\int (1-\sin^2(2x))\cos(2x)\,dx \\ &= \frac{5}{16}x+\frac{3}{16}\sin(2x)+\frac{3}{64}\sin(4x)+\frac{1}{16}\sin(2x)-\frac{1}{48}\sin^3(2x)+C \\ &= \frac{5}{16}x+\frac{1}{4}\sin(2x)+\frac{3}{64}\sin(4x)-\frac{1}{48}\sin^3(2x)+C\end{aligned}$$

65. Use $\cos^2(x) = \frac{1}{2}(1+\cos(2x))$ and $\sin^2(x) = \frac{1}{2}(1-\cos(2x))$ twice and proceed as follows. The last integral is evaluated with the substitution $u = \sin(2x)$, $du = 2\cos(2x)\,dx$.

$$\begin{aligned}\int \cos^2(x)\sin^4(x)\,dx &= \int \tfrac{1}{2}(1+\cos(2x))\tfrac{1}{2^2}(1-\cos(2x))^2\,dx \\ &= \frac{1}{8}\int (1+\cos(2x)(1-2\cos(2x)+\cos^2(2x))\,dx \\ &= \frac{1}{8}\int (1-\cos(2x)-\cos^2(2x)+\cos^3(2x))\,dx \\ &= \frac{1}{8}\int (1-\cos(2x)-\tfrac{1}{2}(1+\cos(4x)))\,dx+\frac{1}{8}\int \cos^2(2x)\cos(2x)\,dx \\ &= \frac{1}{8}\left(\frac{1}{2}x-\frac{1}{2}\sin(2x)-\frac{1}{8}\sin(4x)\right)+\frac{1}{8}\int (1-\sin^2(2x))\cos(2x)\,dx \\ &= \frac{1}{16}x-\frac{1}{16}\sin(2x)-\frac{1}{64}\sin(4x)+\frac{1}{16}\sin(2x)-\frac{1}{48}\sin^3(2x)+C \\ &= \frac{1}{16}x-\frac{1}{64}\sin(4x)-\frac{1}{48}\sin^3(2x)+C\end{aligned}$$

67. Use the identity $\sin(x)\cos(x) = \frac{1}{2}\sin(2x)$ twice.

$$\begin{aligned}\int \cos^2(x)\sin^2(x)\,dx &= \int \frac{1}{2^2}\sin^2(2x)\,dx = \frac{1}{4}\int \frac{1}{2}(1-\cos(4x))\,dx \\ &= \frac{1}{8}\left(x-\frac{1}{4}\sin(4x)\right)+C = \frac{1}{8}x-\frac{1}{32}\sin(4x)+C\end{aligned}$$

69. Use the identity $\sin(x)\cos(x) = \frac{1}{2}\sin(2x)$ four times, then $\sin^2(2x) =$

$\frac{1}{2}(1-\cos(4x))$ twice, followed by $\cos^2(4x)=\frac{1}{2}(1+\cos(8x))$.

$$\begin{aligned}\int \cos^4(x)\sin^4(x)\,dx &= \int \frac{1}{2^4}\sin^4(2x)\,dx = \frac{1}{16}\int \frac{1}{2^2}(1-\cos(4x))^2\,dx \\ &= \frac{1}{64}\int (1-2\cos(4x)+\cos^2(4x))\,dx \\ &= \frac{1}{64}\int \left(1-2\cos(4x)+\frac{1}{2}(1+\cos(8x))\right)dx \\ &= \frac{3}{128}x - \frac{1}{128}\sin(4x) + \frac{1}{1024}\sin(8x) + C\end{aligned}$$

Further Theory and Practice

71. With $u=\sin(x)$ and $du=\cos(x)\,dx$, $\int \sin(x)\cos(x)\,dx = \int u\,du = \frac{1}{2}u^2+C=\frac{1}{2}\sin^2(x)+C$. Similarly, with $u=\cos(x)$ and $du=-\sin(x)\,dx$, $\int \sin(x)\cos(x)\,dx = -\int u\,du = -\frac{1}{2}u^2+C = -\frac{1}{2}\cos^2(x)+C$. Moreover, using the identity $\sin(x)\cos(x)=\frac{1}{2}\sin(2x)$, $\int \sin(x)\cos(x)\,dx = \int \frac{1}{2}\sin(2x)\,dx = -\frac{1}{4}\cos(2x)+C$.

    These are not contradictory because the second and third antiderivatives each differ from $\frac{1}{2}\sin^2(x)$ by a constant. In fact, $-\frac{1}{2}\cos^2(x) = -\frac{1}{2}(1-\sin^2(x)) = \frac{1}{2}\sin^2(x)-\frac{1}{2}$, and $-\frac{1}{4}\cos(2x) = -\frac{1}{4}(\cos^2(x)-\sin^2(x)) = -\frac{1}{4}(1-2\sin^2(x)) = \frac{1}{2}\sin^2(x)-\frac{1}{4}$.

73. Split off a cosine and convert the remaining cosines into sines. Then substitute $u=\sin(x)$, $du=\cos(x)\,dx$, followed by $v=2+2u$, $dv=2\,du$.

$$\begin{aligned}\int_{-\pi/2}^{\pi/2}\sqrt{2+2\sin(x)}\cos^3(x)\,dx &= \int_{-\pi/2}^{\pi/2}\sqrt{2+2\sin(x)}\cos^2(x)\cos(x)\,dx \\ &= \int_{-\pi/2}^{\pi/2}\sqrt{2+2\sin(x)}(1-\sin^2(x))\cos(x)\,dx = \int_{-1}^{1}\sqrt{2+2u}(1-u^2)\,du \\ &= \int_0^4 \sqrt{v}\left(1-\frac{1}{4}(v-2)^2\right)\cdot\frac{1}{2}\,dv = \frac{1}{2}\int_0^4 v^{1/2}\left(v-\frac{1}{4}v^2\right)dv \\ &= \frac{1}{8}\int_0^4 (4v^{3/2}-v^{5/2})\,dv = \frac{1}{8}\left(\frac{8}{5}v^{5/2}-\frac{2}{7}v^{7/2}\right)\Big|_0^4 = \frac{32}{5}-\frac{32}{7}=\frac{64}{35}\end{aligned}$$

75. Since $\sin^3(x)\le\sin^2(x)$ for $0\le x\le\pi/2$, the area is

$$\int_0^{\pi/2}(\sin^2(x)-\sin^3(x))\,dx = \int_0^{\pi/2}\left(\frac{1}{2}(1-\cos(2x))-(1-\cos^2(x))\cdot\sin(x)\right)dx$$

$$= \left(\frac{1}{2}x-\frac{1}{4}\sin(2x)+\cos(x)-\frac{1}{3}\cos^3(x)\right)\Big|_0^{\pi/2} = \frac{\pi}{4}-\left(1-\frac{1}{3}\right)=\frac{\pi}{4}-\frac{2}{3}.$$

77. Apply integration by parts with $u = \sec^{n-2}(x)$ and $dv = \sec^2(x)\,dx$ so $du = (n-2)\sec^{n-3}(x) \cdot \sec(x)\tan(x)\,dx$ and $v = \tan(x)$.

$$\int \sec^n(x)\,dx = \sec^{n-2}(x)\tan(x) - (n-2)\int \sec^{n-2}(x)\tan^2(x)\,dx$$

Replace $\tan^2(x)$ with $\sec^2(x) - 1$ and rearrange to obtain the following equation.

$$\begin{aligned}\int \sec^n(x)\,dx &= \sec^{n-2}(x)\tan(x) - (n-2)\int \sec^{n-2}(x)(\sec^2(x) - 1)\,dx \\ &= \sec^{n-2}(x)\tan(x) - (n-2)\int \sec^n(x)\,dx + (n-2)\int \sec^{n-2}(x)\,dx\end{aligned}$$

Now move the $\sec^n(x)$ integral on the right side of the equation to the left side and simplify to obtain

$$(n-1)\int \sec^n(x)\,dx = \sec^{n-2}(x)\tan(x) + (n-2)\int \sec^{n-2}(x)\,dx\,.$$

Finally, divide both sides by $n-1$ to obtain formula (6.2.13).

79. Split off one sine and use it to integrate the cosine product in the following integration by parts scheme: $u = \sin^{n-1}(x)$, $dv = \cos^m(x)\sin(x)\,dx$ and $du = (n-1)\sin^{n-2}(x)\cos(x)\,dx$, $v = -\frac{1}{m+1}\cos^{m+1}(x)$.

$$\int \sin^n(x)\cos^m(x)\,dx = -\frac{\sin^{n-1}(x)\cos^{m+1}(x)}{m+1} + \frac{n-1}{m+1}\int \sin^{n-2}(x)\cos^{m+2}(x)\,dx$$

Split off two cosines in the integral on the right hand side, replace $\cos^2(x)$ with $1 - \sin^2(x)$, and divide the integral into two to obtain the following equation.

$$\begin{aligned}\int \sin^n(x)\cos^m(x)\,dx = -\frac{\sin^{n-1}(x)\cos^{m+1}(x)}{m+1} &+ \frac{n-1}{m+1}\int \sin^{n-2}(x)\cos^m(x)\,dx \\ &- \frac{n-1}{m+1}\int \sin^n(x)\cos^m(x)\,dx\end{aligned}$$

Now move the second integral on the right hand side over to the left side and then simplify the left side to obtain

$$\frac{m+n}{m+1}\int \sin^n(x)\cos^m(x)\,dx = -\frac{\sin^{n-1}(x)\cos^{m+1}(x)}{m+1} + \frac{n-1}{m+1}\int \sin^{n-2}(x)\cos^m(x)\,dx\,.$$

Finally, multiply both sides by $m+1$ and divide by $m+n$ to obtain the desired reduction formula.

81. If $k = 1$, then

$$\int \frac{1}{A\cos(x) + B\sin(x)}\,dx = \int \frac{1}{\sqrt{A^2+B^2}\cos(x+\phi)}\,dx = \frac{1}{\sqrt{A^2+B^2}}\int \sec(x+\phi)\,dx$$
$$= \frac{1}{\sqrt{A^2+B^2}}\ln\left(|\sec(x+\phi) + \tan(x+\phi)|\right) + C\,.$$

If $k = 2$, then

$$\int \frac{1}{(A\cos(x) + B\sin(x))^2}\,dx = \int \frac{1}{\sqrt{A^2+B^2}\cos^2(x+\phi)}\,dx = \frac{1}{\sqrt{A^2+B^2}}\int \sec^2(x+\phi)\,dx$$
$$= \frac{1}{\sqrt{A^2+B^2}}\tan(x+\phi) + C\,.$$

CALCULATOR/COMPUTER EXERCISES

83. The graphs of $y = \cos^3(x)$ and $y = \sin^2(x)$ are displayed on the right. Using *Maple*'s *fsolve* procedure the abscissa of the point of intersection on the positive side of the x-axis is $a \approx 0.7153287499$ and the desired area is

$$\int_{-a}^{a} (\cos^3(x) - \sin^2(x))\,dx \approx 0.9034161512\,.$$

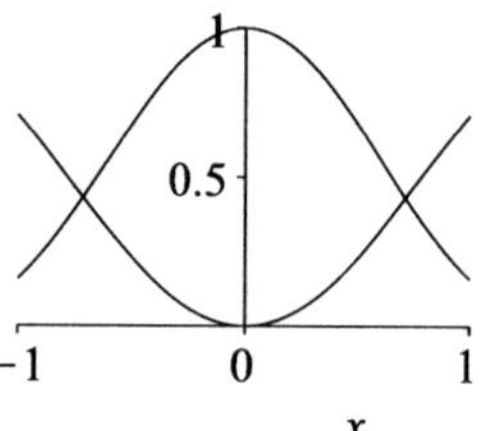

85. The graphs of $f(x) = \sin^2(x)\cos^3(x)$ and $g(x) = \sin^5(x)\cos^2(x)$ are displayed on the right. Using *Maple*'s *fsolve* procedure the abscissa of the point of intersection in $[0.5, 1]$ is $b \approx 0.9720296215$. The desired area is

$$\int_0^b (f(x) - g(x))\,dx \approx 0.07428138337\,.$$

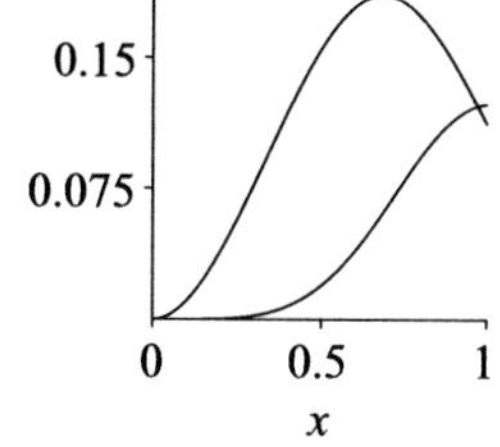

87. The tangent line equation at $(x_0, \sin^4(x_0))$ is $y = 4\sin^3(x_0)\cos(x_0)(x - x_0) + \sin^4(x_0)$. It will pass through the origin when x_0 satisfies $0 = 4\sin^3(x_0)\cos(x_0)\cdot(-x_0) + \sin^4(x_0)$. That is, $\sin(x_0) = 4x_0\cos(x_0)$. Using *Maple*, $x_0 \approx 1.393249075$. See the picture.

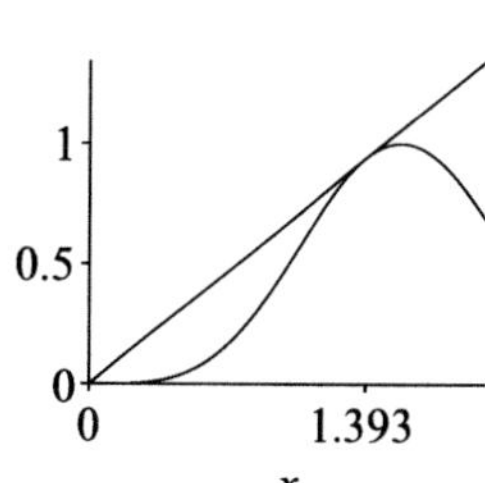

The desired area is

$$\int_0^{x_0} (T(x) - f(x))\,dx \approx 0.2386681149$$

where $f(x) = \sin^4(x)$ and $T(x) = f'(x_0)(x - x_0) + f(x_0)$.

6.3 Trigonometric Substitution

Problems for Practice

1. Let $x = 3\sin(\theta)$, $dx = 3\cos(\theta)\,d\theta$. Note that $\sin(\theta) = x/3$ implies that $\cos(\theta) = \frac{1}{3}\sqrt{9-x^2}$ (draw a right triangle).

$$\int \frac{x}{\sqrt{9-x^2}}\,dx = \int \frac{3\sin(\theta)}{\sqrt{9-9\sin^2(\theta)}} \cdot 3\cos(\theta)\,d\theta = \frac{3\cdot 3}{3}\int \frac{\sin(\theta)\cos(\theta)}{\sqrt{1-\sin^2(\theta)}}\,d\theta$$
$$= 3\int \sin(\theta)\,d\theta = -3\cos(\theta) + C = -\sqrt{9-x^2} + C$$

3. Let $x = 2\sin(\theta)$, $dx = 2\cos(\theta)\,d\theta$. Note that $\sin(\theta) = x/2$ implies that $\cos(\theta) = \frac{1}{2}\sqrt{4-x^2}$ (draw a right triangle).

$$\int \frac{4}{\sqrt{4-x^2}}\,dx = \int \frac{4}{\sqrt{4-4\sin^2(\theta)}} \cdot 2\cos(\theta)\,d\theta = \frac{4\cdot 2}{2}\int \frac{\cos(\theta)}{\sqrt{1-\sin^2(\theta)}}\,d\theta$$
$$= 4\int d\theta = 4\theta + C = 4\arcsin(x/2) + C$$

5. Let $x = \sin(\theta)$, $dx = \cos(\theta)\,d\theta$. Note that $\sin(\theta) = x$ implies that $\cos(\theta) = \sqrt{1-x^2}$ (draw a right triangle).

$$\int \frac{x^2}{\sqrt{1-x^2}}\,dx = \int \frac{\sin^2(x)}{\sqrt{1-\sin^2(\theta)}} \cdot \cos(\theta)\,d\theta = \int \frac{\sin^2(\theta)\cos(\theta)}{\cos(\theta)}\,d\theta$$
$$= \int \sin^2(\theta)\,d\theta = \frac{1}{2}(\theta - \sin(\theta)\cos(\theta)) + C = \frac{1}{2}\left(\arcsin(x) - x\sqrt{1-x^2}\right) + C$$

7. Let $x = \sin(\theta)$, $dx = \cos(\theta)\,d\theta$. Note that $\sin(\theta) = x$ implies that $\tan(\theta) = \frac{x}{\sqrt{1-x^2}}$ (draw a right triangle).

$$\int \frac{x^2}{(1-x^2)^{3/2}}\,dx = \int \frac{\sin^2(x)}{(1-\sin^2(\theta))^{3/2}} \cdot \cos(\theta)\,d\theta = \int \frac{\sin^2(\theta)\cos(\theta)}{\cos^3(\theta)}\,d\theta$$
$$= \int \tan^2(\theta)\,d\theta = \int (\sec^2(\theta) - 1)\,d\theta = \tan(\theta) - \theta + C = \frac{x}{\sqrt{1-x^2}} - \arcsin(x) + C$$

9. Let $x = \frac{1}{2}\sin(\theta)$, $dx = \frac{1}{2}\cos(\theta)\,d\theta$. Note that $\sin(\theta) = 2x$ implies that $\cos(\theta) = \sqrt{1-4x^2}$ (draw a right triangle).

$$\int \sqrt{1-4x^2}\,dx = \int \sqrt{1-\sin^2(\theta)} \cdot \frac{1}{2}\cos(\theta)\,d\theta = \frac{1}{2}\int \cos^2(\theta)\,d\theta$$
$$= \frac{1}{4}(\theta + \sin(\theta)\cos(\theta)) + C = \frac{1}{4}\arcsin(2x) + \frac{1}{2}x\sqrt{1-4x^2} + C$$

11. Let $x = \tan(\theta)$, $dx = \sec^2(\theta)\,d\theta$. Note that $\tan(\theta) = x$ implies that $\sec(\theta) = \sqrt{1+x^2}$ (draw a right triangle).

$$\int \frac{4x}{x^2+1}\,dx = \int \frac{4\tan(\theta)}{\tan^2(\theta)+1} \cdot \sec^2(\theta)\,d\theta = 4\int \frac{\tan(\theta)\sec^2(\theta)}{\sec^2(\theta)}\,d\theta$$
$$= 4\int \tan(\theta)\,d\theta = 4\ln(|\sec(\theta)|) + C = 4\ln(\sqrt{1+x^2}) + C = 2\ln(1+x^2) + C$$

13. Let $x = \frac{1}{3}\tan(\theta)$, $dx = \frac{1}{3}\sec^2(\theta)\,d\theta$. Note that $\tan(\theta) = 3x$ implies that $\sec(\theta) = \sqrt{9x^2+1}$ (draw a right triangle).

$$\int \frac{6}{9x^2+1}\,dx = \int \frac{6}{\tan^2(\theta)+1} \cdot \frac{1}{3}\sec^2(\theta)\,d\theta = 2\int \frac{\sec^2(\theta)}{\sec^2(\theta)}\,d\theta$$
$$= 2\int d\theta = 2\theta + C = 2\arctan(3x) + C$$

15. Let $x = \frac{1}{2}\tan(\theta)$, $dx = \frac{1}{2}\sec^2(\theta)\,d\theta$. Note that $\tan(\theta) = 2x$ implies that $\sec(\theta) = \sqrt{1+4x^2}$ (draw a right triangle). The integral of $\sec^3(\theta)$ is obtained using the reduction formula (6.2.13).

$$\int 4\sqrt{1+4x^2}\,dx = \int 4\sqrt{1+\tan^2(\theta)} \cdot \frac{1}{2}\sec^2(\theta)\,d\theta = 2\int \sec(\theta)\sec^2(\theta)\,d\theta$$
$$= 2\int \sec^3(\theta)\,d\theta = \sec(\theta)\tan(\theta) + \ln(|\sec(\theta)+\tan(\theta)|) + C$$
$$= 2x\sqrt{1+4x^2} + \ln\left(2x + \sqrt{1+4x^2}\right) + C$$

The absolute value signs can be dropped from the ln term because $2x + \sqrt{1+4x^2} > 0$ for all x.

17. Let $x = \tan(\theta)$, $dx = \sec^2(\theta)\,d\theta$. Note that $\tan(\theta) = x$ implies that $\sin(\theta) = \frac{x}{\sqrt{1+x^2}}$ (draw a right triangle).

$$\int \frac{1}{(1+x^2)^{3/2}}\,dx = \int \frac{1}{(1+\tan^2(\theta))^{3/2}} \cdot \sec^2(\theta)\,d\theta = \int \frac{\sec^2(\theta)}{\sec^3(\theta)}\,d\theta$$
$$= \int \cos(\theta)\,d\theta = \sin(\theta) + C = \frac{x}{\sqrt{1+x^2}} + C$$

19. Let $x = \tan(\theta)$, $dx = \sec^2(\theta)\,d\theta$. Note that $\tan(\theta) = x$ implies that $\sec(\theta) = \sqrt{1+x^2}$ (draw a right triangle). The integral of $\sec^3(\theta)$ is obtained using the reduction formula (6.2.13).

$$\int \frac{2x^2}{\sqrt{1+x^2}}\,dx = 2\int \frac{\tan^2(\theta)}{\sqrt{1+\tan^2(\theta)}} \cdot \sec^2(\theta)\,d\theta = 2\int \frac{\tan^2(\theta)\sec^2(\theta)}{\sec(\theta)}\,d\theta$$
$$= 2\int \tan^2(\theta)\sec(\theta)\,d\theta = 2\int (\sec^2(\theta)-1)\sec(\theta)\,d\theta = 2\int (\sec^3(\theta)-\sec(\theta))\,d\theta$$
$$= \sec(\theta)\tan(\theta) + \ln(|\sec(\theta)+\tan(\theta)|) - 2\ln(|\sec(\theta)+\tan(\theta)|) + C$$
$$= x\sqrt{1+x^2} - \ln(x + \sqrt{1+x^2}) + C$$

21. Let $x = 2\sec(\theta)$, $dx = 2\sec(\theta)\tan(\theta)\,d\theta$.

$$\int_{2\sqrt{2}}^{4} \frac{4}{x\sqrt{x^2-4}}\,dx = \int_{\pi/4}^{\pi/3} \frac{4}{2\sec(\theta)\sqrt{4\sec^2(\theta)-4}} \cdot 2\sec(\theta)\tan(\theta)\,d\theta$$
$$= 2\int_{\pi/4}^{\pi/3} \frac{\sec(\theta)\tan(\theta)}{\sec(\theta)\sqrt{\sec^2(\theta)-1}}\,d\theta = 2\int_{\pi/4}^{\pi/3} d\theta = \frac{\pi}{6}$$

23. Let $x = \frac{1}{2}\sec(\theta)$, $dx = \frac{1}{2}\sec(\theta)\tan(\theta)\,d\theta$. Note that $\tan\left(\operatorname{arcsec}(\sqrt{5}\,)\right) = 2$. Draw a right triangle.

$$\int_{1/\sqrt{2}}^{\sqrt{5}/2} \frac{2}{\sqrt{4x^2-1}}\,dx = \int_{\pi/4}^{\operatorname{arcsec}(\sqrt{5}\,)} \frac{2}{\sqrt{\sec^2(\theta)-1}} \cdot \frac{1}{2}\sec(\theta)\tan(\theta)\,d\theta$$
$$= \int_{\pi/4}^{\operatorname{arcsec}(\sqrt{5}\,)} \sec(\theta)\,d\theta = \ln(|\sec(\theta)+\tan(\theta)|)\Big|_{\pi/4}^{\operatorname{arcsec}(\sqrt{5}\,)}$$
$$= \ln(\sqrt{5}+2) - \ln(\sqrt{2}+1) = \ln\left(\frac{2+\sqrt{5}}{1+\sqrt{2}}\right)$$

25. Let $x = \sec(\theta)$, $dx = \sec(\theta)\tan(\theta)\,d\theta$. Note that $\tan\left(\operatorname{arcsec}(\sqrt{5}\,)\right) = 2$ and $\tan\left(\operatorname{arcsec}(\sqrt{2}\,)\right) = 1$. Draw right triangles. The integral of $\sec^3(\theta)$ is obtained using the reduction formula (6.2.13).

$$\int_{\sqrt{2}}^{\sqrt{5}} \frac{2x^2}{\sqrt{x^2-1}}\,dx = \int_{\operatorname{arcsec}(\sqrt{2}\,)}^{\operatorname{arcsec}(\sqrt{5}\,)} \frac{2\sec^2(\theta)}{\sqrt{\sec^2(\theta)-1}} \cdot \sec(\theta)\tan(\theta)\,d\theta$$
$$= 2\int_{\operatorname{arcsec}(\sqrt{2}\,)}^{\operatorname{arcsec}(\sqrt{5}\,)} \frac{\sec^3(\theta)\tan(\theta)}{\tan(\theta)}\,d\theta = 2\int_{\operatorname{arcsec}(\sqrt{2}\,)}^{\operatorname{arcsec}(\sqrt{5}\,)} \sec^3(\theta)\,d\theta$$
$$= 2\left(\frac{1}{2}\sec(\theta)\tan(\theta) + \frac{1}{2}\ln(|\sec(\theta)+\tan(\theta)|\right)\Big|_{\operatorname{arcsec}(\sqrt{2}\,)}^{\operatorname{arcsec}(\sqrt{5}\,)}$$
$$= 2\sqrt{5} + \ln(\sqrt{5}+2) - \left(\sqrt{2} + \ln(\sqrt{2}+1)\right) = 2\sqrt{5} - \sqrt{2} + \ln\left(\frac{2+\sqrt{5}}{1+\sqrt{2}}\right)$$

27. Let $x = \sec(\theta)$, $dx = \sec(\theta)\tan(\theta)\,d\theta$. Note that $\sin\left(\operatorname{arcsec}(\sqrt{10}\,)\right) =$

$3/\sqrt{10}$ and $\sin\left(\operatorname{arcsec}(\sqrt{5}\,)\right) = 2/\sqrt{5}$. Draw right triangles.

$$\int_{\sqrt{5}}^{\sqrt{10}} \frac{6}{(x^2-1)^{3/2}}\,dx = \int_{\operatorname{arcsec}(\sqrt{5}\,)}^{\operatorname{arcsec}(\sqrt{10}\,)} \frac{6}{(\sec^2(\theta)-1)^{3/2}} \cdot \sec(\theta)\tan(\theta)\,d\theta$$

$$= 6\int_{\operatorname{arcsec}(\sqrt{5}\,)}^{\operatorname{arcsec}(\sqrt{10}\,)} \frac{\sec(\theta)\tan(\theta)}{\tan^3(\theta)}\,d\theta = 6\int_{\operatorname{arcsec}(\sqrt{5}\,)}^{\operatorname{arcsec}(\sqrt{10}\,)} \frac{\sec(\theta)}{\tan^2(\theta)}\,d\theta$$

$$= 6\int_{\operatorname{arcsec}(\sqrt{5}\,)}^{\operatorname{arcsec}(\sqrt{10}\,)} \frac{\cos(\theta)}{\sin^2(\theta)}\,d\theta \overset{(u=\sin(\theta))}{=} 6\int_{2/\sqrt{5}}^{3/\sqrt{10}} u^{-2}\,du$$

$$= -6u^{-1}\Big|_{2/\sqrt{5}}^{3/\sqrt{10}} = 3\sqrt{5} - 2\sqrt{10}$$

29. Let $x = \sec(\theta)$, $dx = \sec(\theta)\tan(\theta)\,d\theta$. The integral of $\sec^3(\theta)$ is obtained using the reduction formula (6.2.13).

$$\int_1^{\sqrt{2}} \sqrt{x^2-1}\,dx = \int_0^{\pi/4} \sqrt{\sec^2(\theta)-1} \cdot \sec(\theta)\tan(\theta)\,d\theta$$

$$= \int_0^{\pi/4} \sec(\theta)\tan^2(\theta)\,d\theta = \int_0^{\pi/4} \sec(\theta)(\sec^2(\theta)-1)\,d\theta$$

$$= \int_0^{\pi/4} (\sec^3(\theta) + \sec(\theta))\,d\theta = \left(\frac{1}{2}\sec(\theta)\tan(\theta) - \frac{1}{2}\ln(|\sec(\theta)+\tan(\theta)|)\right)\Big|_0^{\pi/4}$$

$$= \frac{1}{2}\sqrt{2} - \frac{1}{2}\ln(\sqrt{2}+1)$$

31. Let $x = \sin(\theta)$, $dx = \cos(\theta)\,d\theta$.

$$\int \frac{2x+3}{\sqrt{1-x^2}}\,dx = \int \frac{2\sin(\theta)+3}{\sqrt{1-\sin^2(\theta)}} \cdot \cos(\theta)\,d\theta = \int (2\sin(\theta)+3)\,d\theta$$

$$= -2\cos(\theta) + 3\theta + C = 3\arcsin(x) - 2\sqrt{1-x^2} + C$$

33. This integrates directly.

$$\int \frac{2x+3}{x^2+1}\,dx = \ln(x^2+1) + 3\arctan(x) + C$$

35. Let $x = \tan(\theta)$, $dx = \sec^2(\theta)\,d\theta$. The integral of $\sec^3(\theta)$ is obtained using

recursion formula (6.2.13).

$$\int \frac{2x^2+1}{\sqrt{x^2+1}}\,dx = \int \frac{2\tan^2(\theta)+1}{\sqrt{\tan^2(\theta)+1}} \cdot \sec^2(\theta)\,d\theta = \int (2\tan^2(\theta)\sec(\theta)+\sec(\theta))\,d\theta$$
$$= \int (2(\sec^2(\theta)-1)\sec(\theta)+\sec(\theta))\,d\theta = \int (2\sec^3(\theta)-\sec(\theta))\,d\theta$$
$$= 2\left(\frac{1}{2}\sec(\theta)\tan(\theta)+\frac{1}{2}\ln(|\sec(\theta)+\tan(\theta)|)\right) - \ln(|\sec(\theta)+\tan(\theta)|)+C$$
$$= x\sqrt{x^2+1}+C$$

37. Let $x = 3\sin(\theta)$, $dx = 3\cos(\theta)\,d\theta$. Note that when $\sin(\theta) = x/3$, $\tan(\theta) = x/\sqrt{9-x^2}$. Draw a right triangle.

$$\int \frac{x^2+1}{(9-x^2)^{3/2}}\,dx = \int \frac{9\sin^2(\theta)+1}{(9-9\sin^2(\theta))^{3/2}} \cdot 3\cos(\theta)\,d\theta = \frac{1}{9}\int \frac{(9\sin^2(\theta)+1)\cos(\theta)}{(1-\sin^2(\theta))^{3/2}}\,d\theta =$$
$$= \frac{1}{9}\int \frac{9\sin^2(\theta)+1}{\cos^2(\theta)}\,d\theta = \frac{1}{9}\int (9\tan^2(\theta)+\sec^2(\theta))\,d\theta$$
$$= \frac{1}{9}\int (10\sec^2(\theta)-9)\,d\theta = \frac{1}{9}(10\tan(\theta)-9\theta)+C$$
$$= \frac{10x}{9\sqrt{9-x^2}} - \arcsin(x/3)+C$$

Further Theory and Practice

39. Let $x = (1/\sqrt{2})\tan(\theta)$ and $dx = (1/\sqrt{2})\sec^2(\theta)\,d\theta$.

$$\int_0^{1/\sqrt{2}} \frac{1}{16x^2+8}\,dx = \int_0^{\pi/4} \frac{1}{8\tan^2(\theta)+8} \cdot \frac{1}{\sqrt{2}}\sec^2(\theta)\,d\theta = \frac{\sqrt{2}}{16}\int_0^{\pi/4} \frac{\sec^2(\theta)}{\tan^2(\theta)+1}\,d\theta$$
$$= \frac{\sqrt{2}}{16}\int_0^{\pi/4} d\theta = \frac{\sqrt{2}}{16}\cdot\frac{\pi}{4} = \frac{\sqrt{2}\,\pi}{64}$$

41. Let $x = (4/3)\tan(\theta)$ and $dx = (4/3)\sec^2(\theta)\,d\theta$. Note that if $\tan(\theta) = 3/4$, then $\sec(\theta) = 5/4$. Draw a right triangle.

$$\int_0^1 \frac{3}{\sqrt{16+9x^2}}\,dx = \int_0^{\arctan(3/4)} \frac{3}{\sqrt{16+16\tan^2(\theta)}} \cdot \frac{4}{3}\sec^2(\theta)\,d\theta$$
$$= \int_0^{\arctan(3/4)} \frac{\sec^2(\theta)}{\sqrt{1+\tan^2(\theta)}}\,d\theta = \int_0^{\arctan(3/4)} \sec(\theta)\,d\theta$$
$$= \ln(|\sec(\theta)+\tan(\theta)|)\Big|_0^{\arctan(3/4)} = \ln(5/4+3/4) = \ln(2)$$

43. Let $x = (3/2)\tan(\theta)$ and $dx = (3/2)\sec^2(\theta)\,d\theta$. Note that if $\tan(\theta) = 4/3$, then $\sin(\theta) = 4/5$. Draw a right triangle.

$$\int_0^2 \frac{6}{(9+4x^2)^{3/2}}\,dx = \int_0^{\arctan(4/3)} \frac{6}{(9+9\tan^2(\theta))^{3/2}} \cdot \frac{3}{2}\sec^2(\theta)\,d\theta$$
$$= \frac{1}{3}\int_0^{\arctan(4/3)} \frac{\sec^2(\theta)}{(1+\tan^2(\theta))^{3/2}}\,d\theta = \frac{1}{3}\int_0^{\arctan(3/4)} \cos(\theta)\,d\theta$$
$$= \frac{1}{3}\sin(\theta)\bigg|_0^{\arctan(4/3)} = \frac{4}{15}$$

45. Let $x = (4/5)\sec(\theta)$ and $dx = (4/5)\sec(\theta)\tan(\theta\,d\theta$. Note that if $\sec(\theta) = 5/4$, then $\tan(\theta) = 3/4$ and if $\sec(\theta) = 5/2$, then $\tan(\theta) = \sqrt{21}/2$. Draw right triangles.

$$\int_1^2 \frac{10}{\sqrt{25x^2-16}}\,dx = \int_{\operatorname{arcsec}(5/4)}^{\operatorname{arcsec}(5/2)} \frac{10}{\sqrt{16\sec^2(\theta)-16}} \cdot \frac{4}{5}\sec(\theta)\tan(\theta)\,d\theta$$
$$= 2\int_{\operatorname{arcsec}(5/4)}^{\operatorname{arcsec}(5/2)} \frac{\sec(\theta)\tan(\theta)}{\sqrt{\sec^2(\theta)-1}}\,d\theta = 2\int_{\operatorname{arcsec}(5/4)}^{\operatorname{arcsec}(5/2)} \sec(\theta)\,d\theta$$
$$= 2(\ln(|\sec(\theta)+\tan(\theta)|)\bigg|_{\operatorname{arcsec}(5/4)}^{\operatorname{arcsec}(5/2)} = 2\ln\left(\frac{5}{2}+\frac{\sqrt{21}}{2}\right) - 2\ln\left(\frac{5}{4}+\frac{3}{4}\right)$$
$$= 2\ln\left(\frac{5+\sqrt{21}}{4}\right)$$

47. Let $u = \operatorname{arcsec}(x)$ and $dv = dx$, so $du = \frac{1}{x\sqrt{x^2-1}}\,dx$ and $v = x$. In the integral that remains make the substitution $x = \sec(\theta)$, $dx = \sec(\theta)\tan(\theta)\,d\theta$.

$$\int_{\sqrt{2}}^2 \operatorname{arcsec}(x)\,dx = x\operatorname{arcsec}(x)\bigg|_{\sqrt{2}}^2 - \int_{\sqrt{2}}^2 \frac{1}{\sqrt{x^2-1}}\,dx$$
$$= 2\cdot\frac{\pi}{3} - \sqrt{2}\cdot\frac{\pi}{4} - \int_{\pi/4}^{\pi/3} \frac{1}{\sqrt{\sec^2(\theta)-1}} \cdot \sec(\theta)\tan(\theta)\,d\theta$$
$$= \frac{(8-3\sqrt{2})\pi}{12} - \int_{\pi/4}^{\pi/3} \sec(\theta)\,d\theta = \frac{(8-3\sqrt{2})\pi}{12} - \ln(|\sec(\theta)+\tan(\theta)|)\bigg|_{\pi/4}^{\pi/3}$$
$$= \frac{(8-3\sqrt{2})\pi}{12} - \ln\left(2+\sqrt{3}\right) + \ln\left(\sqrt{2}+1\right)$$

49. Let $u = \arcsin(x)$ and $dv = x^2\,dx$, so $du = \frac{1}{\sqrt{1-x^2}}\,dx$ and $v = \frac{1}{3}x^3$. In the integral that remains make the substitution $x = \sin(\theta)$, $dx = \cos(\theta)\,d\theta$.

Then split off a sine term and convert the remaining sines into cosines.

$$\int_0^{1/2} x^2 \arcsin(x)\, dx = \frac{1}{3}x^3 \arcsin(x)\Big|_0^{1/2} - \frac{1}{3}\int_0^{1/2} \frac{x^3}{\sqrt{1-x^2}}\, dx$$

$$= \frac{1}{24}\cdot\frac{\pi}{6} - \frac{1}{3}\int_0^{\pi/6} \frac{\sin^3(\theta)}{\sqrt{1-\sin^2(\theta)}}\cdot\cos(\theta)\, d\theta = \frac{\pi}{144} - \frac{1}{3}\int_0^{\pi/6} \sin^3(\theta)\, d\theta$$

$$= \frac{\pi}{144} - \frac{1}{3}\int_0^{\pi/6} (1-\cos^2(\theta))\sin(\theta)\, d\theta = \frac{\pi}{144} - \frac{1}{3}\left(-\cos(\theta) + \frac{1}{3}\cos^3(\theta)\right)\Big|_0^{\pi/6}$$

$$= \frac{\pi}{144} - \frac{1}{3}\left(-\frac{\sqrt{3}}{2} + \frac{1}{3}\cdot\frac{3\sqrt{3}}{8}\right) + \frac{1}{3}\left(-1 + \frac{1}{3}\right) = \frac{\pi}{144} + \frac{\sqrt{3}}{8} - \frac{2}{9}$$

51. Let $x = \tan(\theta)$, $dx = \sec^2(\theta)\, d\theta$. Note that if $\tan(\theta) = x$, then $\sin(\theta) = x/\sqrt{x^2+1}$ and $\cos(\theta) = 1/\sqrt{x^2+1}$.

$$\int \frac{2}{(x^2+1)^2}\, dx = \int \frac{2}{(\tan^2(\theta)+1)^2}\cdot\sec^2(\theta)\, d\theta = 2\int \frac{1}{\sec^2(\theta)}\, d\theta$$

$$= 2\int \cos^2(\theta)\, d\theta = \theta + \sin(\theta)\cos(\theta) + C$$

$$= \arctan(x) + \frac{x}{x^2+1} + C$$

53. Let $x = \tan(\theta)$, $dx = \sec^2(\theta)\, d\theta$. Note that if $\tan(\theta) = x$, then $\csc(\theta) = \sqrt{x^2+1}/x$ and $\sec(\theta) = \sqrt{x^2+1}$, and $\cot(\theta) = 1/x$.

$$\int \frac{\sqrt{1+x^2}}{x}\, dx = \int \frac{\sqrt{1+\tan^2(\theta)}}{\tan(\theta)}\cdot\sec^2(\theta)\, d\theta = \int \frac{\sec^3(\theta)}{\tan(\theta)}\, d\theta$$

$$= \int \frac{\sec^2(\theta)}{\tan(\theta)}\cdot\sec(\theta)\, d\theta = \int \frac{\tan^2(\theta)+1}{\tan(\theta)}\cdot\sec(\theta)\, d\theta = \int (\sec(\theta)\tan(\theta) + \csc(\theta))\, d\theta$$

$$= \sec(\theta) - \ln(|\csc(\theta) + \cot(\theta)|) + C = \sqrt{x^2+1} - \ln\left(\frac{1+\sqrt{x^2+1}}{|x|}\right) + C$$

55. Let $x = \tan(\theta)$, $dx = \sec^2(\theta)\, d\theta$. For the last step, consider the effect of the substitution $u = \sin(\theta)$ in the second part of the integrand.

$$\int \frac{\sqrt{1+x^2}}{x^2}\, dx = \int \frac{\sqrt{1+\tan^2(\theta)}}{\tan^2(\theta)}\cdot\sec^2(\theta)\, d\theta = \int \frac{\sec^3(\theta)}{\tan^2(\theta)}\, d\theta$$

$$= \int \frac{\sec^2(\theta)}{\tan^2(\theta)}\cdot\sec(\theta)\, d\theta = \int \frac{\tan^2(\theta)+1}{\tan^2(\theta)}\cdot\sec(\theta)\, d\theta = \int \left(\sec(\theta) + \frac{\cos(\theta)}{\sin^2(\theta)}\right) d\theta$$

$$= \ln(|\sec(\theta) + \tan(\theta)|) - \frac{1}{\sin(\theta)} + C = \ln(\sqrt{1+x^2} + x) - \frac{\sqrt{1+x^2}}{x} + C$$

57. Substitute $x = \tan(\theta)$, $dx = \sec^2(\theta)\,d\theta$.

$$\int \frac{x^2}{(1+x^2)^{3/2}}\,dx = \int \frac{\tan^2(\theta)}{(1+\tan^2(\theta))^{3/2}} \cdot \sec^2(\theta)\,d\theta = \int \frac{\tan^2(\theta)}{\sec(\theta)}\,d\theta$$
$$= \int \frac{\sec^2(\theta)-1}{\sec(\theta)}\,d\theta = \int (\sec(\theta)-\cos(\theta))\,d\theta$$
$$= \ln(|\sec(\theta)+\tan(\theta)|) - \sin(\theta) + C = \ln(\sqrt{x^2+1}+x) - \frac{x}{\sqrt{x^2+1}} + C$$

59. Complete the square in the denominator then substitute $x-1=\sin(\theta)$, $dx = \cos(\theta)\,d\theta$.

$$\int \frac{1}{\sqrt{2x-x^2}}\,dx = \int \frac{1}{\sqrt{1-(x-1)^2}}\,dx = \int \frac{1}{\sqrt{1-\sin^2(\theta)}} \cdot \cos(\theta)\,d\theta$$
$$= \int d\theta = \theta + C = \arcsin(x-1) + C$$

61. Complete the square in the denominator then substitute $x+1=\tan(\theta)$, $dx = \sec^2(\theta)\,d\theta$. Note that if $\tan(\theta) = x+1$, then $\sec(\theta) = \sqrt{x^2+2x+2}$. The integral of $\sec^3(\theta)$ is obtained using the reduction formula (6.2.13).

$$\int \frac{x}{\sqrt{x^2+2x+2}}\,dx = \int \frac{x}{\sqrt{(x+1)^2+1}}\,dx = \int \frac{\tan(\theta)-1}{\sqrt{\tan^2(\theta)+1}} \cdot \sec^2(\theta)\,d\theta$$
$$= \int (\sec(\theta)\tan(\theta) - \sec(\theta))\,d\theta = \sec(\theta) - \ln(|\sec(\theta)+\tan(\theta)|) + C$$
$$= \sqrt{x^2+2x+2} - \ln(x+1+\sqrt{x^2+2x+2}\,) + C$$

63. Complete the square in the denominator then substitute $x-1=\tan(\theta)$, $dx = \sec^2(\theta)\,d\theta$. If $\tan(\theta) = x+1$, then $\sin(\theta) = (x-1)/\sqrt{x^2-2x+2}$ and $\cos(\theta) = 1/\sqrt{x^2-2x+2}$.

$$\int \frac{4}{(x^2-2x+2)^2}\,dx = \int \frac{4}{((x-1)^2+1)^2}\,dx = \int \frac{4}{(\tan^2(\theta)+1)^2} \cdot \sec^2(\theta)\,d\theta$$
$$= 4\int \frac{\sec^2(\theta)}{\sec^4(\theta)}\,d\theta = 4\int \cos^2(\theta)\,d\theta = 2(\theta + \sin(\theta)\cos(\theta)) + C$$
$$= 2\left(\arctan(x-1) + \frac{x-1}{x^2-2x+2}\right) + C$$

65. Complete the square in the denominator then substitute $x-7 = 3\tan(\theta)$,

$dx = 3\sec^2(\theta)\,d\theta$. If $\tan(\theta) = (x-7)/3$, then $\sec(\theta) = \frac{1}{3}\sqrt{x^2-14x+58}$.

$$\int \frac{6x}{x^2-14x+58}\,dx = \int \frac{6x}{(x-7)^2+9}\,dx = 6\int \frac{3\tan(\theta)+7}{9\tan^2(\theta)+9}\cdot 3\sec^2(\theta)\,d\theta$$
$$= 2\int \frac{3\tan(\theta)+7}{\tan^2(\theta)+1}\sec^2(\theta)\,d\theta = 2\int (3\tan(\theta)+7)\,d\theta$$
$$= 6\ln(|\sec(\theta)|) + 14\theta + C = 3\ln(x^2-14x+58) + 14\arctan\left(\frac{x-7}{3}\right) + C$$

67. Manipulate the numerator to simplify the integral.

$$\int \frac{x^2-6x+8}{x^2+4}\,dx = \int \frac{x^2+4-6x+4}{x^2+4}\,dx = \int \left(1 - 3\cdot\frac{2x-4/3}{x^2+4}\right)dx$$
$$= \int \left(1 - 3\cdot\frac{2x}{x^2+4} + \frac{4}{x^2+4}\right)dx = x - 3\ln(x^2+4) + 2\arctan(x/2) + C$$

69. Factor a 2 out of the numerator and denominator and complete the square in the denominator. Then substitute $x + \frac{3}{2} = \frac{1}{2}\tan(\theta)$, $dx = \frac{1}{2}\sec^2(\theta)\,d\theta$. If $\tan(\theta) = 2x+3$, then $\sec(\theta) = \sqrt{4x^2+12x+10}$. Draw a right triangle.

$$\int \frac{12x}{2x^2+6x+5}\,dx = \int \frac{6x}{x^2+3x+5/2}\,dx = 6\int \frac{x}{\left(x+\frac{3}{2}\right)^2+\frac{1}{4}}\,dx$$
$$= 6\int \frac{\frac{1}{2}\tan(\theta)-\frac{3}{2}}{\frac{1}{4}\tan^2(\theta)+\frac{1}{4}}\cdot\frac{1}{2}\sec^2(\theta)\,d\theta = 6\int (\tan(\theta)-3)d\theta$$
$$= 6\ln(|\sec(\theta)|) - 18\theta + C = 3\ln(2x^2+6x+5) - 18\arctan(2x+3) + C$$

71. Complete the square in the denominator and substitute $x+1 = \tan(\theta)$, $dx = \sec^2(\theta)\,d\theta$. The integral of $\cos^4(\theta)$ uses the integration formula in Exercise 23 of Section 6.2. Note that if $\tan(\theta) = x+1$, then $\sec(\theta) = \sqrt{x^2+2x+1}$, $\cos(\theta) = 1/\sqrt{x^2+2x+2}$, and $\sin(\theta) = (x+1)/\sqrt{x^2+2x+2}$.

$$\int \frac{8x}{(x^2+2x+2)^3}\,dx = \int \frac{8x}{((x+1)^2+1)^3}\,dx = 8\int \frac{\tan(\theta)-1}{(\tan^2(\theta)+1)^3}\cdot\sec^2(\theta)\,d\theta$$
$$= 8\int \frac{(\tan(\theta)-1)\sec^2(\theta)}{\sec^6(\theta)}\,d\theta = 8\int \left(\frac{\sec(\theta)\tan(\theta)}{\sec^5(\theta)} - \cos^4(\theta)\right)d\theta$$
$$= 8\left(-\frac{1}{4\sec^4(\theta)} - \frac{1}{4}\cos^3(\theta)\sin(\theta) - \frac{3}{8}\cos(\theta)\sin(\theta) - \frac{3}{8}\theta\right) + C$$
$$= -\frac{2}{(x^2+2x+2)^2} - \frac{2(x+1)}{(x^2+2x+2)^2} - \frac{3(x+1)}{x^2+2x+2} - 3\arctan(x+1) + C$$
$$= -\frac{2(x+2)}{(x^2+2x+2)^2} - \frac{3(x+1)}{x^2+2x+2} - 3\arctan(x+1) + C$$

73. Complete the square in the denominator and substitute $x - 5 = 5\sin(\theta)$, $dx = 5\cos(\theta)$. Note that if $\sin(\theta) = (x-5)/5$, then $\cos(\theta) = \frac{1}{5}\sqrt{10x - x^2}$.

$$\int \frac{x+1}{\sqrt{10x - x^2}}\,dx = \int \frac{x+1}{\sqrt{25-(x-5)^2}}\,dx = \int \frac{5\sin(\theta)+6}{\sqrt{25-25\sin^2(\theta)}} \cdot 5\cos(\theta)\,d\theta$$
$$= \int (5\sin(\theta)+6)\,d\theta = -5\cos(\theta) + 6\theta + C$$
$$= -\sqrt{10x - x^2} + 6\arcsin\left(\frac{x-5}{5}\right) + C$$

75. Do a long division, then substitute $x = 2\tan(\theta)$, $dx = 2\sec^2(\theta)\,d\theta$.

$$\int \frac{1+x^2}{4+x^2}\,dx = \int \left(1 - \frac{3}{4+x^2}\right)dx = x - 3\int \frac{1}{4+x^2}\,dx$$
$$= x - 3\int \frac{1}{4+4\tan^2(\theta)}\,2\sec^2(\theta)\,d\theta = x - \frac{3}{2}\int d\theta$$
$$= x - \frac{3}{2}\theta + C = x - \frac{3}{2}\arctan(x/2) + C$$

77. Do a long division, then manipulate the expression to simplify the integration. Finish by completing the square and substituting $u = x + 2$.

$$\int \frac{1+x^2}{5+4x+x^2}\,dx = \int \left(1 - \frac{4x-4}{x^2+4x+5}\right)dx$$
$$= \int \left(1 - 2\cdot\frac{2x+4}{x^2+4x+5} + 4\cdot\frac{1}{x^2+4x+5}\right)dx$$
$$= \int \left(1 - 2\cdot\frac{2x+4}{x^2+4x+5} + 4\cdot\frac{1}{(x+2)^2+1}\right)dx$$
$$= x - 2\ln(x^2+4x+5) + 4\arctan(x+2) + C$$

79. Let $x = u^2$, $dx = 2u\,du$. Then make a long division.

$$\int \frac{1}{1+\sqrt{x}}\,dx = \int \frac{1}{1+u}\cdot 2u\,du = 2\int\left(1 - \frac{1}{1+u}\right)du$$
$$= 2(u - \ln(|1+u|)) + C = 2\sqrt{x} - 2\ln(1+\sqrt{x}\,) + C$$

81. Let $x = u^2$, $dx = 2u\,du$. Then make a long division and integrate.

$$\int \frac{1-\sqrt{x}}{1+\sqrt{x}}\,dx = \int \frac{1-u}{1+u}\cdot 2u\,du = 2\int \frac{u-u^2}{1+u}\,du$$
$$= 2\int\left(2 - u - \frac{2}{1+u}\right)du = 4u - u^2 - 4\ln(|1+u|) + C$$
$$= 4\sqrt{x} - x - 4\ln(1+\sqrt{x}\,) + C$$

83. Let $u = \sqrt{1+x}$ so $x = u^2 - 1$ and $dx = 2u\,du$. Finish with the substitution $u = \sec(\theta)$. Note that if $\sec(\theta) = u$, then $\csc(u) = u/\sqrt{u^2-1}$ and $\cot(u) = 1/\sqrt{u^2-1}$.

$$\begin{aligned}
\int \frac{\sqrt{1+x}}{x}\,dx &= \int \frac{u}{u^2-1}\cdot 2u\,du = 2\int \frac{u^2-1+1}{u^2-1}\,du \\
&= 2\int \left(1+\frac{1}{u^2-1}\right)du = 2\int \left(1+\frac{1}{\sec^2(\theta)-1}\right)\cdot \sec(\theta)\tan(\theta)\,d\theta \\
&= 2\int (\sec(\theta)\tan(\theta)+\csc(\theta))\,d\theta = 2\sec(\theta) - 2\ln(|\csc(\theta)+\cot(\theta)|) + C \\
&= 2u - 2\ln\left(\frac{u+1}{\sqrt{u^2-1}}\right) + C = 2u - 2\ln\left(\frac{u+1}{\sqrt{u+1}\sqrt{u-1}}\right) + C \\
&= 2\sqrt{1+x} - 2\ln\left(\frac{\sqrt{u+1}}{\sqrt{u-1}}\right) + C = 2\sqrt{1+x} + \ln\left(\frac{\sqrt{1+x}-1}{\sqrt{1+x}+1}\right) + C
\end{aligned}$$

85. Let $x = a\sin(\theta)$, then $dx = a\cos(\theta)\,d\theta$. Use the fact that if $\sin(\theta) = x/a$, then $\sec(\theta) = a/\sqrt{a^2-x^2}$ and $\tan(\theta) = x/\sqrt{a^2-x^2}$.

$$\begin{aligned}
\int \frac{1}{a^2-x^2}\,dx &= \int \frac{1}{a^2-a^2\sin^2(\theta)}\cdot a\cos(\theta)\,d\theta = \frac{1}{a}\int \frac{\cos(\theta)}{1-\sin^2(\theta)}\,d\theta \\
&= \frac{1}{a}\int \sec(\theta)\,d\theta = \frac{1}{a}\ln(|\sec(\theta)+\tan(\theta)|) + C \\
&= \frac{1}{a}\ln\left(\frac{a+x}{\sqrt{a^2-x^2}}\right) + C
\end{aligned}$$

Let $x = a\sec(\theta)$, then $dx = a\sec(\theta)\tan(\theta)\,d\theta$. Use the fact that if $\sec(\theta) = x/a$, then $\csc(\theta) = x/\sqrt{x^2-a^2}$ and $\cot(\theta) = a/\sqrt{x^2-a^2}$.

$$\begin{aligned}
\int \frac{1}{a^2-x^2}\,dx &= \int \frac{1}{a^2-a^2\sec^2(\theta)}\cdot a\sec(\theta)\tan(\theta)\,d\theta = \frac{1}{a}\int \frac{\sec(\theta)\tan(\theta)}{1-\sec^2(\theta)}\,d\theta \\
&= -\frac{1}{a}\int \csc(\theta)\,d\theta = \frac{1}{a}\ln(|\csc(\theta)+\cot(\theta)|) + C \\
&= \frac{1}{a}\ln\left(\frac{x+a}{\sqrt{x^2-a^2}}\right) + C
\end{aligned}$$

Using the first formula,

$$\alpha = \int_0^{1/2} \frac{1}{1-x^2}\,dx = \ln\left(\frac{1+x}{\sqrt{1-x^2}}\right)\Bigg|_0^{1/2} = \ln\left(\frac{3/2}{\sqrt{3/4}}\right) = \frac{1}{2}\ln(3).$$

Using the second formula,

$$\begin{aligned}
\beta = \int_{\sqrt{2}}^{2} \frac{1}{1-x^2}\,dx = \ln\left(\frac{x+1}{\sqrt{x^2-1}}\right)\Bigg|_{\sqrt{2}}^{2} &= \ln\left(\frac{3}{\sqrt{3}}\right) - \ln\left(\sqrt{2}+1\right) \\
&= \frac{1}{2}\ln(3) - \ln(\sqrt{2}+1).
\end{aligned}$$

Using the identity,

$$\begin{aligned}\int \frac{1}{a^2 - x^2}\,dx &= \frac{1}{2a}\int\left(\frac{1}{x+a} - \frac{1}{x-a}\right)dx\\ &= \frac{1}{2a}\big(\ln(|x+a|) - \ln(|x-a|)\big) + C\\ &= \frac{1}{2a}\ln\left(\left|\frac{x+a}{x-a}\right|\right) + C\,.\end{aligned}$$

Using the integral formula from the identity,

$$\alpha = \int_0^{1/2} \frac{1}{1-x^2}\,dx = \frac{1}{2}\ln\left(\left|\frac{x+1}{x-1}\right|\right)\Bigg|_0^{1/2} = \frac{1}{2}\ln\left(\left|\frac{3/2}{1/2}\right|\right) = \frac{1}{2}\ln(3)\,.$$

Using the integral formula from the identity,

$$\begin{aligned}\beta = \int_{\sqrt{2}}^{2} \frac{1}{1-x^2}\,dx &= \frac{1}{2}\ln\left(\left|\frac{x+1}{x-1}\right|\right)\Bigg|_{\sqrt{2}}^{2} = \frac{1}{2}\ln\left(\left|\frac{3}{1}\right|\right) - \frac{1}{2}\ln\left(\left|\frac{\sqrt{2}+1}{\sqrt{2}-1}\right|\right)\\ &= \frac{1}{2}\ln(3) - \frac{1}{2}\ln\left(\frac{\sqrt{2}+1}{\sqrt{2}-1}\cdot\frac{\sqrt{2}+1}{\sqrt{2}+1}\right)\\ &= \frac{1}{2}\ln(3) - \ln(\sqrt{2}+1)\,.\end{aligned}$$

CALCULATOR/COMPUTER EXERCISES

87. The graphs of $f(x) = (1+x^2)^{-2/3}$ and $g(x) = x^2/\sqrt{1-x^2}$ are shown on the right. Using *Maple* the points of intersection are $a = -0.6604768777\ldots$ and $b = 0.6604768777\ldots$. Also, using *Maple*, the area of the region between the curves is

$$\int_a^b (f(x) - g(x))\,dx = 0.8767016878\ldots\,.$$

89. The graphs of $f(x) = 2x - x^2$ and $g(x) = x^2/(2 + 2x + x^2)$ are shown on the right. Using *Maple* the point of intersection is $b = 1.796321903\ldots$. Also, using *Maple*, the area of the region between the curves is

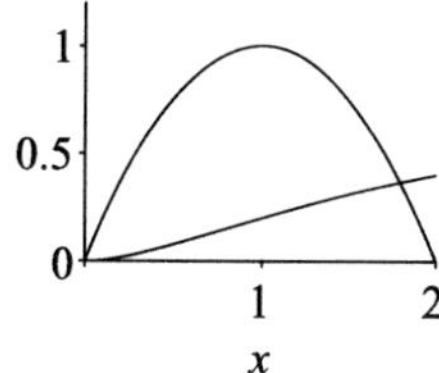

$$\int_0^b (f(x) - g(x))\,dx = 0.9821516711\ldots\,.$$

6.4 Partial Fractions—Linear Factors

Problems for Practice

1. $\dfrac{x+2}{x(x+1)} = \dfrac{A}{x} + \dfrac{B}{x+1}$

3. Make a long division and then factor the denominator.

$$\frac{3x^2+1}{x^2-4} = A + \frac{B}{x-2} + \frac{C}{x+2}$$

5. Long division first to obtain

$$\frac{x^3}{(2x-5)(x+4)} = Ax + B + \frac{C}{2x-5} + \frac{D}{x+4}.$$

7. $\dfrac{x^3+7}{x^2(x+1)(x-2)} = \dfrac{B}{x} + \dfrac{C}{x^2} + \dfrac{D}{x+1} + \dfrac{E}{x-2}$

9. Long division first.

$$\frac{x^6-17}{(x-5)^2(x+3)^2} = Ax^2 + Bx + C + \frac{D}{x-5} + \frac{E}{(x-5)^2} + \frac{F}{x+3} + \frac{G}{(x+3)^2}$$

11. The decomposition is $\dfrac{1}{x(x+1)} = \dfrac{A}{x} + \dfrac{B}{x+1}$ implying that

$$1 = A(x+1) + Bx.$$

If $x = 0$ then $A = 1$, and if $x = -1$ then $B = -1$, so $\dfrac{1}{x(x+1)} = \dfrac{1}{x} - \dfrac{1}{x+1}$.

13. The decomposition is $\dfrac{x+6}{x^2-4} = \dfrac{A}{x-2} + \dfrac{B}{x+2}$ implying that

$$x + 6 = A(x+2) + B(x-2).$$

If $x = 2$ then $A = 2$, and if $x = -2$ then $B = -1$, so

$$\frac{x+6}{x^2-4} = \frac{2}{x-2} - \frac{1}{x+2}.$$

15. The decomposition is $\dfrac{11x}{(2x-5)(x+3)} = \dfrac{A}{2x-5} + \dfrac{B}{x+3}$ implying that

$$11x = A(x+3) + B(2x-5).$$

If $x = 5/2$ then $55/2 = A \cdot 11/2$ and $A = 5$. If $x = -3$ then $B = 3$, so

$$\frac{11x}{(2x-5)(x+3)} = \frac{5}{2x-5} + \frac{3}{x+3}.$$

17. The decomposition is $\dfrac{2x^3+x^2-5x+2}{x^2(x+1)(x-2)} = \dfrac{A}{x} + \dfrac{B}{x^2} + \dfrac{C}{x+1} + \dfrac{D}{x-2}$ implying that

$$2x^3+x^2-5x+2 = Ax(x+1)(x-2)+B(x+1)(x-2)+Cx^2(x-2)+Dx^2(x+1)\,.$$

If $x=0$ then $B=-1$, if $x=-1$ then $C=-2$, and if $x=2$ then $D=1$. To find A observe that the coefficient of the x^3 term on the right hand side of the equation above is $A+C+D$. Since the x^3 coefficient on the left hand side is 2, $A+C+D=2$, and $A=2-C-D=3$. Therefore,

$$\frac{2x^3+x^2-5x+2}{x^2(x+1)(x-2)} = \frac{3}{x} - \frac{1}{x^2} - \frac{2}{x+1} + \frac{1}{x-2}\,.$$

19. Since $\dfrac{2x^3-4x^2-13x+76}{(x-2)^2(x+3)^2} = \dfrac{A}{x-2} + \dfrac{B}{(x-2)^2} + \dfrac{C}{x+3} + \dfrac{D}{(x+3)^2}$,

$$2x^3-4x^2-13x+76 = A(x-2)(x+3)^2+B(x+3)^2+C(x-2)^2(x+3)+D(x-2)^2\,.$$

If $x=2$ then $25B=50$ and $B=2$. If $x=-3$ then $25D=25$ and $D=1$. To find A and C observe that the coefficient of the x^3 term on the right hand side of the equation above is $A+C$. Since the coefficient on the left hand side is 2, $A+C=2$. Both A and C also appear in the constant term on the right. Expand the right hand side to see that the constant is $-18A+9B+12C+4D \overset{(B=2,D=1)}{=} -18A+12C+22$. Since the constant on the left hand side is 76, $-18A+12C=54$. Combining this with $A=2-C$ yields $-36+30C=54$ and $C=3$, $A=-1$. Therefore,

$$\frac{2x^3-4x^2-13x+76}{(x^2-4x+4)(x^2+6x+9)} = -\frac{1}{x-2} + \frac{2}{(x-2)^2} + \frac{3}{x+3} + \frac{1}{(x+3)^2}\,.$$

21. The decomposition is $\dfrac{x+1}{(2x+7)(2x+9)} = \dfrac{A}{2x+7} + \dfrac{B}{2x+9}$ implying that

$$x+1 = A(2x+9)+B(2x+7)\,.$$

If $x=-7/2$ then $-5/2 = A\cdot 2$ and $A=-5/4$. If $x=-9/2$ then $-7/2 = B\cdot(-2)$ and $B=7/4$. Therefore,

$$\frac{x+1}{(2x+7)(2x+9)} = -\frac{5/4}{2x+7} + \frac{7/4}{2x+9}\,.$$

23. The decomposition is $\dfrac{5x^2+3x+1}{(x-2)(x+3)(x+4)} = \dfrac{A}{x-2} + \dfrac{B}{x+3} + \dfrac{C}{x+4}$ implying that

$$5x^2+3x+1 = A(x+3)(x+4)+B(x-2)(x+4)+C(x-2)(x+3)\,.$$

If $x = 2$ then $27 = A \cdot 30$ and $A = 9/10$. If $x = -3$ then $37 = B \cdot (-5)$ and $B = -37/5$. If $x = -4$ then $69 = C \cdot 6$ and $C = 23/2$. Therefore,

$$\frac{5x^2 + 3x + 1}{(x-2)(x+3)(x+4)} = \frac{9/10}{x-2} - \frac{37/5}{x+3} + \frac{23/2}{x+4}.$$

25. The decomposition is $\dfrac{7x^2 + 9x + 5}{x(2x-1)(2x+1)} = \dfrac{A}{x} + \dfrac{B}{2x-1} + \dfrac{C}{2x+1}$ implying that

$$7x^2 + 9x + 5 = A(2x-1)(2x+1) + Bx(2x+1) + Cx(2x-1).$$

If $x = 0$ then $A = -5$. If $x = 1/2$ then $B = 45/4$. If $x = -1/2$ then $C = 9/4$. Therefore,

$$\frac{7x^2 + 9x + 5}{4x^3 - x} = -\frac{5}{x} + \frac{45/4}{2x-1} + \frac{9/4}{2x+1}.$$

27. The integrand decomposes as $\dfrac{3x+1}{(x-1)(x+1)} = \dfrac{A}{x-1} + \dfrac{B}{x+1}$. This implies that $3x+1 = A(x+1) + B(x-1)$. If $x = 1$ then $A = 2$ and if $x = -1$ then $B = 1$ so $\dfrac{3x+1}{x^2-1} = \dfrac{2}{x-1} + \dfrac{1}{x+1}$ and

$$\int \frac{3x+1}{x^2-1}\,dx = \int \left(\frac{2}{x-1} + \frac{1}{x+1}\right) dx = 2\ln(|x-1|) + \ln(|x+1|) + C.$$

29. The integrand decomposes as $\dfrac{9x+18}{(x-3)(x+6)} = \dfrac{A}{x-3} + \dfrac{B}{x+6}$. This implies that $9x + 18 = A(x+6) + B(x-3)$. If $x = 3$ then $A = 5$ and if $x = -6$ then $B = 4$ so $\dfrac{9x+18}{(x-3)(x+6)} = \dfrac{5}{x-3} + \dfrac{4}{x+6}$ and

$$\int \frac{9x+18}{(x-3)(x+6)}\,dx = \int \left(\frac{5}{x-3} + \frac{4}{x+6}\right) dx = 5\ln(|x-3|) + 4\ln(|x+6|) + C.$$

31. Begin with a long division, then decompose to obtain

$$\frac{3x^2}{x^2 + 2x + 1} = 3 + \frac{-6x - 3}{(x+1)^2} = 3 + \frac{A}{x+1} + \frac{B}{(x+1)^2}.$$

This implies that $-6x - 3 = A(x+1) + B$. Therefore, $A = -6$ and if $x = -1$ then $B = 3$ so $\dfrac{3x^2}{x^2+2x+1} = 3 - \dfrac{6}{x+1} + \dfrac{3}{(x+1)^2}$ and

$$\begin{aligned}\int \frac{3x^2}{x^2 + 2x + 1}\,dx &= \int \left(3 - \frac{6}{x+1} + \frac{3}{(x+1)^2}\right) dx \\ &= 3x - 6\ln(|x+1|) - \frac{3}{x+1} + C.\end{aligned}$$

33. The integrand is $\dfrac{x^2+14x+39}{(x+2)(x+3)(x+5)} = \dfrac{A}{x+2} + \dfrac{B}{x+3} + \dfrac{C}{x+5}$. This implies that $x^2+14x+39 = A(x+3)(x+5)+B(x+2)(x+5)+C(x+2)(x+3)$. If $x=-2$ then $A=5$, if $x=-3$ then $B=-3$, and if $x=-5$ then $C=-1$ so $\dfrac{x^2+14x+39}{(x+2)(x+3)(x+5)} = \dfrac{5}{x+2} - \dfrac{3}{x+3} - \dfrac{1}{x+5}$ and

$$\begin{aligned}\int \frac{x^2+14x+39}{(x+2)(x+3)(x+5)}\,dx &= \int \left(\frac{5}{x+2} - \frac{3}{x+3} - \frac{1}{x+5}\right) dx \\ &= 5\ln(|x+2|) - 3\ln(|x+3|) - \ln(|x+5|) + C\,.\end{aligned}$$

35. The integrand decomposes as $\dfrac{x^2+1}{x^2(x+1)} = \dfrac{A}{x} + \dfrac{B}{x^2} + \dfrac{C}{x+1}$. This implies that

$$x^2+1 = Ax(x+1) + B(x+1) + Cx^2\,.$$

If $x=0$ then $B=1$ and if $x=-1$ then $C=2$. To find A observe that the coefficient of x^2 on the right hand side of the displayed equation is $A+C$ and the x^2 coefficient on the left hand side is 1 so $A+C=1$ and $A=1-C=-1$. Therefore, $\dfrac{x^2+1}{x^2(x+1)} = -\dfrac{1}{x} + \dfrac{1}{x^2} + \dfrac{2}{x+1}$ and

$$\begin{aligned}\int \frac{x^2+1}{x^3+x^2}\,dx &= \int \left(-\frac{1}{x} + \frac{1}{x^2} + \frac{2}{x+1}\right) dx \\ &= -\ln(|x|) - \frac{1}{x} + 2\ln(|x+1|) + C\,.\end{aligned}$$

37. The integrand decomposes as $\dfrac{5x-6}{(x-2)^2(x+2)} = \dfrac{A}{x-2} + \dfrac{B}{(x-2)^2} + \dfrac{C}{x+2}$. This implies that

$$5x-6 = A(x-2)(x+2) + B(x+2) + C(x-2)^2\,.$$

If $x=2$ then $B=1$ and if $x=-2$ then $C=-1$. To find A observe that the coefficient of x^2 on the right hand side of the displayed equation is $A+C$ and the x^2 coefficient on the left hand side is 0 so $A+C=0$ and $A=-C=1$. Therefore, $\dfrac{5x-6}{(x-2)^2(x+2)} = \dfrac{1}{x-2} + \dfrac{1}{(x-2)^2} - \dfrac{1}{x+2}$ and

$$\begin{aligned}\int \frac{5x-6}{(x-2)^2(x+2)}\,dx &= \int \left(\frac{1}{x-2} + \frac{1}{(x-2)^2} - \frac{1}{x+2}\right) dx \\ &= \ln(|x-2|) - \frac{1}{x-2} - \ln(|x+2|) + C\,.\end{aligned}$$

39. The integrand decomposes as $\dfrac{4x-7}{(x-1)(x-4)} = \dfrac{A}{x-1} + \dfrac{B}{x-4}$. This implies that $4x-7 = A(x-4) + B(x-1)$. If $x=1$ then $A=1$ and if $x=4$

then $B = 3$ so $\dfrac{4x-7}{(x-1)(x-4)} = \dfrac{1}{x-1} + \dfrac{3}{x-4}$ and

$$\int_2^3 \frac{4x-7}{(x-1)(x-4)}\,dx = \int_2^3 \left(\frac{1}{x-1} + \frac{3}{x-4}\right)dx$$
$$= \big(\ln(|x-1|) + 3\ln(|x-4|)\big)\Big|_2^3 = \ln(2) - 3\ln(2) = -2\ln(2)\,.$$

41. The integrand decomposes as $\dfrac{x+6}{(x-2)(x+2)} = \dfrac{A}{x-2} + \dfrac{B}{x+2}$. This implies that $x+6 = A(x+2) + B(x-2)$. If $x = 2$ then $A = 2$ and if $x = -2$ then $B = -1$ so $\dfrac{x+6}{(x-2)(x+2)} = \dfrac{2}{x-2} - \dfrac{1}{x+2}$ and

$$\int_{-1}^1 \frac{x+6}{(x-2)(x+2)}\,dx = \int_{-1}^1 \left(\frac{2}{x-2} - \frac{1}{x+2}\right)dx$$
$$= \big(2\ln(|x-2|) - \ln(|x+2|)\big)\Big|_{-1}^1 = -\ln(3) - (2\ln(3) - 0) = -3\ln(3)\,.$$

43. The integrand decomposes as $\dfrac{2x^2+5x+4}{x(x+1)(x+2)} = \dfrac{A}{x} + \dfrac{B}{x+1} + \dfrac{C}{x+2}$. This implies that $2x^2+5x+4 = A(x+1)(x+2) + Bx(x+2) + Cx(x+1)$. If $x = 0$ then $A = 2$, if $x = -1$ then $B = -1$ and if $x = -2$ then $C = 1$ so $\dfrac{2x^2+5x+4}{x(x+1)(x+2)} = \dfrac{2}{x} - \dfrac{1}{x+1} + \dfrac{1}{x+2}$ and

$$\int_1^2 \frac{2x^2+5x+4}{x(x+1)(x+2)}\,dx = \int_1^2 \left(\frac{2}{x} - \frac{1}{x+1} + \frac{1}{x+2}\right)dx$$
$$= \big(2\ln(|x|) - \ln(|x+1|) + \ln(|x+2|)\big)\Big|_1^2$$
$$= 2\ln(2) - \ln(3) + \ln(4) - (-\ln(2) + \ln(3))$$
$$= 5\ln(2) - 2\ln(3)\,.$$

Further Theory and Practice

45. The integrand decomposes as $\dfrac{4x^2+x+2}{x^3(x+2)} = \dfrac{A}{x} + \dfrac{B}{x^2} + \dfrac{C}{x^3} + \dfrac{D}{x+2}$. This implies that

$$4x^2+x+2 = Ax^2(x+2) + Bx(x+2) + C(x+2) + Dx^3\,.$$

If $x = 0$ then $C = 1$ and if $x = -2$ then $D = -2$. To find A observe that the x^3 coefficient on the right hand side of the displayed equation is $A+D$ and the x^3 coefficient on the left hand side is 0. Therefore, $A + D = 0$

and $A = -D = 2$. To find B, observe that the x coefficient on the right is $2B + C$ and on the left it is 1 so $2B + C = 1$ and $B = 0$. Consequently, $\dfrac{4x^2 + x + 2}{x^3(x+2)} = \dfrac{2}{x} + \dfrac{1}{x^3} - \dfrac{2}{x+2}$ so

$$\begin{aligned}\int \frac{4x^2 + x + 2}{x^3(x+2)}\,dx &= \int \left(\frac{2}{x} + \frac{1}{x^3} - \frac{2}{x+2}\right) dx \\ &= 2\ln(|x|) - \frac{1}{2x^2} - 2\ln(|x+2|) + C\,.\end{aligned}$$

47. From $\dfrac{3x^3 - 16x^2 + 26x - 14}{(x-1)^2(x-2)^2} = \dfrac{A}{x-1} + \dfrac{B}{(x-1)^2} + \dfrac{C}{x-2} + \dfrac{D}{(x-2)^2}$ we infer that

$$3x^3 - 16x^2 + 26x - 14 = A(x-1)(x-2)^2 + B(x-2)^2 + C(x-1)^2(x-2) + D(x-1)^2\,.$$

If $x = 1$ then $B = -1$ and if $x = 2$ then $D = -2$. To find A and C observe that equating the x^3 coefficients on the right and left hand side of the displayed equation yields $A + C = 3$ and equating the constant terms yields $-4A + 4B - 2C + D = -14$ or $-4A - 2C = -8$. That is, $2A + C = 4$. Therefore, $A = 1$, $C = 2$ and $\dfrac{3x^3 - 16x^2 + 26x - 14}{(x-1)^2(x-2)^2} = \dfrac{1}{x-1} - \dfrac{1}{(x-1)^2} + \dfrac{2}{x-2} - \dfrac{2}{(x-2)^2}$. Consequently,

$$\begin{aligned}\int \frac{3x^3 - 16x^2 + 26x - 14}{(x-1)^2(x-2)^2}\,dx &= \int \left(\frac{1}{x-1} - \frac{1}{(x-1)^2} + \frac{2}{x-2} - \frac{2}{(x-2)^2}\right) dx \\ &= \ln(|x-1|) + \frac{1}{x-1} + 2\ln(|x-2|) + \frac{2}{x-2} + C\,.\end{aligned}$$

49. The integrand is $\dfrac{2x^3 + 6x^2 + 9x + 2}{x(x+1)^3} = \dfrac{A}{x} + \dfrac{B}{x+1} + \dfrac{C}{(x+1)^2} + \dfrac{D}{(x+1)^3}$ from which we infer that

$$2x^3 + 6x^2 + 9x + 2 = A(x+1)^3 + Bx(x+1)^2 + Cx(x+1) + Dx\,.$$

If $x = 0$ then $A = 2$ and if $x = -1$ then $D = 3$. To find B and C observe that equating the x^3 coefficients on the right and left hand side of the displayed equation yields $A + B = 2$ so $B = 0$ and equating the x^2 coefficients yields $3A + 2B + C = 6$. Since $A = 2$ and $B = 0$, $C = 0$ also so $\dfrac{2x^3 + 6x^2 + 9x + 2}{x(x+1)^3} = \dfrac{2}{x} + \dfrac{3}{(x+1)^3}$. Consequently,

$$\begin{aligned}\int_1^2 \frac{2x^3 + 6x^2 + 9x + 2}{x(x+1)^3}\,dx &= \int_1^2 \left(\frac{2}{x} + \frac{3}{(x+1)^3}\right) dx = \left(2\ln(|x|) - \frac{3}{2(x+1)^2}\right)\Bigg|_1^2 \\ &= \left(2\ln(2) - \frac{1}{6}\right) - \left(0 - \frac{3}{8}\right) = 2\ln(2) + \frac{5}{24}\,.\end{aligned}$$

51. The integrand is $\dfrac{3x^2+8x+6}{(x+1)^2(x+2)} = \dfrac{A}{x+1} + \dfrac{B}{(x+1)^2} + \dfrac{C}{x+2}$ which implies that

$$3x^2+8x+6 = A(x+1)(x+2) + B(x+2) + C(x+1)^2 .$$

If $x=-1$ then $B=1$ and if $x=-2$ then $C=2$. To find A observe that equating the x^2 coefficients on the right and left hand side of the displayed equation yields $A+C=3$ so $A=1$ and $\dfrac{3x^2+8x+6}{(x+1)^2(x+2)} = \dfrac{1}{x+1} + \dfrac{1}{(x+1)^2} + \dfrac{2}{x+2}$. Consequently,

$$\begin{aligned}\int_0^1 \frac{3x^2+8x+6}{(x+1)^2(x+2)}\,dx &= \int_0^1 \left(\frac{1}{x+1} + \frac{1}{(x+1)^2} + \frac{2}{x+2}\right) dx \\ &= \left(\ln(|x+1|) - \frac{1}{x+1} + 2\ln(|x+2|)\right)\Bigg|_0^1 \\ &= \ln(2) - \frac{1}{2} + 2\ln(3) - (0 - 1 + 2\ln(2)) = \frac{1}{2} + \ln(9/2) .\end{aligned}$$

53. Make the substitution $u = \exp(x)$, $du = \exp(x)\,dx$.

$$\int \frac{\exp(x)}{\exp(2x)-1}\,dx = \int \frac{1}{u^2-1}\,du$$

Now apply partial fractions: $\frac{1}{(u-1)(u+1)} = \frac{A}{u-1} + \frac{B}{u+1}$ implies that $1 = A(u+1) + B(u-1)$. Let $u=1$ to see that $A = 1/2$ and let $u=-1$ to see that $B=-1/2$ and the integration can be completed as follows.

$$\begin{aligned}\int \frac{\exp(x)}{\exp(2x)-1}\,dx &= \frac{1}{2}\int \left(\frac{1}{u-1} - \frac{1}{u+1}\right) du = \frac{1}{2}(\ln(|u-1|) - \ln(|u+1|) + C \\ &= \frac{1}{2}\big(\ln(|\exp(x)-1|) - \ln(\exp(x)+1)\big) + C\end{aligned}$$

55. Make the substitution $u = \sin(x)$, $du = \cos(x)\,dx$.

$$\int \frac{\cos(x)}{\sin^2(x) - 5\sin(x) + 6}\,dx = \int \frac{1}{u^2-5u+6}\,du$$

Now apply partial fractions: $\frac{1}{(u-3)(u-2)} = \frac{A}{u-3} + \frac{B}{u-2}$ implies that $1 = A(u-2) + B(u-3)$. Let $u=3$ to see that $A=1$ and let $u=2$ to see that $B=-1$ and the integration can be completed as follows.

$$\begin{aligned}&\int \frac{\cos(x)}{\sin^2(x)-5\sin(x)+6}\,dx = \int \left(\frac{1}{u-3} - \frac{1}{u-2}\right) du \\ &\quad = \ln(|u-3|) - \ln(|u-2|) + C = \ln(3-\sin(x)) - \ln(2-\sin(x)) + C\end{aligned}$$

57. Make the substitution $u = x^{1/2}$, $du = \frac{1}{2}x^{-1/2}\,dx$, and $dx = 2u\,du$.

$$\int \frac{1}{x^{3/2} - x}\,dx = \int \frac{2u}{u^3 - u^2}\,du = 2\int \frac{1}{u^2 - u}\,du$$

Now apply partial fractions. The decomposition $\frac{1}{u(u-1)} = \frac{A}{u} + \frac{B}{u-1}$ implies that $1 = A(u-1) + Bu$. Let $u = 0$ to see that $A = -1$ and let $u = 1$ to see that $B = 1$ and the integration can be completed as follows.

$$\begin{aligned}\int \frac{1}{x^{3/2} - x}\,dx &= 2\int \left(-\frac{1}{u} + \frac{1}{u-1}\right)du = 2\big(\ln(|u-1|) - \ln(|u|)\big) + C \\ &= 2\ln(|x^{1/2} - 1|) - \ln(x) + C\end{aligned}$$

59. Make the substitution $u = \sqrt{x+2}$, $du = \frac{1}{2\sqrt{x+2}}\,dx$, and $x = u^2 - 2$.

$$\int \frac{1}{\sqrt{x+2}\,(x+1)}\,dx = 2\int \frac{1}{u^2 - 1}\,du$$

Now apply partial fractions: $\frac{1}{(u-1)(u+1)} = \frac{A}{u-1} + \frac{B}{u+1}$ implies that $1 = A(u+1) + B(u-1)$. Let $u = 1$ to see that $A = 1/2$ and let $u = -1$ to see that $B = -1/2$ and the integration can be completed as follows.

$$\begin{aligned}\int \frac{1}{\sqrt{x+2}\,(x+1)}\,dx &= \int \left(\frac{1}{u-1} - \frac{1}{u+1}\right)du = \big(\ln(|u-1|) - \ln(|u+|)\big) + C \\ &= \ln(|\sqrt{x+2} - 1|) - \ln(\sqrt{x+2} + 1) + C\end{aligned}$$

61. Make the substitution $u = x^{1/6}$, $du = \frac{1}{6}x^{-5/6}\,dx$, and $dx = 6u^5$.

$$\int \frac{1}{x^{1/2} - x^{1/6}}\,dx = 6\int \frac{u^5}{u^3 - u}\,du = 6\int \frac{u^4}{u^2 - 1}\,du$$

Do a long division and then make a partial fraction decomposition: $\frac{u^4}{u^2-1} = u^2 + 1 + \frac{1}{u^2-1} = u^2 + 1 + \frac{A}{u-1} + \frac{B}{u+1}$ implies that $1 = A(u+1) + B(u-1)$. Let $u = 1$ to see that $A = 1/2$ and let $u = -1$ to see that $B = -1/2$ and the integration can be completed as follows.

$$\begin{aligned}\int \frac{1}{x^{1/2} - x^{1/6}}\,dx &= 6\int \left(u^2 + 1 + \frac{1/2}{u-1} - \frac{1/2}{u+1}\right)du \\ &= 2u^3 + 6u + 3\ln(|u-1|) - 3\ln(|u+1|) + C \\ &= 2x^{1/2} + 6x^{1/6} + 3\ln(|x^{1/6} - 1|) - 3\ln(x^{1/6} + 1) + C\end{aligned}$$

63. The partial fraction decomposition $\dfrac{1}{x^2-a^2} = \dfrac{1}{2a}\left(\dfrac{1}{x-a} - \dfrac{1}{x+a}\right)$ implies that

$$\int \frac{dx}{a^2-x^2}\,dx = \frac{1}{2a}\int\left(\frac{1}{a+x}+\frac{1}{a-x}\right)dx = \frac{1}{2a}\big(\ln(|a+x|)-\ln(|a-x|)\big)+C$$
$$= \frac{1}{2a}\ln\left(\frac{a+x}{a-x}\right)+C\,.$$

65. The assumption that $\dfrac{5x^2+18x-1}{(x+4)^2(x-3)} = \dfrac{A_1}{x+4}+\dfrac{B}{x-3}$ implies, that for all x, $5x^2+18x-1 = A_1(x+4)(x-3)+B(x+4)^2$. But this is not possible for if $x=-4$, then the right hand side is 0 and the left hand side is 7.

67. (a) This method of solution is incorrect because the decomposition has only two constants to satisfy three conditions (one for each coefficient of the polynomial in the numerator).

(b) The incorrect method yeilded the correct answer because, for the particular quadratic in the numerator, the other two terms in the *correct* partial decomposition happen to be 0.

(c) Suppose the numerator is $5x^2-4x+1$. Then the displayed partial fraction decomposition implies that, for all x,

$$5x^2-4x+1 = A(x-1)^2+Bx^2\,.$$

If $x=0$ then $A=1$ and $x=1$ implies that $B=2$. But then the coefficient of x^2 on the right hand side of the displayed equation is 3 and on the left hand side it is 5 so the procedure fails.

The correct partial fraction decomposition is $\frac{5x^2-4x+1}{x^2(x-1)^2} = \frac{A}{x}+\frac{B}{x^2}+\frac{C}{x-1}+\frac{D}{(x-1)^2}$. This implies that, for all x,

$$5x^2-4x+1 = Ax(x-1)^2+B(x-1)^2+Cx^2(x-1)+Dx^2\,.$$

If $x=0$ then $B=1$, and if $x=1$ then $D=2$. To find A observe that the coefficient of x on the right hand side of the displayed equation is $A-2B$ and the left hand side coefficient is -4 implying that $A-2B=-4$ so $A=-4+2B=-2$. To obtain C note that the coefficient of x^3 on the right hand side is $A+C$ and the x^3 coefficient on the left is 0 so $A+C=0$ and $C=-A=2$. Therefore, the partial fraction decomposition is

$$\frac{5x^2-4x+1}{x^2(x-1)^2} = -\frac{2}{x}+\frac{1}{x^2}+\frac{2}{x-1}+\frac{2}{(x-1)^2}\,.$$

CALCULATOR/COMPUTER EXERCISES

69. The graphs of $f(x) = (2x^2+5)/q(x)$ and $g(x) = (x^3+2)/q(x)$ are displayed on the right. The point of intersection is $b = 2.485583998$ and the area between the curves is

$$\int_0^b (f(x) - g(x))\,dx = 0.1174351425\ldots .$$

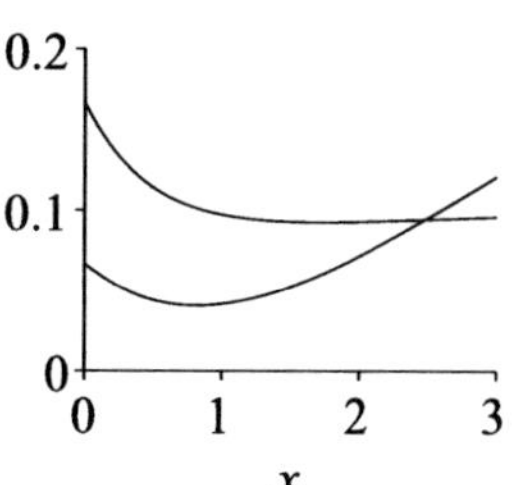

71. The graphs of $f(x) = 4 + 2x - x^2$ and $g(x) = (2x^3 + x + 2)/(x^4 - 5x^3 + 6x - 1)$ are displayed on the right. The point of intersection in $[0.25, 0.5]$ is $a = 0.2679491924$. Using *Maple*'s numerical integration procedure the area between the curves is

$$\int_a^1 (f(x) - g(x))\,dx = 1.681321663\ldots .$$

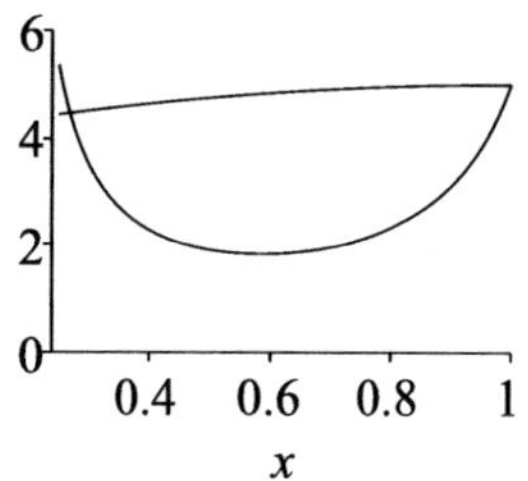

Also using *Maple*, the partial fraction decomposition of $g(x)$ is approximately

$$g(x) = \frac{0.0943501935016854}{x + 1.06900248016489} + \frac{0.390582339725722}{x - 0.170667875021293} - \frac{0.794227179122753}{x - 1.15571663183678} + \frac{2.30929464589535}{x - 4.74261797330682}.$$

The integral approximation using the partial fraction decomposition is the same as the one obtained above.

73. Using *Maple*,

$$\frac{x^3 + 10x^2 + 27x + 24}{x^4 + 10x^3 + 35x^2 + 50x + 24} = \frac{1}{x+1} - \frac{1}{x+2} + \frac{3}{x+3} - \frac{2}{x+4}.$$

75. Using *Maple*,

$$\frac{3x^4 + 12x^3 + 17x^2 + 14x + 2}{x^6 + 2x^5 - x^4 - 4x^3 - x^2 + 2x + 1} = \frac{3}{(x-1)^2} - \frac{1}{(x+1)^4}.$$

6.5 Partial Fractions—Irreducible Quadratic Factors

Problems for Practice

1. No long division is required. Both quadratics in the denominator are irreducible.

$$\frac{2x^3+x+1}{(x^2+1)(x^2+4)} = \frac{Ax+B}{x^2+1} + \frac{Cx+D}{x^2+4}$$

3. No long division is required. Both quadratics in the denominator are irreducible.

$$\frac{2x^3+x+1}{(x^2+1)(x^2+x+1)^2} = \frac{Ax+B}{x^2+1} + \frac{Cx+D}{x^2+x+1} + \frac{Ex+F}{(x^2+x+1)^2}$$

5. Long division is not required. The quadratic in the denominator is irreducible.

$$\frac{2x+1}{(x^2+x+3)(x-4)} = \frac{Ax+B}{x^2+x+3} + \frac{C}{x-4}$$

7. Long division is not required. The quadratic in the denominator is irreducible.

$$\frac{2x^6}{(x^2+4)^3(x-2)} = \frac{Ax+B}{x^2+4} + \frac{Cx+D}{(x^2+4)^2} + \frac{Ex+F}{(x^2+4)^3} + \frac{G}{x-2}$$

9. Long division is not required. The first two quadratics in the denominator are reducible, the third one is irreducible.

$$\frac{3x^5+x+1}{x^2(x-1)(x+1)(x^2+1)} = \frac{A}{x} + \frac{B}{x^2} + \frac{C}{x-1} + \frac{D}{x+1} + \frac{Ex+F}{x^2+1}$$

11. Long division is not required. The quadratic in the denominator is irreducible so $\dfrac{3x^2-5x+4}{(x-1)(x^2+1)} = \dfrac{A}{x-1} + \dfrac{Bx+C}{x^2+1}$. This implies that

$$\begin{aligned} 3x^2-5x+4 &= A(x^2+1)+(Bx+C)(x-1) \\ &= (A+B)x^2+(-B+C)x+A-C\,. \end{aligned}$$

Like powers of x have the same coefficients so $A+B=3$, $-B+C=-5$, and $A-C=4$. Add the first two equations to obtain $A+C=-2$ then add this to the third equation to obtain $2A=2$. Consequently, $A=1$, $B=2$, $C=-3$, and the partial fraction decomposition is

$$\frac{3x^2-5x+4}{(x-1)(x^2+1)} = \frac{1}{x-1} + \frac{2x-3}{x^2+1}\,.$$

13. Long division is not required. Both quadratics in the denominator are irreducible so $\dfrac{7x^3+9x-3x^2-6}{(x^2+2)(x^2+1)} = \dfrac{Ax+B}{x^2+2} + \dfrac{Cx+D}{x^2+1}$. This implies that

$$\begin{aligned}7x^3+9x-3x^2-6 &= (Ax+B)(x^2+1)+(Cx+D)(x^2+2)\\ &= (A+C)x^3+(B+D)x^2+(A+2C)x+B+2D\,.\end{aligned}$$

Comparing like powers of x the fact that $A+C=7$ and $A+2C=9$ implies that $C=2$ and $A=5$. Moreover, $B+D=-3$ and $B+2D=-6$ implies that $D=-3$ and $B=0$. The partial fraction decomposition is

$$\frac{7x^3+9x-3x^2-6}{(x^2+2)(x^2+1)} = \frac{5x}{x^2+2} + \frac{2x-3}{x^2+1}\,.$$

15. Long division is not required. The quadratic in the denominator is irreducible so $\dfrac{x^3-x}{(x^2+1)^2} = \dfrac{Ax+B}{x^2+1} + \dfrac{Cx+D}{(x^2+1)^2}$. This implies that

$$\begin{aligned}x^3-x &= (Ax+B)(x^2+1)+Cx+D\\ &= Ax^3+Bx^2+(A+C)x+B+D\,.\end{aligned}$$

Comparing like powers of x it is clear that $A=1$ and $B=0$. Moreover, because $A+C=-1$, $C=-2$, and because $B+D=0$, $D=0$. The partial fraction decomposition is

$$\frac{x^3-x}{(x^2+1)^2} = \frac{x}{x^2+1} - \frac{2x}{(x^2+1)^2}\,.$$

17. Start with a long division: $\dfrac{x^3+12x^2-9x+48}{(x-3)(x^2+4)} = 1 + \dfrac{15x^2-13x+60}{(x-3)(x^2+4)}$. The quadratic in the denominator of the fractional part is irreducible so $\dfrac{15x^2-13x+60}{(x-3)(x^2+4)} = \dfrac{A}{x-3} + \dfrac{Bx+C}{x^2+4}$. This implies that

$$\begin{aligned}15x^2-13x+60 &= A(x^2+4)+(Bx+C)(x-3)\\ &= (A+B)x^2+(-3B+C)x+4A-3C\,.\end{aligned}$$

Comparing like powers of x, $A+B=15$, $-3B+C=-13$, and $4A-3C=60$. The value of A can be found by substituting $x=3$ above to get $13A=156$ and $A=12$. Therefore $B=3$, and $C=-4$. The partial fraction decomposition is

$$\frac{x^3+12x^2-9x+48}{(x-3)(x^2+4)} = 1 + \frac{12}{x-3} + \frac{3x-4}{x^2+4}\,.$$

19. Long division is not required. The quadratics in the denominator are irreducible so $\dfrac{3x^3-5x^2+10x-19}{(x^2+4)(x^2+3)} = \dfrac{Ax+B}{x^2+4} + \dfrac{Cx+D}{(x^2+3)}$. This implies that

$$\begin{aligned} 3x^3-5x^2+10x-19 &= (Ax+B)(x^2+3)+(Cx+D)(x^2+4) \\ &= (A+C)x^3+(B+D)x^2+(3A+4C)x+3B+4D\,. \end{aligned}$$

Comparing like powers of x, $A+C=3$, $B+D=-5$, $3A+4C=10$, and $3B+4D=-19=0$. Subtract 3 times the first equation from the third to obtain $C=1$ and $A=2$. Subtract 3 times the second equation from the fourth to obtain $D=-4$ and $B=-1$. The partial fraction decomposition is

$$\frac{3x^3-5x^2+10x-19}{(x^2+4)(x^2+3)} = \frac{2x-1}{x^2+4} + \frac{x-4}{(x^2+3)}\,.$$

21. The integrand is in the proper form for a partial fraction decomposition. It is $\frac{3x^2-5x+4}{(x-1)(x^2+1)} = \frac{A}{x-1} + \frac{Bx+C}{x^2+1}$ implying that

$$\begin{aligned} 3x^2-5x+4 &= A(x^2+1)+(Bx+C)(x-1) \\ &= (A+B)x^2+(-B+C)x+A-C\,. \end{aligned}$$

Equating coefficients of like powers of x: $A+B=3$, $-B+C=-5$, and $A-C=4$. Add the first two equations to obtain $A+C=-2$, then add this to the third equation to see that $A=1$ so $B=2$ and $C=-3$. Consequently

$$\begin{aligned} \int \frac{3x^2-5x+4}{(x-1)(x^2+1)}\,dx &= \int \left(\frac{1}{x-1} + \frac{2x-3}{x^2+1}\right) dx \\ &= \int \left(\frac{1}{x-1} + \frac{2x}{x^2+1} - \frac{3}{x^2+1}\right) dx \\ &= \ln(|x-1|) + \ln(x^2+1) - 3\arctan(x) + C\,. \end{aligned}$$

23. The integrand is in the proper form for a partial fraction decomposition. It is $\frac{7x^3+9x-3x^2-6}{(x^2+2)(x^2+1)} = \frac{Ax+B}{x^2+2} + \frac{Cx+D}{x^2+1}$ implying that

$$\begin{aligned} 7x^3+9x-3x^2-6 &= (Ax+B)(x^2+1)+(Cx+D)(x^2+2) \\ &= (A+C)x^3+(B+D)x^2+(A+2C)x+B+2D\,. \end{aligned}$$

Equating coefficients of like powers of x, $A+C=7$ and $A+2C=9$, implying that $C=2$ and $A=5$. Moreover, $B+D=-3$ and $B+2D=-6$, implying $D = -3$ and $B = 0$. Consequently

$$\begin{aligned}\int \frac{7x^3 + 9x - 3x^2 - 6}{(x^2+2)(x^2+1)}\,dx &= \int \left(\frac{5x}{x^2+2} + \frac{2x-3}{x^2+1}\right) dx \\ &= \int \left(\frac{5}{2}\cdot\frac{2x}{x^2+2} + \frac{2x}{x^2+1} - \frac{3}{x^2+1}\right) dx \\ &= \frac{5}{2}\ln(x^2+2) + \ln(x^2+1) - 3\arctan(x) + C\,.\end{aligned}$$

25. The integrand is in the proper form for a partial fraction decomposition. It is $\frac{x^3-x}{(x^2+1)^2} = \frac{Ax+B}{x^2+1} + \frac{Cx+D}{(x^2+1)^2}$ implying that

$$\begin{aligned}x^3 - x &= (Ax+B)(x^2+1) + Cx + D \\ &= Ax^3 + Bx^2 + (A+C)x + B + D\,.\end{aligned}$$

Equating coefficients of like powers of x, $A = 1$, $B = 0$ and $A + C = -1$, $B + D = 0$, implying that $C = -2$ and $D = 0$. Consequently

$$\begin{aligned}\int \frac{x^2+2x}{(x^2+1)^2}\,dx &= \int \left(\frac{x}{x^2+1} - \frac{2x}{(x^2+1)^2}\right) dx \\ &= \frac{1}{2}\ln(x^2+1) + \frac{1}{x^2+1} + C\,.\end{aligned}$$

Both terms in the integrand are easily evaluated using the substitution $u = x^2 + 1$.

27. Start with a long division: $\frac{x^3+12x^2-9x+48}{(x-3)(x^2+4)} = 1 + \frac{15x^2-13x+60}{(x-3)(x^2+4)}$. The partial fraction decomposition of the fractional part is $\frac{15x^2-13x+60}{(x-3)(x^2+4)} = \frac{A}{x-3} + \frac{Bx+C}{x^2+4}$. This implies that

$$\begin{aligned}15x^2 - 13x + 60 &= A(x^2+4) + (Bx+C)(x-3) \qquad (1)\\ &= (A+B)x^2 + (-3B+C)x + 4A - 3C\,.\end{aligned}$$

Comparing like powers of x, $A+B = 15$, $-3B+C = -13$, and $4A-3C = 60$. The value of A can be found by substituting $x = 3$ in Equation (1) above to get $13A = 156$ and $A = 12$. Therefore $B = 3$, and $C = -4$. Consequently

$$\begin{aligned}\int \frac{x^3+12x^2-9x+48}{(x-3)(x^2+4)}\,dx &= \int \left(1 + \frac{12}{x-3} + \frac{3x-4}{x^2+4}\right) dx \\ &= \int \left(1 + \frac{12}{x-3} + \frac{3}{2}\cdot\frac{2x}{x^2+4} - \frac{4}{x^2+4}\right) dx \\ &= x + 12\ln(|x-3|) + \frac{3}{2}\ln(x^2+4) - 2\arctan\left(\frac{x}{2}\right) + C\,.\end{aligned}$$

The last term in the integrand is evaluated using the substitution $x = 2\tan(\theta)$ to obtain $\int 2\,d\theta$.

29. The integrand is in the proper form for a partial fraction decomposition. It is $\frac{3x^3-5x^2+10x-19}{(x^2+4)(x^2+3)} = \frac{Ax+B}{x^2+4} + \frac{Cx+D}{x^2+3}$ implying that

$$\begin{aligned}3x^3 - 5x^2 + 10x - 19 &= (Ax+B)(x^2+3) + (Cx+D)(x^2+4) \\ &= (A+C)x^3 + (B+D)x^2 + (3A+4C)x + 3B + 4D\,.\end{aligned}$$

Comparing like powers of x, $A+C=3$, $B+D=-5$, $3A+4C=10$, and $3B+4D=-19$. Subtract 3 times the first equation from the third to obtain $C=1$ and $A=2$. Subtract 3 times the second equation from the fourth to obtain $D=-4$ and $B=-1$. Consequently

$$\begin{aligned}\int \frac{3x^3-5x^2+10x-19}{(x^2+4)(x^2+3)}\,dx &= \int \left(\frac{2x-1}{x^2+4} + \frac{x-4}{x^2+3}\right) dx \\ &= \int \left(\frac{2x}{x^2+4} - \frac{1}{x^2+4} + \frac{1}{2}\cdot\frac{2x}{x^2+3} - \frac{4}{x^2+3}\right) dx\,.\end{aligned}$$

The first and third terms integrate easily to logarithms. The second and fourth terms integrate via the substitutions $x = 2\tan(\theta)$ and $x = \sqrt{3}\tan(\theta)$ yielding

$$\begin{aligned}\int \frac{3x^3-5x^2+10x-19}{(x^2+4)(x^2+3)}\,dx &= \ln(x^2+4) - \frac{1}{2}\arctan\left(\frac{x}{2}\right) \\ &\quad + \frac{1}{2}\ln(x^2+3) - \frac{4}{\sqrt{3}}\arctan\left(\frac{x}{\sqrt{3}}\right) + C\,.\end{aligned}$$

31. The integrand is in the proper form for a partial fraction decomposition. It is $\frac{x^2+3}{(x^2+1)(x^2+2)} = \frac{Ax+B}{x^2+1} + \frac{Cx+D}{x^2+2}$ implying that

$$\begin{aligned}x^2+3 &= (Ax+B)(x^2+2) + (Cx+D)(x^2+1) \\ &= (A+C)x^3 + (B+D)x^2 + (2A+C)x + 2B + D\,.\end{aligned}$$

Comparing like powers of x, $A+C=0$, $B+D=1$, $2A+C=0$, and $2B+D=3$. Subtract the first equation from the third to obtain $A=0$ and $C=0$. Subtract the second from the fourth to obtain $B=2$ and $D=-1$. Consequently

$$\begin{aligned}\int_0^1 \frac{x^2+3}{(x^2+1)(x^2+2)}\,dx &= \int_0^1 \left(\frac{2}{x^2+1} - \frac{1}{x^2+2}\right) dx \\ &= \left(2\arctan(x) - \frac{\sqrt{2}}{2}\arctan\left(\frac{x}{\sqrt{2}}\right)\right)\Bigg|_0^1 \\ &= \frac{\pi}{2} - \frac{\sqrt{2}}{2}\arctan(\left(\frac{1}{\sqrt{2}}\right)\,.\end{aligned}$$

The second term in the integrand evaluates via $x = \sqrt{2}\tan(\theta)$.

33. The integrand is in the proper form for a partial fraction decomposition. It is $\frac{7x^2+4x+6}{(x+1)(x^2+2)} = \frac{A}{x+1} + \frac{Bx+C}{x^2+2}$ implying that

$$\begin{aligned} 7x^2 + 4x + 6 &= A(x^2+2) + (Bx+C)(x+1) \qquad (1) \\ &= (A+B)x^2 + (B+C)x + 2A + C\,. \end{aligned}$$

Let $x = -1$ in Equation (1) to see that $A = 3$ so, comparing coefficients of like powers of x, $B = 4$ and $C = 0$. Consequently

$$\begin{aligned} \int_0^1 \frac{7x^2+4x+6}{(x+1)(x^2+2)}\,dx &= \int_0^1 \left(\frac{3}{x+1} + \frac{4x}{x^2+2}\right) dx \\ &= \left(3\ln(|x+1|) + 2\ln(x^2+2)\right)\Big|_0^1 \\ &= 3\ln(2) + 2\ln(3) - 2\ln(2) = \ln(18) \end{aligned}$$

35. The integrand is in the proper form for a partial fraction decomposition. It is $\frac{2x^2-x+2}{x(x^2+1)} = \frac{A}{x} + \frac{Bx+C}{x^2+1}$ implying that

$$\begin{aligned} 2x^2 - x + 2 &= A(x^2+1) + (Bx+C)x \\ &= (A+B)x^2 + Cx + A\,. \end{aligned}$$

Comparing coefficients of like powers of x, $A = 2$, $B = 0$, and $C = -1$. Consequently

$$\begin{aligned} \int_1^2 \frac{2x^2-x+2}{x(x^2+1)}\,dx &= \int_1^2 \left(\frac{2}{x} - \frac{1}{x^2+1}\right) dx \\ &= \left(2\ln(|x|) - \arctan(x)\right)\Big|_1^2 \\ &= 2\ln(2) - \arctan(2) + \frac{\pi}{4}\,. \end{aligned}$$

37. Rearrange the numerator to simplify the integrand.

$$\begin{aligned} \int_0^1 \frac{x^2+x+1}{(x^2+1)^2}\,dx &= \int_0^1 \frac{x^2+1+x}{(x^2+1)^2}\,dx \\ &= \int_0^1 \left(\frac{1}{x^2+1} + \frac{1}{2}\cdot\frac{2x}{(x^2+1)^2}\right) dx \\ &= \left(\arctan(x) - \frac{1}{2}\cdot\frac{1}{x^2+1}\right)\Big|_0^1 \\ &= \frac{\pi}{4} - \frac{1}{4} - \left(0 - \frac{1}{2}\right) = \frac{\pi}{4} + \frac{1}{4}\,. \end{aligned}$$

Further Theory and Practice

39. The integrand is in the proper form for a partial fraction decomposition. It is $\frac{3x^2+6x+4}{(x+1)(x^2+2x+2)} = \frac{A}{x+1} + \frac{Bx+C}{x^2+2x+2}$ implying that

$$\begin{aligned} 3x^2 + 6x + 4 &= A(x^2 + 2x + 2) + (Bx + C)(x + 1) \\ &= (A + B)x^2 + (2A + B + C)x + 2A + C\,. \end{aligned}$$

Comparing coefficients of like powers of x, $A + B = 3$, $2A + B + C = 6$, and $2A + C = 4$. Subtract the third equation from the second to see that $B = 2$, so $A = 1$ and $C = 2$. Consequently

$$\begin{aligned} \int \frac{3x^2 + 6x + 4}{(x + 1)(x^2 + 2x + 2)}\,dx &= \int \left(\frac{1}{x + 1} + \frac{2x + 2}{x^2 + 2x + 2}\right) dx \\ &= \ln(|x + 1|) + \ln(x^2 + 2x + 2) + C\,. \end{aligned}$$

41. Factor the denominator so that the integrand is in the proper form for a partial fraction decomposition. It is $\frac{2x^2+4x+9}{(x-1)(x^2+x+1)} = \frac{A}{x-1} + \frac{Bx+C}{x^2+x+1}$ implying that

$$\begin{aligned} 2x^2 + 4x + 9 &= A(x^2 + x + 1) + (Bx + C)(x - 1) \\ &= (A + B)x^2 + (A - B + C)x + A - C\,. \end{aligned}$$

Comparing coefficients of like powers of x, $A + B = 2$, $A - B + C = 4$, and $A - C = 9$. Add all three equations to see that $A = 5$ implying that $B = -3$ and $C = -4$. Consequently

$$\begin{aligned} \int \frac{2x^2 + 4x + 9}{x^3 - 1}\,dx &= \int \left(\frac{5}{x - 1} - \frac{3x + 4}{x^2 + x + 1}\right) dx \\ &= 5\ln(|x - 1|) - \frac{3}{2}\int \frac{2x + 1 + 5/3}{x^2 + x + 1}\,dx \\ &= 5\ln(|x - 1|) - \frac{3}{2}\ln(x^2 + x + 1) - \frac{5}{2}\int \frac{1}{x^2 + x + 1}\,dx\,. \end{aligned}$$

Now, in the last integral, complete the square: $x^2+x+1 = (x+1/2)^2+3/4$, and make the substitution $x+1/2 = \frac{\sqrt{3}}{2}\tan(\theta)$ to obtain $\frac{2}{\sqrt{3}}\arctan\left(\frac{2x+1}{\sqrt{3}}\right)$ and

$$\int \frac{2x^2 + 4x + 9}{x^3 - 1}\,dx = 5\ln(|x-1|) - \frac{3}{2}\ln(x^2+x+1) - \frac{5}{\sqrt{3}}\arctan\left(\frac{2x + 1}{\sqrt{3}}\right) + C\,.$$

43. Both 1 and -1 are roots of the denominator of the integrand. Make a long division (divide by $x^2 - 1$) to obtain its partial fraction form: $\frac{8x}{(x+1)(x-1)^2(x^2+1)} = \frac{A}{x+1} + \frac{B}{x-1} + \frac{C}{(x-1)^2} + \frac{Dx+E}{x^2+1}$ implying that

$$\begin{aligned} 8x &= A(x - 1)^2(x^2 + 1) + B(x + 1)(x - 1)(x^2 + 1) \\ &\qquad + C(x + 1)(x^2 + 1) + (Dx + E)(x + 1)(x - 1)^2 \\ &= (A + B + D)x^4 + (-2A + C - D + E)x^3 + (2A + C - D - E)x^2 \\ &\qquad + (-2A + C + D - E)x + A - B + C + E\,. \end{aligned}$$

Let $x = -1$ to see that $A = -1$ and let $x = 1$ to see that $C = 2$. Comparing coefficients of x^3, $-D + E = -4$, and comparing coefficients of x^2, $D + E = 0$. This implies that $D = 2$, $E = -2$ and $B = -1$. Consequently

$$\begin{aligned}\int \frac{8x}{x^5 - x^4 - x + 1}\,dx &= \int \left(-\frac{1}{x+1} - \frac{1}{x-1} + \frac{2}{(x-1)^2} + \frac{2x-2}{x^2+1}\right) dx\\ &= -\ln(|x+1|) - \ln(|x-1|) - \frac{2}{x-1} + \ln(x^2+1) - 2\arctan(x) + C\\ &= \ln\left(\frac{x^2+1}{|x^2-1|}\right) - \frac{2}{x-1} - 2\arctan(x) + C\,.\end{aligned}$$

45. The proposed decomposition is $\frac{3x^2+11x+10}{(x+1)(x^2+4x+3)} = \frac{A}{x+1} + \frac{Bx+C}{x^2+4x+3}$. If this is an identity then so is

$$\begin{aligned}3x^2 + 11x + 10 &= A(x^2 + 4x + 3) + (Bx + C)(x + 1)\\ &= (A + B)x^2 + (4A + B + C)x + 3A + C\,,\end{aligned}$$

implying that $A + B = 3$, $4A + B + C = 11$, and $3A + C = 10$. Add the first and third equations to obtain $4A + B + C = 13$ contradicting the second equation. There is no combination of A, B, and C that will make the decomposition an identity.

CALCULATOR/COMPUTER EXERCISES

47. The graphs of $f(x) = x^4+2$ and $g(x) = \frac{5x^3+3x+2}{x^3+x^2+x+1}$ are shown on the right for $0 \le x \le 1$. The abscissa of the point of intersection is $b = 0.7585244270\ldots$ and the area of the region between the two curves is

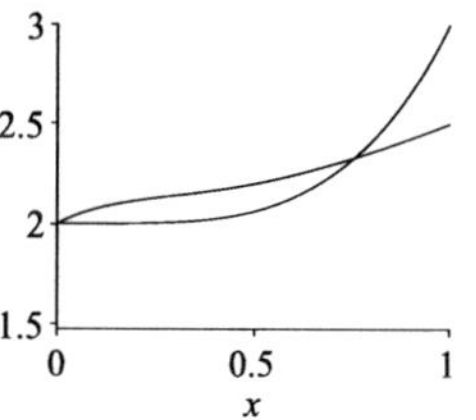

$$\int_0^b (g(x) - f(x))\,dx = 0.07744458064\ldots.$$

We note that $g(x)$ has the partial fraction decomposition $5 - \frac{3}{x+1} - \frac{2x}{x^2+1}$ and the "exact" area of the region is $3b - \frac{1}{5}b^5 - 3\ln(b+1) - \ln(b^2+1)$.

49. The graphs of $f(x) = 1-x^2$ and $g(x) = \frac{3x^2+4x+5}{x^3+3x^2+7x+5}$ are shown on the right for $0 \le x \le 0.5$. The abscissas of the points of intersection are $a = 0$ and $b = 0.3843270419\ldots$ and the area of the region between the two curves is

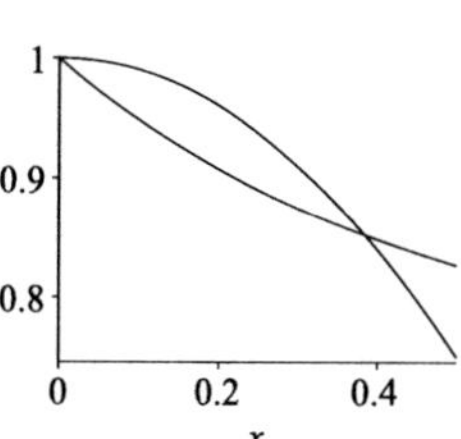

$$\int_0^b (f(x) - g(x))\,dx = 0.01370615330\ldots.$$

We note that $g(x)$ has the partial fraction decomposition $\frac{x}{x+1} + \frac{2x}{x^2+2x+5}$

and the "exact" area of the region is $\frac{1}{3}b^3 - b + \arctan\left(\frac{1}{2}\right) - \arctan\left(\frac{b+1}{2}\right) + \ln(b^2 + 2b + 5) + \ln(b + 1)$.

6.6 Improper Integrals—Unbounded Integrands

Problems for Practice

1. The integrand is infinite at the right endpoint. Let $0 < \epsilon < 1$. Then

$$\begin{aligned}\int_1^{5-\epsilon} (x-5)^{-4/3}\,dx &= \left(-3(x-5)^{-1/3}\right)\Big|_1^{5-\epsilon} \\ &= 3\epsilon^{-1/3} - 3\cdot 4^{-1/3} \xrightarrow[\epsilon\to 0^+]{} \infty\,.\end{aligned}$$

The integral diverges.

3. The integrand is infinite at the right endpoint. Let $0 < \epsilon < 1$. Then

$$\begin{aligned}\int_2^{4-\epsilon} (4-x)^{-0.9}\,dx &= -10(4-x)^{0.1}\Big|_2^{4-\epsilon} \\ &= -10\epsilon^{0.1} + 10\cdot 2^{0.1} \xrightarrow[\epsilon\to 0^+]{} 10\cdot 2^{0.1}\,.\end{aligned}$$

The integral converges, $\int_2^4 (4-x)^{-0.9}\,dx = 10\cdot 2^{0.1}$.

5. The integrand is infinite at the right endpoint. Let $0 < \epsilon < 1$. Then

$$\begin{aligned}\int_0^{\pi/2-\epsilon} \tan(x)\,dx &= \ln(|\sec(x)|)\Big|_0^{\pi/2-\epsilon} \\ &= \ln(\sec(\pi/2) - \epsilon) \xrightarrow[\epsilon\to 0^+]{} \infty\,.\end{aligned}$$

The integral diverges.

7. The integrand is infinite at the right endpoint. Let $0 < \epsilon < 1$. Then

$$\begin{aligned}\int_0^{1-\epsilon} \frac{x}{(1-x^2)^{1/4}}\,dx &= -\frac{1}{2}\int_0^{1-\epsilon} \frac{-2x}{(1-x^2)^{1/4}}\,dx \\ &= -\frac{1}{2}\left(\frac{4}{3}(1-x^2)^{3/4}\right)\Big|_0^{1-\epsilon} \\ &= -\frac{2}{3}(1-(1-\epsilon)^2)^{3/4} + \frac{2}{3} \xrightarrow[\epsilon\to 0^+]{} \frac{2}{3}\,.\end{aligned}$$

The integral converges, $\int_0^1 \frac{x}{(1-x^2)^{1/4}}\,dx = 2/3$.

9. The integrand is infinite at the right endpoint. Let $0 < \epsilon < 1$. Then, using a partial fraction decomposition,

$$\begin{aligned}\int_0^{2-\epsilon} \frac{1}{4-x^2}\,dx &= \frac{1}{4}\int_0^{2-\epsilon}\left(\frac{1}{2+x}+\frac{1}{2-x}\right)dx\\ &= \frac{1}{4}\big(\ln(2+x)-\ln(2-x)\big)\Big|_0^{2-\epsilon}\\ &= \frac{1}{4}\big(\ln(4-\epsilon)-\ln(\epsilon)\big) \xrightarrow[\epsilon\to 0^+]{} \infty\,.\end{aligned}$$

The integral diverges.

11. The integrand is infinite at the left endpoint. Let $0 < \epsilon < 1$. Then

$$\begin{aligned}\int_{-3+\epsilon}^{2} (x+3)^{-1.1}\,dx &= -10(x+3)^{-0.1}\Big|_{-3+\epsilon}^{2}\\ &= -10\cdot 5^{-0.1} + 10\,\epsilon^{-0.1} + \xrightarrow[\epsilon\to 0^+]{} \infty\,.\end{aligned}$$

The integral diverges.

13. The integrand is infinite at the left endpoint. Let $0 < \epsilon < 1$. Then

$$\begin{aligned}\int_{0+\epsilon}^{8} x^{-1/3}\,dx &= \frac{3}{2}x^{2/3}\Big|_{0+\epsilon}^{8}\\ &= 6 - \frac{3}{2}\epsilon^{2/3} \xrightarrow[\epsilon\to 0^+]{} 6\,.\end{aligned}$$

The integral converges, $\int_0^8 x^{-1/3}\,dx = 6$.

15. The integrand is infinite at the left endpoint. Let $0 < \epsilon < 1$. The integral evaluates using the substitution $u = \ln(x)$, $du = \frac{1}{x}\,dx$.

$$\begin{aligned}\int_{0+\epsilon}^{1} \frac{\ln(x)}{x}\,dx &= \int_{\ln(\epsilon)}^{0} u\,du = \frac{1}{2}u^2\Big|_{\ln(\epsilon)}^{0}\\ &= -\frac{1}{2}\ln^2(\epsilon) \xrightarrow[\epsilon\to 0^+]{} -\infty\end{aligned}$$

The integral diverges.

17. The integrand is infinite at the left endpoint. Let $0 < \epsilon < 1$. Then

$$\begin{aligned}\int_{0+\epsilon}^{3} (x^{-1/2}+x^{1/2})\,dx &= \left(2x^{1/2}+\frac{2}{3}x^{3/2}\right)\Big|_{0+\epsilon}^{3}\\ &= 2\cdot 3^{1/2} + \frac{2}{3}\cdot 3^{3/2} - \left(2\epsilon^{1/2}+\frac{2}{3}\epsilon^{3/2}\right) \xrightarrow[\epsilon\to 0^+]{} 4\sqrt{3}\,.\end{aligned}$$

The integral converges, $\int_0^3 x^{-1/2}(1+x)\,dx = 4\sqrt{3}$.

19. The integrand is infinite at the left endpoint. Let $0 < \epsilon < 1$. The integral evaluates via integration by parts, $u = \ln(x)$, $dv = dx$, $du = \frac{1}{x}\,dx$, $v = x$. Note that $\lim_{\epsilon\to 0^+} \epsilon\ln(\epsilon) = 0$ by L'Hópital's Rule.

$$\begin{aligned}\int_{0+\epsilon}^{1} \ln(x)\,dx &= x\ln(x)\Big|_{\epsilon}^{1} - \int_{\epsilon}^{1} dx \\ &= -\epsilon\ln(\epsilon) - (1-\epsilon) \xrightarrow[\epsilon\to 0^+]{} -1\end{aligned}$$

The integral converges, $\int_0^1 \ln(x)\,dx = -1$.

21. The integrand is infinite at the interior point $x = 1$. Let $0 < \epsilon < 1$. Then

$$\int_0^{1-\epsilon} \frac{1}{x-1}\,dx = \ln(|x-1|)\Big|_0^{1-\epsilon} = \ln(\epsilon) \xrightarrow[\epsilon\to 0^+]{} -\infty\,.$$

The integral diverges.

23. The integrand is infinite at the interior point $x = -3$. Let $0 < \epsilon < 1$. Then

$$\int_{-5}^{-3-\epsilon} 3(x+3)^{-2/5}\,dx = 5(x+3)^{3/5}\Big|_{-5}^{-3-\epsilon} = -5\epsilon^{3/5} + 5\cdot 2^{3/5} \xrightarrow[\epsilon\to 0^+]{} 5\cdot 2^{3/5}$$

and

$$\int_{-3+\epsilon}^{-1} 3(x+3)^{-2/5}\,dx = 5(x+3)^{3/5}\Big|_{-3+\epsilon}^{-1} = 5\cdot 2^{3/5} - 5\epsilon^{3/5} \xrightarrow[\epsilon\to 0^+]{} 5\cdot 2^{3/5}\,.$$

The integral converges, $\int_{-5}^{-1} \frac{3}{(x+3)^{2/5}}\,dx = 10\cdot 2^{3/5}$.

25. The integrand is infinite at the interior point $x = -1$. Let $0 < \epsilon < 1$. Then

$$\int_{-2}^{-1-\epsilon} (x+1)^{-2/3}\,dx = 3(x+1)^{1/3}\Big|_{-2}^{-1-\epsilon} = -3\epsilon^{1/3} + 3 \xrightarrow[\epsilon\to 0^+]{} 3$$

and

$$\int_{-1+\epsilon}^{4} (x+1)^{-2/3}\,dx = 3(x+1)^{1/3}\Big|_{-1+\epsilon}^{4} = 3\cdot 5^{1/3} - 3\epsilon^{1/3} \xrightarrow[\epsilon\to 0^+]{} 3\cdot 5^{1/3}\,.$$

The integral converges, $\int_{-2}^{4}(x+1)^{-2/3}\,dx = 3(1+5^{1/3})$.

27. The integrand is infinite at the interior point $x = \sqrt{2}$. Let $0 < \epsilon < 1$. The integrals is easy via $u = x^2 - 2$, $du = 2x\,dx$.

$$\begin{aligned}\frac{1}{2}\int_0^{\sqrt{2}-\epsilon} \frac{2x}{x^2-2}\,dx &= \frac{1}{2}\ln(|x^2-2|)\Big|_0^{\sqrt{2}-\epsilon} \\ &= \frac{1}{2}\ln(|\epsilon^2 - 2\sqrt{2}\,\epsilon|) - \frac{1}{2}\ln(2) \xrightarrow[\epsilon\to 0^+]{} -\infty\end{aligned}$$

The integral diverges.

29. The integrand is infinite at the interior point $x = 4$. Let $0 < \epsilon < 1$. The integral is evaluated using the substitution $u = \sqrt{x}$, $du = \frac{1}{2\sqrt{x}}\,dx$.

$$\int_1^{4-\epsilon} \frac{1}{\sqrt{x}}\left(\frac{1}{2-\sqrt{x}}\right)^{1/3} dx = 2\int_1^{\sqrt{4-\epsilon}} (2-u)^{-1/3}\,du = -3(2-u)^{2/3}\Big|_1^{\sqrt{4-\epsilon}}$$
$$= -3(2-\sqrt{4-\epsilon}\,)^{2/3} + 3 \xrightarrow[\epsilon\to 0^+]{} 3$$

and

$$\int_{4+\epsilon}^{16} \frac{1}{\sqrt{x}}\left(\frac{1}{2-\sqrt{x}}\right)^{1/3} dx = 2\int_{\sqrt{4+\epsilon}}^{4} (2-u)^{-1/3}\,du = -3(2-u)^{2/3}\Big|_{\sqrt{4+\epsilon}}^{4}$$
$$= -3\cdot 2^{2/3} + 3(2-\sqrt{4+\epsilon}\,)^{2/3} \xrightarrow[\epsilon\to 0^+]{} -3\cdot 2^{2/3}$$

The integral converges, $\int_1^{16} \frac{1}{\sqrt{x}}\left(\frac{1}{2-\sqrt{x}}\right)^{1/3} dx = 3(1-2^{2/3})$.

31. The integrand is infinite at both endpoints. We split it into two at $x = 0$. Let $0 < \epsilon < 1$. The integral is evaluated via partial fractions.

$$\int_{-1+\epsilon}^{0} \frac{1}{1-x^2}\,dx = \frac{1}{2}\int_{-1+\epsilon}^{0}\left(\frac{1}{1+x} + \frac{1}{1-x}\right)dx$$
$$= \frac{1}{2}\big(\ln(|1+x|) - \ln(|1-x|)\big)\Big|_{-1+\epsilon}^{0} = -\frac{1}{2}\big(\ln(\epsilon) - \ln(2-\epsilon)\big) \xrightarrow[\epsilon\to 0^+]{} \infty$$

The integral diverges.

33. The integrand is infinite at both endpoints. We split it into two at $x = 1/2$. Let $0 < \epsilon < 1/2$. The integral is evaluated via the substitution $x = u^2$, $dx = 2u\,du$ and a partial fraction decomposition.

$$\int_{0+\epsilon}^{1/2} \frac{1}{\sqrt{x}\,(1-x)}\,dx = \int_{\sqrt{\epsilon}}^{\sqrt{1/2}} \frac{2u}{u(1-u^2)}\,du = \int_{\sqrt{\epsilon}}^{\sqrt{1/2}}\left(\frac{1}{1+u} + \frac{1}{1-u}\right)du$$
$$= \big(\ln(|1+u|) - \ln(|1-u|)\big)\Big|_{\sqrt{\epsilon}}^{\sqrt{1/2}} = \ln\left(\frac{1+\sqrt{1/2}}{1-\sqrt{1/2}}\right) - \ln\left(\frac{1+\sqrt{\epsilon}}{1-\sqrt{\epsilon}}\right) \xrightarrow[\epsilon\to 0^+]{} \ln\left(\frac{1+\sqrt{1/2}}{1-\sqrt{1/2}}\right)$$

and

$$\int_{1/2}^{1-\epsilon} \frac{1}{\sqrt{x}\,(1-x)}\,dx = \int_{\sqrt{1/2}}^{\sqrt{1-\epsilon}} \frac{2u}{u(1-u^2)}\,du = \int_{\sqrt{1/2}}^{\sqrt{1-\epsilon}}\left(\frac{1}{1+u} + \frac{1}{1-u}\right)du$$
$$= \big(\ln(|1+u|) - \ln(|1-u|)\big)\Big|_{\sqrt{1/2}}^{\sqrt{1-\epsilon}} = \ln\left(\frac{1+\sqrt{1-\epsilon}}{1-\sqrt{1-\epsilon}}\right) - \ln\left(\frac{1+\sqrt{1/2}}{1-\sqrt{1/2}}\right) \xrightarrow[\epsilon\to 0^+]{} \infty$$

The integral diverges.

35. The integrand is infinite at both endpoints. We split it into two at $x = 0$. Let $0 < \epsilon < 1$. The integral is evaluated easily by making the substitution $u = x^2$, $du = 2x\,dx$.

$$\int_{-1+\epsilon}^{0} \frac{x}{(1-x^2)^{1/4}}\,dx = \int_{(-1+\epsilon)^2}^{0} \frac{1/2}{(1-u)^{1/4}}\,du = -\frac{2}{3}(1-u)^{3/4}\Big|_{(-1+\epsilon)^2}^{0}$$
$$= -\frac{2}{3} + \frac{2}{3}\left(1-(-1+\epsilon)^2\right) \xrightarrow[\epsilon\to 0^+]{} -\frac{2}{3}$$

and

$$\int_{0}^{1-\epsilon} \frac{x}{(1-x^2)^{1/4}}\,dx = \int_{0}^{(1-\epsilon)^2} \frac{1/2}{(1-u)^{1/4}}\,du = -\frac{2}{3}(1-u)^{3/4}\Big|_{0}^{(1-\epsilon)^2}$$
$$= -\frac{2}{3}\left(1-(1-\epsilon)^2\right) + \frac{2}{3} \xrightarrow[\epsilon\to 0^+]{} \frac{2}{3}$$

The integral converges, $\int_{-1}^{1} \frac{x}{(1-x^2)^{1/4}}\,dx = 0$.

37. The integrand is infinite at both endpoints. We split it into two at $x = 3/2$. Let $0 < \epsilon < 1$. The integral is evaluated by completing the square and making the substitution $x - 3/2 = \frac{5}{2}u$, $dx = \frac{5}{2}\,du$.

$$\int_{-1+\epsilon}^{3/2} \frac{1}{\sqrt{(1+x)(4-x)}}\,dx = \int_{-1+\epsilon}^{3/2} \frac{1}{\sqrt{25/4-(x-3/2)^2}}\,dx = \int_{-1+\frac{2}{5}\epsilon}^{0} \frac{5/2}{\sqrt{25/4-25u^2/4}}\,du$$
$$= \int_{-1+\frac{2}{5}\epsilon}^{0} \frac{1}{\sqrt{1-u^2}}\,du = \arcsin(u)\Big|_{-1+\frac{2}{5}\epsilon}^{0} = -\arcsin(-1+\tfrac{2}{5}\epsilon) \xrightarrow[\epsilon\to 0^+]{} \frac{\pi}{2}$$

and

$$\int_{3/2}^{4-\epsilon} \frac{1}{\sqrt{(1+x)(4-x)}}\,dx = \int_{3/2}^{4-\epsilon} \frac{1}{\sqrt{25/4-(x-3/2)^2}}\,dx = \int_{0}^{1-\frac{2}{5}\epsilon} \frac{5/2}{\sqrt{25/4-25u^2/4}}\,du$$
$$= \int_{0}^{1-\frac{2}{5}\epsilon} \frac{1}{\sqrt{1-u^2}}\,du = \arcsin(u)\Big|_{0}^{1-\frac{2}{5}\epsilon} = \arcsin(1-\tfrac{2}{5}\epsilon) \xrightarrow[\epsilon\to 0^+]{} \frac{\pi}{2}$$

The integral converges, $\int_{-1}^{4} \frac{1}{\sqrt{1+x}\sqrt{4-x}}\,dx = \pi$.

39. The integrand is infinite at both endpoints. We split it into two at $x = 1$. Let $0 < \epsilon < 1$. The integral is evaluated by making the substitution $u = \sqrt{x}$, $du = \frac{1}{2\sqrt{x}}\,dx$.

$$\int_{0+\epsilon}^{1} \frac{1}{\sqrt{x}}\left(\frac{1}{2-\sqrt{x}}\right)^{1/3} dx = 2\int_{\sqrt{\epsilon}}^{1} (2-u)^{-1/3}\,du$$
$$= -3(2-u)^{2/3}\Big|_{\sqrt{\epsilon}}^{1} = -3 + 3(2-\sqrt{\epsilon})^{2/3} \xrightarrow[\epsilon\to 0^+]{} 3(2^{2/3}-1)$$

and

$$\int_1^{4-\epsilon} \frac{1}{\sqrt{x}}\left(\frac{1}{2-\sqrt{x}}\right)^{1/3} dx = 2\int_1^{\sqrt{4-\epsilon}} (2-u)^{-1/3}\, du$$
$$= -3(2-u)^{2/3}\Big|_1^{\sqrt{4-\epsilon}} = -3(2-\sqrt{4-\epsilon}\,)^{2/3} + 3 \xrightarrow[\epsilon\to 0^+]{} 3$$

The integral converges, $\int_0^4 \frac{1}{\sqrt{x}}\left(\frac{1}{2-\sqrt{x}}\right)^{1/3} dx = 3\cdot 2^{2/3}$.

Further Theory and Practice

41. The integrand is infinite at the left endpoint. Let $0 < \epsilon < 1$. The integral is evaluated using integration by parts, $u = \ln(x+5)$, $dv = dx$, $du = \frac{1}{x+5}\, dx$, $v = x$ and continuing with a long division. Note that $\lim_{\epsilon\to 0^+} \epsilon\ln(\epsilon) = 0$ by L'Hópital's Rule.

$$\int_{-5+\epsilon}^{2} \ln(x+5)\, dx = x\ln(x+5)\Big|_{-5+\epsilon}^{2} - \int_{-5+\epsilon}^{2} \frac{x}{x+5}\, dx$$
$$= x\ln(x+5)\Big|_{-5+\epsilon}^{2} - \int_{-5+\epsilon}^{2} \left(1 - \frac{5}{x+5}\right) dx$$
$$= \big((x+5)\ln(x+5) - x\big)\Big|_{-5+\epsilon}^{2}$$
$$= 7\ln(7) - 2 - \big(\epsilon\ln(\epsilon) + 5 - \epsilon\big) \xrightarrow[\epsilon\to 0^+]{} 7\ln(7) - 7$$

The integral converges, $\int_{-5}^{2} \ln(x+5)\, dx = 7(\ln(7) - 1)$.

43. The integrand is infinite at the left endpoint. Let $0 < \epsilon < 1$. The integral is evaluated using the substitution $u = \ln(x)$, $du = \frac{1}{x}\, dx$.

$$\int_{0+\epsilon}^{1} \frac{\ln^{1/3}(x)}{x}\, dx = \int_{\ln(\epsilon)}^{0} u^{1/3}\, du = \frac{3}{4}u^{4/3}\Big|_{\ln(\epsilon)}^{0} = -\frac{3}{4}\ln^{4/3}(\epsilon) \xrightarrow[\epsilon\to 0^+]{} \infty$$

The integral diverges.

45. The integrand is infinite at the left endpoint. Let $0 < \epsilon < 1$. The integral is evaluated using integration by parts, $u = \ln(x)$, $dv = x^{-1/2}\, dx$, $du = \frac{1}{x}\, dx$, $v = 2x^{1/2}$. Note that $\lim_{\epsilon\to 0^+} \epsilon^{1/2}\ln(\epsilon) = 0$ by L'Hópital's Rule.

$$\int_{0+\epsilon}^{4} x^{-1/2}\ln(x)\, dx = 2x^{1/2}\ln(x)\Big|_{\epsilon}^{4} - \int_{\epsilon}^{4} 2x^{-1/2}\, dx$$
$$= \left(2x^{1/2}\ln(x) - 4x^{1/2}\right)\Big|_{\epsilon}^{4}$$
$$= 4\ln(4) - 8 - \left(2\epsilon^{1/2}\ln(\epsilon) - 4\epsilon^{1/2}\right) \xrightarrow[\epsilon\to 0^+]{} 8(\ln(2) - 1)$$

The integral converges, $\int_0^4 x^{-1/2}\ln(x)\, dx = 8(\ln(2) - 1)$.

47. The integrand is infinite at the left endpoint. Let $0 < \epsilon < 1$. The integral is evaluated using integration by parts, $u = \ln(x)$, $dv = -\frac{1}{3}x^{-1/3}\,dx$, $du = \frac{1}{x}\,dx$, $v = -\frac{1}{2}x^{2/3}$. Note that $\lim_{\epsilon\to 0^+} \epsilon^{2/3}\ln(\epsilon) = 0$ by L'Hópital's Rule.

$$\begin{aligned}
\int_{0+\epsilon}^{8} \frac{1}{x^{1/3}} \ln\left(\frac{1}{x^{1/3}}\right) dx &= \int_{\epsilon}^{8} -\frac{1}{3}x^{-1/3}\ln(x)\,dx \\
&= -\frac{1}{2}x^{2/3}\ln(x)\Big|_{\epsilon}^{8} + \frac{1}{2}\int_{\epsilon}^{8} x^{-1/3}\,dx \\
&= \left(-\frac{1}{2}x^{2/3}\ln(x) + \frac{3}{4}x^{2/3}\right)\Big|_{\epsilon}^{8} \\
&= -2\ln(8) + 3 - \left(-\frac{1}{2}\epsilon^{2/3}\ln(\epsilon) + \frac{3}{4}\epsilon^{2/3}\right) \xrightarrow[\epsilon\to 0^+]{} 3 - 6\ln(2)
\end{aligned}$$

The integral converges, $\int_0^8 \frac{1}{x^{1/3}} \ln\left(\frac{1}{x^{1/3}}\right) dx = 3 - 6\ln(2)$.

49. The integrand is infinite at both endpoints. Let $0 < \epsilon < 1$. The integral is evaluated using integration by parts, $u = \ln(1-x^2)$, $dv = dx$, $du = \frac{-2x}{1-x^2}\,dx$, $v = x$, followed by a long division and partial fraction decomposition. Note that $\lim_{\epsilon\to 0^+} \epsilon\ln(\epsilon) = 0$ by L'Hópital's Rule.

$$\begin{aligned}
\int_{-1+\epsilon}^{0} \ln(1-x^2)\,dx &= x\ln(1-x^2)\Big|_{-1+\epsilon}^{0} + 2\int_{-1+\epsilon}^{0} \frac{x^2}{1-x^2}\,dx \\
&= x\ln(1-x^2)\Big|_{-1+\epsilon}^{0} + 2\int_{-1+\epsilon}^{0} \left(-1 + \frac{1}{2}\cdot\left(\frac{1}{1+x} + \frac{1}{1-x}\right)\right) dx \\
&= \left(x\ln(1-x^2) - 2x + \ln\left(\frac{1+x}{1-x}\right)\right)\Big|_{-1+\epsilon}^{0} \\
&= -(\epsilon-1)\ln(1-(\epsilon-1)^2) + 2(\epsilon-1) - \ln\left(\frac{\epsilon}{2-\epsilon}\right) \\
&= -(\epsilon-1)\big(\ln(\epsilon) + \ln(2+\epsilon)\big) + 2(\epsilon-1) - \ln(\epsilon) + \ln(2-\epsilon) \\
&= -\epsilon\ln(\epsilon) - \epsilon\ln(2+\epsilon) + 2(\epsilon-1) + 2\ln(2-\epsilon) \xrightarrow[\epsilon\to 0^+]{} 2\ln(2) - 2
\end{aligned}$$

and

$$\begin{aligned}
\int_0^{1-\epsilon} \ln(1-x^2)\,dx &= x\ln(1-x^2)\Big|_0^{1-\epsilon} + 2\int_0^{1-\epsilon} \frac{x^2}{1-x^2}\,dx \\
&= x\ln(1-x^2)\Big|_0^{1-\epsilon} + 2\int_0^{1-\epsilon} \left(-1+\frac{1}{2}\cdot\left(\frac{1}{1+x}+\frac{1}{1-x}\right)\right)dx \\
&= \left(x\ln(1-x^2) - 2x + \ln\left(\frac{1+x}{1-x}\right)\right)\Big|_0^{1-\epsilon} \\
&= (1-\epsilon)\ln(1-(1-\epsilon)^2) - 2(1-\epsilon) + \ln\left(\frac{2-\epsilon}{\epsilon}\right) \\
&= (1-\epsilon)\big(\ln(\epsilon)+\ln(2+\epsilon)\big) + 2(1-\epsilon) + \ln(2-\epsilon) - \ln(\epsilon) \\
&= -\epsilon\ln(\epsilon) - \epsilon\ln(2+\epsilon) + 2(\epsilon-1) + 2\ln(2-\epsilon) \xrightarrow[\epsilon\to 0^+]{} 2\ln(2) - 2
\end{aligned}$$

The integral converges, $\int_{-1}^{1} \ln(1-x^2)\,dx = 4(\ln(2)-1)$.

51. The integrand is infinite at the left endpoint. Let $0<\epsilon<1$. The integral is evaluated using the substitutions $u=\sin(x)$, $du=\cos(x)\,dx$, $\cos^2(x) = 1-u^2$.

$$\begin{aligned}
\int_{0+\epsilon}^{\pi/2} \frac{\cos^3(x)}{\sin^{1/3}(x)}\,dx &= \int_{\sin(\epsilon)}^{1} \frac{1-u^2}{u^{1/3}}\,du = \int_{\sin(\epsilon)}^{1} (u^{-1/3}-u^{5/3})\,du \\
&= \left(\frac{3}{2}u^{2/3} - \frac{3}{8}u^{8/3}\right)\Big|_{\sin(\epsilon)}^{1} = \frac{9}{8} - \left(\frac{3}{2}\sin^{2/3}(\epsilon) - \frac{3}{8}\sin^{8/3}(\epsilon)\right) \xrightarrow[\epsilon\to 0^+]{} \frac{9}{8}
\end{aligned}$$

The integral converges, $\int_{0+\epsilon}^{\pi/2} \frac{\cos^3(x)}{\sin^{1/3}(x)}\,dx = \frac{9}{8}$.

53. The integrand is infinite at the right endpoint. Let $0<\epsilon<1$. The integral is evaluated using the substitution $u=\arcsin(x)$, $du = \frac{1}{\sqrt{1-x^2}}\,dx$.

$$\begin{aligned}
\int_{0+\epsilon}^{1} \frac{\arcsin(x)}{\sqrt{1-x^2}}\,dx &= \int_{\arcsin(\epsilon)}^{\pi/2} u\,du = \frac{1}{2}u^2\Big|_{\arcsin(\epsilon)}^{\pi/2} \\
&= \frac{1}{2}\cdot\frac{\pi^2}{4} - \frac{1}{2}\arcsin^2(\epsilon) \xrightarrow[\epsilon\to 0^+]{} \frac{\pi^2}{8}
\end{aligned}$$

The integral converges, $\int_0^1 \frac{\arcsin(x)}{\sqrt{1-x^2}}\,dx = \frac{\pi^2}{8}$.

55. If $p=1$, then $\int_0^b \frac{1}{x}\,dx = \lim_{\epsilon\to 0^+}\int_\epsilon^b \frac{1}{x}\,dx = \lim_{\epsilon\to 0^+}\big(\ln(b)-\ln(\epsilon)\big) = \infty$ and the integral diverges.

If $p\neq 1$, then

$$\int_0^b \frac{1}{x^p}\,dx = \lim_{\epsilon\to 0^+}\int_\epsilon^b x^{-p}\,dx = \lim_{\epsilon\to 0^+}\left(\frac{b^{1-p}}{1-p} - \frac{\epsilon^{1-p}}{1-p}\right) = \begin{cases} \dfrac{b^{1-p}}{1-p} & , \quad p<1 \\ \infty & , \quad p>1 \end{cases}.$$

57. Make the substitution, $u = \ln(x)$, $du = \frac{1}{x}\,dx$, to obtain

$$\int_1^e \frac{1}{x\ln^p(x)}\,dx = \lim_{\epsilon\to 0^+}\int_{1+\epsilon}^e \frac{1}{x\ln^p(x)}\,dx = \lim_{\epsilon\to 0^+}\int_{\ln(1+\epsilon)}^1 \frac{1}{u^p}\,du = \int_0^1 \frac{1}{u^p}\,du\,.$$

Therefore, the integral converges for $0 < p < 1$ and diverges for $p \geq 1$ (see Exercise 55.)

59. If $0 < x \leq 1$, then $0 \leq e^x \leq e$ so $0 \leq \frac{e^x}{\sqrt{x}} \leq \frac{e}{\sqrt{x}}$. Since $\int_0^1 \frac{e}{\sqrt{x}}\,dx$ converges ($p = 1/2$, see Exercise 55), the Comparison Theorem implies that $\int_0^1 \frac{e^x}{\sqrt{x}}\,dx$ also converges.

61. If $0 < x \leq 3$, then $0 \leq \sqrt{1+x} \leq 2$ so $0 \leq \frac{\sqrt{1+x}}{x^{2/3}} \leq \frac{2}{x^{2/3}}$. Since $\int_0^3 \frac{2}{x^{2/3}}\,dx$ converges ($p = 2/3$, see Exercise 55), the Comparison Theorem implies that $\int_0^3 \frac{\sqrt{1+x}}{x^{2/3}}\,dx$ also converges.

63. If $1/2 \leq x < 1$, then $0 \leq 1/x \leq 2$ so $0 \leq \frac{1/x}{(1-x)^{1/3}} \leq \frac{2}{(1-x)^{1/3}}$. Since $\int_{1/2}^1 \frac{2}{(1-x)^{1/3}}\,dx$ converges ($p = 1/3$, see Exercise 56), the Comparison Theorem implies that $\int_{1/2}^1 \frac{1}{x(1-x)^{1/3}}\,dx$ also converges.

65. If $0 \leq x < 1/3$, then $0 < 1 \leq \sec(\pi x)$ so $0 \leq \frac{1}{1-3x} \leq \frac{\sec(\pi x)}{1-3x}$. Since $\int_0^{1/3} \frac{1}{1-3x}\,dx$ diverges ($p = 1$ after the substitution $u = 3x$, see Exercise 56), the Comparison Theorem implies that $\int_0^{1/3} \frac{\sec(\pi x)}{1-3x}\,dx$ also diverges.

67. The integrand is infinite at the left endpoint.

Assume that $0 < x \leq 2$. Then $1 \leq x^3$ and $x^3 + x^2 - 2 \geq 1 + x^2 - 2 = x^2 - 1$. Since $x^2 - 1 = (x-1)(x+1) \geq 2(x-1)$, $x^3 + x^2 - 2 \geq 2(x-1) > 0$, implying that $\sqrt{x^3+x^2-2} \geq \sqrt{2(x-1)} > 0$. Consequently, $\frac{1}{\sqrt{x^3+x^2-2}} \leq \frac{1/\sqrt{2}}{\sqrt{x-1}}$ and, since $\int_1^2 \frac{1/\sqrt{2}}{\sqrt{x-1}}\,dx$ converges ($p = 1/2$, see Exercise 56), $\int_1^2 \frac{1}{\sqrt{x^3+x^2-2}}\,dx$ also converges by the Comparison Test.

69. The integrand is infinite at both endpoints. Split it into two parts at $x = 2$.

If $1 < x \leq 2$, then $1 \leq \frac{1}{\sqrt{3-x}}$ implying that $0 < \frac{1}{(x-1)^{5/3}} \leq \frac{1}{(x-1)^{5/3}\sqrt{3-x}}$. Since $\int_0^2 \frac{1}{(x-1)^{5/3}}\,dx$ diverges ($p = 5/3$, see Exercise 55), the Comparison Test implies that the integral $\int_0^2 \frac{1}{(x-1)^{5/3}\sqrt{3-x}}\,dx$ also diverges. Therefore, $\int_0^3 \frac{1}{(x-1)^{5/3}\sqrt{3-x}}\,dx$ diverges.

71. (a) Because $p > 0$ and $q = 0$, the integrand is infinite at the left endpoint.

Let $0 < \epsilon < 1$, then

$$B(p,0) = \lim_{\epsilon\to 0^+} \int_{0+\epsilon}^{1} \frac{1-x}{x^{1-p}}\,dx = \lim_{\epsilon\to 0^+} \int_{\epsilon}^{1} \left(x^{p-1} - x^p\right)dx$$

$$= \lim_{\epsilon\to 0^+} \left(\frac{1}{p}x^p - \frac{1}{p+1}x^{p+1}\right)\Big|_{\epsilon}^{1}$$

$$= \lim_{\epsilon\to 0^+} \left(\frac{1}{p} - \frac{1}{p+1} - \frac{1}{p}\epsilon^p + \frac{1}{p+1}\epsilon^{p+1}\right) = \frac{1}{p(p+1)}.$$

(b) Apply integration by parts, $u = (1-x)^{1+q}$, $dv = x^{-1+p}\,dx$, $u = -(1+q)(1-x)^q$, $v = \frac{1}{p}x^p$.

$$B(p,q) = \int_0^1 x^{-1+p}(1-x)^{1+q}\,dx$$

$$= \frac{1}{p}x^p(1-x)^{1+q}\Big|_0^1 + \frac{1+q}{p}\int_0^1 x^p(1-x)^q\,dx = \frac{1+q}{p}B(p+1,q-1).$$

(c) Let $q = 1$ in part b to obtain $B(p,1) = \frac{2}{p}B(p+1,0)$. Now use part a to see that $B(p,1) = \frac{2}{p}\cdot\frac{1!}{(p+1)(p+2)} = \frac{2!}{p(p+1)(p+2)}$.

Let $q = 2$ in part b to obtain $B(p,2) = \frac{3}{p}B(p+1,1)$. Now use the formula just obtained (with p replaced by $p+1$) to see that $B(p,2) = \frac{3}{p}\cdot\frac{2!}{(p+1)(p+2)(p+3)} = \frac{3!}{p(p+1)(p+2)(p+3)}$.

CALCULATOR/COMPUTER EXERCISES

73. Observe that $f(x) = \frac{\sqrt{x+1}}{\sqrt{x}}$. Since $0 \le \sqrt{x+1} \le \sqrt{2}$ for $0 < x \le 1$, $0 \le \frac{\sqrt{x+1}}{x^{1/2}} \le \frac{\sqrt{2}}{x^{1/2}}$ and $\int_0^1 \sqrt{1+1/x}\,dx$ converges because $\int_0^1 \frac{\sqrt{2}}{x^{1/2}}\,dx$ converges ($p = 1/2$, see Exercise 55).

    Given ϵ such that $0 < \epsilon < 1$, $\int_0^1 \sqrt{1+1/x}\,dx = \int_0^{\epsilon} \sqrt{1+1/x}\,dx + \int_{\epsilon}^1 \sqrt{1+1/x}\,dx$, so if $\int_0^{\epsilon} \sqrt{1+1/x}\,dx \le 5\times 10^{-4}$, then $\int_{\epsilon}^1 \sqrt{1+1/x}\,dx$ will be a three decimal approximation to $\int_0^1 \sqrt{1+1/x}\,dx$. Since $\int_0^{\epsilon} \sqrt{1+1/x}\,dx \le \int_0^{\epsilon} \frac{\sqrt{2}}{x^{1/2}}\,dx = 2\sqrt{2\epsilon} = 5\times 10^{-4}$ when $\epsilon = a = 25/8 \times 10^{-8}$, $\int_a^1 \sqrt{1+1/x}\,dx \approx 2.295$ is the desired three decimal approximation.

75. Observe that $f(x) = \frac{\sqrt{x}}{\sin(x)} = \frac{x/\sin(x)}{\sqrt{x}}$. Since $0 \le \frac{x}{\sin(x)} \le \frac{3}{2}$ for $0 < x \le 1$, $0 \le \frac{\sqrt{x}}{\sin(x)} \le \frac{3/2}{x^{1/2}}$ and $\int_0^1 \frac{\sqrt{x}}{\sin(x)}\,dx$ converges because $\int_0^1 \frac{3/2}{x^{1/2}}\,dx$ converges ($p = 1/2$, see Exercise 55).

    Given ϵ such that $0 < \epsilon < 1$, $\int_0^1 \frac{\sqrt{x}}{\sin(x)}\,dx = \int_0^{\epsilon} \frac{\sqrt{x}}{\sin(x)}\,dx + \int_{\epsilon}^1 \frac{\sqrt{x}}{\sin(x)}\,dx$, so if $\int_0^{\epsilon} \frac{\sqrt{x}}{\sin(x)}\,dx \le 5\times 10^{-4}$, then $\int_{\epsilon}^1 \frac{\sqrt{x}}{\sin(x)}\,dx$ will be a three decimal approximation to $\int_0^1 \frac{\sqrt{x}}{\sin(x)}\,dx$. Since $\int_0^{\epsilon} \frac{\sqrt{x}}{\sin(x)}\,dx \le \int_0^{\epsilon} \frac{3/2}{x^{1/2}}\,dx = 3\sqrt{\epsilon} =$

5×10^{-4} when $\epsilon = a = 25/9 \times 10^{-8}$, $\int_a^1 \frac{\sqrt{x}}{\sin(x)}\,dx \approx 2.071$ is the desired three decimal approximation.

77. Since $0 \le \ln(2+x) \le \ln(3)$ for $0 < x \le 1$, $0 \le \frac{\ln(2+x)}{x^{\ln(2)}} \le \frac{\ln(3)}{x^{\ln(2)}}$ and $\int_0^1 \frac{\ln(2+x)}{x^{\ln(2)}}\,dx$ converges because $\int_0^1 \frac{\ln(3)}{x^{\ln(2)}}\,dx$ converges ($p = \ln(2)$, see Exercise 55).

Given ϵ such that $0 < \epsilon < 1$, $\int_0^1 \frac{\ln(2+x)}{x^{\ln(2)}}\,dx = \int_0^\epsilon \frac{\ln(2+x)}{x^{\ln(2)}}\,dx + \int_\epsilon^1 \frac{\ln(2+x)}{x^{\ln(2)}}\,dx$, so if $\int_0^\epsilon \frac{\ln(2+x)}{x^{\ln(2)}}\,dx \le 5 \times 10^{-4}$, then $\int_\epsilon^1 \frac{\ln(2+x)}{x^{\ln(2)}}\,dx$ will be a three decimal approximation to $\int_0^1 \frac{\ln(2+x)}{x^{\ln(2)}}\,dx$. Since $\int_0^\epsilon \frac{\ln(2+x)}{x^{\ln(2)}}\,dx \le \int_0^\epsilon \frac{\ln(3)}{x^{\ln(2)}}\,dx = \frac{\ln(3)}{1-\ln(2)}\epsilon^{1-\ln(2)} = 5\times 10^{-4}$ when $\epsilon = a = (5(1-\ln(2))/\ln(3)\times 10^{-4})^{(1-\ln(2))^{-1}}$, $\int_a^1 \frac{\ln(2+x)}{x^{\ln(2)}}\,dx \approx 2.597$ is the desired three decimal approximation.

6.7 Improper Integrals—Unbounded Intervals

Problems for Practice

1. Since

$$\int_3^\infty x^{-3/2}\,dx = \lim_{N\to\infty} \int_3^N x^{-3/2}\,dx = \lim_{N\to\infty} \left(-2x^{-1/2}\Big|_3^N \right)$$
$$= \lim_{N\to\infty} \left(-N^{-1/2} + 2\cdot 3^{-1/2} \right) = \frac{2}{\sqrt{3}},$$

the integral converges.

3. Since

$$\int_{-1}^\infty (3+x)^{-3/2}\,dx = \lim_{N\to\infty} \int_{-1}^N (3+x)^{-3/2}\,dx = \lim_{N\to\infty} \left(-2(3+x)^{-1/2}\Big|_{-1}^N \right)$$
$$= \lim_{N\to\infty} \left(-2(3+N)^{-1/2} + 2\cdot 2^{-1/2} \right) = \sqrt{2},$$

the integral converges.

5. Since

$$\int_0^\infty \frac{1}{1+x^2}\,dx = \lim_{N\to\infty} \int_0^N \frac{1}{1+x^2}\,dx = \lim_{N\to\infty} \left(\arctan(x)\Big|_0^N \right)$$
$$= \lim_{N\to\infty} \arctan(N) = \pi/2,$$

the integral converges.

7. The integral evaluates using the substitution $u = 1 + x^2$. Since

$$\int_0^\infty \frac{x}{(1+x^2)^2}\,dx = \lim_{N\to\infty} \int_0^N \frac{1}{2}\cdot\frac{2x}{(1+x^2)^2}\,dx = \lim_{N\to\infty}\left(-\frac{1}{2}(1+x^2)^{-1}\Big|_0^N\right)$$
$$= \lim_{N\to\infty}\left(-\frac{1}{2}(1+N^2)^{-1}+\frac{1}{2}\right) = 1/2\,,$$

the integral converges.

9. Evaluate the integral with a partial fraction decomposition. Since

$$\int_2^\infty \frac{1}{x(x-1)}\,dx = \lim_{N\to\infty}\int_2^N \left(\frac{1}{x-1}-\frac{1}{x}\right)dx = \lim_{N\to\infty}\left(\ln\left(\frac{x-1}{x}\right)\Big|_2^N\right)$$
$$= \lim_{N\to\infty}\left(\ln\left(1-\frac{1}{N}\right)-\ln\left(\frac{1}{2}\right)\right) = \ln(2)\,,$$

the integral converges.

11. The integral evaluates using the substitution $u = e^x + 1$, $du = e^x\,dx$. Since

$$\int_0^\infty \frac{e^x}{(e^x+1)^3}\,dx = \int_2^\infty \frac{1}{u^3}\,du = \lim_{N\to\infty}\int_2^N \frac{1}{u^3}\,du = \lim_{N\to\infty}\left(-\frac{1}{2}u^{-2}\Big|_2^N\right)$$
$$= \lim_{N\to\infty}\left(-\frac{1}{2}N^{-2}+\frac{1}{8}\right) = \frac{1}{8}\,,$$

the integral converges.

13. Evaluate the integral with a partial fraction decomposition. Substitute $x = \sqrt{2}\tan(\theta)$ in the second fraction. Since

$$\int_0^\infty \frac{6x^2+8}{(x^2+1)(x^2+2)}\,dx = \lim_{N\to\infty}\int_0^N \left(\frac{2}{x^2+1}+\frac{4}{x^2+2}\right)dx$$
$$= \lim_{N\to\infty}\left(2\arctan(x)+2\sqrt{2}\arctan(x/\sqrt{2})\Big|_0^N\right)$$
$$= \lim_{N\to\infty}\left(2\arctan(N)+2\sqrt{2}\arctan(N/\sqrt{2}\,)\right) = (1+\sqrt{2}\,)\pi\,,$$

the integral converges.

15. Use the substitution $u = 3x^2$, $du = 6x\,dx$. Since

$$\int_1^\infty xe^{-3x^2}\,dx = \int_3^\infty e^{-u}\cdot\frac{1}{6}\,du = \lim_{N\to\infty}\int_3^N e^{-u}\cdot\frac{1}{6}\,du = \lim_{N\to\infty}\left(-\frac{1}{6}e^{-u}\Big|_3^N\right)$$
$$= \lim_{N\to\infty}\left(-\frac{1}{6}e^{-N}+\frac{1}{6}e^{-3}\right) = e^{-3}/6\,,$$

the integral converges.

17. Use the substitution $u = \ln(x)$, $du = \frac{1}{x}\,dx$. Then

$$\int_e^\infty \frac{1}{x\ln(x)}\,dx = \int_1^\infty \frac{1}{u}\,du = \lim_{N\to\infty}\int_1^N \frac{1}{u}\,du = \lim_{N\to\infty}\left(\ln(u)\Big|_1^N\right) = \infty\,,$$

the integral diverges.

19. Use the substitution $u = \arctan(x)$, $du = \frac{1}{x^2+1}\,dx$. Then

$$\int_1^\infty \frac{\arctan(x)}{x^2+1}\,dx = \lim_{b\to(\pi/2)^-}\int_{\pi/4}^b u\,du = \lim_{b\to(\pi/2)^-}\left(\frac{1}{2}u^2\Big|_{\pi/4}^b\right)$$
$$= \lim_{b\to(\pi/2)^-}\left(\frac{1}{2}b^2 - \frac{1}{2}\cdot\frac{\pi^2}{16}\right) = \frac{\pi^2}{8} - \frac{\pi^2}{32} = \frac{3}{32}\pi^2\,,$$

the integral converges.

21. Since

$$\int_{-\infty}^{-2} x^{-3}\,dx = \lim_{M\to-\infty}\int_M^{-2} x^{-3}\,dx = \lim_{M\to-\infty}\left(-\frac{1}{2}x^{-2}\Big|_M^{-2}\right)$$
$$= \lim_{M\to-\infty}\left(-\frac{1}{8} + \frac{1}{2}M^{-2}\right) = -\frac{1}{8}\,,$$

the integral converges.

23. Since

$$\int_{-\infty}^{-2} \frac{1}{(1+x)^{4/3}}\,dx = \lim_{M\to-\infty}\int_M^{-2} \frac{1}{(1+x)^{4/3}}\,dx = \lim_{M\to-\infty}\left(-3(1+x)^{-1/3}\Big|_M^{-2}\right)$$
$$= \lim_{M\to-\infty}\left(3 + 3(1+M)^{-1/3}\right) = 3\,,$$

the integral converges.

25. Since

$$\int_{-\infty}^{2} \frac{1}{(3-x)^{3/2}}\,dx = \lim_{M\to-\infty}\int_M^{2} (3-x)^{-3/2}\,dx = \lim_{M\to-\infty}\left(2(3-x)^{-1/2}\Big|_M^{2}\right)$$
$$= \lim_{M\to-\infty}\left(2 - 2(3-M)^{-1/2}\right) = 2\,,$$

the integral diverges.

27. Evaluate the integral using the substitution $x = \tan(\theta)$, $dx = \sec^2(\theta)\,d\theta$.

$$\int_{-\infty}^{0} \frac{1}{(1+x^2)^{3/2}}\,dx = \int_{-\pi/2}^{0} \frac{1}{(1+\tan^2(\theta))^{3/2}}\cdot\sec^2(\theta)\,d\theta = \int_{-\pi/2}^{0}\cos(\theta)\,d\theta$$
$$= \sin(\theta)\Big|_{-\pi/2}^{0} = 0 - (-1) = 1\,.$$

The integral converges.

29. Since

$$\int_{-\infty}^{4} e^{x/3}\,dx = \lim_{M\to-\infty}\int_{M}^{4} e^{x/3}\,dx = \lim_{M\to-\infty}\left(3e^{x/3}\Big|_{M}^{4}\right)$$
$$= \lim_{M\to-\infty}\left(3e^{4/3} - 3e^{M/3}\right) = 3e^{4/3},$$

the integral converges.

31. Evaluate the integral using integration by parts, $u = x$, $dv = \exp(x/2)\,dx$, $du = dx$, $v = 2\exp(x/2)$. Note that $\lim_{M\to-\infty} 2M\exp(M/2) = 0$ (L'Hópital).

$$\int_{-\infty}^{0} xe^{x/2}\,dx = \lim_{M\to-\infty}\int_{M}^{0} xe^{x/2}\,dx = \lim_{M\to-\infty}\left(2x\exp(x/2)\Big|_{M}^{0} - \int_{M}^{0} 2\exp(x/2)\,dx\right)$$
$$= \lim_{M\to-\infty}\left(2x\exp(x/2) - 4\exp(x/2)\Big|_{M}^{0}\right)$$
$$= \lim_{M\to-\infty}\left(-4 - 2M\exp(M/2) + 4\exp(M/2)\right) = -4.$$

The integral converges.

33. Evaluate the integral using the substitution $u = \pi/x$, $du = -\pi/x^2\,dx$.

$$\int_{-\infty}^{-1} \frac{\sin(\pi/x)}{x^2}\,dx = \int_{0}^{-\pi} -\frac{1}{\pi}\sin(u)\,du = \frac{1}{\pi}\cos(u)\Big|_{0}^{-\pi} = \frac{1}{\pi}(-1-1) = -\frac{2}{\pi}.$$

The integral converges.

35. Evaluate the integral using the substitution $u = \ln(2-x)$, $du = -\frac{1}{2-x}\,dx$.

$$\int_{-\infty}^{0} \frac{1}{(2-x)\ln^2(2-x)}\,dx = \int_{\infty}^{\ln(2)} -\frac{1}{u^2}\,du = \lim_{N\to\infty}\int_{N}^{\ln(2)} -\frac{1}{u^2}\,du$$
$$= \lim_{N\to\infty}\left(\frac{1}{u}\Big|_{N}^{\ln(2)}\right) = \lim_{N\to\infty}\left(\frac{1}{\ln(2)} - \frac{1}{N}\right) = \frac{1}{\ln(2)}.$$

The integral converges.

37. Split the integral into two at $x = 0$. Evaluate the integrals using the substitution $u = x^2$, $du = 2x\,dx$. We do the right side first. What happens there will be reflected on the left side because the integrand is an odd function.

$$\int_{0}^{\infty} xe^{-x^2}\,dx = \int_{0}^{\infty} \frac{1}{2}e^{-u}\,du = \lim_{N\to\infty}\int_{0}^{N} \frac{1}{2}e^{-u}\,du$$
$$= \lim_{N\to\infty}\left(-\frac{1}{2}e^{-u}\Big|_{0}^{N}\right) = \lim_{N\to\infty}\left(-\frac{1}{2}e^{-N} + \frac{1}{2}\right) = \frac{1}{2}.$$

Since the integrand is an odd function, $\int_{-\infty}^{0} xe^{-x^2}\,dx = -1/2$, and the complete integral converges: $\int_{-\infty}^{\infty} xe^{-x^2}\,dx = -1/2 + 1/2 = 0$.

39. Split the integral into two at $x = 0$. Evaluate the integrals using the substitution $x = 2\tan(\theta)$, $dx = \sec^2(\theta)\,d\theta$. We do the right side first. Note that the integrand is an even function.

$$\int_0^\infty \frac{1}{4+x^2}\,dx = \int_0^\pi \frac{1}{4+4\tan^2(\theta)} \cdot \sec^2(\theta)\,d\theta = \frac{1}{4}\int_0^\pi d\theta = \frac{1}{4}\theta\Big|_0^\pi = \frac{\pi}{4}.$$

Since the integrand is an even function, $\int_{-\infty}^0 \frac{1}{4+x^2}\,dx = \pi/4$ also and $\int_{-\infty}^\infty \frac{1}{4+x^2}\,dx = \pi/2$.

41. Split the integral into two at $x = 0$. Evaluate the integrals using the substitution $u = 1 + x^2$, $du = 2x\,dx$. We do the right side first. What happens there will be reflected on the left side because the integrand is an odd function.

$$\int_0^\infty \frac{x}{(1+x^2)^2}\,dx = \int_1^\infty \frac{1}{2}u^{-2}\,du = \lim_{N\to\infty}\int_1^N \frac{1}{2}u^{-2}\,du$$
$$= \lim_{N\to\infty}\left(-\frac{1}{2}u^{-1}\Big|_1^N\right) = \lim_{N\to\infty}\left(-\frac{1}{2}u^{-N} + \frac{1}{2}\right) = \frac{1}{2}.$$

Since the integrand is an odd function, $\int_{-\infty}^0 \frac{x}{(1+x^2)^2}\,dx = -1/2$, and the complete integral converges: $\int_{-\infty}^\infty \frac{x}{(1+x^2)^2}\,dx = -1/2 + 1/2 = 0$.

43. Split the integral into two at $x = 0$. Evaluate the integrals by making a partial fraction decomposition and then substituting $x = \sqrt{2}\tan(\theta)$, $dx = \sqrt{2}\sec^2(\theta)\,d\theta$ in the second fraction. We do the right side first. Note that the integrand is an even function.

$$\int_0^\infty \frac{1}{(x^2+1)(x^2+2)}\,dx = \lim_{N\to\infty}\int_0^N \left(\frac{1}{x^2+1} - \frac{1}{x^2+2}\right)dx$$
$$= \lim_{N\to\infty}\left(\arctan(x) - \frac{\sqrt{2}}{2}\arctan\left(\frac{x}{\sqrt{2}}\right)\Big|_0^N\right)$$
$$= \lim_{N\to\infty}\left(\arctan(N) - \frac{\sqrt{2}}{2}\arctan\left(\frac{N}{\sqrt{2}}\right)\right) = \frac{\pi}{2}\cdot\left(1 - \frac{\sqrt{2}}{2}\right).$$

Since the integrand is even, $\int_{-\infty}^0 \frac{1}{(x^2+1)(x^2+2)}\,dx = \frac{\pi}{2}\cdot\left(1-\frac{\sqrt{2}}{2}\right)$ also, and $\int_{-\infty}^\infty \frac{1}{(x^2+1)(x^2+2)}\,dx = \pi\cdot\left(1-\frac{\sqrt{2}}{2}\right)$

45. $\text{PV} = \int_0^\infty 1000e^{-0.05t}\,dt = \lim_{N\to\infty}\left(-\frac{1000}{0.05}e^{-0.05t}\Big|_0^N\right) = \frac{10^3}{5\cdot 10^{-2}} = \$20,000$

47. The integral evaluates using integration by parts, $u = 1000 + 50t$, $dv =$

$e^{-0.05t}$, $du = 50\,dt$, $v = -20e^{-0.05t}$.

$$\begin{aligned}
\text{PV} &= \int_0^{\infty} (1000 + 50t)e^{-0.05t}\,dt = \lim_{N\to\infty} \int_0^N (1000 + 50t)e^{-0.05t}\,dt \\
&= \lim_{N\to\infty} \left(-20(1000 + 50t)e^{-0.05t}\Big|_0^N + 1000 \int_0^N e^{-0.05t}\,dt \right) \\
&= 20,000 + 20,000 = \$40,000\,.
\end{aligned}$$

See Exercise 45 for the evaluation of the integral in line 2 above.

Further Theory and Practice

49. Evaluate the integral using integration by parts, $u = \arctan(x)$, $dv = 4x^{-2}$, $du = \frac{1}{x^2+1}\,dx$, $v = -4x^{-1}$, followed by a partial fraction decomposition.

$$\begin{aligned}
\int_1^{\infty} 4x^{-2}\arctan(x)\,dx &= \lim_{N\to\infty} \int_1^N 4x^{-2}\arctan(x)\,dx \\
&= \lim_{N\to\infty} \left(-4x^{-1}\arctan(x)\Big|_1^N + \int_1^N \frac{4}{x(x^2+1)}\,dx \right) \\
&= \lim_{N\to\infty} \left(-4x^{-1}\arctan(x)\Big|_1^N + \int_1^N \left(\frac{4}{x} - \frac{4x}{x^2+1} \right) dx \right) \\
&= \lim_{N\to\infty} \left(-4x^{-1}\arctan(x) + 4\ln(x) - 2\ln(x^2+1)\Big|_1^N \right) \\
&= \lim_{N\to\infty} \left(-4x^{-1}\arctan(x) + \ln\left(\frac{x^4}{(x^2+1)^2} \right)\Big|_1^N \right)
\end{aligned}$$

$$= \lim_{N\to\infty} \left(-4\frac{\arctan(N)}{N} + \ln\left(\frac{N^4}{(N^2+1)^2} \right) + 4\cdot\frac{\pi}{4} - \ln\left(\frac{1}{4}\right) \right) = \pi + 2\ln(2)\,.$$

51. Evaluate the integral using integration by parts, $u = x$, $dv = 2^x\,dx$, $du = dx$, $v = 2^x / \ln(2)$.

$$\int_{-\infty}^0 x2^x\,dx = \lim_{M\to-\infty} \int_M^0 x2^x\,dx = \lim_{M\to-\infty} \left(\frac{x2^x}{\ln(2)}\Big|_M^0 - \int_M^0 \frac{2^x}{\ln(2)}\,dx \right)$$

$$= \lim_{M\to-\infty} \left(\frac{x2^x}{\ln(2)} - \frac{2^x}{\ln^2(2)}\Big|_M^0 \right) = \lim_{M\to-\infty} \left(-\frac{1}{\ln^2(2)} - \frac{M2^M}{\ln(2)} + \frac{2^M}{\ln^2(2)} \right) = -\frac{1}{\ln^2(2)}\,.$$

53. Using double integration by parts as in Example 5 of Section 6.1 the

integral evaluates as follows.

$$\int_0^\infty e^{-x}\cos(x)\,dx = \lim_{N\to\infty}\int_0^N e^{-x}\cos(x)\,dx = \lim_{N\to\infty}\left(\frac{1}{2}(e^{-x}\sin(x) - e^{-x}\cos(x))\Big|_0^N\right)$$

$$= \lim_{N\to\infty}\left(\frac{1}{2}(e^{-N}\sin(N) - e^{-N}\cos(N)) + \frac{1}{2}\right) = \frac{1}{2}.$$

55. The domain is unbounded and the integrand is unbounded at the left endpoint. Split it into two parts at $x = 1$. The integrations can be carried out with the substitution $u = -1/x$, $du = \frac{1}{x^2}\,dx$.

$$\int_0^1 \frac{\exp(-1/x)}{x^2}\,dx = \lim_{\epsilon\to0^+}\int_{0+\epsilon}^1 \frac{\exp(-1/x)}{x^2}\,dx = \lim_{\epsilon\to0^+}\int_{-1/\epsilon}^{-1} \exp(u)\,du$$

$$\lim_{\epsilon\to0^+}\left(\exp(u)\Big|_{-1/\epsilon}^{-1}\right) = \lim_{\epsilon\to0^+}(\exp(-1) - \exp(-1/\epsilon)) = 1/e\,,$$

and

$$\int_1^\infty \frac{\exp(-1/x)}{x^2}\,dx = \lim_{N\to\infty}\int_1^N \frac{\exp(-1/x)}{x^2}\,dx = \lim_{N\to\infty}\int_{-1}^{-1/N} \exp(u)\,du$$

$$\lim_{N\to\infty}\left(\exp(u)\Big|_{-1}^{-1/N}\right) = \lim_{N\to\infty}(\exp(-1/N) - \exp(-1)) = 1 - 1/e\,.$$

Therefore, the integral converges, $\int_0^\infty \frac{\exp(-1/x)}{x^2}\,dx = 1$.

57. The domain is unbounded and the integrand is unbounded at the left endpoint. Split it into two parts at $x = 2$. The integrations can be carried out with the substitution $u = \ln x$, $du = \frac{1}{x}\,dx$.

$$\int_1^2 \frac{1}{x\ln^2(x)}\,dx = \lim_{\epsilon\to0^+}\int_{1+\epsilon}^2 \frac{1}{x\ln^2(x)}\,dx = \lim_{\epsilon\to0^+}\int_{\ln(1+\epsilon)}^{\ln(2)} u^{-2}\,du$$

$$\lim_{\epsilon\to0^+}\left(-u^{-1}\Big|_{\ln(1+\epsilon)}^{\ln(2)}\right) = \lim_{\epsilon\to0^+}\left(-\frac{1}{\ln(2)} + \frac{1}{\ln(1+\epsilon)}\right) = \infty\,.$$

The integral diverges.

59. The domain is unbounded and the integrand is unbounded at the left endpoint. Split it into two parts at $x = 2$.

$$\int_1^2 \frac{2}{x\sqrt{x^2-1}}\,dx = \lim_{\epsilon\to0^+}\int_{1+\epsilon}^2 \frac{2}{x\sqrt{x^2-1}}\,dx = \lim_{\epsilon\to0^+}\left(2\text{arcsec}(x)\Big|_{1+\epsilon}^2\right)$$

$$= \lim_{\epsilon\to0^+}(2\text{arcsec}(2) - 2\text{arcsec}(1+\epsilon)) = \frac{2\pi}{3}\,,$$

and

$$\int_2^\infty \frac{2}{x\sqrt{x^2-1}}\,dx = \lim_{N\to\infty}\int_2^N \frac{2}{x\sqrt{x^2-1}}\,dx = \lim_{N\to\infty}\left(2\text{arcsec}(x)\Big|_2^N\right)$$
$$= \lim_{N\to\infty}\left(2\text{arcsec}(N) - 2\text{arcsec}(2)\right) = \pi - \frac{2\pi}{3}\,.$$

The integral converges, $\int_1^\infty \frac{2}{x\sqrt{x^2-1}}\,dx = \pi$.

61. The domain is unbounded.

If $p = 0$, the integral $\int_0^\infty 1\,dx$ clearly diverges to infinity.

If $p < 0$, the integral also diverges, as shown below.

$$\int_0^\infty x^{-p}\,dx = \lim_{N\to\infty}\int_0^N x^{-p}\,dx = \lim_{N\to\infty}\left(\frac{x^{1-p}}{1-p}\Big|_0^N\right) = \lim_{N\to\infty}\frac{N^{1-p}}{1-p} = \infty\,.$$

If $p > 0$ the integrand is infinite at the left endpoint. Split it into two parts at $x = 1$.

If $0 < p < 1$, then

$$\int_1^\infty x^{-p}\,dx = \lim_{N\to\infty}\int_1^N x^{-p}\,dx = \lim_{N\to\infty}\left(\frac{x^{1-p}}{1-p}\Big|_1^N\right) = \lim_{N\to\infty}\left(\frac{N^{1-p}}{1-p} - \frac{1}{1-p}\right) = \infty\,.$$

If $p = 1$, then

$$\int_1^\infty x^{-1}\,dx = \lim_{N\to\infty}\int_1^N x^{-1}\,dx = \lim_{N\to\infty}\left(\ln(x)\Big|_1^N\right) = \lim_{N\to\infty}\ln(N) = \infty\,.$$

If $p > 1$, then

$$\int_0^1 x^{-p}\,dx = \lim_{\epsilon\to0^+}\int_\epsilon^1 x^{-p}\,dx = \lim_{\epsilon\to0^+}\left(\frac{x^{1-p}}{1-p}\Big|_\epsilon^1\right) = \lim_{\epsilon\to0^+}\left(\frac{1}{1-p} - \frac{\epsilon^{1-p}}{1-p}\right) = \infty\,.$$

63. Since $0 \le \frac{x}{1+x^3} \le \frac{1}{x^2}$ for $1 \le x < \infty$, and $\int_1^\infty \frac{1}{x^2}\,dx$ converges (see Example 3), the Comparison Theorem implies that $\int_1^\infty \frac{x}{1+x^3}\,dx$ converges.

65. If $1 \le x < \infty$, then $1 + x^{5/2} > x^{5/2} > 0$ so $\sqrt{1+x^{5/2}} > x^{5/4} > 0$ and $0 < \frac{1}{\sqrt{1+x^{5/2}}} \le \frac{1}{x^{5/4}}$. Since $\int_1^\infty \frac{1}{x^{5/4}}\,dx$ converges (see Example 3), the Comparison Theorem implies that $\int_1^\infty \frac{1}{\sqrt{1+x^{5/2}}}\,dx$ converges.

67. If $x \ge 1$, then $\frac{1}{\sqrt{x}} \le 1$ and $0 < \frac{e^{-x}}{\sqrt{x}} \le e^{-x}$. Since $\int_1^\infty e^{-x}\,dx$ converges (direct integral calculation) the Comparison Theorem implies that $\int_1^\infty \frac{e^{-x}}{\sqrt{x}}\,dx$ converges.

69. If $x \geq 1$, then $0 < 1 + x^2 \leq 2x^2$ so $0 < \sqrt{1+x^2} \leq \sqrt{2}x$ implying that $\frac{1/\sqrt{2}}{x} \leq \frac{1}{\sqrt{1+x^2}}$. Since $\int_1^\infty \frac{1/\sqrt{2}}{x}\,dx$ diverges (direct integral calculation) the Comparison Theorem implies that $\int_1^\infty \frac{1}{\sqrt{1+x^2}}\,dx$ diverges.

71. If $x \geq 1$, then $\frac{\sin^2(x)+x}{x^{3/2}} \geq \frac{x}{x^{3/2}} = \frac{1}{x^{1/2}}$. Since $\int_1^\infty \frac{1}{x^{1/2}}\,dx$ diverges (direct integral calculation) the Comparison Theorem implies that $\int_1^\infty \frac{\sin^2(x)+x}{x^{3/2}}\,dx$ diverges.

73. If $x \geq 1$, then $\frac{\exp(x)}{x\exp(x)-1} = \frac{1}{x-1/\exp(x)} \geq \frac{1}{x}$. Since $\int_1^\infty \frac{1}{x}\,dx$ diverges (direct integral calculation) the Comparison Theorem implies that $\int_1^\infty \frac{\exp(x)}{x\exp(x)-1}\,dx$ diverges.

75. Make the substitution $u = \sqrt{x}$ so $x = u^2$ and $dx = 2u\,du$. This yields

$$\int_0^\infty \frac{1}{\sqrt{x}}e^{-x}\,dx = \int_0^\infty \frac{1}{u}e^{-u^2}\cdot 2u\,du = \int_0^\infty 2e^{-u^2}\,du = \sqrt{\pi}\,.$$

77. Make the substitution $u = \frac{x-\mu}{\sqrt{2}\sigma}$, $du = \frac{1}{\sqrt{2}\sigma}\,dx$. Then

$$\begin{aligned}\int_{-\infty}^\infty x\exp\left(-\frac{1}{2}\left(\frac{x-\mu}{\sigma}\right)^2\right)dx &= \sqrt{2}\,\sigma\int_{-\infty}^\infty (\sqrt{2}\,\sigma u + \mu)e^{-u^2}\,du \\ &= 2\sigma^2\int_{-\infty}^\infty ue^{-u^2}\,du + \sqrt{2}\,\sigma\mu\int_{-\infty}^\infty e^{-u^2}\,du \\ &= \sigma\mu\sqrt{2\pi}\,.\end{aligned}$$

Note that $\int_0^\infty ue^{-u^2}\,du = 1/2$ (integrate by parts) so $\int_{-\infty}^\infty ue^{-u^2}\,du = 0$ because the integrand is an odd function.

79. Apply integration by parts to $\Gamma(s+1) = \int_0^\infty x^s e^{-x}\,dx$ where $u = x^s$, $dv = e^{-x}\,dx$, $du = sx^{s-1}\,dx$, $v = -e^{-x}$. Then, since $s > 0$,

$$\Gamma(s+1) = -x^s e^{-x}\Big|_{x=0}^{x\to\infty} + s\int_0^\infty x^{s-1}e^{-x}\,dx = s\Gamma(s)\,.$$

Since

$$\Gamma(1) = \int_0^\infty e^{-x}\,dx = \lim_{N\to\infty}\int_0^N e^{-x}\,dx = \lim_{N\to\infty}\left(-e^{-x}\Big|_0^N\right) = \lim_{N\to\infty}\left(-e^{-N}+1\right) = 1\,,$$

$$\begin{aligned}\Gamma(2) &= \Gamma(1+1) = 1\cdot\Gamma(1) = 1\\ \Gamma(3) &= \Gamma(2+1) = 2\cdot\Gamma(2) = 2\cdot 1\\ \Gamma(4) &= \Gamma(3+1) = 3\cdot\Gamma(3) = 3\cdot 2\cdot 1\end{aligned}$$

and, in general, $\Gamma(n+1) = n!$ for each natural number n.

CALCULATOR/COMPUTER EXERCISES

81. Since $0 \le \frac{1}{\sqrt{1+x^5}} \le \frac{1}{x^{5/2}}$ for $1 \le x < \infty$, $\int_1^\infty \frac{1}{\sqrt{1+x^5}}\,dx$ converges because $\int_1^\infty \frac{1}{x^{5/2}}\,dx$ converges.

Given $\epsilon > 0$, $\int_1^\infty \frac{1}{\sqrt{1+x^5}}\,dx = \int_1^\epsilon \frac{1}{\sqrt{1+x^5}}\,dx + \int_\epsilon^\infty \frac{1}{\sqrt{1+x^5}}\,dx$, so if $\int_\epsilon^\infty \frac{1}{\sqrt{1+x^5}}\,dx \le 5\times10^{-3}$, then $\int_1^\epsilon \frac{1}{\sqrt{1+x^5}}\,dx$ will be a two decimal approximation to $\int_1^\infty \frac{1}{\sqrt{1+x^5}}\,dx$. Since $\int_\epsilon^\infty \frac{1}{\sqrt{1+x^5}}\,dx \le \int_\epsilon^\infty \frac{1}{x^{5/2}}\,dx = \frac{2}{3}\epsilon^{-3/2} = 5\times10^{-3}$ when $\epsilon = b = \left(\frac{15}{2}\times10^{-3}\right)^{-2/3}$, $\int_1^b \frac{1}{\sqrt{1+x^5}}\,dx \approx 0.61$ is the desired two decimal approximation.

83. Since $0 \le \frac{1}{x^{3x}} \le \frac{1}{x^3}$ for $1 \le x < \infty$, $\int_1^\infty \frac{1}{x^{3x}}\,dx$ converges because $\int_1^\infty \frac{1}{x^3}\,dx$ converges.

Given $\epsilon > 0$, $\int_1^\infty \frac{1}{x^{3x}}\,dx = \int_1^\epsilon \frac{1}{x^{3x}}\,dx + \int_\epsilon^\infty \frac{1}{x^{3x}}\,dx$, so if $\int_\epsilon^\infty \frac{1}{x^{3x}}\,dx \le 5\times10^{-3}$, then $\int_1^\epsilon \frac{1}{x^{3x}}\,dx$ will be a two decimal approximation to $\int_1^\infty \frac{1}{x^{3x}}\,dx$. Since $\int_\epsilon^\infty \frac{1}{x^{3x}}\,dx \le \int_\epsilon^\infty \frac{1}{x^3}\,dx = \frac{1}{2}\epsilon^{-1/2} = 5\times10^{-3}$ when $\epsilon = b = \left(10\times10^{-3}\right)^{-2}$, $\int_1^b \frac{1}{x^{3x}}\,dx \approx 0.28$ is the desired two decimal approximation.

REVIEW EXERCISES FOR CHAPTER 6

1. Use integration by parts, $u = 16x$, $dv = e^{4x}\,dx$, $du = 16\,dx$, $v = \frac{1}{4}e^{4x}$.

$$\int_0^1 16xe^{4x}\,dx = 4xe^{4x}\Big|_0^1 - \int_0^1 4e^{4x}\,dx = \left(4xe^{4x} - e^{4x}\right)\Big|_0^1 = 3e^4 + 1$$

3. Use integration by parts, $u = 4x$, $dv = \sin(x/3)\,dx$, $du = 4\,dx$, $v = -3\cos(x/3)$.

$$\int_0^{\pi/2} 4x\sin(x/3)\,dx = -12x\cos(x/3)\Big|_0^{\pi/2} + \int_0^{\pi/2} 12\cos(x/3)\,dx$$
$$= \left(-12x\cos(x/3) + 36\sin(x/3)\right)\Big|_0^{\pi/2} = 18 - 3\sqrt{3}\pi$$

5. Use integration by parts, $u = \ln(3x)$, $dv = 2x\,dx$, $du = \frac{1}{x}\,dx$, $v = x^2$.

$$\int_{1/3}^1 2x\ln(3x)\,dx = x^2\ln(3x)\Big|_{1/3}^1 - \int_{1/3}^1 x\,dx = \left(x^2\ln(3x) - \frac{1}{2}x^2\right)\Big|_{1/3}^1$$
$$= \ln(3) - \frac{1}{2} + \frac{1}{18} = \ln(3) - \frac{4}{9}$$

7. Use integration by parts, $u = x$, $dv = e^{-x}\,dx$, $du = dx$, $v = -e^{-x}$.

$$\int_{-1}^1 xe^{-x}\,dx = -xe^{-x}\Big|_{-1}^1 + \int_{-1}^1 e^{-x}\,dx = \left(-xe^{-x} - e^{-x}\right)\Big|_{-1}^1 = -2/e$$

9. Use integration by parts, $u = x$, $dv = \sin(x)\,dx$, $du = dx$, $v = -\cos(x)$.

$$\int_0^\pi x\sin(x)\,dx = -x\cos(x)\Big|_0^\pi + \int_0^\pi \cos(x)\,dx = (-x\cos(x) + \sin(x))\Big|_0^\pi = \pi$$

11. Use integration by parts, $u = \ln(x/3)$, $dv = dx$, $du = \frac{1}{x}\,dx$, $v = x$.

$$\int_1^{3e} \ln(x/3)\,dx = x\ln(x/)\Big|_1^{3e} - \int_1^{3e} dx = \big(x\ln(x/3) - x\big)\Big|_1^{3e} = \ln(3) + 1$$

13. Use integration by parts, twice. Start with $u = x^2$, $dv = e^x\,dx$, $du = 2x\,dx$, $v = e^x$. Finish with $u = x$, $dv = e^x$, $du = dx$, $v = e^x$.

$$\int_0^1 x^2e^x\,dx = x^2e^x\Big|_0^1 - 2\int_0^1 xe^x\,dx = x^2e^x\Big|_0^1 - 2\left(xe^x\Big|_0^1 - \int_0^1 e^x\,dx\right)$$
$$= \big(x^2e^x - 2xe^x + 2e^x\big)\Big|_0^1 = e - 2$$

15. Use integration by parts, twice. Start with $u = x^2$, $dv = \sin(x)\,dx$, $du = 2x\,dx$, $v = -\cos(x)$. Finish with $u = x$, $dv = \cos(x)$, $du = dx$, $v = \sin(x)$.

$$\int_0^\pi x^2\sin(x)\,dx = -x^2\cos(x)\Big|_0^\pi + 2\int_0^\pi x\cos(x)dx$$
$$= -x^2\cos(x)\Big|_0^\pi + 2\left(x\sin(x)\Big|_0^\pi - \int_0^\pi \sin(x)dx\right)$$
$$= \big(-x^2\cos(x) + 2x\sin(x) + 2\cos(x)\big)\Big|_0^\pi = \pi^2 - 4$$

17. Expand the integrand. Use the reduction formula (6.2.13) for the integral of $\sec^3(x)$.

$$\int_0^{\pi/6} 12(1+\sec(x))^3\,dx = 12\int_0^{\pi/6} (1 + 3\sec(x) + 3\sec^2(x) + \sec^3(x))\,dx$$
$$= 12\left(x + 3\ln(|\sec(x) + \tan(x)|) + 3\tan(x) + \frac{1}{2}\sec(x)\tan(x) + \frac{1}{2}\ln(|\sec(x) + \tan(x)|)\right)\Big|_0^{\pi/6}$$
$$= 2\pi + 42\ln\left(\frac{2}{\sqrt{3}} + \frac{1}{\sqrt{3}}\right) + 36\cdot\frac{1}{\sqrt{3}} + 6\cdot\frac{2}{3} = 2\pi + 21\ln(3) + 12\sqrt{3} + 4$$

19. Use a half-angle identity.

$$\int_0^{\pi/4} \sin^2(2x)\,dx = \int_0^{\pi/4} \frac{1}{2}(1 - \cos(4x))\,dx = \frac{1}{2}\left(x - \frac{1}{4}\sin(4x)\right)\Big|_0^{\pi/4} = \frac{\pi}{8}$$

21. Split off a sine and convert the other sines to cosines.

$$\int_{\pi/4}^{\pi/2} 12\sin^3(x)\,dx = \int_{\pi/4}^{\pi/2} 12\sin^2(x)\sin(x)\,dx = \int_{\pi/4}^{\pi/2} 12\big(1-\cos^2(x)\big)\sin(x)\,dx$$

$$= 12\left(-\cos(x) + \frac{1}{3}\cos^3(x)\right)\Bigg|_{\pi/4}^{\pi/2} = 12\left(\frac{\sqrt{2}}{2} - \frac{1}{3}\cdot\frac{2\sqrt{2}}{8}\right) = 5\sqrt{2}$$

23. Split off a cosine and convert the other cosines to sines.

$$\int_0^{\pi/2} \sin^2(x)\cos^3(x)\,dx = \int_0^{\pi/2} \sin^2(x)\cos^2(x)\cos(x)\,dx$$

$$= \int_0^{\pi/2} \sin^2(x)(1-\sin^2(x))\cos(x)\,dx = \int_0^{\pi/2} (\sin^2(x) - \sin^4(x))\cos(x)\,dx$$

$$= \left(\frac{1}{3}\sin^3(x) - \frac{1}{5}\sin^5(x)\right)\Bigg|_0^{\pi/2} = \frac{1}{3} - \frac{1}{5} = \frac{2}{15}$$

25. Expand the integrand and simplify.

$$\int_0^{\pi} (\cos(x)+\sin(x))^2\,dx = \int_0^{\pi} (\cos^2(x) + 2\cos(x)\sin(x) + \sin^2(x))\,dx$$

$$= \int_0^{\pi} (1 + 2\cos(x)\sin(x))\,dx = \big(x - \cos^2(x)\big)\Big|_0^{\pi} = \pi$$

27. Split off a cosine and convert the other cosines to sines.

$$\int_0^{\pi} \sin^3(x)\cos^4(x)\,dx = \int_0^{\pi} \sin^2(x)\cos^4(x)\sin(x)\,dx$$

$$= \int_0^{\pi} (1-\cos^2(x))\cos^4(x)\sin(x)\,dx = \int_0^{\pi} (\cos^4(x) - \cos^6(x))\sin(x)\,dx$$

$$= \left(-\frac{1}{5}\cos^5(x) + \frac{1}{7}\cos^7(x)\right)\Bigg|_0^{\pi} = \frac{2}{5} - \frac{2}{7} = \frac{4}{35}$$

29. Split off a cosine and convert the other cosines to sines.

$$\int_0^{\pi/2} \sin^{3/2}(x)\cos^3(x)\,dx = \int_0^{\pi/2} \sin^{3/2}(x)\cos^2(x)\cos(x)\,dx$$

$$= \int_0^{\pi/2} \sin^{3/2}(x)(1-\sin^2(x))\cos(x)\,dx = \int_0^{\pi/2} (\sin^{3/2}(x) - \sin^{7/2}(x))\cos(x)\,dx$$

$$= \left(\frac{2}{5}\sin^{5/2}(x) - \frac{2}{9}\sin^{9/2}(x)\right)\Bigg|_0^{\pi/2} = \frac{2}{5} - \frac{2}{9} = \frac{8}{45}$$

31. Split off a cosine and convert the other cosines to sines.

$$\int_{\pi/2}^{\pi} \sin^2(x)\cos^7(x)\,dx = \int_{\pi/2}^{\pi} \sin^2(x)\cos^6(x)\cos(x)\,dx$$
$$= \int_{\pi/2}^{\pi} \sin^2(x)(1-\sin^2(x))^3\cos(x)\,dx$$
$$= \int_{\pi/2}^{\pi} \sin^2(x)(1-3\sin^2(x)+3\sin^4(x)-\sin^6(x))\cos(x)\,dx$$
$$= \int_{\pi/2}^{\pi} (\sin^2(x)-3\sin^4(x)+3\sin^6(x)-\sin^8(x))\cos(x)\,dx$$
$$= \left(\frac{1}{3}\sin^3(x)-\frac{3}{5}\sin^5(x)+\frac{3}{7}\sin^7(x)-\frac{1}{9}\sin^9(x)\right)\Bigg|_{\pi/2}^{\pi} = -\frac{1}{3}+\frac{3}{5}-\frac{3}{7}+\frac{1}{9} = -\frac{16}{315}$$

33. Use a half-angle identity, twice.

$$\int_0^{2\pi} \sin^4(x)\,dx = \int_0^{2\pi} \frac{1}{4}(1-\cos(2x))^2\,dx = \int_0^{2\pi} \frac{1}{4}(1-2\cos(2x)+\cos^2(2x))\,dx$$
$$= \int_0^{2\pi} \frac{1}{4}\left(1-2\cos(2x)+\frac{1}{2}(1+\cos(4x))\right)dx$$
$$= \left(\frac{3}{8}x-\frac{1}{4}\sin(2x)+\frac{1}{32}\sin(4x)\right)\Bigg|_0^{2\pi} = \frac{3\pi}{4}$$

35. Use half-angle identities.

$$\int_0^{2\pi} \cos^6(x/4)\,dx = \int_0^{2\pi} \frac{1}{8}(1+\cos(x/2))^3\,dx$$
$$= \int_0^{2\pi} \frac{1}{8}(1+3\cos(x/2)+3\cos^2(x/2)+\cos^3(x/2))\,dx$$
$$= \int_0^{2\pi} \frac{1}{8}\left(1+3\cos(x/2)+\frac{3}{2}(1+\cos(x))+(1-\sin^2(x/2))\cos(x/2)\right)dx$$
$$= \left(\frac{5}{16}x+\sin(x/2)+\frac{3}{16}\sin(x)-\frac{2}{3}\sin^3(x/2)\right)\Bigg|_0^{2\pi} = \frac{5\pi}{8}$$

37. Use half-angle identities.

$$\int_0^{\pi/3} \cos^2(x)\sin^4(x)\,dx = \int_0^{\pi/3} \frac{1}{8}(1+\cos(2x)(1-\cos(2x))^2\,dx$$
$$= \int_0^{\pi/3} \frac{1}{8}(1-\cos^2(2x))(1-\cos(2x))\,dx$$
$$= \int_0^{\pi/3} \frac{1}{8}(1-\cos(2x)-\cos^2(2x)+\cos^3(2x))\,dx$$
$$= \int_0^{\pi/3} \frac{1}{8}\left(1-\cos(2x)-\frac{1}{2}(1+\cos(4x))+(1-\sin^2(2x))\cos(2x)\right)dx$$
$$= \left(\frac{1}{16}x-\frac{1}{64}\sin(4x)-\frac{1}{48}\sin^3(2x)\right)\bigg|_0^{\pi/3} = \frac{\pi}{48}$$

39. Split off a secant squared and convert to tangents.

$$\int_0^{\pi/3} \sec^4(x)\,dx = \int_0^{\pi/3} (\tan^2(x)+1)\sec^2(x)\,dx = \left(\frac{1}{3}\tan^3(x)+\tan(x)\right)\bigg|_0^{\pi/3} = 2\sqrt{3}$$

41. This integrates directly.

$$\int_0^{\pi/3} \sec^2(x)\tan^2(x)\,dx = \frac{1}{3}\tan^3(x)\bigg|_0^{\pi/3} = \sqrt{3}$$

43. Convert the tangents to secants and use the reduction formula (6.2.13) to integrate $\sec^3(x)$.

$$\int_0^{\pi/4} 2\sec(x)\tan^2(x)\,dx = \int_0^{\pi/4} 2\sec(x)(\sec^2(x)-1)\,dx$$
$$= \int_0^{\pi/4} (2\sec^3(x)-2\sec(x))\,dx$$
$$= \big(\sec(x)\tan(x)-\ln(|\sec(x)+\tan(x)|)\big)\bigg|_0^{\pi/4} = \sqrt{2}-\ln(1+\sqrt{2})$$

45. Split off two secants and convert the other two to tangents.

$$\int_0^{\pi/4} \tan(x)\sec^4(x)\,dx = \int_0^{\pi/4} \tan(x)(\tan^2(x)+1)\sec^2(x)\,dx$$
$$= \int_0^{\pi/4} (\tan^3(x)+\tan(x))\sec^2(x)\,dx$$
$$= \left(\frac{1}{4}\tan^4(x)+\frac{1}{2}\tan^2(x)\right)\bigg|_0^{\pi/4} = \frac{1}{4}+\frac{1}{2} = \frac{3}{4}$$

47. Convert to sines and cosines, then to cosines and simplify.

$$\int_0^{\pi/4} 2\tan(x)\sin(x)\,dx = 2\int_0^{\pi/4} \frac{\sin^2(x)}{\cos(x)}\,dx = 2\int_0^{\pi/4} \frac{1-\cos^2(x)}{\cos(x)}\,dx$$

$$= 2\int_0^{\pi/4} (\sec(x) - \cos(x))\,dx = 2\left(\ln(|\sec(x)+\tan(x)|) - \sin(x)\right)\Big|_0^{\pi/4}$$

$$= 2\ln(1+\sqrt{2}) - \sqrt{2}$$

49. Split off two cosecants and convert the other two to cotangents.

$$\int_{\pi/4}^{\pi/2} 4\cot(x)\csc^4(x)\,dx = \int_{\pi/4}^{\pi/2} 4\cot(x)(\cot^2(x)+1)\csc^2(x)\,dx$$

$$= \int_{\pi/4}^{\pi/2} (4\cot^3(x) + 4\cot(x))\csc^2(x)\,dx$$

$$= \left(-\cot^4(x) - 2\cot^2(x)\right)\Big|_{\pi/4}^{\pi/2} = 3$$

51. Convert to sines and cosines, then split off a sine.

$$\int_0^{\pi/4} 4\sin^2(x)\tan(x)\,dx = 4\int_0^{\pi/4} \frac{\sin^3(x)}{\cos(x)}\,dx = 4\int_0^{\pi/4} \frac{1-\cos^2(x)}{\cos(x)}\cdot\sin(x)\,dx$$

$$= 4\int_0^{\pi/4} (\tan(x) - \cos(x)\sin(x))\,dx = 4\left(\ln(|\sec(x)|) - \frac{1}{2}\sin^2(x)\right)\Big|_0^{\pi/4}$$

$$= 2\ln(2) - 1$$

53. Use the substitution $u = \sin(x)$, $du = \cos(x)\,dx$, then make a partial fraction decomposition.

$$\int_0^{\pi/4} \frac{3\sin(x)+1}{1-\sin^2(x)}\cos(x)\,dx = \int_0^{\sqrt{2}/2} \frac{3u+1}{1-u^2}\,du = \int_0^{\sqrt{2}/2} \left(\frac{2}{1-u} - \frac{1}{1+u}\right)du$$

$$= \left(-2\ln(|1-u|) - \ln(|1+u|)\right)\Big|_0^{\sqrt{2}/2} = -2\ln\left(1-\frac{\sqrt{2}}{2}\right) - \ln\left(1+\frac{\sqrt{2}}{2}\right)$$

$$= -\ln\left(\left(1-\frac{1}{2}\right)\left(\frac{2-\sqrt{2}}{2}\right)\right) = \ln\left(\frac{4}{2-\sqrt{2}}\right) = \ln(4+2\sqrt{2})$$

55. Use the substitutions $u = x^{1/4}$, $x = u^4$, $dx = 4u^3\,du$, then make a long

long division.

$$\int_0^1 \frac{1+x^{1/2}}{1+x^{1/4}}\,dx = \int_0^1 \frac{1+u^2}{1+u}\cdot 4u^3\,du = \int_0^1 \frac{4u^5+4u^3}{1+u}\,du$$
$$= \int_0^1 \left(4u^4 - 4u^3 + 8u^2 - 8u + 8 - \frac{8}{u+1}\right) du$$
$$= \left(\frac{4}{5}u^5 - u^4 + \frac{8}{3}u^3 - 4u^2 + 8u - 8\ln(u+1)\right)\Bigg|_0^1 = \frac{97}{15} - 8\ln(2)$$

57. Make a long division, then a partial fraction decomposition.

$$\int_0^{1/\sqrt{2}} \frac{2x^2}{1-x^2}\,dx = \int_0^{1/\sqrt{2}} \left(-2 - \frac{2}{x^2-1}\right) dx = \int_0^{1/\sqrt{2}} \left(-2 + \frac{1}{x+1} - \frac{1}{x-1}\right) dx$$
$$= \left(-2x + \ln\left(\left|\frac{x+1}{x-1}\right|\right)\right)\Bigg|_0^{1/\sqrt{2}} = \ln\left(\left|\frac{1/\sqrt{2}+1}{1/\sqrt{2}-1}\right|\right) - \sqrt{2} = \ln(3+2\sqrt{2}) - \sqrt{2}$$

59. This integrates directly.

$$\int_{\sqrt{2}}^{2/\sqrt{3}} \frac{1}{x\sqrt{x^2-1}}\,dx = \operatorname{arcsec}(x)\Bigg|_{\sqrt{2}}^{2/\sqrt{3}} = \frac{\pi}{6} - \frac{\pi}{4} = -\frac{\pi}{12}$$

61. Substitute $x = 4\tan(\theta)$, $dx = 4\sec^2(\theta)\,d\theta$.

$$\int_0^3 \frac{1}{\sqrt{16+x^2}}\,dx = \int_0^{\arctan(3/4)} \frac{1}{\sqrt{16+16\tan^2(\theta)}}\cdot 4\sec^2(\theta)\,d\theta = \int_0^{\arctan(3/4)} \sec(\theta)\,d\theta$$
$$= \ln(|\sec(\theta) + \tan(\theta)|)\Bigg|_0^{\arctan(3/4)} = \ln(2)$$

63. Substitute $x = \frac{1}{2}\sin(\theta)$, $dx = \frac{1}{2}\cos(\theta)\,d\theta$.

$$\int_0^{1/2} \sqrt{1-4x^2}\,dx = \int_0^{\pi/2} \sqrt{1-\sin^2(\theta)}\cdot\frac{1}{2}\cos(\theta)\,d\theta = \int_0^{\pi/2} \frac{1}{2}\cos^2(\theta)\,d\theta$$
$$= \int_0^{\pi/2} \frac{1}{4}(1+\cos(2\theta))\,d\theta = \left(\frac{1}{4}\theta + \frac{1}{4}\sin(\theta)\cos(\theta)\right)\Bigg|_0^{\pi/2} = \frac{\pi}{8}$$

65. Substitute $x = 2\sin(\theta)$, $dx = 2\cos(\theta)\,d\theta$, then use half-angle identities,

twice.

$$\int_0^{\sqrt{3}} 4(4-x^2)^{3/2}\,dx = \int_0^{\pi/3} 4(4-4\sin^2(\theta))^{3/2}\cdot 2\cos(\theta)\,d\theta = 64\int_0^{\pi/3}\cos^4(\theta)\,d\theta$$
$$= 16\int_0^{\pi/3}(1+\cos(2\theta))^2\,d\theta = 16\int_0^{\pi/3}(1+2\cos(2\theta)+\cos^2(2\theta))\,d\theta$$
$$= 16\int_0^{\pi/3}\left(1+2\cos(2\theta)+\frac{1}{2}(1+\cos(4\theta))\right)d\theta =$$
$$= \left(24\theta+16\sin(2\theta)+2\sin(4\theta)\right)\Big|_0^{\pi/3} = 8\pi+7\sqrt{3}$$

67. Make a partial fraction decomposition.

$$\int_{\sqrt{10}}^5 \frac{3}{x^2-9}\,dx = \frac{1}{2}\int_{\sqrt{10}}^5\left(\frac{1}{x-3}-\frac{1}{x+3}\right)dx = \frac{1}{2}\ln\left(\frac{x-3}{x+3}\right)\Big|_{\sqrt{10}}^5$$
$$= \frac{1}{2}\ln\left(\frac{2}{8}\right)-\frac{1}{2}\ln\left(\frac{\sqrt{10}-3}{\sqrt{10}+3}\right) = -\ln(2)-\ln(\sqrt{10}-3)$$
$$= \ln\left(\frac{\sqrt{10}+3}{2}\right)$$

69. Substitute $x = 3\tan(\theta)$, $dx = 3\sec^2(\theta)\,d\theta$. Integrate $\sec^3(\theta)$ using the reduction formula (6.2.13).

$$\int_0^4 (9+x^2)^{-3/2}\,dx = \int_0^{\arctan(4/3)}(9+9\tan^2(\theta))^{-3/2}\cdot 3\sec^2(\theta)\,d\theta$$
$$= \frac{1}{9}\int_0^{\arctan(4/3)}\cos(\theta)\,d\theta = \frac{1}{9}\sin(\theta)\Big|_0^{\arctan(4/3)} = \frac{4}{45}$$

71. Substitute $x = \sec(\theta)$, $dx = \sec(\theta)\tan(\theta)\,d\theta$. Then split off two secants and convert the remaining secants into tangents.

$$\int_2^3 \frac{3x^3}{\sqrt{x^2-1}}\,dx = \int_{\pi/3}^{\operatorname{arcsec}(3)}\frac{3\sec^3(\theta)}{\sqrt{\sec^2(\theta)-1}}\cdot\sec(\theta)\tan(\theta)\,d\theta$$
$$= \int_{\pi/3}^{\operatorname{arcsec}(3)} 3\sec^4(\theta)\,d\theta = \int_{\pi/3}^{\operatorname{arcsec}(3)} 3(\tan^2(\theta)+1)\sec^2(\theta)\,d\theta$$
$$= \left(\tan^3(\theta)+3\tan(\theta)\right)\Big|_{\pi/3}^{\operatorname{arcsec}(3)}$$
$$= \ln(3)-\ln\left(\frac{\sqrt{10}+1}{3}\right) = 22\sqrt{2}-6\sqrt{3}$$

73. Make a long division followed by a partial fraction decomposition.

$$\int_0^1 \frac{x^2}{2-x^2}\,dx = \int_0^1 \left(-1+\frac{2}{2-x^2}\right)dx = \int_0^1 \left(-1+\frac{\sqrt{2}/2}{\sqrt{2}-x}+\frac{\sqrt{2}/2}{\sqrt{2}+x}\right)dx$$
$$= \left(-x+\frac{\sqrt{2}}{2}\ln\left(\frac{\sqrt{2}+x}{\sqrt{2}-x}\right)\right)\Bigg|_0^1 = \frac{\sqrt{2}}{2}\ln\left(\frac{\sqrt{2}+1}{\sqrt{2}-1}\right)-1 = \sqrt{2}\ln(\sqrt{2}+1)-1$$

75. The denominator factors, $\frac{1+x}{1-2x+x^2} = \frac{1+x}{(x-1)^2}$, yielding the partial fraction decomposition $\frac{x+1}{(x-1)^2} = \frac{A}{x-1}+\frac{B}{(x-1)^2}$. Therefore, $x+1 = A(x-1)+B$ so $A = 1$, $B = 2$ and

$$\int_{-1}^0 \frac{1+x}{1-2x+x^2}\,dx = \int_{-1}^0 \left(\frac{1}{x-1}+\frac{2}{(x-1)^2}\right)dx$$
$$= \left(\ln(|x-1|)-\frac{2}{x-1}\right)\Bigg|_{-1}^0 = 1-\ln(2)\,.$$

77. The denominator factors and $\frac{3x^2-2x+3}{x(x^2+1)} = \frac{A}{x}+\frac{Bx+C}{x^2+1}$. Therefore,

$$3x^2-2x+3 = A(x^2+1)+(Bx+C)x\,,$$

so $A = 3$, $B = 0$, $C = -2$, and

$$\int_1^2 \frac{3x^2-2x+3}{x(x^2+1)}\,dx = \int_1^2 \left(\frac{3}{x}-\frac{2}{x^2+1}\right)dx = \big(3\ln(x)-2\arctan(x)\big)\Big|_1^2$$
$$= 3\ln(2)-2\arctan(2)+\frac{\pi}{2}\,.$$

79. The denominator is irreducible so complete the square after simplifying the integrand. The second term integrates easily using the substitution $u = x + 1$.

$$\int_{-1}^0 \frac{2x+4}{x^2+2x+2}\,dx = \int_{-1}^0 \frac{2x+2+2}{x^2+2x+2}\,dx = \int_{-1}^0 \left(\frac{2x+2}{x^2+2x+2}+\frac{2}{(x+1)^2+1}\right)dx$$
$$= \big(\ln(x^2+2x+2)+2\arctan(x+1)\big)\Big|_{-1}^0 = \ln(2)+\frac{\pi}{2}\,.$$

81. The denominator factors and $\frac{2x^3+2x+2}{(x^2+1)^2} = \frac{Ax+B}{x^2+1}+\frac{Cx+D}{(x^2+1)^2}$. Therefore,

$$2x^3+2x+2 = (Ax+B)(x^2+1)+Cx+D$$
$$= Ax^3+Bx^2+(A+C)x+B+D\,.$$

Therefore, $A = 2$, $B = 0$, $C = 0$, $D = 2$ and

$$\int_0^1 \frac{2x^3+2x+2}{(x^2+1)^2}\,dx = \int_0^1 \left(\frac{2x}{x^2+1} + \frac{2}{(x^2+1)^2}\right) dx$$
$$= \ln(x^2+1)\Big|_0^1 + \int_0^1 \frac{2}{(x^2+1)^2}\,dx\,.$$

Evaluate the remaining term with the substitution $x = \tan(\theta)$, $dx = \sec^2(\theta)\,d\theta)$ to obtain

$$\int_0^1 \frac{2x^3+2x+2}{(x^2+1)^2}\,dx = \ln(2) + \int_0^{\pi/4} \frac{2}{(\tan^2(\theta)+1)^2}\cdot \sec^2(\theta)\,d\theta$$
$$= \ln(2) + \int_0^{\pi/4} 2\cos^2(\theta)\,d\theta = \ln(2) + \int_0^{\pi/4} (1+\cos(2\theta))\,d\theta$$
$$= \ln(2) + \left(\theta + \frac{1}{2}\sin(2\theta)\right)\Big|_0^{\pi/4} = \ln(2) + \frac{\pi}{4} + \frac{1}{2}\,.$$

83. The denominator factors and $\frac{2x^3+x^2+2x+2}{(x^2+1)(x^2+2)} = \frac{Ax+B}{x^2+1} + \frac{Cx+D}{x^2+2}$. Therefore,

$$2x^3+x^2+2x+2 = (Ax+B)(x^2+2) + (Cx+D)(x^2+1)$$
$$= (A+C)x^3 + (B+D)x^2 + (2A+C)x + 2B + D\,.$$

The four constants satisfy the following four equations.

$$A + C = 2 \qquad B + D = 1$$
$$2A + C = 2 \qquad 2B + D = 2$$

Therefore, $A = 0$, $C = 2$, $B = 1$, $D = 0$ and

$$\int_0^1 \frac{2x^3+x^2+2x+2}{(x^2+1)(x^2+2)}\,dx = \int_0^1 \left(\frac{1}{x^2+1} + \frac{2x}{x^2+2}\right) dx$$
$$= \big(\arctan(x) + \ln(x^2+2)\big)\Big|_0^1 = \frac{\pi}{4} + \ln(3/2)\,.$$

85. Substitute $u = x^{1/2}$, $x = u^2$, $dx = 2u\,du$ and simplify.

$$\int_1^4 \frac{1}{x+x^{1/2}}\,dx = \int_1^2 \frac{1}{u^2+u}\cdot 2u\,du = \int_1^2 \frac{2}{u+1}\,du = 2\ln(u+1)\Big|_1^2 = 2\ln(3/2)$$

87. The integrand is unbounded at the left endpoint.

$$\int_0^4 \frac{1}{x^{1/2}}\,dx = \lim_{\epsilon\to 0^+} \int_\epsilon^4 \frac{1}{x^{1/2}}\,dx = \lim_{\epsilon\to 0^+} \left(2x^{1/2}\Big|_\epsilon^4\right) = \lim_{\epsilon\to 0^+} (4 - 2\sqrt{\epsilon}) = 4$$

89. The integrand is unbounded at the right endpoint. Begin with the substitution $u = x^2$, $du = 2x\,dx$.

$$\int_0^{2\sqrt{2}} \frac{x}{(8-x^2)^{1/3}}\,dx = \int_0^8 \frac{1}{(8-u)^{1/3}} \cdot \frac{1}{2}\,du = \lim_{\epsilon \to 0^+} \int_0^{8-\epsilon} \frac{1}{2} \cdot (8-u)^{-1/3}\,du$$

$$= \lim_{\epsilon \to 0^+} \left(-\frac{3}{4} \cdot (8-u)^{2/3} \Big|_0^{8-\epsilon} \right) = \lim_{\epsilon \to 0^+} \left(-\frac{3}{4}\epsilon^{2/3} + 3 \right) = 3$$

91. The integral is improper. Use integration by parts, $u = x$, $dv = \exp(-x/3)\,dx$, $du = dx$, $v = -3\exp(-x/3)$.

$$\int_0^\infty x\exp(-x/3)\,dx = \lim_{N\to\infty} \int_0^N x\exp(-3x)\,dx$$

$$= \lim_{N\to\infty} \left(-3x\exp(-x/3)\Big|_0^N + \int_0^N 3\exp(-x/3)\,du \right)$$

$$= \lim_{N\to\infty} \left(\big(-3x\exp(-x/3) - 9\exp(-x/3)\big)\Big|_0^N \right)$$

$$= \lim_{N\to\infty} \big(-3N\exp(-N/3) - 9\exp(-N/3) + 9 \big) = 9$$

93. The integral is improper. Use the substitution $u = x^4$, $du = 4x^{-3}\,dx$.

$$\int_0^\infty x^3\exp(-x^4)\,dx = \lim_{N\to\infty} \int_0^N x^3\exp(-x^4)\,dx = \lim_{N\to\infty} \int_0^{N^4} \frac{1}{4}\exp(-u)\,du$$

$$= \lim_{N\to\infty} \left(-\frac{1}{4}\exp(-u)\Big|_0^{N^4} \right) = \lim_{N\to\infty} \left(-\frac{1}{4}\exp(-N^4) + \frac{1}{4} \right) = \frac{1}{4}$$

95. The integral is improper. The substitution $x = \tan(\theta)$, $du = \sec^2(\theta)\,d\theta$, and the appropriate cancellations, makes it proper

$$\int_1^\infty \frac{1}{x\sqrt{1+x^2}}\,dx = \int_{\pi/4}^{\pi/2} \frac{1}{\tan(\theta)\sqrt{1+\tan^2(\theta)}} \cdot \sec^2(\theta)\,d\theta = \int_{\pi/4}^{\pi/2} \frac{\sec(\theta)}{\tan(\theta)}\,d\theta$$

$$= \int_{\pi/4}^{\pi/2} \csc(\theta)\,d\theta = \ln(|\csc(\theta) - \cot(\theta)|)\Big|_{\pi/4}^{\pi/2} = -\ln(\sqrt{2}-1) = \ln(\sqrt{2}+1)$$

97. The area is

$$A = \int_0^{\pi/2} (1 - \sin^3(x) - \cos^2(x))\,dx$$

$$= \int_0^{\pi/2} \left(1 - (1-\cos^2(x))\sin(x) - \frac{1}{2}(1+\cos(2x)) \right) dx$$

$$= \left(\frac{1}{2}x + \cos(x) - \frac{1}{3}\cos^3(x) - \frac{1}{4}\sin(2x) \right)\Big|_0^{\pi/2} = \frac{\pi}{4} - \frac{2}{3}.$$

99. The area is

$$A = \int_0^1 (\ln(1+x) - x\ln(1+x))\,dx = \int_0^1 (1-x)\ln(1+x)\,dx\,.$$

Integrate by parts, $u = \ln(1+x)$, $dv = (1-x)\,dx$, $du = \frac{1}{x+1}\,dx$, $v = x - \frac{1}{2}x^2$, then make a long division.

$$\begin{aligned}
A &= \left(x - \frac{1}{2}x^2\right)\ln(1+x)\Big|_0^1 - \int_0^1 \frac{x - x^2/2}{x+1}\,dx \\
&= \frac{1}{2}\ln(2) - \int_0^1 \left(\frac{3}{2} - \frac{1}{2}x - \frac{3/2}{x+1}\right)\,dx \\
&= \frac{1}{2}\ln(2) - \left(\frac{3}{2}x - \frac{1}{4}x^2 - \frac{3}{2}\ln(x+1)\right)\Big|_0^1 \\
&= 2\ln(2) - \frac{5}{4}
\end{aligned}$$

Chapter 7

Applications of the Integral

7.1 Volumes

Problems for Practice

1. The disk obtained by rotating the rectangle with base $[x, x + \Delta x]$ and height $\sqrt{x}$ about the x-axis has volume $\pi x \Delta x$. The volume of the solid is $V = \int_1^3 \pi x\, dx = \frac{1}{2}\pi x^2\Big|_1^3 = 4\pi$.

3. The disk obtained by rotating the rectangle with base $[x, x + \Delta x]$ and height $\sqrt{\sin(x)}$ about the x-axis has volume $\pi \sin(x)\Delta x$. The volume of the solid is $V = \int_0^\pi \pi \sin(x)\, dx = -\pi \cos(x)\Big|_0^\pi = 2\pi$.

5. The disk obtained by rotating the rectangle with base $[x, x + \Delta x]$ and height $\exp(x)$ about the x-axis has volume $\pi \exp^2(x)\Delta x$. The volume of the solid is $V = \int_0^{\ln(2)} \pi \exp^2(x)\, dx = \pi \int_0^{\ln(2)} \exp(2x)\, dx = \frac{1}{2}\pi \exp(2x)\Big|_0^{\ln(2)} = 3\pi/2$.

7. The disk obtained by rotating the rectangle with base $[y, y + \Delta y]$ and height $\sin(y)$ about the y-axis has volume $\pi \sin^2(y)\Delta y$. The volume of the solid is

$$\begin{aligned} V &= \int_0^{\pi/2} \pi \sin^2(y)\, dy = \frac{\pi}{2}\int_0^{\pi/2} (1 - \cos(2y)\, dy \\ &= \frac{\pi}{2}\left(y - \frac{1}{2}\sin(2y)\right)\Big|_0^{\pi/2} = \frac{\pi^2}{4}. \end{aligned}$$

9. The disk obtained by rotating the rectangle with base $[y, y+\Delta y]$ and height e^y about the y-axis has volume $\pi e^{2y}\Delta y$. The volume of the solid is $V = \int_0^1 \pi e^{2y}\,dy = \frac{\pi}{2}e^{2y}\big|_0^1 = \frac{\pi}{2}(e^2-1)$.

11. The disk obtained by rotating the rectangle with base $[y, y+\Delta y]$ and height y^2 about the y-axis has volume $\pi y^4\Delta y$. The volume of the solid is $V = \int_0^2 \pi y^4\,dy = \frac{\pi}{5}y^5\big|_0^2 = 32\pi/5$.

13. The washer obtained by rotating the rectangle with base $[x, x+\Delta x]$ outer radius x and inner radius x^2 about the x-axis has volume $\pi(x^2-x^4)\Delta x$. The volume of the solid is $V = \int_0^1 \pi(x^2-x^4)\,dx = \pi\left(\frac{1}{3}x^3 - \frac{1}{5}x^5\right)\big|_0^1 = 2\pi/15$.

15. The region $\mathcal{R}$ lies above the interval $-2 \le x \le 1$. The washer obtained by rotating the rectangle with base $[x, x+\Delta x]$ outer radius $4-x^2$ and inner radius $x+2$ about the x-axis has volume $\pi\big((4-x^2)^2-(x+2)^2\big)\Delta x$. The volume of the solid is

$$V = \int_{-2}^{1} \pi\big((4-x^2)^2-(x+2)^2\big)\,dx = \pi\int_{-2}^{1}\left(12-4x-9x^2+x^4\right)dx$$
$$= \pi\left(12x-2x^2-3x^3+\frac{1}{5}x^5\right)\Big|_{-2}^{1} = \frac{108\pi}{5}.$$

17. The region $\mathcal{R}$ lies above the interval $0 \le x \le 1$. The washer obtained by rotating the rectangle with base $[x, x+\Delta x]$ outer radius $x+1$ and inner radius 2^x about the x-axis has volume $\pi\big((x+1)^2-2^{2x}\big)\Delta x$. The volume of the solid is

$$V = \int_0^1 \pi\big((x+1)^2-2^{2x}\big)\,dx = \pi\left(\frac{1}{3}(x+1)^3 - \frac{4^x}{\ln(4)}\right)\Big|_0^1$$
$$= \pi\left(\frac{7}{3} - \frac{3}{2\ln(2)}\right) = \frac{\pi(14\ln(2)-9)}{6\ln(2)}.$$

19. The washer obtained by rotating the rectangle with base $[y, y+\Delta y]$ outer radius $2y$ and inner radius y^2 about the y-axis has volume $\pi(4y^2-y^4)\Delta y$. The volume of the solid is $V = \int_0^2 \pi(4y^2-y^4)\,dy = \pi\left(\frac{4}{3}y^3 - \frac{1}{5}y^5\right)\big|_0^2 = \pi\left(\frac{32}{3}-\frac{32}{5}\right) = \frac{64\pi}{15}$.

21. The region lies to the right of the interval $-2 \le y \le 1$. The washer obtained by rotating the rectangle with base $[y, y+\Delta y]$ outer radius $4-y$ and inner radius y^2+2 about the y-axis has volume $\pi\big((4-y)^2-(y^2+2)^2\big)\Delta y$. The volume of the solid is

$$V = \int_{-2}^{1} \pi\big((4-y)^2-(y^2+2)^2\big)\,dy = \pi\int_{-2}^{1}\left(12-8y-3y^2-y^4\right)dy$$
$$= \pi\left(12y-4y^2-y^3-\frac{1}{5}y^5\right)\Big|_{-2}^{1} = \frac{162\pi}{5}.$$

23. The region lies to the right of the interval $0 \leq y \leq 4$. The washer obtained by rotating the rectangle with base $[y, y + \Delta y]$ outer radius $\sqrt{y}$ and inner radius $y/4$ about the y-axis has volume $\pi(y - y^2/16)\Delta y$. The volume of the solid is $V = \int_0^4 \pi(y - y^2/16)\,dy = \pi\left(\frac{1}{2}y^2 - \frac{1}{48}y^3\right)\Big|_0^4 = \pi\left(8 - \frac{4}{3}\right) = \frac{20\pi}{3}$.

25. The cylindrical shell obtained by rotating a rectangle having base $[x, x + \Delta x]$ and height $\exp(x^2)$ about the y-axis has the approximate volume $2\pi x \exp(x^2)\Delta x$. The volume of the solid is $V = \int_0^1 2\pi x \exp(x^2)\,dx = \pi \exp(x^2)\Big|_0^1 = \pi(e - 1)$.

27. The cylindrical shell obtained by rotating a rectangle having base $[x, x + \Delta x]$ and height $x^2 + 1$ about the y-axis has the approximate volume $2\pi x(x^2 + 1)\Delta x$. The volume of the solid is $V = \int_2^4 2\pi x(x^2 + 1)\,dx = 2\pi \int_2^4 (x^3 + x)\,dx = 2\pi\left(\frac{1}{4}x^4 + \frac{1}{2}x^2\right)\Big|_2^4 = 132\pi$.

29. The cylindrical shell obtained by rotating a rectangle having base $[x, x + \Delta x]$ and height $\sin(x)/x$ about the y-axis has the approximate volume $2\pi x(\sin(x)/x)\Delta x$. The volume of the solid is $V = \int_0^\pi 2\pi x \cdot (\sin(x)/x)\,dx = 2\pi \int_0^\pi \sin(x)\,dx = -2\pi \cos(x)\Big|_0^\pi = 4\pi$.

31. The cylindrical shell obtained by rotating a rectangle having base $[y, y + \Delta y]$ and width $y + 2$ about the x-axis has the approximate volume $2\pi y(y + 2)\Delta y$. The volume of the solid is $V = \int_0^2 2\pi y(y + 2)\,dy = 2\pi \int_0^2 (y^2 + 2y)\,dy = 2\pi\left(\frac{1}{3}y^3 + y^2\right)\Big|_0^2 = \frac{40\pi}{3}$.

33. The cylindrical shell obtained by rotating a rectangle having base $[y, y + \Delta y]$ and width $y^2 + 1$ about the x-axis has the approximate volume $2\pi y \cdot (y^2 + 1)\Delta y$. The volume of the solid is $V = \int_0^1 2\pi y(y^2 + 1)\,dy = 2\pi \int_0^1 (y^3 + y)\,dy = 2\pi\left(\frac{1}{4}y^4 + \frac{1}{2}y^2\right)\Big|_0^1 = \frac{3\pi}{2}$.

35. The cylindrical shell obtained by rotating a rectangle having base $[y, y + \Delta y]$ and width $y \exp(y^3)$ about the x-axis has the approximate volume $2\pi y \cdot (y \exp(y^3))\Delta y$. The volume of the solid is $V = \int_0^1 2\pi y \cdot (y \exp(y^3))\,dy = 2\pi \int_0^1 y^2 \exp(y^3)\,dy = \frac{2}{3}\pi \exp(y^3)\Big|_0^1 = \frac{2\pi(e-1)}{3}$.

37. The region lies above the interval $0 \leq x \leq 1$. The cylindrical shell obtained by rotating about the y-axis a rectangle having base $[x, x + \Delta x]$ and extending vertically from x^2 to x has the approximate volume $2\pi x \cdot (x - x^2)\Delta x$. The volume of the solid is $V = \int_0^1 2\pi x \cdot (x - x^2)\,dx = 2\pi \int_0^1 (x^2 - x^3)\,dx = 2\pi\left(\frac{1}{3}x^3 - \frac{1}{4}x^4\right)\Big|_0^1 = \frac{\pi}{6}$.

39. The cylindrical shell obtained by rotating about the y-axis a rectangle having base $[x, x + \Delta x]$ and extending vertically from x to $x^2 + 1$ has the approximate volume $2\pi x \cdot (x^2 + 1 - x)\Delta x$. The volume of the solid is $V = \int_1^3 2\pi x \cdot (x^2 + 1 - x)\,dx = 2\pi \int_1^3 (x^3 + x - x^2)\,dx = 2\pi\left(\frac{1}{4}x^4 + \frac{1}{2}x^2 - \frac{1}{3}x^3\right)\Big|_1^3 = \frac{92\pi}{3}$.

41. The region lies above the interval $0 \le x \le 4$. The cylindrical shell obtained by rotating about the y-axis a rectangle having base $[x, x+\Delta x]$ and extending vertically from $x+1$ to $\sqrt{9+x^2}$ has the approximate volume $2\pi x \cdot (\sqrt{9+x^2} - (x+1))\Delta x$. The volume of the solid is

$$V = \int_0^4 2\pi x \cdot \left(\sqrt{9+x^2} - (x+1)\right) dx = 2\pi \int_0^4 \left(x\sqrt{9+x^2} - x^2 + x)\right) dx$$
$$= 2\pi \left(\frac{1}{3}(9+x^2)^{3/2} - \frac{1}{3}x^3 + \frac{1}{2}x^2\right)\Big|_0^4 = \frac{20\pi}{3}.$$

43. The region lies to the right of the interval $1 \le y \le 4$. The cylindrical shell obtained by rotating about the x-axis a rectangle having base $[y, y+\Delta y]$ and extending horizontally from $\sqrt{y}$ to 3 has the approximate volume $2\pi y \cdot (3 - \sqrt{y}\,)\Delta y$. The volume of the solid is $V = \int_1^4 2\pi y(3-\sqrt{y}\,)\,dy = 2\pi\left(\frac{3}{2}y^2 - \frac{2}{5}y^{5/2}\right)\Big|_1^4 = \frac{101\pi}{5}$.

45. The region lies to the left and right of the interval $1 \le y \le 4$. The cylindrical shell obtained by rotating about the x-axis a rectangle having base $[y, y+\Delta y]$ and extending horizontally from $-\sqrt{5-y}$ to $\sqrt{5-y}$ has the approximate volume $2\pi y \cdot (\sqrt{5-y} - (-\sqrt{5-y}\,)\Delta y$. The volume of the solid is $V = \int_1^4 2\pi y \cdot 2\sqrt{5-y}\,dy = 4\pi \int_1^4 y\sqrt{5-y}\,dy$. Make the substitution $u = 5-y$ and $V = 4\pi \int_1^4 (5-u)u^{1/2}\,du = 4\pi\left(\frac{10}{3}u^{3/2} - \frac{2}{5}u^{5/2}\right)\Big|_1^4 = \frac{656\pi}{15}$.

47. The region lies to the right of the interval $1 \le y \le 3$. The cylindrical shell obtained by rotating about the x-axis a rectangle having base $[y, y+\Delta y]$ and width $4y - y^2 - 3$ has the approximate volume $2\pi y \cdot (4y - y^2 - 3)\Delta y$. The volume is $V = \int_1^3 2\pi y \cdot (4y - y^2 - 3)\,dy = 2\pi \int_1^3 (4y^2 - y^3 - 3y)\,dy = 2\pi\left(\frac{4}{3}y^3 - \frac{1}{4}y^4 - \frac{3}{2}y^2\right)\Big|_1^3 = \frac{16\pi}{3}$.

49. The region lies above the interval $0 \le x \le 1$. The cylindrical shell obtained by rotating about the line $x = 2$ a rectangle having base $[x, x+\Delta x]$ and height $\sqrt{x} - x^2$ has the approximate volume $2\pi(2-x) \cdot (\sqrt{x} - x^2)\Delta x$. Therefore, the volume is $V = \int_0^1 2\pi(2-x)(\sqrt{x} - x^2)\,dx$ $= 2\pi \int_0^1 (2x^{1/2} - x^{3/2} - 2x^2 + x^3)\,dx = 2\pi\left(\frac{4}{3}x^{3/2} - \frac{2}{5}x^{5/2} - \frac{2}{3}x^3 + \frac{1}{4}x^4\right)\Big|_0^1 = \frac{31\pi}{30}$.

51. The region lies to the right of the interval $0 \le y \le 1$. The cylindrical shell obtained by rotating about the line $y = 1$ a rectangle having base $[y, y+\Delta y]$ and width $\sqrt{y} - y^2$ has the approximate volume $2\pi(1-y) \cdot (\sqrt{y} - y^2)\Delta y$. Therefore, the volume is $V = \int_0^1 2\pi(1-y)(\sqrt{y} - y^2)\,dy$ $= 2\pi \int_0^1 (y^{1/2} - y^{3/2} - y^2 + y^3)\,dx = 2\pi\left(\frac{2}{3}y^{3/2} - \frac{2}{5}y^{5/2} - \frac{1}{3}y^3 + \frac{1}{4}y^4\right)\Big|_0^1 = \frac{11\pi}{30}$.

53. The region lies above the interval $1 \le x \le 2$. The cylindrical shell obtained by rotating about the line $x = -1$ a rectangle having base $[x, x+\Delta x]$ and

height $4 - 4/x^2$ has the approximate volume $2\pi(x+1)\cdot(4-4/x^2)\Delta x$. Therefore, the volume is $V = \int_1^2 2\pi(x+1)(4-4x^{-2})\,dx$
$= 2\pi\int_1^2 (4+4x-4x^{-1}-4x^{-2})\,dx = 2\pi\left(4x+2x^2-4\ln(x)+4x^{-1}\right)\Big|_1^2 = 8\pi(2-\ln(2))$.

55. The region lies above the interval $1 \le x \le 2$. The disk obtained by rotating about the line $y = 4$ a rectangle having as its base the line segment from $(x, 4)$ to $(x+\Delta x, 4)$ and extending from $y = 4$ to $y = 4/x^2$ has the volume $\pi(4-4/x^2)^2\Delta x$. Therefore, the volume is $V = \int_1^2 \pi(4-4x^{-2})^2\,dx$
$= 16\pi\int_1^2(1-2x^{-2}-x^{-4})\,dx = 16\pi\left(x+2x^{-1}+\frac{1}{3}x^{-3}\right)\Big|_1^2 = \frac{14\pi}{3}$.

Further Theory and Practice

57. The disk obtained by rotating about the x-axis a rectangle having as its base the interval $[x, x+\Delta x]$ and height $\sin(x)$ has volume $\pi\sin^2(x)\Delta x$. Therefore, the volume is $V = \int_0^\pi \pi\sin^2(x)\,dx = \frac{\pi}{2}\int_0^\pi(1-\cos(2x))\,dx = \frac{\pi}{2}\left(x-\frac{1}{2}\cos(2x)\right)\Big|_0^\pi = \frac{\pi^2}{2}$.

59. The disk obtained by rotating about the x-axis a rectangle having base $[x, x+\Delta x]$ and height $\sqrt{x}\exp(x)$ has volume $\pi x\exp(2x)\Delta x$. Therefore, the volume is $V = \int_0^1 \pi x\exp(2x)\,dx$ which can be evaluated using integration by parts: $u = \pi x$, $dv = \exp(2x)\,dx$, $du = \pi\,dx$, $v = \frac{1}{2}\exp(2x)$.

$$V = \frac{\pi}{2}x\exp(2x)\Big|_0^1 - \frac{\pi}{2}\int_0^1 \exp(2x)\,dx = \frac{\pi}{2}\left(x\exp(2x)-\frac{1}{2}\exp(2x)\right)\Big|_0^1$$
$$= \frac{\pi}{2}\left(\frac{1}{2}e^2+\frac{1}{2}\right) = \frac{\pi(e^2+1)}{4}$$

61. The region lies to the right of the interval $1 \le y \le e$. The disk obtained by rotating about the y-axis a rectangle having base $[y, y+\Delta y]$ and width $\ln(y)$ has volume $\pi\ln^2(y)\Delta y$. Therefore, the volume is $V = \int_1^e \pi\ln^2(y)\,dy$ which can be evaluated starting with integration by parts: $u = \pi\ln^2(y)$, $dv = dy$, $du = 2\pi\ln(y)\cdot\frac{1}{y}\,dy$, $v = y$, and finishing with integration by parts: $u = \ln(y)$, $dv = dy$, $du = \frac{1}{y}\,dy$, $v = y$.

$$V = \pi y\ln^2(y)\Big|_1^e - 2\pi\int_1^e \ln(y)\,dy = \pi\left(y\ln^2(y) - 2\left(y\ln(y)-y\right)\right)\Big|_1^e = \pi(e-2)$$

63. If $y = x^2/(1-x^2)$ and $y \ge 0$, then $x = \sqrt{\frac{y}{y+1}}$. The disk obtained by rotating about the y-axis a rectangle having base $[y, y+\Delta y]$ and width $\sqrt{\frac{y}{y+1}}$ has volume $\pi\left(\frac{y}{y+1}\right)\Delta y$. Therefore, the volume is
$V = \int_0^{1/3}\pi\left(\frac{y}{y+1}\right)dy = \pi\int_0^{1/3}\left(1-\frac{1}{y+1}\right)dy = \pi(y-\ln(y+1))\Big|_0^{1/3} = \pi\left(\frac{1}{3}-2\ln(2)+\ln(3)\right)$.

65. The curves intersect at the points $(-1,0)$ and $(5,0)$ so the region lies above and below the interval $-1 \le x \le 5$. The cylindrical shell obtained by rotating about the line $x = -3$ a rectangle having base $[x, x+\Delta x]$ and height $(-x^2+4x+5)-(x^2-4x-5)$ has volume $2\pi(x+3)(-2x^2+8x+25)\Delta x$. Therefore, the volume is $V = \int_{-1}^{5} 2\pi(x+3)(-2x^2+8x+25)\,dx$ $= 2\pi\int_{-1}^{5}(-2x^3+2x^2+49x+30)\,dx = 2\pi\left(-\frac{1}{2}x^4+\frac{2}{3}x^3+\frac{49}{2}x^2+30x\right)\Big|_{-1}^{5}$ $= 1620\pi$.

67. The region lies above the interval $-2 \le x \le 3$. The cylindrical shell obtained by rotating about the line $x = 5$ a rectangle having base $[x, x+\Delta x]$ and height $(-x^2+x+7)-(x^2-x-5)$ has volume $2\pi(5-x)(-2x^2+2x+12)\Delta x$. Therefore, the volume is $V = \int_{-2}^{3} 2\pi(5-x)(-2x^2+2x+12)\,dx$ $= 2\pi\int_{-2}^{3}(2x^3-12x^2-2x+60)\,dx = 2\pi\left(\frac{1}{2}x^4-4x^3-x^2+60x\right)\Big|_{-2}^{3}$ $= 375\pi$.

69. The region lies above and below the interval $-2 \le x \le 3$. The cylindrical shell obtained by rotating about the line $x = -4$ a rectangle having base $[x, x+\Delta x]$ and height $6-x^2-(-x)$ has volume $2\pi(x+4)(6-x^2+x)\Delta x$. The volume is $V = \int_{-2}^{3} 2\pi(x+4)(6-x^2+x)\,dx = 2\pi\int_{-2}^{3}(24+10x-3x^2-x^3)\,dx = 2\pi\left(24x+5x^2-x^3-\frac{1}{4}x^4\right)\Big|_{-2}^{3} = \frac{375\pi}{2}$.

71. The region lies below the interval $\pi/2 \le x \le 3\pi/2$. The cylindrical shell obtained by rotating about the line $x = 3\pi$ a rectangle having base $[x, x+\Delta x]$ and height $0-\cos(x)$ has volume $2\pi(3\pi-x)(-\cos(x))\Delta x$. The volume is $V = \int_{\pi/2}^{3\pi/2} 2\pi(x-3\pi)\cos(x)\,dx$ which can be evaluated using integration by parts, $u = x-3\pi$, $dv = \cos(x)\,dx$, $du = dx$, $v = \sin(x)$.

$$V = 2\pi\left((x-3\pi)\sin(x)\Big|_{\pi/2}^{3\pi/2} - \int_{\pi/2}^{3\pi/2}\sin(x)\,dx\right) = 8\pi^2$$

73. The region is to the right of the interval $0 \le y \le \pi/2$. The cylindrical shell obtained by rotating about the line $y = 2$ a rectangle having base $[y, y+\Delta y]$ and width $\sin(y)$ has volume $2\pi(2-y)\sin(y)\Delta y$. The volume of the solid is $V = \int_0^{\pi/2} 2\pi(2-y)\sin(y)\,dy$ which can be integrated by parts, $u = 2-y$, $dv = \sin(y)$, $du = -dy$, $v = -\cos(y)$.

$$V = 2\pi\left(-(2-y)\cos(y)\Big|_0^{\pi/2} - \int_0^{\pi/2}\cos(y)\,dy\right)$$
$$= 2\pi\big((y-2)\cos(y) - \sin(y)\big)\Big|_0^{\pi/2} = 2\pi$$

75. The cylindrical shell obtained by rotating about the line $x = -1$ a rectangle having base $[x, x+\Delta x]$ and height x^3+x has volume $2\pi(x+$

$1)(x^3 - x)\Delta x$. The volume of the solid is $V = \int_0^1 2\pi(x+1)(x^3+x)\,dx = 2\pi\int_0^1 (x + x^2 + x^3 + x^4)\,dx = 2\pi\left(\frac{1}{2}x^2 + \frac{1}{3}x^3 + \frac{1}{4}x^4 + \frac{1}{5}x^5\right)\Big|_0^1 = \frac{77\pi}{30}$

77. We will double the volume of the top half of the torus (using cylindrical shells). The top half of the rotated region lies above the interval $-r \le x \le r$. The cylindrical shell obtained by rotating about the line $x = R$ a rectangle having base $[x, x+\Delta x]$ and height $\sqrt{r^2 - x^2}$ has volume $2\pi(R - x)\sqrt{r^2 - x^2}\Delta x$. The volume of the entire solid is

$$V = 2{\cdot}2\pi\int_{-r}^{r}(R-x)\sqrt{r^2 - x^2}\,dx = 4\pi R\int_{-r}^{r}\sqrt{r^2 - x^2}\,dx - 4\pi R\int_{-r}^{r} x\sqrt{r^2 - x^2}\,dx\,.$$

The first integral evaluates to $\frac{1}{2}\pi r^2$ (half the area of a circle of radius r). The second one evaluates to 0 because the integrand is odd. Therefore, $V = 4\pi R\cdot\frac{1}{2}\pi r^2 = 2\pi R\cdot \pi r^2$.

79. The solid generated by the rotating the circle is a torus. According to the solution to Exercise 77 its volume is $(2\pi\cdot 3)\cdot(\pi\cdot 2^2) = 24\pi^2$. Using cylindrical shells, the volume of the "hole" inside of the torus is

$$\int_{-1}^{1} 2\pi(3 - y)(2)\,dy = 4\pi\left(3y - \frac{1}{2}y^2\right)\Big|_{-1}^{1} = 24\pi\,.$$

Therefore, the desired volume is $V = 24\pi^2 - 24\pi = 24\pi(\pi - 1)$.

81. Partition the interval $-5 \le x \le 5$ into subintervals of equal width Δx. Let $[x, x+\Delta x]$ be a subinterval of the partition and ΔV be the volume of the portion of the solid that lies above the vertical strip $[x, x+\Delta x]\times(-\infty,\infty)$. Since the vertical plane at $(x, 0)$ interesects the solid region in a square of area $A(x) = 4(25 - x^2)$, $\Delta V \approx A(x)\Delta x = 4(25 - x^2)\,\Delta x$. Therefore, the volume of the solid is $V = \int_{-5}^{5} 4(25 - x^2)\,dx = 4\left(25x - \frac{1}{3}x^3\right)\Big|_{-5}^{5} = \frac{2000}{3}$.

83. Slice the sphere through its center with the center line of the hole along the y-axis. Let r and R be the radii of the hole and the sphere, respectively. Since the hole is 10 inches long the point $(r, 5)$ is on the circle that rotates to generate the surface of the sphere. That is, $r^2 + 5^2 = R^2$. Using the method of cylindrical shells to calculate the volume V of the sphere minus the hole,

$$\begin{aligned} V &= \int_r^R 2\pi x\cdot 2\sqrt{R^2 - x^2}\,dx = -2\pi\cdot\frac{2}{3}(R^2 - x^2)^{3/2}\Big|_r^R \\ &= \frac{4\pi}{3}\cdot(R^2 - r^2)^{3/2} = \frac{4\pi}{3}\cdot 5^3 = \frac{500\pi}{3}\,. \end{aligned}$$

CALCULATOR/COMPUTER EXERCISES

85. Using *Maple* the curves intersect at $b = 0.4263027510\ldots$. The volume integral, using cylindrical shells, is $V = \int_0^b 2\pi x(\sqrt{x} - x\exp(x))\,dx$ which evaluates to $V = 0.0741$ using Simpson's Rule, $N = 6$.

87. Using *Maple* the volume integral evaluates as

$$V = \int_0^{2.8} \pi((2.05\sqrt{x}+0.546)^2-(2.05\sqrt{x}+0.496)^2)\,dx = 2.469932104\ldots\ \text{cm}^3$$

and the mass is $M = V \cdot 3.74 = 9.237546069\ldots$ grams.

7.2 Arc Length and Surface Area

Problems for Practice

1. The length of the graph is $L = \int_1^4 \sqrt{1+3^2}\,dx = \sqrt{10}\cdot 3$. The endpoints of the graph are $P = (1,3)$ and $Q = (4,12)$ so $|\overline{PQ}| = \sqrt{3^2+9^2} = \sqrt{90} = 3\sqrt{10}$.

3. The length of the graph is $L = \int_{-1}^{2} \sqrt{1+1}\,dx = \sqrt{2}\cdot 3$. The endpoints of the graph are $P = (-1,4)$ and $Q = (2,1)$ so $|\overline{PQ}| = \sqrt{3^2+3^2} = \sqrt{18} = 3\sqrt{2}$.

5. Since $f'(x) = \frac{3}{2}x^{1/2}$, $L = \int_1^4 \sqrt{1+\frac{9}{4}x}\,dx$. Use the substitution $u = 1+\frac{9}{4}x$, $du = \frac{9}{4}\,dx$ to obtain $L = \int_{13/4}^{10} \frac{4}{9}u^{1/2}\,du = \frac{8}{27}u^{3/2}\Big|_{13/4}^{10} = \frac{80}{27}\sqrt{10} - \frac{13}{27}\sqrt{13}$.

7. Since $f'(x) = 3(2x+1)^{1/2}$, $L = \int_0^4 \sqrt{1+9(2x+1)}\,dx = \int_0^4 \sqrt{18x+10}\,dx = \frac{1}{27}(18x+10)^{3/2}\Big|_0^4 = \frac{1}{27}\left(82\sqrt{82} - 10\sqrt{10}\right)$.

9. Since $f'(x) = -\tan(x)$, $L = \int_0^{\pi/3} \sqrt{1+\tan^2(x)}\,dx = \int_0^{\pi/3} \sec(x)\,dx = \ln(|\sec(x)+\tan(x)|)\Big|_0^{\pi/3} = \ln(2+\sqrt{3})$.

11. Since $f'(x) = 2x(x^2+1)^{1/2}$, $L = \int_0^1 \sqrt{1+4x^2(x^2+1)}\,dx = \int_0^1 \sqrt{4x^4+4x^2+1}\,dx$ $\int_0^1 \sqrt{(2x^2+1)^2}\,dx = \int_0^1 (2x^2+1)\,dx = \left(\frac{2}{3}x^3+x\right)\Big|_0^1 = \frac{5}{3}$.

13. Since $g'(y) = \frac{3}{2}y^{1/2}$, $L = \int_0^1 \sqrt{1+\frac{9}{4}y}\,dy = \frac{8}{27}\left(1+\frac{9}{4}y\right)^{3/2}\Big|_0^1 = \frac{13\sqrt{13}-8}{27}$.

15. Solve for x to obtain $x = g(y) = 2y^{3/2}$. Since $g'(y) = 3y^{1/2}$, $L = \int_{1/3}^{7} \sqrt{1+9y}\,dy = \frac{2}{27}\left(1+9y\right)^{3/2}\Big|_{1/3}^{7} = \frac{2}{27}(8^3-8) = \frac{112}{3}$.

17. $\frac{dx}{dt} = 3$, $\frac{dy}{dt} = -4$, and $L = \int_{-1}^{1} \sqrt{3^2+4^2}\,dt = 5\cdot 2 = 10$.

19. $\frac{dx}{dt} = -8\sin(2t)$, $\frac{dy}{dt} = 8\cos(2t)$, and $L = \int_{-\pi/6}^{\pi/6} \sqrt{64\left(\sin^2(2t)+\cos^2(2t)\right)}\,dt = 8\cdot 2\cdot \frac{\pi}{6} = \frac{8\pi}{3}$.

21. $\frac{dx}{dt} = 6t$, $\frac{dy}{dt} = 6t^2$, and $L = \int_0^1 \sqrt{36t^2(1+t^2)}\,dt = \int_0^1 6t\sqrt{1+t^2}\,dt = 2(1+t^2)^{3/2}\Big|_0^1 = 4\sqrt{2}-2$.

23. $\frac{dx}{dt} = e^t \cos(t) - e^t \sin(t)$, $\frac{dy}{dt} = e^t \sin(t) + e^t \cos(t)$. When these are squared and added the cross products cancel and $L = \int_0^1 \sqrt{2e^{2t}\cos^2(t) + 2e^{2t}\sin^2(2t)}\, dt = \int_0^1 \sqrt{2}e^t\, dt = \sqrt{2}e^t\big|_0^1 = \sqrt{2}(e-1)$.

25. Since $f'(x) = 3$, rotating the graph of $y = f(x)$ about the x-axis produces a surface of area $S = \int_1^4 2\pi(3x-1)\sqrt{1+9}\, dx = 2\pi\sqrt{10}\left(\frac{3}{2}x^2 - x\right)\big|_1^4 = 39\pi\sqrt{10}$.

27. Since $f'(x) = \frac{3}{2}x^{-1/2}$, rotating the graph of $y = f(x)$ about the x-axis produces a surface of area

$$S = \int_{7/4}^{4} 2\pi \cdot 3x^{1/2}\sqrt{1 + \frac{9}{4}x^{-1}}\, dx = 6\pi \int_{7/4}^{4} \sqrt{x + \frac{9}{4}}\, dx$$
$$= 4\pi\left(x + \frac{9}{4}\right)^{3/2}\Bigg|_{7/4}^{4} = \frac{61\pi}{2}.$$

29. Since $f'(x) = e^x/2 - e^{-x}/2$, rotating the graph of $y = f(x)$ about the x-axis produces a surface of area

$$S = \int_{-1}^{1} 2\pi \cdot \left(\frac{e^x}{2} + \frac{e^{-x}}{2}\right)\sqrt{1 + \left(\frac{e^x}{2} - \frac{e^{-x}}{2}\right)^2}\, dx$$
$$= \pi \int_{-1}^{1} \left(e^x + e^{-x}\right)\sqrt{\frac{(e^x + e^{-x})^2}{4}}\, dx = \frac{\pi}{2}\int_{-1}^{1} \left(e^x + e^{-x}\right)^2 dx$$
$$= \frac{\pi}{2}\int_{-1}^{1} \left(e^{2x} + 2 + e^{-2x}\right) dx = \frac{\pi}{2}\left(\frac{1}{2}e^{2x} + 2x - \frac{1}{2}e^{-2x}\right)\Bigg|_{-1}^{1} = \frac{\pi}{2}\left(e^2 + 4 - e^{-2}\right).$$

31 Since $y = x/4 - 2$ implies that $x = 4(y+2)$, the surface is obtained by rotating the graph of $g(y) = 4y + 8$ over the interval $0 \le y \le 1$ about the y-axis. The area is $S = \int_0^1 2\pi \cdot (4y+8)\sqrt{1+4^2}\, dy = 2\pi\sqrt{17}\left(2y^2 + 8y\right)\big|_0^1 = 20\pi\sqrt{17}$.

33. Since $y = x^{1/3}$ implies that $x = y^3$, the surface is obtained by rotating the graph of $g(y) = y^3$ over the interval $0 \le y \le 2$ about the y-axis. Since $g'(y) = 3y^2$, the area is $S = \int_0^2 2\pi \cdot y^3\sqrt{1+9y^4}\, dy = \frac{\pi}{18}\int_0^2 36y^3\sqrt{1+9y^4}\, dy = \frac{\pi}{27}\left(1+9y^4\right)^{3/2}\big|_0^2 = \frac{\pi}{27}\left(145\sqrt{145} - 1\right)$.

Further Theory and Practice

35. Since $\frac{dy}{dx} = x$, $L = \int_0^1 \sqrt{1+x^2}\, dx$. This evaluates using the substitution $x = \tan(\theta)$, $dx = \sec^2(\theta)\, d\theta$. The integral of $\sec^3(\theta)$ is handled with the

reduction formula (6.2.13).

$$L = \int_0^{\pi/4} \sqrt{1+\tan^2(\theta)}\sec^2(\theta)\,d\theta = \int_0^{\pi/4} \sec^3(\theta)\,d\theta$$

$$= \frac{1}{2}\big(\sec(\theta)\tan(\theta) + \ln(|\sec(\theta)+\tan(\theta)|)\big)\Big|_0^{\pi/4} = \frac{1}{2}(\sqrt{2} + \ln(\sqrt{2}+1))$$

37. Using $\frac{dy}{dx} = x^{-1/2}$, $L = \int_{1/3}^{1} \sqrt{1+\frac{1}{x}}\,dx$. The length can be calculated more easily using $x = \frac{1}{4}y^2$, $\frac{dx}{dy} = \frac{1}{2}y$, and $L = \int_{2/\sqrt{3}}^{2} \sqrt{1+\frac{1}{4}y^2}\,dy$. Continue with the substitution $y = 2\tan(\theta)$, $dy = 2\sec^2(\theta)\,d\theta$. The $\sec^3(\theta)$ integral is evaluated using reduction formula (6.2.13).

$$L = \int_{\pi/6}^{\pi/4} \sqrt{1+\tan^2(\theta)}\cdot 2\sec^2(\theta)\,d\theta = 2\int_{\pi/6}^{\pi/4} \sec^3(\theta)\,d\theta$$

$$= \big(\sec(\theta)\tan(\theta) + \ln(|\sec(\theta)+\tan(\theta)|)\big)\Big|_{\pi/6}^{\pi/4}$$

$$= \sqrt{2} + \ln(\sqrt{2}+1) - \frac{2}{3} - \ln\left(\frac{2}{\sqrt{3}} + \frac{1}{\sqrt{3}}\right) = \sqrt{2} - \frac{2}{3} + \ln(\sqrt{2}+1) - \frac{1}{2}\ln(3)$$

39. Since $f'(x) = x(x^2+2)^{1/2}$, $L = \int_0^1 \sqrt{1+x^2(x^2+2)}\,dx = \int_0^1 \sqrt{(x^2+1)^2}\,dx = \int_0^1 (x^2+1)\,dx = \left(\frac{1}{3}x^3 + x\right)\Big|_0^1 = \frac{4}{3}$.

41. Since $f'(x) = \frac{1}{2}e^x - \frac{1}{2}e^{-x}$, $L = \int_0^1 \sqrt{1+\frac{1}{4}(e^x-e^{-x})^2}\,dx = \int_0^1 \sqrt{\frac{1}{4}(e^x+e^{-x})^2}\,dx = \int_0^1 \frac{1}{2}(e^x+e^{-x})\,dx = \frac{1}{2}\left(e^x - e^{-x}\right)\Big|_0^1 = \frac{1}{2}\left(e - 1/e\right)$.

43. Since $f'(x) = 8x^3 - \frac{1}{32x^3}$, $L = \int_{1/2}^{1} \sqrt{1+(8x^3-\frac{1}{32x^3})^2}\,dx = \int_{1/2}^{1} \sqrt{\left(8x^3+\frac{1}{32}x^{-3}\right)^2}\,dx = \int_{1/2}^{1} \left(8x^3+\frac{1}{32}x^{-3}\right)\,dx = \left(2x^4 - \frac{1}{64}x^{-2}\right)\Big|_{1/2}^{1} = \frac{123}{64}$.

45. Since $\frac{dx}{dt} = \cos(t) - t\sin(t)$ and $\frac{dy}{dt} = \sin(t) + t\cos(t)$,

$$L = \int_0^{3\pi} \sqrt{\cos^2(t)+\sin^2(t)+t^2(\cos^2(t)+\sin^2(t))}\,dt$$

(the cross product terms in each squared derivative cancel). Therefore,

$$L = \int_0^{3\pi} \sqrt{1+t^2}\,dt \overset{(t=\tan(\theta))}{=} \int_0^{\arctan(3\pi)} \sqrt{1+\tan^2(\theta)}\cdot\sec^2(\theta)\,d\theta$$

$$= \int_0^{\arctan(3\pi)} \sec^3(\theta)\,d\theta \overset{(6.2.13)}{=} \frac{1}{2}\big(\sec(\theta)\tan(\theta) + \ln(|\sec(\theta)+\tan(\theta)|)\big)\Big|_0^{\arctan(3\pi)}$$

$$= \frac{3}{2}\pi\sqrt{9\pi^2+1} + \frac{1}{2}\ln(3\pi + \sqrt{9\pi^2+1}).$$

47. Since $\frac{dx}{dt} = t - 1$ and $\frac{dy}{dt} = t + 1$,

$$L = \int_0^1 \sqrt{(t-1)^2 + (t+1)^2}\, dt = \int_0^1 \sqrt{2t^2 + 2}\, dt$$
$$\overset{(t=\tan(\theta))}{=} \int_0^{\pi/4} \sqrt{2}\sqrt{\tan^2(\theta) + 1} \cdot \sec^2(\theta)\, d\theta = \sqrt{2} \int_0^{\pi/4} \sec^3(\theta)\, d\theta$$
$$\overset{(6.2.13)}{=} \frac{\sqrt{2}}{2}\big(\sec(\theta)\tan(\theta) + \ln(|\sec(\theta) + \tan(\theta)|)\big)\Big|_0^{\pi/4}$$
$$= 1 + \frac{\sqrt{2}}{2} \ln(\sqrt{2} + 1)\,.$$

49. Since $\frac{dx}{dt} = 1 - \cos(t)$ and $\frac{dy}{dt} = \sin(t)$, $L = \int_0^{2\pi} \sqrt{(1-\cos(t))^2 + \sin^2(t)}\, dt =$ $\int_0^{2\pi} \sqrt{2 - 2\cos(t)}\, dt = \int_0^{2\pi} \sqrt{4\sin^2\left(\frac{1}{2}t\right)}\, dt = \int_0^{2\pi} 2\sin\left(\frac{1}{2}t\right)\, dt = -4\cos\left(\frac{1}{2}t\right)\Big|_0^{2\pi}$ $= 8\,.$

51. Since $\frac{dx}{dt} = -\sin(t) + \frac{\sec^2(t/2)}{2\tan(t/2)} = -\sin(t) + \frac{1}{2\sin(t/2)\cos(t/2)}$ $= -\sin(t) + \frac{1}{\sin(t)}$ and $\frac{dy}{dt} = \cos(t)$, $L = \int_{\pi/3}^{\pi/2} \sqrt{\sin^2(t) - 2 + \csc^2(t) + \cos^2(t)}\, dt$ $= \int_{\pi/3}^{\pi/2} \sqrt{\csc^2(t) - 1}\, dt = \int_{\pi/3}^{\pi/2} \cot(t)\, dt = \ln(|\sin(\theta)|)\Big|_{\pi/3}^{\pi/2} = \ln(2) - \frac{1}{2}\ln(3)\,.$

53. This is the graph of $g(y) = \frac{1}{16}y^4 + \frac{1}{2}y^{-2}$ for $1 \le y \le 2$. Since $g'(y) =$ $\frac{1}{4}y^3 - y^{-3}$, $L = \int_1^2 \sqrt{1 + \left(\frac{1}{4}y^3 - y^{-3}\right)^2}\, dy = \int_1^2 \sqrt{\left(\frac{1}{4}y^3 + y^{-3}\right)^2}\, dy$ $= \int_1^2 \left(\frac{1}{4}y^3 + y^{-3}\right)\, dy = \left(\frac{1}{16}y^4 - \frac{1}{2}y^{-2}\right)\Big|_1^2 = \frac{21}{16}\,.$

55. Let L and L' denote the lengths of the curves $\mathcal{C}$ and $\mathcal{C}'$ respectively. Then $L = \int_a^b \sqrt{1 + f'(x)^2}\, dx$ and, since $\mathcal{C}'$ is the graph of $g(x) = f(x-h) + \nu$ over the interval $a + h \le x \le b + h$,

$$L' = \int_{a+h}^{b+h} \sqrt{1 + f'(x-h)^2}\, dx \overset{(u=x-h)}{=} \int_a^b \sqrt{1 + f'(u)^2}\, du = L\,.$$

57. Since $\frac{dx}{dt} = a\cos(t)$ and $\frac{dy}{dt} = -b\sin(t)$,

$$\mathcal{L}(\tau) = \int_0^\tau \sqrt{a^2\cos^2(t) + b^2\sin^2(t)}\, dt$$
$$= \int_0^\tau \sqrt{a^2(\cos^2(t) + \sin^2(t)) - (a^2 - b^2)\sin^2(t)}\, dt$$
$$= \int_0^\tau a\sqrt{1 - \varepsilon^2\sin^2(t)}\, dt = aE(\varepsilon, \tau)\,.$$

59. Since $f'(x) = 4x^3 - \frac{1}{16}x^{-3}$, the surface area is

$$S = \int_{1/2}^{1} 2\pi \left(x^4 + \frac{1}{32}x^{-2}\right)\sqrt{1 + \left(4x^3 - \frac{1}{16}x^{-3}\right)^2}\,dx$$
$$= 2\pi \int_{1/2}^{1} \left(x^4 + \frac{1}{32}x^{-2}\right)\sqrt{\left(4x^3 + \frac{1}{16}x^{-3}\right)^2}\,dx$$
$$= 2\pi \int_{1/2}^{1} \left(x^4 + \frac{1}{32}x^{-2}\right)\left(4x^3 + \frac{1}{16}x^{-3}\right)dx$$
$$= 2\pi \int_{1/2}^{1} \left(4x^7 + \frac{3}{16}x + \frac{1}{512}x^{-5}\right)dx = 2\pi\left(\frac{1}{2}x^8 + \frac{3}{32}x^2 - \frac{1}{2048}x^{-4}\right)\Big|_{1/2}^{1} = \frac{1179\pi}{1024}.$$

61. Since $f'(x) = \frac{3}{2}x^2 - \frac{1}{6}x^{-2}$, the surface area is

$$S = \int_{1}^{2} 2\pi \left(\frac{1}{2}x^3 + \frac{1}{6}x^{-1}\right)\sqrt{1 + \left(\frac{3}{2}x^2 - \frac{1}{6}x^{-2}\right)^2}\,dx$$
$$= 2\pi \int_{1}^{2} \left(\frac{1}{2}x^3 + \frac{1}{6}x^{-1}\right)\sqrt{\left(\frac{3}{2}x^2 + \frac{1}{6}x^{-2}\right)^2}\,dx$$
$$= 2\pi \int_{1}^{2} \left(\frac{1}{2}x^3 + \frac{1}{6}x^{-1}\right)\left(\frac{3}{2}x^2 + \frac{1}{6}x^{-2}\right)dx$$
$$= 2\pi \int_{1}^{2} \left(\frac{3}{4}x^5 + \frac{1}{3}x + \frac{1}{36}x^{-3}\right)dx = 2\pi\left(\frac{1}{8}x^6 + \frac{1}{6}x^2 - \frac{1}{72}x^{-2}\right)\Big|_{1}^{2} = \frac{805\pi}{48}.$$

63. Rotate the curve $y = \frac{R}{h}x$, $0 \le x \le h$, about the x-axis to obtain the surface of a right circular cone of radius R and height h. Its surface area is

$$S = \int_0^h 2\pi\cdot\frac{R}{h}x\cdot\sqrt{1 + \frac{R^2}{h^2}}\,dx = 2\pi\cdot\frac{R\sqrt{h^2+R^2}}{h^2}\int_0^h x\,dx = 2\pi\cdot\frac{R\sqrt{h^2+R^2}}{h^2}\cdot\frac{1}{2}h^2.$$

This simplifies to $S = \pi R\sqrt{h^2+R^2} = \pi R\ell$, where ℓ is the slant height.

65. The volume is finite: $V = \lim_{N\to\infty}\int_1^N \pi\cdot\frac{1}{x^2}\,dx = \pi\lim_{N\to\infty}\left(-\frac{1}{x}\right)\Big|_1^N$ $= \pi\lim_{N\to\infty}\left(-\frac{1}{N}+1\right) = \pi$.

Since $\sqrt{1+\frac{1}{x^4}} \ge 1$, $S_N = \int_1^N 2\pi\cdot\frac{1}{x}\cdot\sqrt{1+\frac{1}{x^4}}\,dx \ge \int_1^N 2\pi\cdot\frac{1}{x}\,dx = 2\pi\ln(N) > \ln(N)$, and $\lim_{N\to\infty} S_N = \infty$.

CALCULATOR/COMPUTER EXERCISES

67. By symmetry the arc length is $L = 4\int_0^{\pi/2}\sqrt{1+\cos^2(x)}\,dx$. The Simpson's Rule approximations for $N = 2, 4, 6$ all round to 7.6404.

69. Since $\frac{dy}{dx} = 1.025/\sqrt{x}$, $S = \int_{0.02}^{2.80} 2\pi \cdot (2.05\sqrt{x} + 0.496)\sqrt{1 + 1.025^2/x}\,dx$. The Simpson's Rule approximations for $N = 2, 4, 6$ all round to 69 cm^2.

71. Let f denote the function defining the Arch. Then $f(x) = A - B\cosh(kx)$ where $A = 693.8597$ and $B = 2 \cdot 34.38365$. Restricting its domain to $0 \le x \le b$, there $b = 299.2239$, f has an inverse function g defined as $g(y) = \frac{1}{k}\text{arccosh}\left(\frac{A-y}{B}\right)$, $0 \le y \le c$, where $c = f(0) = 625.0914$. Using *Maple*, the area of the surface obtained by rotating the Arch about the y-axis is

$$S = \int_0^c 2\pi g(y)\sqrt{1 + g'(y)^2}\,dy \approx 9.385 \times 10^5 .$$

7.3 The Average Value of a Function

Problems for Practice

1. $f_{\text{ave}} = \frac{1}{\frac{\pi}{2}-0}\int_0^{\pi/2} \cos(x)\,dx = \frac{2}{\pi} \cdot \sin(x)\Big|_0^{\pi/2} = \frac{2}{\pi}$

3. $f_{\text{ave}} = \frac{1}{4-1}\int_1^4 \frac{1}{x}\,dx = \frac{1}{3} \cdot \ln(x)\Big|_1^4 = \frac{2}{3}\ln(2)$

5. $f_{\text{ave}} = \frac{1}{\pi-\frac{\pi}{3}}\int_{\pi/3}^{\pi} \sin(x)\,dx = \frac{3}{2\pi} \cdot \big(-\cos(x)\big)\Big|_{\pi/3}^{\pi} = \frac{9}{4\pi}$

7. $f_{\text{ave}} = \frac{1}{5-2}\int_2^5 (x-1)^{1/2}\,dx = \frac{1}{3} \cdot \frac{2}{3}(x-1)^{3/2}\Big|_2^5 = \frac{14}{9}$

9. $f_{\text{ave}} = \frac{1}{3-1}\int_1^3 60x^{-2}\,dx = \frac{1}{2} \cdot \big(-60x^{-1}\big)\Big|_1^3 = 20$

11. $f_{\text{ave}} = \frac{1}{e-1}\int_1^e \ln(x)\,dx \overset{(u=\ln(x),\,dv=dx)}{=} \frac{1}{e-1} \cdot \big(x\ln(x) - x\big)\Big|_1^e = \frac{1}{e-1}$

13. The average value is $f_{\text{ave}} = \frac{1}{3-1}\int_1^3 \frac{1}{x}\,dx = \frac{1}{2}\ln(3)$. Therefore, c must be a solution to the equation $\frac{1}{c} = \frac{1}{2}\ln(3)$. That is, $c = \frac{2}{\ln(3)}$.

15. The average value is $f_{\text{ave}} = \frac{1}{3-1}\int_1^3 \left(x^2 - \frac{10}{3}x\right)\,dx = \frac{1}{2}\left(\frac{1}{3}x^3 - \frac{5}{3}x^2\right)\Big|_2^3 = -\frac{7}{3}$. Therefore, c must be a solution to the equation $c^2 - \frac{10}{3}c = -\frac{7}{3}$. That is, $3c^2 - 10c + 7 = 0$ or $(3c-7)(c-1) = 0$, so $c = \frac{7}{3}$.

17. $P(1 \le X \le 2) = \int_1^2 \frac{3}{8}x^2\,dx = \frac{1}{8}x^3\Big|_1^2 = \frac{7}{8}$

19. $P\left(\frac{\pi}{4} \le X \le \frac{\pi}{2}\right) = \int_{\pi/4}^{\pi/2} \frac{1}{2}\sin(x)\,dx = -\left(\frac{1}{2}\cos(x)\right)\Big|_{\pi/4}^{\pi/2} = \frac{\sqrt{2}}{4}$

21. $P\left(0 \le X \le \frac{1}{2}\right) = \int_0^{1/2} \frac{e^{1-x}}{e-1}\,dx = -\left(\frac{e^{1-x}}{e-1}\right)\Big|_0^{1/2} = \frac{e-\sqrt{e}}{e-1} = \frac{\sqrt{e}(\sqrt{e}-1)}{e-1} = \frac{\sqrt{e}}{\sqrt{e}+1}$

23. Since $\int_0^{\pi/3} \sec^2(x)\,dx = \tan(x)\Big|_0^{\pi/3} = \sqrt{3}$, $f(x) = \frac{1}{\sqrt{3}} \cdot \sec^2(x) = \frac{\sqrt{3}}{3}\sec^2(x)$ is a probability density on $[0, \pi/3]$.

25. Since $\int_{-1}^{3}(9-x^2)\,dx = \left(9x-\frac{1}{3}x^3\right)\Big|_{-1}^{3} = \frac{80}{3}$, $f(x) = \frac{3}{80}\cdot(9-x^2)$ is a probability density on $[-1,3]$.

27. Since $\int_0^{\infty} e^{-2x}\,dx = \lim_{N\to\infty}\int_0^N e^{-2x}\,dx = \lim_{N\to\infty}\left(-\frac{1}{2}e^{-2x}\Big|_0^N\right)$ $= \lim_{N\to\infty}\left(\frac{1}{2}-\frac{1}{2}e^{-N}\right) = \frac{1}{2}$, $f(x) = 2\cdot e^{-2x}$ is a probability density on $[0,1]$.

29. The mean is $\mu = \int_0^1 x\cdot 3x^2\,dx = \int_0^1 3x^3\,dx = \frac{3}{4}x^4\Big|_0^1 = \frac{3}{4}$.

31. The mean is $\mu = \int_{1/4}^{4} x\cdot\frac{1}{3\sqrt{x}}\,dx = \int_{1/4}^{4}\frac{1}{3}x^{1/2}\,dx = \frac{2}{9}x^{3/2}\Big|_{1/4}^{4} = \frac{7}{4}$.

33. The mean is $\mu = \int_{-1}^{2} x\cdot\frac{1}{3}x^2\,dx = \int_{-1}^{2}\frac{1}{3}x^3\,dx = \frac{1}{12}x^4\Big|_{-1}^{2} = \frac{5}{4}$.

35. The mean is $\mu = \int_1^e x\cdot\frac{2\ln(x)}{x}\,dx = 2\int_1^e \ln(x)\,dx = 2(x\ln(x)-x)\Big|_1^e = 2$.

37. The population is $P = 2\pi\int_0^6 5860x\exp(-0.148x)\,dx$. Integrating by parts, $u = x$, $dv = \exp(-0.148)\,dx$, $du = dx$, $v = -\frac{1}{0.148}\exp(-0.148x)\,dx$,

$$P = 2\pi\cdot 5860\left(-\frac{x}{0.148}\exp(-0.148x) - \frac{1}{(0.148)^2}\exp(-0.148x)\right)\Bigg|_0^6 = 375,069\,.$$

Further Theory and Practice

39. In 1950 the central core of Akron had a population of $P = 2\pi\int_0^4 13000x\exp(-0.38x)\,dx$. Integrating by parts, $u = x$, $dv = \exp(-0.38x)\,dx$, $du = dx$, $v = -\frac{1}{0.38}\exp(-0.38x)\,dx$, $P = 253,895$.

In 1970 the central core had a population of $Q = 2\pi\int_0^4 10000x\exp(-0.29x)\,dx$. Integrating by parts, $u = x$, $dv = \exp(-0.29x)\,dx$, $du = dx$, $v = -\frac{1}{0.29}\exp(-0.29x)\,dx$, $Q = 241,219$.

The population decreased by approximately 12,676.

41. The average temperature is

$$T_{\text{ave}} = \frac{1}{3-0}\int_0^3 (99.6 - t + 0.8t^2)\,dt = \frac{1}{3}\left(99.6t - \frac{1}{2}t^2 + \frac{0.8}{3}t^3\right)\Bigg|_0^3 = 100.5^\circ\text{F}\,.$$

43. The average value is $f_{\text{ave}} = \frac{1}{2-0}\int_0^2 x\exp(x^2)\,dx = \frac{1}{4}\exp(x^2)\Big|_0^2 = \frac{1}{4}(e^4-1)$.

45. The average value is $f_{\text{ave}} = \frac{1}{\frac{\pi}{2}-0}\int_0^{\pi/2} 3\pi\cos(x)\sqrt{1+\sin(x)}\,dx$. This integrates easily to $f_{\text{ave}} = 6\cdot\frac{2}{3}(1+\sin(x))^{3/2}\Big|_0^{\pi/2} = 4(2\sqrt{2}-1)$.

47. The average value is

$$f_{\text{ave}} = \frac{1}{\pi - 0}\int_0^{\pi} \cos^2(x)\sin^3(x)\,dx = \frac{1}{\pi}\int_0^{\pi} \cos^2(x)(1-\cos^2(x))\sin(x)\,dx$$
$$= \frac{1}{\pi}\left(-\frac{1}{3}\cos^3(x) + \frac{1}{5}\cos^5(x)\right)\Bigg|_0^{\pi} = \frac{4}{15\pi}\,.$$

49. The average value is $f_{\text{ave}} = \frac{1}{e-1}\int_1^e 4x\ln(x)\,dx$. Use integration by parts, $u = \ln(x)$, $dv = x\,dx$, $du = \frac{1}{x}\,dx$, $v = \frac{1}{2}x^2$.

$$f_{\text{ave}} = \frac{4}{e-1}\left(\frac{1}{2}x^2\ln(x) - \frac{1}{4}x^2\right)\Bigg|_1^e = \frac{e^2+1}{e-1}$$

51. The average value is $f_{\text{ave}} = \frac{1}{3-1}\int_1^3 \frac{5x+2}{x^2+x}\,dx$. Use a partial fraction decomposition, $\frac{5x+2}{x(x+1)} = \frac{A}{x} + \frac{B}{x+1}$, implying that $5x+2 = A(x+1) + Bx$ so $A = 2$ and $B = 3$.

$$f_{\text{ave}} = \frac{1}{2}\int_1^3\left(\frac{2}{x} + \frac{3}{x+1}\right)dx = \frac{1}{2}(2\ln(x) + 3\ln(x+1))\Bigg|_1^3 = \ln(3) + \frac{3}{2}\ln(2)$$

53. The average value of $f(x) = (x-c)\sin(x)$ is $f_{\text{ave}} = \frac{1}{\frac{\pi}{3}-0}\int_0^{\pi/3}(x-c)\sin(x)\,dx$. Integrate by parts, $u = x - c$, $dv = \sin(x)\,dx$, $du = dx$, $v = -\cos(x)\,dx$, to obtain $f_{\text{ave}} = \frac{3}{\pi}\big(-(x-c)\cos(x) + \sin(x)\big)\big|_0^{\pi/3} = \frac{3(\sqrt{3}-c)-\pi}{2\pi}$. This is $-1/2$ when $c = \sqrt{3}$.

55. The average temperature for days one, two, and three were $T_1 = \frac{1}{24}\int_0^{24} T(t)\,dt = 34.315$, $T_2 = \frac{1}{24}\int_{24}^{48} T(t)\,dt = 39.355$, and $T_3 = \frac{1}{24}\int_{48}^{72} T(t)\,dt = 38.635$. At \$0.30 per degree day, it cost her \$9.21 on day one, \$7.69 on day two, and \$7.91 on day three.

57. The average of f over $[a, a+\pi]$ is $F(a) = \frac{1}{\pi}\int_a^{a+\pi}(\sin(x) - \cos(x))\,dx = \frac{1}{\pi}\big(-\cos(x) - \sin(x)\big)\big|_a^{a+\pi} = \frac{1}{\pi}\big(\cos(a) - \cos(a+\pi) + \sin(a) - \sin(a+\pi)\big)$. This simplifies to $F(a) = \frac{2}{\pi}(\cos(a) + \sin(a))$. The maximum value of F on the interval $[0,\pi]$ is attained where $F'(a) = \frac{2}{\pi}(\cos(a) - \sin(a)) = 0$. That is, $a = \pi/4$.

59. Integrate by parts, $u = \ln(x)$, $dv = x\,dx$, $du = \frac{1}{x}\,dx$, $v = \frac{1}{2}x^2$, to obtain $\int_1^e x\ln(x)\,dx = \left(\frac{1}{2}x^2\ln(x) - \frac{1}{4}x^2\right)\big|_1^e = \frac{1}{4}(e^2+1)$. Therefore, the function $f(x) = \frac{4}{e^2+1}\cdot g(x)$ is a probability density function.

61. Since $\int_0^1 x^2(1-x^3)^{-1/2}\,dx = -\frac{2}{3}(1-x^3)^{1/2}\big|_0^1 = \frac{2}{3}$, the function $f(x) = \frac{3}{2}\cdot g(x)$ is a probability density function.

63. Integrate by parts, $u = \arcsin(x)$, $dv = dx$, $du = \frac{1}{\sqrt{1-x^2}}\,dx$, $v = x$, to obtain $\int_0^1 \arcsin(x)\,dx = x\arcsin(x)\big|_0^1 - \int_0^1 \frac{x}{\sqrt{1-x^2}}\,dx = \frac{\pi}{2} + \left(\sqrt{1-x^2}\,\big|_0^1\right) =$

$\frac{\pi}{2} - 1$. Therefore, the function $f(x) = \frac{2}{\pi-2} \cdot g(x)$ is a probability density function.

65. Integrate by parts, $u = x$, $dv = \sin(x)\,dx$, $du = dx$, $v = -\cos(x)$, to obtain $E(X) = \int_0^{\pi} x \cdot \frac{1}{2}\sin(x)\,dx = \frac{1}{2}\big(-x\cos(x) + \sin(x)\big)\big|_0^{\pi} = \frac{\pi}{2}$.

67. $E(X) = \int_0^1 x \cdot \frac{4x}{(x^2+1)^2}\,dx = 4\int_0^1 \frac{x^2+1-1}{(x^2+1)^2}\,dx = 4\int_0^1 \left(\frac{1}{x^2+1} - \frac{1}{(x^2+1)^2}\right)dx$. Now make the substitutions $x = \tan(\theta)$, $dx = \sec^2(\theta)\,d\theta$ to obtain

$$E(X) = 4\int_0^{\pi/4} \left(\frac{1}{\sec^2(\theta)} - \frac{1}{\sec^4(\theta)}\right) \cdot \sec^2(\theta)\,d\theta = 4\int_0^{\pi/4} (1 - \cos^2(\theta))\,d\theta$$

$$= 4\int_0^{\pi/4} \sin^2(\theta)\,d\theta = 2\int_0^{\pi/4} (1 - \cos(2\theta))\,d\theta = \big(2\theta - \sin(2\theta)\big)\Big|_0^{\pi/4} = \frac{\pi}{2} - 1\,.$$

69. $E(X) = \int_0^4 x \cdot \frac{3}{16}\sqrt{4-x}\,dx = \frac{3}{16}\int_0^4 x\sqrt{4-x}\,dx$. Substitute $u = 4 - x$, $x = 4 - u$, $dx = -du$ to obtain $E(X) = \frac{3}{16}\int_4^0 (4-u)u^{1/2} \cdot (-1)\,du$ $= \frac{3}{16}\int_0^4 (4u^{1/2} - u^{3/2})\,du = \frac{3}{16}\left(\frac{8}{3}u^{3/2} - \frac{2}{5}u^{5/2}\right)\big|_0^4 = \frac{8}{5}$.

71. $E(X) = \int_0^{\infty} x \cdot 2e^{-2x}\,dx = 2\lim_{N\to\infty}\int_0^N xe^{-2x}\,dx$. Integrate by parts, $u = x\,dx$, $dv = 2e^{-2x}\,dx$, $du = dx$, $v = -e^{-2x}$, to obtain

$$E(X) = \lim_{N\to\infty}\left(\left(-xe^{-2x} - \frac{1}{2}e^{-2x}\right)\Big|_0^N\right)$$

$$= \lim_{N\to\infty}\left(-Ne^{-2N} - \frac{1}{2}e^{-2N} + \frac{1}{2}\right) = \frac{1}{2}\,.$$

73. $E(X) = \int_0^{\infty} x \cdot \frac{4}{\pi(1+x^2)^2}\,dx = \frac{2}{\pi}\lim_{N\to\infty}\int_0^N \frac{2x}{(1+x^2)^2}\,dx$ $= \frac{2}{\pi}\lim_{N\to\infty}\left(-\frac{1}{1+x^2}\Big|_0^N\right) = \frac{2}{\pi}\lim_{N\to\infty}\left(-\frac{1}{1+N^2} - 1\right) = \frac{2}{\pi}$.

75. $P(X \le m) = \int_0^m \cos(x)\,dx = \sin(m)$. Therefore, $P(X \le \pi/6) = 1/2$ and the median is $m = \pi/6$.

77. $P(X \le m) = \int_0^m \frac{e^{1-x}}{e-1}\,dx = \frac{e}{e-1}\int_0^m e^{-x}\,dx = \frac{e}{e-1} \cdot (1 - e^{-m}) = \frac{e-e^{1-m}}{e-1}$. This is 1/2 when $e - e^{1-m} = \frac{1}{2}(e-1)$ or $e^{1-m} = \frac{1}{2}e + \frac{1}{2}$. Therefore, $1 - m = \ln(e+1) - \ln(2)$ and $m = 1 + \ln(2) - \ln(e+1)$.

79. The average for X is $\overline{X} = \int_0^1 x(p+1)x^p\,dx = (p+1)\frac{x^{p+2}}{p+2}\Big|_0^1 = \frac{p+1}{p+2}$. The average value for f is $f_{\text{ave}} = \int_0^1 (p+1)x^p\,dx = x^{p+1}\big|_0^1 = 1$.

81. It is appropriate to make this definition. If the limit is finite, then f_{ave} can be interpreted at the "long-term" average value of f.

83. The mean is $\overline{X} = \frac{1}{\sqrt{2\pi}\sigma}\int_{-\infty}^{\infty} x\exp\left(-\frac{1}{2}\left(\frac{x-\mu}{\sigma}\right)^2\right)\,dx$. Make the substitution $u = \frac{x-\mu}{\sqrt{2}\sigma}$, $du = \frac{1}{\sqrt{2}\sigma}\,dx$ to obtain

$$\begin{aligned}\overline{X} &= \frac{1}{\sqrt{2\pi}\sigma}\int_{-\infty}^{\infty} (\mu + \sqrt{2}\sigma u)\exp(-u^2)\cdot\sqrt{2}\sigma\,du \\ &= \frac{\mu}{\sqrt{\pi}}\int_{-\infty}^{\infty}\exp(-u^2)\,du + \sigma\sqrt{\frac{2}{\pi}}\int_{-\infty}^{\infty} u\exp(-u^2)\,du\,.\end{aligned}$$

The second integral evaluates to 0 because the integrand is odd (split it into two integrals at $u = 0$ and the improper integrals are finite with opposite sign). The first integral evaluates to $\sqrt{\pi}$ (see Equation (6.2.7)) implying that $\overline{X} = \mu$.

CALCULATOR/COMPUTER EXERCISES

85. The graphs of f and A_f are displayed on the right. The smooth curve is the graph of $y = A_f(x)$.

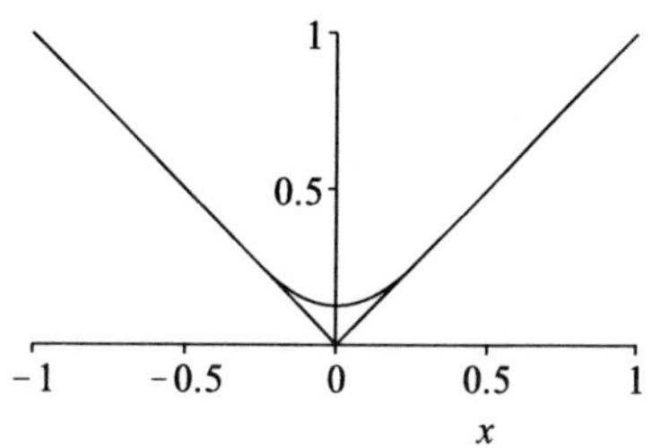

87. The graphs of f and A_f are displayed on the right. The continuous broken line segment is the graph of $y = A_f(x)$.

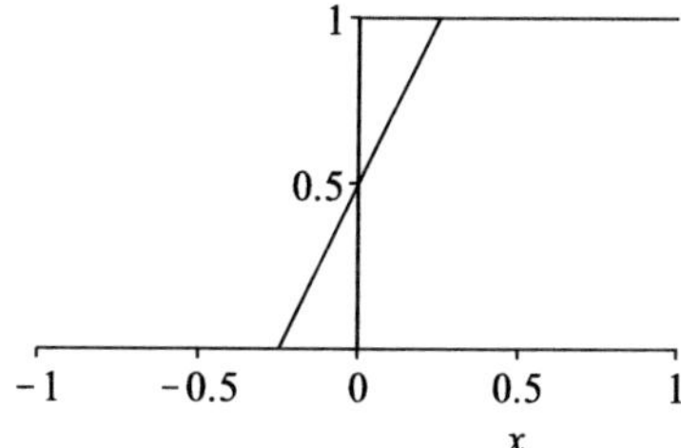

89. Using *Maple*'s numerical integration, $f_{\text{ave}} = \frac{1}{4}\int_0^4 \sqrt{x}\exp(-x)\,dx = 0.211$ and, using the *fsolve* procedure, the value of c such that $f(c) = f_{\text{ave}}$ is $c = 1.866$.

91. Using *Maple*'s numerical integration, $f_{\text{ave}} = \frac{1}{2}\int_2^4 \frac{x}{16+x^3}\,dx = 0.069$ and, using the *fsolve* procedure, the value of c such that $f(c) = f_{\text{ave}}$ is $c = 3.041$.

93. According to Exercise 84, and using *Maple*,

$$P(\mu - \sigma \le X \le \mu + \sigma) = \frac{1}{\sqrt{2\pi}}\int_{-1}^{1}\exp(-x^2/2)\,dx = 0.683\ .$$

7.4 Center of Mass

Problems for Practice

1. The region is all (x, y) such that $0 \leq x \leq 6$, $0 \leq y \leq 2 - x/3$. Therefore, $M_{x=3} = \int_0^6 (x-3)(2-x/3)\,dx = \int_0^6 (3x-6-x^2/3)\,dx = \left(\frac{3}{2}x^2 + 6x - \frac{1}{9}x^3\right)\Big|_0^6$ $= -6$.

3. The region is all (x, y) such that $0 \leq x \leq 6$, $0 \leq y \leq 2 - x/3$. Therefore, $M_{x=2} = \int_0^6 (x-2)(2-x/3)\,dx = \int_0^6 (8x/3-4-x^2/3)\,dx = \left(\frac{4}{3}x^2 - 4x - \frac{1}{9}x^3\right)\Big|_0^6$ $= 0$.

5. The region is all (x, y) such that $0 \leq x \leq 2$, $0 \leq y \leq 4x - x^3$. Therefore, $M_{x=3} = \int_0^2 (x-3)(4x-x^3)\,dx = \int_0^2 (4x^2 - x^4 - 12x + 3x^3)\,dx = \left(\frac{4}{3}x^3 - \frac{1}{5}x^5 - 6x^2 + \frac{3}{4}x^4\right)\Big|_0^2 = -\frac{116}{15}$.

7. The region is all (x, y) such that $1 \leq x \leq 2$, $0 \leq y \leq 1/x$. Therefore, $M_{x=-3} = \int_1^2 (x+3)(1/x)\,dx = \int_1^2 (1+3/x)\,dx = (x + 3\ln(x))\Big|_1^2$ $= 1 + 3\ln(2)$.

9. The region is all (x, y) such that $0 \leq x \leq 6$, $0 \leq y \leq 2 - x/3$. Therefore, $M_{y=0} = \frac{1}{2}\int_0^6 (2 - x/3)^2\,dx = -\frac{1}{2}\left(2 - \frac{1}{3}x\right)^3\Big|_0^6 = 4$.

11. The region is all (x, y) such that $1 \leq x \leq 2$, $0 \leq y \leq 1/x$. Therefore, $M_{y=0} = \frac{1}{2}\int_1^2 (1/x)^2\,dx = -\frac{1}{2}(1/x)\Big|_1^2 = 1/4$.

13. The region is all (x, y) such that $-2 \leq x \leq 2$, $0 \leq y \leq \sqrt{4-x^2}$. Therefore, $M_{y=0} = \frac{1}{2}\int_{-2}^2 (\sqrt{4-x^2}\,)^2\,dx = \frac{1}{2}\left(4x - \frac{1}{3}x^3\right)\Big|_{-2}^2 = \frac{16}{3}$.

15. The region is all (x, y) such that $0 \leq x \leq 1$, $0 \leq y \leq \sqrt{x}(1-x^2)^2$. Therefore, $M_{y=0} = \frac{1}{2}\int_0^1 (\sqrt{x}(1-x^2)^2)^2\,dx = \frac{1}{2}\int_0^1 x(1-x^2)^4\,dx = -\frac{1}{20}\left(1-x^2\right)^5\Big|_0^1$ $= \frac{1}{20}$.

17. The region is $0 \leq x \leq 6$, $0 \leq y \leq 2 - x/3$. Assuming that $\delta = 1$, $M_{x=0} = \int_0^6 x(2-x/3)\,dx = 12$, $M_{y=0} = \int_0^6 \frac{1}{2}(2-x/3)^2\,dx = 4$, $M = \int_0^6 (2-x/3)\,dx = 6$, and the center of mass is $(\overline{x}, \overline{y}) = \left(2, \frac{2}{3}\right)$.

19. The region is $0 \leq x \leq 2$, $0 \leq y \leq \sqrt{4-x^2}$. It is a quarter circle of radius 2. Assuming that $\delta = 1$, $M = \pi$ and, by symmetry, $M_{x=0} = M_{y=0}$. Since $M_{x=0} = \int_0^2 x\sqrt{4-x^2}\,dx = \frac{8}{3}$, the center of mass is $(\overline{x}, \overline{y}) = \left(\frac{8}{3\pi}, \frac{8}{3\pi}\right)$.

21. The region is $-4 \leq x \leq 2$, $0 \leq y \leq 8 - 2x - x^2$. Assuming that $\delta = 1$, $M_{x=0} = \int_{-4}^2 x(8-2x-x^2)\,dx = -36$, $M_{y=0} = \int_{-4}^2 \frac{1}{2}(8-2x-x^2)^2\,dx = \frac{648}{5}$, $M = \int_{-4}^2 (8-2x-x^2)\,dx = 36$, and the center of mass is $(\overline{x}, \overline{y}) = \left(-1, \frac{18}{5}\right)$.

23. The region is $1 \leq x \leq 2$, $0 \leq y \leq 1/x$. Assuming that $\delta = 1$, $M_{x=0} = \int_1^2 x(1/x)\,dx = 1$, $M_{y=0} = \int_1^2 \frac{1}{2}(1/x)^2\,dx = \frac{1}{4}$, $M = \int_1^2 1/x\,dx = \ln(2)$, and the center of mass is $(\overline{x}, \overline{y}) = \left(\frac{1}{\ln(2)}, \frac{1}{4\ln(2)}\right)$.

25. The region is $0 \le x \le 1$, $0 \le y \le e^x$. Assuming that $\delta = 1$, $M_{x=0} = \int_0^1 xe^x\,dx = 1$, $M_{y=0} = \int_0^1 \frac{1}{2}(e^x)^2\,dx = \frac{1}{4}(e^2-1)$, $M = \int_0^1 e^x\,dx = e-1$, and the center of mass is $(\overline{x},\overline{y}) = \left(\frac{1}{e-1}, \frac{e+1}{4}\right)$.

27. The region is $0 \le x \le 1/\sqrt{2}$, $0 \le y \le 1/\sqrt{1-x^2}$. Assuming that $\delta = 1$, $M_{x=0} = \int_0^{1/\sqrt{2}} x(1/\sqrt{1-x^2})\,dx = 1-\frac{1}{2}\sqrt{2}$, $M_{y=0} = \int_0^{1/\sqrt{2}} \frac{1}{2}(1/\sqrt{1-x^2}\,)^2\,dx = \frac{1}{4}\ln(3+2\sqrt{2})$, $M = \int_0^{1/\sqrt{2}} 1/\sqrt{1-x^2}\,dx = \pi/4$, and the center of mass is $(\overline{x},\overline{y}) = \left(\frac{2(2-\sqrt{2})}{\pi}, \frac{\ln(3+2\sqrt{2})}{\pi}\right)$.

29. The region is $-1 \le x \le 1$, $1-x^2 \le y \le 4(1-x^2)$. By symmetry, $\overline{x} = 0$. Since $\delta = 1$, $M_{y=0} = \int_{-1}^1 \frac{1}{2}(16(1-x^2)^2 - (1-x^2)^2)\,dx = 8$, $M = \int_{-1}^1 (4(1-x^2)-(1-x^2))\,dx = 4$, and the center of mass is $(\overline{x},\overline{y}) = (0,2)$.

31. The region is $0 \le x \le 4$, $x \le y \le 2\sqrt{x}$. Since $\delta = 1$, $M_{x=0} = \int_0^4 x(2\sqrt{x}-x)\,dx = \frac{64}{15}$, $M_{y=0} = \int_0^4 \frac{1}{2}((2\sqrt{x}\,)^2 - x^2)\,dx = \frac{16}{3}$, $M = \int_0^4 (2\sqrt{x}-x)\,dx = \frac{8}{3}$, and the center of mass is $(\overline{x},\overline{y}) = \left(\frac{8}{5}, 2\right)$.

33. The region is $-1 \le x \le 2$, $4-x^2 \le y \le x+5$. Since $\delta = 1$, $M_{x=0} = \int_{-1}^1 x(x+5-(4-x^2))\,dx = \frac{33}{4}$, $M_{y=0} = \int_{-1}^1 \frac{1}{2}((x+5)^2-(4-x^2)^2)\,dx = \frac{156}{5}$, $M = \int_{-1}^1 (x+5-(4-x^2))\,dx = \frac{15}{2}$, and the center of mass is $(\overline{x},\overline{y}) = \left(\frac{11}{10}, \frac{104}{25}\right)$.

35. The region is $0 \le x \le 3$, $\sqrt{4-x} \le y \le 2$. Since $\delta = 1$, $M_{x=0} = \int_0^3 x(2-\sqrt{4-x}\,)\,dx = \frac{41}{15}$, $M_{y=0} = \int_0^3 \frac{1}{2}(2^2 - (\sqrt{4-x}\,)^2))\,dx = \frac{9}{4}$, $M = \int_0^3 (2-\sqrt{4-x}\,)\,dx = \frac{35}{6}$, and the center of mass is $(\overline{x},\overline{y}) = \left(\frac{41}{20}, \frac{27}{16}\right)$.

Further Theory and Practice

37. The region is $0 \le x \le \pi/2$, $0 \le y \le \cos(x)$. Since $\delta = 1$,
$M_{x=0} = \int_0^{\pi/2} x\cos(x)\,dx = \left(x\sin(x)+\cos(x)\right)\Big|_0^{\pi/2} = \frac{\pi}{2}-1$,
$M_{y=0} = \int_0^{\pi/2} \frac{1}{2}\cos^2(x)\,dx = \frac{1}{4}\left(x + \frac{1}{2}\sin(2x)\right)\Big|_0^{\pi/2} = \frac{\pi}{8}$,
$M = \int_0^{\pi/2} \cos(x)\,dx = 1$, and the center of mass is $(\overline{x},\overline{y}) = \left(\frac{\pi}{2}-1, \frac{\pi}{8}\right)$.

39. The region is $0 \le x \le 1$, $0 \le y \le 1/(1+x^2)$. Since $\delta = 1$,
$M_{x=0} = \int_0^1 x\cdot\frac{1}{1+x^2}\,dx = \frac{1}{2}\ln(1+x^2)\Big|_0^1 = \frac{1}{2}\ln(2)$,
$M_{y=0} = \int_0^1 \frac{1}{2}\left(\frac{1}{1+x^2}\right)^2\,dx \overset{(x=\tan(\theta))}{=} \frac{1}{4}\left(\theta + \frac{1}{2}\sin(2\theta)\right)\Big|_0^{\pi/4} = \frac{\pi}{16}+\frac{1}{8}$,
$M = \int_0^1 \frac{1}{1+x^2}\,dx = \arctan(x)\Big|_0^1 = \frac{\pi}{4}$, and the center of mass is $(\overline{x},\overline{y}) = \left(\frac{2\ln(2)}{\pi}, \frac{\pi+2}{4\pi}\right)$.

41. The region is $-1 \le x \le 1$, $0 \le y \le (1-x^2)^{3/2}$. Since $\delta = 1$,
$M_{x=0} = \int_{-1}^1 x\cdot(1-x^2)^{3/2}\,dx = 0$ (by symmetry, the integrand is odd),
$M_{y=0} = \int_{-1}^1 \frac{1}{2}\left((1-x^2)^{3/2}\right)^2\,dx = \frac{1}{2}\int_{-1}^1 (1-x^2)^3\,dx = \frac{16}{35}$, and, by symme-

try and the reduction formula for $\cos^4(\theta)$, $M = 2\int_0^1 (1-x^2)^{3/2}\,dx \overset{(x=\sin(\theta))}{=} 2\int_0^{\pi/2}\cos^4(\theta)\,d\theta = \left(\frac{1}{2}\cos^3(\theta)\sin(\theta) + \frac{3}{4}\cos(\theta)\sin(\theta) + \frac{3}{4}\theta\right)\Big|_0^{\pi/2} = \frac{3\pi}{8}$, so the center of mass is $(\overline{x},\overline{y}) = \left(0, \frac{128}{105\pi}\right)$.

43. The region is $1 \le x \le 2$, $0 \le y \le x - 1/x$. Since $\delta = 1$, $M_{x=0} = \int_1^2 x\cdot(x-1/x)\,dx = \frac{4}{3}$, $M_{y=0} = \int_1^2 \frac{1}{2}\left((x-1/x)\right)^2 dx = \frac{5}{12}$, $M = \int_1^2 (x-1/x)\,dx = \frac{3}{2} - \ln(2)$, and the center of mass is $(\overline{x},\overline{y}) = \left(\frac{8}{3(3-2\ln(2))}, \frac{5}{6(3-2\ln(2))}\right)$.

45. The region is $-1 \le x \le 1$, $0 \le y \le \begin{cases} 1 & x \le 0 \\ 1-x & x \ge 0 \end{cases}$. Since $\delta = 1$,
$M_{x=0} = \int_{-1}^0 x\cdot 1\,dx + \int_0^1 x\cdot(1-x)\,dx = -\frac{1}{3}$,
$M_{y=0} = \int_{-1}^0 \frac{1}{2}\cdot 1^2\,dx + \int_0^1 \frac{1}{2}\cdot(1-x)^2\,dx = \frac{2}{3}$,
$M = \int_{-1}^0 1\,dx + \int_0^1 (1-x)\,dx = \frac{3}{2}$, the center of mass is $(\overline{x},\overline{y}) = \left(-\frac{2}{9}, \frac{4}{9}\right)$.

47. The region is $-2 \le x \le 0$, $0 \le y \le 4 - x^2$ and $0 \le x \le 1$, $3x \le y \le 4 - x^2$. Since $\delta = 1$, $M_{x=0} = \int_{-2}^0 x\cdot(4-x^2)\,dx + \int_0^1 x\cdot(4-x^2-3x)\,dx = -\frac{13}{4}$,
$M_{y=0} = \int_{-2}^0 \frac{1}{2}\cdot(4-x^2)^2\,dx + \int_0^1 \frac{1}{2}\cdot((4-x^2)^2 - (3x)^2)\,dx = \frac{69}{5}$,
$M = \int_{-2}^0 (4-x^2)\,dx + \int_0^1 (4-x^2-3x)\,dx = \frac{15}{2}$, the center of mass is $(\overline{x},\overline{y}) = \left(-\frac{13}{30}, \frac{46}{25}\right)$.

49. The region is $-1 \le x \le 0$, $-x \le y \le 2 - x^2$ and $0 \le x \le 1$, $\sqrt{x} \le y \le 2 - x^2$. Since $\delta = 1$,
$M_{x=0} = \int_{-1}^0 x\cdot(2-x^2-(-x))\,dx + \int_0^1 x\cdot(2-x^2-\sqrt{x})\,dx = -\frac{1}{15}$,
$M_{y=0} = \int_{-1}^0 \frac{1}{2}\cdot((2-x^2)^2 - (-x)^2)\,dx + \int_0^1 \frac{1}{2}\cdot((2-x^2)^2 - (\sqrt{x})^2)\,dx = \frac{49}{20}$,
$M = \int_{-1}^0 (2-x^2-(-x))\,dx + \int_0^1 (2-x^2-\sqrt{x})\,dx = \frac{13}{6}$, the center of mass is $(\overline{x},\overline{y}) = \left(-\frac{2}{65}, \frac{147}{130}\right)$.

51. The region is $-1 \le x \le 4$, $(x-1)^2 \le y \le \begin{cases} 2(1-x) & x \le 0 \\ 2(1+x) & x \ge 0 \end{cases}$. Since $\delta = 1$, $M_{x=0} = \int_{-1}^0 x\cdot(2(1-x)-(x-1)^2)\,dx + \int_0^4 x\cdot(2(1+x)-(x-1)^2)\,dx = \frac{349}{12}$,
$$M_{y=0} = \int_{-1}^0 \frac{1}{2}\cdot((2(1-x))^2 - ((x-1)^2)^2)\,dx + \int_0^4 \frac{1}{2}\cdot((2(1+x))^2 - ((x-1)^2)^2)\,dx = \frac{359}{6},$$
$M = \int_{-1}^0 (2(1-x)-(x-1)^2)\,dx + \int_0^4 (2(1+x)-(x-1)^2)\,dx = \frac{46}{3}$, the center of mass is $(\overline{x},\overline{y}) = \left(\frac{349}{184}, \frac{359}{92}\right)$.

53. The mass of the region $\mathcal{R}$ is $M = \int_a^b f(x)\,dx = 1$. Therefore, $\overline{x} = \int_a^b xf(x)\,dx = \overline{X}$.

55. Second moment: $\int_1^2 (x+1)^2\cdot x^2\,dx = \left(\frac{1}{3}x^3 + \frac{1}{2}x^4 + \frac{1}{5}x^5\right)\Big|_1^2 = \frac{481}{30}$

57. Second moment: (integration by parts, twice)

$$\int_0^\pi (x-2\pi)^2 \sin(x)\,dx \overset{(u=(x-2\pi)^2)}{\underset{(dv=\sin(x)dx)}{=}} -(x-2\pi)^2\cos(x)\Big|_0^\pi + 2\int_0^\pi (x-2\pi)\cos(x)\,dx$$

$$\overset{(u=(x-2\pi))}{\underset{(dv=\cos(x)dx)}{=}} -(x-2\pi)^2\cos(x)\Big|_0^\pi + 2\left((x-2\pi)\sin(x)\Big|_0^\pi - \int_0^\pi \sin(x)\,dx\right)$$

$$= 5\pi^2 - 4$$

59. The mean is $\mu_X = \int_0^1 x\cdot 3x^2\,dx = \frac{3}{4}x^4\Big|_0^1 = \frac{3}{4}$. Therefore, $\mathrm{Var}(X) = \int_0^1 \left(x-\frac{3}{4}\right)^2\cdot 3x^2\,dx = \left(\frac{3}{5}x^5 - \frac{9}{8}x^4 + \frac{9}{16}x^3\right)\Big|_0^1 = \frac{3}{80}$.

61. The mean is $\mu_X = \int_a^b x\cdot\frac{1}{b-a}\,dx = \frac{x^2}{2(b-a)}\Big|_a^b = \frac{a+b}{2}$. Therefore, $\mathrm{Var}(X) = \int_a^b \left(x-\frac{a+b}{2}\right)^2\cdot\frac{1}{b-a}\,dx = \frac{1}{b-a}\cdot\frac{1}{3}\cdot\left(x-\frac{a+b}{2}\right)^3\Big|_a^b$
$= \frac{1}{b-a}\cdot\frac{1}{3}\cdot\frac{1}{8}\cdot\left((b-a)^3-(a-b)^3\right) = \frac{1}{24}\cdot 2(b-a)^2 = \frac{(b-a)^2}{12}$.

63. Let $\mu = E(X) = \int_a^b xf(x)\,dx$. Then

$$\begin{aligned}\mathrm{Var}(X) &= \int_a^b (x-\mu)^2 f(x)\,dx = \int_a^b (x^2-2\mu x+\mu^2)f(x)\,dx \\ &= \int_a^b x^2 f(x)\,dx - 2\mu\int_a^b xf(x)\,dx + \mu^2\int_a^b f(x)\,dx \\ &= E(X^2) - 2\mu\cdot\mu + \mu^2 = E(X^2) - \mu^2 = E(X^2) - E(X)^2.\end{aligned}$$

65. Make the substitution $z = \frac{x-\mu}{\sqrt{2}\sigma}$, $x = \mu + \sqrt{2}\,\sigma z$, $dx = \sqrt{2}\,\sigma\,dz$ in the integral for $E(X)$: $E(X) = \frac{1}{\sqrt{2\pi}\sigma}\int_{-\infty}^{\infty} x\exp\left(-\frac{1}{2}\left(\frac{x-\mu}{\sigma}\right)^2\right)dx$, to obtain

$$\begin{aligned}E(X) &= \frac{1}{\sqrt{2\pi}\sigma}\int_{-\infty}^{\infty}(\mu+\sqrt{2}\,\sigma z)\exp(-z^2)\cdot\sqrt{2}\sigma\,dz \\ &= \frac{\mu}{\sqrt{\pi}}\int_{-\infty}^{\infty}\exp(-z^2)\,dz + \sqrt{\frac{2}{\pi}}\,\sigma\int_{-\infty}^{\infty} z\exp(-z^2)\,dz.\end{aligned}$$

This simpifies to $E(X) = \mu$ because $\int_{-\infty}^{\infty}\exp(-z^2)\,dz = \sqrt{\pi}$ (see (6.7.2)) and $\int_{-\infty}^{\infty} z\exp(-z^2)\,dz = 0$ by symmetry (the integrand is odd and the improper integral $\int_0^{\infty} z\exp(-z^2)\,dz$ is finite).

For $\mathrm{Var}(X) = \frac{1}{\sqrt{2\pi}\sigma}\int_{-\infty}^{\infty}(x-\mu)^2\exp\left(-\frac{1}{2}\left(\frac{x-\mu}{\sigma}\right)^2\right)dx$ make the same substitution $z = \frac{x-\mu}{\sqrt{2}\sigma}$ to obtain $\mathrm{Var}(X) = \frac{1}{\sqrt{2\pi}\sigma}\int_{-\infty}^{\infty} 2\sigma^2 z^2\exp(-z^2)\cdot\sqrt{2}\sigma\,dz$. Now apply integration by parts: $u = z$, $dv = 2z\exp(-z^2)\,dz$, $du = dz$, $v = -\exp(-z^2)$, to obtain

$$\mathrm{Var}(X) = \frac{\sigma^2}{\sqrt{\pi}}\left(-z\exp(-z^2)\Big|_{z\to-\infty}^{z\to\infty} + \int_{-\infty}^{\infty}\exp(-z^2)\,dz\right) = \sigma^2.$$

Note that (6.7.2) was used to evaluate the last integral.

CALCULATOR/COMPUTER EXERCISES

67. Assume that $\delta = 1$. The region is $0 \le x \le b$, $\exp(x^2) \le y \le 1 + x$ where $b = 0.7468\ldots$ (*Maple*). Using *Maple*'s numerical integrator,

$$M = 0.1133\ldots \quad , \quad M_{x=0} = 0.0443\ldots \quad , \quad M_{y=0} = 0.1473\ldots$$

and the center of mass is $(0.3913\ldots, 1.3004\ldots)$.

69. Assume that $\delta = 1$. The region is $a \le x \le b$, $x^2 - 3 \le y \le \ln(1 + x + x^2)$ where $a = -2.0324\ldots$ and $b = 2.2643\ldots$ (*Maple*). Using *Maple*'s numerical integrator,

$$M = 9.2662\ldots \quad , \quad M_{x=0} = 2.1930\ldots \quad , \quad M_{y=0} = -6.5232\ldots$$

and the center of mass is $(0.2366\ldots, -0.7039\ldots)$.

71. Assume that $\delta = 1$. The region is $a \le x \le b$, $0 \le y \le A - B(\exp(kx) + \exp(-kx))$ where $A = 693.8597$, $B = -34.38365$, $k = 0.0100333$, and $a = -299.2239$, $b = 299.2239$. Using *Maple*'s numerical integrator,

$$M = 277611 \quad , \quad M_{x=0} = 0 \quad , \quad M_{y=0} = 73145100$$

and the center of mass is $(0, 263.48)$.

7.5 Work

Problems for Practice

1. Measuring time in seconds, at time t there are $500-t$ pounds of sand in the shovel. Since the shovel is $y = \frac{80}{60} \cdot t$ feet above the ground, the force applied to the shovel is $F(y) = 500 - \frac{60}{80} \cdot y$ pounds. Therefore, the work done to lift the shovel from $y = 0$ to $y = 80$ is $W = \int_0^{80} (500 - \frac{3}{4}y)\, dy = 37\,600$ foot-pounds.

3. Since there are 5280 feet in a mile,

$$\begin{aligned} W &= \int_0^{100\cdot 5280} 15697444 \cdot \frac{100}{(3962 + y/5280)^2}\, dy \\ &= -5280 \left(15697444 \cdot \frac{100}{3962 + y/5280} \right) \Bigg|_0^{528000} = 51\,500\,135 \text{ ft-lb.} \end{aligned}$$

5. Using formula (7.5.4), in the mks system, an object of mass m kilograms weighs $F(r) = 3.98621 \times 10^{14} \frac{m}{r^2}$ newtons when it is r meters from the center of the Earth. Using 6375580 meters for the radius of the Earth, the rocket's mass must satisfy the equation

$$\int_{6375580}^{6400580} 3.98621 \cdot 10^{14} \cdot \frac{m}{r^2}\, dr = 58\,610\,091 \text{ joules}.$$

That is, $2.442087121 \cdot 10^5\, m = 58610091$ and $m = 240$ kg.

7. When the rocket is at height y feet it weighs $F(y) = 7000 - 30 \cdot \frac{y}{5280}$ pounds. Consequently, the work done to lift the rocket 20 miles into space is $W = \int_0^{20\cdot 5280} \left(7000 - 30 \cdot \frac{y}{5280}\right) dy = 707\,520\,000$ ft-lbs.

9. When the bottom of the chain has been pulled up y feet, the hanging portion weighs $F(y) = 3 \cdot (50 - y)$ pounds. This is approximately the force that must be exerted by the man to lift the chain from that point to the point where the bottom of the chain is at $y + \Delta y$ (Δy is very small). The approximate work done is $F(y)\,\Delta y$. Summing these the total work is approximately $\sum F(y)\,\Delta y$, and the actual work to pull the chain from $y = 0$ to $y = 30$ is $\int_0^{30} 3 \cdot (50 - y)\, dy = 3150$ ft-lb.

11. The load weighs $300 \cdot 9.80665 = 2941.995$ newtons. When it has been pulled up y meters, the hanging portion of the cable and the load weighs $F(y) = 220 \cdot (40 - y) + 2941.995$ newtons. This is approximately the force that must be exerted to lift the load from that point to the point where the load is pulled to $y + \Delta y$ (Δy is very small). The approximate work done is $F(y)\,\Delta y$. Summing these the total work is approximately $\sum F(y)\,\Delta y$, and the actual work to lift the load to a height of 24 meters above ground level is $W = \int_0^{24} \big(220 \cdot (40 - y) + 2941.995\big)\, dy = 218\,448$ N-m.

13. When it has been pulled up y meters, the hanging portion of the cable and the load weighs $F(y) = 20 \cdot (100 - y) + 300$ pounds. This is approximately the force that must be exerted to lift the load from that point to the height $y + \Delta y$ (Δy is very small). The approximate work done is $F(y)\,\Delta y$. Summing these the total work is approximately $\sum F(y)\,\Delta y$, and the actual work is $W = \int_0^{100} \big(20 \cdot (100 - y) + 300\big)\, dy = 130\,000$ ft-lb.

15. When it has been pulled up y meters, the hanging portion of the cable and the load weighs $F(y) = 20 \cdot (100 - y) + 300$ pounds. This is approximately the force that must be exerted to lift the load from that point to the height $y + \Delta y$ (Δy is very small). The approximate work done is $F(y)\,\Delta y$. Summing these the total work is approximately $\sum F(y)\,\Delta y$, and the actual work is $W = \int_{30}^{80} \big(20 \cdot (100 - y) + 300\big)\, dy = 60\,000$ ft-lb.

17. Working with feet and pounds, the spring constant is $k = 8 \cdot 12 = 96$ pounds per foot. Since the spring is stretched 7/12 of a foot, the work done is $W = \int_0^{7/12} 96x\, dx = \frac{49}{3}$ ft-lb.

19. Working in the mks system, the spring constant is $k = \frac{240}{0.1} = 2400$ newtons per meter. Since the spring is stretched 1/10 of a meter, the work done is $W = \int_0^{1/10} 2400x\, dx = 12$ N-m.

21. Working with feet and pounds, the spring constant is $k = \frac{280}{4/12} = 840$ pounds per foot. Since the spring is compressed from 1/3 of a foot to 2/3 of a foot, the work done is $W = \int_{1/3}^{2/3} 840x\, dx = 140$ ft-lb.

23. Working in the mks system let k denote the spring constant. Since it takes 40 joules of work to stretch the spring 0.08 meters, $\int_0^{0.08} kx\,dx = 40$ implying that $\frac{1}{2}k \cdot 0.08^2 = 40$ and $k = 12500$ newtons per meter. The additional work required to stretch the spring from $x = 0.08$ meters to $x = 0.12$ meters is $W = \int_{0.08}^{0.12} 12500x\,dx = 50$ joules.

25. Working with feet and pounds let k denote the spring constant. Since it takes 2/3 ft-lb of work to stretch the spring 1/6 feet, $\int_0^{1/6} kx\,dx = 2/3$ implying that $\frac{1}{2}k \cdot \frac{1}{36} = \frac{2}{3}$ and $k = 48$ pounds per foot. The force required to maintain the spring stretched $x = 1/6$ feet is $F = kx = 48 \cdot \frac{1}{6} = 8$ pounds.

27. Let the y-axis point downward with its origin at the top of the pool. A thin slice of water at y with thickness Δy has volume $15^2 \Delta y$ and weighs $62.428 \cdot 15^2 \cdot \Delta y$ pounds. It will take approximately $62.428 \cdot 15^2 \cdot y\,\Delta y$ foot-pounds of work to lift it to the top of the pool. Therefore, the total work is approximately $\sum 62.428 \cdot 15^2 \cdot y\,\Delta y$ implying that the exact amount of work ks $W = \int_0^{10} 62.428 \cdot 15^2 \cdot y\,dy = 702\,315$ ft-lb.

29. Let the y-axis point downward with its origin at the top of the pool. A thin slice of water at y with thickness Δy has volume $6 \cdot 10\Delta y$ and weighs $9806.65 \cdot 60 \cdot \Delta y$ newtons. It will take approximately $9806.65 \cdot 60 \cdot y\,\Delta y$ joules of work to lift it to the top of the pool. Therefore, the total work is approximately $\sum 9806.65 \cdot 60 \cdot y\,\Delta y$ implying that the exact amount of work is $W = \int_1^3 9806.65 \cdot 60 \cdot y\,dy = 2\,353\,596$ joules.

31. When half of the water is pumped there is 4 feet of water left in the pool. The work to get the top 4 feet of water out of the pool is $W = \int_2^6 62.428 \cdot 480 \cdot y\,dy = 479\,447$ ft-lb.

33. Let the y-axis point downward with its origin at the top of the tank. A thin slice of water at y with thickness Δy is approximately a rectangular box with width $2\sqrt{36-y^2}$, length 16, and height Δy, so its volume is $32\sqrt{36-y^2}\,\Delta y$. The slice weighs $62.428 \cdot 32\sqrt{36-y^2}\,\Delta y$ pounds so it will take approximately $62.428 \cdot 32\sqrt{36-y^2} \cdot y\,\Delta y$ ft-lb of work to lift it to the top of the tank. Therefore, the total work is approximately $\sum 62.428 \cdot 32\sqrt{36-y^2} \cdot y\,\Delta y$ implying that the exact amount of work is $W = \int_0^6 62.428 \cdot 32\sqrt{36-y^2} \cdot y\,dy$ ft-lb. The integral can be evaluated with the substitution $u = 36 - y^2$, $du = -2y\,dy$ yielding

$$W = 62.428 \int_{36}^{0} 32\sqrt{u} \cdot \left(-\frac{1}{2}\right)\,du = -62.428 \cdot 16 \cdot \frac{2}{3} u^{3/2}\bigg|_{36}^{0} = 143\,834 \text{ ft-lb}.$$

35. Let the y axis point downward with the origin at the center of the top surface of the tank. A thin slice of water at y with thickness Δy is approximately a cylinder with base radius $50 + y$ (draw a picture). Its volume is $\pi(50+y)^2\,\Delta y$ and it weighs approximately $62.428\pi(50+y)^2\,\Delta y$

pounds. It will take about $62.428\pi(50+y)^2 \cdot y\,\Delta y$ foot pounds of work to lift this cylinder to the top of the tank. This implies that the work required to lift all of the water to the top is given by the following integral: $W = \int_0^{50} 62.428\pi(50+y)^2 \cdot y\,dy = 1\,736\,508\,794$ ft-lb.

Further Theory and Practice

37. Since the satellite weighs 800 pounds when $y = 0$, its mass M satisfies the equation $800 = \frac{1.4077 \cdot 10^{16} M}{(2.0917 \cdot 10^7)^2}$ so $M = 24.864$ slugs. Therefore, the work required to lift the satellite from $y = 0$ to $y = 200 \cdot 5280$ feet is

$$W = \int_0^{200 \cdot 5280} \frac{1.4077 \cdot 10^{16} \cdot 24.864}{(2.0917 \cdot 10^7 + y)^2}\,dy = 8.0419 \times 10^8 \text{ ft-lb}\,.$$

39. Let the length of the stretch be 2 units and the spring constant be k. Then the work to stretch the spring 2 units is $\int_0^2 kx\,dx = \frac{1}{2}k \cdot 4 = 2k$. The work to stretch the spring 1 unit is $\int_0^1 kx\,dx = \frac{1}{2}k \cdot 1 = \frac{1}{2}k$ so 25% of the work is expended by the first half of the stretch.

41. Let $x_1 < s_j < x_2$ where s_j is the jth point of a uniform partition of the interval $[x_1, x_2]$ (partition width $\Delta x = (x_2 - x_1)/N$). If the piston is s_j inches from the cylinder head, then the pressure of the gas on the cylinder walls is $p(As_j)$ and the force required to compress the gas by moving the piston is $A \cdot p(As_j)$. Since p is a continuous function, the work required to compress the gas from volume As_j to volume As_{j-1} is approximately $A \cdot p(As_j) \cdot \Delta x$ and the total work to compress the gas from volume Ax_2 to Ax_1 is approximately $\sum_{j=1}^{N} A \cdot p(As_j) \cdot \Delta x$. Allowing $N \to \infty$ yields the integral formula $W = A\int_{x_1}^{x_2} p(Ax)\,dx$.

43. A cross-sectional slice of the tank at a point y, $0 \le y \le 18$, is a circlular disk with radius $x = \sqrt{y/2}$ so its area is $A(y) = \pi \cdot \frac{y}{2}$. The volume of water between the slices at y and $y + \Delta y$ (Δy positive and small) is approximately $\pi \cdot \frac{y}{2} \cdot \Delta y$. Since it weighs $62.428\pi \cdot \frac{y}{2} \cdot \Delta y$ pounds it will take approximately $62.428\pi \cdot \frac{y}{2} \cdot (18 - y)\,\Delta y$ ft-lbs of work to lift the water to the top of the tank. This implies that the work required to lift the water between $y = 3$ and $y = 18$ is $\int_3^{18} 62.428\pi \cdot \frac{y}{2} \cdot (18 - y)\,dy = 88\,256$ ft-lb.

45. This is $\int_0^4 62.428\pi \cdot \frac{y}{2} \cdot (18 - y)\,dy = 12\,029$ ft-lb. See the solution to Exercise 43.

47. According to Newton's Law,

$$F = ma = m\frac{dv}{dt} = m\frac{dv}{dx}\frac{dx}{dt} = mv\frac{dv}{dx} = \frac{d}{dx}\left(\tfrac{1}{2}mv^2\right).$$

Consequently, because the mass is at rest at $x = 0$,

$$W(b) = \int_0^b F(x)\,dx = \int_0^b \frac{d}{dx}\left(\tfrac{1}{2}mv^2\right)dx = \tfrac{1}{2}mv^2\Big|_{x=0}^{x=b} = \tfrac{1}{2}mv(b)^2\,.$$

49. The parametric plot of $F = kx$ and $W = \int_0^x kz\,dz = \frac{1}{2}kx^2$ in the FW-plane is a portion of a parabola. (Observe that $W = \frac{1}{2k}F^2$.)

CALCULATOR/COMPUTER EXERCISES

51. Let $b = 10(1 - 1/e^2)$ so the tank extends from $y = 0$ to $y = b$. For $0 \le y \le b$, the water in the slice between y and $y + \Delta y$ (Δy positive and small) has a volume that is approximately equal to the volume of a circular disk with base radius x where $x^2 = -2\ln(1 - y/10)$ and height Δy. The disk's volume is $\pi x^2 \Delta y = -2\pi \ln(1 - y/10)\,\Delta y$ so the water in the slice weighs approximately $-62.428 \cdot 2\pi \ln(1 - y/10)\,\Delta y$ pounds. It will take $-62.428 \cdot 2\pi \ln(1 - y/10)(b - y)\,\Delta y$ ft-lb of work to lift the slice to the top of the tank. Consequently, the work required to pump out a volume of water that leaves 4 feet of water in the tank is $W = -\int_4^b 62.428 \cdot 2\pi \ln(1 - y/10)(b - y)\,dy = 3582$ ft-lb (*Maple*).

53. The work required to pump all of the water out of the tank is $W = -\int_0^b 62.428 \cdot 2\pi \ln(1 - y/10)(b - y)\,dy = 5754.9$ ft-lb (*Maple*).

7.6 First Order Differential Equations–Separable Equations

Problems for Practice

1. Substitute into the left side: $\frac{dy}{dx} = C \cdot xe^{x^2/2}$. Substitute into the right side: $xy = x \cdot Ce^{x^2/2}$. The left side equals the right side for all x.

3. Substitute into the left side: $\frac{dy}{dx} = \frac{1}{3} - 3Ce^{-3x}$. Substitute into the right side: $x - 3y = x - 3(x/3 - 1/9 + Ce^{-3x}) = \frac{1}{3} - 3Ce^{-3x}$. The left side equals the right side for all x.

5. Substitute into the left side: $\frac{dy}{dx} = Ce^x - 1$. Substitute into the right side: $x + y = x + Ce^x - x - 1 = Ce^x - 1$. The left side equals the right side for all x.

7. Substitute into the left side: $\frac{dy}{dx} = Ce^x - 2x - 2$. Substitute into the right side: $y + x^2 = Ce^x - x^2 - 2x - 2 + x^2 = Ce^x - 2x - 2$. The left side equals the right side for all x.

9. Assuming x and y are both positive, the variables separate: $\frac{dy}{\sqrt{y}} = 6\sqrt{x}\,dx$. Integrate to obtain $\int y^{-1/2}\,dy = \int 6x^{1/2}\,dx + C$. That is, $2y^{1/2} = 4x^{3/2} + C$ or, equivalently, $y = (2x^{3/2} + C)^2$.

11. The variables separate: $\frac{dy}{y^2} = \frac{dx}{2+x}$. Integrate: $\int y^{-2}\,dy = \int \frac{1}{2+x}\,dx + C$. That is, $-\frac{1}{y} = \ln(|2 + x|) + C$ or, equivalently, $y = \frac{1}{C - \ln(|2+x|)}$.

13. The variables separate: $\frac{dy}{y} = 3x^2\,dx$. Integrate: $\int \frac{dy}{y} = \int 3x^2\,dx + C$. That is, $\ln(|y|) = x^3 + C$ or $|y| = e^C \cdot e^{x^3}$. Equivalently, $y = \pm Ke^{x^3}$ where K is a positive constant. Because $y = 0$ is also a solution, this can be expressed in the form $y = Ce^{x^3}$ where the constant C is arbitrary.

15. Write the differential equation in the form $e^{3y}\frac{dy}{dx} = e^{-2x}$. Then $e^{3y}\,dy = e^{-2x}\,dx$ which integrates to $\frac{1}{3}e^{3y} = -\frac{1}{2}e^{-2x} + C$. Equivalently, $e^{3y} = -\frac{3}{2}e^{-2x} + C$, or $y = \frac{1}{3}\ln\left(C - \frac{3}{2}e^{-2x}\right)$.

17. Separate the variables to $\cos(y)\,dy = x\,dx$ which integrates to the implicit solution $\sin(y) = \frac{1}{2}x^2 + C$. That is, $y = \arcsin\left(\frac{1}{2}x^2 + C\right)$.

19. Integrate to obtain the general solution $y(x) = \int 2x\,dx + C = x^2 + C$. Since $y(1) = 3$, the constant C must satisfy the equation $1 + C = 3$ and $C = 2$. The solution to the initial value problem is $y(x) = x^2 + 2$.

21. Integrate to obtain the general solution $y(x) = \int \cos(x)\,dx + C = \sin(x) + C$. Since $y(0) = 2$, the constant C must satisfy the equation $0 + C = 2$ and $C = 2$. The solution to the initial value problem is $y(x) = \sin(x) + 2$.

23. The variables separate to $y\,dy = x\,dx$ which integrates to yield the following implicit solution $\frac{1}{2}y^2 = \frac{1}{2}x^2 + C$. Since $y(0) = 1$, the constant C must be chosen so that $\frac{1}{2} = 0 + C$ and $C = \frac{1}{2}$. Therefore, the implicit solution is $y^2 = x^2 + 1$ which, since $y > 0$, can be written in the equivalent form $y = \sqrt{x^2 + 1}$.

25. The variables separate to $y^{-2}\,dy = \sin(x)\,dx$ which integrates to yield the implicit solution $-\frac{1}{y} = -\cos(x) + C$ or, equivalently, $y = \frac{1}{\cos(x)+C}$. Since $y(0) = 2$, the constant C must be chosen so that $2 = \frac{1}{1+C}$ and $C = -\frac{1}{2}$. Therefore, the solution to the initial value problem is $y = \frac{1}{\cos(x)-\frac{1}{2}}$.

27. The variables separate to $y\sqrt{1+y^2}\,dy = 2x\,dx$ which integrates to yield the implicit solution $\frac{1}{3}(1+y^2)^{3/2} = x^2 + C$. Since $y(0) = 0$, the constant C must be chosen to satisfy the equation $\frac{1}{3} = 0 + C$ and $C = \frac{1}{3}$. The solution is given implicitly by the equation $\frac{1}{3}(1+y^2)^{3/2} = x^2 + \frac{1}{3}$ which implies that $1 + y^2 = (3x^2+1)^{2/3}$ and $y^2 = (3x^2+1)^{2/3} - 1$.

29. The variables separate to $\frac{y}{y^2+1}\,dy = dx$ which integrates to $\frac{1}{2}\ln(y^2+1) = x + C$. Since $y(0) = 3$, the constant C must satisfy the equation $\frac{1}{2}\ln(10) = 0 + C$ and $C = \frac{1}{2}\ln(10)$. The implicit solution is $\frac{1}{2}\ln(y^2+1) = x + \frac{1}{2}\ln(10)$. This is equivalent to $y^2 + 1 = 10e^{2x}$ or, since $y > 0$, $y = \sqrt{10e^{2x} - 1}$.

31. The variables separate to $\frac{dy}{y} = \frac{1}{x^2}\sin(1/x)\,dx$ which integrates to $\ln(|y|) = \cos(1/x) + C$. Since $y(2/\pi) = 3$, the constant C must satisfy the equation $\ln(3) = 0 + C$ and $C = \ln(3)$. Because $y > 0$, the implicit solution is $\ln(y) = \cos(1/x) + \ln(3)$. Explicitly, $y = 3\exp(\cos(1/x))$.

33. Since $\frac{dy}{dx} = 4x^2y$, the variables separate to $\frac{dy}{y} = 4x^2\,dx$ which integrates to $\ln(|y|) = \frac{4}{3}x^3 + C$. Since $y(0) = 1$, the constant C must satisfy the

equation $0 = 0 + C$ and $C = 0$. Because $y > 0$, the implicit solution is $\ln(y) = \frac{4}{3}x^3$. Explicitly, $y = \exp\left(\frac{4}{3}x^3\right)$.

35. The variables separate to $y\,dy = \frac{dx}{1+x^2}$ which integrates to $\frac{1}{2}y^2 = \arctan(x) + C$. Since $y(1) = \sqrt{\pi}$, the constant C must satisfy the equation $\frac{\pi}{2} = \arctan(1) + C$ and $C = \frac{\pi}{4}$. The implicit solution is $\frac{1}{2}y^2 = \arctan(x) + \frac{\pi}{4}$. Explicitly, because $y > 0$, $y = \sqrt{2\arctan(x) + \frac{\pi}{2}}$.

Further Theory and Practice

37. $F_1(x, y) = x$ must match with a slope field that has horizontal tangents on the y-axis and constant slopes on vertical lines. This is Figure 9e.

    $F_2(x, y) = y$ must match with a slope field that has horizontal tangents on the x-axis and constant slopes on horizontal lines. This is Figure 9c.

    $F_3(x, y) = xy$ must match with a slope field that has horizontal tangents on both axes. This is Figure 9a.

    $F_4(x, y) = x/y$ must match with a slope field that has horizontal tangents on the y-axis and vertical tangents on the x-axis. This is Figure 9f.

    $F_5(x, y) = x^2 - y^2$ must match with a slope field that has horizontal tangents on the lines $y = \pm x$. This is Figure 9d.

    $F_6(x, y) = y^2$ must match with a slope field where all tangent lines have positive slope except the ones on the x-axis where the slope is 0. This is Figure 9b.

39. The variables separate: $e^y\,dy = xe^{-x}\,dx$. This integrates (by parts on the right side) to $e^y = -xe^{-x} - e^{-x} + C$ or $y = \ln(C - (x+1)e^{-x})$.

41. The variables separate: $\frac{dy}{y(y-1)} = 3x^2\,dx$. The left side integrates by partial fractions $\int\left(\frac{1}{y-1} - \frac{1}{y}\right)dy = x^3 + C$. Consequently, $\ln\left(\left|\frac{y-1}{y}\right|\right) = x^3 + C$. This is equivalent to $\left|\frac{y-1}{y}\right| = Ke^{x^3}$ where K is a positive constant or $\frac{y-1}{y} = Ce^{x^3}$, C arbitrary. Note that $y = 1$ is a solution, as is $y = 0$. Replace C with $-C$ and solve for y to get the explicit formula $y = \frac{1}{1+Ce^{x^3}}$.

43. The variables separate: $\frac{dy}{y} = \frac{e^x}{4+e^{2x}}\,dx$. The right side integrates via the substitution $u = e^x$, $du = e^x\,dx$, and $u = 2\tan(\theta)$, $du = 2\sec^2(\theta)\,d\theta$, to yield

$$\begin{aligned}\ln(|y|) &= \int \frac{e^x}{4+e^{2x}}\,dx = \int \frac{1}{4+u^2}\,du = \int \frac{1}{4+4\tan^2(\theta)} \cdot 2\sec^2(\theta)\,d\theta \\ &= \frac{1}{2}\int d\theta = \frac{1}{2}\theta + C = \frac{1}{2}\arctan(e^x/2) + C.\end{aligned}$$

Therefore, the solution is $y = C\exp\left(\frac{1}{2}\arctan\left(\frac{1}{2}e^x\right)\right)$. (Note that $y = 0$ is a solution).

45. Integrate both sides with respect to x: $y(x) = f^4(x) + C$. To determine C substitute $x = 1$ to obtain $y(1) = f^4(1) + C$ or $24 = 2^4 + C$, so $C = 8$. The solution is $y(x) = f^4(x) + 8$. For example, if $f(x) = 2\sqrt{x}$, then the differential equation is $y'(x) = 4 \cdot 8x^{3/2} \cdot x^{-1/2} = 32x$, and the solution satisfying $y(1) = 24$ is $y(x) = (2\sqrt{x})^4 + 8 = 16x^2 + 8$.

47. Integrate both sides with respect to x: $y(x) = g(f(x)) + C$. To determine C substitute $x = 0$ to obtain $y(0) = g(3) + C$ or $6 = 2 + C$, so $C = 4$. The solution is $y(x) = g(f(x)) + 4$. For example, if $f(x) = 3\cos(x)$ and $g(x) = \sqrt{1+x}$, then the differential equation is $y'(x) = -\frac{3\sin(x)}{2\sqrt{1+3\cos(x)}}$, and the solution satisfying $y(0) = 6$ is $y(x) = \sqrt{1 + 3\cos(x)} + 4$.

49. Let $A(t)$ be the area of the surface of the water at time t. Observe that $\frac{dV}{dy} = A(y)$ so $\frac{dV}{dt} = \frac{dV}{dy}\frac{dy}{dt} = A(y)\frac{dy}{dt}$ and Torrecelli's Law can be expressed in the form $A(y)\frac{dy}{dt} = -a \cdot \sqrt{2gy}$.

51. Referring to Exercise 49, $A(y) = \pi(1+y)^2$ (draw a picture). Therefore, the height $y(t)$ satisfies the differential equation $(1 + y)^2\frac{dy}{dt} = -\alpha\sqrt{y}$ where $\alpha = \frac{1}{\pi} \cdot \frac{32}{15} \cdot (\pi \cdot 0.01^2) \cdot \sqrt{2 \cdot 9.80665}$. Separate variables to obtain the equation $(1 + 2y + y^2) \cdot y^{-1/2}\, dy = -\alpha\, dt$ or $\left(y^{-1/2} + 2y^{1/2} + y^{3/2}\right) dy = -\alpha\, dt$. This integrates to

$$2y^{1/2} + \frac{4}{3}y^{3/2} + \frac{2}{5}y^{5/2} = -\alpha t + C\,.$$

Since $y(0) = 1$, the constant C equals 56/15. The tank is empty when $y = 0$. That is, in $t = \frac{56}{15\alpha} \approx 3952$ seconds (about 1.1 hours).

53. Assuming that $\frac{dP}{dt} = \frac{1}{2} \cdot P(t) \cdot (2P_0 - P(t))$, and $P(0) = P_0$, formula (7.6.10) yields the solution

$$P(t) = \frac{2P_0^2}{P_0 + P_0\exp(-P_0 t)} = \frac{2P_0}{1 + \exp(-P_0 t)}\,.$$

$P(t) = \frac{2}{3}P0$ when $\frac{3}{2} = \frac{2}{1+\exp(-P_0 t)}$. That is, $1 + \exp(-P_0 t) = \frac{4}{3}$, so $\exp(-P_0 t) = \frac{1}{3}$ implying that $P_0 t = \ln(3)$ and $t = \frac{\ln(3)}{P_0}$.

55. Write (7.6.8) in the form $\frac{dP}{dt} = -k \cdot \left(P^2 - P_\infty P\right)$ and complete the square to obtain $\frac{dP}{dt} = -k \cdot \left((P - \frac{1}{2}P_\infty)^2 - \frac{1}{4}P_\infty^2\right)$. Consequently, $P'(t)$ is maximum at the time t_I when $P(t_I) = \frac{1}{2}P_\infty$.

57. According to Exercise 56, $\exp(k \cdot P_\infty \cdot t_I) = \frac{P_\infty - P_0}{P_0}$. This allows us to replace $P_\infty - P_0$ in formula (7.6.10) with $P_0 \cdot \exp(k \cdot P_\infty \cdot t_I)$. Doing so yields

$$\begin{aligned} P(t) &= \frac{P_0 \cdot P_\infty}{P_0 + P_0 \cdot \exp(k \cdot P_\infty \cdot t_I) \cdot \exp(-k \cdot P_\infty \cdot t)} \\ &= \frac{P_\infty}{1 + \exp(-k \cdot P_\infty \cdot (t - t_I))}\,. \end{aligned}$$

59. Let $t = 0$ correspond to the year 1990 and $\tau = 10$ years. Using the formulas developed in Exercise 58 with $A = P(-\tau) = 36$, $P_0 = P(0) = 60$, and $P(\tau) = B = 90$, first calculate $P_\infty = \frac{(A+B)P_0^2 - 2ABP_0}{P_0^2 - AB} = 180$. Then, to obtain the population in the year 2020, 30 years after 1990, calculate

$$P(30) = \frac{60 \cdot 180}{60 + (180 - 60)\left(\frac{60}{36} \cdot \frac{180-36}{180-60}\right)^{-30/10}} = 144 \text{ thousand people}.$$

61. Assuming that the inflection point is at $t_I = 9$, and $P(t_I) = 25$, the formula $P(t_I) = \frac{1}{2}P_\infty$ derived in Exercise 55 implies that $P_\infty = 2P(t_I) = 50$.

The formula $t_I = \frac{1}{kP_\infty}\ln\left(\frac{P_\infty - P_0}{P_0}\right)$ from Exercise 56 implies that

$$k = \frac{1}{t_I \cdot P_\infty}\ln\left(\frac{P_\infty - P_0}{P_0}\right) = \frac{1}{9 \cdot 50}\ln\left(\frac{50 - 3.6}{3.6}\right) = 0.00568.$$

63. Since C, the concentration of sucrose, satisfies the equation $C(t) = C_0e^{-kt}$, where $k = 5.7 \cdot 10^{-5}$, $C(t) = \frac{1}{3}C_0$ when $e^{-kt} = \frac{1}{3}$. That is, $t = \frac{\ln(3)}{k} = 19274$ time units.

65. Let C denote the concentration of dinitrogen pentoxide. Then $\frac{dC}{dt} = -kC$ and $C(t) = C_0e^{-kt}$ where $C_0 = C(0) = 2.32$ moles per liter. If $C(3000) = 0.37$ moles per liter, then $0.37 = 2.32e^{-k \cdot 3000}$ and $k = -\frac{1}{3000} \cdot \ln\left(\frac{0.37}{2.32}\right) = 0.000612$. Therefore, $C(t) = 2.32\exp(-0.000612t)$ moles per liter.

67. Assuming that $y(x) = x \cdot w(x)$ implies that $y'(x) = w(x) + x \cdot w'(x)$. Therefore, substituting $y = x \cdot w$ into the differential equation $\frac{dy}{dx} = \phi(y/x)$ yields $w + x \cdot \frac{dw}{dx} = \phi(w)$. Consequently, $w(x)$ is the solution to the separable differential equation $\frac{dw}{dx} = \frac{\phi(w) - w}{x}$.

For example, the substitution $y = x \cdot w$ in the differential equation $\frac{dy}{dx} = \frac{2xy}{x^2+y^2} = \frac{2y/x}{1+(y/x)^2}$ yields $w + x\frac{dw}{dx} = \frac{2w}{1+w^2}$ or

$$x\frac{dw}{dx} = \frac{2w}{1+w^2} - w = \frac{2w - w(1+w^2)}{1+w^2} = \frac{w - w^3}{1+w^2}.$$

The variables separate to $\frac{w^2+1}{w^3-w}\,dw = -\frac{dx}{x}$ implying that

$$\begin{aligned}\ln(|x|) + C &= -\int \frac{w^2+1}{w(w+1)(w-1)} \\ &= -\int\left(\frac{a}{w} + \frac{b}{w+1} + \frac{c}{w-1}\right)\end{aligned}$$

where $w^2 + 1 = a(w+1)(w-1) + bw(w-1) + cw(w+1)$. Consequently, $a = -1$, $b = 1$, $c = 1$, and $\ln(|x|) + C = \ln(|w|) - \ln(|w+1|) - \ln(|w-1|)$.

Therefore, $\ln\left(\frac{|w|}{|w^2-1|}\right) = \ln(|x|) + C$ and $\frac{|w|}{|w^2-1|} = K|x|$ where K is a positive constant. This is equivalent to $\frac{w}{w^2-1} = Cx$ where C is arbitrary. (Note that $w = 0$ is a solution to the ode displayed above.) Replace w with y/x to obtain $\frac{y/x}{\frac{y^2}{x^2}-1} = Cx$ or $\frac{xy}{y^2-x^2} = Cx$. Two explicit solution formulas can be obtained by writing the implicit solution as $xy = Cxy^2 - Cx^3$ or, assuming $x \neq 0$, $Cy^2 - y - Cx^2 = 0$, and using the quadratic formula:

$$y = \frac{1 \pm \sqrt{1 + 4C^2x^2}}{2C}.$$

69. Separate the variables to $P\,dP = CT^{7.5}\,dT$ and integrate to obtain $\frac{1}{2}P^2 = \frac{C}{8.5}T^{8.5} + D$. Since $P = 0$ when $T = 0$, $D = 0$, and $P(T) = \left(\frac{2C}{8.5}T^{8.5}\right)^{1/2}$. Therefore, $P(T) = \sqrt{\frac{4C}{17}}\,T^{17/4} = 2\sqrt{\frac{C}{17}}\,T^{17/4}$.

71. Since the mass density is ρ, a constant, $m(r) = \frac{4}{3}\pi r^3 \cdot \rho$. Consequently, the pressure P satisfies the differential equation $\frac{dP}{dr} = -\frac{G\cdot\frac{4}{3}\pi r^3\cdot\rho\cdot\rho}{r^2} = -\frac{4}{3}\pi G\rho^2 r$. Integrate to obtain $P(r) = -\frac{2}{3}\pi G\rho^2 r^2 + C$. Because $P(R) = 0$, $C = \frac{2}{3}\pi G\rho^2 R^2$, and $P(r) = \frac{2}{3}\pi G\rho^2(R^2 - r^2)$. The pressure at the center of the star is $P(0) = 2\pi G\rho^2 R^2/3$. In terms of the stellar mass M, because $M = \frac{4}{3}\pi R^3\rho$, $R^3 = \frac{3M}{4\pi\rho}$ so $R^2 = \left(\frac{3M}{4\pi\rho}\right)^{2/3}$ and $P(0) = \frac{2}{3}\pi G\rho^2\left(\frac{3M}{4\pi\rho}\right)^{2/3}$. This simplifies to $P(0) = \frac{\pi^{1/3}G\rho^{4/3}M^{2/3}}{3^{1/3}2^{1/3}} = \left(\frac{\pi}{6}\right)^{1/3}G\rho^{4/3}M^{2/3}$.

73. Let $x(t)$ be the distance travelled by the bullet in t seconds. Since $x'(t) = (4 - 100t)^5$ and $x(0) = 0$, the distance traveled by the bullet is $x(0.04) - x(0) = \int_0^{0.04} x'(t)\,dt = \int_0^{0.04}(4 - 100t)^5\,dt = -\frac{1}{600}(4 - 100t)^6\Big|_0^{0.04} = \frac{4^6}{600} = \frac{512}{75} \approx 6.83$ feet.

75. The variables separate to $\frac{cy-d}{y}\,dy = \frac{a-bx}{x}\,dx$ so $\int\left(c - \frac{d}{y}\right)dy = \int\left(\frac{a}{x} - b\right)dx$ and, because x and y are both positive, $cy - d\ln(y) = a\ln(x) - bx + C$. Substitute the values for the constants a, b, c, d and the general solution is $4y - 7\ln(y) = 6\ln(x) - 2x + C$. The initial conditions for the predator and prey populations imply that the solution curve passes through the point $(2, 3/2)$ in the xy-plane so the integration constant C satisfies the equation $4\cdot\frac{3}{2} - 7\ln\left(\frac{3}{2}\right) = 6\ln(2) - 2\cdot 2 + C$ and $C = 10 - 7\ln(3) + \ln(2)$, which is slightly more than 3. The predator-prey relationship is approximately $4y - 7\ln(y) = 6\ln(x) - 2x + 3$.

77. Let $(0, M(x))$ be the position of the Man when the boat is at the point (x, y) as in Figure 12. The slope of the Chord is equal to $\frac{dy}{dx}$ and also equal to $-\frac{M(x)-y}{x}$. That is, $\frac{dy}{dx} = -\frac{M(x)-y}{x}$. The Pythagorean Theorem, applied to the right triangle with vertices $(0, y)$, (x, y), and $(0, M(x))$, implies that $(M(x) - y)^2 + x^2 = L^2$ so, since $M(x) - y > 0$, $M(x) - y = \sqrt{L^2 - x^2}$ and $\frac{dy}{dx} = -\frac{\sqrt{L^2-x^2}}{x}$ as desired.

79. Let $y = L(x)$. Makeham's assumption is $-\frac{1}{y}\frac{dy}{dx} = M + Bg^x$, so $\frac{dy}{y} = -(M + Bg^x)\,dx$ and, since $y > 0$, $\ln(y) = -\left(Mx + Bg^x \cdot \frac{1}{\ln(g)}\right) + C$. Therefore, $L(x) = Ce^{-\left(Mx + \frac{Bg^x}{\ln(g)}\right)}$.

81. Divide the second equation by the first to obtain $\frac{x}{y} \cdot \frac{dy}{dx} = \frac{\beta}{\alpha}$. Consequently, $\frac{dy}{y} = p \cdot \frac{dx}{x}$, where $p = \frac{\beta}{\alpha}$. Since both x and y are positive, this implies that $\ln(y) = p\ln(x) + C$ and $y = kx^p$.

83. Clearly $y(0) = 0$ so Dawson's integral satisfies the initial condition. Applying the product rule and Fundamental Theorem of Calculus,

$$\begin{aligned} y'(x) &= -2x\exp(-x^2)\int_0^x \exp(t^2)\,dt + \exp(-x^2)\exp(x^2) \\ &= 1 - 2x\exp(-x^2)\int_0^x \exp(t^2)\,dx = 1 - 2xy(x)\,. \end{aligned}$$

85. Divide both sides of the differential equation by m and write it in the form $\frac{dv}{dt} = -\frac{k}{m} \cdot (v^2 + \alpha^2)$ where $\alpha = \sqrt{mg/k}$. Separate the variables, $\frac{dv}{v^2+\alpha^2} = -\frac{k}{m}\,dt$, and integrate: $\int \frac{dv}{v^2+\alpha^2} = -\int \frac{k}{m}\,dt + C$. The dv integral evaluates via substitution, $v = \alpha\tan(\theta)$, $dv = \alpha\sec^2(\theta)\,d\theta$, yielding $\frac{1}{\alpha}\arctan\left(\frac{v}{\alpha}\right) = -\frac{k}{m}t + C$. This is equivalent to $\arctan\left(\frac{v}{\alpha}\right) = -\frac{\alpha k}{m}t + C$ or $v = \alpha\tan\left(C - \frac{\alpha k}{m}t\right)$. The initial condition implies that $\tan(C) = \frac{v_0}{\alpha}$ so $v(t) = \alpha\tan\left(\arctan\left(\frac{v_0}{\alpha}\right) - \frac{\alpha k}{m}t\right)$. Replace α with $\sqrt{\frac{mg}{k}}$ to obtain

$$v(t) = \sqrt{\frac{mg}{k}}\tan\left(\arctan\left(v_0\sqrt{\frac{k}{mg}}\right) - \sqrt{\frac{kg}{m}}\,t\right).$$

87. Divide both sides of the differential equation by m and write it in the form $\frac{dv}{dt} = -\frac{k}{m} \cdot (v^2 - \alpha^2)$ where $\alpha = \sqrt{mg/k}$. Separate the variables, $\frac{dv}{v^2-\alpha^2} = -\frac{k}{m}\,dt$, and integrate: $\int \frac{dv}{v^2-\alpha^2} = -\int \frac{k}{m}\,dt + C$. The dv integral can be evaluated be via substitution, $v = \alpha\tanh(\theta)$, $dv = \alpha\,\mathrm{sech}^2(\theta)\,d\theta$, yielding $\frac{1}{\alpha}\mathrm{arctanh}\left(\frac{v}{\alpha}\right) = -\frac{k}{m}t + C$. This is equivalent to $\mathrm{arctanh}\left(\frac{v}{\alpha}\right) = -\frac{\alpha k}{m}t + C$ or $v = \alpha\tanh\left(C - \frac{\alpha k}{m}t\right)$. The initial condition $v(0) = 0$ implies that $\tanh(C) = 0$ so $C = 0$ and $v(t) = \alpha\tanh\left(-\frac{\alpha k}{m}t\right)$. Replace α with $\sqrt{\frac{mg}{k}}$ to obtain

$$v(t) = \sqrt{\frac{mg}{k}}\tanh\left(-\sqrt{\frac{kg}{m}}\,t\right).$$

89. Let $\rho(y)$ be the density (weight/volume) of air y cm above sea level. Since $\rho(y) = kp(y)$ for some positive constant k, the weight $p(y)$ of the column of air from level y to ∞ is $p(y) = \int_y^\infty A(s) \cdot \rho(s)\,ds$ where $A(y)$ is the cross-sectional area of the column at level y. That is, $p(y) = -\int_\infty^y kp(s)\,ds$, implying that $p'(y) = -kp(y)$ (Fundamental Theorem of Calculus). This is the model for exponential decay: $p(y) = p(0)e^{-ky}$. The physical dimensions of k are cm^{-1}.

91. $E'(t) = E(t) \cdot \frac{d}{dt}\left(\frac{\alpha T}{T+\beta}\right) = E(t) \cdot \left(\frac{(T+\beta)\cdot\alpha - \alpha T}{(T+\beta)^2}\right) = E(t) \cdot \frac{\alpha\beta}{(T+\beta)^2}$, and $E''(t) = E'(t) \cdot \frac{\alpha\beta}{(T+\beta)^2} - E(t) \cdot \frac{2\alpha\beta}{(T+\beta)^3}$. Therefore,

$$E''(t) = E(t) \cdot \left(\frac{\alpha^2\beta^2}{(T+\beta)^4} - \frac{2\alpha\beta}{(T+\beta)^3}\right) = E(t) \cdot \frac{\alpha^2\beta^2 - 2\alpha\beta(T+\beta)}{(T+\beta)^4}$$
$$= E(t) \cdot \frac{2\alpha\beta\big((\alpha/2-1)\beta - T\big)}{(T+\beta)^4}.$$

Since $E'(t)$ is positive, E is an increasing function of T with the concavity changing from up to down at the point $(T, E(T))$, where the temperature is $T = (\alpha/2 - 1)\beta$. The graph is sigmoidal.

CALCULATOR/COMPUTER EXERCISES

93. Separate variables, $(1+y^4)\,dy = (1+x^2)\,dx$, and integrate to obtain $y + \frac{1}{5}y^5 = x + \frac{1}{3}x^3 + C$ or, equivalently, $15y + 3y^5 = 15x + 5x^3 + C$. Since $y(0) = 0$, $C = 0$, and the implicit solution is $15y + 3y^5 = 15x + 5x^3$.

The implicit plot (*Maple*) of the solution curve displayed on the right indicates that $y(2)$ is approximately 1.5. Using *Maple*'s *fsolve* procedure we find that $y(2) = 1.71338505\ldots$.

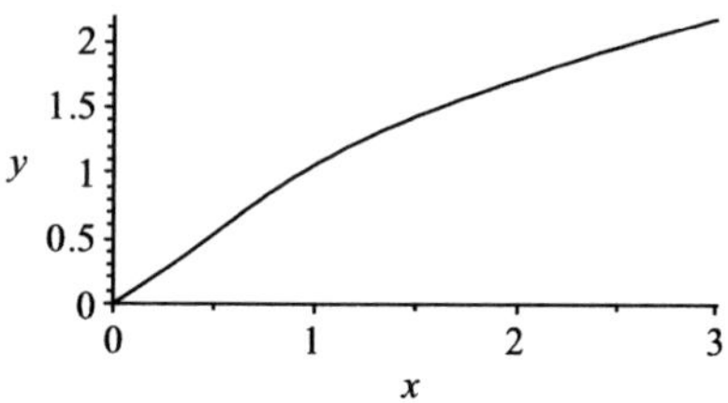

95. Separate variables, $\frac{y^2}{1+y}\,dy = \frac{dx}{1+2x}$, and integrate to obtain the relation $\int \frac{y^2}{1+y}\,dy = \frac{1}{2}\ln(1+2x) + C$ or, after a long division, $\int \left(y - 1 + \frac{1}{1-y}\right) dy = \frac{1}{2}\ln(1+2x) + C$. Since $y(0) = 0$, this is $\frac{1}{2}y^2 - y - \ln(1-y) = \frac{1}{2}\ln(1+2x) + C$, and $C = 0$. The implicit solution is $\frac{1}{2}y^2 - y - \ln(1-y) = \frac{1}{2}\ln(1+2x)$.

The implicit plot (*Maple*) of the solution curve displayed on the right indicates that $y(2)$ is approximately 1.5. Using *Maple*'s *fsolve* procedure we find that $y(2) = 1.760613117\ldots$.

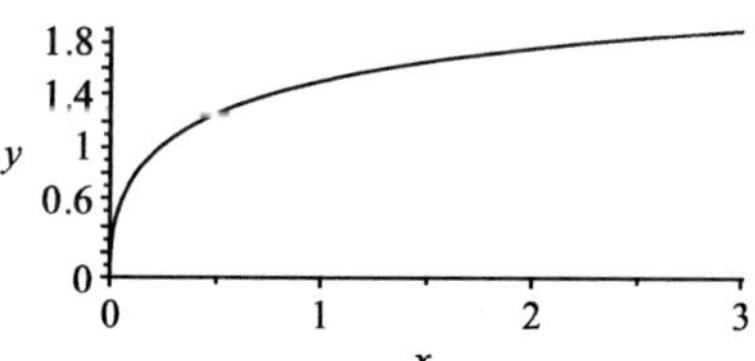

97. Separate variables, $\frac{dy}{5000-y} = 0.09t^{1/2}\,dt$, implying that $-\ln(5000-y) = 0.06t^{3/2} + C$ and $5000 - y = Ke^{-0.06t^{3/2}}$, where $K = e^C$. Since $y(0) = 175$, $K = 5000 - 175 = 4825$ and $y(t) = 5000 - 4825e^{-0.06t^{3/2}}$. Since $\lim_{t\to\infty} y(t) = 5000$, the goose reaches half of its mature weight when t satisfies the equation $y(t) = 2500$. According to *Maple*, $t \approx 4.93$ weeks.

99. (a) The central difference quotient approximations for $P'(1984)$ through $P'(1988)$ are 950, 1650, 3000, 2450, 2900 respectively. These yield the

following approximations for the constant k:

$6.21891{\cdot}10^{-6}, 8.17803{\cdot}10^{-6}, 1.14508{\cdot}10^{-5}, 6.66921{\cdot}10^{-6}, 7.55208{\cdot}10^{-6}$.

The average of these approximations is $k = 8.01381 \times 10^{-6}$.

(b) Using the approxmation for k found in part a, and the given values for $P(0)$, P_∞,

$$P(t) = \frac{3.52 \cdot 10^7}{800 + 43200 \exp(-0.35261t + 697.81049)}.$$

The predicted values for $P(1990)$, $P(1991)$, $P(1992)$, and $P(1993)$ are $20788, 24652, 28358$, and 31707, respectively. The observed values are $20700, 26200, 29300$, and 32100.

(c) The logistic growth curve for the time interval $1979 \le t \le 1993$ is shown on the right along with the population data points. Both compare favorably to the picture displayed in Figure 13.

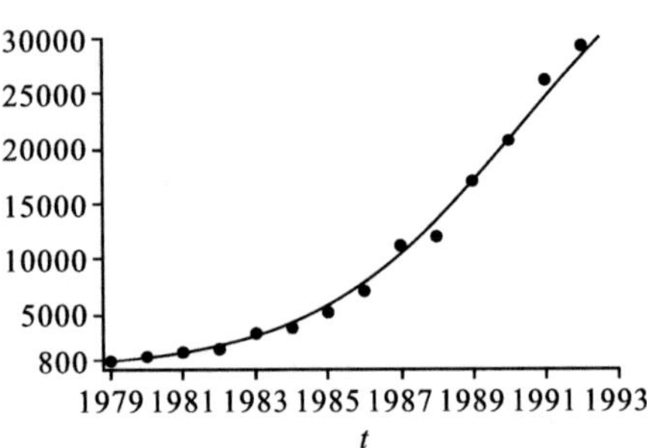

(d) According to the logistic model defined in part b, $P(t) = 43000$ when $t = 2001$. In fact, $P(2001) = 43007$.

7.7 First Order Differential Equations–Linear Equations

Problems for Practice

1. The standard form equation is $\frac{dy}{dx} - 3y = 0$ so the integrating factor is $e^{\int -3\,dx} = e^{-3x}$. Multiply the standard form equation by e^{-3x} and rearrange to $\frac{d}{dx}\left(e^{-3x}y\right) = 0$. Therefore, $e^{-3x}y = C$ and $y = Ce^{3x}$.

3. The equation is in standard form so the integrating factor is $e^{\int 4\,dx} = e^{4x}$. Multiply the standard form equation by e^{4x} and rearrange to $\frac{d}{dx}\left(e^{4x}y\right) = 0$. Therefore, $e^{4x}y = C$ and $y = Ce^{-4x}$.

5. The standard form equation is $\frac{dy}{dx} - \frac{1}{2}y = 1$ so the integrating factor is $e^{\int -\frac{1}{2}\,dx} = e^{-\frac{1}{2}x}$. Multiply the standard form equation by $e^{-\frac{1}{2}x}$ and

rearrange to $\frac{d}{dx}\left(e^{-\frac{1}{2}x}y\right) = e^{-\frac{1}{2}x}$. Therefore, $e^{-\frac{1}{2}x}y = -2e^{-\frac{1}{2}x} + C$ and $y = -2 + Ce^{\frac{1}{2}x}$.

7. The standard form equation is $\frac{dy}{dx} + \frac{1}{x}y = 1$ so the integrating factor is $e^{\int \frac{1}{x}\,dx} = e^{\ln(x)} = x$. Multiply the standard form equation by x and rearrange to $\frac{d}{dx}(xy) = x$. Therefore, $xy = \frac{1}{2}x^2 + C$ and $y = \frac{1}{2}x + \frac{C}{x}$.

9. The equation is in standard form so the integrating factor is $e^{\int x\,dx} = e^{\frac{1}{2}x^2}$. Multiply the standard form equation by $e^{\frac{1}{2}x^2}$ and rearrange to $\frac{d}{dx}\left(e^{\frac{1}{2}x^2}y\right) = xe^{\frac{1}{2}x^2}$. Therefore, $e^{\frac{1}{2}x^2}y = e^{\frac{1}{2}x^2} + C$ and $y = 1 + Ce^{-\frac{1}{2}x^2}$.

11. The equation is in standard form so the integrating factor is $e^{\int -3\,dx} = e^{-3x}$. Multiply the standard form equation by e^{-3x} and rearrange to $\frac{d}{dx}\left(e^{-3x}y\right) = 1$. Therefore, $e^{-3x}y = x + C$ and $y = xe^{3x} + Ce^{3x}$.

13. The equation is in standard form so the integrating factor is $e^{\int 1\,dx} = e^x$. Multiply the standard form equation by e^x and rearrange to $\frac{d}{dx}\left(e^x y\right) = 6e^{3x}$. Therefore, $e^x y = 2e^{3x} + C$ and $y = 2e^{2x} + Ce^{-x}$.

15. The equation is in standard form so the integrating factor is $e^{\int 2x\,dx} = e^{x^2}$. Multiply the standard form equation by e^{x^2} and rearrange to $\frac{d}{dx}\left(e^{x^2}y\right) = 8x^3e^{x^2}$. Therefore, $e^{x^2}y = \int 8x^3e^{x^2}\,dx + C$. The integral evaluates using integration by parts, $u = x^2$, $dv = 2xe^{x^2}\,dx$, $du = 2x\,dx$, $v = e^{x^2}$ yielding

$$e^{x^2}y = 4\left(x^2e^{x^2} - \int 2xe^{x^2}\,dx\right) + C = 4x^2e^{x^2} - 4e^{x^2} + C\,.$$

Therefore, $y = 4x^2 - 4 + Ce^{-x^2}$.

17. The equation is in standard form so the integrating factor is $e^{\int 2\,dx} = e^{2x}$. Multiply the standard form equation by e^{2x} and rearrange to $\frac{d}{dx}\left(e^{2x}y\right) = 3e^{2x}$. Therefore, $e^{2x}y = \frac{3}{2}e^{2x} + C$. The initial condition implies that $C = -1$ and the solution is $y = \frac{3}{2} - e^{-2x}$.

19. The standard form equation is $\frac{dy}{dx} + \frac{1}{x}y = 6\sqrt{3+x^2}$ so the integrating factor is $e^{\int \frac{1}{x}\,dx} = e^{\ln(x)} = x$. Multiply the standard form equation by x and rearrange to $\frac{d}{dx}(xy) = 2x\cos(\pi x^2)$. Therefore, $xy = 2(3+x^2)^{3/2} + C$. The initial condition implies that $7 = 16 + C$ so $C = -9$ and the solution is $y = \frac{2(3+x^2)^{3/2}-9}{x}$.

21. The equation is in standard form so the integrating factor is $e^{\int \frac{2}{x}\,dx} = e^{2\ln(x)} = x^2$. Multiply the standard form equation by x^2 and rearrange to $\frac{d}{dx}(x^2y) = 7x^{5/2}$. Therefore, $x^2y = 2x^{7/2} + C$. The initial condition implies that $16 \cdot 17 = 256 + C$ so $C = 17\cdot 16 - 16\cdot 16 = 16$ and the solution is $y = \frac{2x^{7/2}+16}{x^2}$.

23. The equation is in standard form so the integrating factor is $e^{\int -1\,dx} = e^{-x}$. Multiply the standard form equation by e^{-x} and rearrange to $\frac{d}{dx}\left(e^{-x}y\right) =$

1. Therefore, $e^{-x}y = x + C$. The initial condition implies that $C = -3$ and the solution is $y = (x-3)e^x$.

25. The equation is in standard form so the integrating factor is $e^{\int 2\,dx} = e^{2x}$. Multiply the standard form equation by e^{2x} and rearrange to $\frac{d}{dx}(e^{2x}y) = 4xe^{2x}$. Therefore, $e^{2x}y = \int \underbrace{4x}_{u}\, \underbrace{e^{2x}\,dx}_{dv} + C = 2xe^{2x} - e^{2x} + C$. The initial condition implies that $C = 4$ and the solution is $y = 2x - 1 + 4e^{-2x}$.

27. The standard form equation is $\frac{dy}{dx} - y = x^2$ so the integrating factor is $e^{\int -1\,dx} = e^{-x}$. Multiply the standard form equation by e^{-x} and rearrange to $\frac{d}{dx}(e^{-x}y) = x^2e^{-x}$. Therefore, $e^{-x}y = \int \underbrace{x^2}_{u}\, \underbrace{e^{-x}\,dx}_{dv} + C = -x^2e^{-x} + \int \underbrace{2x}_{u}\, \underbrace{e^{-x}\,dx}_{dv} + C = -x^2e^{-x} - 2xe^{-x} - 2e^{-x} + C$. The initial condition implies that $C = 3$ and the solution is $y = -x^2 - 2x - 2 + 3e^x$.

29. The rate that salt enters the tank is $\frac{1\text{kg}}{10\,\text{L}} \cdot \frac{20\text{ L}}{1\text{ min}} = 2$ kg per min. It leaves the tank at the rate of $\frac{m\text{ kg}}{200\,\text{L}} \cdot \frac{20\text{ L}}{1\text{ min}} = \frac{m}{10}$ kg per min. Therefore, $\frac{dm}{dt} = 2 - \frac{m}{10}$. The standard form equation is $\frac{dm}{dt} + \frac{1}{10}m = 2$ so the integrating factor is $e^{\int \frac{1}{10}\,dt} = e^{\frac{t}{10}}$. Multiply the standard form equation by this factor and rearrange to $\frac{d}{dt}(e^{\frac{t}{10}}m) = 2e^{\frac{t}{10}}$. Therefore, $e^{\frac{t}{10}}m = 20e^{\frac{t}{10}} + C$. Since $m(0) = 10$, $C = -10$, and $m(t) = 20 - 10e^{-\frac{t}{10}}$. Consequently, $\lim_{t\to\infty} m(t) = 20$.

31. Let m be the amount of salt in the tank at time t, t in minutes and m in pounds, $m(0) = 60$. There is no salt entering the tank and salt leaves the tank at the rate of $\frac{m\text{ lb}}{100\,\text{gal}} \cdot \frac{2.5\text{ gal}}{1\text{ min}} = \frac{m}{40}$ lb per min. Therefore, $\frac{dm}{dt} = -\frac{m}{40}$, and the amount of salt in the tank decreases exponentially: $m(t) = m(0)e^{-\frac{t}{40}} = 60e^{-\frac{t}{40}}$. Consequently, $\lim_{t\to\infty} m(t) = 0$.

33. Let T denote the temperature of the bar at time t, t in minutes, $T(0) = 150°\text{F}$. Newton's Law of Cooling with $T_\infty = 70$ is $\frac{dT}{dt} = k\cdot(70 - T)$. The standard form equation is $\frac{dT}{dt} + kT = 70k$. The integrating factor is $e^{\int k\,dt} = e^{kt}$ so $\frac{d}{dt}(e^{kt}T) = 70ke^{kt}$ and $e^{kt}T = 70e^{kt} + C$. The initial condition implies that $C = 80$ so $T(t) = 70 + 80e^{-kt}$. Since $T(4) = 125$, the constant k satisfies the equation $70 + 80e^{-k\cdot 4} = 125$ so $e^{-4k} = \frac{55}{80} = \frac{11}{16}$ and $k = \frac{1}{4}\ln(16/11)$. Thus $T(t) = 70 + 80\cdot\left(\frac{11}{16}\right)^{t/4}$ and the temperature after 10 minutes is $T(10) = 70 + 80\cdot\left(\frac{11}{16}\right)^{5/2} = 70 + 80\cdot\left(\frac{11}{16}\right)^2\cdot\frac{\sqrt{11}}{4} = 70 + \frac{605}{64}\sqrt{11} \approx 101.35°\text{F}$.

Further Theory and Practice

35. Using formula (7.7.13) developed in Example 7: $v(t) = -\frac{mg}{k}(1 - e^{-kt/m})$. When $t = 10$, $v(t) = 0.99\cdot\left(-\frac{mg}{k}\right)$ implying that $0.99 = 1 - e^{-kt/m}$. Since $k = 2$, $e^{-20/m} = 0.01$ so $-\frac{20}{m} = \ln(0.01)$ and $m = -\frac{20}{\ln(0.01)} \approx 4.3429$ kg.

37. Let V denote the volume of the lake and W the weight of the pollutants. Let $W(0) = W_0$. We wish to determine the time t when $W(t) = \frac{1}{20}W_0$. There are no pollutants flowing into the lake. Pollutants flow out of the lake at the rate of $\frac{W \text{ lbs}}{V \text{ cu ft}} \cdot \frac{V/3000 \text{ cu ft}}{1 \text{ day}} = \frac{W}{3000}$ lbs per day. Therefore, $\frac{dW}{dt} = -\frac{W}{3000}$ and the amount of pollutants in the lake decreases exponentially: $W(t) = W_0 e^{-t/3000}$. The weight of the pollutants will be $\frac{1}{20}W_0$ in t days where $e^{-t/3000} = \frac{1}{20}$. That is, $-\frac{t}{3000} = \ln(1/20)$, so $t = -3000\ln(1/20) \approx 8987$ days.

39. According to Newton's Law of Cooling $T(t) = T_\infty + (T_0 - T_\infty)e^{-kt}$. Therefore, t minutes after the thermometer is brought outdoors, its temperature is $T(t) = T_\infty + (72 - T_\infty)e^{-kt}$. Since $T(1) = 50$ and $T(3/2) = 44$, the constants k and T_∞ satisfy the following two equations.

$$(72 - T_\infty)e^{-k} = 50 - T_\infty \quad \text{and} \quad (72 - T_\infty)e^{-3k/2} = 44 - T_\infty\,.$$

Solve the left equation for e^{-k} and substitute into the equation on the right to obtain $(72 - T_\infty) \cdot \left(\frac{50-T_\infty}{72-T_\infty}\right)^{3/2} = 44 - T_\infty$. This simplifies to the equation $(50 - T_\infty)^{3/2} = (44 - T_\infty)(72 - T_\infty)^{1/2}$. Now square both sides and expand to obtain the quadratic equation $5T_\infty^2 - 386T_\infty + 7196 = 0$ implying that $T_\infty = \frac{193 \pm 3\sqrt{141}}{5} \approx 45.7$ or 31.5. The outside temperature is approximately 31.5°F.

41. Since $L\frac{dI}{dt} + RI = E(t)$, the ode is $12\frac{dI}{dt} + 60I = 24e^{-5t}$ or $\frac{dI}{dt} + 5I = 2e^{-5t}$. This is linear with integrating factor $e^{\int 5\,dt} = e^{5t}$ implying that $\frac{d}{dt}(e^{5t}I) = 2$ and $e^{5t}I = 2t + C$. Therefore, $I(t) = (2t + C)e^{-5t}$.

43. Since $L\frac{dI}{dt} + RI = E(t)$, the ode is $5\frac{dI}{dt} + 15I = 20\sin(t)$ or $\frac{dI}{dt} + 3I = 4\sin(t)$. This is linear with integrating factor $e^{\int 3\,dt} = e^{3t}$ implying that $\frac{d}{dt}(e^{3t}I) = 4e^{3t}\sin(t)$ and, using the integration formula in Exercise 86 of Section 6.1, $e^{3t}I = \int 4e^{3t}\sin(t)\,dt + C = \frac{1}{5}e^{3t}\big(6\sin(t) - 2\cos(t)\big) + C$. Therefore, $I(t) = \frac{1}{5}\big(6\sin(t) - 2\cos(t)\big) + Ce^{-3t}$.

45. Let m and V denote the amount of salt in the tank and the volume of solution in the tank at time t. Then $m(0) = 24$, $V(0) = 200$, and $V(t) = 200 - 4t$ liters. The tank is drained in 50 minutes. There is no salt entering the tank, but salt leaves the tank at the rate of $\frac{m \text{ kg}}{V \text{ L}} \cdot \frac{16 \text{ L}}{1 \text{ min}} = \frac{16m}{200-4t}$ kg per min. Therefore, $\frac{dm}{dt} = -\frac{16m}{200-4t}$, or $\frac{dm}{dt} + \frac{4}{50-t}m = 0$. The integrating factor is $e^{\int \frac{4}{50-t}\,dt} = e^{-4\ln(50-t)} = (50-t)^{-4}$. Multiply the standard form equation by this factor and rearrange to $\frac{d}{dt}\big((50-t)^{-4}m\big) = 0$. Therefore, $(50-t)^{-4}m = C$. Since $m(0) = 24$, $C = 24 \cdot 50^{-4}$, and $m(t) = \frac{24}{50^4} \cdot (50-t)^4$. After 25 minutes there are $m(25) = \frac{24}{50^4} \cdot 25^4 = \frac{24}{16} = 1.5$ kg of salt in the tank.

47. Differentiate $c \cdot \lambda^t = p - F(t)$ to obtain $c \cdot \lambda^t \cdot \ln(\lambda) = -F'(t)$. Therefore, $F'(t) = -(p - F(t)) \cdot \ln(\lambda)$. Equivlently, $\frac{dF}{dt} - \ln(\lambda)F = -p\ln(\lambda)$, so $\alpha = -p\ln(\lambda)$ and $\beta = -\ln(\lambda)$.

49. Since $\frac{dm}{dt} = \alpha - \beta m$, $\frac{dm}{dt} + \beta m = \alpha$ and $m(t) = \frac{\alpha}{\beta} + \big(m(0) - \frac{\alpha}{\beta}\big)e^{-\beta t}$ (formula (7.7.7)). When t is large there will be approximately $\frac{\alpha}{\beta}$ units of drug in the patient's bloodsrtream.

51. Equation (7.7.16) has the form $\frac{dI}{dt} + (\alpha + \lambda)I = \alpha$ implying that the potassium current at time t is $I(t) = \frac{\alpha}{\alpha+\lambda} + \big(I(0) - \frac{\alpha}{\alpha+\lambda}\big)e^{-(\alpha+\lambda)t}$.

53. (a) Let $V(t)$ be the volume of solution in the tank at time t. Then $V(0) = L$, $V(t) = L + (\rho - r)t$, and $m(0) = m_0$. There is no salt entering the tank, but salt leaves the tank at the rate of $\frac{m\text{ kg}}{V\text{ L}} \cdot \frac{r\text{ L}}{1\text{ hr}} = \frac{rm}{L+(\rho-r)t}$ kg per hr. Therefore, $\frac{dm}{dt} = -\frac{rm}{L+(\rho-r)t}$.

(b) If $\rho = r$, then $V(t) = L$, a constant, and $\frac{dm}{dt} = -\frac{r}{L}m$. This is the model for exponential decay and $m(t) = m(0)e^{-rt/L}$.

(c) If $\rho \neq r$, then $\frac{dm}{dt} + \frac{r}{L+(\rho-r)t}m = 0$. The integrating factor is $e^{\int \frac{r}{L+(\rho-r)t}\,dt} = e^{\frac{r}{\rho-r}\ln(L+(\rho-r)t)} = (L+(\rho-r)t)^{r/(\rho-r)}$. Consequently, $\frac{d}{dt}\big((L+(\rho-r)t)^{r/(\rho-r)}\,m\big) = 0$ and $m(t) = C(L+(\rho-r)t)^{-r/(\rho-r)}$.

55. Since $\frac{d\sigma}{dt} + \frac{\kappa}{m}\cdot\sigma = -g$, equation (7.7.5) can be used to obtain the general solution formula $\sigma(t) = -\frac{mg}{\kappa} + Ce^{-\kappa t/m}$. The initial condition $\sigma(\tau) = v_\tau$ implies that $C = e^{\kappa\tau/m}\big(v_\tau + \frac{mg}{\kappa}\big)$. Therefore,

$$\sigma(t) = -\frac{mg}{\kappa} + \Big(\frac{mg}{\kappa} + v_\tau\Big)\exp\Big(-\frac{\kappa}{m}(t-\tau)\Big)\,.$$

By the Fundamental Theorem of Calculus, $\eta(t) - \eta(\tau) = \int_\tau^t \sigma(u)\,du$. Therefore, $\eta(t) = y_\tau + \int_\tau^t \sigma(u)\,du$. Explicitly,

$$\begin{aligned}\eta(t) &= y_\tau + \int_\tau^t \Big(-\frac{mg}{\kappa} + \Big(\frac{mg}{\kappa} + v_\tau\Big)\exp\Big(-\frac{\kappa}{m}(u-\tau)\Big)\Big)\,du\\ &= y_\tau - \Big(\frac{mg}{\kappa}u + \Big(\frac{mg}{\kappa} + v_\tau\Big)\cdot\frac{m}{\kappa}\exp\Big(-\frac{\kappa}{m}(u-\tau)\Big)\Big)\Big|_{u=\tau}^{u=t}\\ &= H - \frac{mg\tau}{k} + \frac{m^2g}{k^2}\Big(1 - e^{-k\tau/m}\Big) - \frac{mg}{\kappa}(t-\tau)\\ &\quad + \frac{m}{\kappa^2}(mg + v_\tau\kappa)\Big(1 - e^{-(t-\tau)\kappa/m}\Big)\,.\end{aligned}$$

57. The height $y(t)$ of the projectile at any time $t \in [0, T_u + T_d]$ satisfies the

equation $y(t) - y(0) = \int_0^t v(u)\,du$. Therefore, since $y(0) = 0$,

$$\begin{aligned} y(t) &= \int_0^t \frac{mg}{k}\left(\left(1+\frac{kv_0}{mg}\right)e^{-ku/m} - 1\right) du \\ &= -\left(\frac{mg}{k}\cdot\frac{m}{k}\left(1+\frac{kv_0}{mg}\right)e^{-ku/m} + \frac{mg}{k}u\right)\Bigg|_{u=0}^{u=t} \\ &= \frac{mg}{k^2}\left(\left(m+\frac{kv_0}{g}\right)(1-e^{-kt/m}) - kt\right). \end{aligned}$$

The height H reached by the projectile before it begins to fall is $y(T_u)$. Therefore, since $(1+\frac{kv_0}{mg})e^{-kT_u/m} = 1$ and $(m+\frac{kv_0}{g})e^{-kT_u/m} = m$,

$$\begin{aligned} H &= \frac{mg}{k^2}\left(\left(m+\frac{kv_0}{g}\right)(1-e^{-kT_u/m}) - kT_u\right) \\ &= \frac{mg}{k^2}\left(m+\frac{kv_0}{g} - m - kT_u\right) = \frac{mv_0}{k} - \frac{mg}{k}T_u \\ &= \frac{mv_0}{k} - \frac{m^2 g}{k^2}\ln\left(1+\frac{k}{mg}v_0\right). \end{aligned}$$

Finally, since $e^{kT_u/m} = 1 + \frac{kv_0}{mg}$, $me^{kT_u/m} = m + \frac{kv_0}{g}$. Therefore,

$$\begin{aligned} y(t) &= \frac{mg}{k^2}\left(\left(m+\frac{kv_0}{g}\right)(1-e^{-kt/m}) - kt\right) \\ &= \frac{mg}{k^2}\left(me^{kT_u/m}(1-e^{-kt/m}) - kt\right). \end{aligned}$$

59. Let $t = T_u + T_d$ in the formula for $y(t)$ in Exercise 57 to obtain the equation

$$0 = \frac{mg}{k^2}\left(me^{kT_u/m}(1-e^{-k(T_u+T_d)/m}) - k(T_u+T_d)\right).$$

This is easily rearranged to $e^{kT_u/m} - e^{-kT_d/m} = \frac{kT_u}{m} + \frac{kT_d}{m}$.

To verify that $T_u < T_d$ suppose that a and b are positive numbers with $e^a - e^{-b} = a + b$. We will show that $a < b$. First rearrange the assumption to $e^a - a = e^{-b} + b$. Because $e^{-b} + b < e^b - b$ (an inequality that is equivalent to $0 < b < \sinh(b)$), we see that $e^a - a < e^b - b$. From this it follows that $a < b$ because the function $f(x) = e^x - x$ is strictly increasing for x positive.

61. Since $v(t) = P\tau(1 - e^{-t/\tau})$, $\lim_{t\to\infty} v(t) = P\tau$.

CALCULATOR/COMPUTER EXERCISES

63. The model is $m\frac{dv}{dt} = -mg - kv$, $v(0) = v_0$ and, using equation (7.7.7), $v(t) = -\frac{mg}{k} + (v_0 + \frac{mg}{k})e^{-kt/m}$. The initial data: $v_0 = 0$ and $v(1) = -9.4$, implies that $-9.4 = -\frac{9.80665}{k}(1 - e^{-k})$. Using *Maple*'s *fsolve* procedure, $k = 0.08531$, and, since $v(t) = -\frac{mg}{k}(1 - e^{-kt/m})$, the terminal velocity is $v_\infty = -\frac{mg}{k} = -114.96$ m/sec.

65. (a) The relevant formula is $y(t) = H - \frac{mg}{k}t + \frac{m^2 g}{k^2}\left(1 - e^{-kt/m}\right)$ from Exercise 54. Substitute $t = 26$, $y(26) = 0$, $H = 3170$, $m = 800$, and $g = 9.80665$, then solve for k (*Maple*) to obtain $k = 4.1655$. Since kv is a force it has the units of mass times acceleration implying that k is in units of kg/sec.

(b) Since $v(t) = -\frac{mg}{k}\left(1 - e^{-kt/m}\right)$, the terminal velocity of the bomb was $v_\infty = -\frac{mg}{k} = -1883$ m/s.

(c) The actual velocity was $v(26) = -238$ m/s.

(d) The plot of height $y(t)$ for $0 \le t \le 26$ is displayed on the right.

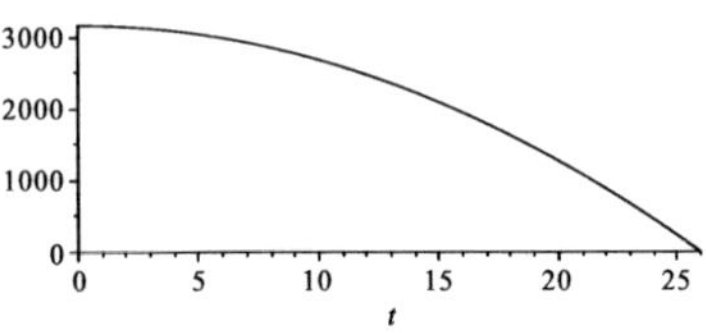

(e) With no air resistance, $m\frac{dv}{dt} = -mg$ so $v(t) = -gt$ and $y(t) = -\frac{1}{2}gt^2 + H$. The time of fall would be $t = \sqrt{\frac{2H}{g}} = 25.43$ seconds, and the velocity at impact would be $v(25.43) = -249.34$ m/s.

67. The projectile rises for $T_u = \frac{mg}{k}\ln\left(1 + \frac{k}{mg}v_0\right) \approx 1.59$ seconds to reach a height of $H = y(T_u) \approx 14.65$ meters. Solve the equation $y(t) = 0$ to find that the projectile hits the ground in $T_u + T_d = 3.48$ seconds, implying that $T_d \approx 1.89$ seconds. The velocity at impact is $v(3.48) \approx -14.16$ m/s.

69. (a) The distance traveled at time t is $x(t) = v_\infty\left(t - \tau\left(1 - e^{-t/\tau}\right)\right)$.

For Johnson, $x(9.83) = 100$ so $100 = 11.8\left(9.83 - \tau\left(1 - e^{-9.83/\tau}\right)\right)$ implying that $\tau = 1.356$ and $P = \frac{v_\infty}{\tau} = 8.702$.

For Lewis, $x(9.93) = 100$ so $100 = 11.8\left(9.93 - \tau\left(1 - e^{-9.93/\tau}\right)\right)$ implying that $\tau = 1.457$ and $P = \frac{v_\infty}{\tau} = 8.099$.

Johnson's victory can be attributed to a greater propulsive force.

(b) The plots are shown on the right. The two velocity curves are almost identical but there is clearly more area under the solid curve (Johnson) than under the dotted curve (Lewis).

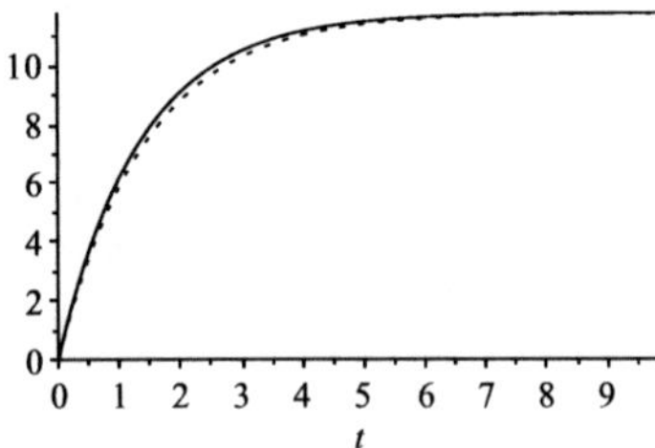

(c) In 9.83 seconds Lewis has traveled

$$x(9.83) = 11.8\left(9.83 - 1.457\left(1 - e^{-9.83/1.457}\right)\right) = 98.82 \text{ meters.}$$

Johnson's winning margin was 1.18 meters.

71. Let $v(0) = v_0$. Since $\frac{dv}{dt} = -kv^{0.05}$, $v^{-0.05}\,dv = -k\,dt$, and $\frac{1}{0.95}v^{0.95} = -kt+C$. This implies that that $C = \frac{v_0^{0.95}}{0.95}$ and $v(t) = (v_0^{0.95} - 0.95kt)^{1/0.95}$. Since $v(10) = 0$, $k = \frac{v_0^{0.95}}{9.5}$. Substitute this expression for k into the equation $\int_0^{10} v(u)\,du = 93$ and solve for v_0 (*Maple*) to obtain $v_0 = \frac{3627}{190} \approx 19.1$ ft/sec.

REVIEW EXERCISES FOR CHAPTER 7

1. Use circular disks, each one of volume $\pi(\sqrt{x-1}\,)^2\,\Delta x$. The total volume is $V = \int_1^2 \pi(x-1)\,dx = \pi \cdot \frac{1}{2}(x-1)^2\Big|_1^2 = \frac{\pi}{2}$.

3. Use circular disks, each one of volume $\pi(1+x^2)^2\,\Delta x$. The total volume is $V = \int_0^1 \pi(1+x^2)^2\,dx = \pi \cdot \left(x + \frac{2}{3}x^3 + \frac{1}{5}x^5\right)\Big|_0^1 = \frac{28\pi}{15}$.

5. Use cylindrical shells, base radius $1-x$ and height $1+x^2$, with thickness Δx. Each one has approximate volume $2\pi(1-x)(1+x^2)\,\Delta x$. The total volume is $V = \int_0^1 2\pi(1-x)(1+x^2)\,dx = 2\pi\left(x - \frac{1}{2}x^2 + \frac{1}{3}x^3 - \frac{1}{4}x^4\right)\Big|_0^1 = \frac{7\pi}{6}$.

7. Use circular disks, each one of volume $\pi\left(\frac{5x}{2+x^3}\right)^2\Delta x$. The total volume is $V = \int_0^2 \pi\left(\frac{5x}{2+x^3}\right)^2 dx = \frac{25\pi}{3}\int_0^2 \frac{3x^2}{(2+x^3)^2}\,dx = -\frac{25\pi}{3} \cdot \frac{1}{2+x^3}\Big|_0^2 = \frac{10\pi}{3}$.

9. The upper boundary curve is the graph of $y = -(x-1)(x-3)$ which is nonnegative for $1 \le x \le 3$. Use cylindrical shells, each one of approximate volume $2\pi x(4x - x^2 - 3)\,\Delta x$. The volume is $V = \int_1^3 2\pi x(4x - x^2 - 3)\,dx = 2\pi\left(\frac{4}{3}x^3 - \frac{1}{4}x^4 - \frac{3}{2}x^2\right)\Big|_1^3 = \frac{16\pi}{3}$.

11. The planar region lies above the interval $1 \le x \le 2$. Use circular washers of outer radius 1 and inner radius $\frac{1}{x}$. Each one has volume $\pi\left(1^2 - \frac{1}{x^2}\right)\Delta x$. The volume of the solid is $V = \int_1^2 \pi\left(1 - \frac{1}{x^2}\right)dx = \pi\left(x + \frac{1}{x}\right)\Big|_1^2 = \frac{\pi}{2}$.

13. The planar region lies above the interval $1 \le x \le 2$. Use cylindrical shells, each one with base radius $2-x$, height $1 - \frac{1}{x}$, and thickness Δx, for an approximate volume of $2\pi(2-x)\left(1 - \frac{1}{x}\right)\Delta x$. The volume of the entire solid is $V = \int_1^2 2\pi(2-x)\left(1 - \frac{1}{x}\right)dx = \pi(3 - 4\ln(2))$.

15. The planar region lies above the interval $-2 \le x \le 1$. Use circular washers of outer radius $9 - x^2$ and inner radius $x + 7$. Each one has volume $\pi\left((9-x^2)^2 - (x+7)^2\right)\Delta x$. The volume of the solid of revolution is $V = \int_{-2}^1 \pi\left((9-x^2)^2 - (x+7)^2\right)dx = \frac{333\pi}{5}$.

17. Use circular disks. The radius is $x\sqrt{\ln(x)}$ so they have volume $\pi x^2\ln(x)\,\Delta x$. The volume of the solid region is $V = \int_1^e \pi x^2 \ln(x)\,dx$. Using Integral Formula 231, $V = \frac{\pi}{9}\left(1 + 2e^3\right)$.

19. Use circular washers. The outer radius is $1 + \sin(x)$, the inner radius is $\cos(x)$, and they have volume $\pi((1+\sin(x))^2 - \cos^2(x))\,\Delta x$. The volume

of the solid region is $V = \int_0^{\pi/2} \pi((1+\sin(x))^2 - \cos^2(x))\,dx = \frac{1}{2}\pi(\pi+4)$.

21. Since $y' = 2(4x+1)^{1/2}$, the arc length is

$$L = \int_0^{1/4} \sqrt{1+4(4x+1)}\,dx = \int_0^{1/4} \sqrt{5+16x}\,dx$$
$$= \frac{1}{24}(5+16x)^{3/2}\Big|_0^{1/4} = \frac{27-5\sqrt{5}}{24}.$$

23. Since $y' = \frac{1}{2x^{1/2}}\left(\frac{2}{3}x^{1/2} + \frac{1}{36}\right)^{1/2}$, the arc length is

$$L = \int_{1/4}^{1} \sqrt{1 + \frac{1}{4x}\left(\frac{2}{3}x^{1/2} + \frac{1}{36}\right)}\,dx = \int_{1/4}^{1} \sqrt{\frac{1}{144x} + \frac{1}{6x^{1/2}} + 1\,dx}$$
$$= \int_{1/4}^{1} \sqrt{\left(\frac{1}{12x^{1/2}} + 1\right)^2}\,dx = \int_{1/4}^{1} \frac{1}{12}x^{-1/2} + 1\,dx =$$
$$= \left(\frac{1}{6}x^{1/2} + x\right)\Big|_{1/4}^{1} = \frac{5}{6}.$$

25. Since $y' = x^{3/2} - \frac{1}{4}x^{-3/2}$, the arc length is

$$L = \int_{1/4}^{1} \sqrt{1 + \left(x^{3/2} - \frac{1}{4}x^{-3/2}\right)^2}\,dx = \int_{1/4}^{1} \sqrt{x^3 + \frac{1}{2} + \frac{1}{16}x^{-3}}\,dx$$
$$= \int_{1/4}^{1} \sqrt{\left(x^{3/2} + \frac{1}{4}x^{-3/2}\right)^2}\,dx = \int_{1/4}^{1} x^{3/2} + \frac{1}{4}x^{-3/2}\,dx$$
$$= \left(\frac{2}{5}x^{5/2} - \frac{1}{2}x^{-1/2}\right)\Big|_{1/4}^{1} = \frac{71}{80}.$$

27. Since $f'(x) = -\sin(x)$, and using symmetry, the surface area is

$$S = 2\int_0^{\pi/2} 2\pi\cos(x)\sqrt{1+\sin^2(x)}\,dx \overset{u=\sin(x)}{=} 4\pi \int_0^1 \sqrt{1+u^2}\,du$$
$$\overset{\text{Int Table, 89}}{=} 2\pi\left(u\sqrt{1+u^2} + \ln\left(u + \sqrt{1+u^2}\right)\right)\Big|_0^1$$
$$= 2\pi\left(\sqrt{2} + \ln\left(1+\sqrt{2}\right)\right)$$

29. Since $f'(x) = 2/\sqrt{x}$, the surface area is

$$S = \int_0^5 2\pi \cdot 4\sqrt{x}\sqrt{1+\frac{4}{x}}\,dx = 8\pi\int_0^5 \sqrt{x}\sqrt{\frac{x+4}{x}}\,dx$$
$$= 8\pi\int_0^5 \sqrt{x+4}\,dx = \frac{16}{3}\pi(x+4)^{3/2}\Big|_0^5 = \frac{304\pi}{3}.$$

31. Since $\frac{dx}{dt} = 2t$ and $\frac{dy}{dt} = -6t$, the length of the curve is

$$\begin{aligned} L &= \int_0^{\sqrt{2}} \sqrt{4t^2 + 36t^2}\, dt = \int_0^{\sqrt{2}} \sqrt{40t^2}\, dt \\ &= 2\sqrt{10} \int_0^{\sqrt{2}} t\, dt = 2\sqrt{10} \cdot \frac{1}{2} t^2 \Big|_0^{\sqrt{2}} = 2\sqrt{10}\,. \end{aligned}$$

33. Since $\frac{dx}{dt} = 3t^2$ and $\frac{dy}{dt} = 6t$, the length of the curve is

$$\begin{aligned} L &= \int_0^1 \sqrt{9t^4 + 36t^2}\, dt = \int_0^1 3t\sqrt{t^2 + 4}\, dt \\ &= (t^2 + 4)^{3/2} \Big|_0^1 = 5\sqrt{5} - 8\,. \end{aligned}$$

35. $f_{\text{avg}} = \frac{1}{2} \int_1^3 (4x - x^2 - 3)\, dx = \frac{1}{2} \left(2x^2 - \frac{1}{3}x^3 - 3x \right) \Big|_1^3 = \frac{2}{3}$

37. $f_{\text{avg}} = \frac{1}{6} \int_1^7 \frac{3}{1 + x}\, dx = \frac{1}{2} \ln(1 + x) \Big|_1^7 = \ln(2)$

39. $f_{\text{avg}} = \frac{1}{2} \int_0^2 \frac{x}{(1 + 2x^2)^2}\, dx = -\frac{1}{8} \cdot \frac{1}{1 + 2x^2} \Big|_0^2 = \frac{1}{9}$

41. $f_{\text{avg}} = \frac{2}{3} \int_{1/2}^2 \ln(x)\, dx \overset{\text{Int Tab 228}}{=} \frac{2}{3} \cdot (x \ln(x) - x) \Big|_{1/2}^2 = \frac{5}{3} \ln(2) - 1$

43. $\mu_X = \int_0^{\pi/2} x \sin(x)\, dx \overset{\text{Int Tab 125}}{=} (\sin(x) - x \cos(x)) \Big|_0^{\pi/2} = 1$

45. $\mu_X = \int_1^3 x \cdot \frac{3}{4}(4x - x^2 - 3)\, dx = \frac{3}{4} \left(\frac{4}{3}x^3 - \frac{1}{4}x^4 - \frac{3}{2}x^2 \right) \Big|_1^3 - 2$

47. $\mu_X = \int_0^{1/\sqrt{2}} x \cdot \frac{4}{\pi}(1 - x^2)^{-1/2}\, dx = -\frac{4}{\pi}(1 - x^2)^{1/2} \Big|_0^{1/\sqrt{2}} = \frac{2(2 - \sqrt{2}\,)}{\pi}$

49. $\mu_X = \int_1^e x \cdot \ln(x)\, dx \overset{\text{Int Tab 230}}{=} \left(\frac{1}{2}x^2 \ln(x) - \frac{1}{4}x^2 \right) \Big|_1^e = \frac{1}{4}(e^2 + 1)$

51. The mass is $M = \int_1^2 \frac{1}{1+x}\, dx = \ln(1 + x)\big|_1^2 = \ln(3/2)$. Since $\mathcal{M}_{x=0} = \int_1^2 \frac{x}{1+x}\, dx = \int_1^2 \left(1 - \frac{1}{1+x}\right)\, dx = (x - \ln(1 + x))\big|_1^2 = 1 - \ln(3/2)$ and $\mathcal{M}_{y=0} = \int_1^2 \frac{1}{2} \cdot \frac{1}{(1+x)^2}\, dx = -\frac{1}{2} \cdot \frac{1}{1+x}\big|_1^2 = \frac{1}{12}$, the center of mass is $(\overline{x}, \overline{y}) = \left(\frac{1 - \ln(3/2)}{\ln(3/2)}, \frac{1}{12 \ln(3/2)} \right)$.

53. The mass is $M = \int_0^{\pi/2} 2\sin(x)\,dx = -2\cos(x)\Big|_0^{\pi/2} = 2$. Since $\mathcal{M}_{x=0} = \int_0^{\pi/2} x \cdot 2\sin(x)\,dx \stackrel{\text{Int Tab 125}}{=} 2(\sin(x) - x\cos(x))\Big|_0^{\pi/2} = 2$ and $\mathcal{M}_{y=0} = \int_0^{\pi/2} \frac{1}{2} \cdot 4\sin^2(x)\,dx = \frac{\pi}{2}$, the center of mass is $(\overline{x}, \overline{y}) = \left(1, \frac{\pi}{4}\right)$.

55. The region lies above and below the interval $-4 \le x \le 1$. The mass is $M = \int_{-4}^{1} (4 - x^2 - 3x)\,dx = \frac{125}{6}$. Since $\mathcal{M}_{x=0} = \int_{-4}^{1} x \cdot (4 - x^2 - 3x)\,dx = -\frac{125}{4}$ and $\mathcal{M}_{y=0} = \int_{-4}^{1} \frac{1}{2} \cdot ((4 - x^2)^2 - 9x^2)\,dx = -\frac{125}{3}$, the center of mass is $(\overline{x}, \overline{y}) = \left(-\frac{3}{2}, -2\right)$.

57. The work done is $W = \int_0^{160} 15697440 \cdot 1600 \cdot (3962 + y)^{-2}\,dy = 246\,063.0762$ mile-lbs. Multiply by 5280 to obtain $W = 1\,299\,213\,042$ ft-lbs of work.

59. Adapting the integral from Exercise 58, we solve the equation

$$\int_{6375580}^{D} 3.98621 \cdot 10^{14} \cdot 100 \cdot r^{-2}\,dr = 3 \cdot 10^8$$

for D. This yields $D = 6\,696\,913$ meters from the center of the Earth or 321.333 km above the surface of the Earth.

61. The weight of the load is $200 \cdot 9.80665$ newtons. If it is y meters above the ground, then the force required to lift it is $F(y) = 200 \cdot 9.80665 + 160 \cdot (20 - y)$ newtons. The work required to lift the load and cable from $y = 0$ to $y = 20$ is

$$W = \int_0^{20} 200 \cdot 9.80665 + 160 \cdot (20 - y)\,dy = 71\,226.6 \text{ N-m}.$$

63. The force required to hold the spring x feet from equilibrium is $F(x) = 120x$ pounds. The work needed to pull the spring from $x = 0$ to $x = \frac{1}{3}$ ft is

$$W = \int_0^{1/3} 120x\,dx = \frac{20}{3} \text{ ft-lb}.$$

65. The force required to hold the spring x meters from equilibrium is $F(x) = \frac{360}{12} \cdot 100x = 3000x$ newtons. The work needed to pull the spring from $x = 0$ to $x = 0.12$ meters is

$$W = \int_0^{0.12} 3000x\,dx = 21.6 \text{ N-m}.$$

67. The force required to compress the spring x meters from its natural length is $F(x) = \frac{864}{0.12} \cdot x = 7200x$ newtons. The work needed to compress the spring from $x = 0.12$ to $x = 0.18$ meters is

$$W = \int_{0.12}^{0.18} 7200x\,dx = 64.8 \text{ N-m}.$$

69. The spring constant k newtons/meter satisfies the equation $\int_0^{0.05} kx\,dx = 2$. That is, $0.00125k = 2$. Therefore, $k = 1600$, and the force required maintain the spring 5 cm beyond equilibrium is $F(0.05) = 0.05k = 80$ newtons.

71. A thin layer of water of thickness Δx that is x meters from the bottom of the tank weighs $W = 9806.65 \cdot \pi \cdot 0.3^2 \cdot \Delta x$ newtons. It will take approximately $W \cdot (2-x)$ newton-meters of work to lift it to the top of the tank. Therefore, the work required to pump the top half of the water out of the tank is $\int_1^2 9806.65 \cdot \pi \cdot 0.3^2 \cdot (2-x)\,dx = 1386.4$ N-m.

73. This equation is separable: $\frac{dy}{y^2} = (x+1)\,dx$. Integrate to obtain the relation $-\frac{1}{y} = \frac{1}{2}(x+1)^2 + C$ implying that the explicit solution has the form $y = \frac{2}{C-(x+1)^2}$ or $y = \frac{2}{C-2x-x^2}$.

75. This equation is first-order linear: $y' - y = 2x$, with integrating factor $e^{\int(-1)\,dx} = e^{-x}$. Multiply both sides of the equation by this factor and rearrange to $(e^{-x}y)' = 2xe^{-x}$, implying that $e^{-x}y = \int 2xe^{-x}\,dx + C$. Using formula 215 in the Table of Integrals, $e^{-x}y = -2(x+1)e^{-x} + C$, and $y = -2x - 2 + Ce^{x}$.

77. This equation is separable: $e^y\,dy = 2e^{2x}\,dx$. Integrate to obtain the relation $e^y = e^{2x} + C$ implying that the explicit solution has the form $y = \ln(e^{2x} + C)$.

79. This equation is separable: $\frac{dy}{y^2} = \frac{dx}{1+x^2}$. Integrate to obtain the relation $-\frac{1}{y} = \arctan(x) + C$ implying that the explicit solution has the form $y = \frac{1}{C-\arctan(x)}$.

81. This equation is first-order linear: $y' - 2y = x$, with integrating factor $e^{\int(-2)\,dx} = e^{-2x}$. Multiply both sides of the equation by this factor and rearrange to $\left(e^{-2x}y\right)' = xe^{-2x}$, implying that $e^{-2x}y = \int xe^{-2x}\,dx + C$. Using the Table of Integrals, formula 215, $e^{-2x}y = -\frac{1}{2}\left(x+\frac{1}{2}\right)e^{-2x} + C$, and the general solution is $y(x) = -\frac{1}{2}x - \frac{1}{4} + Ce^{2x}$. Since $y(0) = 1/2$, the constant C must be chosen so that $\frac{1}{2} = -\frac{1}{4} + C$. That is, $C = \frac{3}{4}$, and the solution to the initial value problem is $y(x) = -\frac{1}{2}x - \frac{1}{4} + \frac{3}{4}e^{2x}$.

83. This equation is first-order linear: $y' - 2y = 3$, with integrating factor $e^{\int(-2)\,dx} = e^{-2x}$. Multiply both sides of the equation by this factor and rearrange to $\left(e^{-2x}y\right)' = 3e^{-2x}$, implying that $e^{-2x}y = \int 3e^{-2x}\,dx + C = -\frac{3}{2}e^{-2x} + C$. The general solution is $y(x) = -\frac{3}{2} + Ce^{2x}$. Since $y(0) = -1$, the constant C must be chosen so that $-1 = -\frac{3}{2} + C$. That is, $C = \frac{1}{2}$, and the solution to the initial value problem is $y(x) = -\frac{3}{2} + \frac{1}{2}e^{2x}$.

85. This equation is first-order linear: $y' + y = 2e^{x}$, with integrating factor $e^{\int dx} = e^{x}$. Multiply both sides of the equation by this factor and rearrange to $(e^{x}y)' = 2e^{2x}$, implying that $e^{x}y = \int 2e^{2x}\,dx + C = e^{2x} + C$. The general solution is $y(x) = e^{x} + Ce^{-x}$. Since $y(0) = 5$, the constant C

must be chosen so that $5 = 1 + C$. That is, $C = 4$, and the solution to the initial value problem is $y(x) = e^x + 4e^{-x}$.

87. Salt enters the tank at the rate of $\frac{1}{10} \cdot 32 = 3.2$ kg/min. It exits the tank at the rate of $\frac{m(t)}{400} \cdot 32 = 0.8m(t)$ kg/min. Consequently,

$$m'(t) = 3.2 - 0.8m(t) \quad \text{and} \quad m(0) = 25\,.$$

This is a first order linear differential equation: $m'(t) + 0.8m(t) = 3.2$, with integrating factor $e^{\int 0.8\,dt} = e^{0.8t}$. Multiply the ode by this factor and simplify to $\left(e^{0.8t}m(t)\right)' = 3.2e^{0.8t}$ implying that $e^{0.8t}m(t) = \int 3.2e^{0.8t}\,dt + C = 4e^{0.8t} + C$. Consequently, $m(t) = 4 + Ce^{-0.8t}$, where the constant C must satisfy the equation $25 = 4 + C$. Consequently, $C = 21$, and $m(t) = 4 + 21e^{-0.8t}$. Clearly, $\lim_{t\to\infty} m(t) = 4$ kg.

89. There is no salt entering the tank. Salt exits the tank at the rate of $\frac{m(t)}{200} \cdot 5 = \frac{1}{40}m(t)$ lb/min. Consequently,

$$m'(t) = -\frac{1}{40}m(t) \quad \text{and} \quad m(0) = 100\,.$$

By inspection, $m(t) = 100e^{-t/40}$. Clearly, $\lim_{t\to\infty} m(t) = 0$ lb.

Chapter 8

Infinite Series

8.1 Series

Problems for Practice

1. Using (8.1.10), with $p_1 = q_1 = 3$, $\lim_{n\to\infty} a_n = \frac{1}{5}$.

3. Divide the numerator and the denominator by 2^n.

$$\lim_{n\to\infty} a_n = \lim_{n\to\infty} \frac{2 + 5/2^n}{1 + 3/2^n} = 2$$

5. Divide the numerator and the denominator by 5^n.

$$\lim_{n\to\infty} a_n = \lim_{n\to\infty} \frac{(3/5)^n + 2}{(2/5)^n + 3} = \frac{2}{3}$$

7. Divide the numerator and the denominator by n.

$$\lim_{n\to\infty} a_n = \lim_{n\to\infty} \frac{6 + \cos(n)/n}{3 + 2/n - \sin(n^2)/n} = 2$$

9. Divide the numerator and the denominator by 3^n and use the fact that $\lim_{x\to\infty} (x \cdot (2/3)^x) = 0$ by L'Hópital's Rule.

$$\lim_{n\to\infty} a_n = \lim_{n\to\infty} \frac{n(2/3)^n + 1 + 1/3^n}{n(2/3)^n + 1 + 1/2^n} = 1$$

11. By L'Hópital's Rule, applied three times, $\lim_{x\to\infty} \dfrac{\ln^2(x)}{\sqrt{x}} = 0$. Therefore, $\lim_{n\to\infty} \dfrac{\ln^2(n)}{\sqrt{n}} = 0$.

13. By L'Hópital's Rule, applied twice, $\lim_{x\to\infty} \dfrac{x^2}{e^x} = 0$. Therefore, $\lim_{n\to\infty} \dfrac{n^2}{e^n} = 0$.

15. Observe that

$$\sqrt{n^2+3n} - n = \frac{\sqrt{n^2+3n}-n}{1} \cdot \frac{\sqrt{n^2+3n}+n}{\sqrt{n^2+3n}+n} = \frac{3n}{\sqrt{n^2+3n}+n}.$$

Therefore, upon dividing the numerator and the denominator by n,

$$\lim_{n\to\infty} a_n = \lim_{n\to\infty} \frac{3}{\sqrt{1+3/n}+1} = \frac{3}{2}.$$

17. Since $\lim_{x\to\infty} \arctan(x) = \pi/2$, $\lim_{n\to\infty} \arctan(n) = \pi/2$.

19. Observe that $\left(\frac{x+1}{x}\right)^x = \exp\left(x \cdot \ln(1+1/x)\right)$ and, as $x \to \infty$, the expression $x \cdot \ln(1+1/x)$ is indeterminate of type $\infty \cdot 0$. L'Hópital's Rule (applied just once) will verify that $\lim_{x\to\infty} \frac{\ln(1+1/x)}{x^{-1}} = 1$, implying that $\lim_{n\to\infty} \left(\frac{n+1}{n}\right)^n = e$.

21. $S_N = \sum_{n=1}^{N} \frac{1}{n}$

N	1	2	3	4	5
S_N	1	$\frac{3}{2}$	$\frac{11}{6}$	$\frac{25}{12}$	$\frac{137}{60}$

23. $S_N = \sum_{n=1}^{N} \left(\frac{1}{2^n} + 1\right)$

N	1	2	3	4	5
S_N	$\frac{3}{2}$	$\frac{11}{4}$	$\frac{31}{8}$	$\frac{79}{16}$	$\frac{191}{32}$

25. $S_N = \sum_{n=1}^{N} \frac{2^n}{3^n}$

N	1	2	3	4	5
S_N	$\frac{2}{3}$	$\frac{10}{9}$	$\frac{38}{27}$	$\frac{130}{81}$	$\frac{422}{243}$

27. $S_N = \sum_{n=1}^{N} \frac{(-1)^{n+1}}{n^2}$

N	1	2	3	4	5
S_N	1	$\frac{3}{4}$	$\frac{31}{36}$	$\frac{115}{144}$	$\frac{3019}{3600}$

29. $\sum_{n=1}^{\infty} \left(\frac{3}{7}\right)^n = \frac{3/7}{1-3/7} = \frac{3}{4}$

31. $\sum_{n=0}^{\infty} \left(\frac{-2}{3}\right)^n = \frac{1}{1+2/3} = \frac{3}{5}$

33. $\sum_{n=0}^{\infty} 9\cdot(0.1)^n = 9\cdot\frac{1}{1-0.1} = 9\cdot\frac{10}{9} = 10$

35. $\sum_{n=-3}^{\infty}\left(\frac{1}{5}\right)^n = \frac{(1/5)^{-3}}{1-1/5} = \frac{5^3}{4/5} = \frac{625}{4}$

37. $\sum_{n=2}^{\infty}\frac{1}{2^{n/2}} = \sum_{n=2}^{\infty}\left(\frac{1}{\sqrt{2}}\right)^n = \frac{1/2}{1-1/\sqrt{2}} = \frac{1}{2}\cdot\frac{\sqrt{2}}{\sqrt{2}-1} = \frac{2+\sqrt{2}}{2}$

39. $\sum_{n=3}^{\infty}\frac{(0.1)^n}{(0.2)^{n+2}} = \sum_{n=3}^{\infty}\left(\frac{1}{(0.2)^2}\cdot\left(\frac{1}{2}\right)^n\right) = 25\cdot\frac{1/8}{1-1/2} = \frac{25}{4}$

41. $\sum_{n=1}^{\infty}(2^{-n}\cdot 3^{-n} + 7^{-n}) = \sum_{n=1}^{\infty}\left(\frac{1}{6}\right)^n + \sum_{n=1}^{\infty}\left(\frac{1}{7}\right)^n = \frac{1/6}{1-1/6} + \frac{1/7}{1-1/7}$
$= \frac{1}{5} + \frac{1}{6} = \frac{11}{30}$

43. $\sum_{n=1}^{\infty}\frac{2^{n+1}}{5^{n-1}} = \sum_{n=1}^{\infty}10\cdot\left(\frac{2}{5}\right)^n = \frac{4}{1-2/5} = \frac{20}{3}$

45. $8\cdot\left(0.1 + (0.1)^2 + (0.1)^3 + \cdots\right) = 8\cdot\frac{0.1}{1-0.1} = \frac{8}{9}$

47. $17\cdot\left((0.1)^3 + (0.1)^6 + (0.1)^9 + \cdots\right) = 17\cdot\frac{0.001}{1-0.001} = \frac{17}{999}$

49. $\frac{122}{1000} + 12\cdot\left((0.1)^5 + (0.1)^7 + (0.1)^9 + \cdots\right) = \frac{122}{1000} + \frac{12}{1000}\cdot\frac{(0.1)^2}{1-(0.1)^2}$
$= \frac{122}{1000} + \frac{12}{1000}\cdot\frac{1}{99} = \frac{403}{3300}$

51. $\sum_{n=1}^{\infty}a_n = \lim_{N\to\infty}S_N = \lim_{N\to\infty}\left(2-\frac{1}{N^2}\right) = 2$

53. $\sum_{n=1}^{\infty}a_n = \lim_{N\to\infty}S_N = \lim_{N\to\infty}\left(\frac{3N+2}{N+1}\right) = 3$

55. $\sum_{n=1}^{\infty}a_n = \lim_{N\to\infty}S_N = \lim_{N\to\infty}\left(\frac{3N+1}{2N+4}\right) = \frac{3}{2}$

Further Theory and Practice

57. Since $(1.01)^n \geq 1$ for all $n \geq 1$, $S_N = \sum_{n=1}^{N}(1.01)^n \geq \sum_{n=1}^{N}1 = N$.

59. Since $\frac{n}{2n+3} \geq \frac{1}{5}$ for all $n \geq 1$ (the function $x \mapsto \frac{x}{2x+3}$ is increasing on $[1,\infty)$), $S_N = \sum_{n=1}^{N}\frac{n}{2n+3} \geq \sum_{n=1}^{N}\frac{1}{5} = \frac{1}{5}\cdot N$.

61. Since $\frac{4^n}{4^n+2^n} \geq \frac{2}{3}$ for all $n \geq 1$, (the function $x \mapsto \frac{4^x}{4^x+2^x}$ is increasing on $[1,\infty)$), $S_N = \sum_{n=1}^{N}\frac{4^n}{4^n+2^n} \geq \sum_{n=1}^{N}\frac{2}{3} = \frac{2}{3}\cdot N$.

63. The Nth partial sum is

$$S_N = \left(1-\frac{1}{2}\right) + \left(\frac{1}{2}-\frac{1}{3}\right) + \cdots + \left(\frac{1}{N}-\frac{1}{N+1}\right) = 1 - \frac{1}{N+1} \xrightarrow[N\to\infty]{} 1$$

implying that the series sums to 1.

65. The Nth partial sum is

$$S_N = \left(0-\frac{2}{8}\right) + \left(\frac{2}{8}-\frac{4}{27}\right) + \cdots + \left(\frac{2(N-1)}{N^3}-\frac{2N}{(N+1)^3}\right) = 0 - \frac{2N}{(N+1)^3} \xrightarrow[N\to\infty]{} 0$$

implying that the series sums to 0.

67. The Nth partial sum is

$$S_N = \left(1 - \frac{1}{\sqrt{2}}\right) + \left(\frac{1}{\sqrt{2}} - \frac{1}{\sqrt{3}}\right) + \cdots + \left(\frac{1}{\sqrt{N}} - \frac{1}{\sqrt{N+1}}\right) = 1 - \frac{1}{\sqrt{N+1}} \xrightarrow[N\to\infty]{} 1$$

implying that the series sums to 1.

69. Observe that $\frac{1}{n(n+2)} = \frac{A}{n} + \frac{B}{n+2}$ so $1 = A(n+2) + Bn$. This implies that $A = \frac{1}{2}$ and $B = -\frac{1}{2}$. Therefore, the Nth partial sum is

$$\begin{aligned} S_N &= \frac{1}{2}\sum_{n=1}^{N}\left(\frac{1}{n} - \frac{1}{n+2}\right) \\ &= \frac{1}{2}\left(\left(1 - \frac{1}{3}\right) + \left(\frac{1}{2} - \frac{1}{4}\right) + \left(\frac{1}{3} - \frac{1}{5}\right) + \cdots + \left(\frac{1}{N} - \frac{1}{N+1}\right)\right) \\ &= \frac{1}{2}\left(1 + \frac{1}{2} - \frac{1}{N+1}\right) \xrightarrow[N\to\infty]{} \frac{3}{4} \end{aligned}$$

implying that the series sums to $\frac{3}{4}$.

71. Observe that $\frac{1}{(2n+1)(2n+3)} = \frac{A}{2n+1} + \frac{B}{2n+3}$ so $1 = A(2n+3) + B(2n+1)$. This implies that $A = \frac{1}{2}$ and $B = -\frac{1}{2}$. Therefore, the Nth partial sum is

$$\begin{aligned} S_N &= \frac{1}{2}\sum_{n=1}^{N}\left(\frac{1}{2n+1} - \frac{1}{2n+3}\right) \\ &= \frac{1}{2}\left(\left(\frac{1}{3} - \frac{1}{5}\right) + \left(\frac{1}{5} - \frac{1}{7}\right) + \left(\frac{1}{7} - \frac{1}{9}\right) + \cdots + \left(\frac{1}{2N+1} - \frac{1}{2N+3}\right)\right) \\ &= \frac{1}{2}\left(\frac{1}{3} - \frac{1}{2N+3}\right) \xrightarrow[N\to\infty]{} \frac{1}{6} \end{aligned}$$

implying that the series converges to $\frac{1}{6}$.

73. Since $\ln\left(\frac{n}{n+1}\right) = \ln(n) - \ln(n+1)$, the Nth partial sum is

$$\begin{aligned} S_N &= \sum_{n=1}^{N}(\ln(n) - \ln(n+1)) \\ &= (\ln(1) - \ln(2)) + (\ln(2) - \ln(3)) + (\ln(3) - \ln(4)) + \cdots + (\ln(N) - \ln(N+1)) \\ &= -\ln(N+1) \xrightarrow[N\to\infty]{} -\infty \end{aligned}$$

implying that the series diverges.

75. Observe that $\dfrac{1}{r^n - 1} - \dfrac{1}{r^{n+1} - 1} = \dfrac{r^{n+1} - r^n}{(r^n - 1)(r^{n+1} - 1)} = \dfrac{r^n(r-1)}{(r^n - 1)(r^{n+1} - 1)}$, implying that $\dfrac{r^n}{(r^n - 1)(r^{n+1} - 1)} = \dfrac{1}{r-1} \cdot \left(\dfrac{1}{r^n - 1} - \dfrac{1}{r^{n+1} - 1}\right)$. Con-

sequently, the Nth paratial sum of the series is

$$\begin{aligned} S_N &= \frac{1}{r-1}\sum_{n=1}^{N}\left(\frac{1}{r^n-1}-\frac{1}{r^{n+1}-1}\right) \\ &= \frac{1}{r-1}\left(\left(\frac{1}{r-1}-\frac{1}{r^2-1}\right)+\left(\frac{1}{r^2-1}-\frac{1}{r^3-1}\right)+\cdots+\left(\frac{1}{r^N-1}-\frac{1}{r^{N+1}-1}\right)\right) \\ &= \frac{1}{r-1}\left(\frac{1}{r-1}-\frac{1}{r^{N+1}-1}\right) \xrightarrow[N\to\infty]{} \frac{1}{(r-1)^2}\,. \end{aligned}$$

Therefore, the series converges to $\frac{1}{(r-1)^2}$.

77. The first series converges to $\pi^2/24$ because

$$\sum_{n=1}^{\infty}\frac{1}{(2n)^2}=\sum_{n=1}^{\infty}\left(\frac{1}{4}\cdot\frac{1}{n^2}\right)=\frac{1}{4}\sum_{n=1}^{\infty}\frac{1}{n^2}=\frac{1}{4}\cdot\frac{\pi^2}{6}=\frac{\pi^2}{24}\,.$$

Since

$$\begin{aligned} \frac{\pi^2}{6} &= 1+\frac{1}{2^2}+\frac{1}{3^2}+\frac{1}{4^2}+\frac{1}{5^2}+\frac{1}{6^2}+\ldots \\ &= \left(1+0+\frac{1}{3^2}+0+\frac{1}{5^2}+0+\cdots\right)+\left(0+\frac{1}{2^2}+0+\frac{1}{4^2}+0+\cdots\right) \\ &= \left(1+0+\frac{1}{3^2}+0+\frac{1}{5^2}+0+\cdots\right)+\frac{\pi^2}{24}, \end{aligned}$$

the second series converges to $\frac{\pi^2}{6}-\frac{\pi^2}{24}=\frac{\pi^2}{8}$.

79. Observe that

$$f(x,y)=\sum_{n=1}^{\infty}\left(x\cdot\frac{1}{n^4}+\pi^2y\cdot\frac{1}{n^2}\right)=x\sum_{n=1}^{\infty}\frac{1}{n^4}+\pi^2y\sum_{n=1}^{\infty}\frac{1}{n^2}=x\cdot\frac{\pi^4}{90}+y\cdot\frac{\pi^4}{6}\,.$$

Consequently, the equation $f(x,y)=0$ is equivalent to $\frac{x}{15}+y=0$, implying that the solution set is the line $y=-\frac{x}{15}$.

81. If $n\geq 3$, then $\ln(n)\geq\ln(e)=1$. Therefore, for such n, $\ln(1/n)=-\ln(n)\leq -1$ and $\sum_{n=-3}^{N}\ln(1/n)\leq\sum_{n=3}^{N}(-1)=-(N-2)$. Consequently,

$$S_N=\sum_{n=1}^{N}\ln(1/n)=-\ln(2)+\sum_{n=3}^{N}<-(N-2)\xrightarrow[N\to\infty]{}-\infty\,,$$

and the series diverges.

83. After falling 20 cm to the floor, the ball will bounce up $\frac{2}{3}\cdot 20$ cm and then fall $\frac{2}{3}\cdot 20$ cm traveling $2\cdot\frac{2}{3}\cdot 20$ cm in the first bounce. During the next

bounce the ball will travel $2 \cdot \left(\frac{2}{3}\right)^2 \cdot 20$ cm more, and so on. The total distance traveled by the ball is

$$D = 20 + 40 \cdot \frac{2}{3} + 40 \cdot \left(\frac{2}{3}\right)^2 + 40 \cdot \left(\frac{2}{3}\right)^3 + \cdots = 20 + 40 \sum_{n=1}^{\infty} \left(\frac{2}{3}\right)^n .$$

Therefore, $D = 20 + 40 \cdot \frac{2/3}{1-2/3} = 20 + 40 \cdot 2 = 100$ cm $= 1$ meter.

85. The tax cut will immediately inject $10^{11} \cdot 0.95$ dollars into the economy. Of this amount, $10^{11} \cdot (0.95)^2$ dollars will be spent again, then $10^{11} \cdot (0.95)^3$ more, and so on. The total amount of spending will be

$$T = 10^{11} \cdot 0.95 + 10^{11} \cdot (0.95)^2 + 10^{11} \cdot (0.95)^3 + \cdots = 10^{11} \sum_{n=1}^{\infty} (0.95)^n .$$

Therefore, $T = 10^{11} \cdot \frac{0.95}{1-0.95} = 1.9 \times 10^{12}$ dollars.

87. At time $t = 0$ there is M_0 pollutant. At $t = T$ there is $M_0 + M_0 e^{-kT}$. At $t = 2T$ there is $M_0 + (M_0 + M_0 e^{-kT})e^{-kT} = M_0(1 + e^{-kT} + e^{-2kT})$ and so on. The total amount P will continue to grow towards a limiting value of

$$P_\infty = M_0(1 + e^{-kT} + e^{-2kT} + \ldots) = M_0 \cdot \frac{1}{1 - e^{-kT}} .$$

Consequently, M_0 must be chosen so that $M_0 \cdot \frac{1}{1-e^{-kT}} \le Q$. That is, $M_0 \le Q(1 - e^{-kT})$.

89. Compute its derivative to see that the function $f(x) = (1 - 1/x)^x$ is increasing on $[2, \infty)$. Therefore, $(1 - 1/n)^n \ge 1/4$ for all $n \ge 2$ implying that $S_N = \sum_{n=1}^{N} (1-1/n)^n \ge \frac{1}{4} \cdot (N-1) \xrightarrow[N \to \infty]{} \infty$, and the series diverges.

91. Observe that

$$1 \cdot 3 \cdot 5 \cdot 7 \cdots (2N-3) \cdot (2N-1) \cdot 2 \cdot 4 \cdot 6 \cdot 8 \cdots (2N-2) \cdot 2N = (2N)! .$$

Consequently, using the identity in Exercise 90,

$$1 \cdot 3 \cdot 5 \cdot 7 \cdots (2N-3) \cdot (2N-1) = \frac{(2N)!}{N! 2^N} .$$

CALCULATOR/COMPUTER EXERCISES

93. Using *Maple*, S_8, S_9, and S_{10} round to 0.4587.

95. Using *Maple*, S_6, S_7, and S_8 round to 1.4581.

97. Using *Maple*, S_{15}, S_{16}, and S_{17} round to 1.1875.

8.2 The Divergence Test and the Integral Test

Problems for Practice

1. Since $\lim_{n\to\infty} ne^{-n} = 0$, we can make no conclusion.

3. Since $\lim_{n\to\infty} \frac{n^2}{n^2+1} = 1$, the series diverges.

5. Since $\lim_{n\to\infty} \frac{1}{1+1/n} = 1$, the series diverges.

7. Since $\lim_{n\to\infty} \frac{3^n}{4^n+1} = \lim_{n\to\infty} \frac{(3/4)^n}{1+(1/4)^n} = 0$, we can make no conclusion.

9. Since $\lim_{n\to\infty} \frac{3^n+5^n}{8^n} = \lim_{n\to\infty}((3/8)^n + (5/8)^n) = 0$, we can make no conclusion.

11. Since $\lim_{n\to\infty} \frac{2^n+1}{2^n+n^2} = \lim_{n\to\infty} \frac{1+1/2^n}{1+n^2/2^n} = 1$, the series diverges.

13. Since $\lim_{n\to\infty} \frac{2n}{3n^2+1} = \lim_{n\to\infty} \frac{2/n}{3+1/n^2} = 0$, we can make no conclusion.

15. Since $\lim_{n\to\infty} n^{1/n} = \lim_{n\to\infty} \exp\left(\frac{\ln(n)}{n}\right) = \exp\left(\lim_{n\to\infty}\left(\frac{1/n}{1}\right)\right) = \exp(0) = 1$, the series diverges.

17. Since $\lim_{n\to\infty}(\pi/2 - \arctan(n)) = 0$, we can make no conclusion.

19. Since $\lim_{n\to\infty} \sin(1/n) = 0$, we can make no conclusion.

21. "may", $\sum_{n=1}^{\infty} \frac{1}{2^n} = 1$

23. "may", $\sum_{n=1}^{\infty} \frac{n}{n+1}$

25. "may", $\sum_{n=1}^{\infty}(-1)^n$

27. The function $f(x) = e^{-x}$ is positive, continuous, and decreases on $[1,\infty)$. Since $\int_1^\infty e^{-x}\,dx = \lim_{N\to\infty}\left(-e^{-x}\Big|_1^N\right) = 1$, the series converges.

29. The function $f(x) = \frac{1}{x^2+4}$ is positive, continuous, and decreases on $[1,\infty)$. Since $\int_1^\infty \frac{1}{x^2+4}\,dx = \lim_{N\to\infty}\left(\frac{1}{2}\arctan\left(\frac{x}{2}\right)\Big|_1^N\right) = \frac{\pi}{4} - \frac{1}{2}\arctan\left(\frac{1}{2}\right)$, the series converges.

31. The function $f(x) = \frac{1}{x+3}$ is positive, continuous, and decreases on $[1,\infty)$. Since $\int_1^\infty \frac{1}{x+3}\,dx = \lim_{N\to\infty}\left(\ln(x+3)\Big|_1^N\right) = \infty$, the series converges.

33. The function $f(x) = \frac{2x^2}{x^3+4}$ is positive, continuous, and decreases on $[1,\infty)$. Since $\int_1^\infty \frac{2x^2}{x^3+4}\,dx = \lim_{N\to\infty}\left(\frac{2}{3}\ln(x^3+4)\Big|_1^N\right) = \infty$, the series diverges.

35. The function $f(x) = \frac{1}{(x+3)^{5/4}}$ is positive, continuous, and decreases on $[1,\infty)$. Since $\int_1^\infty \frac{1}{(x+3)^{5/4}}\,dx = \lim_{N\to\infty}\left(-\frac{4}{(x+3)^{1/4}}\Big|_1^N\right) = 2\sqrt{2}$, the series converges.

37. The function $f(x) = \frac{e^x}{(1+e^x)^2}$ is positive, continuous, and decreases on $[1, \infty)$. Since $\int_1^\infty \frac{e^x}{(1+e^x)^2}\,dx = \lim_{N\to\infty}\left(-\frac{1}{1+e^x}\Big|_1^N\right) = \frac{1}{1+e}$, the series converges.

39. The function $f(x) = \frac{3}{x^2+x} = \frac{3}{x} - \frac{3}{x+1}$ is positive, continuous, and decreases on $[1, \infty)$. Since $\int_1^\infty \left(\frac{3}{x} - \frac{3}{x+1}\right)dx = \lim_{N\to\infty}\left(3\ln\left(\frac{x}{x+1}\right)\Big|_1^N\right) = 3\ln(2)$, the series converges.

41. The function $f(x) = \frac{\ln(x)}{x}$ is positive, continuous, and decreases on $[3, \infty)$. Since $\int_3^\infty \frac{\ln(x)}{x}\,dx = \lim_{N\to\infty}\left(\frac{1}{2}\ln^2(x)\Big|_3^N\right) = \infty$, the series $\sum_{n=3}^\infty \frac{\ln(n)}{n}$ diverges implying that $\sum_{n=1}^\infty \frac{\ln(n)}{n}$ also diverges.

43. The function $f(x) = xe^{-2x}$ is positive, continuous, and decreases on $[1, \infty)$. Since $\int_1^\infty xe^{-2x}\,dx = \lim_{N\to\infty}\left(-\frac{1}{4}(1+2x)e^{-2x}\Big|_1^N\right) = \frac{3}{4}e^{-2}$, the series converges.

45. The function $f(x) = \frac{1}{x\sqrt{x^2-1}}$ is positive, continuous, and decreases on $[2, \infty)$. Since $\int_2^\infty \frac{1}{x\sqrt{x^2-1}}\,dx = \lim_{N\to\infty}\left(\operatorname{arcsec}(x)\Big|_2^N\right) = \frac{\pi}{2} - \frac{\pi}{3} = \frac{\pi}{6}$, the series converges.

47. This is a p-series, $p = 1/2 \le 1$, so it diverges.

49. This is $\sqrt{3}$ times a p-series, $p = \sqrt{2} > 1$, so the series converges.

51. This series converges because

$$\sum_{n=1}^\infty \frac{\sqrt{n}+5}{n^2} = \sum_{n=1}^\infty \left(\frac{1}{n^{3/2}} + \frac{5}{n^2}\right) = \sum_{n=1}^\infty \frac{1}{n^{3/2}} + 5\sum_{n=1}^\infty \frac{1}{n^2}.$$

Both series are convergent p-series, ($p = 3/2$ and $p = 2$).

53. This series converges because

$$\sum_{n=1}^\infty \frac{n^2+11}{n^4} = \sum_{n=1}^\infty \left(\frac{1}{n^2} + \frac{11}{n^4}\right) = \sum_{n=1}^\infty \frac{1}{n^2} + 11\sum_{n=1}^\infty \frac{1}{n^4}.$$

Both series are convergent p-series, ($p = 2$ and $p = 4$).

Further Theory and Practice

55. Since $\lim_{n\to\infty} n^{1/n} = 1$ (see the solution to Exercise 15), $\lim_{n\to\infty}\left(\frac{1}{n}\right)^{1/n} = \lim_{n\to\infty}\frac{1}{n^{1/n}} = 1$, and the series diverges.

57. Since $\lim_{n\to\infty}(1+1/n)^n = e$ (see Chapter 5 Section 5), the series diverges.

59. Since $\lim_{n\to\infty}\frac{\ln(n^3)}{n} = 0$, no conclusion may be drawn.

61. Since $\lim_{n\to\infty}\left((1+1/n)^n - n\right) = -\infty$ (see Exercise 55), the series diverges.

63. Since $\lim_{n\to\infty} \frac{\sin(1/n)}{1/n} = 1$, the series diverges.

65. Since $\lim_{n\to\infty} \frac{1}{\arctan(n)} = \frac{1}{\pi/2} = \frac{2}{\pi}$, the series diverges.

67. Since $\lim_{n\to\infty}(\operatorname{arcsec}(n) - \arctan(n)) = \frac{\pi}{2} - \frac{\pi}{2} = 0$, no conclusion may be drawn.

69. The function $f(x) = \frac{x^2}{2^x}$ is positive, continuous, and decreases on $[3, \infty)$. Using the Table of Integrals, formula 216,

$$\begin{aligned}
\int_3^\infty \frac{x^2}{2^x}\,dx &= \int_3^\infty \frac{x^2}{e^{x\ln(2)}}\,dx \\
&= \lim_{N\to\infty}\left(-\frac{1}{\ln^3(2)}\left(\ln^2(2)x^2 + 2\ln(2)x + 2\right)e^{-x\ln(2)}\Big|_3^N\right) \\
&= \frac{1}{8\ln^3(2)}\left(9\ln^2(2) + 6\ln(2) + 2\right),
\end{aligned}$$

so the series converges.

71. The function $f(x) = \frac{1}{1+\sqrt{x}}$ is positive, continuous, and decreases on $[1, \infty)$. Using the substitutions $u = 1 + \sqrt{x}$, $x = (u-1)^2$, $dx = 2(u-1)\,du$,

$$\begin{aligned}
\int_1^\infty \frac{1}{1+\sqrt{x}}\,dx &= \int_2^\infty \frac{1}{u}\cdot 2(u-1)\,du = 2\int_2^\infty \left(1 - \frac{1}{u}\right)du \\
&= \lim_{N\to\infty}\left(2(u - \ln(u))\Big|_1^N\right) \\
&= 2\lim_{N\to\infty}\left(u\left(1 - \frac{\ln(u)}{u}\right)\Big|_1^N\right) \\
&= \infty\,, \qquad\qquad \left(\lim_{u\to\infty}\frac{\ln(u)}{u} = 0\right)
\end{aligned}$$

so the series diverges.

73. The function $f(x) = \ln\left(\frac{x^2+1}{x^2}\right)$ is positive, continuous, and decreases on $[1, \infty)$. Using integration by parts, $u = \ln(1+1/x^2)$, $dv = dx$, $du = \frac{-2x^{-3}}{1+x^{-2}}$,

$v = x$, the integral evaluates as follows.

$$\begin{aligned}\int_1^{\infty} \ln\left(1+x^{-2}\right) dx &= \lim_{N\to\infty} \left(x\ln(1+x^{-2})\Big|_1^N + 2\int_1^N \frac{x^{-2}}{1+x^{-2}}\, dx \right) \\ &= \lim_{N\to\infty} \left(x\ln(1+x^{-2})\Big|_1^N + 2\int_1^N \frac{1}{x^2+1}\, dx \right) \\ &= \lim_{N\to\infty} \left(\left(x\ln(1+x^{-2}) + 2\arctan(x)\right)\Big|_1^N \right) \\ &= \pi - \left(\ln(2) + \frac{\pi}{2}\right) \qquad \left(\lim_{x\to\infty} \frac{\ln(1+x^{-2})}{x^{-1}} \overset{\text{(L'Hópital)}}{=} 0\right) \\ &= \frac{\pi}{2} - \ln(2)\,,\end{aligned}$$

so the series converges.

75. The function $f(x) = \frac{x}{(3+x^2)^{5/2}}$ is positive, continuous, and decreases on $[1,\infty)$. Since $\int_1^{\infty} \frac{x}{(3+x^2)^{5/2}}\, dx = \lim_{N\to\infty} \left(-\frac{1}{3}\cdot \frac{1}{(3+x^2)^{3/2}}\Big|_1^N\right) = \frac{1}{24}$, the series converges. Its sum lies in the open interval $\left(\frac{1}{24}, f(1)+\frac{1}{24}\right) = \left(\frac{1}{24}, \frac{7}{96}\right)$.

77. The function $f(x) = \frac{e^x}{(1+e^x)^3}$ is positive, continuous, and decreases on $[1,\infty)$. Since $\int_1^{\infty} \frac{e^x}{(1+e^x)^3}\, dx = \lim_{N\to\infty} \left(-\frac{1}{2}\cdot \frac{1}{(1+e^x)^2}\Big|_1^N\right) = \frac{1}{2(1+e)^2}$, the series converges. Its sum lies in the open interval $\left(\frac{1}{2(1+e)^2}, f(1)+\frac{1}{2(1+e)^2}\right) = \left(\frac{1}{2(1+e)^2}, \frac{1+3e}{2(1+e)^3}\right)$.

79. The function $f(x) = (1/2)^{x^2}$ is positive, continuous, and decreases on $[1,\infty)$. Since

$$\int_1^{\infty} (1/2)^{x^2}\, dx = \int_1^{\infty} 2^{-x^2}\, dx < \int_0^{\infty} e^{-x^2}\, dx \overset{(6.7.2)}{=} \frac{\sqrt{\pi}}{2}\,,$$

the series converges.

81. Since $n! > 2^{n-1}$ when $n \geq 3$,

$$\sum_{n=1}^{\infty} \frac{1}{n!} < 1 + \frac{1}{2} + \sum_{n=3}^{\infty} \left(\frac{1}{2}\right)^{n-1} = \frac{3}{2} + \frac{1/4}{1-1/2} = 2\,.$$

83. Since $\int_n^{2n} e^{-x}\, dx = e^{-n} - e^{-2n}$ and $f(x) = e^{-x} - e^{-2x}$ is positive, continuous, and decreases on $[1,\infty)$, the infinite series converges if, and only if, the improper integral $\int_1^{\infty} (e^{-x} - e^{-2x})\, dx$ is finite, which it is (easy calculation).

85. Since $\int_n^{n+1} \sin^2(\pi x)\,dx = \frac{1}{2}$, given an integer $N \geq 2$,

$$\int_1^N f(x)\,dx > \sum_{n=1}^{N-1} \int_n^{n+1} \sin^2(\pi x)\,dx = \sum_{n=1}^{N-1} \frac{1}{2} = \frac{1}{2} \cdot (N-1) \xrightarrow[N\to\infty]{} \infty\,,$$

and $\int_1^\infty f(x)\,dx$ diverges. On the other hand, since $f(n) = 1/n^2$ for $n = 1, 2, 3, \dots$, $\sum_{n=1}^\infty f(n)$ is a p-series, $p = 2$, so it converges.

The conclusion of the Integral Test is not necessarily valid because its hypotheses are not satisfied. The function $f(x)$ is positive and continuous, but it is not decreasing on $[1, \infty)$. For example, if n is a positive integer, then $n < n + 1/2$, but $f(n) = \frac{1}{n^2} < 1 + \frac{1}{(n+1/2)^2} = f(n + 1/2)$.

87. Since $0 < a_n \leq \left(\frac{1}{2}\right)^n$ for $n = 1, 2, 3, \dots$,

$$\sum_{n=1}^\infty a_n \leq \sum_{n=1}^\infty \left(\frac{1}{2}\right)^n = \frac{1/2}{1 - 1/2} = 1\,,$$

and $\sum_{n=1}^\infty a_n$ converges.

89. Let $d_n = \frac{1}{2} \cdot c_n$. Then $\sum_{n=1}^\infty d_n$ must diverge, for otherwise, $\sum_{n=1}^\infty c_n = 2\sum_{n=1}^\infty d_n$ would converge.

91. The sketch below illustrates the situation for $N = 5$. Note the 4 "triangular" regions and a rectangular region of area $\frac{1}{5}$.

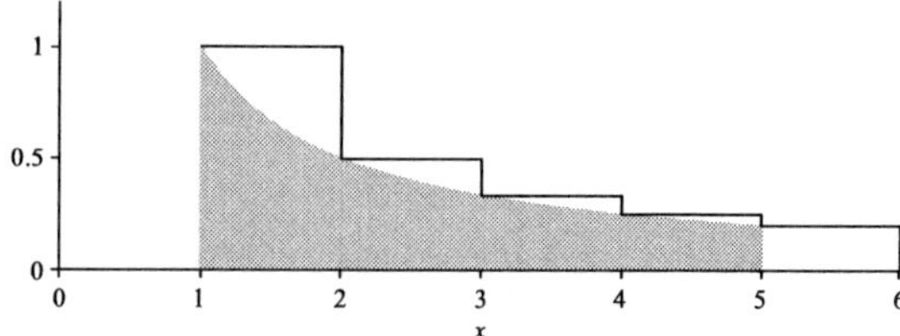

The shaded region has area $\int_1^5 \frac{1}{x}\,dx = \ln(5)$ and the area enclosed in the rectangles is $\sum_{n=1}^5 \frac{1}{n} = H_5$. Thus $A_5 = H_5 - \ln(5)$ can also be expressed as $A_5 = \sum_{n=1}^4 a_n + \frac{1}{5}$, where a_n denotes the area of the nth "triangle".

Since the base of the nth "triangle" is 1 and its "height" is $\frac{1}{n} - \frac{1}{n+1}$ it is enclosed in a rectangle with the same base and height. Consequently, $0 < a_n < \left(\frac{1}{n} - \frac{1}{n+1}\right)$.

93. The derivative is $f'(x) = 1 - \frac{1}{1-x} = \frac{x}{x-1}$, which is negative when $0 < x < 1$. Since its derivative is negative, f is strictly decreasing on $[0, 1)$. Because $f(0) = 0$, it must be the case that $f(x) < 0$ for all $0 < x \leq 1$.

Using the formula for A_N in Exercise 91,

$$A_{N+1} - A_N = \sum_{n=1}^N a_n + \frac{1}{N+1} - \left(\sum_{n=1}^{N-1} a_n + \frac{1}{N}\right) = \frac{1}{N+1} - \left(\frac{1}{N} - a_N\right),$$

and, since $1 - \frac{1}{N+1} = \frac{N}{N+1}$,

$$f\left(\frac{1}{N+1}\right) = \frac{1}{N+1} + \ln\left(\frac{N}{N+1}\right) = \frac{1}{N+1} - \left(\ln(N+1) - \ln(N)\right).$$

These two expressions evaluate to the same number because $\frac{1}{N} - a_N$ is the area under the graph of $\frac{1}{x}$ and above the interval $[N, N+1]$ (draw a picture), and so is $\ln(N+1) - \ln(N)$ $\left(= \int_N^{N+1} \frac{1}{x}\, dx\right)$.

Thus $A_{N+1} - A_N < 0$ for all N implying that $A_N > A_{N+1} > 0$.

CALCULATOR/COMPUTER EXERCISES

95. Using *Maple*, when $M = 13$, $U(M) - L(M) = 0.000455166$. The average of $L(13)$ and $U(13)$ is 1.202.

97. Using *Maple*, when $M = 7$, $U(M) - L(M) = 0.00035899606$. The average of $L(7)$ and $U(7)$ is 0.058.

8.3 The Comparison Tests

Problems for Practice

1 - 27. (odd) See the solutions in the back of the book.

29. This series diverges. Limit Compare to $\sum_{n=1}^{\infty} \frac{1}{n}$ (which diverges).

$$\frac{3n/(2n^2-1)}{1/n} = \frac{3n^2}{2n^2-1} \xrightarrow[n\to\infty]{} \frac{3}{2}$$

31. This series diverges. Limit Compare to $\sum_{n=1}^{\infty} \frac{1}{n}$ (which diverges).

$$\frac{\sqrt{n^2+1}/n^2}{1/n} = \frac{\sqrt{n^2+1}}{n} = \sqrt{1+1/n^2} \xrightarrow[n\to\infty]{} 1$$

33. This series converges. Limit Compare to $\sum_{n=1}^{\infty} \frac{1}{n^2}$ (which converges, $p = 2$).

$$\frac{(10n+3)/(100n^3-99)}{1/n^2} = \frac{10n^3+3n^2}{100n^3-99} \xrightarrow[n\to\infty]{} \frac{1}{10}$$

35. This series converges. Limit Compare to $\sum_{n=1}^{\infty} \frac{1}{n^{3/2}}$ (which converges, $p = 3/2$).

$$\frac{(\sqrt{n}+1)/(2n^2-n)}{1/n^{3/2}} = \frac{n^2+n^{3/2}}{2n^2-n} \xrightarrow[n\to\infty]{} \frac{1}{2}$$

37. This series converges. Limit Compare to $\sum_{n=1}^{\infty}\left(\frac{2}{3}\right)^n$ (which converges, geometric, $r = 2/3$).

$$\frac{(2^n+11)/(3^n-1)}{(2/3)^n} = \frac{1+11/2^n}{1-1/3^n} \xrightarrow[n\to\infty]{} 1$$

39. This series diverges. Limit Compare to $\sum_{n=1}^{\infty}\frac{1}{n}$ (which diverges).

$$\frac{\sqrt{1+n^2}/(1+n^8)^{1/4}}{1/n} = \frac{n\sqrt{1+n^2}}{(1+n^8)^{1/4}} = \frac{n^2\sqrt{1/n^2+1}}{n^2(1/n^8+1)^{1/4}} = \frac{\sqrt{1/n^2+1}}{(1/n^8+1)^{1/4}} \xrightarrow[n\to\infty]{} 1$$

41. This series converges. Limit Compare to $\sum_{n=1}^{\infty}\left(\frac{1}{2}\right)^n$ (which converges, geometric, $r = 1/2$).

$$\frac{2^n/(1+2^{2n})}{(1/2)^n} = \frac{2^{2n}}{1+2^{2n}} = \frac{1}{\dfrac{1}{2^{2n}}+1} \xrightarrow[n\to\infty]{} 1$$

43. This series converges. Limit Compare to $\sum_{n=1}^{\infty}\frac{1}{n^2}$ (which converges, $p = 2$).

$$\frac{\sin(1/n)/n}{1/n^2} = \frac{\sin(1/n)}{1/n} \xrightarrow[n\to\infty]{} 1$$

45. This series diverges. Limit Compare to $\sum_{n=1}^{\infty}\frac{1}{n}$ (which diverges).

$$\frac{1/(n+\ln(n))}{1/n} = \frac{n}{n+\ln(n)} \xrightarrow[n\to\infty]{\text{(L'Hòpital)}} 1$$

47. This series converges. Limit Compare to $\sum_{n=1}^{\infty}\left(\frac{7}{10}\right)^n$ (which converges, geometric, $r = 7/10$).

$$\frac{7^n/(10^n+n^{10})}{(7/10)^n} = \frac{1}{1+\dfrac{n^{10}}{10^n}} \xrightarrow[n\to\infty]{} 1$$

Further Theory and Practice

49. This series diverges by Limit Comparison to $\sum_{n=1}^{\infty}\frac{1}{n}$ (which diverges): $\frac{(2n+1)/n^2}{1/n} = \frac{2n+1}{n} \xrightarrow[n\to\infty]{} 2$.

51. This series diverges by Limit Comparison to $\sum_{n=1}^{\infty}\frac{1}{n^{1/2}}$ (which diverges, $p = 1/2$): $\frac{(n+2)/(2n^{3/2}+3)}{1/n^{1/2}} = \frac{n^{3/2}+2n^{1/2}}{2n^{3/2}+3} \xrightarrow[n\to\infty]{} \frac{1}{2}$.

53. This series converges by Limit Comparison to $\sum_{n=1}^{\infty}\frac{1}{n^2}$ (which converges, $p = 2$): $\frac{((n+2)/(n^2+1))^2}{1/n^2} = \frac{n^2(n+2)^2}{(n^2+1)^2} = \frac{n^4+4n^3+4n^2}{n^4+2n^2+2} \xrightarrow[n\to\infty]{} 1$.

55. This series diverges by Limit Comparison to $\sum_{n=1}^{\infty} \frac{1}{n}$ (which diverges): $\frac{(1+1/n)/n}{1/n} = 1 + \frac{1}{n} \xrightarrow[n\to\infty]{} 1$.

57. This series diverges. Apply L'Hòpital's Rule to see that $\lim_{n\to\infty} \frac{n}{\ln(n)^3} = \infty$. Therefore, the series diverges by the Divergence Test. One can also use comparison to the divergent series $\sum_{n=2}^{\infty} 1$ since, for n sufficiently large, $\frac{n}{\ln(n)^3} > 1$.

59. The series converges by the Comparison Test for Convergence. If $n \geq 101$, then $\frac{100^n}{n^n} \leq \frac{100^n}{101^n} = \left(\frac{100}{101}\right)^n$. The series $\sum_{n=1}^{\infty} \left(\frac{100}{101}\right)^n$ converges because it is geometric, $r = 100/101$.

61. The series converges by the Comparison Test for Convergence. This is because for $n = 1, 2, 3, \ldots$, $0 < \sin\left(\frac{1}{n^2}\right) < \frac{1}{n^2}$, and the series $\sum_{n=1}^{\infty} \frac{1}{n^2}$ converges (p-series, $p = 2$).

63. The series converges by the Comparison Test for Convergence. Observe that for $n = 1, 2, 3, \ldots$, $0 < \frac{1}{2} - \frac{1}{3n} < \frac{1}{2}$ so $0 < \left(\frac{1}{2} - \frac{1}{3n}\right)^n < \left(\frac{1}{2}\right)^n$ and the series $\sum_{n=1}^{\infty} \left(\frac{1}{2}\right)^n$ converges (geometric, $r = 1/2$).

65. The series converges by the Comparison Test for Convergence. Observe that for $n = 1, 2, 3, \ldots$, $0 < \frac{2^n+3^n}{7^n+5^n} < \frac{2^n+3^n}{5^n} = \left(\frac{2}{5}\right)^n + \left(\frac{3}{5}\right)^n < 2 \cdot \left(\frac{3}{5}\right)^n$, and the series $\sum_{n=1}^{\infty} \left(2 \cdot \left(\frac{3}{5}\right)^n\right)$ converges (twice a geometric, $r = 1/2$).

67. The series converges by the Comparison Test for Convergence. Observe that for $n = 1, 2, 3, \ldots$, $0 < \frac{n^3}{(n^3+n^2+1)^3} < \frac{n^3}{(n^3)^3} = \frac{1}{n^6}$, and the series $\sum_{n=1}^{\infty} \frac{1}{n^6}$ converges (p-series, $r = 6$).

69. The series converges by the Comparison Test for Convergence. Observe that for $n = 1, 2, 3, \ldots$,

$$\begin{aligned} 0 < \sqrt{n^4 + 2\sqrt{n}} - \sqrt{n^4 + \sqrt{n}} &= \frac{\sqrt{n^4 + 2\sqrt{n}} - \sqrt{n^4 + \sqrt{n}}}{1} \cdot \frac{\sqrt{n^4 + 2\sqrt{n}} + \sqrt{n^4 + \sqrt{n}}}{\sqrt{n^4 + 2\sqrt{n}} + \sqrt{n^4 + \sqrt{n}}} \\ &= \frac{\sqrt{n}}{\sqrt{n^4 + 2\sqrt{n}} + \sqrt{n^4 + \sqrt{n}}} \\ &< \frac{\sqrt{n}}{\sqrt{n^4}} = \frac{1}{n^{3/2}}, \end{aligned}$$

and the series $\sum_{n=1}^{\infty} \frac{1}{n^{3/2}}$ converges (p-series, $r = 3/2$).

71. The series converges by Limit Comparison to the series $\sum_{n=2}^{\infty} \frac{1}{n\ln(n)^2}$ which converges by the Integral Test: $\int_2^{\infty} \frac{1}{x\ln(x)^2}\,dx = \frac{1}{\ln(2)}$. Observe that

$$\frac{\sqrt{n^2+1}/(n\ln(n))^2}{1/(n\ln(n)^2)} = \frac{\sqrt{n^2+1}}{n} = \sqrt{1 + 1/n^2} \xrightarrow[n\to\infty]{} 1.$$

73. Because $\lim_{n\to\infty} \frac{a_n}{b_n} = 0$, if n is sufficiently large, then $0 < \frac{a_n}{b_n} < \frac{1}{2}$ implying that $0 < a_n < \frac{1}{2} \cdot b_n$. Since $\sum_{n=1}^{\infty} \left(\frac{1}{2} \cdot b_n\right)$ converges, so does $\sum_{n=1}^{\infty} a_n$ by the Comparison Test for Convergence.

75. Because $\sum_{n=1}^{\infty} a_n$ converges, $\lim_{n\to\infty} a_n = 0$ and, for all n sufficiently large, $0 < a_n < 1$, implying that $0 < a_n^2 < a_n$. Thus $\sum_{n=1}^{\infty} a_n^2$ converges by the Comparison Test for Convergence.

77. Let $0 < r < 1$ and $s = \frac{r+1}{2}$. Then $0 < r < s < 1$ and we will show that $\sum_{n=1}^{\infty} n^p r^n$ converges by limit comparison to $\sum_{n=1}^{\infty} s^n$ (with the help of Exericse 73). The series $\sum_{n=1}^{\infty} s^n$ converges because it is geometric with ratio s. Let $b = s/r$, note that $b > 1$.

   Consider then the ratio of the terms in the two series in question.

$$\frac{n^p r^n}{s^n} = n^p \cdot \left(\frac{r}{s}\right)^n = \frac{n^p}{(s/r)^n} = \frac{n^p}{b^n} \xrightarrow[n\to\infty]{\text{(L'Hòpital)}} 0$$

   Now apply Exercise 75.

79. If n is any positive integer, then

$$0 < \frac{n!}{n^n} = \left(\frac{n}{n}\right) \cdot \left(\frac{n-1}{n}\right) \cdot \left(\frac{n-2}{n}\right) \cdots \left(\frac{2}{n}\right) \cdot \left(\frac{1}{n}\right) \leq \frac{2}{n^2}.$$

   Since $\sum_{n=1}^{\infty} \frac{2}{n^2}$ converges (p-series, $p = 2$), $\sum_{n=1}^{\infty} \frac{n!}{n^n}$ converges by the Comparison Test for Convergence.

CALCULATOR/COMPUTER EXERCISES

81. Using a plot of the the graphs of $x^{10} \cdot 0.99^x$ and 0.995^x and *fsolve* shows that the expressions are equal when $x \approx 19620.33$. Therefore, $n^{10} \cdot 0.99^n < 0.995^n$ for all integers $n \geq 19621$.

   Since $\sum_{n=1}^{\infty} 0.995^n$ converges, the series in question converges by the Comparison Test for Convergence (see the solution to Exercise 73 for the general case).

8.4 Alternating Series

Problems for Practice

1. The terms of the sequence $\left\{\frac{1}{n^3+1}\right\}$ are positive and decreasing to 0.

3. The terms of the sequence $\left\{\frac{1}{\sqrt{n^2+1}}\right\}$ are positive and decreasing to 0.

5. This series, which is $\sum_{n=1}^{\infty}(-1)^{n+1}\left(\frac{3}{2}\right)^n$, diverges by the Divergence Test.

7. Since the sequence $\{(4/5)^n\}$ decreases to 0 and the sequence $\{n+2\}$ increases without bound, the sequence $\left\{\frac{(4/5)^n}{n+2}\right\}$ decreases monotonically to 0.

9. Let $f(x) = x^2e^{-x}$. Then $f'(x) = 2xe^{-x} - x^2e^{-x} = x(2-x)e^{-x}$ so f is positive and decreasing (to 0) on $[2, \infty)$. Consequently, the sequence $\left\{\frac{n^2}{e^n}\right\}$ is positive and decreasing monotonically to 0 also.

11. Because $\cos(n\pi) = (-1)^n$ and $\sin(\pi/n)$ is positive when $n \geq 2$, the series alternates. Observe that $\sin(x)$ is strictly increasing on $[0, \pi/2]$, so $\sin(1/x)$ is strictly decreasing (to 0) on $[1, \infty)$. Consequently, the sequence $\{\sin(\pi/n)\}_{n=2}^{\infty}$ is positive and decreasing monotonically to 0 also.

13. This series converges absolutely because $\sum_{n=1}^{\infty} \frac{1}{n^3+1}$ converges by the Comparison Test for Convergence: $0 < \frac{1}{n^3+1} < \frac{1}{n^3}$.

15. This series converges conditionally. It converges by the Alternating Series Test but $\sum_{n=1}^{\infty} \frac{n}{n^2+1}$ diverges by Limit Comparison to $\sum_{n=1}^{\infty} \frac{1}{n}$.

17. This series converges conditionally. It converges by the Alternating Series Test but the series $\sum_{n=2}^{\infty} \frac{1}{\ln(n)}$ diverges by the Comparison Test for Divergence: $0 < \frac{1}{n} < \frac{1}{\ln(n)}$ for $n \geq 2$.

19. This series converges absolutely. Apply the Limit Comparison Test to $\sum_{n=1}^{\infty} \frac{\sqrt{\sqrt{n}+1}}{\sqrt{n^3+1}}$ compared to $\sum_{n=1}^{\infty} \frac{1}{n^{5/4}}$ (which converges, $p = 5/4$).

$$\frac{\sqrt{\sqrt{n}+1}/\sqrt{n^3+1}}{1/n^{5/4}} = \frac{n^{5/4}\sqrt{n^{1/2}+1}}{\sqrt{n^3+1}} = \frac{\sqrt{n^3+n^{5/2}}}{\sqrt{n^3+1}} = \sqrt{\frac{1+1/\sqrt{n}}{1+1/n^3}} \xrightarrow[n\to\infty]{} 1$$

21. This series converges conditionally. It converges by the Alternating Series Test (the function $f(x) = \frac{x+\ln(x)}{x^{3/2}}$ is strictly decreasing to 0 on $[2, \infty)$). However, the series $\sum_{n=1}^{\infty} \frac{n+\ln(n)}{n^{3/2}}$ diverges by the Comparison Test for Divergence: $0 < \frac{1}{n^{1/2}} \leq \frac{n+\ln(n)}{n^{3/2}}$ for $n = 1, 2, 3, \ldots$.

23. This series converges absolutely by the Comparison Test for Convergence: $0 < |\sin(n)e^{-n}| \leq e^{-n}$ for $n = 1, 2, 3, \ldots$, and $\sum_{n=1}^{\infty} \left(\frac{1}{e}\right)^n$ is geometric, $r = 1/e$.

25. This series converges absolutely by the Comparison Test for Convergence. Since $\ln(n)^3 \leq n$ for n sufficiently large (the derivative of $f(x) = x - \ln^3(x)$ is positive for large x), for such n,

$$0 < \left(\frac{\ln(n)}{n}\right)^3 \leq \frac{n}{n^3} = \frac{1}{n^2}.$$

27. This series converges absolutely by the Limit Comparison Test, compare to $\sum_{n=1}^{\infty}\left(\frac{1}{2}\right)^n$.

$$\frac{((1+1/n)/2)^n}{1/2^n} = 2^n\left(\frac{1+1/n}{2}\right)^n = \left(1+\frac{1}{n}\right)^n \xrightarrow[n\to\infty]{(5.5.1)} e$$

29. This series diverges by the Divergence Test: $\lim_{n\to\infty}\left(1+\frac{1}{n}\right) = 1$.

31. This series diverges by the Divergence Test: $\lim_{n\to\infty}\left(1-\frac{1}{n}\right)^n = e^{-1}$.

33. This series converges absolutely by the Comparison Test for Convergence. Since $5^{(n^2)} \geq 5^n$ for $n = 1, 2, 3 \ldots$, $0 < \frac{3^n}{5^{(n^2)}} \leq \frac{3^n}{5^n} = \left(\frac{3}{5}\right)^n$, and $\sum_{n=1}^{\infty}\left(\frac{3}{5}\right)^n$ converges (geometric, $r = 3/5$).

35. This series converges conditionally. It converges by the Alternating Series Test (the function $f(x) = \frac{1}{x\sqrt{\ln(x}}$ is strictly decreasing to 0 on $[3,\infty)$). However, the infinite series $\sum_{n=3}^{\infty}\frac{1}{n\sqrt{\ln(n)}}$ diverges by the Integral Test: $\int_3^N (\ln(x))^{-1/2}\cdot\frac{1}{x}\,dx = \left(2\sqrt{\ln(x)}\right)\Big|_3^N = 2\sqrt{\ln(N)} - 2\sqrt{\ln(3)} \xrightarrow[N\to\infty]{} \infty$.

37. $a_{N+1} \leq 0.01$ when $\frac{1}{\sqrt{N+1}} \leq 0.01$. That is, $\sqrt{N+1} \geq 100$, so $N+1 \geq 10^4$ and $N = 9999$.

39. $a_{N+1} \leq 0.01$ when $\frac{4}{2(N+1)-1} \leq 0.01$. That is, $2N+1 \geq 400$, so $N \geq \frac{399}{2} = 199.5$ and $N = 200$.

41. $a_{N+1} \leq 0.01$ when $\frac{1}{(N+1)^2+15(N+1)} \leq 0.01$. That is, $(N+1)(N+16) \geq 100$, so $N \geq 4$ and $N = 4$.

Further Theory and Practice

43. The quadratic $f(x) = x^2 - 20x + 101$ attains its minumum value when $2x - 20 = 0$, that is, $x = 10$. Note that $f(10) = 1$. Consequently, $1/f(x)$ is maximum and positive when $x = 10$ and decreases monotonically to 0 on $[10,\infty)$. The Alternating Series test applies to $\sum_{n=10}^{\infty}\frac{(-1)^n}{n^2-20x+101}$.

45. The function $f(x) = \dfrac{x^5}{2^x}$ attains its maximum value on the interval $[1,\infty)$ when

$$\frac{2^x\cdot 5x^4 - x^5\cdot 2^x\cdot\ln(2)}{2^{2x}} = \frac{x^4(5-x\ln(2))}{2^x} = 0,$$

that is, $x = 5/\ln(2) \approx 7.2$. Consequently, $f(x)$ decreases monotonically to 0 on $[8,\infty)$. The Alternating Series test applies to $\sum_{n=8}^{\infty}(-1)^n\frac{n^5}{2^n}$.

47. Since $1/6! \approx 0.0014$ and $1/7! \approx 0.0002$, $S_6 = -0.632$ is a three decimal approximation to S.

49. Since $4!/8! \approx 0.0006$ and $5!/10! \approx 0.00003$, $S_4 = -0.424$ is a three decimal approximation to S.

CALCULATOR/COMPUTER EXERCISES

51. Since $a_{200} = 1/200 = 0.005$, $S_{199} = 0.70$ is our two decimal approximation to $\ln(2) = 0.69$.

53. Since $a_9 = 1/9! \approx 0.000002$, $S_8 = 0.63212$ is our five decimal approximation to $1 - 1/e = 0.63212$.

55. Since $a_5 = 1/(2 \cdot 5)! \approx 2 \cdot 10^{-7}$, $S_4 = 0.54030$ is our five decimal approximation to $\cos(1) = 0.54030$.

57. A plot of $f(x) = \dfrac{9x^2 + 13}{x^3 + 55x + 60}$ will show a maximum value of approximately 0.6 near $x = 8$. *Maple*'s *Maximize* procedure tells us the max is at $x = 8.02$, so $M = 9$.

59. A plot of $f(x) = \dfrac{100x^{9/4} + x}{150 + x^{5/2}}$ will show a maximum value of approximately 45 near $x = 15$. *Maple*'s *Maximize* procedure tells us the max is at $x = 17.86$, so $M = 18$.

8.5 The Root and Ratio Tests

Problems for Practice

1. $$\frac{a_{n+1}}{a_n} = \frac{(n+1)/e^{n+1}}{n/e^n} = \frac{n+1}{n} \cdot \frac{1}{e} \xrightarrow[n\to\infty]{} \frac{1}{e} < 1, \text{ converges}$$

3. $$\frac{a_{n+1}}{a_n} = \frac{2^{n+1}/(n+1)^3}{2^n/n^3} = 2 \cdot \left(\frac{n}{n+1}\right)^3 \xrightarrow[n\to\infty]{} 2 > 1, \text{ diverges}$$

5. $$\frac{a_{n+1}}{a_n} = \frac{(n+1)^{100}/(n+1)!}{n^{100}/n!} = \left(\frac{n+1}{n}\right)^{100} \cdot \frac{1}{n+1} \xrightarrow[n\to\infty]{} 0 < 1, \text{ converges}$$

7. $$\frac{a_{n+1}}{a_n} = \frac{(n+1)!/((n+1)3^{n+1})}{n!/(n3^n)} = (n+1) \cdot \frac{n}{n+1} \cdot \frac{1}{3} \xrightarrow[n\to\infty]{} \infty, \text{ diverges}$$

9. $$\frac{a_{n+1}}{a_n} = \frac{(3^{n+1} + n + 1)/(2^{n+1} + (n+1)^3)}{(3^n + n)/(2^n + n^3)} = \frac{3^{n+1} + n + 1}{3^n + n} \cdot \frac{2^n + n^3}{2^{n+1} + (n+1)^3}$$
$$\xrightarrow[n\to\infty]{} \frac{3}{2} > 1, \text{ diverges}$$

11. $$\frac{a_{n+1}}{a_n} = \frac{2^{n+1}\sqrt{n+1}/3^{n+1}}{2^n\sqrt{n}/3^n} = 2 \cdot \sqrt{\frac{n+1}{n}} \cdot \frac{1}{3} \xrightarrow[n\to\infty]{} \frac{2}{3} < 1, \text{ converges}$$

13. $$\left|\frac{a_{n+1}}{a_n}\right| = \frac{\ln(n+1)/(n+1)}{\ln(n)/n} = \frac{\ln(n+1)}{\ln(n)} \cdot \frac{n}{n+1} \xrightarrow[n\to\infty]{\text{(L'Hòpital)}} 1, \text{ test fails}$$

Since $f(x) = \ln(x)/x$ is decreasing to 0 on $[2, \infty)$, the Alternating Series Test implies that the series converges. The series does not converge absolutely because $\sum_{n=2}^{\infty} \frac{\ln(n)}{n}$ diverges by the Integral Test: $\int_2^N \frac{\ln(x)}{x}\,dx = \left(\frac{1}{2}\ln^2(x)\right)\Big|_2^N = \frac{1}{2}(\ln^2(N) - \ln(2)) \xrightarrow[N\to\infty]{} \infty$. So, the series converges conditionally.

15. $$\left|\frac{a_{n+1}}{a_n}\right| = \frac{\ln(n+1)/(n+1)^2}{\ln(n)/n^2} = \frac{\ln(n+1)}{\ln(n)} \cdot \frac{n^2}{(n+1)^2} \xrightarrow[n\to\infty]{\text{(L'Hòpital)}} 1\,,\ \text{fails}$$

The series converges absolutely because, for n sufficiently large, $\ln(n) \leq \sqrt{n}$ (check the derivative of $f(x) = \sqrt{x} - \ln(x)$). Consequently, for such n, $0 < \frac{\ln(n)}{n^2} \leq \frac{\sqrt{n}}{n^2} = \frac{1}{n^{3/2}}$, and $\sum_{n=1}^{\infty} \frac{\ln(n)}{n^2}$ converges by the Comparison Test for Convergence.

17. $$\left|\frac{a_{n+1}}{a_n}\right| = \frac{1 + 1/(n+1)}{1 + 1/n} = \frac{n+2}{n+1} \cdot \frac{n}{n+1} \xrightarrow[n\to\infty]{\text{(L'Hòpital)}} 1\,,\ \text{fails}$$

The series diverges by the The Divergence Test: $\lim_{n\to\infty}\left(1 + \frac{1}{n}\right) = 1$.

19. $$\left|\frac{a_{n+1}}{a_n}\right| = \frac{(e^{n+1} + \ln(n+1))/(e^{n+1}\cdot(n+1)^2)}{(e^n + \ln(n))/(e^n \cdot n^2)}$$

$$= \frac{e^{n+1} + \ln(n+1)}{e^n + \ln(n)} \cdot \frac{1}{e} \cdot \left(\frac{n}{n+1}\right)^2$$

$$= \frac{e + \ln(n+1)/e^n}{1 + \ln(n)/e^n} \cdot \frac{1}{e} \cdot \left(\frac{1}{1+1/n}\right)^2 = \frac{e}{1} \cdot \frac{1}{e} \cdot 1 \xrightarrow[n\to\infty]{} 1\,,\ \text{test fails}$$

The series converges absolutely because

$$\sum_{n=1}^{\infty} \frac{e^n + \ln(n)}{e^n \cdot n^2} = \sum_{n=1}^{\infty}\left(\frac{1}{n^2} + \frac{\ln(n)}{e^n \cdot n^2}\right) = \sum_{n=1}^{\infty}\frac{1}{n^2} + \sum_{n=1}^{\infty}\frac{\ln(n)}{e^n \cdot n^2}\,.$$

Both the series on the right converge, the first being a p-series, $p = 2$, and the second is bounded above, term-by-term, by the first. $0 < \frac{\ln(n)}{e^n \cdot n^2} \leq \frac{1}{n^2}$.

21. $$\left|\frac{a_{n+1}}{a_n}\right| = \frac{3^{n+1}/((n+1)^3 + 3^{n+1})}{3^n/(n^3 + 3^n)} = \frac{3}{1} \cdot \frac{n^3 + 3^n}{(n+1)^3 + 3^{n+1}}$$

$$= \frac{3}{1} \cdot \frac{n^3/3^n + 1}{(n+1)^3/3^n + 3} \xrightarrow[n\to\infty]{} \frac{3}{1} \cdot \frac{1}{3} = 1\,,\ \text{test fails}$$

The series diverges by the Divergence Test:

$$\lim_{n\to\infty} \frac{3^n}{n^3 + 3^n} = \lim_{n\to\infty} \frac{1}{1 + n^3/3^n} = 1\,.$$

23. $$\left|\frac{a_{n+1}}{a_n}\right| = \frac{\arctan(n+1)}{\arctan(n)} \xrightarrow[n\to\infty]{} \frac{\pi/2}{\pi/2} = 1\,,\ \text{test fails}$$

The series diverges by the Divergence Test: $\lim_{n\to\infty} \arctan(n) = \pi/2$.

25. Since $a_n^{1/n} = (n^{-n/2})^{1/n} = n^{-1/2} = \frac{1}{\sqrt{n}} \xrightarrow[n\to\infty]{} 0 < 1$, the series converges.

27. Since $a_n^{1/n} = \left(\frac{n}{2^n}\right)^{1/n} = \frac{n^{1/n}}{2} \xrightarrow[n\to\infty]{} \frac{1}{2} < 1$, the series converges.

29. Since $a_n^{1/n} = \left(\frac{n^{100}}{(1+n)^n}\right)^{1/n} = \frac{(n^{1/n})^{100}}{1+n} \xrightarrow[n\to\infty]{} 0 < 1$, the series converges.

31. Since $a_n^{1/n} = \frac{\ln(n)}{\ln(n^2-4)} \xrightarrow[n\to\infty]{\text{(L'Hòpital)}} \frac{1}{2} < 1$, the series converges.

33. Since $a_n^{1/n} = \frac{n^2+7n+13}{2n^2+1} \xrightarrow[n\to\infty]{} \frac{1}{2} < 1$, the series converges.

Further Theory and Practice

35. $$\frac{a_{n+1}}{a_n} = \frac{((2(n+1))!/(3(n+1))!}{(2n)!/(3n)!} = \frac{(2n+2)(2n+1)}{(3n+3)(3n+2)(3n+1)}$$
$$= \frac{2+2/n}{3+3/n} \cdot \frac{2+1/n}{3+2/n} \cdot \frac{1}{3n+3} \xrightarrow[n\to\infty]{} 0 < 1,$$
converges

37. $$\frac{a_{n+1}}{a_n} = \frac{((2(n+1))!/((n+1)! \cdot 2^{n+1})}{(2n)!/(n! \cdot 2^n)} = \frac{(2n+2)(2n+1)}{n+1} \cdot \frac{1}{2}$$
$$= 2n+1 \xrightarrow[n\to\infty]{} \infty\text{, diverges}$$

39. $$\frac{a_{n+1}}{a_n} = \frac{(n+1+3^{n+1})/((n+1)^3+2^{n+1})}{(n+3^n)/(n^3+2^n)}$$
$$= \frac{n+1+3^{n+1}}{n+3^n} \cdot \frac{n^3+2^n}{(n+1)^3+2^{n+1}}$$
$$= \frac{(n+1)/3^n+3}{n/3^n+1} \cdot \frac{n^3/2^n+1}{(n+1)^3/2^n+2} \xrightarrow[n\to\infty]{} \frac{3}{1} \cdot \frac{1}{2} = \frac{3}{2} > 1\text{, diverges}$$

41. Since $a_n^{1/n} = \frac{4}{n^{1/n}+2} \xrightarrow[n\to\infty]{} \frac{4}{3} > 1$, the series diverges.

43. This series converges. Observe that $0 < \frac{2^n}{1+\ln^n(n)} < \frac{2^n}{\ln^n(n)}$ and the series $\sum_{n=2}^{\infty} \frac{2^n}{\ln^n(n)}$ converges by the Root Test:
$$\left(\frac{2^n}{\ln^n(n)}\right)^{1/n} = \frac{2}{\ln(n)} \xrightarrow[n\to\infty]{} 0 < 1.$$

45. The series converges by the Alternating Series Test. The convergence is conditional because $\sum_{n=1}^{\infty} \frac{1}{\sqrt{n+10}}$ diverges. Use Limit Comparison to $\sum_{n=1}^{\infty} \frac{1}{\sqrt{n}}$.

47. This series diverges by the Divergence Test: $\lim_{n\to\infty} \frac{n!}{3^n} = \infty$.

49. This series converges conditionally. It converges by the Alternating Series Test applied to the tail $\sum_{n=4}^{\infty}(-1)^n \frac{n}{n^2-11}$ but the convergence is not absolute. For example, one can apply the the Comparison Test for Divergence: For n sufficiently large, $0 < \frac{1}{n} < \frac{n}{n^2-11}$.

51. This series converges conditionally. It converges by the Alternating Series Test: The function $f(x) = \frac{x^{1/x}}{\ln(x)}$ is continuous and decreasing to 0 on $[2, \infty)$ (note that $x^{1/x}$ decreases towards 1 and $\ln(x)$ increases without bound). The divergence is conditional by the Comparison Test for Divergence: $0 < \frac{1}{n} < \frac{n^{1/n}}{\ln(n)}$ when $n \geq 2$ (note that $\ln(n) < n$ when $n \geq 2$).

53. This series diverges by the Divergence Test. Both the numerator and the denominator of the fraction converge to 1 as $n \to \infty$.

55. This series converges absolutely by the Comparison Test for Convergence: $0 < \frac{e^{1/n}}{n^e} \leq \frac{e}{n^e}$ for all $n \geq 1$ and the series $\sum_{n=1}^{\infty} \frac{e}{n^e} = e \sum_{n=1}^{\infty} \frac{1}{n^e}$ converges (p-series, $p = e$).

57. This series converges conditionally. It converges by the Alternating Series Test. The function $f(x) = \ln(1 + 1/x)$ is continuous and decreasing to 0 on $[1, \infty)$ (its derivative is negative). The convergence is not absolute by the Integral Test:

$$\begin{aligned}\int_1^N \ln\left(\frac{x+1}{x}\right) dx &= \int_1^N (\ln(x+1) - \ln(x)), dx \\ &= ((x+1)\ln(x+1) - x\ln(x))\Big|_1^N \\ &= (N+1)\ln(N+1) - N\ln(N) - 2\ln(2) \\ &= N\ln\left(\frac{N+1}{N}\right) + \ln(N+1) - 2\ln(2) \xrightarrow[N\to\infty]{} \infty .\end{aligned}$$

Note that $\lim_{N\to\infty} N \ln(1 + 1/N) = 1$ by L'Hòpital's Rule.

59. This series diverges by the Divergence Test. The denominator of the fraction converges to e as $n \to \infty$.

61. Applying the Ratio Test to $\sum_{n=1}^{\infty} n^p a_n$ leads to the following limit.

$$\begin{aligned}\lim_{n\to\infty} \frac{(n+1)^p a_{n+1}}{n^p a_n} &= \lim_{n\to\infty} \left(\frac{n+1}{n}\right)^p \cdot \frac{a_{n+1}}{a_n} \\ &= \lim_{n\to\infty} \left(\frac{n+1}{n}\right)^p \cdot \lim_{n\to\infty} \frac{a_{n+1}}{a_n} = \lim_{n\to\infty} \frac{a_{n+1}}{a_n} .\end{aligned}$$

This observation allows us to delete any factors of the form n^p in the nth term of an infinite series when applying the Ratio Test.

Since a p-series is of the form $\sum_{n\to 1}^{\infty}(n^p \cdot 1)$ the limit obtained from the Ratio Test will always evaluate to 1.

63. Applying the Ratio Test to $\sum_{n=2}^{\infty} a_n \ln^p(n)$ leads to the following limit.

$$\begin{aligned}\lim_{n\to\infty} \frac{a_{n+1}\ln^p(n+1)}{a_n \ln^p(n)} &= \lim_{n\to\infty} \left(\frac{a_{n+1}}{a_n} \cdot \left(\frac{\ln(n+1)}{\ln(n)} \right)^p \right) \\ &= \lim_{n\to\infty} \frac{a_{n+1}}{a_n} \cdot \left(\lim_{n\to\infty} \frac{\ln(n+1)}{\ln(n)} \right)^p = \lim_{n\to\infty} \frac{a_{n+1}}{a_n}.\end{aligned}$$

Note that $\lim_{n\to\infty} \frac{\ln(n+1)}{\ln(n)} = 1$ by L'Hòpital's Rule.

CALCULATOR/COMPUTER EXERCISES

65. The plot below displays the first 40 points and the asymptote. The diamonds are the values of the ratios and the circles are values of the roots.

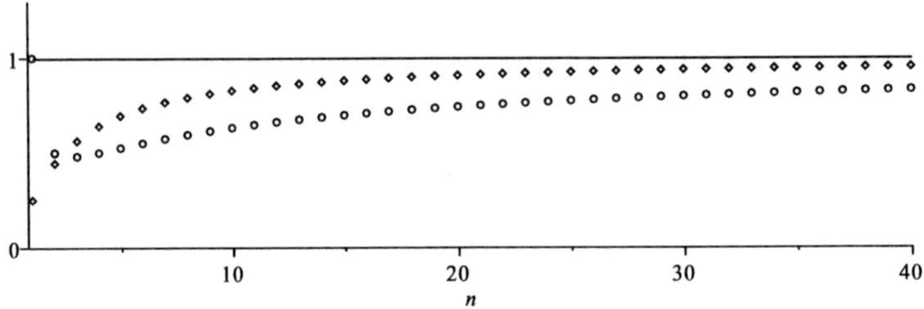

67. The plot below displays the first 40 points and the asymptote. The diamonds are the values of the ratios and the circles are values of the roots.

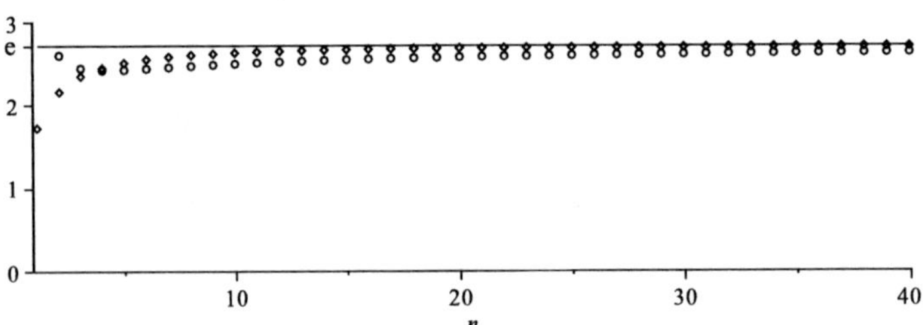

8.6 Introduction to Power Series

Problems for Practice

1. Since $\lim_{n\to\infty} |a_n|^{1/n} = \lim_{n\to\infty} 2 = 2$, the radius of convergence is $R = 1/2$. The series converges absolutely on the open interval $(-1/2, 1/2)$.

   At the left endpoint, $x = -1/2$, the series is $\sum_{n=1}^{\infty}(-1)^n$, which diverges.

   At the right endpoint, $x = 1/2$, the series is $\sum_{n=1}^{\infty} 1$, which also diverges.

   The interval of convergence is $I = (-1/2, 1/2)$.

3. Since $\lim_{n\to\infty}|a_n|^{1/n} = \lim_{n\to\infty} n^{1/n} = 1$, the radius of convergence is $R = 1$. The series converges absolutely on the open interval $(-1, 1)$.

   At the left endpoint, $x = -1$, the series is $\sum_{n=1}^{\infty}(-1)^n n$, which diverges.

   At the right endpoint, $x = 1$, the series is $\sum_{n=1}^{\infty} n$, which also diverges.

   The interval of convergence is $I = (-1, 1)$.

5. Since

$$\lim_{n\to\infty}|a_n|^{1/n} = \lim_{n\to\infty}(n^3)^{1/n} = \lim_{n\to\infty}(n^{1/n})^3 = \left(\lim_{n\to\infty} n^{1/n}\right)^3 = 1\,,$$

   the radius of convergence is $R = 1$. The series converges absolutely on the open interval $(-1, 1)$.

   At the left endpoint, $x = -1$, the series is $\sum_{n=1}^{\infty}(-1)^{2n+1}n^3 = \sum_{n=1}^{\infty}(-n^3)$, which diverges.

   At the right endpoint, $x = 1$, the series is $\sum_{n=1}^{\infty}(-1)^{n+1}n^3$, which also diverges.

   The interval of convergence is $I = (-1, 1)$.

7. Since

$$\lim_{n\to\infty}|a_n|^{1/n} = \lim_{n\to\infty}\left(\frac{n}{5^n}\right)^{1/n} = \lim_{n\to\infty}\frac{n^{1/n}}{5} = \frac{1}{5}\lim_{n\to\infty} n^{1/n} = \frac{1}{5}\,,$$

   the radius of convergence is $R = 5$. The series converges absolutely on the open interval $(-5, 5)$.

   At the left endpoint, $x = -5$, the series is $\sum_{n=1}^{\infty}(-1)^n n$, which diverges.

   At the right endpoint, $x = 5$, the series is $\sum_{n=1}^{\infty} n$, which also diverges.

   The interval of convergence is the open interval $I = (-5, 5)$.

9. Since $\lim_{n\to\infty}|a_n|^{1/n} = \lim_{n\to\infty} n = \infty$, the radius of convergence is $R = 0$. The series converges only at $x = 0$ and the interval of convergence is $I = [0]$.

11. Since

$$\lim_{n\to\infty}\left|\frac{a_{n+1}}{a_n}\right| = \lim_{n\to\infty}\frac{3/\sqrt{(n+1)^2+1}}{3/\sqrt{n^2+1}} = \lim_{n\to\infty}\sqrt{\frac{n^2+1}{(n+1)^2+1}}$$

$$= \sqrt{\lim_{n\to\infty}\frac{1+1/n^2}{(1+1/n)^2+1/n^2}} = 1\,,$$

   the radius of convergence is $R = 1$. The series converges absolutely on the open interval $(-1, 1)$.

At the left endpoint, $x = -1$, the series is $\sum_{n=1}^{\infty}(-1)^{2n}\frac{3}{\sqrt{n^2+1}} = \sum_{n=1}^{\infty}\frac{3}{\sqrt{n^2+1}}$, which diverges by Limit Comparison to the series $\sum_{n=1}^{\infty}\frac{3}{n}$.

At the right endpoint, $x = 1$, the series is $\sum_{n=1}^{\infty}(-1)^n\frac{3}{\sqrt{n^2+1}}$, which converges by the Alternating Series Test.

The interval of convergence is $I = (-1, 1]$.

13. Since

$$\begin{aligned}
\lim_{n\to\infty}\left|\frac{a_{n+1}}{a_n}\right| &= \lim_{n\to\infty}\frac{(n+1)^2/((n+1)^3+1)}{n^2/(n^3+1)} \\
&= \lim_{n\to\infty}\left(\frac{n+1}{n}\right)^2\cdot\left(\frac{n^3+1}{(n+1)^3+1}\right) \\
&= \left(\lim_{n\to\infty}\left(1+\frac{1}{n}\right)\right)^2\cdot\lim_{n\to\infty}\frac{1+1/n^3}{(1+1/n)^3+1/n^3} = 1\,,
\end{aligned}$$

the radius of convergence is $R = 1$. The series converges absolutely on the open interval $(-1, 1)$.

At the left endpoint, $x = -1$, the series is $\sum_{n=1}^{\infty}(-1)^n\frac{n^2}{n^3+1}$, which converges by the Alternating Series Test.

At the right endpoint, $x = 1$, the series is $\sum_{n=1}^{\infty}\frac{n^2}{n^3+1}$, which diverges by Limit Comparison to the Harmonic Series $\sum_{n=1}^{\infty}\frac{1}{n}$.

The interval of convergence is $I = [-1, 1)$.

15. Since

$$\begin{aligned}
\lim_{n\to\infty}\left|\frac{a_{n+1}}{a_n}\right| &= \lim_{n\to\infty}\frac{3^{n+1}/((n+1)^3+1)}{3^n/(n^3+1)} = \lim_{n\to\infty} 3\cdot\frac{n^3+1}{(n+1)^3+1} \\
&= 3\lim_{n\to\infty}\frac{1+1/n^3}{(1+1/n)^3+1/n^3} = 3\,,
\end{aligned}$$

the radius of convergence is $R = 1/3$. The series converges absolutely on the open interval $\left(-\frac{1}{3}, \frac{1}{3}\right)$.

At the left endpoint, $x = -1/3$, the series is $\sum_{n=1}^{\infty}(-1)^n\frac{1}{n^3+1}$, which converges by the Alternating Series Test.

At the right endpoint, $x = 1/3$, the series is $\sum_{n=1}^{\infty}\frac{1}{n^3+1}$, which converges by the Comparison Test for Convergence: $0 < \frac{1}{n^3+1} < \frac{1}{n^3}$, the p-series $\sum_{n=1}^{\infty}\frac{1}{n^3}$ converges.

The interval of convergence is $I = \left[-\frac{1}{3}, \frac{1}{3}\right]$.

17. Since

$$\lim_{n\to\infty}\left|\frac{a_{n+1}}{a_n}\right| = \lim_{n\to\infty}\frac{(n+1)/\sqrt{n+2}}{n/\sqrt{n+1}} = \lim_{n\to\infty}\left(1+\frac{1}{n}\right)\cdot\sqrt{\frac{n+1}{n+2}}$$
$$= \sqrt{\lim_{n\to\infty}\frac{1+1/n}{1+2/n}} = 1\,,$$

the radius of convergence is $R = 1$. The series converges absolutely on the open interval $(-1, 1)$.

At the left endpoint, $x = -1$, the series is $\sum_{n=1}^{\infty}(-1)^{2n}\frac{n}{\sqrt{n+1}}$, which diverges by the Divergence Test: $\lim_{n\to\infty}\frac{n}{\sqrt{n+1}} = \infty$.

The same it true at the right endpoint, $x = 1$, where the series is $\sum_{n=1}^{\infty}(-1)^n\frac{n}{\sqrt{n+1}}$.

The interval of convergence is $I = (-1, 1)$.

19. $\displaystyle\lim_{n\to\infty}\left|\frac{a_{n+1}}{a_n}\right| = \lim_{n\to\infty}\frac{2^{2(n+1)}/\ln(n+1)}{2^{2n}/\ln(n)} = 4\lim_{n\to\infty}\frac{\ln(n)}{\ln(n+1)} \overset{\text{(L'Hòpital)}}{=} 4$, so the radius of convergence is $R = 1/4$. The series converges absolutely on the open interval $\left(-\frac{1}{4}, \frac{1}{4}\right)$.

At the left endpoint, $x = -1/4$, the series is $\sum_{n=2}^{\infty}(-1)^n\frac{1}{\ln(n)}$, which converges by the Alternating Series Test.

At the right endpoint, $x = 1/4$, the series is $\sum_{n=2}^{\infty}\frac{1}{\ln(n)}$, which diverges by the Comparison Test for Divergence: $0 < \frac{1}{n} < \frac{1}{\ln(n)}$ when $n \geq 2$.

The interval of convergence is $I = \left[-\frac{1}{4}, \frac{1}{4}\right)$.

21. The nth coefficient is $a_n = n^{-1/2}$. Using the Root Test, $\lim_{n\to\infty}|a_n|^{1/n} = \lim_{n\to\infty}(n^{-1/2})^{1/n} = \lim_{n\to\infty}(n^{1/n})^{-1/2} = 1$, so the radius of convergence is $R = 1$. Since the base point is $c = -6$, the series converges absolutely on the open interval $(-7, -5)$.

At the left endpoint, $x = -7$, the series is $\sum_{n=1}^{\infty}(-1)^n\frac{1}{\sqrt{n}}$, which converges by the Alternating Series Test.

At the right endpoint, $x = -5$, the series is $\sum_{n=}^{\infty}\frac{1}{\sqrt{n}}$, which, as a p-series, $p = 1/2$, diverges.

The interval of convergence is $I = [-7, -5)$.

23. The nth coefficient is $a_n = \frac{1}{n^5+1}$. Using the Ratio Test, $\lim_{n\to\infty}\left|\frac{a_{n+1}}{a_n}\right| = \lim_{n\to\infty}\frac{1/((n+1)^5+1)}{1/(n^5+1)} = \lim_{n\to\infty}\frac{n^5+1}{(n+1)^5+1} = \lim_{n\to\infty}\frac{1+1/n^5}{(1+1/n)^5+1/n^5} = 1$, so the radius of convergence is $R = 1$. Since the base point is $c = -4$, the series converges absolutely on the open interval $(-5, -3)$.

At the left endpoint, $x = -5$, the series is $\sum_{n=0}^{\infty}(-1)^n\frac{1}{n^5+1}$, which converges by the Alternating Series Test.

At the right endpoint, $x = -3$, the series is $\sum_{n=0}^{\infty} \frac{1}{n^5+1}$, which converges by the Comparison Test for Convergence: $0 < \frac{1}{n^5+1} < \frac{1}{n^5}$.

The interval of convergence is $I = [-5, -3]$.

25. As a power series this series has the form $\sum_{n=0}^{\infty} 2^n \left(x + \frac{1}{2}\right)^n$ so the base point is $c = -1/2$ and the nth coefficient is $a_n = 2^n$. Using the Root Test, $\lim_{n\to\infty} |a_n|^{1/n} = \lim_{n\to\infty} 2 = 2$, so the radius of convergence is $R = 1/2$ and the series converges absolutely on the open interval $(-1, 0)$.

At the left endpoint, $x = -1$, the series is $\sum_{n=0}^{\infty}(-1)^n$, which diverges by the Divertence Test.

At the right endpoint, $x = 0$, the series is $\sum_{n=0}^{\infty} 1$, which also diverges by the Divertence Test.

The interval of convergence is $I = (-1, 0)$.

27. The nth coefficient is $a_n = (-1)^n \frac{n+1}{3n+1}$. Using the Ratio Test,

$$\lim_{n\to\infty} \left|\frac{a_{n+1}}{a_n}\right| = \lim_{n\to\infty} \frac{(n+2)/(3(n+1)+1)}{(n+1)/(3n+1)} = \lim_{n\to\infty} \left(\frac{n+2}{n+1}\right) \cdot \left(\frac{3n+1}{3n+4}\right)$$
$$= \lim_{n\to\infty} \left(\frac{1+2/n}{1+1/n}\right) \cdot \left(\frac{3+1/n}{3+4/n}\right) = 1,$$

so the radius of convergence is $R = 1$. Since the base point is $c = -5$, the series converges absolutely on the open interval $(-6, -4)$.

At the left endpoint, $x = -6$, the series is $\sum_{n=0}^{\infty}(-1)^{2n} \frac{n+1}{3n+1}$, which diverges by the Divergence Test: $\lim_{n\to\infty} \frac{n+1}{3n+1} = \frac{1}{3}$.

At the right endpoint, $x = -4$, the series is $\sum_{n=0}^{\infty}(-1)^n \frac{n+1}{3n+1}$, which also diverges by the Divergence Test.

The interval of convergence is $I = (-6, -4)$.

29. As a power series this series has the form $\sum_{n=1}^{\infty} \frac{1}{n^n}(x+1)^n$ so the base point is $c = -1$ and the nth coefficient is $a_n = 1/n^n$. Using the Root Test, $\lim_{n\to\infty} |a_n|^{1/n} = \lim_{n\to\infty} \frac{1}{n} = 0$, so the radius of convergence is $R = \infty$ and the interval of convergence is $I = (\infty, \infty)$.

31. The nth coefficient is $a_n = 2^n - 2^{-n}$. Using the Ratio Test, $\lim_{n\to\infty} \left|\frac{a_{n+1}}{a_n}\right| = \lim_{n\to\infty} \frac{2^{n+1}-2^{-n-1}}{2^n-2^{-n}} = \lim_{n\to\infty} \frac{2-2^{-2n-1}}{1-2^{-2n}} = 2$, so the radius of convergence is $R = 1/2$. Since the base point is $c = 1$, the series converges absolutely on the open interval $I = \left(\frac{1}{2}, \frac{3}{2}\right)$.

33. The nth coefficient is $a_n = 3^n + 2^n$. Using the Ratio Test, $\lim_{n\to\infty} \left|\frac{a_{n+1}}{a_n}\right| = \lim_{n\to\infty} \frac{3^{n+1}+2^{n+1}}{3^n+2^n} = \lim_{n\to\infty} \frac{3+2\cdot(2/3)^n}{1+(2/3)^n} = 3$, so the radius of convergence is $R = 1/3$. Since the base point is $c = -2$, the series converges absolutely on the open interval $I = \left(-\frac{7}{3}, -\frac{5}{3}\right)$.

35. $f(x) = \sum_{n=1}^{\infty} nx^n$ so $f'(x) = \sum_{n=1}^{\infty} n^2x^{n-1} = \sum_{n=0}^{\infty}(n+1)^2x^n$. The fourth partial sum is

$$\begin{aligned}\sum_{n=0}^{4}(n+1)^2x^n &= 1 + 2^2x + 3^2x^2 + 4^2x^3 + 5^2x^4 \\ &= 1 + 4x + 9x^2 + 16x^3 + 25x^4\,.\end{aligned}$$

Using the Ratio Test for the differentiated series,

$$\lim_{n\to\infty}\left|\frac{a_{n+1}}{a_n}\right| = \lim_{n\to\infty}\frac{(n+2)^2}{(n+1)^2} = \lim_{n\to\infty}\frac{(1+2/n)^2}{(1+1/n)^2} = 1\,,$$

so $R = 1$ and the series converges absolutely on the interval $(-1, 1)$.

At the left endpoint, $x = -1$, and the series is $\sum_{n=0}^{\infty}(-1)^n(n+1)^2$ which diverges by the Divergence Test.

At the right endpoint, $x = 1$, and the series is $\sum_{n=0}^{\infty}(n+1)^2$ which also diverges by the Divergence Test.

The interval of convergence for $f'(x)$ is $I = (-1, 1)$.

37. $f(x) = \sum_{n=1}^{\infty} n^{-3/2}(x-1)^n$ so

$$f'(x) = \sum_{n=1}^{\infty} n^{-1/2}(x-1)^{n-1} = \sum_{n=0}^{\infty}(n+1)^{-1/2}(x-1)^n\,.$$

The fourth partial sum is

$$\sum_{n=0}^{4}(n+1)^{-1/2}(x-1)^n = 1 + \frac{x-1}{\sqrt{2}} + \frac{(x-1)^2}{\sqrt{3}} + \frac{(x-1)^3}{2} + \frac{(x-1)^4}{\sqrt{5}}\,.$$

Using the Ratio Test for the differentiated series,

$$\lim_{n\to\infty}\left|\frac{a_{n+1}}{a_n}\right| = \lim_{n\to\infty}\frac{(n+2)^{-1/2}}{(n+1)^{-1/2}} = \sqrt{\lim_{n\to\infty}\frac{n+1}{n+2}} = 1\,,$$

so $R = 1$. Since the base point is $c = 1$, the series converges absolutely on the interval $(0, 2)$.

At the left endpoint, $x = 0$, and the series is $\sum_{n=0}^{\infty}(-1)^n\frac{1}{\sqrt{n+1}}$ which converges by the Alternating Series Test.

At the right endpoint, $x = 2$, and the series is $\sum_{n=0}^{\infty}\frac{1}{\sqrt{n+1}} = \sum_{n=1}^{\infty}\frac{1}{\sqrt{n}}$ which, as a p-series, $p = 1/2$, diverges.

The interval of convergence for $f'(x)$ is $I = [0, 2)$.

39. $f(x) = \sum_{n=0}^{\infty} \sqrt{n+1}\, x^n$ so $F(x) = \sum_{n=0}^{\infty} \frac{\sqrt{n+1}}{n+1} x^{n+1} = \sum_{n=1}^{\infty} \frac{1}{\sqrt{n}} x^n$. The fourth partial sum is

$$\sum_{n=1}^{4} \frac{1}{\sqrt{n}} x^n = x + \frac{x^2}{\sqrt{2}} + \frac{x^3}{\sqrt{3}} + \frac{x^4}{2}.$$

Using the Root Test for the differentiated series,

$$\lim_{n\to\infty} |a_n|^{1/n} = \lim_{n\to\infty} (n^{-1/2})^{1/n} = \left(\lim_{n\to\infty} n^{1/n}\right)^{-1/2} = 1,$$

so $R = 1$ and the series converges absolutely on the interval $(-1, 1)$.

At the left endpoint, $x = -1$, and the series is $\sum_{n=1}^{\infty} (-1)^n \frac{1}{\sqrt{n}}$ which converges by the Alternating Series Test.

At the right endpoint, $x = 1$, and the series is $\sum_{n=0}^{\infty} \frac{1}{\sqrt{n}}$ which, as a p-series, $p = 1/2$, diverges.

The interval of convergence for $f'(x)$ is $I = [-1, 1)$.

41. $f(x) = \sum_{n=0}^{\infty} \frac{1}{3^n} (x-2)^n$ so

$$F(x) = \sum_{n=0}^{\infty} \frac{1}{(n+1)3^n} (x-2)^{n+1} = \sum_{n=1}^{\infty} \frac{1}{n\, 3^{n-1}} (x-2)^n.$$

The fourth partial sum is

$$\begin{aligned}\sum_{n=1}^{4} \frac{1}{n\, 3^{n-1}} (x-2)^n &= (x-2) + \frac{1}{2\cdot 3}(x-2)^2 + \frac{1}{3\cdot 3^2}(x-2)^3 + \frac{1}{4\cdot 3^3}(x-2)^4 \\ &= (x-2) + \frac{1}{6}(x-2)^2 + \frac{1}{27}(x-2)^3 + \frac{1}{108}(x-2)^4.\end{aligned}$$

Using the Ratio Test for the differentiated series,

$$\lim_{n\to\infty} \left|\frac{a_{n+1}}{a_n}\right| = \lim_{n\to\infty} \frac{1/((n+1)3^n)}{1/(n3^{n-1})} = \frac{1}{3} \lim_{n\to\infty} \frac{n}{n+1} = \frac{1}{3},$$

so $R = 3$. Since the base point is $c = 2$, the series converges absolutely on the interval $(-1, 5)$.

At the left endpoint, $x = -1$, and the series is $\sum_{n=1}^{\infty} 3\frac{(-1)^n}{n}$ which converges by the Alternating Series Test.

At the right endpoint, $x = 5$, and the series is $\sum_{n=1}^{\infty} \frac{3}{n}$ which, as a multiple of the Harmonic Series, diverges.

The interval of convergence for $f'(x)$ is $I = [-1, 5)$.

Further Theory and Practice

43. First put the series into power series form, $\sum_{n=0}^{\infty} n(n+1)(n+2)2^n \left(x+\frac{3}{2}\right)^n$, so $a_n = n(n+1)(n+2)2^n$. We use the Ratio Test,

$$\lim_{n\to\infty}\left|\frac{a_{n+1}}{a_n}\right| = \lim_{n\to\infty}\frac{(n+1)(n+2)(n+3)2^{n+1}}{n(n+1)(n+2)2^n} = 2\lim_{n\to\infty}\frac{n+3}{n} = 2\,,$$

implying that $R = \frac{1}{2}$.

45. We use the Ratio Test,

$$\lim_{n\to\infty}\left|\frac{a_{n+1}}{a_n}\right| = \lim_{n\to\infty}\frac{((n+1)!)^2/(2(n+1))!}{(n!)^2/(2n)!} = \lim_{n\to\infty}\left(\frac{(n+1)!}{n!}\right)^2\left(\frac{(2n)!}{(2(n+1))!}\right)$$
$$= \lim_{n\to\infty}\frac{(n+1)^2}{(2n+2)(2n+1)} = \frac{1}{4}\,.$$

implying that $R = 4$.

47. Put the series into power series form, $\sum_{n=1}^{\infty}\frac{(n!)^2 n^n}{(n^2)!}x^n$, so $a_n = \frac{(n!)^2 n^n}{(n^2)!}$. We use the Ratio Test,

$$\lim_{n\to\infty}\left|\frac{a_{n+1}}{a_n}\right| = \lim_{n\to\infty}\frac{((n+1)!)^2(n+1)^{n+1}/((n+1)^2)!}{(n!)^2 n^n/(n^2)!}$$
$$= \lim_{n\to\infty}(n+1)^2\cdot\left(\frac{n+1}{n}\right)^n\cdot(n+1)\cdot\frac{1}{(n+1)^2((n+1)^2-1)\cdots(n^2+1)}$$
$$= \lim_{n\to\infty}\left(1+\frac{1}{n}\right)^n\cdot\frac{n+1}{(n^2+1)(n^2+2)\cdots(n+1)^2}$$
$$= e\lim_{n\to\infty}\frac{n+1}{(n^2+1)(n^2+2)\cdots(n^2+2n+1)} = 0\,,$$

implying that $R = \infty$.

49. Make the substitution $t = (x+1)^2$ to obtain $\sum_{n=0}^{\infty}(-1)^n\frac{1}{9^n}t^n$, so $a_n = \frac{(-1)^n}{9^n}$. Using the Root Test, $\lim_{n\to\infty}|a_n|^{1/n} = \lim_{n\to\infty}\frac{1}{9} = \frac{1}{9}$, so $R = 9$ and the t-series converges absolutely for $|t| < 9$. This implies that the x-series converges absolutely for $|x+1| < 3$ so its radius of convergence is $R = 3$. Since the base point is $c = -1$ this series converges absolutely on the open interval $(-4, 2)$.

At the left endpoint, $x = -4$ and the series is $\sum_{n=0}^{\infty}(-1)^n\frac{(-3)^{2n}}{9^n} = \sum_{n=0}^{\infty}(-1)^n$, which diverges.

At the right endpoint, $x = 2$ and the series is $\sum_{n=0}^{\infty}(-1)^n\frac{3^{2n}}{9^n} = \sum_{n=0}^{\infty}(-1)^n$, which also diverges.

The interval of convergence is $I = (-4, 2)$.

51. First express the series in power series form: $\sum_{n=0}^{\infty}4^n\cdot 2^{2n}\left(x-\frac{1}{2}\right)^{2n} = \sum_{n=0}^{\infty}16^n\left(x-\frac{1}{2}\right)^{2n}$. Make the substitution $t = \left(x-\frac{1}{2}\right)^2$ to obtain the

series $\sum_{n=0}^{\infty} 16^n t^n$, so $a_n = 16^n$. Using the Root Test, $\lim_{n\to\infty} |a_n|^{1/n} = \lim_{n\to\infty} 16 = 16$, so $R = 1/16$ and the t-series converges absolutely for $|t| < 1/16$. This implies that the x-series converges absolutely for $\left|x - \frac{1}{2}\right| < \frac{1}{4}$ so its radius of convergence is $R = \frac{1}{4}$. Since the base point is $c = \frac{1}{2}$ this series converges absolutely on the open interval $\left(\frac{1}{4}, \frac{3}{4}\right)$.

At the left endpoint, $x = 1/4$ and the series is $\sum_{n=0}^{\infty} 4^n \left(-\frac{1}{2}\right)^{2n} = \sum_{n=0}^{\infty} 1$, which diverges.

At the right endpoint, $x = 3/4$ and the series is $\sum_{n=0}^{\infty} 4^n \left(\frac{1}{2}\right)^{2n} = \sum_{n=0}^{\infty} 1$, which also diverges.

The interval of convergence is $I = \left(\frac{1}{4}, \frac{3}{4}\right)$.

53. First express the series in power series form: $\sum_{n=1}^{\infty} \frac{2^{2n+1}}{n+1} x^{2n+1}$. Then factor out $2x$: $2x \sum_{n=1}^{\infty} \frac{4^n}{n+1} x^{2n}$, and make the substitution $t = x^2$ to obtain the series $\sum_{n=1}^{\infty} \frac{4^n}{n+1} t^n$. Consequently, $a_n = \frac{4^n}{n+1}$. Using the Ratio Test,

$$\lim_{n\to\infty} \left|\frac{a_{n+1}}{a_n}\right| = \lim_{n\to\infty} \frac{4^{n+1}/(n+2)}{4^n/(n+1)} = 4 \lim_{n\to\infty} \frac{n+1}{n+2} = 4,$$

so $R = 1/4$ and the t-series converges absolutely for $|t| < 1/4$. This implies that the x-series converges absolutely for $|x| < \frac{1}{2}$ so its radius of convergence is $R = \frac{1}{2}$. The series converges absolutely on the open interval $\left(-\frac{1}{2}, \frac{1}{2}\right)$.

At the both endpoints, $x = \pm 1/2$, and the series is $\sum_{n=1}^{\infty} \frac{1}{n+1}$, which diverges.

The interval of convergence is $I = \left(-\frac{1}{2}, \frac{1}{2}\right)$.

55. Factor out $x + e$: $(x + e) \sum_{n=0}^{\infty} \frac{(x+e)^{2n}}{3^n}$, and make the substitution $t = (x + e)^2$ to obtain the series $\sum_{n=0}^{\infty} \frac{t^n}{3^n}$. Consequently, $a_n = \frac{1}{3^n}$. Using the Root Test, $\lim_{n\to\infty} |a_n|^{1/n} = \lim_{n\to\infty} \frac{1}{3}$, so $R = 3$ and the t-series converges absolutely for $|t| < 1/3$. This implies that the x-series converges absolutely for $|x + e| < \frac{1}{\sqrt{3}}$ so its radius of convergence is $R = \frac{1}{\sqrt{3}}$. Since the base point is $-e$, the series converges absolutely on the open interval $\left(-e - \sqrt{3}, -e + \sqrt{3}\right)$.

At the both endpoints the series is $\sum_{n=0}^{\infty} 1$, which diverges.

The interval of convergence is $I = \left(-e - \sqrt{3}, \sqrt{3} - e\right)$.

57. Let c denote the midpoint of the interval (α, β), $c = \frac{1}{2}(\alpha + \beta)$. Then the power series $\sum_{n=0}^{\infty} \frac{(x-c)^n}{\left(\frac{1}{2}(\beta-\alpha)\right)^n}$ has, by the Root Test, the radius of convergence $R = \frac{1}{2}(\beta - \alpha)$ and the series converges absolutely on the open interval (α, β).

At the left endpoint $x = \alpha$ the series is $\sum_{n=0}^{\infty}(-1)^n$ which diverges.

At the right endpoint $x = \beta$ the series is $\sum_{n=0}^{\infty} 1$ which also diverges.

The interval of convergence is $I = (\alpha, \beta)$.

59. Let c denote the midpoint of the interval (α, β), $c = \frac{1}{2}(\alpha + \beta)$. Then the power series $\sum_{n=0}^{\infty} \frac{(x-c)^n}{n\left(\frac{1}{2}(\beta-\alpha)\right)^n}$ has, by the Root Test, the radius of convergence $R = \frac{1}{2}(\beta - \alpha)$ and the series converges absolutely on the open interval (α, β).

At the left endpoint $x = \alpha$ the series is $\sum_{n=0}^{\infty} \frac{(-1)^n}{n}$ which converges.

At the right endpoint $x = \beta$ the series is $\sum_{n=0}^{\infty} \frac{1}{n}$ which diverges.

The interval of convergence is $I = [\alpha, \beta)$.

61. This is not possible, what would be the base point?

63. The series $\sum_{k=1}^{\infty} \frac{1}{k^2}$ converges, let S denote its sum. That is, $S = \lim_{n\to\infty} \sum_{k=1}^{n} \frac{1}{k^2}$. Using the Ratio Test,

$$\ell = \lim_{n\to\infty} \frac{\sum_{k=1}^{n+1} \frac{1}{k^2}}{\sum_{k=1}^{n} \frac{1}{k^2}} = \frac{\lim_{n\to\infty} \sum_{k=1}^{n+1} \frac{1}{k^2}}{\lim_{n\to\infty} \sum_{k=1}^{n} \frac{1}{k^2}} = \frac{S}{S} = 1\,,$$

and the radius of convergence is $R = \frac{1}{\ell} = 1$.

65. Applying the Ratio Test to $\sum_{n=1}^{\infty} a_n x^n$, this series converges absolutely for all x such that $\lim_{n\to\infty} \left|\frac{a_{n+1}x^{n+1}}{a_n x^n}\right| < 1$, or, $\lim_{n\to\infty} \left(\left|\frac{a_{n+1}}{a_n}\right| \cdot |x|\right) < 1$. Equivalently, $|x| \lim_{n\to\infty} \left|\frac{a_{n+1}}{a_n}\right| < 1$.

If $\lim_{n\to\infty} \left|\frac{a_{n+1}}{a_n}\right| = \ell > 0$, then absolute convergence is for all x such that $|x| < 1/\ell$, and $R = 1/\ell$.

If $\lim_{n\to\infty} \left|\frac{a_{n+1}}{a_n}\right| = 0$, then absolute convergence is for all x, and $R = \infty$.

67. The radius of convergence is $R = 1$. To see why observe that, for a fixed x positive, the series can be expressed as the sum of two series,

$$\sum_{n=1}^{\infty} a_n x^n = \sum_{n=0}^{\infty} (2n+1)x^{2n+1} + \sum_{n=1}^{\infty} (4n-2)x^{2n}\,.$$

The given series on the left will converge if, and only if, the two series on the right both converge.

By the Ratio Test, as it is applied in Exercise 65, both series on the right converge when $0 < x < 1$ and diverge when $x > 1$.

CALCULATOR/COMPUTER EXERCISES

69. (a) The plot on the right displays the graph of $f(x) = \frac{1}{1-x}$ with a thick curve and the partial sums $S_2(x)$, $S_4(x)$, and $S_6(x)$ as dashed curves. The plot of $S_6(x)$ is only barely visible.

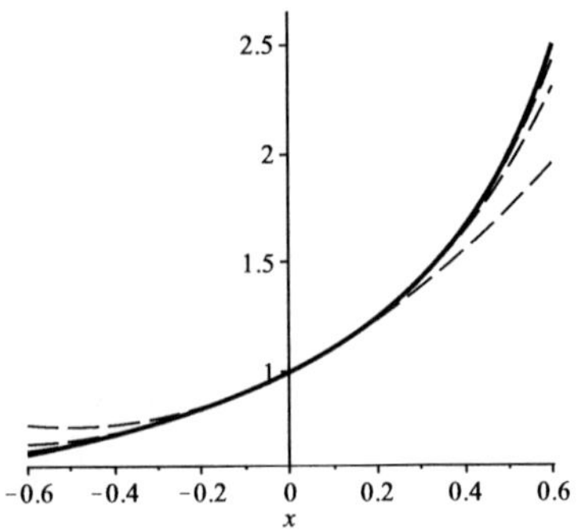

(b) The plot on the right is over the interval $-0.95 \leq x \leq -0.6$. The plot of $S_6(x)$ has been replaced with the a plot of $S_{50}(x)$. It is barely visible.

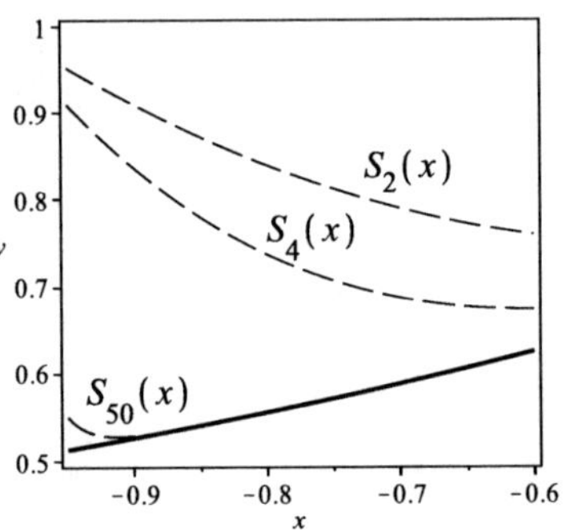

(c) The plot on the right is over the interval $0.6 \leq x \leq 0.98$. The plot of $S_6(x)$ has been restored as well as the plot of $S_{50}(x)$. It is slightly more visible.

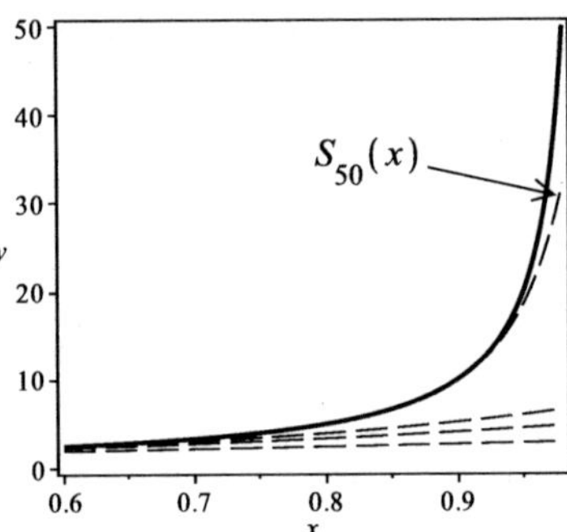

71. (a) The plot on the right displays the graph of $f(x) = \ln(1+x)$ with a thick curve and the partial sums $S_2(x)$, $S_4(x)$, and $S_6(x)$ as dashed curves. All three approximations are clearly visible at the right side of the plot.

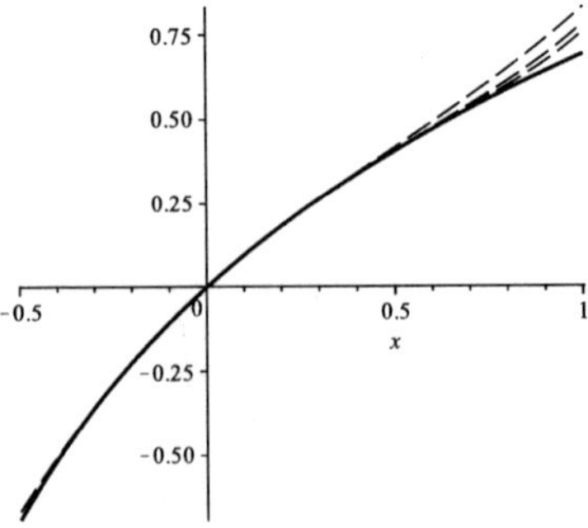

(b) The plot on the right is over the interval $-0.99 \le x \le -0.5$. The plots of $S_2(x)$ and $S_4(x)$ look almost like straight lines. The graph of $S_6(x)$ has been replaced with the the graph of $S_{50}(x)$. It is barely visible.

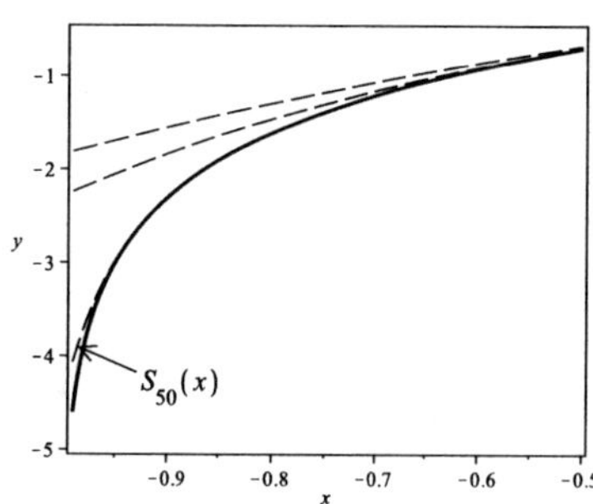

8.7 Representing Functions by Power Series

Problems for Practice

1. As long as $|2x| < 1$, $\frac{1}{1-2x} = 1+2x+(2x)^2+\cdots = \sum_{n=0}^{\infty} 2^n x^n$. Equivalently, this is valid when $|x| < \frac{1}{2}$, so the radius of convergence is $R = \frac{1}{2}$.

3. Observe that $\frac{1}{2-x} = \frac{1}{2} \cdot \frac{1}{1-\frac{x}{2}}$. Therefore, when $\left|\frac{x}{2}\right| < 1$,

$$\frac{1}{2-x} = \frac{1}{2}\left(1 + \left(\frac{x}{2}\right) + \left(\frac{x}{2}\right)^2 + \cdots\right) = \frac{1}{2}\sum_{n=0}^{\infty} \frac{1}{2^n} x^n = \sum_{n=0}^{\infty} \frac{1}{2^{n+1}} x^n .$$

Equivalently, this is valid when $|x| < 2$, so the radius of convergence is $R = 2$.

5. Observe that $\frac{1}{4+x} = \frac{1}{4} \cdot \frac{1}{1-\left(-\frac{x}{4}\right)}$. Therefore, when $\left|\frac{x}{4}\right| < 1$,

$$\frac{1}{4+x} = \frac{1}{4}\left(1 + \left(-\frac{x}{4}\right) + \left(-\frac{x}{4}\right)^2 + \cdots\right) = \frac{1}{4}\sum_{n=0}^{\infty} \frac{(-1)^n}{4^n} x^n = \sum_{n=0}^{\infty} \frac{(-1)^n}{4^{n+1}} x^n .$$

Equivalently, this is valid when $|x| < 4$, so the radius of convergence is $R = 4$.

7. Observe that $\frac{x^2+1}{1-x^2} = (x^2+1) \cdot \frac{1}{1-x^2}$. Therefore, when $\left|x^2\right| < 1$,

$$\begin{aligned}\frac{x^2+1}{1-x^2} &= (x^2+1) \cdot \frac{1}{1-x^2} = (x^2+1)\left(1 + (x^2) + (x^2)^2 + \cdots\right) \\ &= (x^2+1)\sum_{n=0}^{\infty} x^{2n} = \sum_{n=0}^{\infty} x^{2n+2} + \sum_{n=0}^{\infty} x^{2n} \\ &= \sum_{n=1}^{\infty} x^{2n} + \sum_{n=0}^{\infty} x^{2n} = 1 + \sum_{n=1}^{\infty} 2x^{2n} .\end{aligned}$$

Equivalently, this is valid when $|x| < 1$, so the radius of convergence is $R = 1$.

9. Observe that $\frac{x^3}{1+x^4} = x^3 \cdot \frac{1}{1-(-x^4)}$. Therefore, when $|x^4| < 1$,

$$\frac{x^3}{1+x^4} = x^3\left(1 + (-x^4) + (-x^4)^2 + \cdots\right) = \sum_{n=0}^{\infty}(-1)^n x^{4n+3}.$$

Equivalently, this is valid when $|x| < 1$, so the radius of convergence is $R = 1$.

11. Using (8.2.7b), as long as $|4x| < 1$,

$$\ln(1+4x) = 4x - \frac{1}{2}(4x)^2 + \frac{1}{3}(4x)^3 - \frac{1}{4}(4x)^4 + \cdots = \sum_{n=1}^{\infty}\frac{(-1)^{n-1}4^n}{n}x^n.$$

Equivalently, this is valid when $|x| < \frac{1}{4}$, so the radius of convergence is $R = \frac{1}{4}$.

13. Using (8.2.7b), as long as $|x^2| < 1$,

$$\ln(1+x^2) = x^2 - \frac{1}{2}(x^2)^2 + \frac{1}{3}(x^2)^3 - \frac{1}{4}(x^2)^4 + \cdots = \sum_{n=1}^{\infty}\frac{(-1)^{n-1}}{n}x^{2n}.$$

Equivalently, this is valid when $|x| < 1$, so the radius of convergence is $R = 1$.

15. Integrate the series in Exercise 13,

$$\int_0^x \ln(1+t^2)\,dt = \sum_{n=1}^{\infty}\int_0^x \frac{(-1)^{n-1}}{n}t^{2n}\,dt = \sum_{n=1}^{\infty}\frac{(-1)^{n-1}}{n(2n+1)}x^{2n+1}.$$

As for the series in Exercise 13, this is valid when $|x| < 1$, so the radius of convergence is $R = 1$.

17. Using (8.7.3b), as long as $|x^2| < 1$,

$$\arctan(x^2) = x^2 - \frac{1}{3}(x^2)^3 + \frac{1}{5}(x^2)^5 - \frac{1}{7}(x^2)^7 + \cdots = \sum_{n=0}^{\infty}\frac{(-1)^n x^{4n+2}}{2n+1}.$$

Equivalently, this is valid when $|x| < 1$, so the radius of convergence is $R = 1$.

19. Integrate the series in Exercise 17,

$$\int_0^x \arctan(t^2)\,dt = \sum_{n=0}^{\infty}\int_0^x \frac{(-1)^n t^{4n+2}}{2n+1}\,dt = \sum_{n=0}^{\infty}\frac{(-1)^n x^{4n+3}}{(2n+1)(4n+3)}.$$

As for the series in Exercise 17, this is valid when $|x| < 1$, so the radius of convergence is $R = 1$.

21. $\dfrac{1}{6-x}=\dfrac{1}{1-(x-5)}=\sum_{n=0}^{\infty}(x-5)^n$, $|x-5|<1$ and $R=1$

23. $\frac{1}{x+1}=\frac{1}{(x+3)-2}=-\frac{1}{2}\cdot\frac{1}{1-\frac{1}{2}(x+3)}=-\frac{1}{2}\sum_{n=0}^{\infty}\left(\frac{1}{2}(x+3)\right)^n$, for all x such that $\left|\frac{1}{2}(x+3)\right|<1$. Therefore, $\frac{1}{x+1}=\sum_{n=0}^{\infty}\frac{(-1)}{2^{n+1}}(x+3)^n$, provided $|x+3|<2$. The radius of convergence is $R=2$.

25. $\frac{1}{2x+5}=\frac{1}{2(x+1)+3}=\frac{1}{3}\cdot\frac{1}{1+\frac{2}{3}(x+1)}=\frac{1}{3}\sum_{n=0}^{\infty}\left(-\frac{2}{3}(x+1)\right)^n$, for all x such that $\left|\frac{2}{3}(x+1)\right|<1$. Therefore, $\frac{1}{2x+5}=\sum_{n=0}^{\infty}(-1)^n\frac{2^n}{3^{n+1}}(x+1)^n$, provided $|x+1|<\frac{3}{2}$. The radius of convergence is $R=\frac{3}{2}$.

27. The data in the following table

n	$f^{(n)}(x)$	$f^{(n)}(\pi/3)$	$\frac{f^{(n)}(\pi/3)}{n!}$
0	$\cos(x)$	$\frac{1}{2}$	$\frac{1}{2}$
1	$-\sin(x)$	$-\frac{\sqrt{3}}{2}$	$-\frac{\sqrt{3}/2}{1!}$
2	$-\cos(x)$	$-\frac{1}{2}$	$-\frac{1/2}{2!}$
3	$\sin(x)$	$\frac{\sqrt{3}}{2}$	$\frac{\sqrt{3}/2}{3!}$
4	$\cos(x)$	$\frac{1}{2}$	$\frac{1/2}{4!}$

implies that

$$T_4(x)=\frac{1}{2}-\frac{\sqrt{3}}{2}\left(x-\frac{\pi}{3}\right)-\frac{1}{4}\left(x-\frac{\pi}{3}\right)^2+\frac{\sqrt{3}}{12}\left(x-\frac{\pi}{3}\right)^3+\frac{1}{48}\left(x-\frac{\pi}{3}\right)^4.$$

29. The data in the following table

n	$f^{(n)}(x)$	$f^{(n)}(4)$	$\frac{f^{(n)}(4)}{n!}$
0	$\sqrt{x}$	2	2
1	$\frac{1}{2}x^{-1/2}$	$\frac{1}{4}$	$\frac{1/4}{1!}$
2	$-\frac{1}{4}x^{-3/2}$	$-\frac{1}{32}$	$-\frac{1/32}{2!}$
3	$\frac{3}{8}x^{-5/2}$	$\frac{3}{256}$	$\frac{3/265}{3!}$

implies that

$$T_3(x)=2+\frac{1}{4}(x-4)-\frac{1}{64}(x-4)^2+\frac{1}{512}(x-4)^3.$$

31. The data in the following table

n	$f^{(n)}(x)$	$f^{(n)}(1)$	$\frac{f^{(n)}(1)}{n!}$
0	$5x^2 + 1/x$	6	6
1	$10x - x^{-2}$	9	$\frac{9}{1!}$
2	$10 + 2x^{-3}$	12	$\frac{12}{2!}$
3	$-6x^{-4}$	-6	$\frac{-6}{3!}$
4	$24x^{-5}$	24	$\frac{24}{4!}$

implies that

$$T_4(x) = 6 + 9(x-1) + 6(x-1)^2 - (x-1)^3 + (x-1)^4 .$$

33. The data in the following table

n	$f^{(n)}(x)$	$f^{(n)}(2)$	$\frac{f^{(n)}(2)}{n!}$
0	x^{-3}	$\frac{1}{8}$	$\frac{1}{8}$
1	$-3x^{-4}$	$-\frac{3}{16}$	$\frac{-3/16}{1!}$
2	$12x^{-5}$	$\frac{12}{32}$	$\frac{3/8}{2!}$
3	$-60x^{-6}$	$-\frac{60}{64}$	$-\frac{15/16}{3!}$

implies that

$$T_3(x) = \frac{1}{8} - \frac{3}{16}(x-2) + \frac{3}{16}(x-2)^2 - \frac{5}{32}(x-2)^3 .$$

35. The data in the following table

n	$f^{(n)}(x)$	$f^{(n)}(-2)$	$\frac{f^{(n)}(-2)}{n!}$
0	$\ln(3x+7)$	0	0
1	$\frac{3}{3x+7}$	3	$\frac{3}{1!}$
2	$-9(3x+7)^{-2}$	-9	$\frac{-9}{2!}$
3	$54(3x+7)^{-3}$	54	$\frac{54}{3!}$
4	$-486(3x+7)^{-4}$	-486	$\frac{-486}{4!}$

implies that

$$T_4(x) = 3(x+2) - \frac{9}{2}(x+2)^2 + 9(x+2)^3 - \frac{81}{4}(x+2)^4 .$$

37. The data in the following table

n	$f^{(n)}(x)$	$f^{(n)}(0)$	$\frac{f^{(n)}(0)}{n!}$
0	$\tan(x)$	0	0
1	$\sec^2(x)$	1	$\frac{1}{1!}$
2	$2\sec^2(x)\tan(x)$	0	0
3	$4\sec^2(x)\tan^2(x) + 2\sec^4(x)$	2	$\frac{2}{3!}$

implies that

$$T_3(x) = x + \frac{1}{3}x^3 .$$

Further Theory and Practice

39. The data in the following table

n	$f^{(n)}(x)$	$f^{(n)}(1)$	$\frac{f^{(n)}(1)}{n!}$
0	$\frac{x}{x^2+1}$	$\frac{1}{2}$	$\frac{1}{2}$
1	$-\frac{x^2-1}{(x^2+1)^2}$	0	0
2	$\frac{2x(x^2-3)}{(x^2+1))^3}$	$-\frac{1}{2}$	$-\frac{1/2}{2!}$
3	$-\frac{6(x^4-6x^2+1)}{(x^2+1)^4}$	$\frac{3}{2}$	$\frac{3/2}{3!}$

implies that

$$T_3(x) = \frac{1}{2} - \frac{1}{4}(x-1)^2 + \frac{1}{4}(x-1)^3 .$$

41. The data in the following table

n	$f^{(n)}(x)$	$f^{(n)}(3)$	$\frac{f^{(n)}(3)}{n!}$
0	$(x+5)^{1/3}$	2	2
1	$\frac{1}{3}(x+5)^{-2/3}$	$\frac{1}{12}$	$\frac{1/12}{1!}$
2	$-\frac{2}{9}(x+5)^{-5/3}$	$-\frac{1}{144}$	$-\frac{1/144}{2!}$
3	$\frac{10}{27}(x+5)^{-8/3}$	$\frac{5}{3456}$	$\frac{5/3456}{3!}$

implies that

$$T_3(x) = 2 + \frac{1}{12}(x-3) - \frac{1}{288}(x-3)^2 + \frac{5}{20736}(x-3)^3 .$$

43. $$\frac{1}{(1-x)^2} = \frac{d}{dx}\left(\sum_{n=0}^{\infty} x^n\right) = \sum_{n=1}^{\infty} nx^{n-1} = \sum_{n=0}^{\infty}(n+1)x^n$$

45. Using the result in Exercise 43,

$$\frac{2+x}{(1-x)^2} = (2+x)\sum_{n=0}^{\infty}(n+1)x^n = \sum_{n=0}^{\infty}2(n+1)x^n + \sum_{n=0}^{\infty}(n+1)x^{n+1}$$
$$= \sum_{n=0}^{\infty}2(n+1)x^n + \sum_{n=1}^{\infty}nx^n = \sum_{n=0}^{\infty}(3n+2)x^n .$$

47. $\displaystyle\sum_{n=0}^{\infty}x^{n+2} = x^2\sum_{n=0}^{\infty}x^n = \frac{x^2}{1-x}$, provided $|x|<1$, so $R=1$.

49. Let $f(x) = \sum_{n=1}^{\infty} n(n+1)x^n$. Then

$$F(x) = \int_0^x f(t)\,dt = \sum_{n=1}^{\infty} nx^{n+1} = x^2\sum_{n=1}^{\infty}nx^{n-1} = x^2\frac{d}{dx}\left(\sum_{n=0}^{\infty}x^n\right)$$
$$= x^2\frac{d}{dx}\left(\frac{1}{1-x}\right) = \frac{x^2}{(1-x)^2}\,, \quad \text{assuming } |x|<1\,.$$

Consequently, $f(x) = F'(x) = \frac{2x}{(1-x)^3}$. This is valid assuming that $|x|<1$ so $R=1$.

51. Let $f(x) = \sum_{n=1}^{\infty} nx^{2n+1}$. Then $f(x) = x^2\sum_{n=1}^{\infty}nx^{2n-1} = x^2g(x)$, where $g(x) = \sum_{n=1}^{\infty}nx^{2n-1}$. Integrate g to obtain

$$G(x) = \int_0^x g(t)\,dt = \sum_{n=1}^{\infty}\frac{1}{2}x^{2n} = \frac{1}{2}\cdot\frac{x^2}{1-x^2}\,, \quad \text{assuming } |x|<1\,.$$

Consequently, $g(x) = G'(x) = \frac{x}{(x^2-1)^2}$, and $f(x) = \frac{x^3}{(x^2-1)^2}$.

This is valid provided $|x|<1$ so $R=1$.

53. Substitute the series for $y(x)$ in the differential equation to obtain $\frac{d}{dx}\left(\sum_{n=0}^{\infty}a_nx^n\right) = \sum_{n=0}^{\infty}a_nx^n$. That is, $\sum_{n=1}^{\infty}na_nx^{n-1} = \sum_{n=0}^{\infty}a_nx^n$, or, upon replacing n with $n+1$ on the left, $\sum_{n=0}^{\infty}(n+1)a_{n+1}x^n = \sum_{n=0}^{\infty}a_nx^n$.

By the uniqueness of the coefficients in a power series, we may infer that for all $n\geq 0$, $(n+1)a_{n+1} = a_n$. Equivalently, $a_{n+1} = \frac{a_n}{n+1}$, $n = 0,1,2,\ldots$. Since $y(0) = a_0 = 1$,

$$a_1 = \frac{a_0}{1} = 1\,,\ a_2 = \frac{a_1}{2} = \frac{1}{2}\,,\ a_3 = \frac{a_2}{3} = \frac{1}{1\cdot 2\cdot 3}\,,$$

and, in general, $a_n = \frac{1}{n!}$. The solution is $y(x) = \sum_{n=0}^{\infty}\frac{x^n}{n!}$.

55. Substitute the series for $y(x)$ in the differential equation to obtain the equation $\frac{d}{dx}\left(\sum_{n=0}^{\infty} a_n x^n\right) = x + \sum_{n=0}^{\infty} a_n x^n$. That is, $\sum_{n=1}^{\infty} n a_n x^{n-1} = x + \sum_{n=0}^{\infty} a_n x^n$, or, upon replacing n with $n+1$ on the left,

$$\sum_{n=0}^{\infty}(n+1)a_{n+1}x^n = a_0 + (a_1+1)x + \sum_{n=2}^{\infty} a_n x^n .$$

By the uniqueness of the coefficients in a power series, we may infer that

$$a_1 = a_0\,,\ 2a_2 = a_1 + 1\,,\ \text{and for all } n \geq 2\,,\ (n+1)a_{n+1} = a_n\,.$$

Since $a_0 = y(0) = 1$, $a_1 = 1$, $a_2 = \frac{1}{2}(1+1) = 1$, and $a_{n+1} = \frac{a_n}{n+1}$ for $n = 2, 3, 4, \ldots$. That is, $a_3 = \frac{a_2}{3} = \frac{1}{3}$, $a_4 = \frac{a_3}{4} = \frac{1}{3\cdot 4}$, $a_5 = \frac{a_4}{5} = \frac{1}{3\cdot 4\cdot 5}$, and so on. In general, for $n \geq 3$, $a_n = \frac{2}{n!}$. The solution is

$$y(x) = 1 + x + x^2 + \sum_{n=3}^{\infty}\frac{2x^n}{n!} = 1 + x + \sum_{n=2}^{\infty}\frac{2x^n}{n!} .$$

57. Substitute the series for $y(x)$ in the differential equation to obtain the equation $\frac{d}{dx}\left(\sum_{n=0}^{\infty} a_n x^n\right) = 1 + x + \sum_{n=0}^{\infty} a_n x^n$. That is, $\sum_{n=1}^{\infty} n a_n x^{n-1} = 1 + x + \sum_{n=0}^{\infty} a_n x^n$, or, upon replacing n with $n+1$ on the left,

$$\sum_{n=0}^{\infty}(n+1)a_{n+1}x^n = a_0 + 1 + (a_1+1)x + \sum_{n=2}^{\infty} a_n x^n .$$

By the uniqueness of the coefficients in a power series, we may infer that

$$a_1 = a_0 + 1\,,\ 2a_2 = a_1 + 1\,,\ \text{and for all } n \geq 2\,,\ (n+1)a_{n+1} = a_n\,.$$

Since $a_0 = y(0) = 0$, $a_1 = 1$, $a_2 = \frac{1}{2}(1+1) = 1$, and $a_{n+1} = \frac{a_n}{n+1}$ for $n = 2, 3, 4, \ldots$. That is, $a_3 = \frac{a_2}{3} = \frac{1}{3}$, $a_4 = \frac{a_3}{4} = \frac{1}{3\cdot 4}$, $a_5 = \frac{a_4}{5} = \frac{1}{3\cdot 4\cdot 5}$, and so on. In general, for $n \geq 3$, $a_n = \frac{2}{n!}$. The solution is

$$y(x) = x + x^2 + \sum_{n=3}^{\infty}\frac{2x^n}{n!} = x + \sum_{n=2}^{\infty}\frac{2x^n}{n!} .$$

59. Using Exercise 43, $h(x) = \frac{1}{(1-x)^2} = \frac{d}{dx}\left(\sum_{n=0}^{\infty} x^n\right) = \sum_{n=1}^{\infty} n x^{n-1} = \sum_{n=0}^{\infty}(n+1)x^n$. That is, $h(x) = 1 + 2x + 3x^2 + 4x^3 + 5x^4 + 6x^5 + 7x^6 + \cdots$.

Expressing $h(x) = f(x)\cdot g(x) = \frac{1}{1-x}\cdot\frac{1}{1-x}$,

$$h(x) = \left(\sum_{n=0}^{\infty} x^n\right)\left(\sum_{m=0}^{\infty} x^m\right) = \sum_{n=0}^{\infty}\left(\sum_{k=0}^{n} 1\right)x^n = \sum_{n=0}^{\infty}(n+1)x^n ,$$

as above.

61. Using the given series, and expanding as polynomials,

$$\tan(x) = \sin(x)\sec(x) = \left(x - \frac{x^3}{3!} + \frac{x^5}{5!} - \frac{x^7}{7!} + \cdots\right)\left(1 + \frac{1}{2}x^2 + \frac{5}{24}x^4 + \frac{61}{720}x^6 + \cdots\right)$$
$$= x + \left(\frac{1}{2} - \frac{1}{3!}\right)x^3 + \left(\frac{1}{5!} - \frac{1}{2\cdot 3!} + \frac{5}{24}\right)x^5 + \left(\frac{61}{720} - \frac{5}{24\cdot 3!} + \frac{1}{2\cdot 5!} - \frac{1}{7!}\right)x^7 + \cdots$$
$$= x + \frac{1}{3}x^3 + \frac{2}{15}x^5 + \frac{17}{315}x^7 + \cdots .$$

CALCULATOR/COMPUTER EXERCISES

63. (a) The plot on the right displays the graph of the central difference quotient and the derivative of the ninth degree polynomial approximation to f. The difference is barely noticable at the ends of the interval.

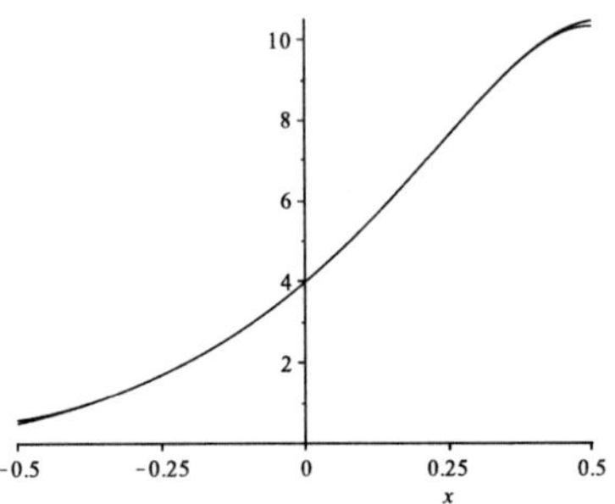

(b) The plot on the right displays the graph of the central difference quotient and the derivative of the ninth degree polynomial approximation to f over a wider interval. The dashed curve is the graph of the polynomial. The difference is quite pronounced near the ends of the interval.

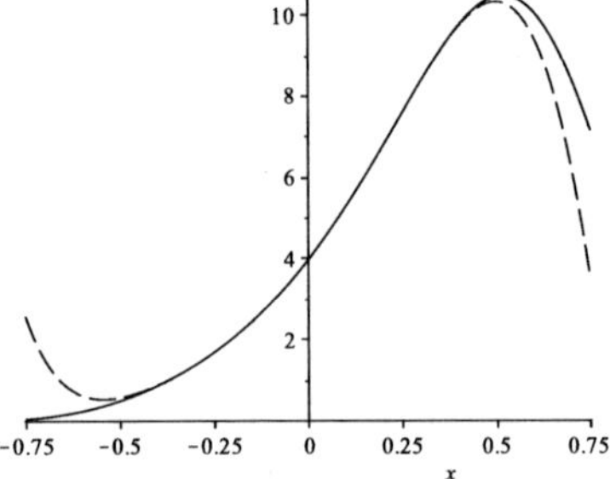

65. (a) Substitute the series for $y(x)$ in the differential equation to obtain the equation $\frac{d}{dx}\left(\sum_{n=0}^{\infty} a_n x^n\right) = x^2 + \sum_{n=0}^{\infty} a_n x^n$. That is, $\sum_{n=1}^{\infty} n a_n x^{n-1} = x^2 + \sum_{n=0}^{\infty} a_n x^n$, or, upon replacing n with $n+1$ on the left,

$$\sum_{n=0}^{\infty}(n+1)a_{n+1}x^n = a_0 + a_1 x + (a_2+1)x^2 + \sum_{n=3}^{\infty} a_n x^n .$$

By the uniqueness of the coefficients in a power series, we may infer that

$$a_1 = a_0\,,\ 2a_2 = a_1\,,\ 3a_3 = a_2 + 1\,,\ \text{and for all } n \ge 3\,,\ (n+1)a_{n+1} = a_n\,.$$

Since $a_0 = y(0) = 1$, $a_1 = 1$, $a_2 = \frac{1}{2}$, $a_3 = \frac{1}{3}\left(\frac{1}{2}+1\right) = \frac{1}{2}$ and $a_{n+1} = \frac{a_n}{n+1}$ for $n = 3, 4, 5, \ldots$. That is, $a_4 = \frac{a_3}{4} = \frac{1}{2\cdot 4}$, $a_5 = \frac{a_4}{5} = \frac{1}{2\cdot 4\cdot 5}$, $a_6 = \frac{a_5}{6} = \frac{1}{2\cdot 4\cdot 5\cdot 6}$, and so on. In general, for $n \ge 3$, $a_n = \frac{3}{n!}$.

The solution is $y(x) = 1 + x + \frac{1}{2}x^2 + \sum_{n=3}^{\infty} \frac{3x^n}{n!}$. The solution, up to the 7th term, is

$$y(x) = 1 + x + \frac{1}{2}x^2 + \frac{1}{2}x^3 + \frac{1}{8}x^4 + \frac{1}{40}x^5 + \frac{1}{240}x^6 + \frac{1}{1680}x^7 + \cdots .$$

(b) The plot on the right displays the graph of $S_3(x) = 1 + x + \frac{1}{2}x^2 + \frac{1}{2}x^3$ in the requested viewing rectangle.

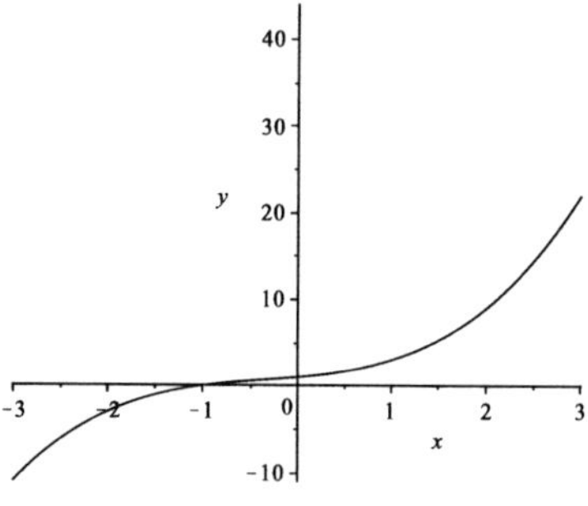

(c) This plot displays the graph of $S_3(x)$ and the solution

$$y(x) = 3e^x - x^2 - 2x - 2$$

in the same viewing rectangle. The solution is the dashed curve.

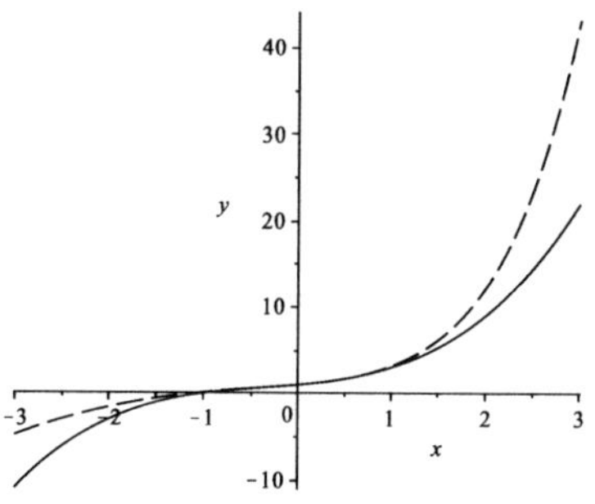

(d) The graph of $S_3(x)$ has been replaced with the graph of the approximation $S_7(x)$. As in part c, the solution is the dashed curve, but it is almost indistinguishable from the approximation over this interval.

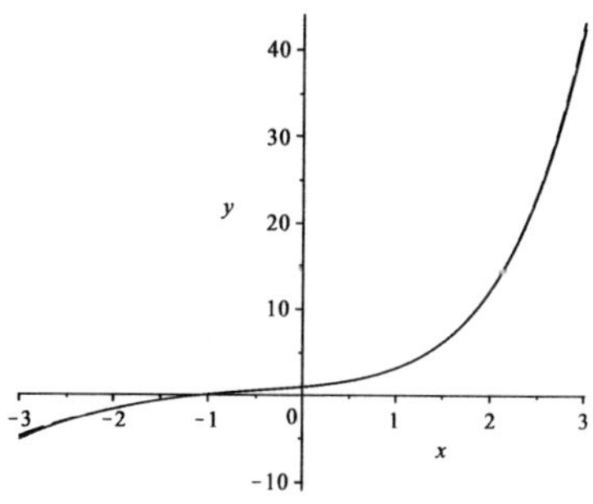

8.8 Taylor Series

Problems for Practice

1. The fourth order Taylor approxmation for $f(0.1) = e^{0.1}$ (base point $c = 0$),

with error term, is

$$\begin{aligned} f(0.1) &= 1 + 0.1 + \frac{1}{2!}(0.1)^2 + \frac{1}{3!}(0.1)^4 + \frac{1}{4!}(0.1)^4 + \frac{f^{(5)}(s)}{5!}(0.1)^5 \\ &= 1.1051708 + \frac{e^s}{5!}(0.1)^5 \quad , \quad \text{where } 0 \le s \le 0.1\,. \end{aligned}$$

Since $f(0.1) = e^{0.1}$, the number s must be a solution of the equation $\frac{e^s}{5!}(0.1)^5 = e^{0.1} - 1.1051708$. Using our solve button, $s = 0.0198026$, which is between 0 and 0.1.

3. The fifth order Taylor approxmation for $f(0.3) = \sin(0.3)$ (base point $c = 0$), with error term, is

$$\begin{aligned} f(0.3) &= 0.3 - \frac{1}{3!}(0.3)^3 + \frac{1}{5!}(0.3)^5 + \frac{f^{(6)}(s)}{6!}(0.3)^6 \\ &= 0.29552025 + \frac{(-\sin(s))}{6!}(0.3)^6 \quad , \quad \text{where } 0 \le s \le 0.3\,. \end{aligned}$$

Since $f(0.3) = \sin(0.3)$, the number s must be a solution of the equation $\frac{(-\sin(s))}{6!}(0.3)^6 = \sin(0.3) - 0.29552025$. Using our solve button, $s = 0.04277848$, which is between 0 and 0.3.

5. Let $f(x) = \cos(x)$. Since $f(\pi/6) = \cos(\pi/6) = \sqrt{3}/2$, $f'(\pi/6) = -\sin(\pi/6) = -1/2$, and $f''(\pi/6) = -\cos(\pi/6) = -\sqrt{3}/2$, the second order Taylor approxmation for $f(0.5)$, (base point $c = \pi/6$), with error term, is

$$\begin{aligned} f(0.5) &= \frac{\sqrt{3}}{2} - \frac{1}{2}\left(0.5 - \frac{\pi}{6}\right) - \frac{\sqrt{3}/2}{2!}\left(0.5 - \frac{\pi}{6}\right)^2 + \frac{f^{(3)}(s)}{3!}\left(0.5 - \frac{\pi}{6}\right)^3 \\ &= 0.8775836 + \frac{\sin(s)}{3!}\left(0.5 - \frac{\pi}{6}\right)^3 \quad , \quad \text{where } 0.5 \le s \le \pi/6\,. \end{aligned}$$

Since $f(0.5) = \cos(0.5)$, the number s must be a solution of the equation $\frac{\sin(s)}{3!}(0.5 - \pi/6)^3 = \cos(0.5) - 0.8775836$. Using our solve button, $s = 0.5178178$, which is between 0.5 and $\pi/6 = 0.5235988$.

7. Let $f(x) = \ln(x)$. Since $f(e) = 1$, $f'(e) = 1/e$, $f''(e) = -1/e^2$, and $f'''(e) = 2/e^3$, the second order Taylor approxmation for $f(3)$, (base point $c = e$), with error term, is

$$\begin{aligned} f(3) &= 1 + \frac{1}{e}(3 - e) - \frac{1/e^2}{2!}(3 - e)^2 + \frac{2/e^3}{3!}(3 - e)^3 + \frac{f^{(4)}(s)}{4!}(3 - e)^4 \\ &= 1.0986389 + \frac{(-6/s^4)}{4!}(3 - e)^4 \quad , \quad \text{where } e \le s \le 3\,. \end{aligned}$$

Since $f(3) = \ln(3)$, the number s must be a solution of the equation $\frac{(-6/s^4)}{4!}(3 - e)^4 = \ln(3) - 1.0986389$. Using our solve button, $s = 2.7727771$, which is between $e = 2.7182818$ and 3.

9. Let $f(x) = \arctan(x)$. Since $f(\sqrt{3}) = \pi/3$, $f'(\sqrt{3}) = \frac{1}{1+3}$, $f''(\sqrt{3}) = -\frac{2\sqrt{3}}{(1+3)^2}$, and $f^{(3)}(s) = \frac{2(3s^2-1)}{(1+s^2)^3}$, the second order Taylor approxmation for $f(1.5)$, (base point $c = \sqrt{3}$), with error term, is

$$\begin{aligned} f(1.5) &= \frac{\pi}{3} + \frac{1}{4}\left(1.5-\sqrt{3}\right) - \frac{2\sqrt{3}/16}{2!}\left(1.5-\sqrt{3}\right)^2 + \frac{f^{(3)}(s)}{3!}\left(1.5-\sqrt{3}\right)^3 \\ &= 0.9833557 + \frac{2(3s^2-1)/(1+s^2)^3}{3!}\left(1.5-\sqrt{3}\right)^3 \quad , \text{ where } 1.5 \le s \le \sqrt{3}. \end{aligned}$$

Since $f(1.5) = \arctan(1.5)$, the number s must be a solution of the equation $\frac{2(3s^2-1)/(1+s^2)^3}{3!}\left(1.5-\sqrt{3}\right)^3 = \arctan(1.5) - 0.9833557$. Using our solve button, $s = 1.6729757$, which is between 1.5 and $\sqrt{3} = 1.7320508$.

11. Let $f(x) = (1+x)^{1/2}$. Since $f(3) = 2$, $f'(3) = \frac{1}{2}(1+3)^{-1/2} = 1/4$, $f''(3) = -\frac{1}{4}(1+3)^{-3/2} = -1/32$, and $f^{(3)}(s) = \frac{3}{8}(1+s)^{-5/2}$, the second order Taylor approxmation for $f(3.2)$, (base point $c = 3$), with error term, is

$$\begin{aligned} f(3.2) &= 2 + \frac{1}{4}(3.2-3) - \frac{1/32}{2!}(3.2-3)^2 + \frac{f^{(3)}(s)}{3!}(3.2-3)^3 \\ &= 2.049375 + \frac{3/(8(1+s)^{5/2})}{3!}(3.2-3)^3 \quad , \text{ where } 3 \le s \le 3.2. \end{aligned}$$

Since $f(3.2) = \sqrt{4.2}$, the number s must be a solution of the equation $\frac{3/(8(1+s)^{5/2})}{3!}(3.2-3)^3 = \sqrt{4.2} - 2.049375$. Using our solve button, $s = 3.049380$, which is between 3 and 3.2.

13. (a) Since $f'(x) = \frac{1}{2}(1+x)^{-1/2}$, $f''(x) = -\frac{1}{4}(1+x)^{-3/2}$, and $f^{(3)}(x) = \frac{3}{8}(1+x)^{-5/2}$, the third order Taylor polynomial for $f(x)$, (base point $c = 0$) is

$$T_3(x) = 1 + \frac{1}{2}x - \frac{1/4}{2!}x^2 + \frac{3/8}{3!}x^3 = 1 + \frac{1}{2}x - \frac{1}{8}x^2 + \frac{1}{16}x^3$$

and $T_3(0.4) = 1.184$.

(b) The remainder is $R_3(0.4) = \frac{f^{(4)}(s)}{4!}(0.4)^4$ where $|f^{(4)}(s)| = \frac{15}{16(1+s)^{7/2}}$ is decreasing on the interval J attaining a maximum value of $\frac{15}{16}$ at $s = c = 0$. Consequently, $|f(0.4) - T_3(0.4)| \le \frac{15}{16 \cdot 4!}(0.4)^4 = 0.001$.

15. (a) Since $f'(x) = 2x\cos(x^2-1)$ and $f''(x) = 2\cos(x^2-1) - 4x^2\sin(x^2-1)$, the second order Taylor polynomial for $f(x)$, (base point $c = 1$) is $T_2(x) = 2(x-1) + \frac{2}{2!}(x-1)^2 = 2(x-1) + (x-1)^2$, and $T_2(0.9) = -0.19$.

(b) The remainder is $R_2(0.9) = \frac{f^{(3)}(s)}{3!}(0.9-1)^3$ where examination of its graph shows that $|f^{(3)}(s)| = |8s^3\cos(s^2-1) + 12s\sin(s^2-1)|$ is increasing on the interval J attaining a maximum value of 8 at $s = c = 1$. Consequently, $|f(0.9) - T_2(0.9)| \le \frac{8}{3!}(0.1)^3 = 0.0013$.

17. (a) Since $f'(x) = 2x + e^{1+x}$ and $f''(x) = 2 + e^{1+x}$, the second order Taylor polynomial for $f(x)$, (base point $c = -1$) is $T_2(x) = 2 - (x+1) + \frac{3}{2!}(x+1)^2$, and $T_2(-1.2) = 2.26$.

    (b) The remainder is $R_2(-1.2) = \frac{f^{(3)}(s)}{3!}(-1.2+1)^3$ where $|f^{(3)}(s)| = e^{1+s}$ is increasing on the interval J attaining a maximum value of 1 at $s = c = -1$. Consequently, $|f(-1.2) - T_2(-1.2)| \le \frac{1}{3!}(0.2)^3 = 0.0013$.

19. (a) Since $f'(x) = 1/x$ and $f''(x) = -1/x^2$, the second order Taylor polynomial for $f(x)$, (base point $c = e$) is $T_2(x) = 1 + \frac{1}{e}(x-e) - \frac{1/e^2}{2!}(x-e)^2$, and $T_2(2.5) = 0.9165$.

    (b) The remainder is $R_2(2.5) = \frac{f^{(3)}(s)}{3!}(2.5-e)^3$ where $|f^{(3)}(s)| = 2/s^3$ is decreasing on the interval J attaining a maximum value of 0.128 at $s = x_0 = 2.5$. Consequently, $|f(2.5) - T_2(2.5)| \le \frac{0.128}{3!}|2.25 - e|^3 = 0.00022$.

21. (a) Since $f'(x) = \frac{1}{1+x^2}$ and $f''(x) = -\frac{2x}{(1+x^2)^2}$, the second order Taylor polynomial for $f(x)$, (base point $c = 0$) is $T_2(x) = x$, and $T_2(0.4) = 0.4$.

    (b) The remainder is $R_2(0.4) = \frac{f^{(3)}(s)}{3!}(0.4-0)^3$ where inspection of its graph shows that $|f^{(3)}(s)| = \frac{2(3s^2-1)}{(1+s^2)^3}$ is decreasing on the interval J attaining a maximum value of 2 at $s = c = 0$. Consequently, $|f(0.4) - T_2(0.4)| \le \frac{2}{3!}(0.4)^3 = 0.021$.

23. We know that $\sin(x) = \sum_{n=0}^{\infty}(-1)^n \frac{x^{2n+1}}{(2n+1)!}$. Therefore,

$$\sin(2x) = \sum_{n=0}^{\infty}(-1)^n \frac{(2x)^{2n+1}}{(2n+1)!} = \sum_{n=0}^{\infty}(-1)^n \frac{2^{2n+1}}{(2n+1)!} x^{2n+1}.$$

25. We know that $\sin(x) = \sum_{n=0}^{\infty}(-1)^n \frac{x^{2n+1}}{(2n+1)!}$. Therefore,

$$x^2\sin(x/2) = x^2 \sum_{n=0}^{\infty}(-1)^n \frac{(x/2)^{2n+1}}{(2n+1)!} = \sum_{n=0}^{\infty}(-1)^n \frac{x^2}{2^{2n+1}(2n+1)!} x^{2n+1}$$
$$= \sum_{n=0}^{\infty}(-1)^n \frac{1}{2^{2n+1}(2n+1)!} x^{2n+3}$$

27. We know that $\cos(2x) = \sum_{n=0}^{\infty}(-1)^n \frac{(2x)^{2n}}{(2n)!}$ and $\sin(3x) = \sum_{n=0}^{\infty}(-1)^n \frac{(3x)^{2n+1}}{(2n+1)!}$. Therefore,

$$5\cos(2x) - 4\sin(3x) = \sum_{n=0}^{\infty}(-1)^n \frac{5 \cdot 2^{2n}}{(2n)!} x^{2n} + \sum_{n=0}^{\infty}(-1)^{n+1} \frac{4 \cdot 3^{2n+1}}{(2n+1)!} x^{2n+1}$$
$$= \sum_{n=0}^{\infty} a_n x^n ,$$

where $a_n = \begin{cases} (-1)^{\frac{n}{2}} \cdot \frac{5 \cdot 2^n}{n!} & , \quad n \text{ even} \\ (-1)^{\frac{n+1}{2}} \cdot \frac{4 \cdot 3^n}{n!} & , \quad n \text{ odd} \end{cases}$.

29. We know that $\exp(x) = \sum_{n=0}^{\infty} \frac{x^n}{n!}$. Therefore,

$$\exp(-x^2) = \sum_{n=0}^{\infty} \frac{(-x^2)^n}{n!} = \sum_{n=0}^{\infty} \frac{(-1)^n}{n!} x^{2n}.$$

31. We know that $\frac{1}{1+x^2} = \sum_{n=0}^{\infty} (-x^2)^n$. Therefore,

$$\frac{x}{1+x^2} = x \sum_{n=0}^{\infty} (-1)^n x^{2n} = \sum_{n=0}^{\infty} (-1)^n x^{2n+1}.$$

33. We know that $(1+x)^{1/2} = \sum_{n=0}^{\infty} \binom{1/2}{n} x^n$. Therefore, $(1+x^3)^{1/2} = \sum_{n=0}^{\infty} \binom{1/2}{n} x^{3n}$.

35. Observe that $f^{(1)}(x) = \frac{1}{2} x^{-1/2}$, $f^{(2)}(x) = -\frac{1}{2^2} x^{-3/2}$, $f^{(3)}(x) = \frac{1 \cdot 3}{2^3} x^{-5/2}$, $f^{(4)}(x) = -\frac{1 \cdot 3 \cdot 5}{2^4} x^{-7/2}$, and, in general, for $n > 1$,

$$f^{(n)}(x) = (-1)^{n-1} \frac{1 \cdot 3 \cdot 5 \cdots (2n-3)}{2^n} x^{-(2n-1)/2}.$$

Therefore, for such n,

$$\begin{aligned} f^{(n)}(1) &= (-1)^{n-1} \frac{1 \cdot 3 \cdot 5 \cdots (2n-3)}{2^n} \\ &= (-1)^{n-1} \frac{1 \cdot 3 \cdot 5 \cdots (2n-3) \cdot 2 \cdot 4 \cdot 6 \cdots (2n-2)}{2 \cdot 4 \cdot 6 \cdots (2n-2) \cdot 2^n} \\ &= (-1)^{n-1} \frac{(2n-2)!}{(n-1)! \cdot 2^{n-1} \cdot 2^n}. \end{aligned}$$

Since this formula is also valid for $n = 1$, and $f(1) = 1$, the Taylor series is

$$T(x) = 1 + \sum_{n=1}^{\infty} (-1)^{n-1} \frac{(2n-2)!}{(n-1)!\, n!\, 2^{2n-1}} (x-1)^n.$$

37. Observe that $f^{(1)}(x) = e^x + xe^x = (1+x)e^x$, $f^{(2)}(x) = e^x + (1+x)e^x = (2+x)e^x$, $f^{(3)}(x) = e^x + (2+x)e^x = (3+x)e^x$, and, in general, for $n \geq 1$, $f^{(n)}(x) = (n+x)e^x$. Therefore, for such n, $f^{(n)}(1) = (n+1)e$ and, since $f(1) = e$, this formula is also valid for $n = 0$. Therefore, the Taylor series is $T(x) = \sum_{n=0}^{\infty} \frac{(n+1)e}{n!} (x-1)^n$.

39. Observe that $f^{(1)}(x) = x^{-1}$, $f^{(2)}(x) = -x^{-2}$, $f^{(3)}(x) = 2x^{-3}$, $f^{(4)}(x) = -3 \cdot 2x^{-4}$, and, in general, for $n \geq 1$, $f^{(n)}(x) = (-1)^{n-1}(n-1)!x^{-n}$. Therefore, for such n, $f^{(n)}(1) = (-1)^{n-1}(n-1)!$ and, since $f(1) = 0$, the Taylor series is $T(x) = \sum_{n=1}^{\infty} \frac{(-1)^{n-1}}{n} (x-1)^n$.

41. Observe that $f^{(1)}(x) = -3^{-x}\ln(3)$, $f^{(2)}(x) = 3^{-x}\ln^2(3)$, $f^{(3)}(x) = -3^{-x}\ln^3(3)$, and, in general, for $n \geq 1$, $f^{(n)}(x) = (-1)^n 3^{-x}\ln^n(3)$. Therefore, for such n, $f^{(n)}(-3) = (-1)^n 3^3 \ln^n(3) = (-1)^n \cdot 27\ln^n(3)$. Since $f(-3) = 27$, this formula is also valid for $n = 0$, and the Taylor series is $T(x) = \sum_{n=0}^{\infty}(-1)^n \frac{27\ln^n(3)}{n!}(x+3)^n$.

43. $$(1+x)^{3/4} = 1 + \frac{3}{4}x + \frac{\frac{3}{4}\cdot\left(-\frac{1}{4}\right)}{2!}x^2 + \frac{\frac{3}{4}\cdot\left(-\frac{1}{4}\right)\cdot\left(-\frac{5}{4}\right)}{3!}x^3 + \cdots$$
$$= 1 + \frac{3}{4}x - \frac{3}{32}x^2 + \frac{5}{128}x^3 + \cdots$$

45. $$(1+x)^{-1/2} = 1 - \frac{1}{2}x + \frac{\left(-\frac{1}{2}\right)\left(-\frac{3}{2}\right)}{2!}x^2 + \frac{\left(-\frac{1}{2}\right)\left(-\frac{3}{2}\right)\left(-\frac{5}{2}\right)}{3!}x^3 + \cdots$$
$$= 1 - \frac{1}{2}x + \frac{3}{8}x^2 - \frac{5}{16}x^3 + \cdots$$

47. $$(1+x)^{-3/2} = 1 - \frac{3}{2}x + \frac{\left(-\frac{3}{2}\right)\left(-\frac{5}{2}\right)}{2!}x^2 + \frac{\left(-\frac{3}{2}\right)\left(-\frac{5}{2}\right)\left(-\frac{7}{2}\right)}{3!}x^3 + \cdots$$
$$= 1 - \frac{3}{2}x + \frac{15}{8}x^2 - \frac{35}{16}x^3 + \cdots$$

49. $$\lim_{x\to 0}\frac{\sin(x) - x}{x^3} = \lim_{x\to 0}\frac{\left(x - \frac{1}{6}x^3 + \frac{1}{120}x^5 - \cdots\right) - x}{x^3}$$
$$= \lim_{x\to 0}\left(-\frac{1}{6} + \frac{1}{120}x^2 - \cdots\right) = -\frac{1}{6}$$

51. $$\lim_{x\to 0}\frac{\ln(1+x) - x}{x^2} = \lim_{x\to 0}\frac{\left(x - \frac{1}{2}x^2 + \frac{1}{3}x^3 - \cdots\right) - x}{x^2}$$
$$= \lim_{x\to 0}\left(-\frac{1}{2} + \frac{1}{3}x - \cdots\right) = -\frac{1}{2}$$

53. $$\lim_{x\to 0}\frac{1 - e^{2x}}{x} = \lim_{x\to 0}\frac{1 - \left(1 + 2x + \frac{1}{2}(2x)^2 + \cdots\right)}{x}$$
$$= \lim_{x\to 0}\left(-2 - 2x - \cdots\right) = -2$$

Further Theory and Practice

55. $$\lim_{x\to 0}\frac{\cos(x) - \exp(-x^2)}{x^2} = \lim_{x\to 0}\frac{\left(1 - \frac{x^2}{2} + \frac{x^4}{24} - \cdots\right) - \left(1 - x^2 + \frac{1}{2}x^4 - \cdots\right)}{x^2}$$
$$= \lim_{x\to 0}\left(\frac{1}{2} - \frac{11}{24}x^2 + \cdots\right) = \frac{1}{2}$$

57. $$\lim_{x\to 0}\frac{\arctan(x) - \sin(x)}{x(1-\cos(x))} = \lim_{x\to 0}\frac{\left(x - \frac{x^3}{3} + \frac{x^5}{5} - \cdots\right) - \left(x - \frac{x^3}{6} + \frac{x^5}{120} - \cdots\right)}{x\left(\frac{x^2}{2} - \frac{x^4}{24} + \cdots\right)}$$

$$= \lim_{x\to 0} \frac{-\frac{x^3}{6} + \frac{23x^5}{120} - \cdots}{\frac{x^3}{2} - \frac{x^5}{24} + \cdots} = \lim_{x\to 0} \frac{-\frac{1}{6} + \frac{23x^2}{120} - \cdots}{\frac{1}{2} - \frac{x^2}{24} + \cdots} = -\frac{1}{3}$$

59. Since $\cos^2(x) = \left(1 - \frac{1}{2}x^2 + \frac{1}{24}x^4 + \cdots\right)\left(1 - \frac{1}{2}x^2 + \frac{1}{24}x^4 + \cdots\right) = 1 - x^2 + \frac{1}{3}x^4 + \cdots$,

$$\lim_{x\to 0} \frac{\cos^2(x) - \exp(x^2)}{x\sin(x)} = \lim_{x\to 0} \frac{\left(1 - x^2 + \frac{1}{3}x^4 + \cdots\right) - \left(1 + x^2 + \frac{1}{2}x^4 + \cdots\right)}{x^2 - \frac{1}{6}x^4 + \cdots}$$
$$= \lim_{x\to 0} \frac{-2x^2 - \frac{1}{6}x^4 + \cdots}{x^2 - \frac{1}{6}x^4 + \cdots} = \lim_{x\to 0} \frac{-2 - \frac{1}{6}x^2 + \cdots}{1 - \frac{1}{6}x^2 + \cdots} = -2\,.$$

61. According the the Half Angle Formula, $\sin^2(x) = \frac{1}{2}(1 - \cos(2x))$. Therefore,

$$\sin^2(x) = \frac{1}{2} - \frac{1}{2}\sum_{n=0}^{\infty}(-1)^n \frac{(2x)^{2n}}{(2n)!} = \sum_{n=1}^{\infty}(-1)^{n+1}\frac{2^{2n-1}}{(2n)!}x^{2n}\,.$$

63. Using the series in Exercise 61, $\sin^2(x) = x^2 - \frac{1}{3}x^4 + \frac{2}{45}x^6 \cdots$. Therefore,

$$\lim_{x\to 0} \frac{\sin^2(x) - \sin(x^2)}{x^4} = \lim_{x\to 0} \frac{\left(x^2 - \frac{1}{3}x^4 + \frac{2}{45}x^6 \cdots\right) - \left(x^2 - \frac{1}{6}x^6 + \frac{1}{120}x^{10} + \cdots\right)}{x^4}$$
$$= \lim_{x\to 0}\left(-\frac{1}{3} + \frac{19}{90}x^2 + \cdots\right) = -\frac{1}{3}\,.$$

65. Since f'' is continuous near c, for $h \neq 0$, sufficiently small,

$$f(c+h) = f(c) + f'(c)h + \frac{f''(s_1)}{2!}h^2$$

and

$$f(c-h) = f(c) - f'(c)h + \frac{f''(s_2)}{2!}h^2$$

where s_1 lies between c and $c+h$ and s_2 lies between c and $c-h$.

Add these two equations to obtain

$$f(c+h) + f(c-h) = 2f(c) + h^2 \cdot \frac{1}{2}(f''(s_1) + f''(s_2))$$

implying that

$$\frac{f(c+h) - 2f(c) + f(c-h)}{h^2} = \frac{1}{2}(f''(s_1) + f''(s_2))\,.$$

Therefore, $\lim_{h\to 0} \frac{f(c+h)-2f(c)+f(c-h)}{h^2} = \lim_{h\to 0} \frac{1}{2}(f''(s_1)+f''(s_2)) = f''(c)$ because $f''(s_1)$ and $f''(s_2)$ both have the limiting value $f''(c)$ as $h \to 0$.

67. This series is the Maclaurin series for $\cos(x)$ evaluated at $x = \pi/3$. Therefore, it converges to $\cos(\pi/3) = \frac{1}{2}$.

69. This series is the Maclaurin series for e^x evaluated at $x = 1$. Therefore, it converges to $e^1 = e$.

71. This series is the Maclaurin series for $e^{2x} - 1$ evaluated at $x = 1$. Therefore, it converges to $e^2 - 1$.

73. With u as defined in the Exercise, $\ln(\cos(x)) = u - \frac{1}{2}u^2 + \frac{1}{3}u^3 - \cdots$. Therefore, replacing u with the first few terms of its series expansion,

$$\begin{aligned}\ln(\cos(x)) &= \left(-\frac{1}{2}x^2 + \frac{1}{24}x^4 - \frac{1}{720}x^6 \cdots\right) - \frac{1}{2}\left(-\frac{1}{2}x^2 + \frac{1}{24}x^4 - \cdots\right)^2 \\ &\qquad + \frac{1}{3}\left(-\frac{1}{2}x^2 + \cdots\right)^3 + \cdots \\ &= -\frac{1}{2}x^2 - \frac{1}{12}x^4 - \frac{1}{45}x^6 + \cdots .\end{aligned}$$

Using this,

$$\tan(x) = -\frac{d}{dx}\ln(\cos(x)) = \frac{d}{dx}\left(\frac{1}{2}x^2 + \frac{1}{12}x^4 + \frac{1}{45}x^6 + \cdots\right) = x + \frac{1}{3}x^3 + \frac{2}{15}x^5 + \cdots .$$

75. Using the definition,

$$\begin{aligned}\binom{-1/2}{n} &= \frac{-\frac{1}{2}\left(-\frac{3}{2}\right)\left(-\frac{5}{2}\right)\cdots\left(-\frac{1}{2} - n + 1\right)}{n!} = \frac{(-1)^n \cdot 1 \cdot 3 \cdot 5 \cdots (2n-1)}{2^n n!} \\ &= \frac{(-1)^n \cdot 1 \cdot 3 \cdot 5 \cdots (2n-1) \cdot 2 \cdot 4 \cdot 6 \cdots (2n)}{2^n n! \cdot 2 \cdot 4 \cdot 6 \cdots (2n)} \\ &= \frac{(-1)^n (2n)!}{2^{2n}(n!)^2} .\end{aligned}$$

Consequently, $(1+u)^{-1/2} = \sum_{n=0}^{\infty}\binom{-1/2}{n}u^n = \sum_{n=0}^{\infty}\frac{(-1)^n(2n)!}{2^{2n}(n!)^2}u^n$, and

$$\arcsin(x) = \int_0^x (1+u^2)^{-1/2}\,du = \sum_{n=0}^{\infty}\int_0^x \binom{-1/2}{n}u^{2n}\,du = \sum_{n=0}^{\infty}\frac{(-1)^n(2n)!}{2^{2n}(2n+1)(n!)^2}x^{2n+1} .$$

77. Using the Maclaurin series for $\tan(x)$ in Exercise 73 and the series for $\sin(x)$,

$$\begin{aligned}f(t) &= \tan(t/3) + 2\sin(t/3) \\ &= \left(\frac{1}{3}t + \frac{1}{3^4}t^3 + \frac{2}{15 \cdot 3^5}t^5 + \cdots\right) + 2\left(\frac{1}{3}t - \frac{1}{3^3 \cdot 3!}t^3 + \frac{1}{3^5 \cdot 5!}t^5 - \cdots\right) \\ &= t + \frac{1}{1620}t^5 + \cdots .\end{aligned}$$

Therefore, $T_5(t) = t + \frac{1}{1620}t^5$. This accounts for the accuracy of the approximation. Note that $|f(\pi/2) - \pi/2| \approx \frac{1}{1620}(\pi/2)^5 = 0.00097$

79. (a) Observe that

$$(n+1)\cdot\binom{\alpha}{n+1}+n\cdot\binom{\alpha}{n}=(n+1)\frac{\alpha(\alpha-1)\cdots(\alpha-n)}{(n+1)!}+n\frac{\alpha(\alpha-1)\cdots(\alpha-n+1)}{n!}$$
$$=\frac{\alpha(\alpha-1)\cdots(\alpha-n+1)}{n!}\cdot(\alpha-n+n)$$
$$=\alpha\cdot\frac{\alpha(\alpha-1)\cdots(\alpha-n+1)}{n!}=\alpha\cdot\binom{\alpha}{n}.$$

(b) Calculate

$$(1+x)\cdot g'(x)=(1+x)\sum_{n=0}^{\infty}n\binom{\alpha}{n}x^{n-1}=\sum_{n=0}^{\infty}n\binom{\alpha}{n}x^{n-1}+\sum_{n=0}^{\infty}n\binom{\alpha}{n}x^{n}$$
$$=\sum_{n=0}^{\infty}(n+1)\binom{\alpha}{n+1}x^{n}+\sum_{n=0}^{\infty}n\binom{\alpha}{n}x^{n}$$
$$=\sum_{n=0}^{\infty}\left((n+1)\binom{\alpha}{n+1}+n\binom{\alpha}{n}\right)x^{n}$$
$$=\sum_{n=0}^{\infty}\alpha\cdot\binom{\alpha}{n}x^{n} \qquad \text{(by part a)}$$
$$=\alpha\cdot g(x)$$

(c) Calculate: $(1+x)\cdot f'(x)=(1+x)\cdot\alpha(1+x)^{\alpha-1}=\alpha\cdot(1+x)^{\alpha}=\alpha\cdot f(x)$.

(d) Duly noted.

(e) Integrate $\frac{y'(x)}{y(x)}=\frac{\alpha}{1+x}$ to obtain $\ln(y(x))=\alpha\ln(1+x)+C$ implying that $y(x)=C\cdot(1+x)^{\alpha}$.

(f) Continuing part e, if $y(0)=1$, then $C=1$, and $y(x)=(1+x)^{\alpha}$.

We conclude that $f=g$ because $y=f(x)$ and $y=g(x)$ are both solutions to this initial value problem.

81. We assume that f is $N+1$ times continuously differentiable and the formula for ρ_N should be $\rho_N(t)=f(x)-\sum_{n=0}^{N}\frac{f^{(n)}(t)}{n!}(x-t)^n$. Note that $\rho_N(c)=R_N(x)$, the Nth Taylor Remainder for base point c. Also observe that

$$\rho_N'(t)=-\sum_{n=0}^{N}\frac{f^{(n+1)}(t)}{n!}(x-t)^n+\sum_{n=0}^{N}\frac{f^{(n)}(t)}{n!}\cdot n(x-t)^{n-1}$$
$$=-\sum_{n=0}^{N}\frac{f^{(n+1)}(t)}{n!}(x-t)^n+\sum_{n=1}^{N}\frac{f^{(n)}(t)}{(n-1)!}(x-t)^{n-1}$$
$$=-\sum_{n=0}^{N}\frac{f^{(n+1)}(t)}{n!}(x-t)^n+\sum_{n=0}^{N-1}\frac{f^{(n+1)}(t)}{n!}(x-t)^n=-\frac{f^{(N+1)}(t)}{N!}(x-t)^N.$$

Therefore,

$$\begin{aligned}\phi'(t) &= \rho_N'(t) + (N+1)\left(\frac{x-t}{x-c}\right)^N \cdot \left(\frac{1}{x-c}\right) \cdot \rho_N(c) \\ &= -\frac{f^{(N+1)}(t)}{N!}(x-t)^N + (N+1)\frac{(x-t)^N}{(x-c)^{N+1}} \cdot R_N(x).\end{aligned}$$

Substitute $t = s$ in the equation above, yielding $\phi'(s) = 0$ on the left side, then solve for $R_N(x)$, to obtain

$$R_N(x) = \frac{f^{(N+1)}(s)}{(N+1)!}(x-c)^{N+1}.$$

CALCULATOR/COMPUTER EXERCISES

83. Since $\sin(0.3) = \sum_{n=0}^{\infty}(-1)^n \frac{(0.3)^{2n+1}}{(2n+1)!}$, an Alternating Series, the absolute error E_N associated with the Nth partial sum approximation satisfies $E_N \leq \frac{(0.3)^{2N+3}}{(2N+3)!}$. Therefore, we look for the smallest positive integer N such that $\frac{(0.3)^{2N+3}}{(2N+3)!} < 5 \times 10^{-5}$. Since $N = 1$ yields 0.0045 and $N = 2$ gives 2.03×10^{-5}, the approximation

$$\sin(0.3) \approx S_2 = 0.3 - \frac{(0.3)^3}{6} + \frac{(0.3)^5}{120} = 0.2955$$

should be accurate to four decimals. According to our calculator, $\sin(0.3) = 0.2955202067$.

85. Since $\arctan(0.15) = \sum_{n=0}^{\infty}(-1)^n \frac{(0.15)^{2n+1}}{2n+1}$, an Alternating Series, the absolute error E_N associated with the Nth partial sum approximation satisfies $E_N \leq \frac{(0.15)^{2N+3}}{2N+3}$. Therefore, we look for the smallest positive integer N such that $\frac{(0.15)^{2N+3}}{2N+3} < 5 \times 10^{-5}$. Since $N = 1$ yields 0.001125 and $N = 2$ gives 1.52×10^{-5}, the approximation

$$\arctan(0.15) \approx S_2 = 0.15 - \frac{(0.15)^3}{3} + \frac{(0.15)^5}{5} = 0.1489$$

should be accurate to four decimals. According to our calculator, $\arctan(0.15) = 0.1488899476$.

87. Differentiate f three times to obtain

$$T_3(x) = 2 + \tfrac{1}{4}(x-1) - \tfrac{1}{64}(x-1)^2 + \tfrac{1}{512}(x-1)^3.$$

The picture on the right displays the horizontal line and the graph of $T_3(x)$. The vertical dashed line meets the t-axis at $s = 1.077$.

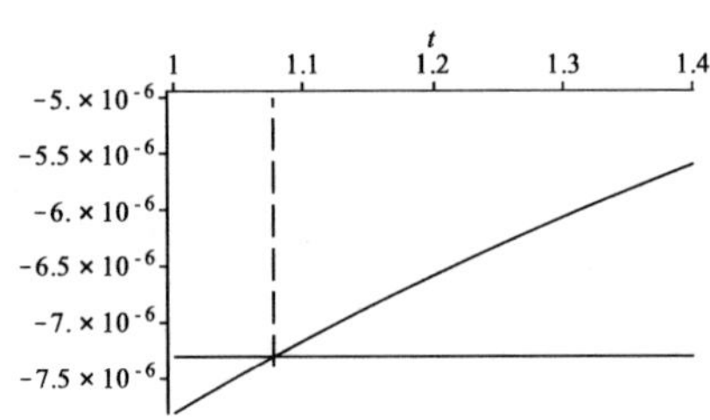

89. Differentiate f three times to obtain

$$T_3(x) = 4 - \tfrac{13}{8}(x - \pi)^2 .$$

The picture on the right displays the horizontal line and the graph of $T_3(x)$. The vertical dashed line meets the t-axis at $s = 3.234$.

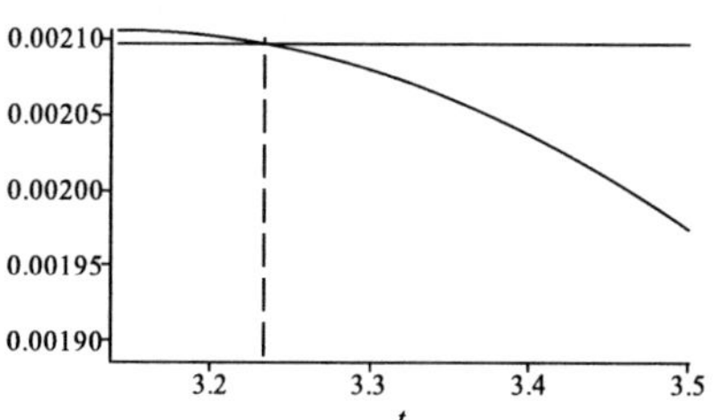

91. 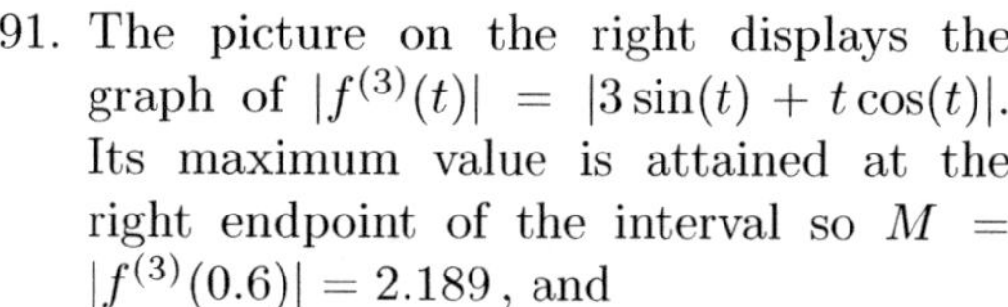
The picture on the right displays the graph of $|f^{(3)}(t)| = |3\sin(t) + t\cos(t)|$. Its maximum value is attained at the right endpoint of the interval so $M = |f^{(3)}(0.6)| = 2.189$, and

$$|R_2(x_0)| \le \frac{M}{3!} \cdot |x_0 - c|^3 = 0.00016 .$$

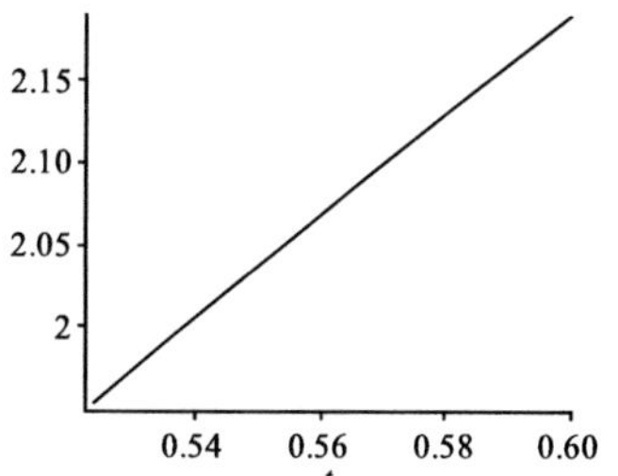

93. 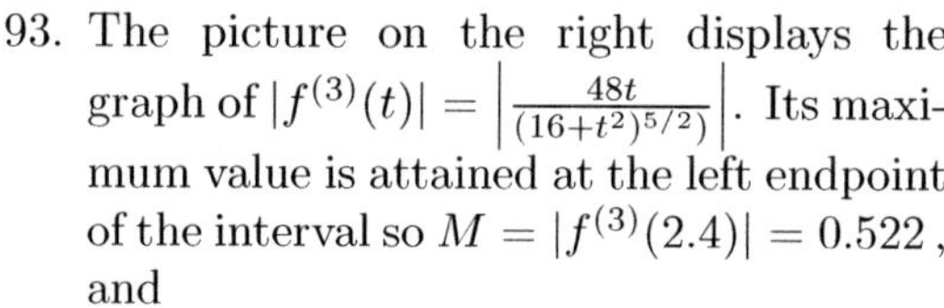
The picture on the right displays the graph of $|f^{(3)}(t)| = \left|\frac{48t}{(16+t^2)^{5/2}}\right|$. Its maximum value is attained at the left endpoint of the interval so $M = |f^{(3)}(2.4)| = 0.522$, and

$$|R_2(x_0)| \le \frac{M}{3!} \cdot |x_0 - c|^3 = 0.001878 .$$

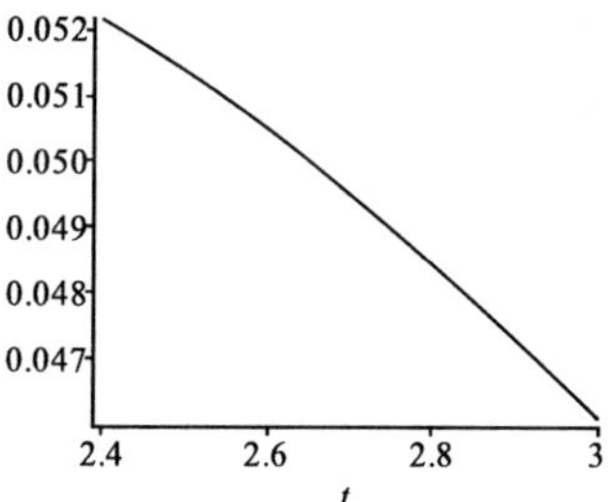

95. Observe that

$$\int_0^{1/2} e^{-x^2}\,dx = \int_0^{1/2} \sum_{n=0}^{\infty} \frac{(-x^2)^n}{n!}\,dx = \int_0^{1/2} \sum_{n=0}^{\infty} (-1)^n \frac{x^{2n}}{n!}\,dx$$

$$= \sum_{n=0}^{\infty} (-1)^n \left(\frac{x^{2n+1}}{(2n+1)\cdot n!} \right)\Bigg|_{x=0}^{x=1/2}$$

$$= \sum_{n=0}^{\infty} (-1)^n \frac{(1/2)^{2n+1}}{(2n+1)\cdot n!} = \sum_{n=0}^{\infty} (-1)^n \frac{1}{2^{2n+1}(2n+1)\cdot n!} .$$

The series alternates, so the absolute error E_N associated with the Nth partial sum approximation satisfies $E_N \le \frac{1}{2^{2N+3}(2N+3)\cdot(N+1)!}$. We look for the first N such that $\frac{1}{2^{2N+3}(2N+3)\cdot(N+1)!} < 5 \times 10^{-6}$. Using *Maple*, this is $N = 4$ for which the upper bound on the error evaluates to 3.7×10^{-7}. This yields the approximation

$$\int_0^{1/2} e^{-x^2}\,dx \approx \sum_{n=0}^{4} (-1)^n \frac{1}{2^{2n+1}(2n+1)\cdot n!} = 0.46128 .$$

Using *Maple*, $\int_0^{1/2} e^{-x^2}\,dx = 0.4612810064$.

97. Observe that

$$\int_0^{1/3} \frac{1}{1+x^5}\,dx = \int_0^{1/3} \sum_{n=0}^{\infty}(-x^5)^n\,dx = \int_0^{1/3} \sum_{n=0}^{\infty}(-1)^n x^{5n}\,dx$$
$$= \sum_{n=0}^{\infty}(-1)^n \left(\frac{x^{5n+1}}{5n+1}\right)\Bigg|_{x=0}^{x=1/3}$$
$$= \sum_{n=0}^{\infty}(-1)^n \frac{(1/3)^{5n+1}}{5n+1} = \sum_{n=0}^{\infty}(-1)^n \frac{1}{3^{5n+1}(5n+1)}\,.$$

The series alternates, so the absolute error E_N associated with the Nth partial sum approximation satisfies $E_N \le \frac{1}{3^{5N+6}(5N+6)}$. We look for the first N such that $\frac{1}{3^{5N+6}(5N+6)} < 5 \times 10^{-6}$. Using *Maple*, this is $N = 1$ for which the upper bound on the error evaluates to 5.1×10^{-7}. This yields the approximation

$$\int_0^{1/3} \frac{1}{1+x^5}\,dx \approx \sum_{n=0}^{1}(-1)^n \frac{1}{3^{5n+1}(5n+1)} = \frac{1}{3} - \frac{1}{4374} = 0.33310\,.$$

Using *Maple*, $\int_0^{1/2} e^{-x^2}\,dx = 0.3331052181$.

99. *Maple*'s plot of f is on the left. Round-off error is rampant. The plot of the quotient of the Maclaurin series approximations is on the right.

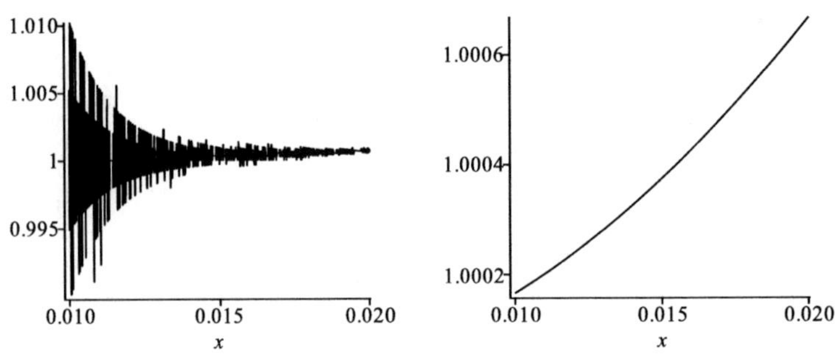

REVIEW EXERCISES FOR CHAPTER 8

1. For $S_N = \sum_{n=1}^{\infty} \frac{6n}{n+1}$,

$$S_1 = 3 \quad , \quad S_2 = 7 \quad , \quad S_3 = \tfrac{23}{2} \quad , \quad S_4 = \tfrac{163}{10} \quad , \quad S_5 = \tfrac{213}{10}\,.$$

3. Geometric, ratio 3/5: $\sum_{n=0}^{\infty} \left(\frac{3}{5}\right)^n = \frac{1}{1-\frac{3}{5}} = \frac{5}{2}$.

5. Almost geometric, ratio 3/7:

$$\sum_{n=2}^{\infty} \frac{3^{n+1}}{7^{n-1}} = 21 \sum_{n=2}^{\infty} \left(\frac{3}{7}\right)^n = 21 \cdot \frac{3^2/7^2}{1-\frac{3}{7}} = 21 \cdot \frac{3^2/7}{4} = \frac{27}{4}\,.$$

7. $0.6313636363\cdots = \frac{631}{1000} + 36 \cdot 10^{-5} \cdot (1 + 10^{-2} + 10^{-4} + \cdots)$
$= \frac{631}{1000} + \frac{36}{100000} \cdot \frac{1}{1-10^{-2}} = \frac{631}{1000} + \frac{36}{100000} \cdot \frac{100}{99} = \frac{1389}{2200}$

9. This series converges by the Alternating Series Test. The convergence is conditional because $\sum_{n=0}^{\infty} \frac{\sqrt{1+\sqrt{n}}}{(1+\sqrt{n})^2}$ diverges by Limit Comparison to the divergent series $\sum_{n=1}^{\infty} \frac{n^{1/4}}{n} = \sum_{n=1}^{\infty} \frac{1}{n^{3/4}}$, ($p$-series, $p = \frac{3}{4}$).

11. This series diverges by the Divergence Test. The nth term has no limit as $n \to \infty$.

13. This series converges by the Alternating Series Test. The convergence is conditional because $\sum_{n=0}^{\infty} \frac{1}{(n^2+1)^{1/3}}$ diverges by Limit Comparison to the divergent series $\sum_{n=1}^{\infty} \frac{1}{n^{2/3}}$, ($p$-series, $p = \frac{2}{3}$).

15. This series diverges by the Divergence Test. The nth term is unbounded as $n \to \infty$ by L'Hòpital's Rule (several differentiations are required).

17. This series converges absolutely by the Comparison Test for Convergence. Observe that $0 < \frac{1}{1+\sqrt{n}+n\sqrt{n}} < \frac{1}{n^{3/2}}$ for all $n \geq 1$ and $\sum_{n=1}^{\infty} \frac{1}{n^{3/2}}$ converges (p-series, $p = \frac{3}{2}$).

19. This series converges absolutely by the Integral Test. The function $f(x) = \frac{1}{x \ln^2(x)}$ is positive and decreases to 0 on $[10, \infty)$ and

$$\int_{10}^{N} \frac{1}{x \ln^2(x)}\, dx = -\frac{1}{\ln(x)}\Bigg|_{10}^{N} = \frac{1}{\ln(10)} - \frac{1}{\ln(N)} \xrightarrow[N\to\infty]{} \frac{1}{\ln(10)}\,.$$

21. This series converges by the Alternating Series Test. The convergence is conditional because $\sum_{n=3}^{\infty} \frac{\sqrt{n}}{n-\sqrt{2}}$ diverges by the Comparison Test for Divergence. Observe that $0 < \frac{1}{\sqrt{n}} < \frac{\sqrt{n}}{n-\sqrt{2}}$ for all $n \geq 3$ and $\sum_{n=3}^{\infty} \frac{1}{n^{1/2}}$ diverges (p-series, $p = \frac{1}{2}$).

23. This series converges absolutely by Limit Comparison to $\sum_{n=1}^{\infty} \frac{1}{n^{19/13}}$, which converges ($p$-series, $p = \frac{19}{13}$). Note that for n large, $(1 + n^6)^{14/13} \approx n^{84/13}$, implying that $\frac{n^5}{(1+n^6)^{14/13}} \approx \frac{1}{n^{19/13}}$.

25. This series converges absolutely by the Integral Test. The function $f(x) = \frac{\ln^2(x)}{x^{3/2}}$ is positive, decreases to 0 on $[4, \infty)$, and $\int_4^{\infty} \frac{\ln^2(x)}{x^{3/2}}\, dx = 4\ln^2(2) + 8\ln(2) + 8$, as can be verified via integration-by-parts, $u = \ln^2(x)$, $dv = x^{-3/2}\, dx$, $du = 2\ln(x) \cdot \frac{1}{x}\, dx$, $v = -2x^{-1/2}$. Then do it again for the $v\,du$ integral to obtain $\int \frac{\ln^2(x)}{x^{3/2}}\, dx = -\frac{2}{\sqrt{x}}(\ln^2(x) + 4\ln(x) + 8) + C$.

27. This series converges by the Alternating Series Test. The function $f(x) = \frac{\sin(1/x)}{1/x} \cdot \sin(1/x)$ decreases monotonically to 0 on the interval $[1, \infty)$. The

convergence is conditional because, for large n, $\frac{\sin(1/n)}{1/n} \approx 1$ and $\sin(1/n) \approx 1/n$. Use Limit Comparison to the Harmonic Series $\sum_{n=1}^{\infty} \frac{1}{n}$.

29. This series converges absolutely by the Comparison Test for Convergence. Observe that $0 < \frac{1}{n+(1.3)^n} \le \frac{1}{(1.3)^n}$ for all $n \ge 0$, and $\sum_{n=1}^{\infty} \frac{1}{(1.3)^n}$ is geometric, $r = \frac{1}{1.3}$.

31. This series converges absolutely by the Comparison Test for Convergence. Observe that $0 < \frac{\arctan(n)}{n\sqrt{n}} \le \frac{\pi/2}{n^{3/2}}$ for all $n \ge 1$, and $\sum_{n=1}^{\infty} \frac{\pi/2}{n^{3/2}}$ is a multiple of a convergent p-series.

33. This series diverges by the Divergence Test. Since $\lim_{n\to\infty} n^{1/n} = 1$, the nth term does not have a limiting value as $n \to \infty$.

35. This series converges absolutely by the root test.

$$|a_n|^{1/n} = \left(\frac{n}{n+1}\right)^n = \frac{1}{(1+1/n)^n} \xrightarrow[n\to\infty]{} \frac{1}{e} < 1\,.$$

37. This series converges by the Alternating Series Test. The function $f(x) = \frac{1}{2\ln(x)}$ decreases monotonically to 0 on the interval $[2, \infty)$. The convergence is conditional by the Comparison Test for Divergence. Observe that $0 < \frac{1}{n} < \frac{1}{2\ln(n)}$ for $n \ge 2$.

39. This series converges absolutely by the root test, $|a_n|^{1/n} = \frac{n+3}{2n+1} \xrightarrow[n\to\infty]{} \frac{1}{2}$.

41. This series converges absolutely by the root test. Observe that

$$|a_n| = \underbrace{\frac{2^{2n-1}}{(n+1)(n+2)\cdots(n+n-1)}}_{n-1 \text{ factors}} < \frac{2^{2n-1}}{(n+1)^{n-1}}\,,$$

implying that

$$|a_n|^{1/n} < \frac{2^{2-1/n}}{(n+1)^{1-1/n}} = \frac{4}{n+1} \cdot \left(\frac{n+1}{2}\right)^{1/n} \xrightarrow[n\to\infty]{} 0\,.$$

L'Hòpital's Rule can be used to verify that

$$\lim_{n\to\infty} \left(\frac{n+1}{2}\right)^{1/n} = \lim_{n\to\infty} \exp\left(\frac{\ln((n+1)/2)}{n}\right) = 1\,.$$

43. This series converges absolutely by the ratio test. Observe that

$$\left|\frac{a_{n+1}}{a_n}\right| = \frac{((n+1)!)^3/(3(n+1))!}{(n!)^3/(3n)!} = \frac{(n+1)^3}{(3n+3)(3n+2))(3n+1)} \xrightarrow[n\to\infty]{} \frac{1}{27}\,.$$

45. (a) The Divergence Test is conclusive. (b) The Ratio Test is inconclusive.

47. (a) The Divergence Test is inconclusive. (b) The Ratio Test is inconclusive.

49. (a) The Divergence Test is conclusive. (b) The Ratio Test is conclusive.

51. Since the series alternates, the estimate will be S_N where $\frac{1}{4(N+1)^3} \leq 0.002$. That is, $N+1 \geq \left(\frac{1}{0.008}\right)^{1/3} = 5$, $S_4 = 0.224\ldots$.

53. Using the geometry of the integral test,

$$\int_1^\infty \frac{1}{x^{3/2}}\,dx < S < 1 + \int_1^\infty \frac{1}{x^{3/2}}\,dx\,.$$

That is, $2 < S < 3$.

55. The series is geometric, it will converge provided $|3x+4| < 1$. That is, $|x+4/3| < 1/3$. The radius of convergence is $R = 1/3$ and, since the base point is $c = -4/3$, the series converges absolutely on the open interval $(-5/3, -1)$.

At the left and right endpoints the series is $\sum_{n=0}^\infty (-1)^n$ and $\sum_{n=0}^\infty 1$, respectively. Both diverge and the interval of convergence is $I = (-5/3, -1)$.

57. Using the Ratio Test,

$$\ell = \lim_{n\to\infty} \frac{3(n+1)/(n+2)^3}{3n/(n+1)^3} = \lim_{n\to\infty} \left(\frac{n+1}{n}\right)\left(\frac{n+1}{n+2}\right)^3 = 1\,,$$

so $R = 1/\ell = 1$. The series converges absolutely on the open interval $(-1, 1)$.

The series converges at both endpoints by the Comparison Test for Convergence: $0 < \frac{n}{(n+1)^3} < \frac{1}{n^2}$, so the interval of convergence is $I = [-1, 1]$.

59. Using the Ratio Test,

$$\ell = \lim_{n\to\infty} \frac{1/\sqrt{(n+1)^{4/3}+2}}{1/\sqrt{n^{4/3}+1}} = \lim_{n\to\infty} \left(\frac{n^{4/3}+1}{(n+1)^{4/3}+1}\right)^{1/2} = 1\,,$$

so $R = 1/\ell = 1$. The series converges absolutely on the open interval $(-1, 1)$.

The series diverges at the left endpoint by Limit Comparison to $\sum_{n=1}^\infty \frac{1}{n^{2/3}}$. It converges at the right endpoint by the Alternating Series Test, so the interval of convergence is $I = (-1, 1]$.

61. Using the Root Test: $\lim_{n\to\infty} \left(\frac{1}{\sqrt{n}}\right)^{1/n} = \lim_{n\to\infty} \frac{1}{\sqrt{n^{1/n}}} = 1$, $R = 1$. Since the base point is $c = -1/2$, the series converges absolutely on the open interval $(-3/2, 1/2)$.

The series diverges at the left endpoint (p-series, $p = 1/2$). It converges at the right endpoint by the Alternating Series Test, so the interval of convergence is $I = (-3/2, 1/2]$.

63. As a power series the series is $\sum_{n=1}^{\infty} \frac{2^n}{3^n \cdot n}(x+3/2)^n$. Using the Root Test, $\ell = \lim_{n\to\infty} a_n^{1/n} = \lim_{n\to\infty} \frac{2/3}{n^{1/n}} = \frac{2}{3}$, so $R = 1/\ell = 3/2$. Since the base point is $c = -3/2$, series converges absolutely on the open interval $(-3, 0)$.

At the left endpoint the series is $\sum_{n=1}^{\infty} \frac{(-1)^n}{n}$, which converges by the Alternating Series Test. The series is the Harmonic Series at the right endpoint, so it diverges. The interval of convergence is $I = [-3, 0)$.

65. As a power series the series is $\sum_{n=2}^{\infty} \frac{(-1)^n}{\ln^2(n)}(x - 1/3)^n$. Using the Ratio Test, $\ell = \lim_{n\to\infty} \left|\frac{a_{n+1}}{a_n}\right| = \lim_{n\to\infty} \frac{1/\ln^2(n+1)}{1/\ln^2(n)} = \lim_{n\to\infty} \left(\frac{\ln(n)}{\ln(n+1)}\right)^2 = 1$, by L'Hòpital's Rule. Therefore, $R = 1/\ell = 1$. Since the base point is $c = 1/3$, series converges absolutely on the open interval $(-2/3, 4/3)$.

At the left endpoint the series is $\sum_{n=2}^{\infty} \frac{1}{\ln^2(n)}$, which diverges by the Comparison Test for Divergence: $0 < \frac{1}{n} < \frac{1}{\ln^2(n)}$ for all n sufficiently large. The series converges at the right endpoint by the Alternating Series Test. The interval of convergence is $I = (-2/3, 4/3]$.

67. Observe that

$$\frac{2}{3+x} = \frac{2}{3} \cdot \frac{1}{1+x/3} = \frac{2}{3}\sum_{n=0}^{\infty}(-x/3)^n = \sum_{n=0}^{\infty}(-1)^n \frac{2}{3^{n+1}}x^n .$$

69. Observe that

$$\frac{27}{81-x^4} = \frac{27}{81} \cdot \frac{1}{1-x^4/81} = \frac{1}{3}\sum_{n=0}^{\infty}(x^4/3^4)^n = \sum_{n=0}^{\infty}\frac{x^{4n}}{3^{4n+1}} .$$

71. Observe that

$$\frac{x^2}{1+x^2} = x^2\sum_{n=0}^{\infty}(-x^2)^n = \sum_{n=0}^{\infty}(-1)^n x^{2n+2} .$$

73. Since $f'(x) = 2^{1-x} - x2^{1-x}\ln(2) = (1 - x\ln(2))2^{1-x}$ and $f''(x) = -\ln(2)2^{1-x} - (1 - x\ln(2))2^{1-x}\ln(2) = (x\ln(2) - 2)\ln(2)2^{1-x}$, $f'(1) = 1 - \ln(2)$ and $f''(1) = (\ln(2) - 2)\ln(2)$. Therefore,

$$T_2(x) = 1 + (1 - \ln(2))(x-1) + \frac{\ln^2(2) - 2\ln(2)}{2}(x-1)^2 ,$$

and $T_2(5/4) = \frac{5}{4} - \frac{5}{16}\ln(2) + \frac{1}{32}\ln^2(2)$.

The associated error is

$$R_2(5/4) = \frac{f^{(3)}(s)}{3!}\left(\frac{5}{4} - 1\right)^3 = \frac{(3 - s\ln(2))\ln^2(2) \cdot 2^{1-s}}{384}, \text{ where } 1 < s < 5/4 .$$

75. Since $f'(x) = \sin(x)$ and $f''(x) = \cos(x)$, $f'(\pi/3) = \sqrt{3}/2$ and $f''(\pi/3) = 1/2$. Therefore,

$$T_2(x) = \frac{5}{2} + \frac{\sqrt{3}}{2}(x - \pi/3) + \frac{1/2}{2}(x - \pi/3)^2\,,$$

and $T_2(\pi/2) = \frac{5}{2} + \frac{\sqrt{3}}{2} \cdot \frac{\pi}{6} + \frac{1}{4} \cdot \left(\frac{\pi}{6}\right)^2$.

The associated error is

$$R_2(\pi/4) = \frac{f^{(3)}(s)}{3!}\left(\frac{\pi}{2} - \frac{\pi}{3}\right)^3 = -\frac{\pi^3 \sin(s)}{6^4}\,, \text{ where } \frac{\pi}{3} < s < \frac{\pi}{2}\,.$$

77. (a) Since $f'(x) = \frac{1}{2}(4+x)^{-1/2}$, $f''(x) = -\frac{1}{4}(4+x)^{-3/2}$, and $f'''(x) = \frac{3}{8}(4+x)^{-5/2}$, $f'(0) = \frac{1}{4}$, $f''(0) = -\frac{1}{32}$, and $f'''(0) = \frac{3}{256}$. Therefore,

$$T_3(x) = 2 + \frac{1}{4}x - \frac{1}{64}x^2 + \frac{1}{512}x^3\,,$$

and $T_3(0.41) = 2.100008049$.

(b) Since $|f^{(4)}(s)| = \frac{15}{16}(4+s)^{-7/2} \le \frac{15}{16 \cdot 2^7}$, the associated error has the following upper bound:

$$|R_3(0.41)| \le \frac{15/16}{2^7 \cdot 4!}\,|0.41 - 0|^4 = 8.6235 \times 10^{-6}\,.$$

79. (a) Since $f'(x) = 1 + 2\exp(2x)$, $f''(x) = 4\exp(2x)$, and $f'''(x) = 8\exp(2x)$, $f'(0) = 3$, $f''(0) = 4$, and $f'''(0) = 8$. Therefore,

$$T_3(x) = 1 + 3x + \frac{4}{2}x^2 + \frac{8}{6}x^3\,,$$

and $T_3(1/2) = 3.166666667$.

(b) Since $|f^{(4)}(s)| = 16\exp(2s) \le 16e$, the associated error has the following upper bound:

$$|R_3(1/2)| \le \frac{16e}{4!}\,|0.5 - 0|^4 = 0.1132617428\,.$$

81. Observe that

$$\int_0^{0.2} \sin(x^2)\,dx = \int_0^{0.2} \sum_{n=0}^{\infty} (-1)^n \frac{(x^2)^{2n+1}}{(2n+1)!}\,dx = \sum_{n=0}^{\infty} \int_0^{0.2} (-1)^n \frac{x^{4n+2}}{(2n+1)!}\,dx$$
$$= \sum_{n=0}^{\infty} (-1)^n \frac{(0.2)^{4n+3}}{(4n+3)(2n+1)!}\,.$$

The series alternates. Therefore, the absolute error associated with the Nth partial sum S_N is no more than $E_N = \frac{(0.2)^{4(N+1)+3}}{(4(N+1)+3)(2(N+1))!}$. We seek the smallest value of N for which $E_N < 10^{-5}$. Since $E_0 = 9.1\times 10^{-7}$, $N = 0$ and $\int_0^{0.2}\sin(x^2)\,dx \approx \frac{(0.2)^3}{3} = 0.002666666667$. Using our calculator, $\int_0^{0.2}\sin(x^2)\,dx = 0.002666361920$.

83. Observe that

$$\int_0^{0.2}\frac{1}{1+x^3}\,dx = \int_0^{0.2}\sum_{n=0}^{\infty}(-x^3)^n\,dx = \sum_{n=0}^{\infty}\int_0^{0.2}(-1)^n x^{3n}\,dx$$
$$= \sum_{n=0}^{\infty}(-1)^n\frac{(0.2)^{3n+1}}{3n+1}.$$

The series alternates. Therefore, the absolute error associated with the Nth partial sum S_N is no more than $E_N = \frac{(0.2)^{3(N+1)+1}}{3(N+1)+1}$. We seek the smallest value of N for which $E_N < 10^{-5}$. Since $E_1 = 1.8\times 10^{-6}$, $N = 1$ and $\int_0^{0.2}\frac{1}{1+x^3}\,dx \approx 0.2 - \frac{(0.2)^4}{4} = 0.1996$. Using our calculator, $\int_0^{0.2}\frac{1}{1+x^3}\,dx = 0.1996018184$.

85. Using the Maclaurin series for the exponential function,

$$\lim_{x\to 0}\frac{\exp(x)+\exp(-x)-2}{x^2} = \lim_{x\to 0}\frac{(1+x+\frac{1}{2}x^2+\frac{1}{6}x^3+\frac{1}{24}x^4+\cdots)+(1-x+\frac{1}{2}x^2-\frac{1}{6}x^3++\frac{1}{24}x^4+\cdots)-2}{x^2}$$
$$= \lim_{x\to 0}\frac{x^2+\frac{1}{12}x^4+\cdots}{x^2} = \lim_{x\to 0}\left(1+\frac{1}{12}x^2+\cdots\right) = 1.$$

87. Using the Maclaurin series for arctan and sine,

$$\lim_{x\to 0}\frac{x\arctan(x^2)}{x-\sin(x)} = \lim_{x\to 0}\frac{x\left(x^2-\frac{1}{3}x^6+\frac{1}{5}x^{10}-\cdots\right)}{x-\left(x-\frac{1}{6}x^3+\frac{1}{120}x^5-\frac{1}{5040}x^7+\cdots\right)}$$
$$= \lim_{x\to 0}\frac{x^3-\frac{1}{3}x^7+\frac{1}{5}x^{11}-\cdots}{\frac{1}{6}x^3-\frac{1}{120}x^5+\frac{1}{5040}x^7-\cdots}$$
$$= \lim_{x\to 0}\frac{1-\frac{1}{3}x^4+\frac{1}{5}x^8-\cdots}{\frac{1}{6}-\frac{1}{120}x^2+\frac{1}{5040}x^4-\cdots} = 6.$$